系统工程手册

系统生命周期流程和活动指南

SYSTEMS ENGINEERING HANDBOOK
A GUIDE FOR SYSTEM LIFE CYCLE PROCESSES AND ACTIVITIES

原书第4版　国际系统工程协会（INCOSE）编著

中英对照版　张新国 译

国际系统工程协会（INCOSE）编写的《系统工程手册》为系统工程师所从事的关键流程活动提供说明。本手册的目标读者是新的系统工程师、需要从事系统工程的非系统工程专业的工程师或需要进行参考的有经验的系统工程师。本手册描述了每个系统工程流程活动在可承受性和性能设计背景中的必需性方面的内容。

本书为中英文对照版。

Copyright © 2015 John Wiley & Sons, Inc. All Rights Reserved. This translation published under license. Authorized translation from the English language edition, entitled Systems engineering handbook : a guide for system life cycle processes and activities, ISBN: 9781118999400, by International Council on Systems Engineering (INCOSE), Published by John Wiley & Sons, Inc. No part of this book may be reproduced in any form without the written permission of the original copyrights holder.

本书中文简体字版由 Wiley 授权机械工业出版社独家出版，未经出版者书面允许，本书的任何部分不得以任何方式复制或抄袭。版权所有，翻印必究。

北京市版权局著作权合同登记 图字：01-2016-0888。

图书在版编目（CIP）数据

系统工程手册：系统生命周期流程和活动指南：原书第 4 版：中英对照版/美国国际系统工程协会（INCOSE）编著；张新国译.—2 版.—北京：机械工业出版社，2017.3（2021.3 重印）

书名原文：INCOSE Systems Engineering Handbook: A Guide for System Life Cycle Processes and Activities

ISBN 978-7-111-56174-3

Ⅰ.①系… Ⅱ.①美…②张… Ⅲ.①系统工程－手册－汉、英 Ⅳ.①N945-62

中国版本图书馆 CIP 数据核字（2017）第 037069 号

机械工业出版社（北京市百万庄大街22号 邮政编码 100037）
策划编辑：廖　岩　责任编辑：廖　岩
责任印制：邹　敏　责任校对：李　伟
北京圣夫亚美印刷有限公司印刷
2021 年 3 月第 2 版第 8 次印刷
170mm×242mm・60.5 印张・3 插页・855 千字
标准书号：ISBN 978-7-111-56174-3
定价：380.00 元

凡购本书，如有缺页、倒页、脱页，由本社发行部调换

电话服务　　　　　　　　　　　　　网络服务
服务咨询热线：（010）88361066　　　机工官网：www.cmpbook.com
读者购书热线：（010）68326294　　　机工官博：weibo.com/cmp1952
　　　　　　　（010）88379203　　　金书网：www.golden-book.com
封面无防伪标均为盗版　　　　　　　教育服务网：www.cmpedu.com

译者序

《系统工程手册》(Systems Engineering Handbook)是国际系统工程协会(INCOSE)在系统工程领域发布的知识体系中的核心,最初由来自美国的几家航空航天/防务领域的会员所大力倡导并创建,在过去的20多年里,先后经历了十多次的版本修订,广泛地采纳INCOSE会员们的建议并扩大作者群体,不断地结合工程应用的发展扩充系统工程流程域的方法。直至2006年3月发布的V3.0,成为基于国际通行的系统工程标准ISO/IEC 15288的首个版本。在此之后的数个版本,作为ISO/IEC的技术报告(TR)提交,其目的是为广泛的工业界遵循国际标准提供丰富的流程和方法的指南。

本次翻译的《系统工程手册》(Systems Engineering Handbook)V4.0,是2015年最新更新的版本。从2006年出版的3.0版本到2015年的SE Handbook 4.0版本,间隔了近9年的时间,而这9年的时间正是传统系统工程向复杂系统工程转型发展最关键的时期,也是相关技术变化最快的时期,全球防务及航空领域正在经历从传统的基于文件的系统工程到基于模型的系统工程(MBSE)的转型期。因此这9年的发展是迅猛的,进步是巨大的,4.0版本与3.0版本自然也会有较大差异,主要体现在以下几个方面。

第一,与开放的《系统工程知识体指南》(SEBoK)接轨。

本手册在可行的范围内与《系统工程知识体指南》(SEBoK, 2014)一致,在很多地方,本手册为读者揭示了SEBoK中相关主题更为详细的适用范围。

SEBoK是一个关于系统工程信息的大纲,是系统工程(SE)知识体的指南。它聚焦于工程系统的背景环境,提供广泛可接受的、基于团体的和周期性更新的SE知识基线,这个基线将加强跨越包括开发和运行系统在内的多学科的共同理解。对于系统的成本、进度和安全性的屡遭挫败的持续研究表明,大多数挫败并非是由领域学科所致,而是缺乏恰当的系统工程(SE)。为了减少这些挫败,SEBoK提供了所需的对SE的共同理解的基础,阐述了SE的边界、术语、内容和结构,以及通用的SE生命周期和流程知识。同时提供SE活动的组织支持的资源,揭示SE与其他学科之间的相互作用,强调系统工程师需要知晓其他学科的哪些内容。

SEBoK采用开源、众创和共享的方式,类似SE领域的维基百科。因此,使得SE知识体的开放、共享和进化达到了空前的程度,《系统工程手册》与SEBoK的接轨,也赋予SE知识更强大的生命力。

第二,加入系统科学和系统思考,回归系统工程的方法论之根。

系统科学,汇集系统各个方面的研究,以识别、探索和理解跨越学科的多

译者序

个领域和应用的诸多范围的复杂性特征为目的，寻求开发能够形成适用于独立于元素类型或应用的所有系统类型的理论基础的跨学科依据。因此，这就为系统工程作为一种使系统能成功实现的跨学科的方法和手段提供了方法论依据。

此外，由于系统科学能够有助于为 SE 提供公共语言，求解复杂问题需要将系统科学、系统思考和 SE 的元素结合起来，跨学科的团队也需要共同语言来进行系统级的对话与沟通。更重要的是，系统科学打破了传统科学中的"解析研究方法"的局限性，即"一个实体在物质上或概念上都可以被分解为多个部分并由其各部分复原整体"——这一"经典"科学的基本原理，从揭示了系统科学——新科学范式，更为关注系统的开放性、涌现性和演进性，并为现在和未来面向模型的系统工程学提供了基础。

系统科学主要研究系统基础知识域，包括系统（工程系统）群、复杂性、涌现性、系统类型等，负责系统表达、概念原理和特征模式的定义与演进，是系统思考的基础，不断探索与发展系统方法的理论和方法论，形成系统方法应用于工程系统的基础。

系统思考是系统工程的思维方式，系统科学是系统工程的方法论基础。系统思考的主要任务是，识别系统主要识别形式和功能，形式是系统是如何组成的（由多少个部分组成的，以及它们的相互关系），功能是系统要做什么，它们是系统最根本的两个要素。在系统思考所基于的知识领域里，系统概念、系统原理与特征模式、系统表达都是很重要的概念。系统概念需要使用系统思考的方法，与系统原理与特征模式建立起相关性，其中，系统原理与特征模式是系统思考形成的基础，而系统表达则是系统思考的应用。而系统的结构化表达成了主要的问题，模型是表达，架构也是表达，是区别于自然语言的表达。因此，系统思考成为用于工程系统的系统方法，而那些以系统方式思考和行为的人们，正是掌握了系统思考的方法。

SE 基于系统思考——对现实的一种独特的视角——塑造我们的整体意识和理解整体内的多个部分如何相互关联。当系统被视为系统元素的组合时，系统思考推崇整体（系统）至上及系统元素之间的相互关系以及与整体之间的关系至上。系统思考来自于发现、学习、诊断和对话引发感知、建模以及探讨真实世界以更好地理解系统、定义系统并与系统一起工作。一个系统思考者应知晓系统如何适配于日常生活中更大的背景环境，系统如何作为以及如何管理系统。

第三，系统工程的起点从利益攸关者需要上升到业务或任务需要。

尽管系统工程覆盖系统的整个生命周期，但是需求工程一直是复杂系统工程最为重要的开端，而且是后续的系统规范、设计、验证、确认的根本和基础。传统的系统工程一般从利益攸关者需要（原理）开始，进而由系统工程师将其转化为系统需求（可验证和确认的需求规范）。这次的新版《系统工程手册》，又将利益攸关者需要上升到其所在组织的业务或任务需要来描述，使需求定义开始于组织或复杂组织体的业务愿景、ConOps（组织的领导层对系统在组织中运行的意图）及其他的组织战略目标和目的，业务管理据此对业务需要（亦作任务需要）进行定义。所以从需要（问题空间）到需求（解决方案空间）的转换图像再次被上升和扩展，也就从一开始看到了"大图像"。特别是从一开始就要了解本系统在其运行环境中与多少其他系统有相关性或相互作用，站在更大的背景环境中来看，最终可能很多系统都处于更复杂的 SoS 之中。

第四，从系统需求规范到系统架构的定义。

传统系统工程通常将系统需求规范作为系统顶层分析与综合的结果并做系统设计的输入。这次的新版《系统工程手册》，将原来的"架构设计流程"分开为"架构定义流程"与"设计定义流程"。在架构定义流程中，凸显了系统架构的顶层决策支持作用。明确其目的是生成系统架构的备选方案，并以一系列一致的视图对备选方案进行表达。

系统架构和设计活动使得能够基于彼此相互逻辑地相关并且协调一致的原则、概念和特性来创造一个全面的解决方案。解决方案架构和设计具有尽可能满足系统需求集合（可追溯到任务/业务和利益攸关者需求）及生命周期概念（如运行、支持）所表达的问题或机会的特征、特性和特点，并且可通过技术（如机械、电子、液压、软件、服务、程序）来实施。

但是架构和设计活动是基于不同且互补的观念的，系统架构是更加抽象的、面向概念化的、全局的，聚焦于达成任务的 OpsCon 和系统及系统元素的高层级结构，有效的架构尽可能独立于设计，以使在设计权衡空间中有最大的灵活性，它要聚焦于"做什么"，而不是"如何做"（属于设计流程的工作）。

系统架构流程中需要开发架构视角，并开发候选架构的模型和视图，通过对系统从需求到功能架构、逻辑架构，再到物理架构，以及对系统功能、行为和结构的映射，就实现了将系统需求和架构实体，向系统元素的划分、对准和分配，并建立系统设计与演进的指导原则。

架构定义期间使用的建模、仿真和原型可大幅度地降低完成后的系统失败的风险。系统架构将是系统工程师最主要的任务之一，有时系统工程师与系统架构师甚至是同义词。

第五，系统工程的应用扩展至众多领域。

系统工程流程可在多种不同的产品行业和领域、产品线、服务乃至复杂组织体得到剪裁应用。在组织层级上，剪裁流程调整组织流程背景环境中的外部标准以满足组织的需要；在项目层级上，剪裁流程调整组织流程以满足项目的独特需要。手册中给出了汽车系统、生物医疗和健康医疗系统、防务和航天系统、基础设施系统、空间系统、（地面）运输系统等典型系统对 SE 流程如何裁剪应用的建议。

用于产品线管理（PLM）的系统工程应用，用于服务的系统工程应用，特别是用于复杂组织体的系统工程应用，阐明 SE 原理在复杂组织体 SE 中应用，用复杂组织体的规划、设计、改进和运行，以指导复杂组织体的转型和持续改进，复杂组织体 SE 是为了应对复杂组织体固有复杂性而聚焦于框架、工具和问题求解途径的新兴学科，而且复杂组织体涉及的不仅仅是求解问题，亦涉及对实现复杂组织体目标的更好方式的机会的探索。

本手册还特别强调指出，复杂组织体与"组织"并不等同，这是常见的对复杂组织体术语的误用。复杂组织体，不仅包括加入其中的组织，还包括人员、知识，以及诸如流程、原则、策略、实践、学说、理论、信仰、设施、土地、知识产权等其他资产。

第六，基于模型的系统工程已成为未来发展的基本趋势。

在计算机出现以后，建模与仿真已经成了区别于理论和实验的第三种方法。它既不同于理论，也没有证明，因为实验就是对理论的证明，它也不同于实验，因为实验有时候不需要理论。建模与仿真同时需要理论和实验，因为它是基于理论而建模，同时仿真实际上等于在做虚拟实验，它在证明理论。所以为什么建模与仿真实际上变成了各个领域的仿真。

在新版《系统工程手册》中，建模和仿真已被定义为跨系统生命周期流程的系统工程共性方法，用以实现对系统概念的完整性表达，支持系统需求、设计、分析、验证与确认活动的开展，涉及建模目的、标准、范围、类型、语言等内容。建模与仿真都是用抽象在应对具体，再用形式描述物理，这是建模与仿真的核心。许多模型和仿真实践已被正规化为 SE 流程，且众多建模语言已成为国际标准，如 SysML、UML 等，在"基于模型的"背景环境下，支持 SE 学科的有关方法论、流程、方法和工具已日趋成熟，且在航空航天/防务领域已被证明。

事实上，基于文件的传统系统工程方法存在着许多难以克服的困难。
- 完整性、一致性以及在需求、设计、工程分析和测试信息之间的关系都是难以评估的。
- 难以理解系统的某一特定方面，并进行必要的跟踪和新变化的影响分析。
- 在系统级需求和设计以及低层硬件/软件之间能否同步。
- 难以为一个持续演进或变化的系统设计维持或复用系统需求和设计信息。

模型将不再像传统系统工程中那些只是文件中的"附属品"，而是成为系统工程的主要"制品"/产物和手段。

而 MBSE 能够使以上这些问题得到很好的解决。不仅能够实现需求/系统架构做早期的验证，而且还能保持系统工程各阶段工作的连续验证，并跨越生命周期，在各个方面支持系统的需求、设计、开发、制造和验证。特别是在工业 4.0 时代，为全系统的开发提供数字线索和"数字双胞胎"，使系统/产品的数字模型与物理试题在全生命周期内保持同步进化。

最后回到国际系统工程协会发布的 2010—2025 年 MBSE 路线图上（见图 0.1），它对 MBSE 成熟度和能力的演进路径进行了表达，涵盖了标准、流程、方法、工具、培训与教育等多个方面，是未来 MBSE 的发展趋势。

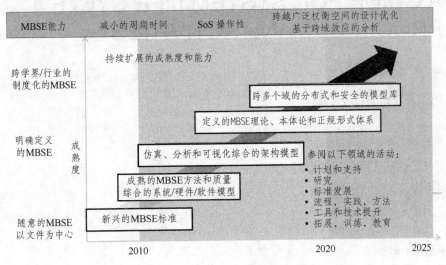

图 0.1　INCOSE MBSE 路线图

总的来说，4.0 版本无疑是更具有全面性、时效性和实用性的。本手册整个翻译过程同样基于协同翻译平台的多学科团队工作模式，通过前几轮翻译过程的知识沉淀，效率及准确性都有了显著提高，这也是 4.0 双语版本可以较短时间内成稿的原因。另外，学习了 3.0 版本的读者应该也可以发现，4.0 版本有一些概念的翻译和 3.0 版本是有差异的。经过几年的学习和实践，理解和认知都在进化。在 4.0 版本的翻译过程中，更注意易读性，在严格遵循原文语义的基础上，尽量减少逻辑的嵌套和句法的转化。

最后，感谢 INCOSE 及北京（中国）分会所给予的信任和支持，对中航工业信息技术中心高星海以及编译工作小组在校对、版本管理、过程控制等工作中付出的努力，在此表示衷心的感谢。

张新国

2016 年 11 月

SYSTEMS ENGINEERING HANDBOOK

A GUIDE FOR SYSTEM LIFE CYCLE PROCESSES AND ACTIVITIES

FOURTH EDITION

INCOSE-TP-2003-002-04
2015

Prepared by:

International Council on Systems Engineering (INCOSE)
7670 Opportunity Rd, Suite 220
San Diego, CA, USA 92111-2222

Compiled and Edited by:
David D. Walden, ESEP
Garry J. Roedler, ESEP
Kevin J. Forsberg, ESEP
R. Douglas Hamelin
Thomas M. Shortell, CSEP

系统工程手册

系统生命周期流程和活动指南

第4版

INCOSE-TP-2003-002-04
2015

编著：

国际系统工程协会（INCOSE）
7670 Opportunity Rd, Suite 220
San Diego, CA, USA 92111-2222

编撰：

David D. Walden, ESEP
Garry J. Roedler, ESEP
Kevin J. Forsberg, ESEP
R. Douglas Hamelin
Thomas M. Shortell, CSEP

WILEY

INCOSE NOTICES

This International Council on Systems Engineering (INCOSE) Technical Product was prepared by the INCOSE Knowledge Management working group. It is approved by INCOSE Technical Operations Leadership for release as an INCOSE Technical Product.

Copyright ©2015 by INCOSE, subject to the following restrictions:

Author Use: Authors have full rights to use their contributions unfettered, with credit to the INCOSE technical source, except as noted in the following text. Abstraction is permitted with credit to the source.

INCOSE Use: Permission to reproduce and use this document or parts thereof by members of INCOSE and to prepare derivative works from this document for INCOSE use is granted, with attribution to INCOSE and the original author(s) where practical, provided this copyright notice is included with all reproductions and derivative works. Content from ISO/IEC/IEEE 15288 and ISO/IEC TR 24748-1 is used by permission, and is not to be reproduced other than as part of this total document.

External Use: This document may not be shared or distributed to any non-INCOSE third party. Requests for permission to reproduce this document in whole or in part, or to prepare derivative works of this document for external and/or commercial use, will be denied unless covered by other formal agreements with INCOSE. Copying, scanning, retyping, or any other form of reproduction or use of the content of whole pages or source documents are prohibited, except as approved by the INCOSE Administrative Office, 7670 Opportunity Road, Suite 220, San Diego, CA 92111-2222, USA.

Electronic Version Use: All electronic versions (e.g., eBook, PDF) of this document are to be used for personal professional use only and are not to be placed on non-INCOSE sponsored servers for general use. Any additional use of these materials must have written approval from the INCOSE Administrative Office.

INCOSE Corporate Advisory Board Use: INCOSE has granted permission to member organizations of the INCOSE Corporate Advisory Board (CAB) to post an electronic (PDF) version of this document on their internal servers for use by their employees, subject to the external use restrictions noted earlier. Additional use of this document by CAB organizations for internal purposes is permitted per INCOSE policy CAB-100.

Notice: Hardcopy versions of this document may not be the most current. The current approved version is always the electronic version posted on the Product Area of the INCOSE website.

General Citation Guidelines: References to this handbook should be formatted as follows, with appropriate adjustments for formally recognized styles:

INCOSE (2015). *Systems Engineering Handbook:A Guide for System Life Cycle Process and Activities* (4th ed.). D. D. Walden, G. J. Roedler, K. J. Forsberg, R. D. Hamelin, and, T. M. Shortell (Eds.). San Diego, CA: International Council on Systems Engineering. Published by John Wiley & Sons, Inc.

INCOSE 声明

本国际系统工程协会（INCOSE）技术产品由 INCOSE 知识管理工作组编制。经 INCOSE 技术运行领导层批准，作为 INCOSE 技术产品发行。

版权©2015 归 INCOSE 所有，但受下列限制：

作者的使用权：说明源自 INCOSE 技术文献，作者有权不受任何约束地使用他们的著作，除非下文另有说明。说明源自 INCOSE，允许摘录。

INCOSE 的使用权：允许 INCOSE 成员复制和使用本文件或其部分内容，允许从本文件中编制衍生文献以便 INCOSE 使用，在可行时，应说明 INCOSE 和原作者的贡献，须将本版权声明包含在复制和衍生文献中。ISO/IEC/IEEE 15288 和 ISO/IEC TR 24748-1 的内容经允许后方可使用，而且除非是该整个文件的一部分，否则不允许复制。

外部使用权：本文件不得共享或分发给任何非 INCOSE 的第三方。除非与 INCOSE 签订的其他正式协议中已涵盖，否则为了外部和/或商业用途而复制全部或部分本文件以及编制本文件的衍生文献的请求均将被拒绝。除非 INCOSE 行政办公室（地址：7670 Opportunity Road，Suite 220，San Diego，CA92111-2222，USA）已批准，否则不得复制、扫描、重新录入或以任何其他形式复制或使用整页或源文件中的内容。

电子版的使用权：本文件的任何电子版（如 eBook、PDF）仅用于个人专业用途，不得为一般用途放置在非 INCOSE 所拥有的服务器中。这些资料的任何额外使用必须经过 INCOSE 行政办公室的书面批准。

INCOSE 公司咨询委员会的使用权：INCOSE 授权允许 INCOSE 公司咨询委员会（CAB）的会员组织在其内部服务器上发布其文件的电子（PDF）版本供其员工使用，内部发布和使用本文件，但受前面提到的外部使用的限制。INCOSE 方针 CAB-100 允许 CAB 组织由于内部用途对本文件进行附加使用。

注意：本文件的文本复印版本可能不是最新版本。当前经批准的版本始终是 INCOSE 网站产品区上发布的电子版本。

一般引用指南：本手册的参考文献按照如下方式编排格式，并按正式认可的风格做适当调整：

INCOSE（2015）. *系统工程手册：系统生命周期流程和活动指南*（第 4 版）. D.D.Walden, G.J.Roedler, K.J.Forsberg, R.D.Hamelin 和 T.M.Shortell（Eds.）。加州圣迭戈：国际系统工程协会。由 John Wiley & Sons 公司出版。

HISTORY OF CHANGES

Revision	Revision date	Change description and rationale
Original	Jun 1994	Draft *Systems Engineering Handbook* (SEH) created by INCOSE members from several defense/aerospace companies—including Lockheed, TRW. Northrop Grumman, Ford Aerospace, and the Center for Systems Management—for INCOSE review
1.0	Jan 1998	Initial SEH release approved to update and broaden coverage of SE process. Included broad participation of INCOSE members as authors. Based on Interim Standards EIA 632 and IEEE 1220
2.0	Jul 2000	Expanded coverage on several topics, such as functional analysis. This version was the basis for the development of the Certified Systems Engineering Professional (CSEP) exam
2.0A	Jun 2004	Reduced page count of SEH v2 by 25% and reduced the US DoD-centric material wherever possible. This version was the basis for the first publically offered CSEP exam
3.0	Jun 2006	Significant revision based on ISO/IEC 15288:2002. The intent was to create a country- and domain-neutral handbook. Significantly reduced the page count, with elaboration to be provided in appendices posted online in the INCOSE Product Asset Library (IPAL)
3.1	Aug 2007	Added detail that was not included in SEH v3, mainly in new appendices. This version was the basis for the updated CSEP exam
3.2	Jan 2010	Updated version based on ISO/IEC/IEEE 15288:2008. Significant restructuring of the handbook to consolidate related topics
3.2.1	Jan 2011	Clarified definition material, architectural frameworks, concept of operations references, risk references, and editorial corrections based on ISO/IEC review
3.2.2	Oct 2011	Correction of errata introduced by revision 3.2.1
4.0	Jan 2015	Significant revision based on ISO/IEC/IEEE 15288:2015. inputs from the relevant INCOSE working groups (WGs), and to be consistent with the Guide to the Systems Engineering Body of Knowledge (SEBoK)

变更历史

修订版本	修订日期	变更说明和原因
原版	1994年6月	系统工程手册（SEH）草案由几家来自防务/航空航天公司的 INCOSE 成员创建——包括 Lockheed、TRW、Northrop Grumman、Ford Aerospace 和系统管理中心——以供 INCOSE 审查
1.0	1998年1月	SEH 的初始发布版本经批准后更新和扩展 SE 流程的范围。INCOSE 成员广泛参与并成为作者。基于暂行标准 EIA 632 和 IEEE 1220
2.0	2000年7月	扩展几个主题的范围，如功能分析。此版本是注册系统工程专业人员(CSEP)考试的发展基础
2.0A	2004年6月	将 SEH v2 的页数减少 25%，并且尽可能减少以美国国防部为中心的材料。此版本是首次公开提供的 CSEP 考试的基础
3.0	2006年6月	基于 ISO/IEC 15288:2002 的重要修订。目的是创造出对于国家和领域都为中立的手册。大幅减少页数，同时在 INCOSE 产品资产库（IPAL）中在线公布的附录中提供详细阐述
3.1	2007年8月	增加在 SEH v3 中没有包含的细节，主要是在新的附录中。此版本是更新的 CSEP 考试的基础
3.2	2010年1月	基于 ISO/IEC/IEEE 15288:2008 的更新版本。对手册大幅度重新建构以巩固相关主题
3.2.1	2011年1月	基于 ISO/IEC 的评审来澄清：定义资料、架构框架、运行方案的引用、风险的引用以及编辑性的更正
3.2.2	2011年10月	修正 3.2.1 版本推出的勘误表
4.0	2015年1月	基于 ISO/IEC/IEEE 15288:2015 进行大幅修订，从相关 INCOSE 工作组（WG）输入，并与《系统工程知识主体指南》相一致

PREFACE

The objective of the International Council on Systems Engineering (INCOSE) Systems Engineering Handbook (SEH) is to describe key process activities performed by systems engineers. The intended audience is the systems engineering (SE) professional. When the term *systems engineer* is used in this handbook, it includes the new systems engineer, a product engineer or an engineer in another discipline who needs to perform SE, or an experienced systems engineer who needs a convenient reference.

The descriptions in this handbook show what each SE process activity entails, in the context of designing for required performance and life cycle considerations. On some projects, a given activity may be performed very informally; on other projects, it may be performed very formally, with interim products under formal configuration control. This document is not intended to advocate any level of formality as necessary or appropriate in all situations. The appropriate degree of formality in the execution of any SE process activity is determined by the following:

1. The need for communication of what is being done (across members of a project team, across organizations, or over time to support future activities)
2. The level of uncertainty
3. The degree of complexity
4. The consequences to human welfare

On smaller projects, where the span of required communications is small (few people and short project life cycle) and the cost of rework is low, SE activities can be conducted very informally and thus at low cost. On larger projects, where the span of required communications is large (many teams that may span multiple geographic locations and organizations and long project life cycle) and the cost of failure or rework is high, increased formality can significantly help in achieving project opportunities and in mitigating project risk.

In a project environment, work necessary to accomplish project objectives is considered "in scope"; all other work is considered "out of scope." On every project, "thinking" is always "in scope." Thoughtful tailoring and intelligent application of the SE processes described in this handbook are essential to achieve the proper balance between the risk of missing project technical and business objectives on the one hand and process paralysis on the other hand. Chapter 8 provides tailoring guidelines to help achieve that balance.

前言

国际系统工程协会（INCOSE）系统工程手册（SEH）的目的是描述系统工程师所实施的关键流程活动。目标读者是系统工程（SE）专业人员。当本手册中使用术语*系统工程师*时，其包括新的系统工程师、产品工程师或需要实施SE的其他学科的工程师，或需要一个便于参考的有经验的系统工程师。

本手册中的描述内容指明在所要求的性能和生命周期考量的设计背景环境中每个 SE 流程活动的必要性。在某些项目中，可能以很不正式的方式实施某一给定活动；而在其他项目中，却非常正式地实施，并且中间过渡产品都处于正式的构型控制之中。本文件的意图不是提倡在所有情况下任何正式等级都是必须或合适的，实施任何 SE 流程活动的适当正式程度取决于：

1. 需要沟通正在做什么（项目团队成员之间、组织之间或随着时间而支持未来活动）
2. 不确定性的等级
3. 复杂性的程度
4. 人类福祉的结果

在比较小的项目中，要求沟通的跨度小（人员少且项目生命周期短），返工成本低，因而 SE 活动可非常不正式地执行，因而成本低。在比较大的项目中，所需沟通的跨度大（许多团队可能跨越多重地理位置和组织且项目生命周期长）且失败或返工成本高，因而提高正式程度可极大地帮助达成项目的机会和减轻项目风险。

在项目环境中，完成项目目标所需的工作属于"范围内"；所有其他工作属于"范围外"。对于每个项目而言，"思考"总是在"范围内"。将本手册所述的 SE 流程进行深思熟虑的剪裁和颇具智慧的应用，对于丧失项目技术和业务目标的风险与流程瘫痪之间的适当平衡是根本性的。第 8 章提供了剪裁准则，以帮助达到这种平衡。

PREFACE

Approved for SEH v4:

Kevin Forsberg, ESEP, Chair, INCOSE Knowledge Management Working Group

Garry Roedler, ESEP, Co-Chair, INCOSE Knowledge Management Working Group

William Miller, INCOSE Technical Director (2013–2014)

Paul Schreinemakers, INCOSE Technical Director (2015–2016)

Quoc Do, INCOSE Associate Director for Technical Review

Kenneth Zemrowski, ESEP, INCOSE Assistant Director for Technical Information

SEH v4 审批：

INCOSE 知识管理工作组主席，Kevin Forsberg，ESEP

INCOSE 知识管理工作组副主席，Garry Roedler，ESEP

INCOSE 技术总监，William Miller（2013—2014）

INCOSE 技术总监，Paul Schreinemakers（2015—2016）

INCOSE 技术评审副总监，Quoc Do

INCOSE 技术信息助理总监，Kenneth Zemrowski, ESEP

CONTENTS

1 Systems Engineering Handbook Scope 2
    1.1 Purpose 2
    1.2 Application 2
    1.3 Contents 4
    1.4 Format 6
    1.5 Definitions of Frequently Used Terms 10
2 Systems Engineering Overview 12
    2.1 Introduction 12
    2.2 Definitions and Concepts of a System 12
    2.3 The Hierarchy *within* a System 18
    2.4 Definition of Systems of Systems 20
    2.5 Enabling Systems 28
    2.6 Definition of Systems Engineering 30
    2.7 Origins and Evolution of Systems Engineering 34
    2.8 Use and Value of Systems Engineering 38
    2.9 Systems Science and Systems Thinking 48
    2.10 Systems Engineering Leadership 62
    2.11 Systems Engineering Professional Development 66
3 Generic Life Cycle Stages 72
    3.1 Introduction 72
    3.2 Life Cycle Characteristics 74
    3.3 Life Cycle Stages 80
    3.4 Life Cycle Approaches 94
    3.5 What is Best for Your Organization, Project, or Team? 108
    3.6 Introduction to Case Studies 114
4 Technical Processes 140
    4.1 Business or Mission Analysis Process 146
    4.2 Stakeholder Needs and Requirements Definition Process ... 154
    4.3 System Requirements Definition Process 172
    4.4 Architecture Definition Process 194
    4.5 Design Definition Process 216
    4.6 System Analysis Process 226

目录

1 系统工程手册的范围 ·· 3
   1.1　目的 ··· 3
   1.2　适用性 ·· 3
   1.3　内容 ··· 5
   1.4　格式 ··· 7
   1.5　常用术语的定义 ··· 11
2 系统工程概览 ··· 13
   2.1　引言 ·· 13
   2.2　系统的定义和概念 ·· 13
   2.3　一个系统内部的层级结构 ··· 19
   2.4　系统之系统的定义 ·· 21
   2.5　使能系统 ··· 29
   2.6　系统工程的定义 ··· 31
   2.7　系统工程的起源和演变 ·· 35
   2.8　系统工程的使用和价值 ·· 39
   2.9　系统科学与系统思考 ·· 49
   2.10　系统工程的领导力 ·· 63
   2.11　系统工程职业发展 ·· 67
3 一般生命周期阶段 ··· 73
   3.1　引言 ·· 73
   3.2　生命周期特性 ·· 75
   3.3　生命周期阶段 ·· 81
   3.4　生命周期方法 ·· 95
   3.5　什么最适合你的组织、项目或团队？ ·· 109
   3.6　案例研究简介 ·· 115
4 技术流程 ·· 141
   4.1　业务或任务分析流程 ··· 147
   4.2　利益攸关者需要和需求定义流程 ·· 155
   4.3　系统需求定义流程 ·· 173
   4.4　架构定义流程 ·· 195
   4.5　设计定义流程 ·· 217
   4.6　系统分析流程 ·· 227

CONTENTS

  4.7 Implementation Process ···236
  4.8 Integration Process ··244
  4.9 Verification Process ···256
  4.10 Transition Process ···270
  4.11 Validation Process ··276
  4.12 Operation Process ···292
  4.13 Maintenance Process ··298
  4.14 Disposal Process ··314
5 Technical Management Processes ··322
  5.1 Project Planning Process ···322
  5.2 Project Assessment and Control Process ··338
  5.3 Decision Management Process ··344
  5.4 Risk Management Process ··358
  5.5 Configuration Management Process ··388
  5.6 Information Management Process ···408
  5.7 Measurement Process ··418
  5.8 Quality Assurance Process ··434
6 Agreement Processes ··444
  6.1 Acquisition Process ··448
  6.2 Supply Process ···456
7 Organizational Project-Enabling Processes ······································464
  7.1 Life Cycle Model Management Process ··466
  7.2 Infrastructure Management Process ···480
  7.3 Portfolio Management Process ···486
  7.4 Human Resource Management Process ···494
  7.5 Quality Management Process ···502
  7.6 Knowledge Management Process ···512
8 Tailoring process and Application of Systems Engineering ················524
  8.1 Tailoring Process ···526
  8.2 Tailoring for Specific Product Sector or Domain Application ··················532
  8.3 Application of Systems Engineering for Product Line Management ·······550
  8.4 Application of Systems Engineering for Services ································556
  8.5 Application of Systems Engineering for Enterprises ·····························570

  4.7 实施流程 ·· 237
  4.8 综合流程 ·· 245
  4.9 验证流程 ·· 257
  4.10 转移流程 ··· 271
  4.11 确认流程 ··· 277
  4.12 运行流程 ··· 293
  4.13 维护流程 ··· 299
  4.14 处置流程 ··· 315
5 技术管理流程 ··· 323
  5.1 项目规划流程 ·· 323
  5.2 项目评估和控制流程 ··· 339
  5.3 决策管理流程 ·· 345
  5.4 风险管理流程 ·· 359
  5.5 构型配置管理流程 ··· 389
  5.6 信息管理流程 ·· 409
  5.7 测度流程 ·· 419
  5.8 质量保证流程 ·· 435
6 协议流程 ·· 445
  6.1 采办流程 ·· 449
  6.2 供应流程 ·· 457
7 组织的项目使能流程 ··· 465
  7.1 生命周期模型管理流程 ·· 467
  7.2 基础设施管理流程 ··· 481
  7.3 项目群管理流程 ·· 487
  7.4 人力资源管理流程 ··· 495
  7.5 质量管理流程 ·· 503
  7.6 知识管理流程 ·· 513
8 系统工程的剪裁流程和应用 ·· 525
  8.1 剪裁流程 ·· 527
  8.2 特定产品行业或领域应用的剪裁 ·· 533
  8.3 用于产品线管理的系统工程应用 ·· 551
  8.4 用于服务的系统工程应用 ··· 557
  8.5 用于复杂组织体的系统工程应用 ·· 571

CONTENTS

  8.6 Application of Systems Engineering for Very Small and Micro Enterprises ⋯580
9 Cross-Cutting Systems Engineering Methods ⋯584
  9.1 Modeling and Simulation ⋯584
  9.2 Model-Based Systems Engineering ⋯614
  9.3 Functions-Based Systems Engineering Method ⋯618
  9.4 Object-Oriented Systems Engineering Method ⋯628
  9.5 Prototyping ⋯640
  9.6 Interface Management ⋯640
  9.7 Integrated Product and Process Development ⋯648
  9.8 Lean Systems Engineering ⋯662
  9.9 Agile Systems Engineering ⋯676
10 Specialty Engineering Activities ⋯686
  10.1 Affordability/Cost-Effectiveness/Life Cycle Cost Analysis ⋯686
  10.2 Electromagnetic Compatibility ⋯710
  10.3 Environmental Engineering/Impact Analysis ⋯714
  10.4 Interoperability Analysis ⋯718
  10.5 Logistics Engineering ⋯718
  10.6 Manufacturing and Producibility Analysis ⋯730
  10.7 Mass Properties Engineering ⋯730
  10.8 Reliability, Availability, and Maintainability ⋯732
  10.9 Resilience Engineering ⋯744
  10.10 System Safety Engineering ⋯752
  10.11 System Security Engineering ⋯762
  10.12 Training Needs Analysis ⋯772
  10.13 Usability Analysis/Human Systems Integration ⋯772
  10.14 Value Engineering ⋯788
APPENDIX A: REFERENCES ⋯800
APPENDIX B: ACRONYMS ⋯824
APPENDIX C: TERMS AND DEFINITIONS ⋯838
APPENDIX D: N^2 DIAGRAM OF SYSTEMS ENGINEERING PROCESSES ⋯856
APPENDIX E: INPUT/OUTPUT DESCRIPTIONS ⋯860
APPENDIX F: ACKNOWLEDGEMENTS ⋯912
APPENDIX G: COMMENT FORM ⋯916
INDEX ⋯918

目录

　　8.6　极小型和微型组织体的系统工程的应用···········581
9　跨领域/学科系统工程方法···········585
　　9.1　建模和仿真···········585
　　9.2　基于模型的系统工程···········615
　　9.3　基于功能的系统工程方法···········619
　　9.4　面向对象的系统工程方法···········629
　　9.5　原型构建···········641
　　9.6　接口管理···········641
　　9.7　综合的产品和流程开发···········649
　　9.8　精益系统工程···········663
　　9.9　敏捷系统工程···········677
10　专业工程活动···········687
　　10.1　可承受性/成本效能/生命周期成本分析···········687
　　10.2　电磁兼容性···········711
　　10.3　环境工程/影响分析···········715
　　10.4　互操作性分析···········719
　　10.5　后勤工程···········719
　　10.6　制造及可生产性分析···········731
　　10.7　质量特性工程···········731
　　10.8　可靠性、可用性和可维护性···········733
　　10.9　可恢复性工程···········745
　　10.10　系统安全性工程···········753
　　10.11　系统安保性工程···········763
　　10.12　培训需要分析···········773
　　10.13　可用性分析/人与系统综合···········773
　　10.14　价值工程···········789

附录A：参考文献···········800
附录B：首字母缩写词···········825
附录C：术语和定义···········839
附录D：系统工程流程的 N^2 图···········857
附录E：输入/输出描述···········861
附录F：致谢···········913
附录G：意见表···········917
索引···········919

LIST OF FIGURES

FIGURE 1.1	System life cycle processes per ISO/IEC/IEEE 15288	6
FIGURE 1.2	Sample of IPO diagram for SE processes	8
FIGURE 2.1	Hierarchy within a system	20
FIGURE 2.2	Example of the systems and systems of systems within a transport system of systems	22
FIGURE 2.3	System of interest, its operational environment, and its enabling systems	30
FIGURE 2.4	Committed life cycle cost against time	40
FIGURE 2.5	Technology acceleration over the past 140 years	42
FIGURE 2.6	Project performance versus SE capability	44
FIGURE 2.7	Cost and schedule overruns correlated with SE effort	46
FIGURE 2.8	Systems science in context	50
FIGURE 2.9	SE optimization system	68
FIGURE 2.10	Professional development system	68
FIGURE 3.1	Generic business life cycle	76
FIGURE 3.2	Life cycle model with some of the possible progressions	84
FIGURE 3.3	Comparisons of life cycle models	84
FIGURE 3.4	Importance of the concept stage	90
FIGURE 3.5	Iteration and recursion	98
FIGURE 3.6	Vee model	100
FIGURE 3.7	Left side of the Vee model	102
FIGURE 3.8	Right side of the Vee model	104
FIGURE 3.9	IID and evolutionary development	108
FIGURE 3.10	The incremental commitment spiral model (ICSM)	110
FIGURE 3.11	Phased view of the generic incremental commitment spiral model process	110
FIGURE 4.1	Transformation of needs into requirements	142
FIGURE 4.2	IPO diagram for business or mission analysis process	146
FIGURE 4.3	Key SE interactions	156
FIGURE 4.4	IPO diagram for stakeholder needs and requirements definition process	158
FIGURE 4.5	IPO diagram for the system requirements definition process	176
FIGURE 4.6	IPO diagram for the architecture definition process	198
FIGURE 4.7	Interface representation	210

图目录

图 1.1　符合 ISO/IEC/IEEE 15288 标准的系统生命周期流程 ················ 7
图 1.2　SE 流程的 IPO 图示例 ·· 9
图 2.1　一个系统内的层级结构 ··· 21
图 2.2　运输系统之系统内的系统及系统之系统的示例 ······················ 23
图 2.3　所感兴趣之系统、其运行环境及其使能系统 ························· 31
图 2.4　所确定的生命周期成本与时间的对比 ································· 41
图 2.5　过去 140 年内技术的加速发展 ··· 43
图 2.6　项目绩效与 SE 能力的对比 ··· 45
图 2.7　超支成本和延期进度与 SE 工作的关系 ······························· 47
图 2.8　背景环境中的系统科学 ··· 51
图 2.9　SE 优化系统 ·· 69
图 2.10　专业的开发系统 ·· 69
图 3.1　一般商业生命周期 ··· 77
图 3.2　具有一些可能的进展的生命周期模型 ································· 85
图 3.3　生命周期模型的比较 ·· 85
图 3.4　概念阶段的重要性 ··· 91
图 3.5　迭代和递归 ··· 99
图 3.6　V 形模型 ··· 101
图 3.7　V 形模型的左侧 ·· 103
图 3.8　V 形模型的右侧 ·· 105
图 3.9　IID 和演进式开发 ·· 109
图 3.10　增量承诺螺旋模型（ICSM） ······································· 111
图 3.11　一般增量承诺螺旋模型流程的阶段视图 ··························· 111
图 4.1　需要到需求的转换 ··· 143
图 4.2　业务或任务分析流程的 IPO 图 ······································· 147
图 4.3　关键的 SE 互动 ··· 157
图 4.4　利益攸关者需要和需求定义流程的 IPO 图 ························· 159
图 4.5　系统需求定义流程的 IPO 图 ·· 177
图 4.6　架构定义流程的 IPO 图 ··· 199
图 4.7　接口表征形式 ··· 211

LIST OF FIGURES

FIGURE 4.8	(a) Initial arrangement of aggregates: (b) final arrangement after reorganization	212
FIGURE 4.9	IPO diagram for the design definition process	218
FIGURE 4.10	IPO diagram for the system analysis process	230
FIGURE 4.11	IPO diagram for the implementation process	238
FIGURE 4.12	IPO diagram for the integration process	246
FIGURE 4.13	IPO diagram for the verification process	256
FIGURE 4.14	Definition and usage of a verification action	264
FIGURE 4.15	Verification level per level	270
FIGURE 4.16	IPO diagram for the transition process	272
FIGURE 4.17	IPO diagram for the validation process	278
FIGURE 4.18	Definition and usage of a validation action	286
FIGURE 4.19	Validation level per level	292
FIGURE 4.20	IPO diagram for the operation process	294
FIGURE 4.21	IPO diagram for the maintenance process	300
FIGURE 4.22	IPO diagram for the disposal process	316
FIGURE 5.1	IPO diagram for the project planning process	326
FIGURE 5.2	IPO diagram for the project assessment and control process	340
FIGURE 5.3	IPO diagram for the decision management process	348
FIGURE 5.4	IPO diagram for the risk management process	364
FIGURE 5.5	Level of risk depends on both likelihood and consequences	370
FIGURE 5.6	Typical relationship among the risk categories	372
FIGURE 5.7	Intelligent management of risks and opportunities	390
FIGURE 5.8	IPO diagram for the configuration management process	392
FIGURE 5.9	Requirements changes are inevitable	398
FIGURE 5.10	IPO diagram for the information management process	412
FIGURE 5.11	IPO diagram for the measurement process	420
FIGURE 5.12	Measurement as a feedback control system	426
FIGURE 5.13	Relationship of technical measures	430
FIGURE 5.14	TPM monitoring	434
FIGURE 5.15	IPO diagram for the quality assurance process	438
FIGURE 6.1	IPO diagram for the acquisition process	450
FIGURE 6.2	IPO diagram for the supply process	460
FIGURE 7.1	IPO diagram for the life cycle model management process	468
FIGURE 7.2	Standard SE process flow	476

图 4.8 （a）聚集的初始排列；（b）重组后的最终排列 ·················· 213
图 4.9 设计定义流程的 IPO 图 ·················· 219
图 4.10 系统分析流程的 IPO 图 ·················· 231
图 4.11 实施流程的 IPO 图 ·················· 239
图 4.12 综合流程的 IPO 图 ·················· 247
图 4.13 验证流程的 IPO 图 ·················· 257
图 4.14 验证措施的定义和使用 ·················· 265
图 4.15 逐层验证 ·················· 271
图 4.16 转移流程的 IPO 图 ·················· 273
图 4.17 确认流程的 IPO 图 ·················· 279
图 4.18 确认措施的定义和使用 ·················· 287
图 4.19 逐层确认 ·················· 293
图 4.20 运行流程的 IPO 图 ·················· 295
图 4.21 维护流程的 IPO 图 ·················· 301
图 4.22 处置流程的 IPO 图 ·················· 317
图 5.1 项目规划流程的 IPO 图 ·················· 327
图 5.2 项目评估和控制流程的 IPO 图 ·················· 341
图 5.3 决策管理流程的 IPO 图 ·················· 349
图 5.4 风险管理流程的 IPO 图 ·················· 365
图 5.5 风险等级取决于可能性和后果两者 ·················· 371
图 5.6 风险类别之间的典型关系 ·················· 373
图 5.7 风险和机会的智能管理 ·················· 391
图 5.8 构型配置管理流程的 IPO 图 ·················· 393
图 5.9 需求变更是不可避免的 ·················· 399
图 5.10 信息管理流程的 IPO 图 ·················· 413
图 5.11 测度流程的 IPO 图 ·················· 421
图 5.12 作为反馈控制系统的测量 ·················· 427
图 5.13 技术测度的关系 ·················· 431
图 5.14 TPM 监控 ·················· 435
图 5.15 质量保证流程的 IPO 图 ·················· 439
图 6.1 采办流程的 IPO 图 ·················· 451
图 6.2 供应流程的 IPO 图 ·················· 461
图 7.1 生命周期模型管理流程的 IPO 图 ·················· 469
图 7.2 标准 SE 流程图 ·················· 477

LIST OF FIGURES

FIGURE 7.3	IPO diagram for the infrastructure management process	480
FIGURE 7.4	IPO diagram for the portfolio management	490
FIGURE 7.5	IPO diagram for the human resource management process	496
FIGURE 7.6	IPO diagram for the quality management process	506
FIGURE 7.7	IPO diagram for the knowledge management process	516
FIGURE 8.1	Tailoring requires balance between risk and process	526
FIGURE 8.2	IPO diagram for the tailoring process	528
FIGURE 8.3	Product line viewpoints	552
FIGURE 8.4	Capitalization and reuse in a product line	552
FIGURE 8.5	Product line return on investment	556
FIGURE 8.6	Service system conceptual framework	564
FIGURE 8.7	Organizations manage resources to create enterprise value	572
FIGURE 8.8	Individual competence leads to organizational, system, and operational capability	576
FIGURE 8.9	Enterprise SE process areas in the context of the entire enterprise	580
FIGURE 9.1	Sample model taxonomy	596
FIGURE 9.2	SysML diagram types	608
FIGURE 9.3	Functional analysis/allocation process	620
FIGURE 9.4	Alternative functional decomposition evaluation and definition	622
FIGURE 9.5	OOSEM builds on established SE foundations	630
FIGURE 9.6	OOSEM activities in context of the system development process	632
FIGURE 9.7	OOSEM activities and modeling artifacts	632
FIGURE 9.8	Sample FFBD and N^2 diagram	644
FIGURE 9.9	Examples of complementary integration activities of IPDTs	656
FIGURE 9.10	Lean development principles	672
FIGURE 10.1	Contextual nature of the affordability trade space	690
FIGURE 10.2	System operational effectiveness	692
FIGURE 10.3	Cost versus performance	698
FIGURE 10.4	Affordability cost analysis framework	698
FIGURE 10.5	Life cycle cost elements (not to scale)	706
FIGURE 10.6	Process for achieving EMC	710
FIGURE 10.7	Supportability analysis	726
FIGURE 10.8	Reliability program plan development	740
FIGURE 10.9	Resilience event model	748
FIGURE 10.10	Sample Function Analysis System Technique (FAST) diagram	796

图 7.3	基础设施管理流程的 IPO 图	481
图 7.4	项目群管理流程的 IPO 图	491
图 7.5	人力资源管理流程的 IPO 图	497
图 7.6	质量管理流程的 IPO 图	507
图 7.7	知识管理流程的 IPO 图	517
图 8.1	剪裁要求风险和流程之间平衡	527
图 8.2	剪裁流程的 IPO 图	529
图 8.3	产品线视角	553
图 8.4	产品线中的资本化和复用	553
图 8.5	产品线投资回报	557
图 8.6	服务系统概念框架	565
图 8.7	组织管理资源以创建复杂组织体价值	573
图 8.8	个人胜任力产生组织能力、系统能力以及运营能力	577
图 8.9	整个复杂组织体背景环境中的复杂组织体 SE 流程领域	581
图 9.1	模型分类法示例	597
图 9.2	SysML 图类型	609
图 9.3	功能分析/分配流程	621
图 9.4	备选功能分解的评价和定义	623
图 9.5	OOSEM 建立在既定的 SE 基础上	631
图 9.6	系统开发流程的背景环境中的 OOSEM 活动	633
图 9.7	OOSEM 活动和建模制品	633
图 9.8	FFBD 和 N^2 图示例	645
图 9.9	IPDT 互补的综合活动的示例	657
图 9.10	精益开发原则	673
图 10.1	可承受性权衡空间的背景环境本质属性	691
图 10.2	系统运行效能	693
图 10.3	成本与性能的对比	699
图 10.4	可承受性成本分析框架	699
图 10.5	生命周期成本元素（不按比例）	707
图 10.6	实现 EMC 的流程	711
图 10.7	维持性分析	727
图 10.8	可靠性项目计划开发	741
图 10.9	一个可恢复性事件模型	749
图 10.10	功能分析系统技巧（FAST）示例图	797

LIST OF TABLES

TABLE 2.1	Important dates in the origins of SE as a discipline	34
TABLE 2.2	Important dates in the origin of SE standards	34
TABLE 2.3	Current significant SE standards and guides	36
TABLE 2.4	SE return on investment	46
TABLE 3.1	Generic life cycle stages, their purposes, and decision gate options	82
TABLE 4.1	Examples of system elements and physical interfaces	210
TABLE 5.1	Partial list of decision situations (opportunities) throughout the life cycle	344
TABLE 8.1	Standardization-related associations and automotive standards	536
TABLE 8.2	Attributes of system entities	562
TABLE 9.1	Types of IPDTs and their focus and responsibilities	652
TABLE 9.2	Pitfalls of using IPDT	664

表目录

表 2.1　SE 学科起源过程中的重要年代 ································ 35
表 2.2　SE 标准起源过程中的重要年代 ································ 35
表 2.3　当前的重要 SE 标准和指南 ··································· 37
表 2.4　SE 投资回报 ··· 47
表 3.1　一般生命周期阶段、其目的和决策门选项 ······················· 83
表 4.1　系统元素和物理接口的示例 ··································· 211
表 5.1　贯穿于生命周期内的决策情景（机会）的部分列表 ················ 345
表 8.1　与标准化有关的协会和汽车标准 ······························· 537
表 8.2　系统实体的属性 ··· 563
表 9.1　IPDT 的类型及其聚焦点和职责 ································ 653
表 9.2　使用 IPDT 的陷阱 ··· 665

1 SYSTEMS ENGINEERING HANDBOOK SCOPE

1.1 PURPOSE

This handbook defines the discipline and practice of systems engineering (SE) for students and practicing professionals alike and provides an authoritative reference to understand the SE discipline in terms of content and practice.

1.2 APPLICATION

This handbook is consistent with ISO/IEC/IEEE 15288:2015, *Systems and software engineering—System life cycle processes* (hereafter referred to as ISO/IEC/ IEEE 15288), to ensure its usefulness across a wide range of application domains—man-made systems and products, as well as business and services.

ISO/IEC/IEEE 15288 is an international standard that provides generic top-level process descriptions and requirements, whereas this handbook further elaborates on the practices and activities necessary to execute the processes. Before applying this handbook in a given organization or project, it is recommended that the tailoring guidelines in Chapter 8 be used to remove conflicts with existing policies, procedures, and standards already in use within an organization. Processes and activities in this handbook do not supersede any international, national, or local laws or regulations.

This handbook is also consistent with the *Guide to the Systems Engineering Body of Knowledge* (SEBoK, 2014) (hereafter referred to as the SEBoK) to the extent practicable. In many places, this handbook points readers to the SEBoK for more detailed coverage of the related topics, including a current and vetted set of references.

For organizations that do not follow the principles of ISO/IEC/IEEE 15288 or the SEBoK to specify their life cycle processes (including much of commercial industry), this handbook can serve as a reference to practices and methods that have proven beneficial to the SE community at large and that can add significant value in new domains, if appropriately selected and applied. Section 8.2 provides top-level guidance on the application of SE in selected product sectors and domains.

INCOSE Systems Engineering Handbook: A Guide for System Life Cycle Processes and Activities, Fourth Edition. Edited by David D. Walden, Garry J. Roedler, Kevin J. Forsberg, R. Douglas Hamelin and Thomas M. Shortell.

© 2015 John Wiley & Sons, Inc. Published 2015 by John Wiley & Sons, Inc.

1 系统工程手册的范围

1.1 目的

本手册为在职专业人士和学生等定义系统工程（SE）的学科和实践，并在以实践和内容来理解 SE 学科方面提供权威性的参考。

1.2 适用性

本手册与 ISO/IEC/IEEE 15288:2015《系统和软件工程——系统生命周期流程》（以下简称为 ISO/IEC/IEEE 15288）一致，以确保其在宽泛的应用领域的有效性——如人造系统和产品以及业务和服务。

ISO/IEC/IEEE 15288 是一个提供通用的顶层流程描述和需求的国际标准，而本手册更进一步地详细阐明执行该流程所需的实践和活动。在特定组织或项目中应用本手册之前，建议采用第 8 章中的剪裁指南来排除与已在组织内使用的现有方针、程序和标准的冲突。本手册中的流程和活动不取代任何国际、国家或地方法律、法规。

本手册在可行的范围内还与《系统工程知识体指南》（SEBoK，2014）（以下简称为 SEBoK）一致。在很多地方，本手册为读者揭示了 SEBoK 中相关主题更详细的适用范围，包括当前的和已审查的参考文献集合。

对于不按照 ISO/IEC/IEEE 15288 或 SEBoK 的原则规定其生命周期流程的组织（包括许多商用行业），可将本手册作为实践和方法的参考，所述实践和方法已被证实对广大的 SE 领域有益，若经适当选择和应用，亦可在新的领域增加重大价值。8.2 节提供关于在所选定的产品行业和领域中应用 SE 的高层级指南。

INCOSE 系统工程手册：系统生命周期流程和活动指南，第 4 版。编撰：David D. Walden, Garry J. Roedler, Kevin J. Forsberg, R. Douglas Hamelin 和 Thomas M. Shortell。

John Wiley & Sons 公司版权所有©2015。由 John Wiley & Sons 公司于 2015 年出版。

SYSTEMS ENGINEERING HANDBOOK SCOPE

1.3 CONTENTS

This chapter defines the purpose and scope of this handbook. Chapter 2 provides an overview of the goals and value of using SE throughout the system life cycle.

Chapter 3 describes an informative life cycle model with six stages: concept, development, production, utilization, support, and retirement.

ISO/IEC/IEEE 15288 identifies four process groups to support SE. Each of these process groups is the subject of an individual chapter. A graphical overview of these processes is given in Figure 1.1:

- *Technical processes* (Chapter 4) include business or mission analysis, stakeholder needs and requirements definition, system requirements definition, architecture definition, design definition, system analysis, implementation, integration, verification, transition, validation, operation, maintenance, and disposal.

- *Technical management processes* (Chapter 5) include project planning, project assessment and control, decision management, risk management, configuration management, information management, measurement, and quality assurance.

- *Agreement processes* (Chapter 6) include acquisition and supply.

- *Organizational project-enabling processes* (Chapter 7) include life cycle model management, infrastructure management, portfolio management, human resource management, quality management, and knowledge management.

This handbook provides additional chapters beyond the process groups listed in Figure 1.1:

- *Tailoring processes and application of systems engineering* (Chapter 8) include information on how to adapt and scale the SE processes and how to apply those processes in various applications. NOT every process will apply universally. Careful selection from the material is recommended. Reliance on process over progress will not deliver a system.

- *Crosscutting systems engineering methods* (Chapter 9) provide insights into methods that can apply across all processes, reflecting various aspects of the iterative and recursive nature of SE.

1.3 内容

本章定义本手册的目的和范围。第 2 章提供贯穿于系统生命周期内使用 SE 目标和价值的概览。

第 3 章按照六个阶段描述翔实的生命周期模型：概念（方案）、开发、生产、使用、保障和退出。

ISO/IEC/IEEE 15288 识别支持 SE 的四个流程组。每个流程组是一个独立章节的主题。图 1.1 给出这些流程的图示化概览：

- *技术流程*（第 4 章）包括业务或任务分析、利益攸关者需要和需求定义、系统需求定义、架构定义、设计定义、系统分析、实施、综合、验证、转移、确认、运行、维护和退出。

- *技术管理流程*（第 5 章）包括项目规划、项目评估和控制、决策管理、风险管理、构型管理、信息管理、测量和质量保证。

- *协议流程*（第 6 章）包括采办和供应。

- *组织的项目—使能流程*（第 7 章）包括生命周期模型管理、基础设施管理、项目组合管理、人力资源管理、质量管理和知识管理。

本手册还提供其他超出图 1.1 所列流程组的章节：

- *系统工程剪裁流程和应用*（第 8 章）包括如何适配并扩展 SE 流程以及如何在各种不同的应用中运用这些流程的信息，并不是每个流程都普遍适用，建议从资料中进行仔细选择，在项目进展中只依赖流程并不能交付系统。

- *跨领域/学科系统工程方法*（第 9 章）提供对跨越所有流程适用的方法的深入理解，并反映 SE 的迭代和递归本质属性的多个方面（视角）。

- *Specialty engineering activities* (Chapter 10) include practical information so systems engineers can understand and appreciate the importance of various specialty engineering topics.

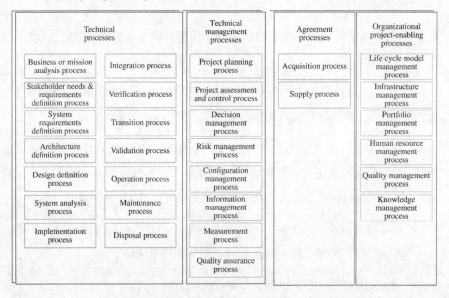

FIGURE 1.1 System life cycle processes per ISO/IEC/IEEE 15288. This figure is excerpted from ISO/IEC/IEEE 15288:2015, Figure 4 on page 17, with permission from the ANSI on behalf of the ISO. © ISO 2015. All rights reserved.

Appendix A contains a list of references used in this handbook. Appendices B and C provide a list of acronyms and a glossary of SE terms and definitions, respectively. Appendix D provides an N^2 diagram of the SE processes showing where dependencies exist in the form of shared inputs or outputs. Appendix E provides a master list of all inputs/outputs identified for each SE process. Appendix F acknowledges the various contributors to this handbook. Errors, omissions, and other suggestions for this handbook can be submitted to the INCOSE using the comment form contained in appendix G.

1.4 FORMAT

A common format has been applied in Chapters 4 through 7 to describe the system life cycle processes found in ISO/IEC/IEEE 15288. Each process is illustrated by an input–process–output (IPO) diagram showing key inputs, process activities, and resulting outputs.

- **专业工程活动**（第 10 章）包括实践的信息，目的是使系统工程师能够理解并领会各种不同专业工程主题的重要性。

技术流程		技术管理流程	协议流程	组织的项目使能流程
业务或任务分析流程	综合流程	项目规划流程	采办流程	生命周期模型管理流程
利益攸关者需要和需求定义流程	验证流程	项目评估和控制流程	供应流程	基础设施管理流程
系统需求定义流程	转移流程	决策管理流程		项目组合管理流程
架构定义流程	确认流程	风险管理流程		人力资源管理流程
设计定义流程	运行流程	构型管理流程		质量管理流程
系统分析流程	维护流程	信息管理流程		知识管理流程
实施流程	退出流程	测量流程		
		质量保证流程		

图 1.1 符合 ISO/IEC/IEEE 15288 标准的系统生命周期流程。经代表 ISO 的 ANSI 许可后，此图摘自 ISO/IEC/IEEE 15288:2015，第 17 页图 4。ISO 版权所有，©2015。版权所有。

附录 A 包含本手册中使用的参考文献列表。附录 B 和 C 分别提供缩写词列表及 SE 术语和定义的词汇。附录 D 提供 SE 流程的 N^2 图，以共享输入或输出的形式，表明 SE 诸多流程之间存在的依赖关系。附录 E 提供针对每个 SE 流程所识别的所有输入/输出的总览表。附录 F 感谢对本手册做出贡献的人们。对于本手册中出现的错误、遗漏和其他建议，请填写在附录 G 给出的意见表中，并提交至 INCOSE。

1.4 格式

第 4 章至第 7 章应用公共的格式来描述 ISO/IEC/IEEE 15288 中给出的系统生命周期流程。每个流程通过输入—流程—输出（IPO）图进行详细阐述，指明关键输入、流程活动和产生的输出。

SYSTEMS ENGINEERING HANDBOOK SCOPE

A sample is shown in Figure 1.2. Note that the IPO diagrams throughout this handbook represent "a" way that the SE processes can be performed, but not necessarily "the" way that they must be performed. The issue is that SE processes produce "results" that are often captured in "documents" rather than producing "documents" simply because they are identified as outputs. To under- stand a given process, readers are encouraged to study the complete information provided in the combination of diagrams and text and not rely solely on the diagrams.

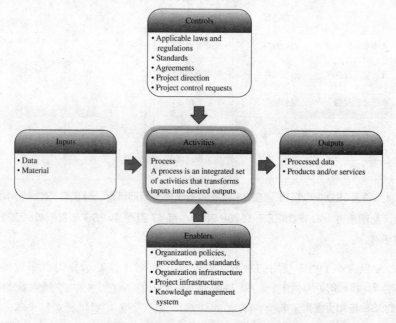

FIGURE 1.2 Sample of IPO diagram for SE processes. INCOSE SEH original figure created by Shortell and Walden. Usage per the INCOSE notices page. All other rights reserved.

The following heading structure provides consistency in the discussion of these processes:

- Process overview
- Purpose
- Description
- Inputs/outputs

图 1.2 给出一个示例。要注意的是，贯穿于本手册的 IPO 图表示 SE 流程能够被执行的"某种"方式，但不一定是 SE 流程必须被执行的"特定"方式。问题是，SE 流程产物通常是在"文件"中捕获到的"结果"，而不是产生"文件"，这仅仅是因为它们被识别为输出。为理解给定的流程，鼓励读者学习以图表和文字结合的方式提供完整的信息，而不只是依赖于图表。

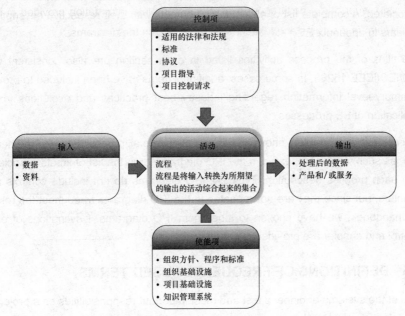

图 1.2 SE 流程的 IPO 图示例。INCOSE SEH 原始图由 Shortell 和 Walden 创建。按照 INCOSE 告知页使用。版权所有。

在论述这些流程中，将采用与如下标题一致的结构：

- 流程概览
- 目的
- 描述
- 输入/输出

- Process activities
- Process elaboration

To ensure consistency with ISO/IEC/IEEE 15288, the purpose statements from the standard are included verbatim for each process described herein. Inputs and outputs are listed by name within the respective IPO diagrams with which they are associated. A complete list of all inputs and outputs with their respective descriptions appears in appendix E.

The titles of the process activities listed in each section are also consistent with ISO/IEC/IEEE 15288. In some cases, additional items have been included to provide summary-level information regarding industry best practices and evolutions in the application of SE processes.

The controls and enablers shown in Figure 1.2 govern all processes described herein and, as such, are not repeated in the IPO diagrams or in the list of inputs associated with each process description. Typically, IPO diagrams do not include controls and enablers, but since they are not repeated in the IPO diagrams throughout the rest of the handbook, we have chosen to label them IPO diagrams. Descriptions of each control and enabler are provided in appendix E.

1.5 DEFINITIONS OF FREQUENTLY USED TERMS

One of the systems engineer's first and most important responsibilities on a project is to establish nomenclature and terminology that support clear, unambiguous communication and definition of the system and its elements, functions, operations, and associated processes. Further, to promote the advancement of the field of SE throughout the world, it is essential that common definitions and understandings be established regarding general methods and terminology that in turn support common processes. As more systems engineers accept and use common terminology, SE will experience improvements in communications, understanding, and, ultimately, productivity.

The glossary of terms used throughout this book (see appendix C) is based on the definitions found in ISO/ IEC/IEEE 15288; ISO/IEC/IEEE 24765, *Systems and Software Engineering—Vocabulary* (2010); and SE VOCAB（2013）.

- 流程活动

- 流程详细阐述

为确保与 ISO/IEC/IEEE 15288 的一致性，将标准的目的申明逐一地列入本文所描述的每个流程中。在与其相关联对应的 IPO 图中，按照名称列出输入和输出。附录 E 给出所有输入和输出的完整列表，并具有对应的描述。

每一节列出的流程活动的标题亦与 ISO/IEC/IEEE 15288 一致。在某些情况下，包括附加项，用以提供关于 SE 流程应用的行业最佳实践及演进方面的综述层级的信息。

图 1.2 表明的控制项和使能项支配着本文所描述的所有流程，正因如此，在 IPO 图或与每个流程描述相关联的输入列表中，它们不再被重复。典型地，IPO 图并不包括控制项和使能项，但由于在贯穿于本手册其余部分的 IPO 图中它们不再被重复，因此我们已选择将它们标识到 IPO 图中。附录 E 提供每个控制项和使能项的描述。

1.5 常用术语的定义

系统工程师首要的也是最重要的项目职责之一就是建立名称及术语，这些名称及术语支撑系统及其元素、功能、运行和相关流程进行清晰的、明确的沟通和定义。此外，为促进 SE 领域在全世界范围内的发展，建立关于支持常用流程的一般方法和术语的常用定义和理解是根本性的基线。随着越来越多的系统工程师们接受并使用常用术语，SE 将会体验沟通、理解和最终的生产率的全面提升。

贯穿于本手册内使用的术语词汇（参见附录 C）以 ISO/ IEC/IEEE 15288；ISO/IEC/IEEE 24765，《系统和软件工程——词汇》（2010）；以及 SE VOCAB（2013）中提供的定义为基础。

2 SYSTEMS ENGINEERING OVERVIEW

2.1 INTRODUCTION

This chapter offers a brief overview of the systems engineering (SE) discipline, beginning with a few key definitions, an abbreviated survey of the origins of the discipline, and discussions on the value of applying SE. Other concepts, such as systems science, systems thinking, SE leadership, SE ethics, and professional development, are also introduced.

2.2 DEFINITIONS AND CONCEPTS OF A SYSTEM

While the concepts of a *system* can generally be traced back to early Western philosophy and later to science, the concept most familiar to systems engineers is often traced to Ludwig von Bertalanffy (1950, 1968) in which a system is regarded as a "whole" consisting of interacting "parts." The ISO/IEC/IEEE definitions provided in this handbook draw from this concept.

2.2.1 General System Concepts

The systems considered in ISO/IEC/IEEE 15288 and in this handbook

> [5.2.1] ... are man-made, created and utilized to provide products or services in defined environments for the benefit of users and other stakeholders.

The definitions cited here and in Appendix C refer to systems in the real world. A system concept should be regarded as a shared "mental representation" of the actual system. The systems engineer must continually distinguish between systems in the real world and system representations. The INCOSE and ISO/IEC/IEEE definitions draw from this view of a system:

> ... an integrated set of elements, subsystems, or assemblies that accomplish a defined objective. These elements include products (hardware, software, firmware), processes, people, information, techniques, facilities, services, and other support elements. (INCOSE)

INCOSE Systems Engineering Handbook: A Guide for System Life Cycle Processes and Activities, Fourth Edition. Edited by David D. Walden, Garry J. Roedler, Kevin J. Forsberg, R. Douglas Hamelin and Thomas M. Shortell.

© 2015 John Wiley & Sons, Inc. Published 2015 by John Wiley & Sons, Inc.

2 系统工程概览

2.1 引言

本章从一些关键定义开始，提供系统工程（SE）学科的简要概览，概要综述该学科的起源，并论述应用 SE 的价值。还引入了其他概念，如系统科学、系统思考、SE 领导力、SE 道德及职业发展。

2.2 系统的定义和概念

尽管系统的概念一般可追溯到早期的西方哲学以及后来的科学，但系统工程师最熟悉的概念往往要追溯到 Ludwig von Bertalanffy（1950, 1968），他将系统看做由相互作用的"多个部分"组成的一个"整体"。本手册中提供的 ISO/IEC/IEEE 定义就是从这个概念中得出的。

2.2.1 一般系统概念

ISO/IEC/IEEE 15288 和本手册中所考虑的系统

> [5.2.1]…是人造的，被创造并使用于定义明确的环境中提供产品或服务，使用户及其他利益攸关者受益。

此处以及附录 C 中引用的定义指的是真实世界中的系统。系统概念应当被看做实际系统所共享的"思维表达"。系统工程师必须不断地区分真实世界中的系统与系统表达。INCOSE 和 ISO/IEC/IEEE 是从系统的视角来提取定义的：

> …一组综合的元素、子系统或组件，以完成一个定义明确的目标。这些元素包括产品（硬件、软件和固件）、流程、人员、信息、技术、设施、服务和其他支持元素。（INCOSE）

INCOSE 系统工程手册：系统生命周期流程和活动指南，第 4 版。编撰：David D. Walden, Garry J. Roedler, Kevin J. Forsberg, R. Douglas Hamelin 和 Thomas M. Shortell。

John Wiley & Sons 公司版权所有©2015。由 John Wiley & Sons 公司于 2015 年出版。

SYSTEMS ENGINEERING OVERVIEW

[4.1.46] ... combination of interacting elements organized to achieve one or more stated purposes. (ISO/IEC/ IEEE 15288)

Thus, the usage of terminology throughout this handbook is clearly an elaboration of the fundamental idea that a system is a purposeful whole that consists of inter- acting parts.

An external view of a system must introduce elements that specifically do not belong to the system but do interact with the system. This collection of elements is called the *operating environment or context* and can include the users (or operators) of the system.

The internal and external views of a system give rise to the concept of a *system boundary*. In practice, the system boundary is a "line of demarcation" between the system itself and its greater context (to include the operating environment). It defines what belongs to the system and what does not. The system boundary is not to be confused with the subset of elements that interact with the environment.

The *functionality* of a system is typically expressed in terms of the interactions of the system with its operating environment, especially the users. When a system is considered as an integrated combination of interacting elements, the functionality of the system derives not just from the interactions of individual elements with the environmental elements but also from how these interactions are influenced by the organization (interrelations) of the system elements. This leads to the concept of *system architecture*, which ISO/IEC/IEEE 42010 (2011) defines as

> the fundamental concepts or properties of a system in its environment embodied in its elements, relationships, and in the principles of its design and evolution.

This definition speaks to both the internal and external views of the system and shares the concepts from the definitions of a system.

2.2.2 Scientific Terminology Related to System Concepts

In general, *engineering* can be regarded as the practice of creating and sustaining services, systems, devices, machines, structures, processes, and products to improve the quality of life—getting things done effectively and efficiently. The repeatability of experiments demanded by science is critical for delivering practical engineering solutions that have commercial value. Engineering in general and SE in particular draw heavily from the terminology and concepts of science.

[4.1.46]…交互作用的元素组织起来的组合,以实现一个或多个特定的目的。(ISO/IEC/IEEE 15288)

因此,贯穿于本手册内的术语的使用显然是对基础理念——系统是由交互作用部分所组成的有目的的整体——的详细阐述。

系统的外部视角必须引入一些不专门属于该系统但与系统交互作用的元素。这些元素的这种集合被称为运行环境或背景环境,且可包括系统的用户(或操作者)。

系统的内部视角和外部视角产生系统边界的概念。实际上,系统边界是系统本身与其更大的背景环境(包括运行环境)之间的"分界线"。它定义了哪些属于系统,哪些不属于系统。不得将系统边界和与环境交互作用的元素的子集合相混淆。

系统的功能性通常按照系统与其运行环境,特别是与用户的交互作用来表达。当系统被视为交互元素的综合组合时,系统的功能性不仅来源于单独元素与环境元素的交互,而且还来源于这些交互如何被系统元素的组织(相互关系)所影响。这导致产生系统架构的概念,ISO/IEC/IEEE 42010(2011)将这些概念定义为

> 系统元素、关系及系统设计和演进原理中所体现的系统环境中的系统的基础概念或特性。

该定义提及了系统的内部视图和外部视图,并共享来源于系统定义的概念。

2.2.2 与系统概念相关的科学术语

一般而言,工程可被看做创造和维持服务、系统、装置、机器、结构、流程和产品的实践,以便改善生活质量——有效地且高效地做好事情。科学所要求的实验重复性对于交付具有商业价值的实际工程解决方法而言是至关重要的。一般的工程和特别是 SE 在很大程度上取自科学的术语和概念。

An *attribute* of a system (or system element) is an observable characteristic or property of the system (or system element). For example, among the various attributes of an aircraft is its air speed. Attributes are represented symbolically by variables. Specifically, a *variable* is a symbol or name that identifies an attribute. Every variable has a domain, which could be but is not necessarily measurable. A *measurement* is the outcome of a process in which the system of interest (SOI) interacts with an observation system under specified conditions. The outcome of a measurement is the assignment of a *value* to a variable. A system is in a *state* when the values assigned to its attributes remain constant or steady for a meaningful period of time (Kaposi and Myers, 2001). In SE and software engineering, the *system elements* (e.g., software objects) have *processes* (e.g., operations) in addition to attributes. These have the binary logical values of being either *idle* or *executing*. A complete description of a system state therefore requires values to be assigned to both attributes and processes. *Dynamic behavior* of a system is the time evolution of the system state. *Emergent behavior* is a behavior of the system that cannot be understood exclusively in terms of the behavior of the individual system elements.

The key concept used for problem solving is the *black box/white box* system representation. The black box representation is based on an external view of the system (attributes). The white box representation is based on an internal view of the system (attributes and structure of the elements). There must also be an understanding of the relationship between the two. A system, then, is represented by the (external) attributes of the system, its internal attributes and structure, and the interrelationships between these that are governed by the laws of science.

2.2.3 General Systems Methodologies

Early pioneers of SE and software engineering, such as Yourdon (1989) and Wymore (1993), long sought to bring discipline and precision to the understanding and management of the dynamic behavior of a system by seeking relations between the external and internal representations of the system. Simply stated, they believed that if the flow of dynamic behavior (the system state evolution) could be mapped coherently into the flow of states of the constituent elements of the system, then emergent behaviors could be better understood and managed.

Klir (1991) complemented the concepts of a system in engineering and science with a general systems methodology. He regarded problem solving in general to rest upon a principle of alternatively using abstraction and interpretation to solve a problem. He considered that his methodology could be used both for system inquiry (i.e., the representation of an aspect of reality) and for system definition (i.e., the representation of purposeful man-made objects).

系统（或系统元素）的属性是系统（或系统元素）的可观察到的特征或特性。例如，在飞机的各种不同属性中有飞机的空速。属性用变量符号化来表示。特定情况下，变量是识别属性的符号或名称。每一个变量都具有一个域，可能但未必是可测量的。测量是在特定情况下所感兴趣之系统（SOI）与观察系统交互的流程的结果。测量的结果是对变量赋值。当赋予系统属性的值在有意义的时间段内保持不变或稳定时，系统处于一种状态（Kaposi 和 Myers, 2001）。在 SE 和软件工程中，系统元素（如软件对象）除了属性以外还具有流程（如运行）。这些流程具有空闲或执行的二进制逻辑值。因此，系统状态的完整描述要求对属性和流程赋值。系统的动态行为是系统状态的时间演进。涌现行为是不能依据单个系统元素的行为来孤立理解的系统行为。

问题求解所用的关键概念是黑盒/白盒的系统表征形式。黑盒表征形式基于系统的外部视图（属性）。白盒表征形式基于系统的内部视图（元素的属性和结构）。还必须理解两者之间的关系。那么系统由系统的（外部）属性、系统的内部属性和结构以及其之间受科学法则支配的相互关系来表示。

2.2.3 一般系统方法论

早期 SE 和软件工程的先行者，如 Yourdon（1989）和 Wymore（1993）长期以来通过寻找系统的外部表达形式和内部表达形式之间的关系，寻求将科律和准确性带入到对系统动态行为的理解和管理中来。简单地说，他们认为如果动态行为流（系统状态的演进）可以被一致地映射到系统构成元素的状态流中，那么涌现行为可以得到更好的理解和管理。

Klir（1991）采用一般系统方法论在工程和科学中补充了系统的概念。他认为一般情况下的问题求解所依赖的原则是使用抽象或解释的方式来求解问题。他认为自己的方法论可以用于系统询问探究（即，一个现实方面的表达形式）和系统定义（即，有目的的人造对象的表达形式）。

2.3 THE HIERARCHY *WITHIN* A SYSTEM

In the ISO/IEC/IEEE usage of terminology, the *system elements* can be *atomic* (i.e., not further decomposed), or they can be *systems on their own merit* (i.e., decomposed into further subordinate system elements). The *integration* of the system elements must establish the relationship between the effects that *organizing* the elements has on their *interactions* and how these effects enable the system to achieve its *purpose*.

One of the challenges of system definition is to understand what level of detail is necessary to define each system element and the interrelations between elements. Because the SOIs are in the real world, this means that the response to this challenge will be domain specific. A system element that needs only a black box representation (external view) to capture its requirements and confidently specify its real-world solution definition can be regarded as atomic. Decisions to make, buy, or reuse the element can be made with confidence without further specification of the element. This leads to the concept of hierarchy within a system.

One approach to defining the elements of a system and their interrelations is to identify a complete set of distinct system elements with regard only to their relation to the whole (system) by suppressing details of their interactions and interrelations. This is referred to as a partitioning of the system. Each element can be either atomic or it can be a much higher level that could be viewed as a system itself. At any given level, the elements are grouped into distinct subsets of elements subordinated to a higher level system, as illustrated in Figure 2.1. Thus, hierarchy within a system is an organizational representation of system structure using a partitioning relation.

The concept of a system hierarchy described in ISO/ IEC/IEEE 15288 is as follows:

> [5.2.2] The system life cycle processes ... are described in relation to a system that is composed of a set of interacting system elements, each of which can be implemented to fulfill its respective specified requirements.

2.3 一个系统内部的层级结构

在 ISO/IEC/IEEE 的术语使用中，系统元素可以是原子的（即，不再被进一步分解）或者它们可以凭借自己的特质是系统（即，被进一步分解成次级系统元素）。系统元素的综合必须建立元素的组织方式对其交互作用产生的影响与这些影响如何促使系统实现其目的之间的关系。

系统定义的挑战之一是理解什么样的详细层级对于定义每一个系统元素以及元素之间的关系而言是必要的。由于 SOI 处于现实世界中，这意味着对这种挑战的回应将是领域特定的。仅需要黑盒表达形式（外部视图）捕获其需求，并有信心地指定其现实解决方案定义的系统元素可被认为是原子的。可以有信心地做出制造、购买或复用元素的决策而无须进一步的元素规范。这导出了系统内部的层级结构的概念。

定义系统元素及其相互关系的一种途径是通过抑制这些元素的交互作用及相互关系的细节来识别仅仅这些元素与整体（系统）之间的关系有关的一整套截然不同的系统元素。这种方法称做系统的划分。每个元素可以是原子的或者可以是被视为系统本身的更高层级。在任意给定层级上，元素被分组成为从属于更高层级系统的截然不同的元素子集，如图 2.1 所示。因此，一个系统内部的层级结构是使用划分关系对系统结构的组织化表达形式。

ISO/IEC/IEEE 15288 中所描述的系统层级结构的概念如下：

> [5.2.2] 描述了与系统有关的系统生命周期流程…，该系统由交互作用的系统元素集组成，每个系统元素可被实现以满足其各自的特定要求。

SYSTEMS ENGINEERING OVERVIEW

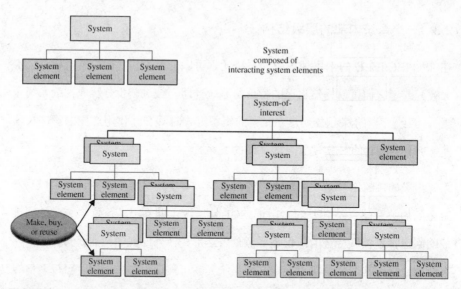

FIGURE 2.1 Hierarchy within a system. This figure is adapted from ISO/IEC/IEEE 15288:2015, Figure 1 on page 11 and Figure 2 on page 12, with permission from the ANSI on behalf of the ISO. © ISO 2015. All rights reserved.

The art of defining a hierarchy within a system relies on the ability of the systems engineer to strike a balance between clearly and simply defining span of control and resolving the structure of the SOI into a complete set of system elements that can be implemented with confidence. Urwick (1956) suggests that a possible heuristic is for each level in the hierarchy to have no more than 7 ± 2 elements subordinate to it. Others have also found this heuristic to be useful (Miller, 1956). A level of design with too few subordinate elements is unlikely to have a distinct design activity. In this case, both design and verification activities may contain redundancy. In practice, the nomenclature and depth of the hierarchy can and should be adjusted to fit the complexity of the system and the community of interest.

2.4 DEFINITION OF SYSTEMS OF SYSTEMS

A "system of systems" (SoS) is an SOI whose elements are managerially and/or operationally independent systems. These interoperating and/or integrated collections of constituent systems usually produce results unachievable by the individual systems alone. Because an SoS is itself a system, the systems engineer may choose whether to address it as either a system or as an SoS, depending on which perspective is better suited to a particular problem.

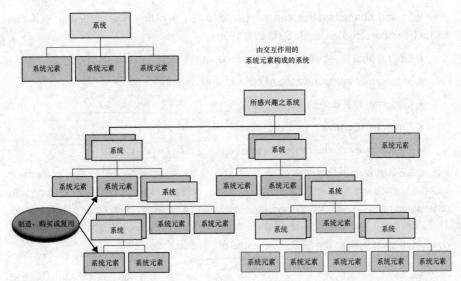

图 2.1 一个系统内的层级结构。经代表 ISO 的 ANSI 许可后，此图改编自 ISO/IEC/IEEE 15288:2015，第 11 页图 1 和第 12 页图 2。ISO 版权所有，©2015。版权所有。

定义一个系统内的层级结构的艺术依赖于系统工程师的能力——在清晰且简单地定义控制跨度与分解 SOI 结构为有信心实施的系统元素完整集之间达成平衡。Urwick（1956）提出，可能的启发法是对于层级结构中的每一层级，隶属于该层级的元素不超过 7±2 个。其他人也发现这种启发法是很有用的（Miller, 1956）。一个具有太少的次级元素的设计层级并非具有截然不同的设计活动。在这种情况下，设计和验证的活动可能会含有冗余。实际上，层级结构的命名和层次能够且应该被调整以适配系统的复杂性和所感兴趣的团体。

2.4 系统之系统的定义

"系统之系统"（SoS）是指其元素在管理上和/或运行上是独立系统的 SOI。这些成员系统的互操作和/或综合的集合通常产生单个系统无法单独达成的结果。由于 SoS 本身是一个系统，因此系统工程师可选择是否将其看做系统或看做 SoS，这取决于哪种视角更适合特定的问题。

SYSTEMS ENGINEERING OVERVIEW

The following characteristics can be useful when deciding if a particular SOI can better be understood as an SoS (Maier, 1998):

- Operational independence of constituent systems
- Managerial independence of constituent systems
- Geographical distribution
- Emergent behavior
- Evolutionary development processes

Figure 2.2 illustrates the concept of an SoS. The air transport system is an SoS comprising multiple aircraft, airports, air traffic control systems, and ticketing systems, which along with other systems such as security and financial systems facilitate passenger transportation. There are equivalent ground and maritime transportation SoS that are all in turn part of the overall transport system (an SoS in the terms of this description).

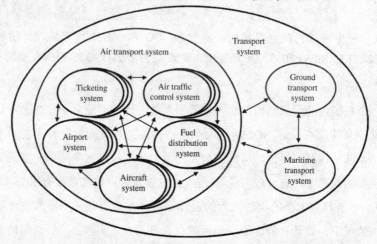

FIGURE 2.2 Example of the systems and systems of systems within a transport system of systems. Reprinted with permission from Judith Dahmann. All other rights reserved.

The SoS usually exhibits complex behaviors, often created by the existence of the aforementioned Maier's characteristics. "Complexity" is essentially different from "complicated". In complicated systems, such as an automobile, the interactions between the many parts are governed by fixed relationships. This allows reasonably reliable prediction of technical, time, and cost issues. In complex systems, such as the air transport system, interactions between the parts complicatedexhibit self-organization, where local interactions give rise to novel, nonlocal, emergent patterns. Complicated systems can often become complex when the behaviors change, but even systems of very few parts can sometimes exhibit surprising complexity.

当判定特定的 SOI 是否能够被更好地理解为 SoS 时，以下特性可能有用（Maier, 1998）：

- 构成诸系统的运行独立性
- 构成诸系统的管理独立性
- 地理分布
- 涌现行为
- 进化式开发流程

图 2.2 详细阐明 SoS 的概念。航空运输系统是一种包括多重的飞机、机场、空中交通管制系统和票务系统的 SoS，它连同例如安保与财务系统等其他系统一起促进旅客的交通运输。存在等效的地面与海上运输 SoS，它们反过来都是总体运输系统的一部分（按此描述的 SoS）。

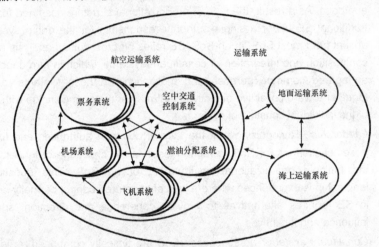

图 2.2　运输系统之系统内的系统及系统之系统的示例。经 Judith Dahmann 许可后转载。版权所有。

SoS 通常呈现出复杂的行为，往往因上述 Maier 的特性的存在而产生。"复杂性"基本上不同于"繁杂的"。在繁杂的系统中，例如汽车，多个部分之间的交互取决于固定关系。这允许对技术问题、时间问题和成本问题进行合理可靠的预测。在复杂系统中，例如航空运输系统，各个部分之间的交互作用呈现出自组织，该自组织中的局部交互作用产生新型模式、非局部模式和涌现模式。繁杂系统在行为发生变化时往往变得复杂，但甚至由很少部分构成的系统有时也可能呈现出惊人的复杂性。

SYSTEMS ENGINEERING OVERVIEW

The best way to understand a complicated system is to break it down into parts recursively until the parts are so simple that we understand them and then to reassemble the parts to understand the whole. However, this approach does not help us to understand a complex system, because the emergent properties that we really care about disappear when we examine the parts in isolation. A fundamentally different approach is required to understand the whole in context through iterative exploration and adaptation. As a result, SE requires a balance of linear, procedural methods for sorting through complicatedness ("systematic activity") and holistic, nonlinear, iterative methods for harnessing complexity ("systemic" or systems thinking and analysis—always required when dealing with SoS). The tension between breaking things apart and keeping them in context must be dynamically managed throughout the SE process.

The following challenges all influence the engineering of an SoS (Dahmann, 2014):

1. *SoS authorities*—In an SoS, each constituent system has its own local "owner" with its stakeholders, users, business processes, and development approach. As a result, the type of organizational structure assumed for most traditional SE under a single authority responsible for the entire system is absent from most SoS. In an SoS, SE relies on crosscutting analysis and on composition and integration of constituent systems, which in turn depend on an agreed common purpose and motivation for these systems to work together toward collective objectives that may or may not coincide with those of the individual constituent systems.

2. *Leadership*—Recognizing that the lack of common authorities and funding poses challenges for SoS, a related issue is the challenge of leadership in the multiple organizational environment of an SoS. This question of leadership is experienced where a lack of structured control normally present in SE requires alternatives to provide coherence and direction, such as influence and incentives.

3. *Constituent systems' perspectives*—SoS are typically composed, at least in part, of in-service systems, which were often developed for other purposes and are now being leveraged to meet a new or different application with new objectives. This is the basis for a major issue facing SoS SE, that is, how to technically address issues that arise from the fact that the systems identified for the SoS may be limited in the degree to which they can support the SoS. These limitations may affect initial efforts at incorporating a system into an SoS, and systems' commitments to other users may mean that they may not be compatible with the SoS over time. Further, because the systems were developed and operate in different situations, there is a risk that there could be a mismatch in understanding the services or data provided by one system to the SoS if the particular system's context differs from that of the SoS.

理解繁杂系统的最佳方式是将其递归地分解成多个部分，直到每个部分简单到我们可以理解为止，然后将各个部分重新组装以便理解整体。然而，这种方法并不能帮助我们理解复杂系统，因为我们真正关心的涌现的特性在我们孤立地检查各个部分时就消失了。需要一种根本上不同的方法以通过迭代式探索和适应以在背景环境中理解整体。结果是，SE 要求在整理繁杂（"系统的活动"）的线性、程序性方法与利用复杂性（"系统的"或系统思考与分析——在应对 SoS 时始终要求）的整体性、非线性、迭代式方法之间取得平衡。"将事物拆分"与"保持其处于背景环境中"之间的紧张关系必须在贯穿于 SE 流程内进行动态的管理。

以下挑战都影响 SoS 的工程（Dahmann, 2014）：

1. *SoS 权限*——在 SoS 中，每个成员系统都具有其自己的局部"所有者"，该"所有者"具有它的利益攸关者、用户、业务流程和开发方法。结果是，在对整个系统负责的单一权限下，针对大多数传统 SE 所假定的组织结构类型，在大多数 SoS 中并不存在。在 SoS 中，SE 依赖跨界分析以及对成员系统的构成和综合，它们反过来取决于协定一致的共同目的和动机，以便这些系统朝着与单独成员系统的目标可能一致或可能不一致的共同目标合作。

2. *领导力*——要意识到缺乏共同的权限和资金对 SoS 提出的挑战，一个相关的问题是在 SoS 的多重组织环境中对领导力的挑战。经历这种领导力问题，通常在于 SE 中对结构化控制的缺乏需要一些备选方案以提供和谐性和方向，如影响力和激励。

3. *成员系统的视角*——SoS 通常至少部分地由正在服役的系统构成，这些在役的系统往往是被开发用于其他目的，并且现在要被充分利用，以便满足具有新目标的新应用或不同的应用。这是面对 SoS SE 的主要问题的基础，即，如何在技术上应对由"针对为 SoS 所识别的系统可能在它们能够支持 SoS 的程度上受限制"这一事实引起的问题。这些局限性可能影响将系统纳入 SoS 中的初始努力，这些系统对其他用户的承诺可能意味着随时间推移可能与 SoS 不兼容。此外，由于系统是过去开发的且在不同情形下运行，因此存在一种风险：如果某特殊系统的背景环境不同于 SoS 的背景环境，则在理解一个系统对 SoS 所提供的服务或数据方面可能存在不匹配。

SYSTEMS ENGINEERING OVERVIEW

4. *Capabilities and requirements*—Traditionally (and ideally), the SE process begins with a clear, complete set of user requirements and provides a disciplined approach to develop a system to meet these requirements. Typically, SoS are comprised of multiple independent systems with their own requirements, working toward broader capability objectives. In the best case, the SoS capability needs are met by the constituent systems as they meet their own local requirements. However, in many cases, the SoS needs may not be consistent with the requirements for the constituent systems. In these cases, SoS SE needs to identify alternative approaches to meeting those needs either through changes to the constituent systems or through the addition of other systems to the SoS. In effect, this is asking the systems to take on new requirements with the SoS acting as the "user."

5. *Autonomy, interdependencies, and emergence*— The independence of constituent systems in an SoS is the source of a number of technical issues facing SE of SoS. The fact that a constituent system may continue to change independently of the SoS, along with interdependencies between that constituent system and other constituent systems, adds to the complexity of the SoS and further challenges SE at the SoS level. In particular, these dynamics can lead to unanticipated effects at the SoS level leading to unexpected or unpredictable behavior in an SoS even if the behavior of the constituent systems is well understood.

6. *Testing, validation, and learning*—The fact that SoS are typically composed of constituent systems that are independent of the SoS poses challenges in conducting end-to-end SoS testing, as is typically done with systems. First, unless there is a clear understanding of the SoS-level expectations and measures of those expectations, it can be very difficult to assess the level of performance as the basis for determining areas that need attention or to ensure users of the capabilities and limitations of the SoS. Even when there is a clear understanding of SoS objectives and metrics, testing in a traditional sense can be difficult. Depending on the SoS context, there may not be funding or authority for SoS testing. Often, the development cycles of the constituent systems are tied to the needs of their owners and original ongoing user base. With multiple constituent systems subject to asynchronous development cycles, finding ways to conduct traditional end-to-end testing across the SoS can be difficult if not impossible. In addition, many SoS are large and diverse, making traditional full end-to-end testing with every change in a constituent system prohibitively costly. Often, the only way to get a good measure of SoS performance is from data collected from actual operations or through estimates based on modeling, simulation, and analysis. Nonetheless, the SoS SE team needs to enable continuity of operation and performance of the SoS despite these challenges.

4. *能力和需求*——传统上（以及理想地），SE 流程开始于清晰、完整的用户需求集合并提供科律化方法开发系统以满足这些需求。典型地，SoS 由具有其各自需求的多重独立系统所组成，它们朝着更广泛的能力目标来工作。在最佳情况下，各成员系统在满足它们各自的局部需求时，也满足 SoS 的能力需要。然而，在许多情况下，SoS 的需要可能与成员系统的需求不一致。在这些情况下，SoS SE 需要识别备选途径以便通过成员系统的改变或通过在 SoS 中增加其他系统来满足那些需要。实际上，这要求系统承担新的需求，并且使 SoS 充当"用户"。

5. *自主性、相互依赖性和涌现性*——SoS 中成员系统的独立性是 SoS 的 SE 面临的许多技术问题的来源。"成员系统可能继续独立于 SoS 而变化"这一事实以及该成员系统与其他成员系统之间的相互依赖增加 SoS 的复杂性，并进一步在 SoS 层级上挑战 SE。特别是，即使成员系统的行为得到很好的理解，这些动态可在 SoS 层级上导致产生意料之外的影响，导致在 SoS 中出现出乎意料的或不可预测的行为。

6. *试验、确认和学习*——"SoS 通常由独立于 SoS 的成员系统构成"这一事实造成在进行端对端 SoS 试验中遭遇挑战，正如各类系统通常所遇。首先，除非对 SoS 层级的预期和这些预期的测度有清晰的理解，否则可能很难将性能水平评估作为确定需要关注的区域的基础或者很难向用户确保 SoS 的能力及限制。即使对 SoS 目标和衡量标准有清晰的理解时，传统意义上的试验也可能很难。根据 SoS 背景环境，也许就没有 SoS 试验的资金授权。通常，成员系统的开发周期与系统所有者的需要以及最初的不断发展的用户基础相关联。由于多重的成员系统经历了异步的开发周期，因此找出跨 SoS 进行传统的端对端试验的方式即使不是不可能，也会很困难。此外，许多 SoS 很大而且多样化，使成员系统中传统的完整端对端试验的每项更改的费用可能高昂得令人却步。通常，使 SoS 性能得到良好测量的唯一方式是来自从实际运行中收集到的数据或通过基于建模、仿真和分析的预估。尽管面临这些挑战，SoS SE 团队需要使 SoS 的运行和性能能够连续。

7. *SoS principles*—SoS is a relatively new area, with the result that there has been limited attention given to ways to extend systems thinking to the issues particular to SoS. Work is needed to identify and articulate the crosscutting principles that apply to SoS in general and to develop working examples of the application of these principles. There is a major learning curve for the average systems engineer moving to an SoS environment and a problem with SoS knowledge transfer within or across organizations.

Beyond these general SE challenges, in today's environment, SoS pose particular issues from a security perspective. This is because constituent system interface relationships are rearranged and augmented asynchronously and often involve commercial off-the-shelf (COTS) elements from a wide variety of sources. Security vulnerabilities may arise as emergent phenomena from the overall SoS configuration even when individual constituent systems are sufficiently secure in isolation.

The SoS challenges cited in this section require SE approaches that combine both the systematic and procedural aspects described in this handbook with holistic, nonlinear, iterative methods.

2.5 ENABLING SYSTEMS

Enabling systems are systems that facilitate the life cycle activities of the SOI. The enabling systems provide services that are needed by the SOI during one or more life cycle stages, although the enabling systems are not a direct element of the operational environment. Examples of enabling systems include collaboration development systems, production systems, logistics support systems, etc. They enable progress of the SOI in one or more of the life cycle stages. The relationship between the enabling system and the SOI may be one where there is interaction between both systems or one where the SOI simply receives the services it needs when it is needed. Figure 2.3 illustrates the relationship of the SOI, enabling systems, and the other systems in the operational environment.

During the life cycle stages for an SOI, it is necessary to concurrently consider the relevant enabling systems and the SOI. All too often, it is assumed that the enabling systems will be available when needed and are not considered in the SOI development. This can lead to significant issues for the progress of the SOI through its life cycle.

7. *SoS 原则*——SoS 是一个相对新的领域，结果是对于系统思考向 SoS 特有问题的延伸方式所给予的关注很有限。需要在识别和明确地表达广义的适用于 SoS 的跨界原理方面的工作并开发这些原理的应用工作示例。有一条主学习曲线，供普通的系统工程师用组织内或跨组织的 SoS 知识转让来转移至 SoS 环境和问题中去。

超越这些一般 SE 挑战之外，在今天的环境中，SoS 从安保视角也提出特殊的问题。这是因为成员系统的接口关系被重新安排和异步增强，并且往往涉及来自广泛而多样的来源的商用货架产品（COTS）元素。安保漏洞可能作为一种涌现现象由总体 SoS 构型产生，即使单独的成员系统在孤立情况下是足够安全的。

本节援引的 SoS 挑战需要的是将本手册中描述的系统方面和程序方面与整体的、非线性的、迭代的方法相结合的 SE 方法。

2.5 使能系统

使能系统是助促 SOI 生命周期活动的系统。使能系统在一个或多个生命周期阶段期间提供 SOI 所需的服务，尽管使能系统并不是运行环境的直接元素。使能系统的示例包括协同开发系统、生产系统、后勤支持系统等。它们使 SOI 能够在一个或多个生命周期阶段中取得进展。使能系统与 SOI 之间的关系可以是两个系统之间存在交互的关系或者是 SOI 仅在需要时接收其所需服务的关系。图 2.3 详细阐明 SOI、使能系统以及运行环境中的其他系统的关系。

在 SOI 的生命周期阶段期间，有必要同时并行考虑 SOI 及相关的使能系统。太多的时候总是假设使能系统在需要时就会现成可用并且在 SOI 开发中不予考虑。这可导致在其生命周期内 SOI 的进展中出现重大问题。

SYSTEMS ENGINEERING OVERVIEW

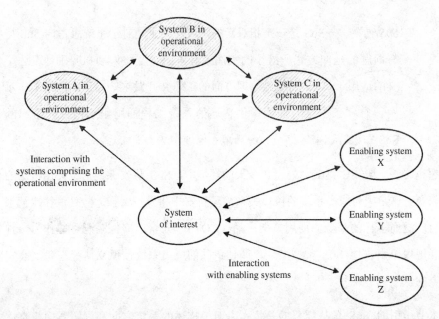

FIGURE 2.3 System of interest, its operational environment, and its enabling systems. This figure is excerpted from ISO/IEC/ IEEE 15288:2015, Figure 3 on page 13, with permission from the ANSI on behalf of the ISO. © ISO 2015. All rights reserved.

2.6 DEFINITION OF SYSTEMS ENGINEERING

SE is a perspective, a process, and a profession, as illustrated by these three representative definitions:

> Systems engineering is an interdisciplinary approach and means to enable the realization of successful systems. It focuses on defining customer needs and required functionality early in the development cycle, documenting requirements, and then proceeding with design synthesis and system validation while considering the complete problem: operations, cost and schedule, performance, training and support, test, manufacturing, and disposal. Systems engineering integrates all the disciplines and specialty groups into a team effort forming a structured development process that proceeds from concept to production to operation. Systems engineering considers both the business and the technical needs of all customers with the goal of providing a quality product that meets the user needs. (INCOSE, 2004)

> Systems engineering is an iterative process of top-down synthesis, development, and operation of a real-world system that satisfies, in a near optimal manner, the full range of requirements for the system. (Eisner, 2008)

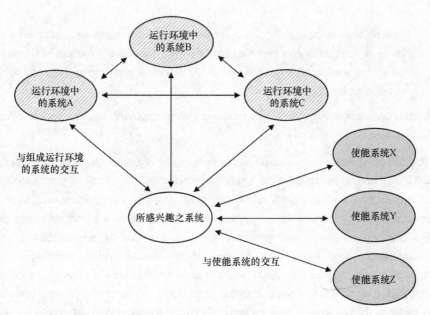

图 2.3 所感兴趣之系统、其运行环境及其使能系统。经代表 ISO 的 ANSI 许可后，此图摘自 ISO/IEC/IEEE 15288:2015，第 13 页图 3。ISO 版权所有，©2015。版权所有。

2.6 系统工程的定义

系统工程（SE）是一个视角、一个流程、一门专业，正如以下三种代表性定义所阐明：

> 系统工程是一种使系统能成功实现的跨学科的方法和手段。系统工程专注于：在开发周期的早期阶段定义客户需要与所要求的功能性，将需求文件化，然后再进行设计综合和系统确认，并同时考虑完整问题，即运行、成本、进度、性能、培训、支持、试验、制造和退出问题时，进行设计综合和系统确认。系统工程把所有学科和专业群体综合为一种团队的努力，形成从概念到生产再到运行的结构化开发流程。系统工程以提供满足用户需要的高质量产品为目的，同时考虑所有客户的业务和技术需要。（INCOSE, 2004）

> 系统工程是一种自上而下的综合、开发和运行一个真实系统的迭代过程，以接近于最优的方式满足系统的全部需求。（Eisner, 2008）

SYSTEMS ENGINEERING OVERVIEW

> Systems engineering is a discipline that concentrates on the design and application of the whole (system) as distinct from the parts. It involves looking at a problem in its entirety, taking into account all the facets and all the variables and relating the social to the technical aspect. (FAA, 2006)

Certain keywords emerge from this sampling—interdisciplinary, iterative, sociotechnical, and wholeness.

The SE perspective is based on systems thinking. Systems thinking (see Section 2.9.2) is a unique perspective on reality—a perspective that sharpens our awareness of wholes and how the parts within those wholes interrelate. When a system is considered as a combination of system elements, systems thinking acknowledges the primacy of the whole (system) and the primacy of the relation of the interrelationships of the system elements to the whole. Systems thinking occurs through discovery, learning, diagnosis, and dialogue that lead to sensing, modeling, and talking about the real world to better understand, define, and work with systems. A systems thinker knows how systems fit into the larger context of day-to-day life, how they behave, and how to manage them.

The SE process has an iterative methodology that supports discovery, learning, and continuous improvement. As the process unfolds, systems engineers gain insights into the relationships between the specified requirements for the system and the emergent properties of the system. Insights into the emergent properties of a system can therefore be gained from understanding the interrelationships of the system elements and the relation of these to the whole (system). Due to circular causation, where one system variable can be both the cause and effect of another, even the simplest of systems can have unexpected and unpredictable emergent properties. Complexity, as discussed in Section 2.4, can further exacerbate this problem; hence, one of the objectives of the SE process is to minimize undesirable consequences. This can be accomplished through the inclusion of and contributions from experts across relevant disciplines coordinated by the systems engineer.

SE includes both technical and management processes, and both processes depend upon good decision making. Decisions made early in the life cycle of a system, whose consequences are not clearly understood, can have enormous implications later in the life of a system. It is the task of the systems engineer to explore these issues and make the critical decisions in a timely manner. The roles of the systems engineer are varied, and Sheard's "Twelve Systems Engineering roles" (1996) provides one description of these variations.

> 系统工程是一门专注于整体（系统）设计和应用的学科而不是各个部分，这涉及从问题的整体性来审视，将问题的所有方面和所有变量都考虑在内，并将社会与技术方面相关联。(FAA, 2006)

该示例中出现的一些关键词——跨学科、迭代性、社会技术和整体性。

SE 视角基于系统思考。系统思考（见 2.9.2 节）是对现实的一种独特的视角——塑造我们的整体意识和理解整体内的各个部分如何互相关联。当系统被视为系统元素的组合时，系统思考推崇整体（系统）至上及系统元素的相互关系以及与整体的关系至上。系统思考来自于发现、学习、诊断和对话引发感知、建模以及探讨真实世界以更好地理解系统、定义系统和与系统一起工作。一个系统思考者知晓系统如何适配于日常生活中更大的背景环境，系统如何行为，以及如何管理系统。

SE 流程具有支持发现、学习和持续改进的迭代式方法论。随着流程的展开，系统工程师深入理解系统的特定需求与系统的涌现性之间的关系。因此，对系统的涌现性的深度透视可以从理解系统元素的相互关系及其与整体（系统）的关系来获得。由于循环的因果关系，其中一个系统变量既可以是另一个变量之因，也可以是其之果，即使最简单的系统也会具有意想不到的、不可预测的涌现性。复杂性，如 2.4 节所述，可使问题进一步恶化；因此，SE 流程的目标之一就是使非预期的后果最小化。系统工程师通过协调跨多个相关学科的专家的融合及其贡献而实现这一目标。

SE 包括技术流程和管理流程，这两种流程均依赖于良好的决策。如果对系统生命周期早期所作决策的后果尚未清晰理解的话，将会在系统生命周期的后期产生巨大的影响。系统工程师的任务正是探索这些议题，并适时地做出关键性的决策。系统工程师的角色是多样性的，Sheard 在"12 种系统工程师角色"（1996）中对这些多样性给出了解释。

SYSTEMS ENGINEERING OVERVIEW

2.7 ORIGINS AND EVOLUTION OF SYSTEMS ENGINEERING

The modern origins of SE can be traced to the 1930s followed quickly by other programs and supporters (Hughes, 1998). Tables 2.1 and 2.2 (Martin, 1996) offer a thumbnail of some important highlights in the origins and standards of SE. A list of current significant SE standards and guides is provided in Table 2.3.

TABLE 2.1 Important dates in the origins of SE as a discipline

1937	British multidisciplinary team to analyze the air defense system
1939–1945	Bell Labs supported NIKE missile project development
1951–1980	SAGE air defense system defined and managed by Massachusetts Institute of Technology (MIT)
1954	Recommendation by the RAND Corporation to adopt the term "systems engineering"
1956	Invention of systems analysis by RAND Corporation
1962	Publication of *A Methodology for Systems Engineering* by Hall
1969	Modeling of urban systems at MIT by Jay Forrester
1990	National Council on Systems Engineering (NCOSE) established
1995	INCOSE emerged from NCOSE to incorporate international view
2008	ISO, IEC, IEEE, INCOSE, PSm, and others fully harmonize SE concepts on ISO/IEC/IEEE 15288:2008

TABLE 2.2 Important dates in the origin of SE standards

1969	Mil-Std 499
1979	Army Field manual 770-78
1994	Perry memorandum urges military contractors to adopt commercial practices. EIA 632 IS (interim standard) and IEEE 1220 (trial version) issued instead of Mil-Std 499B
1998	EIA 632 released
1999	IEEE 1220 released
2002	ISO/IEC 15288 released, adopted by IEEE in 2003
2012	*Guide to the Systems Engineering Body of Knowledge (SEBoK)* released

2.7 系统工程的起源和演变

现代 SE 的起源可追溯到 20 世纪 30 年代，紧随其后，其他计划和支持者相继出现（Hughes, 1998）。表 2.1 和表 2.2（Martin, 1996）提供了 SE 的起源和标准中一些重大事件的概要。表 2.3 提供当前重要的 SE 标准和指南的列表。

表 2.1　SE 学科起源过程中的重要年代

1937	英国多学科团队分析防空系统
1939—1945	贝尔实验室资助的 NIKE 导弹项目的开发
1951—1980	由麻省理工学院（MIT）定义和管理的 SAGE 防空系统
1954	兰德公司建议采用术语"系统工程"
1956	由兰德公司发明系统分析法
1962	Hall 的《系统工程方法论》出版
1969	Jay Forrester 在麻省理工学院对城市系统建模
1990	国家系统工程委员会（NCOSE）成立
1995	为了体现国际化的视角，NCOSE 更名为 INCOSE
2008	ISO、IEC、IEEE、INCOSE、PSM 以及其他机构全面协调 ISO/IEC/IEEE 15288:2008 的 SE 概念

表 2.2　SE 标准起源过程中的重要年代

1969	Mil-Std 499
1979	陆军战场手册 770-78
1994	佩里备忘录呼吁军品承包商采取商用条例。EIA 632 IS（临时标准）和 IEEE 1220（试用版）发行，而不是 Mil-Std 499B
1998	EIA 632 发布
1999	IEEE 1220 发布
2002	ISO/IEC 15288 发布，IEEE 于 2003 年采纳
2012	《系统工程知识体指南》（SEBoK）发布

SYSTEMS ENGINEERING OVERVIEW

TABLE 2.3 Current significant SE standards and guides

ISO/IEC/IEEE 15288	Systems and software engineering—System life cycle processes
ANSI/EIA-632	Processes for engineering a system
ISO/IEC/IEEE 26702	Systems engineering—Application and management of the systems engineering process (replaces IEEE 1220™)
SEBoK	Guide to the systems engineering body of knowledge
ISO/IEC TR 24748	Systems and software engineering—Life cycle management—Part 1, guide for life cycle management; Part 2, guide to the application of ISO/IEC 15288 (system life cycle processes)
ISO/IEC/IEEE 24765	Systems and software engineering—vocabulary
ISO/IEC/IEEE 29148	Software and systems engineering — Life cycle processes — requirements engineering
ISO/IEC/IEEE 42010	Systems and software engineering—Architecture description (replaces IEEE 1471)
ISO 10303-233	Industrial automation systems and integration — Product data representation and exchange—Part 233: Application protocol: Systems engineering
OMG SysML™	Object management group (OMG) systems modeling language (SysML™)
CMMI-DEV v1.3	Capability maturity model integration (CmmI®) for development
ISO/IEC 15504-6	Information technology—Process assessment—Part 6: An exemplar system life cycle process assessment model
ISO/IEC/IEEE 15289	Systems and software engineering—Content of systems and software life cycle process information products (documentation)
ISO/IEC/IEEE 15939	Systems and software engineering—Measurement process
ISO/IEC/IEEE 16085	Systems and software engineering — Life cycle processes — Risk management
ISO/IEC/IEEE 16326	Systems and software engineering—Life cycle processes—Project management
ISO/IEC/IEEE 24748-4	Life cycle management—Part 4: Systems engineering planning
ISO 31000	risk management—Principles and guidelines
TechAmerica/ANSI EIA-649-B	National consensus standard for configuration management
ANSI/AIAA G-043A-2012e	ANSI/AIAA guide to the preparation of operational concept documents
ISO/IEC/IEEE 15026	Systems and software assurance—Part 1, concepts and vocabulary; Part 2, assurance case; Part 4, assurance in the life cycle

表 2.3 当前的重要 SE 标准和指南

标准	描述
ISO/IEC/IEEE 15288	系统和软件工程——系统生命周期流程
ANSI/EIA-632	系统工程设计流程
ISO/IEC/IEEE 26702	系统工程——系统工程流程的应用和管理（取代 IEEE 1220™）
SEBoK	系统工程知识体指南
ISO/IEC TR 24748	系统和软件工程——生命周期管理——第 1 部分，生命周期管理指南；第 2 部分，ISO/IEC 15288 应用指南（系统生命周期流程）
ISO/IEC/IEEE 24765	系统和软件工程——词汇
ISO/IEC/IEEE 29148	软件和系统工程——生命周期流程——需求工程
ISO/IEC/IEEE 42010	软件和系统工程——架构描述（取代 IEEE 1471）
ISO 10303-233	工业自动化系统和综合——产品数据表示和交换——第 233 部分：应用协议：系统工程
OMG SysML™	对象管理组（OMG）的系统建模语言（SysML™）
CMMI-DEV v1.3	用于开发的能力成熟度模型综合（CMMI®）
ISO/IEC 15504-6	信息技术——流程评估——第 6 部分：系统生命周期流程评估模型范例
ISO/IEC/IEEE 15289	系统和软件工程——系统和软件生命周期流程信息产品的内容（文件）
ISO/IEC/IEEE 15939	系统和软件工程——测量流程
ISO/IEC/IEEE 16085	系统和软件工程——生命周期流程——风险管理
ISO/IEC/IEEE 16326	系统和软件工程——生命周期流程——项目管理
ISO/IEC/IEEE 24748-4	生命周期管理——第 4 部分：系统工程规划
ISO 31000	风险管理——原理和指南
TechAmerica/ANSI EIA-649-B	构型管理国家达成一致的标准
ANSI/AIAA G-043A-2012e	ANSI/AIAA 运行方案文档的准备指南
ISO/IEC/IEEE 15026	系统和软件保证——第 1 部分，概念与词汇；第 2 部分，保证案例；第 4 部分，生命周期中的保证

SYSTEMS ENGINEERING OVERVIEW

With the introduction of the international standard ISO/IEC 15288 in 2002, the discipline of SE was formally recognized as a preferred mechanism to establish agreement for the creation of products and services to be traded between two or more organizations — the supplier(s) and the acquirer(s). But even this simple designation is often confused in a web of contractors and subcontractors since the context of most systems today is as a part of an "SoS" (see Section 2.4).

2.8 USE AND VALUE OF SYSTEMS ENGINEERING

Even on its earliest projects, SE emerged as an effective way to manage complexity and change. As both complexity and change continue to escalate in products, services, and society, reducing the risk associated with new systems or modifications to complex systems continues to be a primary goal of the systems engineer. This is illustrated in Figure 2.4. The percentages along the timeline represent the actual life cycle cost (LCC) accrued over time based on a statistical analysis performed on projects in the US Department of Defense (DOD), as reported by the Defense Acquisition University (DAU, 1993). For example, the concept stage of a new system averages 8% of the total LCC. The curve for committed costs represents the amount of LCC committed by project decisions and indicates that when 20% of the actual cost has been accrued, 80% of the total LCC has already been committed. The diagonal arrow under the curve reminds us that errors are less expensive to remove early in the life cycle.

Figure 2.4 also demonstrates the consequences of making early decisions without the benefit of good information and analysis. SE extends the effort performed in concept exploration to reduce the risk of hasty commitments without adequate study. Though shown as linear, the execution of the various life cycle stages associated with modern product development is, in actual application, recursive. Nonetheless, the consequences of ill-formed decisions throughout the life cycle are the same.

Another factor driving the need for SE is that complexity has an ever increasing impact on innovation. Few new products represent the big-bang introduction of new invention; rather, most products and services in today's market are the result of incremental improvement. This means that the life cycle of today's products and services is longer and subject to increasing uncertainty. A well-defined SE process becomes critical to establishing and maintaining a competitive edge in the twenty-first century.

随着 2002 年国际标准 ISO/IEC 15288 的引入，SE 学科被正式认为是在两个或多个组织（供应商和采办方）之间进行所创建产品和服务而建立协议的优选机制。但即便如此，这种简单的指定也常在承包方和分包方网络中引起困惑，因为如今大多数系统的背景环境是作为"SoS"（见 2.4 节）的一个部分。

2.8 系统工程的使用和价值

甚至在 SE 的最早项目上，SE 已经显现为管理复杂性和变化的有效方式。由于产品、服务和社会的复杂性和变化都在不断上升，减少与新系统或复杂系统改进相关的风险仍然是系统工程师的主要目标。图 2.4 阐明了这一点。根据美国国防采办大学（DAU，1993）所报告的美国国防部（DOD）项目中执行的统计分析，沿着时间轴的百分数表示随时间累积的实际生命周期成本（LCC）。例如，一个新系统的概念阶段平均占 LCC 总成本的 8%。所承诺的成本曲线表示由于项目决策所承诺的 LCC 的总和，还表明当累计的实际成本达到 20% 时，80% 的 LCC 总成本已经被承诺。曲线下方的斜箭头提示我们，在生命周期中极早地消除差错所需要的费用并不高。

图 2.4 展现在没有良好的信息和分析之下而做出早期决策所带来的后果。为了减少未经充分研究而轻率承诺所带来的风险，SE 拓展了在概念探索中所付诸的努力。尽管图中所示的各个阶段是线性的，但在实际应用中，与现代产品开发相关的生命周期各个不同阶段的执行却是递归的。尽管如此，在整个生命周期内做出的不良决策所产生的后果是相同的。

需要 SE 的另一个驱动因素是复杂性对创新的影响在前所未有的增加。很少有新产品代表着新发明"大爆炸"式的引入；相反，如今市场上的多数产品和服务都是逐渐改善的结果。这意味着如今的产品和服务的生命周期更长，遭受不断增加的不确定性。良好定义的 SE 流程成为建立和保持 21 世纪竞争优势的关键。

SYSTEMS ENGINEERING OVERVIEW

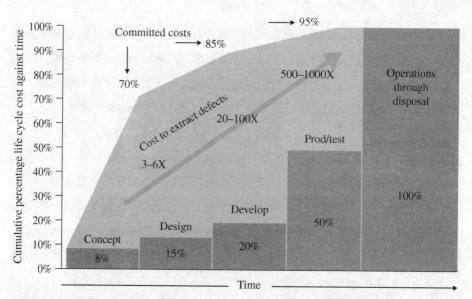

FIGURE 2.4 Committed life cycle cost against time. Reprinted with permission from DAU. All other rights reserved.

As illustrated in Figure 2.5, the development and market penetration of technology has accelerated by more than a factor of four over the past 140 years. In this sample of products, the time it took to achieve 25% market penetration was reduced from about 50 years to below 12 years. On average, development from prototype to 25% market penetration went from 44 to 17 years.

Two studies have demonstrated the value of SE from an effectiveness and return on investment (ROI) perspective. These studies are summarized in the following.

2.8.1 SE Effectiveness

A 2012 study by the National Defense Industrial Association, the Institute of Electrical and Electronic Engineers (IEEE), and the Software Engineering Institute of Carnegie mellon surveyed 148 development projects and found clear and significant relationships between the application of SE activities and the performance of those projects, as seen in Figure 2.6 and explained in the following (Elm and Goldenson, 2012).

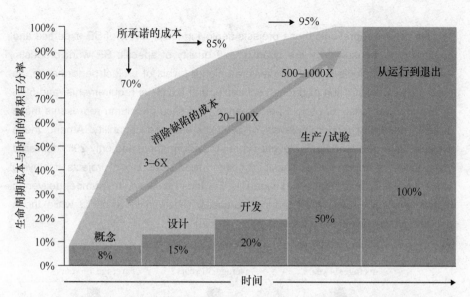

图 2.4 所确定的生命周期成本与时间的对比。经 DAU 许可后转载。版权所有。

如图 2.5 所示,在过去的 140 年内技术的开发和市场渗透加快了四倍以上。在产品的样本中,达到 25%市场渗透率所花费的时间从大约 50 年缩短到 12 年以下。平均而言,从原型机的开发到 25%市场渗透率的开发时间从 44 年缩短到 17 年。

两项研究都已从效能和投资回报(ROI)的视角展示了 SE 的价值。这些研究总结如下。

2.8.1 SE 效能

国家防务工业协会、电气和电子工程师协会(IEEE)和卡内基梅隆大学软件工程研究院 2012 年的研究调查了 148 个开发项目并发现了 SE 活动应用与那些项目的绩效之间清晰且极其重要的关系,正如图 2.6 所示以及如下解释(Elm 和 Goldenson,2012)。

SYSTEMS ENGINEERING OVERVIEW

The left column represents those projects deploying lower levels of SE expertise and capability, as measured by the quantity and quality of specific SE work products. Among these projects, only 15% delivered higher levels of project performance, as measured by satisfaction of budget, schedule, and technical requirements, and 52% delivered lower levels of project performance. The second column represents those projects deploying moderate levels of SE expertise and capability. Among these projects, 24% delivered higher levels of project performance, and only 29% delivered lower levels of performance. The third column represents those projects deploying higher levels of SE expertise and capability. For these projects, the number delivering higher levels of project performance increased substantially to 57%, while those delivering lower levels decreased to 20%.

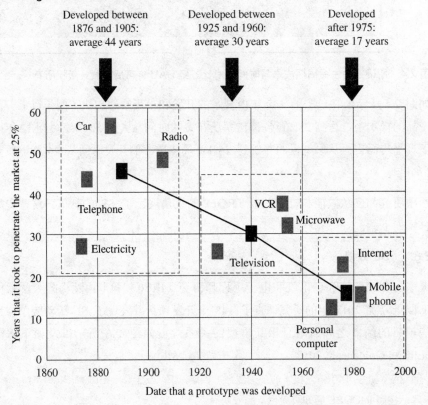

FIGURE 2.5 Technology acceleration over the past 140 years. INCOSE SEH original figure created by Michael Krueger. Usage per the INCOSE Notices page. All other rights reserved.

作为通过特定 SE 工作产物的数量和质量的衡量，左列表示部署较低层级的 SE 专长和能力的那些项目。在这些项目中，按照预算、进度和技术需求的满足情况衡量，仅有 15%交付了较高层级的项目绩效，52%交付了较低层级的项目绩效。第二列表示部署中等层级的 SE 专长和能力的那些项目。在这些项目中，24%已交付了更高层级的项目绩效，仅有 29%交付了较低层级的项目绩效。第三列表示部署较高层级的 SE 专长和能力的那些项目。在这些项目中，交付较高层级的项目绩效的数目明显增加至 57%，而交付较低层级的项目绩效的数目下降到 20%。

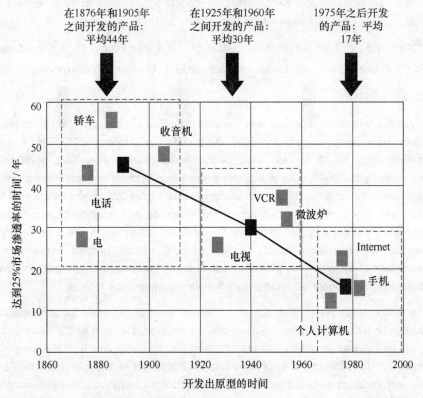

图 2.5　过去 140 年内技术的加速发展。INCOSE SEH 原始图由 Michael Krueger 创建。按照 INCOSE 通知页使用。版权所有。

SYSTEMS ENGINEERING OVERVIEW

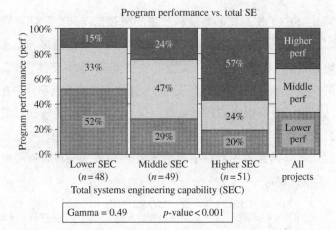

FIGURE 2.6 Project performance versus SE capability. From Elm and goldenson (2012). Reprinted with permission from Joseph Elm. All other rights reserved.

2.8.2 SE ROI

A quantitative research project was completed by Eric Honour and the University of South Australia to quantify the ROI of SE activities (Honour, 2013).This project gathered data on 43 survey points and from 48 detailed interviews, each point representing the total results of a single system development project. Projects had a wide variety of domains, sizes, and success levels. Figure 2.7 compares the total SE effort with cost compliance (top figure) and schedule performance (bottom figure). Both graphs show that SE effort has a significant, quantifiable effect on program success, with correlation factors as high as 80%. In both graphs, increasing the percentage of SE within the project results in better success up to an optimum level, above which additional SE effort results in poorer performance.

The research results showed that the optimum level of SE effort for a normalized program is 14% of the total program cost. In contrast, the median SE effort of the actual interviewed programs was only 7%, showing that programs typically operate at about half the optimum level of SE effort. The optimum level for any program can also be predicted by adjusting for the program characteristics, with levels ranging from 8 to 19% of the total program cost.

The ROI of adding additional SE activities to a project is shown in Table 2.4, and it varies depending on the level of SE activities already in place. If the project is using no SE activities, then adding SE carries a 7:1 ROI; for each cost unit of additional SE, the project total cost will reduce by 7 cost units. At the median level of the programs inter- viewed, additional SE effort carries a 3.5:1 ROI.

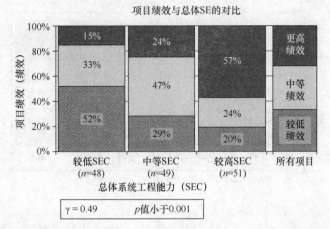

图 2.6 项目绩效与 SE 能力的对比。来自于 Elm 和 Goldenson（2012）。经 Joseph Elm 许可后转载。版权所有。

2.8.2 SE ROI

定量研究项目由 Eric Honour 和南澳大学完成，用以量化 SE 活动的 ROI (Honour, 2013)。该项目从 48 次详细访谈中收集了 43 个调查点的数据，每个调查点表示单一系统开发项目的总体结果。项目具有多种多样的领域、规模和成功水平。图 2.7 将全部 SE 努力与成本符合性（上图）和进度绩效（下图）进行比较。两张图均表明 SE 工作对项目成功具有重要的、可量化的影响，其相关系数高达 80%。在两张图中，增加项目内 SE 的百分比导向更好地成功达到最佳水平，在该层级以上，额外的 SE 努力则导致更差的绩效。

研究结果表明，规范化项目的 SE 努力的最佳水平是全部项目成本的 14%。相比之下，实际访谈项目的 SE 努力的中值仅为 7%时，表明项目通常运行在 SE 工作最佳水平的约二分之一。任何项目的最佳水平还可通过针对项目特点来预测，其水平范围从全部项目成本的 8%到 19%。

在项目中额外增加的 SE 活动的 ROI 如表 2.4 所示，它根据已经到位的 SE 活动的水平而变化。如果项目没有使用 SE 活动，那么增加 SE 会带来 7:1 的 ROI；至于额外增加每单位的 SE 成本，则项目总体成本将下降 7 个单位成本。在访谈的项目的中值水平上，额外的 SE 工作带来 3.5:1 的 ROI。

SYSTEMS ENGINEERING OVERVIEW

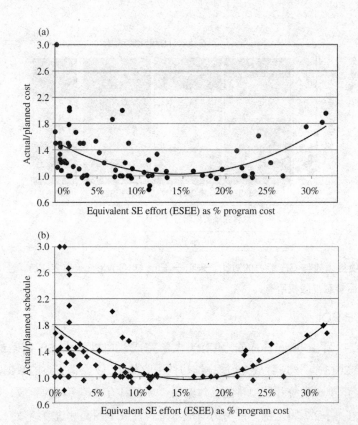

FIGURE 2.7 Cost (a) and schedule (b) overruns correlated with SE effort. From Honour (2013). Reprinted with permission from Eric Honour. All other rights reserved.

TABLE 2.4 SE return on investment

Current SE effort (% of program cost)	Average cost overrun (%)	ROI for additional SE effort (cost reduction $ per $ SE added)
0	53	7.0
5	24	4.6
7.2 (median of all programs)	15	3.5
10	7	2.1
15	3	−0.3
20	10	−2.8

From Honour (2013). Reprinted with permission from Eric Honour. All other rights reserved.

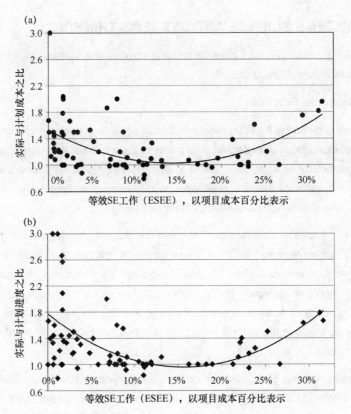

图 2.7 超支成本（a）和延期进度（b）与 SE 工作的关系。来自 Honour（2013）。经 Eric Honour 许可后转载。版权所有。

表 2.4 SE 投资回报

当前的 SE 工作（占项目成本的百分比）	平均成本超支（百分比）	额外的 SE 工作的 ROI（增加每单位 SE 工作带来的成本下降值）
0	53	7.0
5	24	4.6
7.2（所有项目中的中值）	15	3.5
10	7	2.1
15	3	−0.3
20	10	−2.8

来自 Honour（2013）。经 Eric Honour 许可后转载。版权所有。

SYSTEMS ENGINEERING OVERVIEW

2.9 SYSTEMS SCIENCE AND SYSTEMS THINKING

This section summarizes the nature of systems science and systems thinking. It also describes how they relate to SE.

2.9.1 Systems Science

Systems science brings together research into all aspects of systems with the goal of identifying, exploring, and understanding patterns of complexity that cross disciplinary fields and areas of application. It seeks to develop interdisciplinary foundations that can form the basis of theories applicable to all types of systems (e.g., in nature, society, and engineering) independent of element type or application.

Additionally, systems science can help to provide a common language and intellectual foundation for SE and make practical system concepts, principles, patterns, and tools accessible to practitioners of the "systems approach." An integrated systems approach for solving complex problems needs to combine elements of systems science, systems thinking, and SE. As such, systems science can serve as the foundation for a metadiscipline that unifies the traditional scientific specializations.

The information in this section is extracted from the systems science article in Part 2 of the SEBoK (SEBoK, 2014). Figure 2.8 illustrates the relationships between systems science, systems thinking, and general systems approach as applied to engineered systems.

Systems science is an integrative discipline that brings together ideas from a wide range of sources sharing in a common systems theme. Some fundamental concepts now used in systems science have been present in other disciplines for many centuries, while equally fundamental concepts have independently emerged as recently as 40 years ago (Flood and Carson, 1993).

Systems science is both the "science of systems" and the "systems approach to science," covering theories and methods that contrast with those of other sciences, which are generally reductionist in nature. Where it is appropriate, the reductionist approach has been very successful in using the methods of separating and isolating in search of simplicity. However, where those methods are not appropriate, systems science relies on connecting and contextualizing to identify patterns of organized complexity.

2.9 系统科学与系统思考

本节概述系统科学与系统思考的本质属性，并描述它们如何与 SE 相关联。

2.9.1 系统科学

系统科学汇集系统各个方面的研究，以识别、探索和理解跨越学科多个领域和应用的诸多范围的复杂性特征模式为目的。它寻求开发能够形成适用于独立于元素类型或应用的所有系统类型（如在自然、社会和工程中）的理论基础的跨学科依据。

此外，系统科学能够有助于为 SE 提供公共语言和智力根据，并使实际的系统概念、原理、特征模式和工具为"系统法"的实践者所易懂易得。求解复杂问题的综合的系统法需要将系统科学、系统思考和 SE 的元素结合起来。如此一来，系统科学可用做统一传统科学专业的元学科的基础。

本节中的信息提取自 SEBoK（SEBoK，2014）的第 2 部分中的系统科学篇章。图 2.8 详细阐述系统科学、系统思考与应用于工程系统的一般系统途径之间的关系。

系统科学是一门综合性学科，它汇集了来自于公共的系统主题中大范围共享资源的多种构想。现在应用于系统科学中的一些根本性的概念已在其他学科中出现多个世纪了，而同等的根本性概念早在 40 年前已经独立出现（Flood 和 Carson, 1993）。

系统科学既是"系统的科学"，又是"系统到科学的途径"，涵盖着与其他科学（通常本质上是简化论）截然不同的理论和方法。在合适的情形下，简化法在使用分割与隔离方法寻求简单性方面非常成功。然而，在那些方法不合适的情形下，系统科学依赖建立关联并置于背景环境之中来识别有组织的复杂性的特征模式。

Questions about the nature of systems, organization, and complexity are not specific to the modern age. As John Warfield (2006) put it,

> Virtually every important concept that backs up the key ideas emergent in systems literature is found in ancient literature and in the centuries that follow.

It was not until the middle of the twentieth century, however, that there was a growing sense of a need for, and possibility of, a scientific approach to the problems of organization and complexity in a "science of systems" per se.

Biologist Ludwig von Bertalanffy was one of the first to argue for and develop a broadly applicable scientific research approach based on *open system theory* (Bertalanffy, 1950). He explained the scientific need for systems research in terms of the limitations of analytical procedures in science. These limitations are based on the idea that an entity can be resolved into and reconstituted from its parts, either materially or conceptually:

> This is the basic principle of "classical" science, which can be circumscribed in different ways: resolution into isolable causal trains or seeking for "atomic" units in the various fields of science, etc.

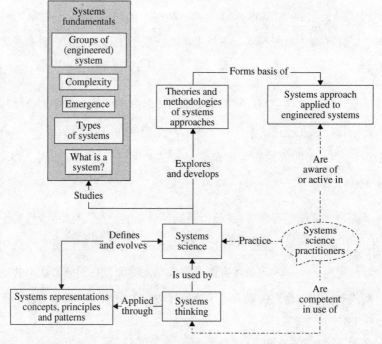

FIGURE 2.8 Systems science in context. From SEBoK (2014). Reprinted with permission from the BKCASE Editorial Board. All other rights reserved.

关于系统、组织和复杂性的本质属性方面的问题并非现代所特有。正如 John Warfield（2006）所说：

> 事实上，在系统文献中出现的支持关键设想的每一个重要概念，在古代及其后续几个世纪中的文献中都能找到。

然而直到 20 世纪中叶，人们才愈发意识到对"系统的科学"本身的组织和复杂性问题的科学途径的需要和可能性。

生物学家 Ludwig von Bertalanffy 是最早支持并基于开放系统理论（Bertalanffy, 1950）开发广泛适用的科学研究方法的学者之一。他以科学中解析程序的局限性解释了对系统研究的科学需求。这些局限性基于一种设想——一个实体在物质上或概念上都可以被分解为多个部分并由其各部分复原整体：

> 这是"经典"科学的基本原理，可采用不同方式对其范围进行限定：在科学等各个领域内分解可隔离的因果链或寻求"原子"单元。

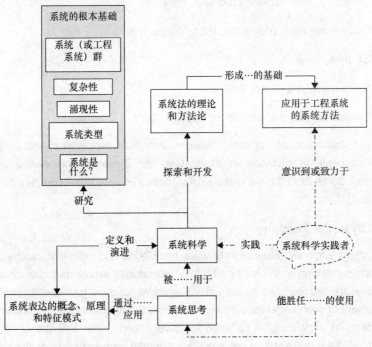

图 2.8 背景环境中的系统科学。来自于 SEBoK (2014)。经 BKCASE 编辑委员会许可后转载。版权所有。

SYSTEMS ENGINEERING OVERVIEW

Research in systems science attempts to compensate for the inherent limitations of classical science, most notably the lack of ways to deal with emergence. Systems science has developed—and continues to develop—hand in hand with practice, each maturing and learning from the other. Various efforts have taken on complementary or overlapping issues of the new "systems approach" as progress has been made over time:

- Cybernetics (Ashby, 1956; Wiener, 1948)
- Open system and general system theory (Bertalanffy, 1950, 1968; Flood, 1999)
- Operations research (Churchman et al., 1950)
- Hard and soft systems thinking (Checkland, 1998; Lewin, 1958)
- Organizational cybernetics (Beer, 1959; Flood, 1999)
- Critical systems thinking (Jackson, 1989)
- System dynamics (Forrester, 1961; Senge, 1990)
- SE (Hall, 1962)
- System analysis (Ryan, 2008)
- Service science and service SE (Katzan, 2008)

A broader discussion of contrasts between analytical procedures and integrative system concepts is provided in *Model-Oriented Systems Engineering Science* (Hybertson, 2009) from the perspective of a traditional versus the complex SE views of systems.

2.9.2 Systems Thinking

As a systems engineer, it is vital to develop knowledge and skills that can be utilized in performing a deep analysis of problem or opportunity situations for which system responses are required. As noted earlier, systems science has contributed to the development of such knowledge. However, during the twentieth century, a number of approaches to performing deep analysis have arisen under the title of "systems thinking." While it is difficult to put a precise boundary around systems thinking and differentiate it from systems science, many systems thinking methods and tools have become popular and have been successfully utilized in multidisciplinary contexts.

在系统科学方面的研究试图弥补经典科学的固有局限性——最显著的是缺乏应对涌现性的方式。系统科学与实践携手并持续发展，各自在成熟并相互借鉴。针对新的"系统法"中的补充或重叠问题做出了多方面努力，随着时间的推移已经取得了进展：

- 控制论（Ashby, 1956；Wiener, 1948）
- 开放系统及一般系统理论（Bertalanffy, 1950, 1968；Flood, 1999）
- 运筹学（Churchman 等，1950）
- 硬与软系统思考（Checkland, 1998；Lewin, 1958）
- 组织控制论（Beer, 1959；Flood, 1999）
- 批判性系统思考（Jackson, 1989）
- 系统动力学（Forrester, 1961；Senge, 1990）
- 系统工程（Hall, 1962）
- 系统分析（Ryan, 2008）
- 服务科学与服务系统工程（Katzan, 2008）

《面向模型的系统工程科学》（Hybertson, 2009）从传统与复杂 SE 的系统观点对比的视角，更广泛地讨论了解析程序与综合系统概念之间的对比。

2.9.2 系统思考

作为一名系统工程师，至关重要的是开发知识和技能，可用于对问题或要求系统响应的机会态势的深度分析。如前所述，系统科学已经对这些知识的开发做出了贡献。然而，在 20 世纪期间出现了在"系统思考"的名义下进行深度分析的大量方法。尽管围绕系统思考划出精确的界限并与系统科学进行区分是很难的，但许多系统思考方法和工具已经非常流行并成功用于多学科背景环境中。

2.9.2.1 System Dynamics Jay Forrester of MIT developed the DYNAMO simulation language and observed that a common means of analyzing complex systems could be used in multiple disciplines (1961). Several students of Forrester refined these ideas into methodologies and tools that have provided a useful basis for analysis. Peter Senge, with his popular book *The Fifth Discipline* (1990), established systems thinking as a discipline. He also developed the link, loop, and delay language as a means of graphically representing system dynamics. Based upon two primary loops (growth and limit), a number of so-called archetypes have been developed to describe a variety of situations. Another student, Barry Richmond, further developed archetypes by adding flow mechanisms in the simulation languages STELLA and iTHINK, which are commercially available.

2.9.2.2 Soft Systems and Action Research Peter Checkland (1975) observed that the use of classical engineering approaches to complex problems falls down since there are many soft factors (attitudes, practices, procedures, etc.) that affect systems. He also observed that the path to improvement must come through the development and analysis of alternative models. Based upon analysis, discussion, and dialogue, a course of action is planned and executed, and the results observed as feedback for further analysis. John Boardman has been inspired by Checkland's work and has developed a version of the soft systems approach supported by a tool called Systemigrams (Boardman and Sauser, 2008).

2.9.2.3 Discovering Patterns Central to systems thinking is the discovery of patterns. A pattern is a representation of similarities in a set of problems, solutions, or systems. Systems thinking captures and exploits what is common in a set of problems and corresponding solutions in the form of patterns of various types, as noted in the archetypes described previously.

Systems engineers use the general information provided by patterns to understand a specific system problem and to develop a specific system solution. Examples include, but are not limited to, the following:

- Software design patterns, such as adapter
- System architecture patterns, such as layered architecture and single sign-on
- Community or urban design patterns, such as ring road, pedestrian street, and food court
- Security and safety patterns, such as fault-tolerant design and role-based access control

2.9.2.1 系统动力学　麻省理工学院的 Jay Forrester 开发了 DYNAMO 仿真语言，并发现一种分析复杂系统的公共手段可用于多种学科（1961）。Forrester 的几个学生将这些理念细化为对分析提供一种有用基础的多种方法论和工具。Peter Senge 在其流行书籍《第五项修炼》（1990）中将系统思考确立为一门学科。他还开发了链接、循环和延迟语言，作为用图形表达系统动力学的手段。基于两种主要循环（增长和限制），开发了大量所谓的基模，用于描述多种多样的情境。另一名学生 Barry Richmond 通过在商品化仿真语言 STELLA 和 iTHINK 中增加流动机制，对基模进行了进一步开发。

2.9.2.2 软系统与行动研究　Peter Checkland（1975）发现，经典工程方法在复杂问题中的使用遭遇挫折是由于系统受很多软因素（态度、实践、程序等）影响。他还发现，改进的路径必须通过备选模型的开发和分析。在分析、讨论和对话的基础上，计划并执行一个行动路线，并将观察到的结果作为反馈以供进一步分析。John Boardman 受到 Checkland 的工作的启发，开发了由名为 Systemigrams（Boardman 和 Sauser，2008）工具所支持的软系统法的版本。

2.9.2.3 发现特征模式　系统思考的核心就是发现特征模式。特征模式是在一系列问题、解决方案或系统中的相似性的一种表达。如前面基模所述，系统思考以各种类型的特征模式的形式捕获并利用一系列问题和相应解决方案中的共性。

系统工程师使用多种特征模式所提供的一般信息来理解特定的系统问题，并开发出特定的系统解决方案。示例包括但不限于以下内容：

- 软件设计特征模式，例如适配器
- 系统架构特征模式，例如分层架构和单点登录
- 社区或城市设计特征模式，例如环形公路、步行街和美食广场
- 安保和安全特征模式，例如容错设计和基于角色的访问权控制

- Interaction patterns, such as publish–subscribe
- Domain-specific patterns, such as the suspension bridge pattern

Pattern categories include domain taxonomies, standards, templates, architecture styles, reference architectures, product lines, abstract data types, and classes in class hierarchies.

Another important category of patterns is associated with complex, counterintuitive systems. An example is "shifting the burden." When a problem appears in a complex system, the intuitive response is to apply a quick short-term fix, rather than take the time to develop a long-term solution. The problem with that approach is that there is often a trade-off between the short-term fix and the long-term solution, such that the initial success of the quick fix reduces the chances the real solution will be developed. This pattern appears in the typical short-term/long-term priority struggle between the systems engineer, project manager, discipline engineering communities, and other stakeholders within a project. Shifting the burden is one of a set of "system archetypes," a class of patterns (sometimes called patterns of failure or antipatterns) that illustrate the underlying systems pathology behind many of the barriers to effective SE. Awareness and understanding of these patterns are important for people trying to embed the systems approach in their organization and take positive action to counter these challenges. Additional examples and references are given in the article "Patterns of Systems Thinking" in (SEBoK, 2014).

2.9.2.4 Habits of a Systems Thinker While there are several additional aspects of systems thinking introduced by multiple contributors, the following list summarizes essential properties of a systems thinker (Waters Foundation, 2013):

- Seeks to understand the big picture
- Observes how elements within the system change over time, generating patterns and trends
- recognizes that a systems' structure (elements and their interactions) generates behavior
- Identifies the circular nature of complex cause-and-effect relationships
- Surfaces and tests assumptions
- Changes perspective to increase understanding
- Considers an issue fully and resists the urge to come to a quick conclusion

- 交互特征模式，例如发布—订阅

- 领域特有的特征模式，例如悬索桥特征模式

特征模式类别包括领域分类学、标准、模板、架构风格、参考架构、产品线、抽象数据类型以及类层级结构中的类。

另一个重要的特征模式类别与复杂的、反直觉的系统相关联。"转移负担"就是一个例子。当一个复杂系统中出现一个问题时，直觉反应是应用快速短期的修补，而不是花时间开发长期的解决方案。该方法带来的问题是，往往需要在短期修补与长期解决方案之间加以权衡，如此一来，快速修补的初步成功便减少了开发真正解决方案的机会。这种特征模式出现在一个项目内的系统工程师、项目经理、学科工程群体及其他利益攸关者之间的典型的短期/长期优先级安排的纠结中。转移负担是一系列"系统基模"之一，是一类特征模式（有时被称为故障特征模式或反特征模式），其详细阐明许多阻碍有效的 SE 背后深层次的系统异常状态。对于试图将系统法嵌入其组织并采取积极行动应对这些挑战的人们而言，知晓并理解这些特征模式是非常重要的。（SEBoK 2014）中的"系统思考的特征模式"篇章给出了附加的示例和参考文献。

2.9.2.4 系统思考者的习惯 多名贡献者引入系统思考的多个附加方面的同时，以下列表概括系统思考者的基本特性（Waters Foundation, 2013）：

- 寻求理解"大图像"

- 观察系统内的元素如何随时间推移发生变化，生成多种特征模式和趋势

- 认识到系统的结构（元素及其交互）产生行为

- 识别复杂因果关系的循环本质属性

- 揭视并对测试假设

- 变换视角以增进理解

- 全面考虑问题并抵制快速结论的冲动

- Considers how mental models affect current reality and the future
- Uses understanding of system structure to identify possible leverage actions
- Considers both short- and long-term consequences of actions
- Finds where unintended consequences emerge
- Recognizes the impact of time delays when exploring cause-and-effect relationships
- Checks results and changes actions if needed: "successive approximation"

People who think and act in a "systems way" are essential to the success of both research and practice (Lawson, 2010). Successful systems research will not only apply systems thinking to the topic being researched but should also consider a systems thinking approach to the way the research is planned and conducted. It would also be of benefit to have people involved in research who have, at a minimum, an awareness of systems practice and ideally are involved in practical applications of the theories they develop.

For a more thorough description of the "discipline" of systems thinking, refer to Part 2 of SEBoK (2014) as well as the popular website *Systems thinking world* (Bellinger, 2013).

2.9.3 Considerations for Systems Engineers

Sillitto (2012) provides a useful digest of concepts from systems science and systems thinking, organized so as to be immediately useful to the SE practitioner.

Properties that are generally true of the sort of systems that systems engineers find themselves involved with are as follows:

1. A system exists within a wider "context" or environment.
   - The context includes an "operational environment," a "threat environment," and a "resource environment" (Hitchens, 2003).
   - The context may also contain collaborating and competing systems.
2. A system is made up of parts that interact with each other and the wider context.
   - The parts may be any or all of hardware, software, information, services, people, organizations, processes, services, etc.

- 考虑心智模型如何影响当前现实及未来
- 利用对系统结构的理解来识别可能的"杠杆"行动
- 考虑行动的短期后果和长期后果
- 寻找非预期后果出现的地方
- 揭示因果关系时,认识到时间延迟的影响
- 检查结果并按需要改变行动:"逐次逼近"

以"系统方式"思考和行动的人们对于研究与实践的成功是根本性的(Lawson, 2010)。成功的系统研究不仅将系统思考应用于所研究的主题,而且还应考虑将系统思考法应用于研究的计划和实施的方式中去。这也有益于使至少有系统实践意识的人员参与到研究中,并且理想情况下是,参与到他们所开发的理论的实际应用中。

关于系统思考"学科"更全面的描述,参考 SEBoK(2014)第 2 部分以及流行的网站 *Systems thinking world*(Bellinger, 2013)。

2.9.3 有关系统工程师的考虑

Sillitto(2012)提供了来自系统科学和系统思考诸多概念的有用文摘,将其组织起来就会立即对 SE 实践者有用。

一般适用于系统工程师自己参与的系统类型的特性如下:

1. 系统存在于更广泛的"背景"或环境内。
   - 背景环境包括"运行环境"、"威胁环境"和"资源环境"(Hitchens, 2003)。
   - 背景环境也许还包含合作系统和竞争系统。
2. 一个系统由彼此交互并与更广泛的背景环境交互的多个部分构成。
   - 这些部分可以是硬件、软件、信息、服务、人员、组织、流程、服务等中的任意一些或全部。

SYSTEMS ENGINEERING OVERVIEW

- Interactions may include exchange of information, energy, and resources.

3. A system has system-level properties ("emergent properties") that are properties of the whole system not attributable to individual parts.

   - Emergent properties depend on the structure (parts and relationships between them) of the whole system and on its interactions with the environment.
   - This structure determines the interactions between functions, behavior, and performance of the parts and interaction of the system with the environment—in ways both intended and unintended.

4. A system has the following:

   - A life cycle
   - Function, which can be characterized following Hitchens as "operate – maintain viability – manage resources" or as "observe – orient – decide – act" (Hitchens, 2003)
   - Structure, including the following:

     –A boundary, which may be static or dynamic and physical or conceptual

     –A set of parts

     –The set of relationships and potential interactions between the parts of the system and across the boundary (interfaces)
   - Behavior, including state change and exchange of information, energy, and resources
   - Performance characteristics associated with function and behavior in given environmental conditions and system states

5. A system both changes and adapts to its environment when it is deployed (inserted into its environment).

6. Systems contain multiple feedback loops with variable time constants, so that cause-and-effect relationships may not be immediately obvious or easy to determine.

Additional properties that are "sometimes true" that systems engineers are likely to encounter are as follows:

- 交互可能包括信息、能量和资源的交换。

3. 系统具有系统级特性("涌现特性"),归属于整体系统的特性而非单个部分。

 - 涌现特性取决于整体系统的结构(各个部分以及其之间的关系)及其与环境的交互。
 - 该结构以意图的和非意图的方式决定各个部分的功能、行为和性能之间的交互以及系统与环境的交互。

4. 系统具有以下内容:

 - 生命周期
 - 功能,能够按照 Hitchens 的观点被特征化为"运行—维持生存—管理资源"或为"观察—定向—决定—行动"(Hitchens, 2003)
 - 结构,包括以下内容:

   — 一个边界,可以是静态的或动态的,以及物理上的或概念上的

   — 各个部分的一个集合

   — 系统各个部分之间以及跨边界(接口)的关系与潜在交互的集合

 - 行为,包括状态变化以及信息、能量和资源的交换
 - 在给定的环境条件和系统状态下与功能和行为相关联的绩效特征

5. 在部署一个系统时(置入它的环境中),系统既改变它的环境,又适应于它的环境。

6. 系统包含时间常数具有的多重反馈回路,从而可能不会立即明显或容易地确定因果关系。

系统工程师可能遇到的"有时适用的"额外特性如下:

SYSTEMS ENGINEERING OVERVIEW

1. A system may exist independent of human intentionality.
2. A system may be part of one or several wider "containing systems."
3. A system may be self-sustaining, self-organizing, and dynamically evolving (such systems include "complex adaptive systems").
4. A system may offer "affordances"—features that provide the potential for interaction by "affording the ability to do something" (Norman, 1990):
   - Affordances will lead to interactions whether planned or not. For example, the affordance of a runway to let planes land and take off also leads to a possibly unintended affordance to drive vehicles across it, which may get in the way of planes, leading to undesirable emergent whole system behavior.
5. A system may be:
   - Clearly bounded and distinct from its context (the solar system, Earth, planes, trains, automobiles, ships, people)
   - Closely coupled with or embedded in its context (a bridge, a town, a runway, the human cardiovascular system, the Internet)
   - Of fluid and dynamic makeup (a club, team, social group, ecosystem, flock of geese, and again the Internet)
6. A system may be technical (requiring one or multiple disciplines to design), social, ecological, environmental, or a compound of any or all of these.

These lists help explain the need for systems engineers to think about the SOI in its wider context, so as to ensure they understand both the properties important to the system's purpose and those that might give rise to undesirable unintended consequences.

2.10 SYSTEMS ENGINEERING LEADERSHIP

Many of the processes in this handbook rightly discuss management (e.g., decision management, risk management, portfolio management, knowledge management), and these are all important aspects of the SE process. However, leadership is an equally important topic to systems engineers. In a paper entitled "What Leaders really Do," J. P. Kotter (2001) states that "leadership is different from management, but not for the reason most people think." Kotter defines the key differences between leaders and managers as:

1. 一个系统可能独立于人的意图而存在。
2. 一个系统可能是一个或若干个更广泛的"包含着众多系统"的一部分。
3. 一个系统可能是自维持、自组织和动态演进的(这些系统包括"复杂的自适应系统")。
4. 一个系统可能提供"可供性"——以通过"提供做事的能力"(Norman, 1990)来提供交互作用的潜力为特征:

   - 可供性将导致交互作用而无论其是计划还是非计划的。例如,跑道使飞机着陆和起飞的可供性也导致产生可能使车辆穿过跑道的非意图的可供性,这可能阻碍飞机,导致不良的涌现性的整体系统行为。

5. 系统可能是:

   - 被清楚地划出边界且与其背景环境(太阳系、地球、飞机、火车、汽车、船舶、人员)完全分开
   - 与其背景环境(一座桥梁、一个城镇、一条跑道、人类心血管系统、互联网)紧密耦合或嵌入其中
   - 由流动和动态内容组成(一个俱乐部、团队、社会群体、生态系统、鹅群和互联网)

6. 一个系统可能是技术的(需要一个学科或多个学科来设计)、社会的、生态的、环境的或者是以上任意几种或全部的组合。

这些列表有助于解释系统工程师在其更广泛的背景环境下去思考 SOI 的需要,以便确保他们既理解对于系统目的而言那些重要的特性,又理解可能引起不合需要的非意图的后果的特性。

2.10 系统工程的领导力

本手册中的许多流程正确地论述了管理(如决策管理、风险管理、项目组合管理、知识管理),这些都是 SE 流程的所有重要方面。然而,领导力对于系统工程师而言是同等重要的主题。在一篇名为"领导者真正要做什么"的文章中,J. P. Kotter(2001)陈述道,"领导力不同于管理,但原因并不是大多数人所想的那样"。Kotter 将领导者与管理者之间的关键区别定义为:

SYSTEMS ENGINEERING OVERVIEW

- Coping with change versus coping with complexity
- Setting a direction versus planning and budgeting
- Aligning people versus organizing and staffing
- Motivating people versus controlling and problem solving

A quote often attributed to Peter Drucker is: "Managers do things right. Leaders do the right things." Compare this to the informal definitions of the SE verification and validation processes: "Verification ensures you built the system right. Validation ensures you built the right system." Both verification and validation are important for the development of systems. Likewise, both management and leadership are important for systems engineers and their teams. Different phases of a project demand emphasis on different aspects of leadership.

Aspects of leadership that are particularly relevant for systems engineers include:

- Thinking strategically and looking at the long-term implications of decisions and actions to set vision and course
- Seeing the "big picture"
- Casting or capturing the vision for the organization and communicating it (the systems engineer may be working in support of the identified leader, or sometimes, it isn't the leader's prerogative to "cast" the vision)
- Defining the journey from the "as is" of today to the "to be" of tomorrow
- Turning ambiguous problem statements into clear, precise solution challenges for the team
- Working with the stakeholders (including customers), representing their points of view to the team and the team's point of view to them
- Maximizing customer value by ensuring a direct tie of all engineering effort to the customer business or mission needs
- Establishing an environment for harmonious teams while working to leverage the potential benefits of diversity (including bridging cultural and communication differences in multidisciplinary teams)
- Challenging conventional wisdom at all levels
- Managing conflicts and facilitating healthy conflict around ideas and alternatives

- 应对变化与应对复杂性的对比

- 设定方向与规划和预算的对比

- 使人们协同一致与组织人员和配备人员的对比

- 激励人员与控制和解决问题的对比

通常引用 Peter Drucker 的话:"管理者正确地做事情。领导者做正确的事情。"将它与 SE 验证与确认流程的非正式定义相比:"验证确保正确地构建系统。确认确保构建正确的系统。"验证和确认对于系统的开发都很重要。同样,管理和领导力对于系统工程师及其团队也都很重要。一个项目的不同阶段需要强调领导的不同方面。

与系统工程师特别相关的领导力方面包括:

- 战略性地思考并着眼于决策和行动的长期影响,以设定愿景和路线

- 看到"大图像"

- 塑造或捕获组织的愿景并去沟通(系统工程师可能正在工作中支持已识别的领导者,或有时,去"塑造"愿景并不是领导者在体会)

- 从今天的"现状"到将来的"目标"的历程进行定义

- 将有含糊不清的问题申明变成团队的清晰、精确的解决方案挑战

- 与利益攸关者(包括客户)一起工作,向团队表达他们的观点并向他们表达团队的观点

- 通过确保所有的工程努力与客户的业务或任务需要的直接联系使客户价值最大化

- 为和谐团队建立一个环境,同时进行工作以撬动多样性的潜在优势(包括桥接多学科团队中的文化和沟通差异)

- 在各个层级挑战传统智慧

- 管理冲突并围绕多种设想和备选方案助促健康性的冲突

- Facilitating decision making
- Demanding and enabling excellence

Leadership is both an opportunity and a critical responsibility of the systems engineer. The SE leader must have a systems view that takes into account the context, boundaries, interrelationships, and scope. They drive better solutions through the holistic understanding of the problem and its context and environment. SE leaders highlight the risks of unintended consequences in a proactive manner. Many times, SE leaders need to move the conversation from "price and cost" to "value and ROI." SE leaders need to serve as a model for the adaptability, agility, and resilience that is sought in both the systems and the teams that develop them. After all, SE leaders have the "best seat in the house" for seeing the broader systems view (Long, 2013).

2.11 SYSTEMS ENGINEERING PROFESSIONAL DEVELOPMENT

To efficiently and cost-effectively deliver differentiated products to the market, an organization needs to know what gaps exist in their overall capability. An individual needs to know what skills would enable them to be more effective, to develop those skills, and to have a standard to demonstrate and communicate their skill levels. The overall system for optimizing SE delivery is shown in Figure 2.9.

Most development, but especially for SE, is achieved through experience and on-the-job training. Typically, 70% of development is achieved through experience, 20% through mentoring, and only 10% through training (Lombardo and Eichinger, 1996). Training creates an understanding of basic concepts, while the mentor helps developing systems engineers absorb the appropriate lessons from practical experience. The model in Figure 2.10 shows how SE development can work for an individual (either pursuing certification or in a development discussion with their manager).

INCOSE has developed an SE Competency Framework based on the work of the INCOSE United Kingdom Chapter. Use of the framework can enable employees to analyze their skills and evaluate the need for training, coaching, or new job assignments to fill any gaps found in the assessment (INCOSE UK, 2010). The framework defines three classes of competencies and aligns relevant competencies to each class. The classes are systems thinking, holistic life cycle view, and SE management. The framework has defined four skill levels: awareness, supervised practitioner, practitioner, and expert. The framework is tailorable to the needs of the organization.

- 助促决策制定

- 要求卓越并使能卓越

领导力是系统工程师的机会和关键职责。SE 领导者必须具有系统视野以将背景环境、边界、相互关系和范围考虑在内。他们通过整体地理解问题及其背景和环境，得到更好的解决方案。SE 领导者以积极主动的方式突显非意图后果的风险。很多时候，SE 领导者需要将谈论从"价格和成本"转移到"价值和 ROI"。SE 领导者需要充当适应性、敏捷性和恢复力的楷模，这是系统及开发系统的团队所追求的。毕竟 SE 领导者具有审视更广泛的系统视野的"最佳席位"（Long, 2013）。

2.11 系统工程职业发展

为了将差异化产品高效率和成本有效益地交付到市场上，组织需要知道这些产品的总体能力中存在哪些差距。个人需要知道哪些技能使他们能够更加高效，以便开发那些技能并具有展示和沟通他们的技能水平的标准。使 SE 交付最优化的总体系统如图 2.9 所示。

大多发展，尤其对于 SE 而言，是通过经验和在岗培训来实现的。典型地，70%的发展是通过经验实现的，20%的发展是通过指导实现的，只有 10%的开发是通过培训实现的（Lombardo 和 Eichinger, 1996）。培训产生对基本概念的理解，而指导有助于进行开发系统的工程师从实际经验中吸取适当的教训。图 2.10 中的模型表明了 SE 开发可以如何为个人（其追求认证或与其管理者进行发展方面的讨论）服务。

INCOSE 已基于 INCOSE 英国分部的工作开发了 SE 能力框架。框架的使用能够使员工分析他们的技能并评估对培训、指导或新岗位分配的需要，以便填充评估中发现的任何差距（INCOSE UK, 2010）。该框架定义了三类能力并将相关能力与每个类别对准。这些类别是系统思考、完整的生命周期视野和 SE 管理。该框架定义了四个技能等级：知晓者、监督下的实践者、实践者和专家。可按照组织的需要对框架进行裁剪。

SYSTEMS ENGINEERING OVERVIEW

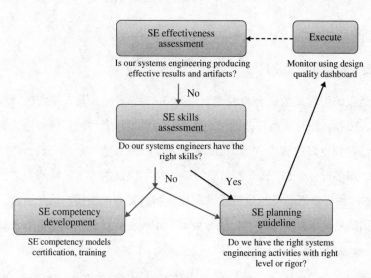

FIGURE 2.9 SE optimization system. Reprinted with permission from Chris Unger. All other rights reserved.

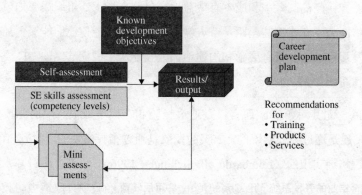

FIGURE 2.10 Professional development system. Reprinted with permission from Chris Unger. All other rights reserved.

2.11.1 SE Professional Ethics

There will always be pressure to cut corners to deliver programs faster or at lower costs, especially for a profession such as SE. As stated in the INCOSE Code of Ethics:

The practice of Systems Engineering can result in significant social and environmental benefits, but only if unintended and undesired effects are considered and mitigated.

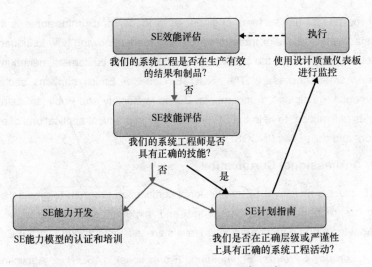

图 2.9 SE 优化系统。经 Chris Unger 许可后转载。版权所有。

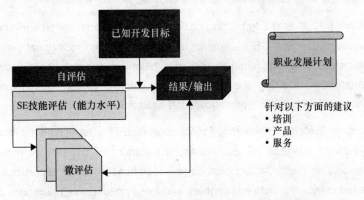

图 2.10 专业的开发系统。经 Chris Unger 许可后转载。版权所有。

2.11.1 SE 职业道德

往往总是在压力下走捷径来加速交付计划和降低成本，尤其是对于诸如 SE 的职业而言是如此。正如 INCOSE 道德规范中所述：

系统工程的实践可导致产生重大的社会效益和环境效益，但只有考虑到非意图的、不合需要的影响并减轻这些影响时才会如此。

Part of the role of the systems engineer as a leader and professional is knowing when unacceptable risks or trade-offs are being made, knowing how to influence key stakeholders, and having the courage to stand up for the customers, community, and profession when necessary. The INCOSE Code of Ethics contains sections on "Fundamental Principles," "Fundamental Duties to Society and Public Infrastructure," and "rules of Practice" to help the SE professional in practical applications of ethics to their work and daily lives (INCOSE, 2006).

2.11.2 Professional Certification

INCOSE offers a multilevel SE professional certification program to provide a formal method for recognizing the knowledge and experience of systems engineers throughout the world. Three certification levels are available through INCOSE:

- *Associate Systems Engineering Professional (ASEP)* — Applicants are required to successfully complete a knowledge examination.

- *Certified Systems Engineering Professional (CSEP)*—requires a minimum of 5 years of practical SE experience, a technical degree (additional years of SE experience can be used in lieu of a technical degree), three professional references covering the candidate's cumulative years of experience, and successful completion of a knowledge examination.

- *Expert Systems Engineering Professional (ESEP)*— requires a minimum of 25 years of practical SE experience, a minimum of 5 years of professional leadership credits, a technical degree (additional years of experience can be used in lieu of a technical degree), and three professional references covering at least the most recent 10 years of experience. The ESEP award is based on panel review and approval.

Additional details on the requirements for SE professional certification are available on the INCOSE website at http://www.incose.org/.

系统工程师作为领导者和专业人员的部分角色知晓何时产生不可接受的风险或进行权衡,知晓如何影响关键利益攸关者,以及在必要时有勇气力挺客户、团体和职业。INCOSE 道德规范包含"基本原理"、"对社会和公共基础设施的基本义务"以及"实践的规则"章节,以便帮助 SE 专业人员将道德实际应用到其工作和日常生活中(INCOSE, 2006)。

2.11.2 专业认证

INCOSE 提供多层级的 SE 专业认证方案,以便提供一种在全世界范围内认可系统工程师的知识和经验的正规方法。通过 INCOSE 可得到三种认证层级:

- 助理级系统工程专业人员(ASEP)——要求申请人成功地完成知识考试。

- 认证的系统工程专业人员(CSEP)——要求最低 5 年的实际 SE 经验、技术学位(如果拥有更多年的 SE 经验,则可不考虑技术学位)、涵盖候选人的多年的累积经验以及知识考试的成功完成的三份专业证明。

- 专家级系统工程专业人员(ESEP)——要求最低 25 年的实际 SE 经验、最低 5 年的职业领导信誉、技术学位(如果拥有更多年的 SE 经验,则可不考虑技术学位)以及涵盖至少最近 10 年经验的三份职业证明。ESEP 的授予是基于委员会的审查和批准。

关于 SE 专业认证要求的附加细节可从 INCOSE 网站 http://www.incose.org/ 上获取。

3 GENERIC LIFE CYCLE STAGES

3.1 INTRODUCTION

Every man-made system has a life cycle, even if it is not formally defined. A life cycle can be defined as the series of stages through which something (a system or manufactured product) passes. In keeping with increased awareness of environmental issues, the life cycle for any system of interest (SOI) must encompass not only the development, production, utilization, and support stages but also provide early focus on the retirement stage when decommissioning and disposal of the system will occur. The needs for each of the subsequent stages must be considered during the earlier stages, especially during the concept and development stages, in order to make the appropriate trades and decisions to accommodate the needs of later stages in an affordable and effective manner.

The role of the systems engineer encompasses the entire life cycle for the SOI. Systems engineers orchestrate the development of a solution from requirements definition through design, build, integration, verification, operations, and ultimately system retirement by assuring that domain experts are properly involved, that all advantageous opportunities are pursued, and that all significant risks are identified and mitigated. The systems engineer works closely with the project manager in tailoring the generic life cycle, including key decision gates, to meet the needs of their specific project. Per ISO/IEC/IEEE 15288,

> 5.4.2—Life cycles vary according to the nature, purpose, use and prevailing circumstances of the system ...

The purpose in defining the system life cycle is to establish a framework for meeting the stakeholders' needs in an orderly and efficient manner for the whole life cycle. This is usually done by defining life cycle stages and using decision gates to determine readiness to move from one stage to the next. Skipping stages and eliminating "time-consuming" decision gates can greatly increase the risks (cost, schedule, and performance) and may adversely affect the technical development as well by reducing the level of the SE effort, as discussed in Section 2.8.

INCOSE Systems Engineering Handbook: A Guide for System Life Cycle Processes and Activities, Fourth Edition. Edited by David D. Walden, Garry J. Roedler, Kevin J. Forsberg, R. Douglas Hamelin and Thomas M. Shortell.

© 2015 John Wiley & Sons, Inc. Published 2015 by John Wiley & Sons, Inc.

3 一般生命周期阶段

3.1 引言

每个人造系统都具有一个生命周期,即使未被正式定义亦是如此。生命周期可以被定义为某事物(系统或制造的产品)所经历的一系列阶段。随着对环境问题意识的增强,任何所感兴趣之系统(SOI)的生命周期不仅必须包含开发、生产、使用和维持阶段,而且在系统即将肢解和处置时还须及早地关注退出阶段。每个后续阶段的需要必须在较早阶段被考虑到,尤其是在概念和开发阶段期间,以便做出恰当的权衡和决策,以可承受的有效方式适应后期阶段的需要。

系统工程师的角色涵盖 SOI 的整个生命周期。通过确保领域专家适当地参与,追求一切有利机会以及识别并减轻所有重大风险,系统工程师们从需求定义到设计、构建、综合、验证、运行以及最终的系统退出来精心策划解决方案的开发。系统工程师与项目经理密切合作对一般通用的生命周期进行剪裁,也包括关键的决策门,以满足他们特定项目的需要。按照 ISO/IEC/IEEE 15288:

> 5.4.2——生命周期根据系统的本质属性、目的、用途和当时环境而变化……

定义系统生命周期的目的是以有序且高效的方式为整个生命周期建立一个满足利益攸关者需要的框架。常常通过定义生命周期阶段并使用决策门来确定成熟度状态,以便从一个阶段进入下一个阶段来实现这一目的。跳过某些阶段和省去一些"耗时的"决策门可能会大幅度增加风险(成本、进度和绩效),减少 SE 投入努力的水平也可能对技术开发造成不利影响,如 2.8 节所述。

INCOSE 系统工程手册:系统生命周期流程和活动指南,第 4 版 编撰:David D. Walden, Garry J. Roedler, Kevin J. Forsberg, R. Douglas Hamelin 和 Thomas M. Shortell。
John Wiley & Sons 公司版权所有©2015。由 John Wiley & Sons 公司于 2015 年出版。

GENERIC LIFE CYCLE STAGES

Systems engineering (SE) tasks are usually concentrated at the beginning of the life cycle, but both industry and government organizations recognize the need for SE throughout the systems' life span, often to modify or change a system product or service after it enters production or is placed in operation. Consequently, SE is an important part of all life cycle stages. During the utilization and support stages, for example, SE executes performance analysis, interface monitoring, failure analysis, logistics analysis, tracking, management, etc. that is essential to ongoing operation and support of the system.

3.2 LIFE CYCLE CHARACTERISTICS

3.2.1 Three Aspects of the Life Cycle

Every system life cycle consists of multiple aspects, including the business aspect (business case), the budget aspect (funding), and the technical aspect (product). The systems engineer creates technical solutions that are consistent with the business case and the funding constraints. System integrity requires that these three aspects are in balance and given equal emphasis at all decision gate reviews. For example, when Motorola's Iridium project started in the late 1980s, the concept of satellite based mobile phones was a breakthrough and would clearly capture a significant market share. Over the next dozen years, the technical reviews ensured a highly successful technical solution. In fact, in the first decade of the twenty-first century, the Iridium project is proving to be a good business venture for all except for the original team who had to sell all the assets—at about 2% of their investment—through the bankruptcy court. The original team lost sight of the competition and changing consumer patterns that substantially altered the original business case. Figure 3.1 highlights two critical parameters that engineers sometimes lose sight of: time to breakeven (indicated by the circle) and return on investment (indicated by the lower curve).

3.2.2 Decision Gates

Decision gates, also known as control gates, are often called "milestones" or "reviews." A decision gate is an approval event in the project cycle, sufficiently important to be defined and included in the schedule by the project manager, executive management, or the customer. Entry and exit criteria are established for each gate at the time they are included into the project management baseline. Decision gates ensure that new activities are not pursued until the previously scheduled activities, on which new activities depend, are satisfactorily completed and placed under configuration control. Proceeding beyond the decision gate before the project is ready entails risk. The project manager may decide to accept that risk, as is done, for instance, with long-lead item procurement.

系统工程（SE）的任务通常集中在生命周期的初期，但工业组织和政府组织都认识到对贯穿整个系统生命跨度的 SE 的需要，因为系统产品或服务进入生产阶段或投入运行后常常要进行修改或变更。由此，SE 成为所有生命周期阶段的重要部分。例如，在使用与保障阶段，SE 执行性能分析、接口监控、失效分析、后勤分析、跟踪和管理等，这对持续的系统运行和维持来说是根本性的。

3.2 生命周期特性

3.2.1 生命周期的三方面

每个系统生命周期都由多方面组成，包括业务方面（业务案例）、预算方面（资金）和技术方面（产品）。系统工程师要创建与业务案例和资金约束一致的技术解决方案。系统完整性要求这三方面达到平衡且在所有决策门评审中受到同等重视。例如，20 世纪 80 年代后期，摩托罗拉公司启动"铱星计划"时，基于卫星的移动电话概念是一个突破，显然会夺取巨大的市场份额。在接下来的十多年内，技术评审确保了技术解决方案取得重大成功。实际上，在 21 世纪的头十年，"铱星计划"正在证明其对于所有投资者来说都是一次良好的业务投资，但原始团队则被排除在外，因为他们早已不得不通过破产法院以其投资的大约 2%出售了所有资产。原始团队忽视了彻底改变原始业务案例的竞争和不断变化的消费者特征模式。图 3.1 强调有时被工程师忽略的两个关键参数：收支平衡时间点（用圆圈表示）和投资回报（用下方的曲线表示）。

3.2.2 决策门

决策门（亦称为控制门）通常被称为"里程碑"或"评审点"。决策门是项目周期内的审批事件，由项目经理、行政管理人员或客户在时间进程中定义并引入是非常重要的。当决策门被包含于项目管理基线中时，每个决策门都需建立进入和退出的准则。决策门确保直到新活动所依赖的预先安排的活动按要求完成并处于构型控制之下时才开展新活动。项目准备就绪前如越过决策门就必须承担风险。项目经理也可能决定接受那种风险，例如，对于长期的前期项采购，就要如此。

GENERIC LIFE CYCLE STAGES

All decision gates are both reviews and milestones; however, not all reviews and milestones are decision gates. Decision gates address the following questions:

- Does the project deliverable still satisfy the business case?
- Is it affordable?
- Can it be delivered when needed?

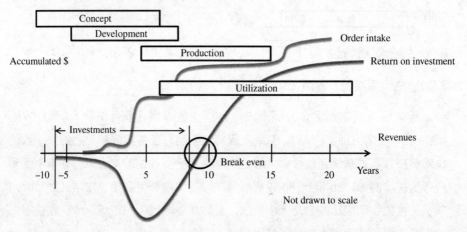

FIGURE 3.1 Generic business life cycle. From Stoewer (2005). Reprinted with permission from Heinz Stoewer. All other rights reserved.

Decision gates represent major decision points in the system life cycle. The primary objectives of decision gates are to:

- Ensure that the elaboration of the business and technical baselines are acceptable and will lead to satisfactory verification and validation (V&V)
- Ensure that the next step is achievable and the risk of proceeding is acceptable
- Continue to foster buyer and seller teamwork
- Synchronize project activities

There are at least two decision gates in any project: authority to proceed and final acceptance of the project deliverable. The project team needs to decide which life cycle stages are appropriate for their project and which decision gates beyond the basic two are needed. Each decision gate must have a beneficial purpose; "proforma" reviews waste everyone's time.

所有决策门既是评审点，又是里程碑；然而，并不是所有评审点和里程碑都是决策门。决策门涉及下列问题：

- 项目交付物仍然满足业务案例吗？
- 是否支付得起？
- 需要时能交付吗？

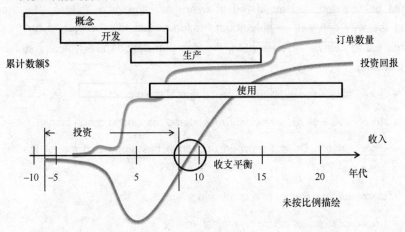

图 3.1 一般商业生命周期。来自于 Stoewer（2005）。经 Heinz Stoewer 许可后转载。版权所有。

决策门表示系统生命周期内的主要决策点。决策门的主要目标是：

- 确保业务基线和技术基线的详细阐述可接受并将引向满意的验证和确认（V&V）
- 确保下一步骤能够达成，继续进行下去的风险可接受
- 继续促进买方和卖方之间的团队工作
- 使项目活动同步

任何项目都具有至少两类决策门：继续进行下去的许可和项目交付物的最终验收。项目团队需要决定：生命周期的哪些阶段对他们的项目是合适的；超出两类基本决策门之外还需要哪些决策门。每个决策门必须具有有益的目的；"形式上的"评审只会浪费大家的时间。

Even in agile development (see Section 9.9), frequent interaction with stakeholders may minimize, but not eliminate, the need for decision gates. The consequences of conducting a superficial review, omitting a critical discipline, or skipping a decision gate are usually long term and costly.

The project business case issues of market demand, affordability, and realistic schedules are important decision criteria influencing concept selection, and they should be updated and evaluated at every decision gate. Inadequate checks along the way can set up subsequent failures—usually a major factor in cost overruns and delays. At each gate, the decision options are typically similar to the following:

- *Acceptable*: Proceed with the next stage of the project.
- *Acceptable with reservations*: Proceed and respond to action items.
- *Unacceptable*: Do not proceed—continue this stage and repeat the review when ready.
- *Unacceptable: Return to a preceding stage*.
- *Unacceptable: Put a hold on project activity*.
- *Unsalvageable: Terminate the project*.

Decision gate descriptions should identify the:

- Purpose and scope of the decision gate
- Entry and exit criteria
- Host and chairperson
- Attendees
- Location
- Agenda and how the decision gate is to be conducted
- Evidence to be evaluated
- Actions resulting from the decision gate
- Method of closing the review, including timing for resolution of open action items

即便在敏捷开发中（见 9.9 节），与利益攸关者频繁的交互也只能将对决策门的需要最小化，但不能将其消除。肤浅的评审、忽略关键规程或跳过决策门所产生的后果通常会是长周期的而且是高代价的。

市场需求、可支付性和现实进度的项目业务案例问题是影响概念选择的重要决策准则，并且这些问题应在每个决策门上被更新和评价。在这一过程中不充分的检查可能引起随后的失败——通常是成本超支和进度延期的主要因素。在每个决策门中，决策选项通常与以下内容类似：

- 可接受——继续进行项目的下一阶段。

- 有保留的接受——继续并回应行动项。

- 不可接受：不继续进行——延续本阶段工作且准备就绪时重做评审。

- 不可接受：返回到前一阶段。

- 不可接受：暂停项目活动。

- 不可挽回：终止项目。

决策门的描述应识别：

- 决策门的目的和范围

- 进入和退出的准则

- 主人和主持人

- 出席者

- 地点

- 议程以及如何实施决策门

- 待评估证据

- 决策门产生的行动

- 关闭评审的方法，包括开放式行动项的解决时序

Decision gate approval follows review by qualified experts and involved stakeholders and is based on hard evidence of compliance to the criteria of the review. Balancing the formality and frequency of decision gates is seen as a critical success factor for all SE process areas. On large or lengthy projects, decisions and their rationale are maintained using an information management process.

Upon successful completion of a decision gate, some artifacts (e.g., documents, models, or other products of a project life cycle stage) have been approved as the basis upon which future work must build. These artifacts are placed under configuration management.

3.3 LIFE CYCLE STAGES

ISO/IEC/IEEE 15288 states:

> 5.4.1—A system progresses through its life cycle as the result of actions, performed and managed by people in organizations, using processes for execution of these actions.

A system "progresses" through a common set of life cycle stages where it is conceived, developed, produced, utilized, supported, and retired. The life cycle model is the framework that helps ensure that the system meets its required functionality throughout its life. For example, to define system requirements and develop system solutions during the concept and development stages, experts from other stages are needed to perform trade-off analyses, help make decisions, and arrive at a balanced solution. This ensures that a system has the necessary attributes as early as possible. It is also essential to have the enabling systems available to perform required stage functions.

Table 3.1 lists six generic life cycle stages (ISO/IEC TR 24748-1, 2010). The purpose of each is briefly identified, and the options from decision gate events are indicated. Note that stages can overlap and the utilization and support stages run in parallel. Note also that the outcome possibilities for decision gates are the same for all decision gates. Although the stages in Table 3.1 are listed as independent, nonoverlapping, and serial, the activities constituting these stages can be in practice interdependent, overlapping, and concurrent.

决策门的批准依据有资质的专家和所涉及的利益攸关者的评审，且以符合评审准则的有力证据为基础。平衡诸多决策门的形式和频度被视为所有 SE 流程领域的关键成功因素。在大规模或长周期的项目中，决策及其依据都是使用信息管理流程来维持的。

当决策门成功完成时，一些工作产物（制品）（如项目生命周期阶段的文件、模型或其他产品）便已经被批准作为开展未来工作必须依据的基础。将这些制品置于构型管理之下。

3.3 生命周期阶段

ISO/IEC/IEEE 15288 陈述了：

> 5.4.1——系统在其生命周期内的进展是组织中的人员通过使用这些行动的执行流程来运行和管理行动的结果。

系统通过其构想、开发、生产、使用、支持和退出的生命周期的各个阶段的一个公共集合而"进展"。生命周期模型是有助于确保系统贯穿于其生命内满足系统所要求的功能性的框架。例如，为了在概念阶段和开发阶段定义系统需求并开发系统解决方案，其他阶段内的专家需要进行权衡分析，帮助做出决策并达成平衡的解决方案。这确保系统尽早具有必要的属性。具有可用于一定方面所要求的阶段功能的使能系统也是非常必要的。

表 3.1 列出六个一般生命周期阶段（ISO/IEC TR 24748-1, 2010）。简要地识别出每个阶段的目的，并指明决策门事件中的选项。需要注意的是，所有阶段可重叠，并且使用阶段和保障阶段并行进行。还需注意的是，对于所有的决策门而言，决策门结果的可能性是相同的。尽管表 3.1 中的阶段被独立、不重叠且序列地列出，但构成这些阶段的活动在实际上可以相互依赖、重叠且并行。

GENERIC LIFE CYCLE STAGES

TABLE 3.1 Generic life cycle stages, their purposes, and decision gate options

Life cycle stages	Purpose	Decision gates
Concept	Define problem space 1. Exploratory research 2. Concept selection Characterize solution space Identify stakeholders' needs Explore ideas and technologies Refine stakeholders' needs Explore feasible concepts Propose viable solutions	Decision options • Proceed with next stage • Proceed and respond to action items • Continue this stage • Return to preceding stage • Put a hold on project activity • Terminate project
Development	Define/refine system requirements Create solution description—architecture and design Implement initial system Integrate, verify, and validate system	
Production	Produce systems Inspect and verify	
Utilization	Operate system to satisfy users' needs	
Support	Provide sustained system capability	
Retirement	Store, archive, or dispose of the system	

This table is excerpted from ISO/IEC TR 24748-1 (2010), Table 1 on page 14, with permission from the ANSI on behalf of the ISO.© ISO 2010. All rights reserved.

Consequently, a discussion of system life cycle stages does not imply that the project should follow a predetermined set of activities or processes unless they add value toward achieving the final goal. Serial time progression is not inherently part of a life cycle model (stages do not necessarily occur serially one after another in time sequence). One possible example of the "progression" of a system through its life cycle is shown in Figure 3.2. When, in this handbook, reference is made to an earlier, prior, next, subsequent, or later stage, this type of model must be kept in mind to avoid confusion by inferring serial time sequencing. Subsequent chapters of this handbook will define processes and activities to meet the objectives of these life cycle stages. Because of the iterative nature of SE, specific processes are not aligned to individual life cycle stages. Rather, the entire set of SE processes is considered and applied at each stage of life cycle development as appropriate to the scope and complexity of the project.

表 3.1 一般生命周期阶段、其目的和决策门选项

生命周期阶段	目的	决策门
概念	定义问题空间 1. 探索性研究 2. 概念选择 特征化解决方案空间 识别利益攸关者的需要 探索构想和技术 细化利益攸关者的需要 探索可行的概念 提出切实可行的解决方案	决策选项 ● 继续进行下一阶段 ● 继续并响应某些行动项 ● 延续本阶段工作 ● 返回前一阶段 ● 暂停项目活动 ● 终止项目
开发	定义/细化系统需求 创建解决方案的描述——架构和设计 实施最初的系统 综合、验证并确认系统	
生产	生产系统 检验和验证	
使用	运行系统以满足用户的需要	
维持	提供持续的系统能力	
退出	系统的封存、归档或处置	

经代表 ISO 的 ANSI 许可后,此表摘自 ISO/IEC TR 24748-1(2010),第 14 页表 1。ISO 版权所有,©2010。版权所有。

因此,系统生命周期阶段的论述并不意味着项目应遵守预定的活动或流程集,除非其增加达成最终目标的价值。连续时间的进展并不是生命周期模型的固有部分(阶段不一定按时序一个接一个地连续出现)。系统贯穿于其生命周期内的"进展"的一个可能示例如图 3.2 所示。在本手册中,当引用早期阶段、先前阶段、下一阶段、后续阶段或后期阶段时,必须牢记通过推断连续时序使这种类型的模型避免出现混淆。本手册的后续章节将定义流程和活动以满足这些生命周期阶段的目标。由于 SE 迭代的本质属性,特定流程与单独的生命周期阶段没有对准。相反,根据项目的范围和复杂性在生命周期开发的每个阶段适当考虑和应用完整的 SE 流程集。

GENERIC LIFE CYCLE STAGES

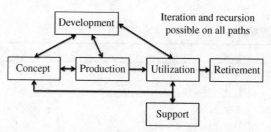

FIGURE 3.2 Life cycle model with some of the possible progressions. This figure is excerpted from ISO/IEC TR 24748-1 (2010), Figure 7 on page 13, with permission from the ANSI on behalf of the ISO.© ISO 2010. All rights reserved.

Figure 3.3 compares the generic life cycle stages to other life cycle viewpoints. For example, the concept stage is aligned with the study period for commercial projects and with the presystem acquisition and the project planning period in the US Departments of Defense and Energy, respectively. Typical decision gates are presented in the bottom line.

FIGURE 3.3 Comparisons of life cycle models. Derived from forsberg et al. (2005), Figure 7.2. Reprinted with permission from Kevin Forsberg. All other rights reserved.

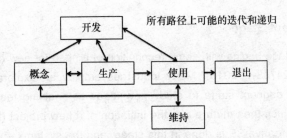

图 3.2 具有一些可能的进展的生命周期模型。经代表 ISO 的 ANSI 许可后，此表摘自 ISO/IEC TR 24748-1（2010），第 13 页图 7。ISO 版权所有，©2010。版权所有。

图 3.3 将一般生命周期阶段与其他的生命周期视角进行比较。例如，将概念阶段分别与商业项目的研究周期、美国国防部的前期系统采办及美国能源部的项目计划周期进行对准。在图最底部的线上给出了典型的决策门。

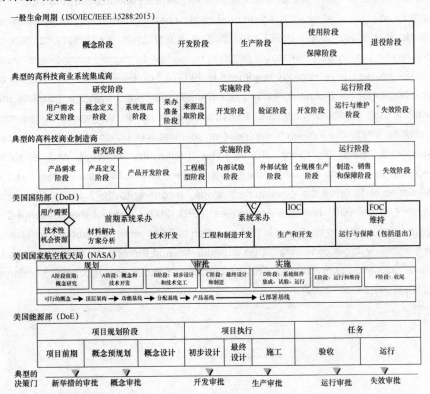

图 3.3 生命周期模型的比较。来自 Forsberg 等（2005），图 7.2。经 Kevin Forsberg 许可后转载。版权所有。

3.3.1 Concept Stage

The concept stage begins with some recognition of a need for new or modified SOI (ISO/IEC TR 24748-1, 2010). Many industries employ an exploratory research activity in the concept stage to study new ideas or enabling technologies and capabilities, which then mature into the initiation of a new project (for the SOI). A great deal of creative SE is done in this stage, and the systems engineer leading these studies is likely to follow a new idea into the concept selection, perhaps as project champion. Often, the exploratory research activity identifies the enabling technologies. If the work is done properly in early stages of the life cycle, it is possible to avoid recalls and rework in later stages.

Many life cycle models show the process beginning with "requirements" or "user requirements." In fact, the process begins earlier with interactions and studies to understand potential new organizational capabilities, opportunities, or stakeholder needs. It is critical that in these early studies, a high-level, preliminary concept be created and explored to whatever depth is necessary to identify technological risks and to assess the technology readiness level (TRL) of the project. The focus is on studying potential technologies and determining the state of what is possible and what is not. In some instances, the project may be an outgrowth of research activities where the research engineer or scientist has no connection to a user-supported need (Forsberg, 1995). The preliminary concept and enabling technologies need to be identified early, and issues arising from the studies need to be addressed during the development stage, according to the National Research Council of the (US) National Academies (NRC, 2008). One of the challenges in developing alternate concepts is that we often build on what has worked well for us in the past, without considering true alternatives, and thereby miss opportunities to make dramatic improvements. This problem has been widely recognized (Adams, 1990; Christensen, 2000).

3.3.1 概念阶段

概念阶段开始于对新的或改进的 SOI（ISO/IEC TR 24748-1，2010）的需要的某种识别。许多行业使用概念阶段中的探索性研究活动来研究诸多新的理念或使能技术和能力，然后使其发展进入到一个新项目（SOI）的启动阶段。大量的创造性 SE 在该阶段中完成，领导这些研究的系统工程师，也许作为项目推动者，有可能将一个新构想引入到概念选择。探索性研究活动常常识别使能技术。如果这项工作在生命周期的早期阶段被恰当地完成，则可能避免后期的召回和返工。

许多生命周期模型均表明流程开始于"需求"或"用户需求"。实际上，流程更早地开始于以理解潜在的新的组织能力、机遇或利益攸关者需要为目的的交互和研究。至关重要的是在这些早期研究中，创建一个高层级的初步概念，并探索识别技术风险和评估项目技术成熟度（TRL）的必要深度。焦点是研究潜在技术并确定可能及不可能的状态。在某些实例中，项目可能是某些研究活动的产物，在这些活动中研究工程师或科学家未与用户支持的需要建立联系（Forsberg，1995）。根据（美国）国家研究院的国家研究委员会的要求（NRC，2008），初步概念和使能技术需要在早期被识别，且在研究中出现的议题需要在开发阶段去应对。备选概念开发中的挑战之一是我们往往依赖于过去做得很好的东西，而不考虑真正的备选方案，因此错失做出根本性改善的机会。此问题已经被广泛认识到（Adams，1990；Christensen，2000）。

The preliminary concept will also be used to generate early cost and schedule projections for the project if it moves ahead. Key activities during exploratory research are to clearly define the problem space, characterize the solution space, identify business or mission requirements and stakeholder needs, and, while avoiding any design work, provide an estimate of the cost and schedule for the full-scale development. Incomplete SE in this stage can lead to poor cost and schedule projections, as well as poor understanding of technical alternatives, resulting in poor trades among the alternatives. For example, the Mars Science Laboratory rover, scheduled for launch in 2009, had to be "delayed because of technical glitches." This resulted in missing the launch window, causing a 2-year delay and a 35% cost growth over the approved development costs. Program critics, however, claimed a 400% cost growth based on the early concept studies, and they threatened the project with cancellation as a result (Achenbach, 2009).

The preliminary concept is a starting point, not an end point, as the project moves into the concept selection activity of the concept stage. The preliminary concept is not put under configuration control, and the key output from exploratory research is a clearer understanding of the business or mission requirements and the stakeholder needs, an assessment of the technology's readiness to move to the next stage, and a rough estimate of the project cost and schedule requirements and technical feasibility to first article delivery.

Concept selection is the second activity of the concept stage. The concept selection activity is a refinement and broadening of the studies, experiments, and engineering models pursued during the exploratory research activity. The first step is to identify, clarify, and document the stakeholders' conceptual operation of the system across the different stages of use and the environments it is to be used in. The operational concept (OpsCon) effort should be undertaken to include any changes caused by changes in the manufacture processes or materials, changes in interface standards, or new feature enhancements being added that can drive various aspects of concept selection of the system.

一般生命周期阶段

如果项目向前推进，初步概念也将被用于生成项目早期的成本和进度预测。探索性研究期间的关键活动是清晰地定义问题空间，特征化解决方案空间，识别业务或使命任务需求以及利益攸关者需要，且在避免任何设计工作的同时，为全规模开发提供成本和进度的估计。本阶段不完整的 SE 可能导致成本和进度的预测欠佳以及对技术备选方案的理解欠佳，导致备选方案之间的权衡欠佳。例如，火星科学实验室漫游车预计 2009 年发射，但"由于技术故障不得不延期"，这导致错过发射窗口，造成两年的延迟且实际成本较批准的开发成本高出 35%。然而该计划的批评者声称，基于早期概念研究成本增长了 400%，结果是他们威胁到该项目导致项目被取消（Achenbach, 2009）。

随着项目进入概念阶段的概念选择活动，初步概念是一个起点而不是终点。初步概念不被置于构型控制下，并且来自探索性研究的关键输出是对业务或使命任务需求及利益攸关者需要的更清晰的理解；是对进入下一阶段的技术成熟度的评估；也是对项目成本和进度需求以及首件交付的技术可行性的粗略估计。

概念选择是概念阶段的第二个活动。概念选择活动是对探索性研究活动期间所开展的研究、实验和工程模型的细化和拓展。第一步是覆盖不同的使用阶段及其将来使用所在的环境来对系统的利益攸关者的概念性运行进行识别、明确并使其文件化。应开展运行概念（OpsCon）工作，包括制造流程或材料中的变更所导致的任何变更、接口标准中的变更或能够驱动系统概念选择的各个不同方面所附加的新特征增强。

GENERIC LIFE CYCLE STAGES

During the concept stage, the team begins in-depth studies that evaluate multiple candidate concepts and eventually provide a substantiated justification for the system concept that is selected. As part of this evaluation, mock-ups may be built (for hardware) or coded (for software), engineering models and simulations may be executed, and prototypes of critical elements may be built and tested. Engineering models and prototypes of critical elements are essential to verify the feasibility of concepts, to aid the understanding of stakeholder needs, to explore architectural trade-offs, and to explore risks and opportunities. These studies expand the risk and opportunity evaluation to include affordability assessment, environmental impact, failure modes, hazard analysis, technical obsolescence, and system disposal. Issues related to integration and verification must also be explored for each alternate system concept, since these can be discriminators in system selection. The systems engineer facilitates these analyses by coordinating the activities of engineers from many disciplines. Key objectives are to provide confidence that the business case is sound and the proposed solutions are achievable.

The concept stage may include system and key system element-level concept and architecture definition and integration, verification, and validation (IV&V) planning. Early validation efforts align requirements with stakeholder expectations. The system capabilities specified by the stakeholders will be met by the combination of system elements. Problems identified for individual system element-level concepts should be addressed early to minimize the risk that they fall short of the required functionality or performance when the elements are finally designed and verified.

FIGURE 3.4 Importance of the concept stage. DILBERT © 1997 Scott Adams. Used with permission from UNIVERSAL UCLICK. All rights reserved.

Many projects are driven by eager project champions who want "to get on with it." They succumb to the temptation to cut short the concept stage, and they use exaggerated projections to support starting development without adequate understanding of the challenges involved, as comically illustrated in Figure 3.4. Many commissions reviewing failed systems after the fact have identified insufficient or superficial study in the concept stage as a root cause of failure.

在概念阶段期间，团队开始深入研究——评估多个候选概念并最终提供所选系统概念的经证实的正当理由。作为这个评估的一部分，初级原型可以被构建（用于硬件）或编码（用于软件），工程模型和仿真可以被执行，以及关键元素的原型机可以被构建和测试。关键元素的工程模型和原型机对于验证概念的可行性而言是必不可少的，以便帮助理解利益攸关者需要，探究架构权衡并探究风险和机会。这些研究扩展了对风险和机会的评估，涵盖可承受性评估、环境影响、失效模式、危险分析、技术淘汰和系统退出。还必须对每个备选系统概念探究出与综合及验证有关的问题，因为这些可在系统选择中作为鉴别项。系统工程师通过协调来自多学科的工程师的活动来促进这些分析工作。关键目的是提供信心，即业务案例是完好的且所建议的解决方案是可达成的。

概念阶段可包括系统和系统关键元素层级的概念和架构的定义，以及综合、验证与确认（IV&V）的规划。早期的确认工作使需求与利益攸关者的期望相一致。利益攸关者指定的系统能力将由系统元素的组合得以满足。在系统个别的元素层级的概念中识别出的问题应尽早解决，以使元素被最终设计和验证时无法达到所需功能性或性能的风险最小化。

图 3.4 概念阶段的重要性。版权归 Scott Adams 的 DILBERT 所有© 1997。经 UNIVERSAL UCLICK 允许后使用。版权所有。

许多项目由试图"抓紧干下去"的急迫的项目推动者驱动。他们经不住诱惑而缩短概念阶段，并在没有充分了解所涉及的挑战的情况下，以浮夸的推测支持开发的启动，正如图 3.4 诙谐的阐明。许多负责审查失败系统的委员会事后确定——概念阶段中不充分的或肤浅的研究是失败的根原因。

GENERIC LIFE CYCLE STAGES

3.3.2 Development Stage

The development stage defines and realizes a SOI that meets its stakeholder requirements and can be produced, utilized, supported, and retired. The development stage begins with the outputs of the concept stage. The primary output of this stage is the SOI. Other outputs can include a SOI prototype, enabling system requirements (or the enabling systems themselves), system documentation, and cost estimates for future stages (ISO/IEC TR 24748-1, 2010).

Business and mission needs, along with stakeholder requirements, are refined into system requirements. These requirements are used to create a system architecture and design. The concept from the previous stage is refined to ensure all system and stakeholder requirements are satisfied. Requirements for production, training, and support facilities are defined. Enabling systems' requirements and constraints are considered and incorporated into the design. System analyses are performed to achieve system balance and to optimize the design for key parameters.

One of the key activities of the development stage is to specify, analyze, architect, and design the system so that the system elements and their interfaces are understood and specified. Hardware and software elements are fabricated and coded.

Operator interfaces are specified, tested, and evaluated during the development stage. Operator and maintainer procedures and training are developed and delivered to ensure humans can interface with the SOI.

Feedback is obtained from both external and internal stakeholders through a series of technical reviews and decision gates. Projects that are not showing acceptable progress may be redirected or even terminated.

The development stage includes detailed planning and execution of IV&V activities. The planning for these activities needs to take place early to ensure that adequate facilities and other resources are available when needed. A source of additional information about IV&V and the significance for project cost and risk when these activities are optimized was the subject of the European Union SysTest program (Engel, 2010).

3.3.2 开发阶段

开发阶段定义和实现满足其利益攸关者需求的 SOI，并且可以被生产、使用、支持和退出。开发阶段开始于概念阶段的输出。该阶段的主要输出是 SOI。其他输出可包括 SOI 原型机、使能系统需求（或使能系统本身）、系统文档以及未来阶段的成本估算（ISO/IEC TR 24748-1, 2010）。

业务和使命任务需要连同利益攸关者需求被细化为系统需求。这些需求用于创建一个系统架构和设计。对前一阶段中的概念进行细化以确保系统和利益攸关者的所有需求得以满足。定义、生产、培训和维持保障设施的需求。考虑使能系统的需求和约束，并将它们纳入到设计中。进行系统分析，以达成系统平衡并优化关键参数的设计。

开发阶段的关键活动之一是规范、分析、构架和设计系统以便对系统元素及其接口进行理解和规范。对硬件和软件元素进行构造和编码。

在开发阶段期间规范、试验并评估操作者的界面。开发并交付操作者和维护者的程序及培训，以确保人能够与 SOI 结合。

通过一系列技术评审和决策门从外部和内部利益攸关者获取反馈。未展示出可接受的进展的项目可能被重新定向或甚至终止。

开发阶段包括 IV&V 活动的详细规划和执行。需要尽早地对这些活动进行规划以确保适当的设施和其他资源在需要时可用。当优化这些活动时，关于 IV&V 以及项目成本与风险重要性的一个额外的信息来源就是欧盟的 SysTest 计划的主题（Engel, 2010）。

3.3.3 Production Stage

The production stage is where the system is produced or manufactured. Product modifications may be required to resolve production problems, to reduce production costs, or to enhance product or system capabilities. Any of these may influence system requirements and may require system reverification or revalidation. All such changes require SE assessment before changes are approved.

3.3.4 Utilization Stage

The utilization stage is where the system is operated in its intended environment to deliver its intended services. Product modifications are often planned for introduction throughout the operation of the system. Such upgrades enhance the capabilities of the system. These changes should be assessed by systems engineers to ensure smooth integration with the operational system.

For large complex systems, midlife upgrades can be substantial endeavors requiring SE effort equivalent to a major program.

3.3.5 Support Stage

The support stage is where the system is provided services that enable continued operation. Modifications may be proposed to resolve supportability problems, to reduce operational costs, or to extend the life of a system. These changes require SE assessment to avoid loss of system capabilities while under operation.

3.3.6 Retirement Stage

The retirement stage is where the system and its related services are removed from operation. SE activities in this stage are primarily focused on ensuring that disposal requirements are satisfied. Planning for retirement is part of the system definition during the concept stage. Experience has repeatedly demonstrated the consequences when system retirement is not considered from the outset. Early in the twenty-first century, many countries have changed their laws to hold the developer of a SOI accountable for proper end-of-life disposal of the system.

3.4 LIFE CYCLE APPROACHES

Various life cycle models, such as the Waterfall (Royce, 1970), Spiral (Boehm, 1986), and Vee (Forsberg and Mooz, 1991), are useful in defining the start, stop, and process activities appropriate to the life cycle stages.

3.3.3 生产阶段

生产阶段是系统被生产或制造的阶段。可能需要产品的修改以解决生产问题，以降低生产成本，或以增强产品或系统的能力。上述任何一点均可能影响系统需求，并且可能要求系统重新验证或重新确认。所有这些变更都要求在变更被批准前进行 SE 评估。

3.3.4 使用阶段

使用阶段是系统在其意图的环境中运行以交付其意图的服务的阶段。通常贯穿于系统运行内有计划地引入产品修改。这样的升级能提高系统的能力。这些变更应由系统工程师评估以确保其与运行的系统顺利综合。

对于大型复杂系统而言，中期升级可能要求大量的 SE 工作，等同于一个重要的计划项目。

3.3.5 支持阶段

维持阶段是为系统提供服务使之能持续运行的阶段。可建议进行修改以解决可维持性问题，以降低运行成本，或延长系统寿命。这些更改需要进行 SE 评估以避免运行时损失系统能力。

3.3.6 退出阶段

退出阶段是系统及其相关服务从运行中移除的阶段。这一阶段中的 SE 活动主要集中于确保退出需求被满足。对退出的规划是概念阶段期间系统定义的一部分。经验反复证明了从一开始不考虑系统退役的后果。21 世纪早期，许多国家已经修改了他们的法律，使 SOI 的开发者负责系统生命终止时恰当的处置。

3.4 生命周期方法

各种不同的生命周期模型，如瀑布模型（Royce, 1970）、螺旋模型（Boehm, 1986）和 V 形模型（Forsberg 和 Mooz, 1991），在定义适合于生命周期阶段的起始、停止和流程活动时均是有用的。

GENERIC LIFE CYCLE STAGES

Graphical representations of life cycle stages tend to be linear, but this hides the true incremental, iterative, and recursive nature of the underlying processes. The approaches that follow imply full freedom to choose a development model and are not restricted to sequential methods.

3.4.1 Iteration and Recursion

Too often, the system definition is viewed as a linear, sequential, single pass through the processes. However, valuable information and insight need to be exchanged between the processes, in order to ensure a good system definition that effectively and efficiently meets the mission or business needs. The application of iteration and recursion to the life cycle processes with the appropriate feedback loops helps to ensure communication that accounts for ongoing learning and decisions. This facilitates the incorporation of learning from further analysis and process application as the technical solution evolves.

Figure 3.5 shows an illustration of iteration and recursion of the processes. Iteration is the repeated application of and interaction between two or more processes at a given level in the system structure or hierarchy. Iteration is needed to accommodate stakeholder decisions and evolving understanding, account for architectural decisions/constraints, and resolve trades for affordability, adaptability, feasibility, resilience, etc. Although the figure only shows a subset of the life cycle technical processes, there can be iteration between any of the processes. For example, there is often iteration between system requirements definition and architecture definition. In this case, there is a concurrent application of the processes with iteration between them, where the evolving system requirements help to shape the architecture through identified constraints and functional and quality requirements. The architecture trades, in turn, may identify requirements that are not feasible, driving further requirements analysis with trades that change some requirements. Likewise, the design definition could identify the need to reconsider decisions and trades in the requirements definition or architecture definition processes. Any of these can invoke additional application of the system analysis and decision management processes.

Recursion is the repeated application of and interaction of processes at successive levels in the system structure. The technical processes are expected to be recursively applied for each successive level of the system structure until the level is reached where the decision is made to make, buy, or reuse a system element. During the recursive application of the processes, the outputs at one level become inputs for the next successive level (below for system definition, above for system realization).

生命周期阶段的图形化的表示形式趋于线性,但这其中隐藏着底层流程的真实的渐进、迭代和递归的本质属性。随之而来的途径意味着完全自由地选择开发模型,并不局限于顺序的方法。

3.4.1 迭代和递归

系统定义太过经常被看做是线性的、顺序的、一次通过流程的。然而,需要在流程之间交换有价值的信息和深入理解,以便确保产生有效且高效地满足使命任务或业务需要的良好的系统定义。迭代和递归在具有适当反馈回路的生命周期流程中的应用,有助于确保产生对持续的学习和决策的沟通。随着技术解决方案的演进,这有助于纳入从进一步分析和流程应用中学习到的知识。

图 3.5 表明流程的迭代和递归的详细阐述。迭代是在系统结构或层级结构的给定层级上两个或更多个流程的重复应用及其之间的交互。迭代需要容纳利益攸关者的决策及持续演进的理解,解释架构决策/约束的原因,并解决可承受性、适应性、可行性、恢复性等方面的权衡。尽管该图仅表明生命周期技术流程的子集,但任何流程之间还可能存在迭代。例如,系统需求定义与架构定义之间往往存在迭代。在这种情况下,它们之间的迭代存在流程的并行应用,在该应用中,持续演进的系统需求通过所识别的约束及功能和质量需求帮助塑形架构。架构权衡可能转而识别出不可行的需求,驱使利用对某些需求进行更改的权衡做出进一步的需求分析。同样地,设计定义可识别决策重新考虑的需要以及在需求定义或架构定义流程中的权衡。任何一种情况都可调用系统分析流程和决策管理流程的再次应用。

递归是在系统结构的相继层级上流程的重复应用及交互。期望技术流程递归地应用于系统结构的每一相继层级,直至对制造、购买或复用系统元素做出决策所在的层级达到为止。在流程的递归应用期间,某一层级上的输出变成下一相继层级的输入(下面是系统定义,上面是系统实现)。

GENERIC LIFE CYCLE STAGES

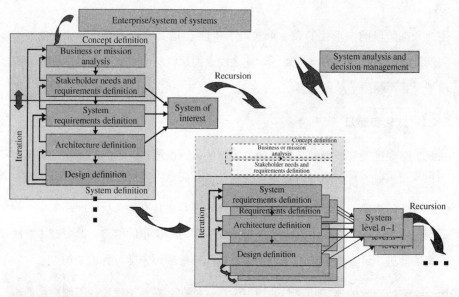

FIGURE 3.5 Iteration and recursion. Reprinted with permission from Garry Roedler. All other rights reserved.

3.4.2 Sequential Methods

On projects where it is necessary to coordinate large teams of people working in multiple companies, sequential approaches provide an underlying framework to provide discipline to the life cycle processes. Sequential methods are characterized by a systematic approach that adheres to specified processes as the system moves through a series of representations from requirements through design to finished product. Specific attention is given to the completeness of documentation, traceability from requirements, and verification of each representation after the fact.

The strengths of sequential methods are predictability, stability, repeatability, and high assurance. Process improvement focuses on increasing process capability through standardization, measurement, and control. These methods rely on the "master plans" to anchor their processes and provide project-wide communication. Historical data is usually carefully collected and maintained as inputs to future planning to make projections more accurate (Boehm and Turner, 2004).

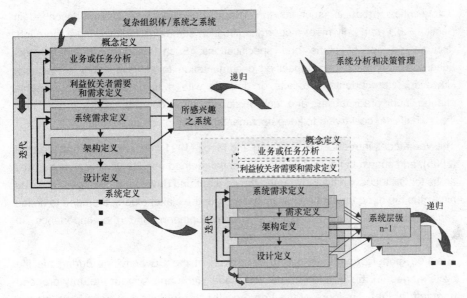

图 3.5　迭代和递归。经 Garry Roedler 许可后转载。版权所有。

3.4.2　顺序方法

在一些需要协调多家公司人员参与的大型团队项目中，顺序法提供一种底层的框架，为生命周期流程提供规程。顺序法的特征在于当系统通过一系列表达形式从需求到设计再到最终产品演进时，始终遵守规定流程的系统化方法。特别关注文档的完整性、需求的可追溯性以及每种表达形式的事后验证。

顺序方法的优势是可预测性、稳定性、可重复性和高保证性。流程改进集中于通过标准化、衡量和控制来提高流程能力。这些方法依靠"主计划"锚定其流程并提供整个项目范围的沟通。历史数据通常被仔细地收集和维护来作为未来计划的输入，以使计划更加精确（Boehm 和 Turner，2004）。

GENERIC LIFE CYCLE STAGES

Safety-critical products, such as the Therac-25 medical equipment described in Section 3.6.1, can only meet modern certification standards by following a thorough, documented set of plans and specifications. Such standards mandate strict adherence to process and specified documentation to achieve safety or security. However, unprecedented projects or projects with a high rate of unforeseeable change, poor predictability, and lack of stability often degrade, and a project may incur significant cost trying to keep documentation and plans up to date.

The Vee model, introduced in Forsberg and Mooz (1991), described in Forsberg et al. (2005), and shown in figure 3.6, is a sequential method used to visualize various key areas for SE focus, particularly during the concept and development stages. The Vee highlights the need for continuous validation with the stakeholders, the need to define verification plans during requirements development, and the importance of continuous risk and opportunity assessment.

The Vee model provides a useful illustration of the SE activities during the life cycle stages. In this version of the Vee model, time and system maturity proceed from left to right. The core of the Vee (i.e., those products that have been placed under configuration control) depicts the evolving baseline from stakeholder requirements agreement to identification of a system concept to definition of elements that will comprise the final system. With time moving to the right, the evolving baseline defines the left side of the core of the Vee, as shown in the shaded portion of figure 3.7.

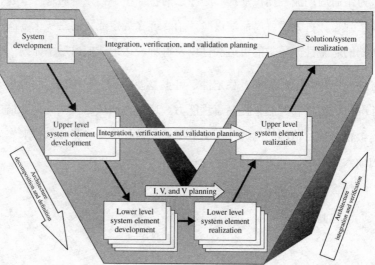

FIGURE 3.6 Vee model. Derived from Forsberg et al. (2005), Figure 7.10. Reprinted with permission from Kevin Forsberg. All other rights reserved.

安全关键产品，如 3.6.1 节所述的 Therac-25 医疗设备，只能通过遵循全面的、文件化的计划和规范集来满足最新的认证标准。这类标准要求严格遵守流程和规定文件，以实现安全或安保。然而，一些前无先例的项目或者某些具有大量无法预见变更的项目，预测性差且稳定性的缺乏通常会降低，并且为了试图保持文件和计划的时效性可能带来巨大的项目成本。

由 Forsberg 和 Mooz（1991）引入、Forsberg 等人（2005）描述的 V 形模型，图 3.6 所示，是用于可视化 SE 焦点的各种关键领域的一种顺序方法，特别是概念阶段和开发阶段。V 形模型强调与利益攸关者持续确认需要、在需求开发期间定义验证计划的需要以及持续进行风险和机会评估的重要性。

V 形模型提供一个在生命周期各个阶段期间 SE 活动的非常有用的图解说明。在此版本的 V 形模型中，时间和系统成熟度从左到右推进。V 形模型的核心（即那些置于构型控制下的产品）描述从利益攸关者需求协议到系统概念识别再到将要构成最终系统的元素定义的演进基线。随着时间向右推移，演进中的基线定义 V 形模型核心的左侧，如图 3.7 中阴影部分所示。

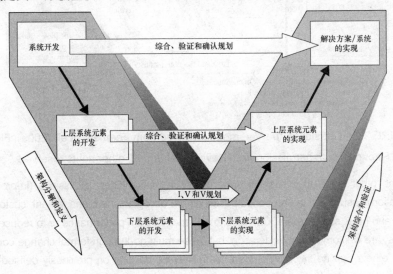

图 3.6 V 形模型。来自 Forsberg 等（2005），图 7.10。经 Kevin Forsberg 许可后转载。版权所有。

GENERIC LIFE CYCLE STAGES

A key attribute of the Vee model is that time and maturity move from the left to the right across the diagram, as shown in Figure 3.7. At any instant of time, the development team then can move their perspective only along the vertical arrow, from the highest level of the system requirements down to the lowest level of detail. The off-core opportunity and risk management investigations going downward are addressing development options to provide assurance that the baseline performance being considered can indeed be achieved and to initiate alternate concept studies at the lower levels of detail to determine the best approach. These downward off-core investigations and development efforts are entirely under control of the development team.

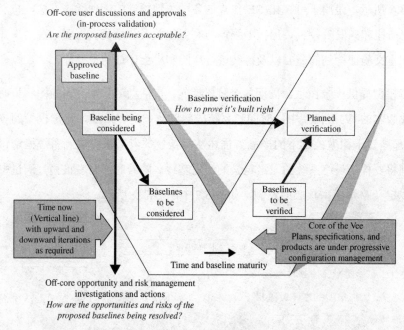

FIGURE 3.7   Left side of the Vee model. Derived from Forsberg et al. (2005), Figure 7.11. Reprinted with permission from Kevin Forsberg. All other rights reserved.

On the other hand, the essential upward off-core stakeholder discussions (in-process validation) ensure that the proposed baselines are acceptable to management, customer, user, and other stakeholders. Changes to enhance system performance or to reduce risk or cost are welcome for consideration, but these must go through formal change control, since others outside the development team may be building on previously defined and released design decisions. The power of understanding the significance of the off-core studies is illustrated in a National Aeronautics and Space Administration (NASA) Jet Propulsion Laboratory (JPL) report (Briedenthal and Forsberg, 2007).

V形模型的关键属性是时间和成熟度跨越该图从左到右推进，如图3.7所示。在任何时间瞬间，开发团队当时可以使他们的视角仅沿着垂直箭头从系统需求的最高层级移动到最低的细节层级。向下的"非核心"机会和风险管理的调查研究是正在应对的开发选项，用以确保所考虑的基线性能确实能够达成并在更低的详细层级上启动备用的概念研究以确定最佳的途径。这些向下的"非核心"调查研究和开发工作完全处于开发团队的控制下。

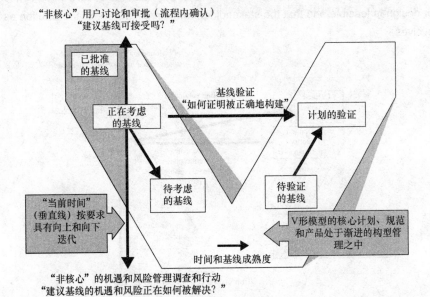

图3.7 V形模型的左侧。来自Forsberg等（2005），图7.11。经Kevin Forsberg许可后转载。版权所有。

另一方面，基本的向上"非核心"利益攸关者的讨论（流程内确认）确保建议基线对于管理者、客户、用户和其他利益攸关者来说是可接受的。考虑为提高系统性能或为减少风险和成本而做出的变更是受欢迎的，但这些变更必须经历正式的变更控制，因为外部的开发团队可能建立在先前定义的已发布的设计决策之上。（美国）国家航空航天管理局（NASA）喷气推进实验室（JPL）报告（Briedenthal和Forsberg, 2007）中详细阐述了理解"非核心"研究重要性的动力。

GENERIC LIFE CYCLE STAGES

As entities are implemented, verified, and integrated, the right side of the core of the Vee is executed. Figure 3.8 illustrates the evolving baseline as system elements are integrated and verified. Since one can never go backward in time, all iterations in the Vee are performed on the vertical "time now" line. Upward iterations involve the stakeholders and are the in-process validation activities that ensure that the proposed baselines are acceptable. The downward vertical iterations are the essential off core opportunity and risk management investigations and actions. In each stage of the system life cycle, the SE processes iterate to ensure that a concept or design is feasible and that the stakeholders remain supportive of the solution as it evolves.

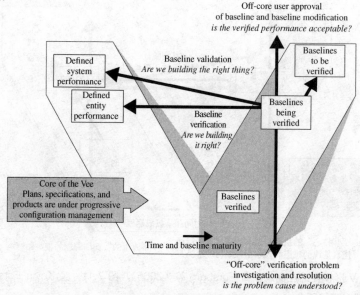

FIGURE 3.8 Right side of the Vee model. Derived from Forsberg et al. (2005), Figure 7.12. Reprinted with permission from Kevin Forsberg. All other rights reserved.

3.4.3 Incremental and Iterative Methods

Incremental and iterative development (IID) methods have been in use since the 1960s (Larman and Basili, 2003). They represent a practical and useful approach that allows a project to provide an initial capability followed by successive deliveries to reach the desired SOI. The goal is to provide rapid value and responsiveness.

随着实体的实施、验证和综合,V形模型核心的右侧被执行。图 3-8 阐明当综合和验证系统元素时演进的基线。因为时间是无法倒退的,V 形模型中的所有迭代都在垂直的"当前时间"线上被执行。向上迭代涉及利益攸关者,且是确保建议基线可接受的流程内的确认活动。向下垂直迭代是基本的"非核心"机遇和风险管理的调查研究与行动。在系统生命周期的每个阶段中,SE 流程的迭代确保概念或设计的可行;并随着解决方案不断演进,确保利益攸关者继续保持对该解决方案的支持。

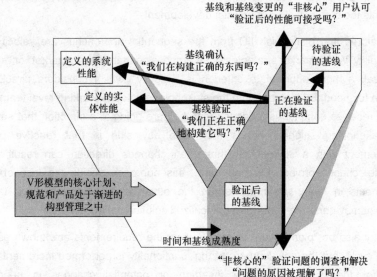

图 3.8　V 形模型的右侧。来自 Forsberg 等(2005),图 7.12。经 Kevin Forsberg 许可后转载。版权所有。

3.4.3　渐进和迭代方法

20 世纪 60 年代以来就已经开始使用渐进和迭代开发(IID)方法(Larman 和 Basili,2003),其代表一种实际的、有用的方法,允许为项目提供一个初始能力,随之提供相继交付以达到期望的 SOI。目标在于提供快速的价值和响应能力。

The IID approach is used when the requirements are unclear from the beginning or the stakeholder wishes to hold the SOI open to the possibilities of inserting new technology. Based on an initial set of assumptions, a candidate SOI is developed and then assessed to determine if it meets the stakeholder needs or requirements. If not, another evolutionary round is initiated, and the process is repeated until a system is delivered to satisfied stakeholders or until the organization decides to terminate the effort.

Most literature agrees that IID methods are best applied to smaller, less complex systems or to system elements. The focus is on flexibility and on allowing selected events to be taken out of sequence when the risk is acceptable. Tailoring in this way highlights the core activities of product development.

The features that distinguish IID from the sequential approaches are velocity and adaptability. While market strategies often emphasize that "time to market" or "speed" is critical, a more appropriate criterion is "velocity," which considers direction in addition to speed. By incorporating the stakeholders into the working-level teams, the project receives continuous feedback that they are going in a direction that satisfies the stakeholders' highest needs first. One downside is that reactive project management with a stakeholder that often changes direction can result in an unstable, chaotic project. On one hand, this approach avoids the loss of large investments in faulty assumptions; on the other hand, emphasis on a tactical viewpoint may generate short-term or localized solution optimizations.

IIDs may also be "plan driven" in nature when the requirements are known early in the life cycle, but the development of the functionality is performed incrementally to allow for the latest technology insertion or potential changes in needs or requirements. A specific IID methodology called evolutionary development (Gilb, 2005) is common in research and development (R&D) environments. Figure 3.9 illustrates how this approach was used in the evolution of the tiles for the NASA space shuttle.

An example of an incremental and iterative method is the Incremental Commitment Spiral Model (ICSM) (Boehm et al., 2014). The ICSM builds on the strengths of current process models, such as early V&V concepts in the Vee model, concurrency concepts in the concurrent engineering model, lighter-weight concepts in the agile and lean models, risk-driven concepts in the spiral model Boehm, 1996, the phases and anchor points in the rational unified process (RUP) (Kruchten, 1999), and recent extensions of the spiral model to address SoS capability acquisition (Boehm and Lane, 2007).

当需求从一开始就不清晰或利益攸关者希望 SOI 对新技术的引入的可能性保持开放时，则使用 IID 方法。基于一系列最初的假设，开发候选的 SOI，然后对其进行评估以确定是否满足利益攸关者需要或需求。若不满足，则启动另一轮演进，并重复该流程，直到交付的系统满足利益攸关者的要求或直到组织决定终止这种努力。

多数文献一致认为 IID 方法最适用于较小的、不太复杂的系统或系统元素。这种方法的重点在于灵活性以及当风险可接受时允许所选事件从序列中排除。以这种方式的剪裁突出了产品开发的核心活动。

IID 方法区别于顺序方法的特征是速度和适应性。当市场战略经常强调"上市时间"或"快速率"至上时，更适合的准则是"速度"，它既考虑了速度的大小，又考虑了方向。通过让利益攸关者加入工作层级团队，项目接收团队工作方向优先满足利益攸关者的最高需要的持续反馈。这种方法的一个缺点是面对经常改变方向的利益攸关者时，这种反应式项目管理可能产生不稳定和混乱的项目。一方面，这种方法避免由于错误假设引起的巨大投资损失；另一方面，对战术观点的强调可能产生短期或局部解决方案最优化。

当需求在生命周期的早期就已知，但为了允许最新技术的引入或需要和需求的潜在变化而渐进地执行功能性开发时，IID 本质上也可以是"计划驱动的"。一种特定的 IID 方法论被称为演进式开发（Gilb, 2005），在研究与开发（R&D）环境中很常用。图 3.9 阐明这种方法如何被用于 NASA 航天飞机隔热瓦的演进之中。

渐进和迭代方法的一个示例是渐进承诺螺旋模型（ICSM）（Boehm 等，2014）。ICSM 建立于当前诸多流程模型的多重优势之上，如 V 形模型中早期的 V&V 概念、并行工程模型中的并行性概念、敏捷模型和精益模型中的更轻负荷的概念、螺旋模型（Boehm, 1996）中的风险驱动的概念、Rational 统一流程（RUP）（Kruchten, 1999）中的阶段和锚点以及螺旋模型的最近扩展，以便应对 SoS 能力采办（Boehm and Lane, 2007）。

GENERIC LIFE CYCLE STAGES

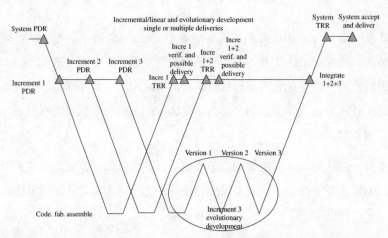

FIGURE 3.9   IID and evolutionary development. Derived from Forsberg et al. (2005), figure 19.18. Reprinted with permission from Kevin Forsberg. All other rights reserved.

A view of the ICSM is shown in figure 3.10. In the ICSM, each increment addresses requirements and solutions concurrently, rather than sequentially. ICSM also considers products and processes; hardware, software, and human factor aspects; and business case analyses of alternative product configurations or product line investments. The stakeholders consider the risks and risk mitigation plans and decide on a course of action. If the risks are acceptable and covered by risk mitigation plans, the project proceeds into the next spiral.

3.5   WHAT IS BEST FOR YOUR ORGANIZATION, PROJECT, OR TEAM?

Conway's law suggests that "organizations which design systems ... are constrained to produce designs which are copies of the communication structures of those organizations" (Conway, 1968). Systems thinking and SE help organizations avoid the pitfall of Conway's law by ensuring that system designs are appropriate to the problem being addressed.

Figure 3.11 presents another view of the ICSM. The top row of activities indicates that a number of system aspects are being concurrently engineered at an increasing level of understanding, definition, and development.

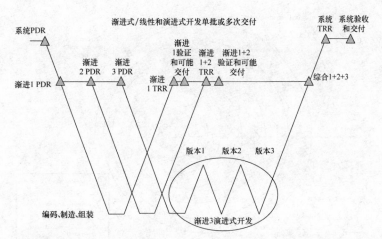

图 3.9 IID 和演进式开发。来自 Forsberg 等（2005），图 19.18。经 Kevin Forsberg 许可后转载。版权所有。

ICSM 的视图如图 3.10 所示。在 ICSM 中，每一轮渐进并行地应对需求和解决方案，而不是按顺序。ICSM 还考虑到产品和流程；硬件、软件和人员因素方面；以及备选的产品构型或产品线投资的业务案例分析。利益攸关者考虑到风险及风险缓解计划并决定行动的路线。如果风险可接受且在风险缓解计划中涵盖，则项目继续进行至下一螺旋。

3.5 什么最适合你的组织、项目或团队？

康威定律建议"设计系统的组织……受其约束而产生的设计正是这些组织沟通结构的复制品"（Conway, 1968）。系统思考和 SE 通过确保系统的设计适合于正在应对的问题，来帮助组织规避陷入康威定律的陷阱。

图 3.11 介绍 ICSM 的另一个视图。最顶行的活动表明正在理解、定义和开发的递增层级上对许多系统方面进行并行地工程设计。

GENERIC LIFE CYCLE STAGES

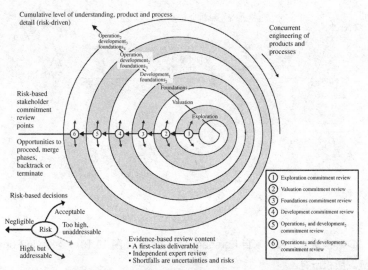

FIGURE 3.10 The incremental commitment spiral model (ICSM). From Boehm et al. (2014). Reprinted with permission from Barry Boehm. All other rights reserved.

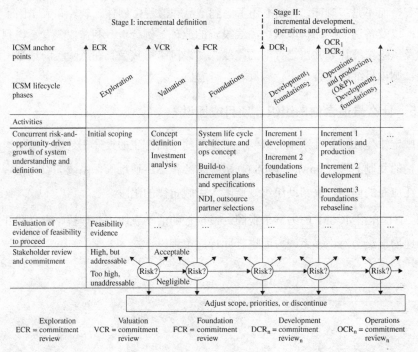

FIGURE 3.11 Phased view of the generic incremental commitment spiral model process. From Boehm et al. (2014). Reprinted with permission from Barry Boehm. All other rights reserved.

一般生命周期阶段

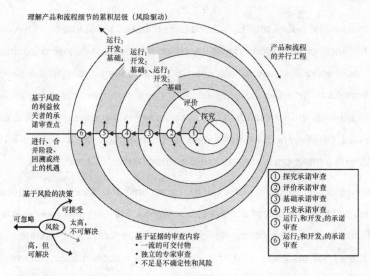

图 3.10 增量承诺螺旋模型（ICSM）。来自 Boehm 等。经 Barry Boehm 许可后转载。版权所有。

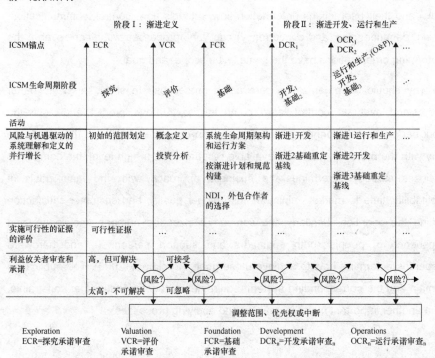

图 3.11 一般增量承诺螺旋模型流程的阶段视图。来自 Boehm 等。经 Barry Boehm 许可后转载。版权所有。

One of the earliest books on SE management (Chase, 1974) identified three simple criteria for such organizations: facilitate communications, streamline controls, and simplify paperwork. The way to effective SE management is not "in the direction of formal, formidable, massive documentation. It does, however, reside in the direction of creating a total environment which is conducive to the emergence and effective utilization of creative and inventive talents oriented toward achieving a system approach with a minimum of management encumbrances" (Chase, 1974).

Whenever someone (be it an individual or a company) wants to reach a desired end, they must perform a series of actions or operations. Further, they must consider the order of those actions, their dependencies, who will perform them, what they require and what they will generate, how long it will take to complete them, and what tools they will employ. Thus, individuals and organizations follow processes, be they predefined or ad hoc. Because process components (activities, products, agents, tools) and their interactions (information flow, artifacts flow, control, communication, timing, dependencies, and concurrency) can vary, processes will differ—even if the performing organizations have the same level, scope, and goal.

So why should an organization care about processes? In short, this is to better understand, evaluate, control, learn, communicate, improve, predict, and certify the work performed (McConnell, 1998). For a given organizational level, the processes vary with the project's goals and available resources. At a high level, the company's business strategy determines the business approach, with the main goals of profitability, time to market, minimum cost, higher quality, and customer satisfaction setting the priorities. Similarly, the company's size; the number, knowledge, and experience of people (both engineers and support personnel); and hardware resources determine how to achieve those goals (Cockburn, 2000). The application domain and the corresponding system requirements, together with other constraints, form another important factor in defining and applying processes.

有关 SE 管理（Chase，1974）方面的最早的书籍之一为这类组织识别了三个简单准则：助促沟通，通畅一致的控制，及简化纸面工作。SE 管理的有效方式不是"指向正式的、令人畏惧的、大规模的文档，而是指向创建一个总体环境，这有助于以实现一种以具有最小管理负担的系统方法为导向的创造性和发明性才能的涌现及有效利用。"

无论何时当有人（可以是个人或公司）想要达到所期望的目标时，他们必须执行一系列行动或运作。而且，他们必须考虑这些行动的顺序、依赖性、谁来执行这些行动、这些行动需要什么、将产生什么、完成这些行动需要花费多长时间以及将要采用哪些工具。因此，个人和组织遵守预先定义的或即时产生的流程。由于流程的组成部分（活动、产品、代理和工具）及其交互（信息流、制品流、控制、沟通、时序、依赖性和并行性）会发生变化，即使执行组织具有相同的层级、范围和目标，则流程也会不同。

那么，组织为什么关心流程？简单地说，是为了更好地理解、评估、控制、学习、沟通、改善、预测和认证所进行的工作（McConnell，1998）。对于给定的组织层级，流程会随着项目的目标和可用资源而变化。在高层级上，公司的业务战略决定了以收益性、上市的时间、最低的成本、较高的质量和客户满意度为主要目标而设定优先级的业务途径。类似的还有：公司的规模，人员（包括工程师和保障人员）的数量、知识和经验，以及决定如何实现这些目标的硬件资源（Cockburn，2000）。应用领域、相应的系统需求连同其他约束一起构成了定义和应用流程的另一重要因素。

So what really is best for my organization? The answer is that it depends on the situation. Depending on the perspective, different processes are defined for entire organizations, teams, or individuals. A "one-size-fits-all" approach does not work when defining processes; thus, organizations must continuously document, define, measure, analyze, assess, compare, and change processes to best meet project goals. One would hardly expect to find the same processes used in a startup e-commerce company as in NASA. The intended goal shapes a process in terms of scope (namely, the stages and activities covered) and organizational level. In any case, the selected processes should help guide people on what to do—how to divide and coordinate the work—and ensure effective communication. Coordination and communication, for example, form the main problems in large projects involving many people, especially in distributed projects where people cannot communicate face to face (Lindvall and Rus, 2000).

3.6 INTRODUCTION TO CASE STUDIES

Real-world examples that draw from diverse industries and types of systems are provided throughout this hand book. Five case studies have been selected to illustrate the diversity of systems to which SE principles and practices can be applied: medical therapy equipment, a bridge, a superhigh-speed train, a breach of a cybersecurity system, and a redesign of a high-tech medical system for low-tech maintenance. They represent examples of failed, successful, and prototype systems that all define(d) the state of the art. These studies may be categorized as medical, infrastructure, and transportation applications; in the manufacturing and construction industry domains; with and without software elements; complex; and subject to scrutiny in the concept, development, utilization, and support stages as all have a need to be safe for humans and are constrained by government regulations.

3.6.1 Case 1: Radiation Therapy—The Therac-25

3.6.1.1 Background Therac-25, a dual-mode medical linear accelerator (LINAC), was developed by the medical division of the Atomic Energy Commission Limited (AECL) of Canada, starting in 1976. A completely computerized system became commercially available in 1982. This new machine could be built at lower production cost, resulting in lower prices for the customers. However, a series of tragic accidents led to the recommended recall and discontinuation of the system.

那么，到底什么对组织是最好的呢？答案是这取决于具体情况。根据这一视角，整个组织、团队或个人定义了不同的流程。"以不变应万变"的方法在定义流程时是行不通的，因此组织必须持续地文件化、定义、测量、分析、评估、比较和更改流程，以最大限度地满足项目目标。几乎没有人期望会发现刚开张的电子商务公司和 NASA 使用的流程是相同的。预期目标是按照范围（即涵盖的阶段和活动）和组织层级形成一个流程。总之，选定的流程应帮助指导人员做什么——如何分解和协调工作——以确保沟通有效。例如，在涉及许多人员的大型项目中协调和沟通就成为主要问题，尤其是分布式项目中，人员不能进行面对面沟通（Lindvall 和 Rus，2000）。

3.6 案例研究简介

本手册提供了取材于不同行业和不同类型系统的真实示例。选择了五个案例研究来阐明 SE 原理和实践可应用于系统的多样性：医疗设备、桥梁、超高速列车、赛博安全系统的攻击及为了低的技术维护而对高技术医疗系统的重新设计。它们代表着全部定义当时最高技术水平的失败的系统、成功的系统和产品原型系统的示例。这些研究可被分类为医疗、基础设施和运输的应用；属于制造和建筑行业领域；具有/不具有软件元素；复杂的；并且由于它们都需要确保人员安全，且受政府法规的约束，所以在概念、开发、使用和保障各个阶段接受仔细审查。

3.6.1 案例1：放射疗法；Therac-25

3.6.1.1 案例背景 Therac-25 双模式医用线性加速器（LINAC）由加拿大原子能有限公司（AECL）医疗部门于 1976 年开发而成。完全计算机化的系统于 1982 年开始在市面上销售。这种新型机器的生产成本较低，从而对客户的销售价格也较低。然而，一系列不幸的事故导致系统被建议召回并且停止生产。

GENERIC LIFE CYCLE STAGES

The Therac-25 was a medical LINAC, or particle accelerator, capable of increasing the energy of electrically charged atomic particles. LINACs accelerate charged particles by introducing an electric field to produce particle beams (i.e., radiation), which are then focused by magnets. Medical LINACs are used to treat cancer patients by exposing malignant cells to radiation. Since malignant tissues are more sensitive than normal tissues to radiation exposure, a treatment plan can be developed that permits the absorption of an amount of radiation that is fatal to tumors but causes relatively minor damage to surrounding tissue.

Six accidents involving enormous radiation overdoses to patients took place between 1985 and 1987. Tragically, three of these accidents resulted in the death of the patients. This case is ranked in the top 10 worst software related incidents on many lists. Details of the accidents and analysis of the case are available from many sources (Jacky, 1989; Leveson and Turner, 1993; Porrello, n.d.).

3.6.1.2 Approach Therac-25 was a revolutionary design compared to its predecessors, Therac-6 and Therac-20, both with exceptional safety records .It was based on a double-pass concept that allowed a more powerful accelerator to be built into a compact and versatile machine. AECL designed Therac-25 to fully utilize the potential of software control. While Therac-6 and Therac-20 were built as stand-alone machines and could be operated without a computer, Therac-25 depended on a tight integration of software and hardware. In the new, tightly coupled system, AECL used software to monitor the state of the machine and to ensure its proper operations and safety. Previous versions had included independent circuits to monitor the status of the beam as well as hardware interlocks that prevented the machine from delivering radiation doses that were too high or from performing any unsafe operation that could potentially harm the patient. In Therac-25, AECL decided not to duplicate these hardware interlocks since the software already performed status checks and handled all the malfunctions. This meant that the Therac-25 software had far more responsibility for safety than the software in the previous models. If, in the course of treatment, the software detected a minor malfunction, it would pause the treatment. In this case, the procedure could be restarted by pressing a single "proceed" key. Only if a serious malfunction was detected was it required to completely reset the treatment parameters to restart the machine.

Therac-25 是一种医用 LINAC 或粒子加速器，能够增加带电原子粒子的能量。LINAC 通过引入电场产生粒子束（即辐射）来加速带电粒子，然后磁铁将带电粒子聚集到一起。医用 LINAC 通过将恶性细胞暴露于放射线下来治疗癌症患者。由于恶性组织对放射性照射比正常组织更敏感，能形成允许吸收适量辐射的治疗计划，这种辐射可以杀死肿瘤，但会对外围组织造成相对轻微的损伤。

1985 年到 1987 年期间，发生六起涉及对患者的放射剂量过多的事故。可悲的是，其中的三起事故导致患者死亡。此案例在多个与事故有关的排行中名列十大最糟软件中。此案例的事故和分析的详细资料可从多个来源获得（Jacky，1989；Leveson 和 Turner，1993；Porrello, n.d.）。

3.6.1.2 实现途径 与前身 Therac-6 和 Therac-20 相比（两者都具有优异的安全记录），Therac-25 是一次革命性设计。Therac-25 基于双程概念，可将功能更强的加速器设计成小型通用机器。AECL 设计了 Therac-25 以全面发挥软件控制的潜力。而 Therac-6 和 Therac-20 被设计成独立的机器，不使用计算机即可运行，Therac-25 依赖于软件和硬件的紧密集成。在全新的紧密耦合系统中，AECL 使用软件监控机器的状态并确保机器的恰当运行和安全性。先前版本包括用于监控射束状态的独立回路，以及硬件互锁装置，用于防止机器发射过大的放射剂量，或防止机器运行任何可能伤害到患者的不安全操作。在 Therac-25 中，AECL 决定不复制这些硬件互锁装置，因为软件已经对所有故障进行了状态检查和处理。这意味着 Therac-25 软件比先前型号中的软件承担更多的安全职责。如果处理过程中软件检测到较小的故障，可暂停处理。在这种情况下，可通过按下一个"继续"键重新启动程序。只有检测到严重的故障时，才需要完全重置治疗参数以重新启动机器。

The software for Therac-25 was developed from the Therac-20's software, which was developed from the Therac-6's software. One programmer, over several years, evolved the Therac-6 software into the Therac-25 software. A stand-alone, real-time operating system was added along with application software written in assembly language and tested as a part of the Therac-25 system operation. In addition, significant adjustments had been made to simplify the operator interface and minimize data entry, since initial operators complained that it took too long to enter a treatment plan.

At the time of its introduction to market in 1982, Therac-25 was classified as a Class II medical device. Since the Therac-25 software was based on software used in the earlier Therac-20 and Therac-6 models, Therac-25 was approved by the federal Drug Administration under Premarket Equivalency.

3.6.1.3 Conclusions The errors were introduced in the concept and early development stages, when the decisions were made to create the software for Therac-25 using the modification of existing software from the two prior machines. The consequences of these actions were difficult to assess at the time, because the starting point (software from Therac-6) was a poorly documented product and no one except the original software developer could follow the logic (Leveson and Turner, 1993).This case illustrates the importance of the off-the-Vee-core studies early in development (see fig. 3.7).

The issues from the Therac case are, unfortunately, still relevant, as evidenced by similar deaths for similar reasons in 2007 upon the introduction of new LINAC based radiation therapy machines (Bogdanich, 2010).

3.6.2 Case 2: Joining Two Countries—The Øresund Bridge

3.6.2.1 Background The Øresund region is composed of eastern Denmark and southern Sweden and since 2000 has been linked by the Øresund Bridge. The area includes two major cities, Copenhagen and Malmö, has a population of 3 million, and counts as Europe's eighth largest economic center. One fifth of the total Danish and Swedish Gross national Product (GNP) is produced in the region. The official name of the bridge is translated "the Øresund Connection" to underscore the full integration of the region. For the first time ever, Sweden is joined permanently to the mainland of Europe by a 10 min drive or train ride. The cost for the entire Øresund Connection construction project was calculated at 30.1 billion DKK (3 billion USD), and the investment is expected to be paid back by 2035.

Therac-20 的软件是从 Therac-6 的软件开发而来的，Therac-25 的软件是从 Therac-20 的软件开发而来的。一个程序经过多年的时间从 Therac-6 软件演进成 Therac-25 软件。单机实时操作系统中添加了以汇编语言编写的应用软件并经测试作为 Therac-25 系统运行的一部分。另外，由于最初的操作者抱怨输入治疗计划需要花费很长时间，因此对系统进行了重大调整以简化操作界面并将数据输入最小化。

1982 年 Therac-25 上市时，被分类为第二类医疗设备。由于 Therac-25 软件是基于早期的 Therac-20 和 Therac-6 模型中使用的软件开发的，按照上市前等价原则，Therac-25 由美国联邦药品管理局批准。

3.6.1.3　结论　决定采用修改先前两类机器原有软件创建 Therac-25 软件时，在概念阶段和早期开发阶段就引入了错误。这些行动的后果在那时很难评估，因为起点（来自 Therac-6 的软件）是文档记录很差的产品，且除了最初的软件开发者没有人能继承这种逻辑（Leveson 和 Turner, 1993）。此案例详细阐明在开发早期 V 形模型中非核心部分研究的重要性（见图 3.7）。

不幸的是，引入新的基于 LINAC 的放射治疗机后类似原因导致的类似死亡的证据表明，Therac 案例中的问题仍然存在（Bogdanich, 2010）。

3.6.2　案例 2：连接两个国家——厄勒海峡大桥

3.6.2.1　案例背景　厄勒海峡区域是由丹麦东部地区和瑞典南部地区组成的，自 2000 年起由厄勒海峡大桥连接起来。该区域包括哥本哈根和马尔默两大城市，拥有 300 万人口，是欧洲第八大经济中心。丹麦和瑞典的全部国民生产总值（GNP）的五分之一来自该区域。大桥的官方名称被翻译为"厄勒海峡连接"，以强调区域的完全综合。瑞典首次实现了与欧洲大陆的永久性连接，开车或乘火车只需 10 分钟即可到达欧洲大陆。整个厄勒海峡连接桥的建造项目的成本预计为 301 亿丹麦克朗（30 亿美元），预期到 2035 年能收回投资。

GENERIC LIFE CYCLE STAGES

The Øresund Bridge is the world's largest composite structure, has the longest cable-stayed bridge span in the world carrying motorway and railway traffic, and boasts the highest freestanding pylons. The 7.9 km (5 miles) long bridge crosses the international navigation route between the Baltic Sea and the North Sea. A cable-stayed high bridge rises 57 m (160 ft) above the surface of the sea, with a main span of 490 m (0.3 miles).Both the main span and the approach bridges are constructed as a two-level composite steel-concrete structure. The upper deck carries a four-lane motorway, and the lower deck carries a two-track railway for both passenger trains and freight trains. The rest of the distance is spanned by the artificial island Peberholm ("Pepper" islet, named to complement the Saltholm islet to the north) and a tunnel on the Danish side that is the longest immersed concrete tunnel in the world. Since completion, Peberholm has become a natural habitat for colonies of rare birds, one of the largest of its kind in Denmark and Sweden.

Nations other than Denmark and Sweden also contributed to this project. Canada provided a floating crane, aptly named Svanen (the swan), to carry prefabricated bridge sections out to the site and place them into position. Forty-nine steel girders for the approach bridges were fabricated in Cádiz, Spain. A specially designed catamaran was built to handle transportation of the foundations for the pylons, which weighed 19,000 tons each.

3.6.2.2 Approach As noted in the many histories of bridge, the development stage of the project began with well-defined time, budget, and quality constraints. The design evolved over more than 7 years, from start to delivery of final documentation and maintenance manuals. More than 4000 drawings were produced. The consortium dealt with changes, as necessary, using a combination of technical competence and stakeholder cooperation. Notably, there were no disputes and no significant claims against the owners at the conclusion, and this has been attributed to the spirit of partnership.

What is not often reported is that the success of the development stage is clearly based on the productive, focused, creative effort in the concept stage that began when the royal families of Denmark and Sweden finally agreed in 1990 to move ahead with a bridge project connecting their two countries. That SE effort shaped the approach to the project with well-defined time, budget, and quality constraints at the transition to the development stage. During the concept stage, the SE team also recognized that the concerns of environmental groups would— and should—impact the approach to the construction of the bridge. The owners took a creative approach by inviting the head of a key environmental group to be part of the board of directors.

厄勒海峡大桥是世界上最大的复合结构，具有世界上最长的斜拉桥跨度，可承载高速公路和铁路运输，拥有最高的独立式索塔。7.9 千米（5 英里）长的桥横跨波罗的海和北海之间的国际航行路线。高斜拉索桥比海平面高 57 米（160英尺），主跨 490 米（0.3 英里）。主跨和引桥被建成双层复合钢混凝土结构。上层桥面承载四车道高速公路，下层桥面承载客运列车和货运列车的双轨铁路。其余的跨度则位处于"佩博霍尔姆"（又名 Pepper 岛，用以在北部地区作为萨尔特岛的补充）人工岛上，以及丹麦一侧的隧道中——世界上最长的浸渍混凝土隧道。佩博霍尔姆岛完工后，成为珍稀鸟类集群的天然栖息地，也是丹麦和瑞典这类最大岛屿之一。

丹麦和瑞典以外的其他国家也对该项目做出了贡献。加拿大提供了名为天鹅（Svanen）的浮式起重机，用于现场承载预制桥段并将其放置于适当位置。在西班牙加迪斯省制造了 49 个引桥钢梁。建造了一艘特殊设计的双体船用于搬运桥塔底座，每个底座重达 19 000 吨。

3.6.2.2 实现途径 正如许多的桥梁历史中所指出，项目的开发阶段从良好定义的时间、预算和质量约束开始。设计从项目启动一直到最终文件和维护手册的交付经历了七年之多，产生了 4 000 多张图样。必要时，联合企业通过结合技术能力和利益攸关者合作来处理各种变更。很明显，项目结束时其与所有者没有任何纠纷和重大索赔，这将归于各方的合作精神。

通常没有被报道的是，开发阶段的成功很明显基于概念阶段中富有成效的、聚焦的、创造性的工作，该工作在丹麦和瑞典两个皇室家族最终于 1990 年推动连接他们两国的桥梁项目上达成一致时开始。该 SE 工作在向开发阶段的转移中形成了具有良好定义的时间、预算和质量约束的项目方法。在概念阶段期间，SE 团队还意识到环境组织的担忧，将会并且应该影响对桥梁施工的方法。所有者通过邀请关键环境组织的领导作为理事会的一员来采取创造性的方法。

From the beginning of the development stage, the owners defined comprehensive requirements and provided definition drawings as part of the contract documents to ensure a project result that not only fulfilled the quality requirements on materials and workmanship but also had the envisioned appearance. The contractor was responsible for the detailed design and for delivering a quality-assured product in accordance with the owners' requirements. The following are representative of the requirements levied at the start of the project:

- Schedule: Design life, 100 years; construction time, 1996–2000
- Railway: rail load, International Union of railways (UIC) 71; train speed, 200 km/h
- Motorway: road axle load, 260 kn; vehicle speed, 120 km/h
- Ambient environment: Wind speed (10 min), 61 m/s; wave height, 2.5 m; ice thickness, 0.6 m; temperature, +/− 27°C
- Ship impact: To pylons, 560 Mn; to girder, 35 Mn

In addition to established requirements, this project crossed national boundaries and was thereby subject to the legislations of each country. Technical requirements were based on the Eurocodes, with project-specific amendments made to suit the national standards of both countries. Special safety regulations were set up for the working conditions, meeting the individual safety standards of Denmark and Sweden.

The railway link introduced yet another challenge. In Denmark, the rail traffic is right handed, as on roadways, whereas the trains in Sweden pass on the left-hand side. The connection needed to ensure a logical transition between the two systems, including safety aspects. In addition, the railway power supply differs between the two countries; thus, it was necessary to develop a system that could accommodate power supply for both railway systems and switch between them on the fly.

The design of a major cable-stayed bridge with approach spans for both road and railway traffic involves several disciplines, including, but not limited to, geotechnical engineering, aerodynamics, foundation engineering, wind tunnel tests, design of piers and pylons, design of composite girders, design of cables and anchorages, design of structural monitoring system, ship impact analysis, earthquake analysis, analysis of shrinkage and creep of concrete, ice load analysis, fatigue analysis, pavement design, mechanical systems, electrical systems, comfort analysis for railway passengers, traffic forecast, operation and maintenance aspects, analysis of construction stages, risk analysis for construction and operation, quality management, and environmental studies and monitoring.

从开发阶段一开始，所有者定义全面的需求并提供定义明确的图样作为合同文件的一部分，以确保项目结果不仅符合材料和工艺上的质量要求，还具有预期的外观。承包商负责按照所有者的需求进行详细设计并交付质量合格的产品。下列内容表示项目开始时制定的需求：

- 进度：设计寿命 100 年；施工时间从 1996 年到 2000 年
- 铁路：轨道载重，国际铁路联盟（UIC）71；列车速度 200 千米/时
- 高速公路：道路轴载 260 千牛；车速 120 千米/时
- 周围环境：风速（10 分钟）61 米/秒；浪高 2.5 米；冰厚 0.6 米；温度 +/-27 摄氏度
- 船舶撞击：索塔 560 兆牛；梁 35 兆牛

除了已确定的需求外，该项目跨越国界，因此同时受每个国家法律的管辖。技术需求基于欧洲标准，项目特有的修正文件符合丹麦和瑞典两个国家的标准。为工作条件设立了特殊的安全规则，从而符合丹麦和瑞典的独特的安全标准。

铁路连接也引入另一个挑战。在丹麦，铁路运输与道路交通一样靠右侧行驶，而瑞典的列车靠左侧行驶。两个国家的铁路连接需要确保两个系统之间的逻辑转换，包括安全方面。另外，两个国家之间的铁路电力供应不同，因此有必要开发一种系统以适应两种铁路系统的电力供应，并且可在高速行驶中互相切换。

用于公路和铁路运输的具有多个引桥的大型斜拉桥的设计涉及多个学科，包括但不限于：岩土工程、空气动力学、基础工程学、风洞试验、桥墩和索塔的设计、组合梁设计、线缆和桥墩的设计、结构监测系统的设计、船舶影响分析、地震分析、混凝土收缩和蠕变的分析、冰负载分析、疲劳分析、路面设计、机械系统、电气系统、铁路乘客舒适度分析、交通预报、运行和维护方面、施工阶段分析、施工和运行的风险分析、质量管理以及环境的研究和监测。

GENERIC LIFE CYCLE STAGES

Comprehensive risk analyses were carried out in connection with the initial planning studies, including specification of requirements to secure all safety aspects. Important examples of the results of these studies for the Øresund Bridge were as follows:

- Navigation span was increased from 330 to 490 m.
- The navigation channel was realigned and deepened to reduce ship groundings.
- Pier protection islands were introduced to mitigate bridge/ship accidents.

Risks were considered in a systematic way, using contemporary risk analysis methods such as functional safety analysis using fault tree and "what-if" techniques.

Three main issues were considered under the design–build contract:

- General identification and assessment of construction risks
- Ship collision in connection with realignment of navigation channel
- Risks in connection with 5-year bridge operation by contractor

A fully quantified risk assessment of the human safety and traffic delay risks was carried out for a comprehensive list of hazards, including fire, explosion, train collisions and derailments, road accidents, ship collisions and groundings, aircraft collisions, environmental loads beyond design basis, and toxic spillages. An example of a consequence of this analysis was the provision of passive fire protection on the tunnel walls and ceilings.

Both Denmark and Sweden are proud of being among the cleanest industrial countries in the world. Their citizens, and therefore the politicians, would not allow for any adverse environmental impact from the construction or operation of a bridge. The Great Belt and Øresund Strait both constitute corridors between the salty Kattegat and the sweeter water of the Baltic Sea. Any reduction in water exchange would reduce the salt content and, therefore, the oxygen content of the Baltic Sea and would alter its ecological balance. The Danish and Swedish authorities decided that the bridge should be designed in such a way that the flow through of water, salt, and oxygen into the Baltic was not affected. This requirement was designated the zero solution. To limit impacts on the local flora and fauna in Øresund during the construction, the Danish and Swedish authorities imposed a restriction that the spillage of seabed material from dredging operations should not exceed 5% of the dredged amounts. The zero solution was obtained by modeling with two different and independent hydrographical models.

结合最初的计划研究,包括保证所有安全方面的需求规范,进行了全面的风险分析。厄勒海峡大桥研究结果的重要示例如下:

- 航道跨度由 330 米增加到 490 米。
- 对航道进行改线和加深,以减少船舶搁浅。
- 引入桥墩保护岛,以减少桥梁/船舶事故。

使用现代风险分析方法,如采用了故障树和"what-if(假设)"技术的功能安全分析,以系统化的方式对风险进行考虑。

按照设计—构建合同考虑以下三个主要问题:

- 施工风险的一般识别和评估
- 关于航道重新调整的船舶碰撞问题
- 承包商五年桥梁运行相关联的风险

项目开展了人身安全风险和交通延误风险的全面量化风险评估,总结出全面的风险列表,包括:火灾、爆炸、列车碰撞和脱轨、公路交通事故、船舶碰撞和搁浅、飞机碰撞、超出设计基础的环境负荷以及有毒物质泄漏。该分析结果的一个示例是在隧道侧壁和隧道顶部提供被动防火设备。

丹麦和瑞典都以位于全世界最洁净的工业国家之列而自豪。两个国家的公民乃至政治家都不准许桥梁的施工或运行带来任何不利于环境的影响。大贝尔特桥和厄勒海峡构成卡特加特海峡咸水和波罗的海淡水之间的走廊。减少水体交换会降低含盐量,因此会改变波罗的海的氧含量即改变其生态平衡。丹麦和瑞典当局决定在桥梁设计时应确保流入波罗的海的水、盐和氧气的流量不受影响。这一要求被指定为"零溶解"。为限制施工期间对厄勒海峡当地的植物群和动物群的影响,丹麦和瑞典的权威机构限制疏浚操作中海底物质泄漏量不应超过疏浚量的 5%。通过用两个不同的独立水道测量模型进行建模得出"零溶解"。

In total, 18 million cubic meters of seabed materials were dredged. All dredged materials were reused for reclamation of the artificial peninsula at Kastrup and the artificial island, Peberholm. A comprehensive and intensive monitoring of the environment was performed to ensure and document the fulfillment of all environmental requirements. In their final status report from 2001, the Danish and Swedish authorities concluded that the zero solution as well as all environmental requirements related to the construction of the link had been fulfilled. Continual monitoring of eel grass and common mussels showed that, after a general but minor decline, populations had recovered by the time the bridge was opened. Overall, the environment paid a low price at both Øresund and the Great Belt because it was given consideration throughout the planning and construction stages of the bridges.

3.6.2.3 Conclusions This award-winning bridge is the subject of numerous articles and a PhD thesis, where details of the construction history and collaboration among all the stakeholders are provided (Jensen, 2014; Nissan, 2006; Skanska, 2013).This project provides a clear example of the benefit of a solid concept stage where the management team was able to resist the customer-driven temptation to jump prematurely into the development stage.

3.6.3 Case 3: Prototype System—The Superhigh-Speed Train in China

3.6.3.1 Background Shanghai Transrapid is the first commercial high-speed commuting system using the state-of-the-art electromagnetic levitation (or maglev) technology. The train runs from Shanghai's financial district to Pudong International Airport, and the total track length is about 30 km (20 miles). The train takes 7 min and 20 s to complete the journey, can reach almost 320 km/h (200 mph) in 2 min, and reaches its maximum speed of 430 km/h (267 mph) within 4 min. The Shanghai Transrapid project cost 10 billion yuan (1.2 billion USD) and took 2.5 years to complete. Construction began in March 2001, and public service commenced on January 1, 2003. Critics argue that the speed over such a short distance is unnecessary and that the line may never recoup this cost. However, the speculation is that this is a prototype to gather operational data assessing the feasibility of a Shanghai to Beijing maglev train route. From this perspective, this project makes perfect sense.

项目总计疏浚海底物质 1 800 万立方米。所有疏浚物质重新用于围垦卡斯楚普的人工半岛和人工岛"佩博霍尔姆"。开展全面深入的环境监测以确保所有环境需求的完成并文件化。在 2001 年的最终状态报告中，丹麦和瑞典的当局总结说"零溶解"和与"连接"施工有关的所有环境需求已被满足。对鳗草和常见贻贝的持续监控表明其种群经过普遍但略微的衰退之后，在大桥开放时得以恢复。总之，由于在大桥的计划阶段和施工阶段都考虑到环境问题，厄勒海峡大桥和大贝尔特桥的环境代价很低。

3.6.2.3 结论 这一获奖大桥是诸多文章和博士学位论文的主题，其中提供了所有利益攸关者之间的施工历史和合作的详细资料（Jensen，2014；Nissan，2006；Skanska，2013）。该项目提供了对完备的概念阶段收益的清晰示例，其中管理团队能够抵制过早地坠入开发阶段的客户驱动的诱惑。

3.6.3 案例3：产品原型系统——中国超高速列车

3.6.3.1 案例背景 "上海磁悬浮"是使用最新电磁悬浮（或磁悬浮）技术的首个商业高速通勤系统。列车在上海金融区和浦东国际机场之间运行，轨道总长约 30 千米（20 英里）。列车完成整个行程花费 7 分 20 秒，2 分钟内可达到约 320 千米/时（200 英里/时），4 分钟内达到 430 千米/时（267 英里/时）的最高车速。上海磁悬浮列车项目耗资 100 亿元人民币（12 亿美元），耗时 2.5 年完成。2001 年 3 月开始施工，2003 年 1 月 1 日开始公共服务。批评者认为，这么短的距离内没有必要达到这么高的速度，运行路线可能永远不会收回成本。然而据推测，这是一种收集运行数据的产品原型，用以评估上海到北京磁悬浮列车路线的可行性。从这种视角来看，该项目是完全有意义的。

Prior to this installation, many countries had argued over the feasibility of maglev trains. They do not have wheels or use a traditional rail. Rather, powerful magnets lift the entire train about 10 mm above the special track, called a guideway, which mainly directs the passage of the train. Electromagnetic force is used to make the train hover and to provide vertical and horizontal stabilization. The frequency, intensity, and direction of the electrical current in the track control the train's movement, while the power for the levitation system is supplied by the train's onboard batteries, which recharge whenever the train is moving. Maglev trains also do not have an onboard motor. The guideway contains a built-in electric motor that generates an electromagnetic field that pulls the train down the track. Putting the propulsion system in the guideway rather than onboard the trains makes the cars lighter, which enables the train to accelerate quickly. The superhigh speeds are attained largely due to the reduction of friction.

Despite the high speed, the maglev system runs more quietly than a typical commuter train, consumes less energy, and is nearly impossible to derail because of the way the train's underside partially wraps around the guideway, like a giant set of arms hugging the train to the elevated platform. Passengers experience a comfortable and quiet ride due to the maglev technology and the specially designed window; noise level is less than 60 decibels at a speed of 300 km/h.

3.6.3.2 Approach The Chinese authorities considered the economical operation, low energy consumption, less environmental impact, and high speed when choosing a solution suitable for ground transport between hubs that range from hundreds to over 1000 km apart. But the same solution also needed to be suitable for modern mass rapid passenger transportation between a center city and adjacent cities. Despite the many advantages, in 1999, the technology was considered to be in an experimental stage—its technological superiority, safety, and economic performance not yet proven by commercialized operation. The current line is the result of a compromise; it was built as a demonstration to verify the maturity, availability, economics, and safety of a high speed maglev transportation system.

The basic technology to create a maglev system has been around since 1979, but until this project, it had never been realized, mostly due to the expense of developing a new train system. Many experts believe that superfast steel-wheel rail systems—such as those in France and Japan—have reached the limits of this technology and cannot go any faster. Maglev proponents describe the system as "the first fundamental innovation in the field of railway technology since the invention of the railway" and are watching proposals for maglev installations in Germany and the United States (BBC, 2002; McGrath, 2003; SMTDC, 2005; Transrapid International, 2003).

建造前，许多国家对磁悬浮列车的可行性争论已久。磁悬浮列车没有车轮，也不使用传统的轨道。相反，功率强大的磁铁使整个列车悬浮于特殊轨道，即导轨之上约 10 毫米，导轨主要用于引导列车的通行。电磁力用于使列车悬停和提供垂直与水平稳定性。轨道上电流的频率、强度和方向控制列车的移动，而悬浮系统的电源由列车的车载电池提供，这种电池在列车移动时为列车重新充电。磁悬浮列车也没有车载电动机。导轨包含内置电动机，可产生沿轨道拉动列车的电磁场。将推进系统置于导轨上而不是车载，使车厢更轻，从而使列车迅速加速。实现超高速很大程度上是由于减少了摩擦。

尽管速度很高，但磁悬浮系统的运行噪声比典型的通勤列车更低，消耗的能量更少，由于列车下侧部分地环绕导轨，在高架平台上就像一系列巨大手臂拥抱列车，因此磁悬浮系统几乎不可能脱轨。由于采用了磁悬浮技术和特殊设计的窗户，乘客乘坐时会感到非常舒服、安静；车速 300 千米/时的噪声级小于 60 分贝。

3.6.3.2 实现途径 中国政府为相距几百乃至几千公里的枢纽之间的地面运输选择合适的解决方案时，考虑到了经济运行、低能耗、低环境影响和高速度。但同样的解决方案亦需要适于中心城市和相邻城市之间现代化大容量快速客运问题。尽管优点很多，1999 年该技术被认为处于试验阶段——其技术优势、安全性和经济表现尚未被商业化运作所证明。当前的路线是折中的结果；这一路线被构建为示范以验证高速磁悬浮运输系统的成熟度、可用性、经济效益和安全性。

创造磁悬浮系统的基本技术自从 1979 年就已经出现了，但直到该项目的启动，这一技术从未被实现，原因多数是开发新列车系统的费用。许多专家相信超高速钢轨铁路系统——如法国和日本的铁路系统已经达到了该技术的极限，速度不能再快。磁悬浮支持者将该系统描述成"自铁路发明以来，在铁路技术领域的首次根本性创新"，并且正在观望德国和美国关于磁悬浮装置的提案（BBC，2002；McGrath，2003；SMTDC，2005；Transrapid International 公司，2003）。

3.6.3.3 Conclusions From an SE perspective, this case illustrates the fact that a project that goes through all the stages from concept to operations may in fact simply be part of the concept stage of a larger effort.

3.6.4 Case 4: Cybersecurity Considerations in Systems Engineering—The Stuxnet Attack on a Cyber-Physical System

3.6.4.1 Background As our world becomes increasingly digital, the issue of cybersecurity is a factor that systems engineers need to take into account. Both hardware and software systems are increasingly at risk for disruption or damage caused by threats taking advantage of digital technologies. Stuxnet, a cyber attack on Iran's nuclear capabilities, illustrates the need for systems engineers to be comprehensive in their assessment of vulnerabilities and rigorous in their mitigation of attack potential (Failliere, 2011; Langner, 2012).

This case study discusses a new degree of attack sophistication previously unseen— a new level of malware complexity at military-grade performance, nearly no side effects, and pinpoint accuracy. However, though the creation and deployment of Stuxnet were expensive undertakings, the strategy, tactical methods, and code mechanisms are now openly available for others to reuse and build upon at much less expense. Cyber-physical system attacks are becoming increasingly prevalent, and SE must consider the implications of cybersecurity to reduce the vulnerabilities.

Iran's natanz nuclear fuel enrichment plant (FEP) is a military-hardened facility, with a security fence surrounding a complex of buildings, which are in turn each protected by a series of concrete walls. The complex contains several "cascade halls" for the production of enriched uranium in gas centrifuges. This facility was further hardened with a roof of several meters of reinforced concrete and covered with a thick layer of earth.

Each of the cascade halls is a cyber-physical system, with an industrial control system (ICS) of programmable logic controllers (PLCs), computers, an internal network with no connections to the outside world, and capacity for thousands of centrifuges. Though the internal network is isolated from the outside world by an "air gap," possible vulnerabilities still include malicious insider collusion, nonmalicious insider insertion of memory devices brought in from the outside, visiting service technicians, and supply chain intervention. It has been suggested that all of these breech vectors may have played a role in the massive centrifuge damage that began occurring in 2009 and continued at least through 2010.

3.6.3.3 结论 从 SE 的视角来看，此案例详细阐明"经历从概念到运行的所有阶段的项目事实上可能仅仅是巨大工作的概念阶段的一部分"这一事实。

3.6.4 案例 4：系统工程中赛博安全的考虑因素——赛博物理系统上的 Stuxnet 病毒攻击

3.6.4.1 案例背景 随着我们的世界变得日益数字化，赛博安全问题是系统工程师需要考虑到的一个因素。硬件和软件系统都日益面临着利用数字技术的威胁所造成的破坏或损坏的风险中。Stuxnet，对伊朗核能力的赛博攻击，详细阐明系统工程师在评估漏洞中的全面性和缓解攻击潜力的严格性上的需要（Failliere, 2011；Langner, 2012）。

此案例研究论述前所未见的新的攻击复杂性程度——军事级表现上的恶意软件复杂性的新水平，几乎没有负面影响并且精确度高。然而，尽管 Stuxnet 的创建和部署是昂贵的负担，但现在策略、战术方法和代码机制对其他人公开可用，能够以很少的费用复用并基于其构建。赛博物理系统攻击变得日益普遍，SE 必须考虑赛博安全的意义以减少脆弱性。

伊朗的纳坦兹核燃料浓缩厂（FEP）是一个军事强化设施，围绕建筑群具有安保防护装置，这些建筑分别依次受一系列混凝土墙的保护。该建筑群包含若干"串联的大厅"，用于在气体离心机中生产浓缩铀。该工厂使用由几米钢筋混凝土制成的屋顶进一步强化，并在上面盖有一层厚厚的土。

诸多串联大厅的每一个都是一个赛博物理系统，其具有可编程逻辑控制器（PLC）的工业控制系统（ICS）、计算机、与外界不连接的内部网络以及成千个离心机的容量。尽管"安全隔离网闸"将内部网络与外界隔离，但可能的脆弱性仍然包括恶意的内部合谋，从外部引入的存储设备的非恶意内部插入，来访的服务技术人员并侵入供应链。有人说所有这些双向导向在 2009 年开始发生的、持续到至少 2010 年的离心机大规模损坏中也许发挥了作用。

Malware, now known as Stuxnet, was introduced into the ICS of at least one of the cascade halls and managed to take surreptitious control of the centrifuges, causing them to spin periodically and repeatedly at rates damaging to sustained physical operation. The net effect of the attack is still unclear, but at a minimum, it ranged from disruption of the production process up to potential permanent damage to the affected centrifuges.

3.6.4.2 Approach Many characteristics of Stuxnet are unprecedented and stand as the inflection point that ushers in a new era of system attack methodology and cyber-physical system targeting. Illuminating forensic analysis of the Stuxnet code was conducted by several well-known cybersecurity firms, with detailed postmortems covered in two documents from the Institute for Science and International Security, "Did Stuxnet Take Out 1000 Centrifuges at the natanz Enrichment Plant?" (Albright et al., 2010) and "Stuxnet Malware and natanz: Update of ISIS December 22, 2010 report" (Albright et al., 2011). This analysis is beneficial in expanding the risk landscape that systems engineers should consider during design. Below are some concepts that are concerned in the context of Stuxnet:

- *Knowing what to do (intelligence)*—To be successful, a threat has to be able to take advantage of the targeted system(s). It is uncertain how the perpetrators knew what specific devices were employed in what configuration at natanz; but after the Stuxnet code was analyzed, natanz was clearly identified as the target. Stuxnet infected many sites other than natanz, but it would only activate if that site was configured to certain specific system specifications. The perpetrators needed specific system configuration information to know how to cause damage and also to know how to single out the target among many similar but not identical facilities elsewhere. Systems engineers need to consider that adversaries will attempt to gain intelligence on a system and must consider methods to prevent this.

- *Crafting the code*—A zero-day attack is one that exploits a previously unknown vulnerability in a computer application, one that developers have had no time to address and patch. Stuxnet attacked Windows systems outside the FEP using a variety of zero-day exploits and stolen certificates to get proper insertion into the operating system and then initiated a multistage propagation mechanism that started with Universal Serial Bus (USB) removable media infected outside the FEP and ended with code insertion into the ICS inside the FEP. Systems engineers need to be prepared for many different attack vectors (including internal threats) and must consider them during system design.

恶意软件，现在被称为 Stuxnet，被引入到至少是多个串联大厅中之一的 ICS 中并设法对离心机采取秘密控制，导致离心机以对持续物理运行不利的速率定期反复旋转。该攻击的净效应尚不清楚，但范围至少从生产流程的破坏直到潜在的永久性损坏再到受影响的离心机。

3.6.4.2 实现途径 Stuxnet 的许多特性是前所未见的，作为开创系统攻击方法论和赛博物理系统设定目标的一个新时代的转折点。几家著名的网络安全公司对 Stuxnet 代码进行了启发性取证分析，并在两份文件中涵盖了来自科学与国际安全研究所的详细事后调查，"Stuxnet 在纳坦兹浓缩厂取出 1 000 台离心机？"（Albright 等，2010）以及"Stuxnet 恶意软件和纳坦兹：ISIS 2010 年 12 月 22 日报告的更新"（Albright 等，2011）。该分析有益于系统工程师在设计期间考虑扩大风险的范围。下面是 Stuxnet 背景环境中涉及的一些概念：

- 知道要做什么（情报）——为取得成功，一种威胁必须能够利用目标系统。不能确定作恶者是如何知晓哪些特定装置用于纳坦兹的何种配置的；但对 Stuxnet 代码进行分析后，纳坦兹被明确地认别为攻击目标。Stuxnet 感染了纳坦兹以外的许多地点，但它仅激活按照某些特定系统规范配置的地点。作恶者需要特定系统配置信息以知晓如何导致破坏，以及知晓如何在其他地方的许多类似但不相同的工厂之间挑选出攻击目标。系统工程师需要考虑到对手会尝试获得系统的情报，且必须考虑到阻止这种情况的方法。

- 周密编制代码——零时差攻击是利用对一个计算机应用程序中先前未知的脆弱性的开发者不曾有时间解决和修补的一种攻击。Stuxnet 使用各种零时差攻击和窃取的证书来攻击 FEP 外部的 Windows 系统，以实现在操作系统中适当的插入，然后启动多级传播机制，该多级传播机制开始于在 FEP 外部感染的通用串行总线（USB）可移动介质，结束于在 FEP 内 ICS 中的代码插入。系统工程师需要为许多不同的攻击载体（包括内部威胁）做好准备，且必须在系统设计期间考虑到这些攻击导向。

- *Jumping the air gap*—It is widely believed that Stuxnet crossed the air gap on a USB removable media device, which had been originally infected on a computer outside of the FEP and carried inside. But it is also suggested that the supply chain for PLCs may have been at least one additional infection vector. Whatever the methods, the air gap was crossed multiple times. USB removable media could have also affected a bidirectional transfer of information, sending out detailed intelligence about device types connected to the FEP network subsequently relayed to remote servers outside of the control of the facility. Systems engineers always need to remember that threats to the system are both inside and outside the system boundary.

- *Dynamic updating*—Analysis shows that the attack code, once inserted, could be updated and changed over time, perhaps to take advantage of new knowledge or to implement new objectives. Stuxnet appears to be continuously updated, with new operational parameters reintroduced as new air gap crossings occur. Systems engineers need to prepare for situations after a successful attack has occurred.

3.6.4.3 Conclusions As the complexity and technology of systems change, the systems engineer's perspective needs to adjust accordingly. The increasing use of digital-based technologies in system design offers enormous benefits to everyone. However, the introduction of digital technologies also brings different risks than previously dealt with by SE. The case study earlier illustrates a point in time behind us, and the adversarial community continues to evolve new methods. The lesson of this case study is that systems engineers need to understand the threats toward their system(s), be cognizant that attacks can and will occur, and be proactive in protecting their system(s). Robust and dynamic system security needs full engagement of SE. A database that systems engineers should be aware of is maintained by the national Institute of Standards and Technology (NIST, 2012).

3.6.5 Case 5: Design for Maintainability— Incubators

Note: This case study is excerpted from "Where Good Ideas Come From: The Natural History of Innovation" (Johnson, 2010).

- 跳过安全隔离网闸——人们普遍相信 Stuxnet 跨越了 USB 可移动介质设备上的安全隔离网闸，该设备最初在 FEP 外部的计算机上被感染并携带到内部。但还有建议说 PLC 供应链至少可能是一个附加感染载体。不管是什么方法，网闸都被跨越了很多次。USB 可移动介质还可能影响双向信息传递，发出被连接到 FEP 网络、后续被中继到工厂控制之外的远程服务器上发送关于设备类型的详细情报。系统工程师始终需要牢记，对系统的威胁在系统边界的内部和外部。

- 动态更新——分析表明，插入后的攻击代码可能被更新并随时间推移而变化，也许利用新知识或实施新目标。Stuxnet 似乎不断更新，在出现新的网闸跨越时会再引入新的运行参数。系统工程师需要为出现成功攻击后的情形做好准备。

3.6.4.3 结论　随着系统复杂性和技术的变化，系统工程师的视角需要相应地调整。系统设计使用中基于数字化的技术的日益增加给每个人都提供了巨大的好处。然而，数字化技术的引入还带来了与 SE 先前所处理的不同的风险。早期的案例研究详细阐明了我们背后的某一时间点，对手群体继续演进新的方法。此案例研究的教训是，系统工程师需要理解他们的系统所面临的威胁，需要认识到攻击能够并且即将出现，需要积极保护他们的系统。鲁棒和动态系统安保需要 SE 的全面参与。系统工程师应当意识到这一数据库由（美国）国家标准与技术研究院（NIST, 2012）来维护。

3.6.5　案例 5：可维护性设计——育婴箱

注：此案例研究摘自《伟大创意的诞生：创新的自然史》（Johnson, 2010）。

3.6.5.1 Background In the late 1870s, a Parisian obstetrician named Stephane Tarnier was visiting the Paris Zoo where they had farm animals. While there, he conceived the idea of adapting a chicken incubator to use for human newborns, and he hired "the zoo's poultry raiser to construct a device that would perform a similar function for human newborns." At the time, infant mortality was staggeringly high "even in a city as sophisticated as Paris. One in five babies died before learning to crawl, and the odds were far worse for premature babies born with low birth weights." Tarnier installed his incubator for newborns at Maternité de Paris and embarked on a quick study of 500 babies. "The results shocked the Parisian medical establishment: while 66 percent of low-weight babies died within weeks of birth, only 38 percent died if they were housed in Tarnier's incubating box.... Tarnier's statistical analysis gave newborn incubation the push that it needed: within a few years the Paris municipal board required that incubators be installed in all the city's maternity hospitals."

"Modern incubators, supplemented with high-oxygen therapy and other advances, became standard equipment in all American hospitals after the end of World War II, triggering a spectacular 75 percent decline in infant mortality rates between 1950 and 1998."... "In the developing world, however, the infant mortality story remains bleak. Whereas infant deaths are below ten per thousand births throughout Europe and the United States, over a hundred infants die per thousand (births) in countries like Liberia and Ethiopia, many of them premature babies that would have survived with access to incubators. But modern incubators are complex, expensive things. A standard incubator in an American hospital might cost more than $40,000 (about €30,000). But the expense is arguably the smaller hurdle to overcome. Complex equipment breaks and when it breaks you need the technical expertise to fix it, and you need replacement parts. In the year that followed the 2004 Indian Ocean tsunami, the Indonesian city of Meulaboh received eight incubators from a range of international relief organizations. By late 2008, when an MIT professor named Timothy Prestero visited the hospital, all eight were out of order, the victims of power surges and tropical humidity, along with the hospital staff 's inability to read the English repair manual. The Meulaboh incubators were a representative sample: some studies suggest that as much as 95% of medical technology donated to developing countries breaks within the first 5 years of use."

3.6.5.1 背景 在 19 世纪 70 年代末，名为 Stephane Tarnier 的巴黎产科医师游览拥有农场动物的巴黎动物园。在那里，他设想了使小鸡孵化器适用于人类新生儿的这一想法，他雇用了"动物园的饲养员来构造一种能够实施人类新生儿类似功能的装置。"那时，"即使在如巴黎一样技术先进的城市中，婴儿死亡率也高得惊人。五分之一的婴儿在学习爬行之前就死亡了，出生时体重低的早产儿情况就更糟。"Tarnier 为巴黎出生的新生儿安装了他的培养箱，并开始快速研究 500 名婴儿。"结果震惊了巴黎医疗机构：66%的低体重婴儿在出生几周之内死亡，如果将他们放在 Tarnier 的培养箱中，仅 38%的婴儿死亡。……Tarnier 的统计分析为新生儿培育提供了所需的推动力：在几年内，巴黎市委员会要求将培养箱安装在城市所有的产科医院中。"

"现代育婴箱，补充有高氧疗法及其他先进技术，在第二次世界大战结束后成为所有美国医院的标准设备，引发 1950 年和 1998 年之间的婴儿死亡率明显下降 75%。"……"然而在发展中国家，婴儿死亡率的故事依然令人感到凄凉。鉴于整个欧洲和美国的新生婴儿死亡低于千分之十，像利比里亚和埃塞俄比亚这样的国家每 1 000 名婴儿（新生儿）中有超过 100 名婴儿死亡，其中许多婴儿是早产儿，如果有机会使用育婴箱就能够得以幸存。但现代育婴箱是复杂、昂贵的东西。美国医院的标准育婴箱可能要花费 40 000 美元（约 30 000 欧元）以上。但该费用是可证实的需要克服的较小障碍。复杂的设备损坏，当它损坏时，你需要技术专家来修理它，你还需要替换零件。在 2004 年印度洋海啸的第二年，印度尼西亚米拉务从一系列的国际救援组织中收到了八个育婴箱。到 2008 年年底，当名为 Timothy Prestero 的 MIT 教授参观医院时，所有八个育婴箱都出现故障，受害者受困于电涌和炎热的湿气，以及医务人员没有阅读英文维修手册的能力，而成为牺牲品。米拉务育婴箱是一个代表性样品：一些研究表明，捐赠给发展中国家的医疗技术，差不多 95%的在使用的前五年内损坏。

3.6.5.2 Approach "Prestero had a vested interest in those broken incubators, because the organization he founded, Design that Matters, had been working for several years on a scheme for a more reliable, and less expensive, incubator, one that recognized complex medical technology was likely to have a very different tenure in a developing world context than it would in an American or European hospital. Designing an incubator for a developing country wasn't just a matter of creating something that worked; it was also a matter of designing something that would break in a non-catastrophic way. You couldn't guarantee a steady supply of spare parts, or trained repair technicians. So instead, Prestero and his team decided to build an incubator out of parts that were already abundant in the developing world. The idea had originated with a Boston doctor named Jonathan Rosen, who had observed that even the smaller towns of the developing world seemed to be able to keep automobiles in working order. The towns might lack air conditioning and laptops and cable television, but they managed to keep their Toyota 4runners on the road. So Rosen approached Prestero with an idea: What if you made an incubator out of automobile parts?"

"Three years after Rosen suggested the idea, the Design that Matters team introduced a prototype device called Neonurture. From the outside, it looked like a streamlined modern incubator, but its guts were automotive. Sealed-beam headlights supplied the crucial warmth; dashboard fans provided filtered air circulation; door chimes sounded alarms. You could power the device via an adapted cigarette lighter, or a standard-issue motorcycle battery. Building the neonurture out of car parts was doubly efficient, because it tapped both the local supply of parts themselves and the local knowledge of automobile repair. These were both abundant resources in the developing world context, as Rosen liked to say. You didn't have to be a trained medical technician to fix the Neonurture; you didn't even have to read the manual. You just needed to know how to replace a broken headlight."

3.6.5.3 Conclusions Systems engineers need to consider issues like maintainability, producibility, and supportability at the project outset in the concept stage. It is too late to add these in during the production stage.

3.6.5.2 实现途径 "Prestero 对于那些损坏的育婴箱有持久的兴趣，因为他发现名为"设计很重要"的组织已多年致力于更可靠且更便宜的育婴箱方案，该组织意识到，与美国或欧洲医院相比，发展中国家的背景环境内的复杂医疗技术可能具有截然不同的使用期。为发展中国家设计育婴箱不只是关系到创造出有用东西的问题；它还关系到设计出以非灾难性方式损坏的东西的问题。你无法保证备件或受过训练的维修技术人员的稳定供应。所以，与之相反，Prestero 和他的团队决定从发展中国家中已经很富有的零件中构建育婴箱。这个想法源自名为 Jonathan Rosen 的波士顿医生，他发现即使是发展中国家的较小城镇似乎也能够使汽车正常工作。这些城镇可能缺乏空调、笔记本电脑和有线电视，但他们能设法驾驶丰田 4Runner 越野车在路上行驶。因此 Rosen 与 Prestero 商量这一想法：如果你用汽车零件制造育婴箱将会怎么样？

"三年后，Rosen 提出这个想法，'设计很重要'团队引入了名为 Neonurture 的产品原型装置。从外表上，它看起来像一个流线型现代育婴箱，但其核心是汽车。密封光束的前灯提供至关重要的热量；仪表板风扇提供过滤空气的循环流通；车门蜂鸣器发出警报。你可以通过适合的点烟器或标准版摩托车电池为该装置供电。从汽车零件中构建 Neonurture 具有双重有效性，因为其既利用了零件本身的本地供应，也利用了汽车修理的本地知识。这些是发展中国家背景环境内的丰富资源，正如 Rosen 喜欢说的。你不必一定是受过培训的医疗技术人员，就能够修理 Neonurture；你甚至不必阅读手册。你只需要知晓如何更换损坏的前灯即可。"

3.6.5.3 结论 系统工程师需要在项目一开始时的概念阶段就考虑到诸如可维护性、可生产性和可保障性这样的问题。在生产阶段再去考虑这些问题已为时太晚。

4 TECHNICAL PROCESSES

The ISO/IEC/IEEE 15288 technical processes and supporting process activities are invoked throughout the life cycle stages of a system. Technical processes are defined in ISO/IEC/IEEE 15288 as follows:

> [6.4] The Technical Processes are used to define the requirements for a system, to transform the requirements into an effective product, to permit consistent reproduction of the product where necessary, to use the product to provide the required services, to sustain the provision of those services and to dispose of the product when it is retired from service.

Technical processes enable systems engineers to coordinate the interactions between engineering specialists, other engineering disciplines, system stakeholders and operators, and manufacturing. They also address conformance with the expectations and legislated requirements of society. These processes lead to the creation of a sufficient set of requirements and resulting system solutions that address the desired capabilities within the bounds of performance, environment, external interfaces, and design constraints. Without the technical processes, the risk of project failure would be unacceptably high. As illustrated in Figure 4.1, the technical processes begin with the development of needs and requirements (Ryan, 2013):

- *Needs*—Per the Oxford English Dictionary, a need is a thing that is wanted or required. For a system, needs are often capabilities or things that are lacking but wanted or desired by one or more stakeholders. These can be viewed in at least three contexts in which SE is performed: (i) projects with customers internal to the enterprise that is doing the engineering, (ii) development under an agreement with an external entity, and (iii) entrepreneurial product development in anticipation of future sales.

- *Requirements*—Requirements are formal structured statements that can be verified and validated. There may be more than one requirement defined for each need.

INCOSE Systems Engineering Handbook: A Guide for System Life Cycle Processes and Activities, Fourth Edition. Edited by David D. Walden, Garry J. Roedler, Kevin J. Forsberg, R. Douglas Hamelin and Thomas M. Shortell.

© 2015 John Wiley & Sons, Inc. Published 2015 by John Wiley & Sons, Inc.

4 技术流程

ISO/IEC/IEEE 15288 技术流程和支持流程活动贯穿于系统生命周期的各个阶段被引用。技术流程在 ISO/IEC/IEEE 15288 中的定义如下：

> [6.4] 技术流程用于定义系统需求，以将需求转换成有效的产品、必要时允许产品进行一致的复制、使用产品来提供所要求服务、持续提供这些服务并且在产品退出时处置产品。

技术流程能使系统工程师协调工程专家、其他工程学科、系统利益攸关者与运行者及制造之间的互动。系统工程师亦涉及与社会的预期和法定要求的一致性。这些流程导向需求及由此产生的系统解决方案的充分集合的创建，以使期望的能力在性能、环境、外部接口和设计约束的边界之内。如果没有技术流程，项目失败的风险将会高得令人无法接受。如图 4.1 所示，技术流程从需要和需求的开发开始（Ryan, 2013）：

- 需要——按照《牛津英语词典》，需要是一种想要或所需的东西。对于系统而言，需要往往是一个或多个利益攸关者所缺乏但想要或期望的能力或东西。这些需要可以在实施 SE 所在的至少三个背景环境中审视：（i）具有复杂组织体内部从事工程的客户，（ii）按照与外部实体的协议开发，以及（iii）期待未来销售复杂组织体的产品开发。

- 需求——需求是能够被验证和确认的正式的结构化说明。为每个需要也许要定义一个以上的需求。

INCOSE 系统工程手册：系统生命周期流程和活动指南，第 4 版。编撰：David D. Walden, Garry J. Roedler, Kevin J. Forsberg, R. Douglas Hamelin 和 Thomas M. Shortell。
John Wiley & Sons 公司版权所有©2015。由 John Wiley & Sons 公司于 2015 年出版。

TECHNICAL PROCESSES

Note: One underlying principle illustrated by Figure 4.1 is that when a decision is made to satisfy a need, that need gives rise to a corresponding requirement or set of requirements.

ISO/IEC/IEEE 15288 includes 14 technical processes, the roles of the first four of which are illustrated in Figure 4.1:

- *Business or mission analysis process*—Requirements definition begins with the business vision of the organization or enterprise, the concept of operations (ConOps), and other organization strategic goals and objectives from which business management define business needs (aka mission needs). These needs are supported by preliminary life cycle concepts—acquisition concept, deployment concept, operational concept (OpsCon), support concept, and retirement concept—see Section 4.1.2.2 for a detailed description of the roles of the ConOps and the OpsCon. Business needs are then elaborated and formalized into business requirements, which are often captured in a Business Requirements Specification (BRS).

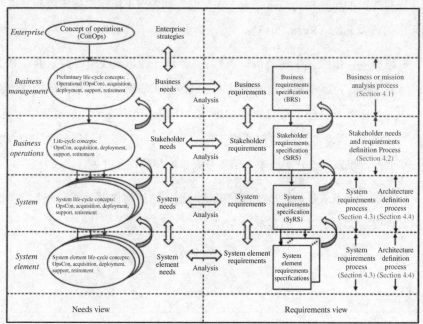

FIGURE 4.1 Transformation of needs into requirements. Reprinted with permission from Mike Ryan. All other rights reserved.

注：图 4.1 详细阐明的一个基本原理是，当满足一个需要做出决策时，该需要会导致一个对应的需求或需求的集合。

ISO/IEC/IEEE 15288 包括 14 个技术流程，其中前四个流程的角色如图 4.1 所示：

- 业务或任务分析流程——需求定义开始于组织或复杂组织体的业务愿景、ConOps（运行意图）及其他的组织战略目标和目的，业务管理据此对业务需要（亦称作任务需要）进行定义。这些需要由初步的生命周期概念——采办概念、部署概念、OpsCon（运行概念）、支持概念和退出概念来支持——关于 ConOps 和 OpsCon 角色的详细描述，参见 4.1.2.2 节。然后，详细说明业务需要并正式化为业务需求，这些需求往往在业务需求规范（BRS）中获取。

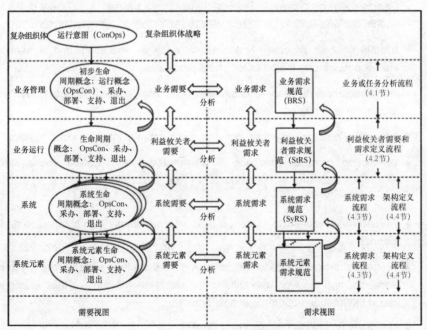

图 4.1 需要到需求的转换。经 Mike Ryan 许可后转载。版权所有。

TECHNICAL PROCESSES

- *Stakeholder needs and requirements definition process*—Using the enterprise-level ConOps from the acquiring enterprise and the system-level preliminary OpsCon from the development enterprise as guidance, requirements engineers lead stakeholders from business operations through a structured process to elicit stakeholder needs (in the form of a refined system-level OpsCon and other life cycle concepts). Stakeholder needs are then transformed by requirements engineers into a formal set of stake-holder requirements, which are often captured in a Stakeholder Requirements Specification (STRS).

- *System requirements definition process*—The stakeholder requirements in the STRS are then transformed by requirements engineers into system requirements, which are often contained in a System Requirements Specification (SYRS).

- *Architecture definition process*—Alternative system architectures are defined and one is selected.

- *Design definition process*—System elements are defined in sufficient detail to enable implementation consistent with the selected system architecture.

- *System analysis process*—Mathematical analysis, modeling, and simulation are used to support the other technical processes.

- *Implementation process*—System elements are realized to satisfy system requirements, architecture, and design.

- *Integration process*—System elements are combined into a realized system.

- *Verification process*—Evidence is provided that the system, the system elements, and the work products in the life cycle meet the specified requirements.

- *Transition process*—The system moves into operations in a planned, orderly manner.

- *Validation process*—Evidence is provided that the system, the system elements, and the work products in the life cycle will achieve their intended use in the intended operational environment.

- *Operation process*—The system is used.

- *Maintenance process*—The system is sustained during operations.

- *Disposal process*—The system or system elements are deactivated, disassembled, and removed from operations.

- *利益攸关者需要和需求定义流程*——使用采办复杂组织体中的复杂组织体级 ConOps 和开发复杂组织体中的系统级初步 OpsCon 作为指南，需求工程师通过结构化的流程促使业务运行中的利益攸关者引出利益攸关者需要（以细化的系统级 OpsCon 及其他生命周期概念的形式）。然后，需求工程师将利益攸关者需要转换成正式的利益攸关者需求集合，这些利益攸关者需求通常在利益攸关者需求规范（STRS）中获取。

- *系统需求定义流程*——然后需求工程师将 STRS 中的利益攸关者需求转换成系统需求，这些系统需求通常包含在系统需求规范（SYRS）中。

- *架构定义流程*——定义备选的系统架构并从中选择一个系统架构。

- *设计定义流程*——充分详细地定义系统元素，使实现与选定的系统架构一致。

- *系统分析流程*——使用数学分析、建模和仿真来支持其他技术流程。

- *实现流程*——实现系统元素以满足系统需求、架构和设计。

- *综合流程*——将系统元素组合到已实现的系统中。

- *验证流程*——提供证据证明生命周期内的系统、系统元素和工作产物满足指定的需求。

- *转移流程*——系统以计划的有序方式投入运行。

- *确认流程*——提供证据证明生命周期内的系统、系统元素和工作产品将实现意图运行环境中的意图使用。

- *运行流程*——使用系统。

- *维护流程*——在运行期间维持系统。

- *处置流程*——系统或系统元素被停用、拆解和从运行中移除。

4.1 BUSINESS OR MISSION ANALYSIS PROCESS

4.1.1 Overview

4.1.1.1 Purpose As stated in ISO/IEC/IEEE 15288,

[6.4.1.1] The purpose of the Business or Mission Analysis process is to define the business or mission problem or opportunity, characterize the solution space, and determine potential solution class(es) that could address a problem or take advantage of an opportunity.

4.1.1.2 Description Business or mission analysis initiates the life cycle of the system of interest (SOI) by defining the problem domain; identifying major stake-holders; identifying environmental conditions and constraints that bound the solution domain; developing preliminary life cycle concepts for acquisition, operations, deployment, support, and retirement; and developing the business requirements and validation criteria. Figure 4.2 is the IPO diagram for the business or mission analysis process.

4.1.1.3 Inputs/Outputs Inputs and outputs for the business or mission analysis process are listed in Figure 4.2. Descriptions of each input and output are provided in Appendix E.

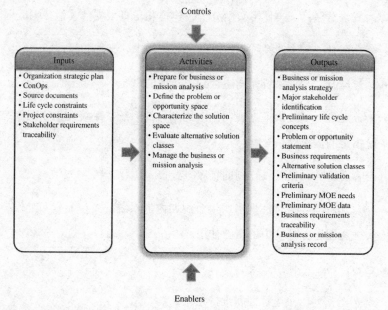

FIGURE 4.2 IPO diagram for business or mission analysis process. INCOSE SEH original figure created by Shortell and Walden. Usage per the INCOSE notices page. All other rights reserved.

4.1 业务或任务分析流程

4.1.1 概览

4.1.1.1 目的 如 ISO/IEC/IEEE 15288 所述，

[6.4.1.1] 业务或任务分析流程的目的是定义业务或任务问题或机会，特征化解决方案空间并确定能够解决问题或利用机会的潜在的解决方案类集。

4.1.1.2 描述 业务或任务分析开启所感兴趣之系统（SOI）的生命周期，通过：定义问题域；识别主要利益攸关者；识别限制解决方案域的环境条件和约束；为采办、运行、部署、支持和退出开发初步生命周期概念；以及开发业务需求和确认准则。图 4.2 是业务或任务分析流程的 IPO 图。

4.1.1.3 输入/输出 图 4.2 列出业务或任务分析流程的输入和输出。附录 E 提供每个输入和输出的描述。

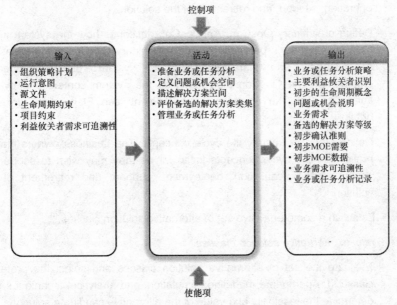

图 4.2 业务或任务分析流程的 IPO 图。INCOSE SEH 原始图由 Shortell 和 Walden 创建。按照 INCOSE 通知页使用。版权所有。

TECHNICAL PROCESSES

4.1.1.4 Process Activities The business or mission analysis process includes the following activities and tasks:

- *Prepare for business or mission analysis.*
  - Establish a strategy for business or mission analysis, including the need for and requirements of any enabling systems, products, or services.
- *Define the problem or opportunity space.*
  - Review identified gaps in the organization strategy with respect to desired organization goals or objectives.
  - Analyze the gaps across the trade space.
  - Describe the problems or opportunities underlying the gaps.
  - Obtain agreement on the problem or opportunity descriptions.
- *Characterize the solution space.*
  - Nominate major stakeholders (individuals or groups). Business owners nominate the major stakeholders who are to be involved in the acquisition, operation, support, and retirement of the solution.
  - Define preliminary OpsCon. An OpsCon describes how the system works from the operator's perspective. The preliminary OpsCon summarizes the needs, goals, and characteristics of the system's user and operator community. The OpsCon also identifies the system context and system interfaces (i.e., the operational environment; see Elaboration for more detail).
  - Define other preliminary life cycle concepts. The business owners identify preliminary life cycle concepts in so far as they may wish to scope any aspect of the acquisition, deployment, support, and retirement of the solution.
  - Establish a comprehensive set of alternative solution classes.
- *Evaluate alternative solution classes.*
  - Evaluate the set of alternative solution classes and select the preferred class(es). Appropriate modeling, simulation, and analytical techniques help determine the feasibility and value of the alternative candidate solutions.
  - Ensure that the preferred alternative solution class(es) has been validated in the context of the proposed business or mission strategy. Feedback on feasibility, market factors, and alternatives is also provided for use in completing the organization strategy and further actions.

4.1.1.4 流程活动　业务或任务分析流程包括以下活动和任务：

- *准备业务或任务分析。*
    - 建立业务或任务分析的策略，包括任何使能系统、产品或服务的需要和需求。
- *定义问题或机会空间。*
    - 评审与期望的组织目标或目的相关的组织战略中已识别到的差距。
    - 跨越权衡空间分析差距。
    - 描述差距底层的问题或机会。
    - 在问题或机会描述上获得一致。
- *描述解决方案空间。*
    - 指定主要利益攸关者（个人或集团）。业务所有者指定将要参与解决方案的采办、运行、保障和退出的主要利益攸关者。
    - 定义初步的 OpsCon。OpsCon 从运行者的视角描述系统如何工作。初步的 OpsCon 概述系统用户和运行者群体的需要、目的和特征。OpsCon 还识别系统背景环境和系统接口（即，运行环境；更多细节参见详细阐述）。
    - 定义其他初步的生命周期概念。业务所有者在他们希望审视解决方案的采办、部署、支持和退出的任何方面的范畴内，识别初步的生命周期概念。
    - 建立备选的解决方案类集的全面集合。
- *评价备选的解决方案类集。*
    - 评价备选的解决方案类集并选择优选的类（集）。适当的建模、仿真和分析技术有助于确定备选的候选解决方案的可行性和价值。
    - 确保在已提出的业务或使命战略的背景环境中确认优选的备选解决方案类（集）。还提供对可行性、市场因素和备选方案的反馈，用于完成组织战略和进一步行动。

- *Manage the business or mission analysis.*
  - Establish and maintain traceability of analysis results, such as requirements and preliminary life cycle concepts.
  - Provide baseline information for configuration management.

4.1.2 Elaboration

4.1.2.1 Nominate Major Stakeholders Although the detailed identification of stakeholders is undertaken in the stakeholder needs and requirements definition process, during business and mission analysis, the business managers are responsible for nominating major stake-holders and often for establishing a stakeholder board.It is fundamentally a business management function to ensure stakeholders are available and willing to contribute to the system development—most stakeholders are heavily occupied in business operations and must be given permission to expend effort and resources on other than their operational tasks.

4.1.2.2 ConOps and OpsCon ANSI/AIAA G-043A-2012 states that the terms " concept of operations " and " operational concept " are often used interchangeably but notes that an important distinction exists in that each has a separate purpose and is used to meet different ends.This handbook uses these terms so that they are consistent with ANSI/AIAA G-043A-2012 and ISO/IEC/IEEE 29148:2011, the same way in which they are used in the US Department of Defense (DoD) and many other defense forces.

ISO/IEC/IEEE 29148 describes the ConOps as

The ConOps, at the organization level, addresses the leadership's intended way of operating the organization. It may refer to the use of one or more systems, as black boxes, to forward the organization's goals and objectives. The ConOps document describes the organization's assumptions or intent in regard to an overall operation or series of operations of the business with using the system to be developed, existing systems, and possible future systems. This document is frequently embodied in long range strategic plans and annual operational plans. The ConOps document serves as a basis for the organization to direct the overall characteristics of the future business and systems, for the project to understand its background, and for the users of [ISO/IEC/IEEE 29148] to implement the stakeholder requirements elicitation.

ISO/IEC/IEEE 29148 describes the OpsCon as

- 管理业务或任务分析。

  – 建立并维护分析结果的可追溯性，如需求和初步生命周期概念。

  – 提供构型配置管理的基线信息。

4.1.2 详细阐述

4.1.2.1 指定主要利益攸关者 尽管在业务和任务分析期间在利益攸关者需要和需求定义流程中进行了利益攸关者的详细识别，而业务管理者负责指定主要利益攸关者且常常负责建立利益攸关者委员会。从根本上讲，业务管理功能是确保利益攸关者可投入并且有意愿贡献于系统开发——大多数利益攸关者很大程度上被其业务运行所占用且必须给予在其运行任务之外扩大精力和资源的许可。

4.1.2.2 ConOps 和 OpsCon ANSI/AIAA G-043A-2012 指出，术语"运行意图"和"运行概念"通常可互换使用，但应注意存在一个重要的区别是各自具有单独的目的且用于满足不同的结果。本手册使用这些术语使其与 ANSI/AIAA G-043A-2012 和 ISO/IEC/IEEE 29148:2011 一致，使用方式与在美国国防部（DoD）和许多其他国防部队中的使用方式相同。

ISO/IEC/IEEE 29148 将 ConOps 描述为

在组织层级，ConOps 指领导层意图的组织运行方式。它可能指一个或多个系统作为黑盒的使用，以促使组织的目标和目的。ConOps 文件描述与业务的总体运行或系列运行有关的组织的假设或意图，并使用待开发系统、现有系统和可能的未来系统。该文件往往体现于长期战略规划和每年的运行计划中。ConOps 文件用作组织指导未来业务和系统总体特征的基础、项目理解其背景的基础以及[ISO/IEC/IEEE 29148]中的使用者实现利益攸关者需求引出的基础。

ISO/IEC/IEEE 29148 将 OpsCon 描述为

A System Operational Concept (OpsCon) document describes what the system will do (not how it will do it) and why (rationale). An OpsCon is a user-oriented document that describes system characteristics of the to-be-delivered system from the user's viewpoint. The OpsCon document is used to communicate overall quantitative and qualitative system characteristics to the acquirer, user, supplier and other organizational elements.

Both the ConOps and the OpsCon are prepared by the organization that has the business need for the SOI. The ConOps is developed by/for the leadership at the enterprise level of the organization using the SOI. In some acquisitions, the ConOps may not be formalized, but rather implied by other business concepts and/or strategies. The OpsCon is prepared at the business level. Business management begins with the preparation of the preliminary OpsCon, which summarizes business needs from an operational perspective for the solution classes that address the problem or opportunity. The preliminary OpsCon is then elaborated and refined by business operations into the OpsCon by engagement with the nominated stakeholders during the stakeholder needs and requirements definition process—the final OpsCon therefore contains both the business needs and the stake-holder needs.The OpsCon may be iteratively refined as a result of feedback obtained through the conduct of the system requirements definition and architecture definition processes.

4.1.2.3 Other Life Cycle Concepts The OpsCon is just one of the life cycle concepts required to address the stakeholder needs across the system life cycle. Preliminary concepts are established in the business or mission analysis process to the extent needed to define the problem or opportunity space and characterize the solution space. These concepts are further refined in the stakeholder needs and requirements definition process. In addition to the operational aspects, other related life cycle concepts are required to address:

- *Acquisition concept*—Describes the way the sys-tem will be acquired including aspects such as stakeholder engagement, requirements definition, design, production, and verification. The supplier enterprise(s) may need to develop more detailed concepts for production, assembly, verification, transport of system, and/or system elements.

- *Deployment concept*—Describes the way the system will be validated, delivered, and introduced into operations, including deployment considerations when the system will be integrated with other systems that are in operation and/or replace any systems in operation.

- *Support concept*—Describes the desired support infrastructure and manpower considerations for supporting the system after it is deployed. A support concept would address operating support, engineering support, maintenance support, supply support, and training support.

系统运行概念（OpsCon）文件描述系统将要做什么（而不是如何做）以及为什么这么做（理由依据）。OpsCon 是从用户的视角描述待交付系统的系统特征的面向用户的文件。OpsCon 文件用于向采办方、用户、供应商及其他组织元素沟通全部定量的和定性的系统特征。

ConOps 和 OpsCon 都是由对 SOI 具有业务需要的组织来准备的。ConOps 由使用 SOI 的组织的复杂组织体层级的领导层来开发。在一些采办中，ConOps 可能不被正式化，而是被其他业务概念和/或策略所隐含。OpsCon 在业务层级准备。业务管理开始于初步 OpsCon 的准备，从运行的视角概述对应对问题或机会的解决方案类集的业务需要。然后，在利益攸关者需要和需求定义流程期间，通过指定的利益攸关者的介入，对初步 OpsCon 进行详细阐述并通过业务运行细化为 OpsCon——因此，最终的 OpsCon 既包含业务需要，也包含利益攸关者需要。作为通过进行系统需求定义流程和架构定义流程所获得的反馈的结果，OpsCon 可能被迭代地细化。

4.1.2.3　其他生命周期概念　OpsCon 仅仅是跨系统生命周期应对利益攸关者需要所要求的生命周期概念之一。初步概念在业务或任务分析流程中得以建立，延伸到定义问题或机会空间和特征化解决方案空间之所需。在利益攸关者需要和需求定义流程中对这些概念进一步细化。除了运行方面外，还需要涉及其他相关的生命周期概念：

- 采办概念——描述系统将被采办的方式，包括诸如利益攸关者介入、需求定义、设计、生产和验证方面。供应商复杂组织体可能需要为生产、组装、验证、系统运输和/或系统元素开发更为详细的概念。

- 部署概念——描述系统将被确认、交付和引入运行的方式，包括系统将与运行中的其他系统综合和/或取代运行中的任何系统时的部署考虑。

- 支持概念——描述为了系统部署后的支持所期望的支持基础设施和人力考虑。支持概念将应对运行支持、工程支持、维护支持、供应支持和培训支持。

- *Retirement concept*—Describes the way the system will be removed from operation and retired, including the disposal of any hazardous materials used in or resulting from the process and any legal obligations—for example, regarding IP rights protection, any external financial/ownership interests, and national security concerns.

4.1.2.4 Business Requirements and Validation

Specify business requirements—It is often helpful to specify business requirements as part of the business and mission analysis process. Business requirements are often contained in a BRS, which the *Guide to the Business Analysis Body of Knowledge* (IIBA, 2009) calls the business requirement document. The term "specification" has some variation in use in various industries, but it is used here to be synonymous with "document" —that is, business requirements are captured in the BRS, stake-holder requirements in the StRS, and system requirements in the SyRS.

Define business validation criteria The business must define how it will know that the solution provided will meet the OpsCon. Validation criteria establish critical and desired system performance—thresholds and objectives for system performance parameters that are critical for system success and those that are desired but may be subject to compromise to meet the critical parameters.

4.2 STAKEHOLDER NEEDS AND REQUIREMENTS DEFINITION PROCESS

4.2.1 Overview

4.2.1.1 Purpose   As stated in ISO/IEC/IEEE 15288,

> [6.4.2.1] The purpose of the Stakeholder needs and Requirements Definition process is to define the stakeholder requirements for a system that can provide the capabilities needed by users and other stakeholders in a defined environment.

4.2.1.2 Description   Successful projects depend on meeting the needs and requirements of the stakeholders throughout the life cycle. A stakeholder is any entity (individual or organization) with a legitimate interest in the system. When nominating stakeholders, business management will take into account all those who may be affected by or able to influence the system—typically, they would consider users, operators, organization decision makers, parties to the agreement, regulatory bodies, developing agencies, support organizations, and society at large (within the context of the business and proposed solution). When direct contact is not possible, systems engineers find agents, such as marketing or nongovernmental organizations, to represent the concerns of a class of stakeholders, such as consumers or future generations.

- 退出概念——描述系统将从运行中移除并退出的方式，包括在流程中使用的或由流程产生的任何危险物料以及任何法律义务的退出——例如，关于知识产权保护、任何外部财务/所有权权益及国家安保的考量。

4.1.2.4 业务需求和确认

规定业务需求——作为业务和任务分析流程的部分，通常有助于规定业务需求。业务需求通常包含在 BRS 中，《业务分析知识体指南》（IIBA, 2009）中称之为业务需求文件。术语"规范"在各种不同的行业中的使用有一些变化，但在这里的使用与"文件"同义——即，业务需求在 BRS 中获取，利益攸关者需求在 StRS 中获取，系统需求在 SyRS 中获取。

定义业务确认准则 业务必须定义其将如何知晓所提供的解决方案将会满足 OpsCon。确认准则建立关键的和期望的系统性能——系统性能参数的阈值和目标值，它们对于系统成功至关重要并为系统所期望，但为满足关键参数可能会被折中。

4.2 利益攸关者需要和需求定义流程

4.2.1 概览

4.2.1.1 目的 如 ISO/IEC/IEEE 15288 所述，

> [6.4.2.1] 利益攸关者需要和需求定义流程的目的是在一个定义明确的环境中，为一个能够提供用户和其他利益攸关者所需能力的系统定义利益攸关者需求。

4.2.1.2 描述 成功的项目取决于贯穿生命周期内满足利益攸关者的需要和需求。利益攸关者是对系统拥有合法权益的任何实体（个人或组织）。当指定利益攸关者时，业务管理将考虑到可能受系统影响或能够影响系统的所有人员——典型地，将考虑到用户、运行者、组织决策者、协议方、法规机构、开发机构、支持组织以及广泛的社会（在业务及提出的解决方案的背景环境内）。当不可能直接接触时，系统工程师找到如市场营销或非政府组织等代理，以表达某一类型利益攸关者（如消费者或新生代）的关注点。

TECHNICAL PROCESSES

After identifying the stakeholders, this process elicits the stakeholder needs that correspond to a new or changed capability or new opportunity. These needs are analyzed and transformed into a set of stakeholder requirements for the operation and effects of the solution and its interaction with the operational and enabling environments. The stakeholder requirements are the primary reference against which the operational capability is validated.

To achieve good results, systems engineers involve themselves in nearly every aspect of a project, pay close attention to interfaces where two or more systems or system elements work together, and establish an interaction network with stakeholders and other organizational units of the organization. Figure 4.3 shows the critical interactions for systems engineers.

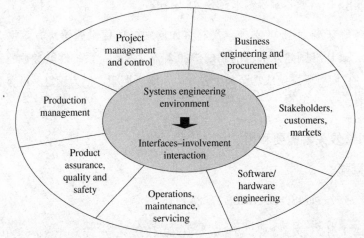

FIGURE 4.3 Key SE interactions. From Stoewer (2005). Figure reprinted with permission from Heinz Stoewer. All other rights reserved.

The stakeholder requirements govern the system's development and are an essential factor in further defining or clarifying the scope of the development project. If an organization is acquiring the system, this process provides the basis for the technical description of the deliverables in an agreement—typically in the form of a system-level specification and defined interfaces at the system boundaries. Figure 4.4 is the IPO diagram for the stakeholder needs and requirements definition process.

4.2.1.3 Inputs/Outputs Inputs and outputs for the stakeholder needs and requirements definition process are listed in Figure 4.4. Descriptions of each input and output are provided in Appendix E.

识别利益攸关者后，该流程引出与新的或改变后的能力或新的机会相对应的利益攸关者需要。对这些需要进行分析并转换成利益攸关者需求集合，用于解决方案的运行和影响及其与运行和使能环境的交互。利益攸关者需求是确认运行能力所依据的主要参考。

为达到良好的结果，系统工程师参与到项目的几乎各个方面，密切关注有两个、更多系统或系统元素共同工作的接口，并建立与组织的利益攸关者和其他组织单元的互动网络。图 4.3 指明系统工程师们的关键的互动。

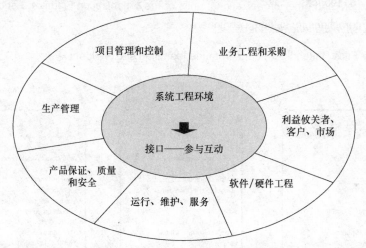

图 4.3 关键的 SE 互动。来自于 Stoewer（2005）。经 Heinz Stoewer 许可后转载图片。版权所有。

利益攸关者的需求支配着系统的开发，并且是进一步定义或明确开发项目范围的一个根本的要素。如果一个组织正在采办系统，该流程在协议中为可交付物的技术描述提供基准——通常采用系统级规范和在系统边界定义接口的形式。图 4.4 是利益攸关者需要和需求定义流程的 IPO 图。

4.2.1.3 输入/输出 图 4.4 列出利益攸关者需要和需求定义流程的输入和输出。附录 E 提供每个输入和输出的描述。

TECHNICAL PROCESSES

4.2.1.4 Process Activities The stakeholder needs and requirements definition process includes the following activities:

- *Prepare for stakeholder needs and requirements definition.*

    - Determine the stakeholders or classes of stake-holders who will participate with systems engineering to develop and define the stakeholder needs and translate these into system requirements, phased throughout the entire life cycle. Capture these results in the ConOps.

    - Determine the need for and requirements of any enabling systems, products, or services.

- *Define stakeholder needs.*

    - Elicit stakeholder needs from the identified stakeholders.

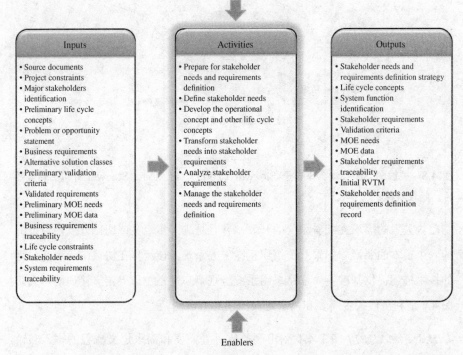

FIGURE 4.4 IPO diagram for stakeholder needs and requirements definition process. INCOSE SEH original figure created by Shortell and Walden. Usage per the INCOSE notices page. All other rights reserved.

4.2.1.4　流程活动　利益攸关者需要和需求定义流程包括以下活动：

- *准备利益攸关者需要和需求定义。*

  – 确定将要参与系统工程贯穿整个生命周期的各个阶段对利益攸关者需要进行开发和定义的利益攸关者或利益攸关者类型，并将这些需要转换成系统需求。在 ConOps 中获取这些结果。

  – 确定对任何使能系统、产品或服务的需要和需求。

- *定义利益攸关者需要。*

  – 从识别的利益攸关者中引出利益攸关者需要。

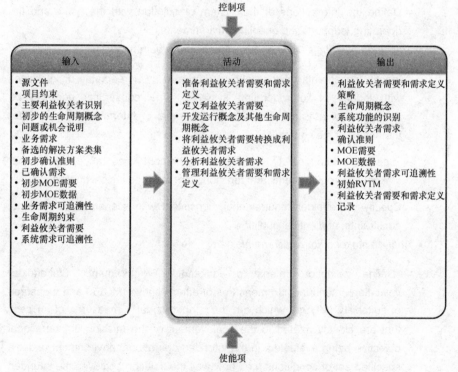

图 4.4　利益攸关者需要和需求定义流程的 IPO 图。INCOSE SEH 原始图由 Shortell 和 Walden 创建。按照 INCOSE 通知页使用。版权所有。

TECHNICAL PROCESSES

- Prioritize the stakeholder needs to identify which to focus on.
- Specify the stakeholder needs.
- *Develop the operational concept and other life cycle concepts.*
  - Identify the expected set of operational scenarios and associated capabilities, behaviors, and responses of the system or solution and environments across the life cycle (in acquisition, deployment, operations, support, and retirement). The scenarios are built to define the life cycle concept documents; the range of anticipated uses of system products; the intended operational environment and the systems' impact on the environment; and interfacing systems, platforms, or products. Scenarios help identify requirements that might otherwise be over-looked. Social and organizational influences also emerge from using scenarios.
  - Define the interactions of the system or solution with the users and the operating, support, and enabling environments.
- *Transform stakeholder needs into stakeholder requirements.*
  - Identify constraints on the solution (imposed by agreements or interfaces with legacy or interoperating systems). The constraints need to be monitored for any interface changes (external or internal) that could alter the nature of the constraint.
  - Specify health, safety, security, environment, assurance, and other stakeholder requirements and functions that relate to critical qualities.
  - Specify stakeholder requirements, consistent with scenarios, interactions, constraints, and critical qualities.
- *Analyze stakeholder requirements.*
  - Define validation criteria for stakeholder requirements. Stakeholder validation criteria include measures of effectiveness (MOEs) and measures of suitability (MOSs), which are the *"operational"* measures of success that are closely related to the achievement of the mission or operational objective being evaluated, in the intended operational environment under a specified set of conditions (i.e., how well the solution achieves the intended purpose). These measures reflect overall customer/user satisfaction (e.g., performance, safety, reliability, availability, maintainability, and workload requirements).

- 对利益攸关者需要进行优先级排序以识别要集中于哪些需要。

- 规定利益攸关者需要。

• *开发运行概念及其他生命周期概念。*

- 跨生命周期（在采办、部署、运行、支持和退出中）识别运行场景集合和系统的相关能力、行为及响应或解决方案和环境。应构建场景为定义生命周期概念文件；系统产品的预期使用范围；预期的运行环境以及系统对环境的影响；以及接口系统、平台或产品。场景有助于识别可能在其他场合被忽略的需求。社会和组织的影响亦从使用这些场景中显现出来。

- 定义系统或解决方案与用户以及运行环境、支持环境和使能环境的互动。

• *将利益攸关者需要转换成利益攸关者需求。*

- 识别对解决方案的约束（由具有遗留系统或互操作系统的协议或接口所施加的）。需要对约束监控以便确定是否具有可能改变约束本质属性的任何接口变化（外部或内部的）。

- 规定健康、安全、安保、环境、保证以及与关键质量相关的其他利益攸关者需求和功能。

- 规定与场景、交互、约束和关键品质一致的利益攸关者需求。

• *分析利益攸关者需求。*

- 定义利益攸关者需求的确认准则。利益攸关者确认准则包括有效性测度（MOE）和适用性测度（MOS），它们是在一系列特定条件集合下意图的运行环境中成功"运行的"测度，与被评估的任务或运行目标的达成密切相关（即解决方案达成意图目的的程度）。这些测度反映客户/用户的总体满意度（如性能、安全性、可靠性、可用性、可维护性和工作负荷需求）。

- Analyze the set of requirements for clarity, completeness, and consistency. Include review of the analyzed requirements to the applicable stakeholders to ensure the requirements reflect their needs and expectations.

- Negotiate modifications to resolve unrealizable or impractical requirements.

- *Manage the stakeholder needs and requirements definition.*

  - Establish with stakeholders that their requirements are expressed correctly.

  - Record stakeholder requirements in a form suitable for maintenance throughout the system life cycle (and beyond for historical or archival purposes).

  - Establish and maintain through the life cycle a traceability of stakeholder needs and requirements (e.g., to the stakeholders, other sources, organizational strategy, and business or mission analysis results).

  - Provide baseline information for configuration management.

4.2.2 Elaboration

Within the context of ISO/IEC/IEEE 15288, requirements (business, stakeholder, and system) are drivers for the majority of the system life cycle processes. Depending on the system development model, stakeholder requirements capture should be conducted nominally once near the beginning of the development cycle or as a continuous activity. Regardless, the reason for eliciting and analyzing requirements is the same—understand the needs of the stakeholders well enough to support the architecture definition and design definition processes.

4.2.2.1 Identify Stakeholders Systems engineers engage with legitimate stakeholders of the system. The major stakeholders at the business management level will have been nominated during the business or mission analysis process—here, systems engineers are interested in identifying stakeholders from the business operations level.

One of the biggest challenges in system development is the identification of the set of stakeholders from whom requirements should be elicited. Customers and eventual end users are relatively easy to identify, but regulatory agencies and other interested parties that may reap the consequences of the deployed system should also be sought out and heard. Stakeholders can include the stake-holders of interoperating systems and enabling systems as these will usually impose constraints that need to be identified and considered. In sustainable development, this includes finding representation for future generations.

- 分析需求集合的清晰性、完整性和一致性。包括对适用的利益攸关者的分析需求进行评审以确保需求反映他们的需要和期望。

- 协商修改以解决无法实现的或不切实际的需求。

- *管理利益攸关者需要和需求定义。*

  - 与利益攸关者确定他们的需求被准确地表达。

  - 贯穿于系统生命周期内（超出历史或归档目的）以适于维护的形式记录利益攸关者需求。

  - 在生命周期内建立和维护利益攸关者需要和需求的可追溯性（如追溯到利益攸关者、其他来源、组织战略以及业务或任务分析结果）。

  - 提供构型管理的基线信息。

4.2.2 详细阐述

在 ISO/IEC/IEEE 15288 背景环境中，需求（业务、利益攸关者和系统）是大多数系统生命周期流程的驱动因素。依赖系统开发模型，利益攸关者需求的捕获应在开发周期将要开始的时候名义地进行一次或作为一个持续的活动。无论如何，引出和分析需求的原因是相同的——理解利益攸关者的需要以足够好地支持架构定义流程和设计定义流程。

4.2.2.1 识别利益攸关者 系统工程师与系统的合法利益攸关者密切联系。业务管理层级的主要利益攸关者将在业务或任务分析流程期间被指定——其中，系统工程师对识别来自业务运行层级的利益攸关者感兴趣。

系统开发中的最大挑战之一是识别引出需求的一系列利益攸关者。客户和最后的最终用户相对容易识别，但亦应找到或听到在已部署的系统中收获结果的管理机构及其他感兴趣的各方。利益攸关者可包括互操作系统和使能系统的利益攸关者，因为这些利益攸关者通常会强加需要被识别和考虑的约束。在可持续的开发中，这包括寻找新生代的代表。

4.2.2.2 Elicit Stakeholder Needs Determining stakeholder needs requires the integration of a number of disparate views, which may not necessarily be harmonious. As the SE process is applied, a common paradigm for examining and prioritizing available information and determining the value of added information should be created. Each of the stakeholder's views of the needed systems can be translated to a common top-level system description that is understood by all participants, and all decision-making activities recorded for future examination. Under some circumstances, it may not be practical to elicit needs from the stakeholder but rather from the marketing organization or other surrogates. There may be stakeholders who oppose the system. These stakeholders or detractors of the system are first considered in establishing consensus needs. Beyond this, they are addressed through the risk management process, the threat analysis of the system, or the system requirements for security, adaptability, or resilience.

Systems engineering should support program and project management in defining what must be done and gathering the information, personnel, and analysis tools to elaborate the business requirements. This includes gathering stakeholder needs, system/project constraints (e.g., costs, technology limitations, and applicable specifications/legal requirements), and system/project "drivers," such as capabilities of the competition, military threats, and critical environments.

The outputs of the stakeholder needs and requirements definition process should be sufficient definition of the business and stakeholder needs and requirements to gain authorization and funding for program initiation through the portfolio management process. The output should also provide necessary technical definition to the acquisition process to generate a request for proposal (RFP) if the system is to be acquired through a contract acquisition process or to gain authorization to develop and market the system if market driven. These outputs can be captured in life cycle concept documents (particularly the OpsCon) and the StRS, which often are used to support the generation of a statement of work (SOW), and/or an RFP, both of which are artifacts of the acquisition process. Contributing users rely on well-defined completion criteria to indicate the successful definition of user and stakeholder needs:

- User *organizations have gained authorization for new system acquisition.*

- *Program development organizations have prepared a SOW, StRS, and gained approval for new system acquisition. If they are going to use support from outside the company, they have issued an RFP and selected a contractor.*

- *Potential contractors have influenced the acquisition needs, submitted a proposal, and have been selected to develop and deliver the system.*

4.2.2.2 引出利益攸关者需要 确定利益攸关者需要要求对未必协调的诸多不同的看法进行综合。随着 SE 流程的应用，应创建一个对可用的信息进行考查和排序并确定增加信息的价值的公共范式。所需系统的每个利益攸关者看法可转换为被所有参与者理解的公共顶层系统的描述，以及记录用于未来考查的所有决策活动。在某些情形下，从利益攸关者引出需要可能是不实际的，但可能从市场组织或其他代理中引出。可能会有反对该系统的利益攸关者。在建立共识的需要时，首先考虑系统的这些利益攸关者或诋毁者。除此之外，为了达到安保性、适应性或复原性，通过风险管理流程、系统的威胁分析或系统需求来应对。

在定义必须做什么和收集用于详述业务需求的信息、人员和分析工具时，系统工程应支持项目群计划和项目管理。这包括收集利益攸关者需要、系统/项目约束（如成本、技术限制和适用规范/法律要求）和系统/项目"驱动因素"，如竞争能力、军事威胁和关键环境。

利益攸关者需要和需求定义流程的输出应充分定义业务和利益攸关者需要和需求，以通过项目群管理流程获取计划启动的授权和资金。如果系统要通过合同采办流程来采办，或在市场驱动下获取开发和营销系统的授权，则输出亦应向采办流程提供必要的技术定义，以产生邀标书（RFP）。这些输出可在生命周期概念文件（特别是 OpsCon）和 StRS 中获取，它们往往用于支持工作说明书（SOW）和/或 RFP 的生成，两者是采办流程的工作制品。做出贡献的用户依靠明确定义的完成准则以表明对用户和利益攸关者需要的成功定义：

- 用户组织已获取采办新系统的授权。

- 项目群计划开发组织已为新系统采办准备了一个 SOW 和 StRS，并获得批准。如果开发组织要使用公司外的支持，则应发布一个 RFP 并选择一个承包商。

- 潜在的承包商已影响了采办需要，提交了投标，并已选定其来开发和交付系统。

TECHNICAL PROCESSES

- *If the system is market driven, the marketing group has learned what consumers want to buy. For expensive items (e.g., aircraft), they have obtained orders for the new systems.*

- *If the system is market and technology driven, the development team has obtained approval to develop the new system from the corporation.*

Since requirements come from multiple sources, eliciting and capturing requirements constitutes a significant effort on the part of the systems engineer. The OpsCon describes the intended operation of the system to be developed and helps the systems engineer understand the context within which requirements need to be captured and defined. Techniques for requirements elicitation include interviews, focus groups, the Delphi method, and soft systems methodology, to name a few. Trade-off analysis and simulation tools can also be used to evaluate mission operational alternatives and select the desired mission alternative. Tools for capturing and managing requirements are many and varied.

The source requirements captured by carrying out this activity are only a portion of the total stakeholder requirements. As such, source requirements will be expanded by a number of activities designed to break down the broad requirement statements and reveal the need for additional clarification, which will lead to either revision of the written source material or additional source documents, such as meeting minutes.

4.2.2.3 Initialize the Requirements Database It is essential to establish a database of baseline requirements traceable to the source needs (and subsequently to system requirements) to serve as a foundation for later refinement and/or revision by subsequent activities in the SE process. The requirements database must first be populated with the source documents that provide the basis for the total set of system requirements that will govern its design.

4.2.2.4 Develop the Life Cycle Concepts The word "scenario" is often used to describe a single thread of behavior; in other cases, it describes a superset of many single threads operating concurrently. Scenarios and what-if thinking are essential tools for planners who must cope with the uncertainty of the future. Scenario thinking can be traced back to the writings of early philosophers, such as Plato and Seneca (Heijden et al., 2002).As a strategic planning tool, scenario techniques have been employed by military strategists throughout history. Building scenarios serves as a methodology for planning and decision making in complex and uncertain environments. The exercise makes people think in a creative way, observations emerge that reduce the chances of overlooking important factors, and the act of creating the scenarios enhances communications within and between organizations. Scenario building is an essentially human activity that may involve interviews with opera- tors of current/similar systems, potential end users, and meetings of an Interface Working Group (IFWG). The results of this exercise can be captured in many graphical forms using modeling tools and simulations.

- 如果系统是市场驱动的，则市场群体已了解到客户想要购买的东西。对于昂贵物品（如飞机），他们已获取新系统的订单。

- 如果系统是市场和技术共同驱动的，则开发团队已从公司得到开发新系统的批准。

由于需求来自多个源头，引出和捕获需求构成系统工程师所要付出的极为重要的努力。OpsCon 描述待开发系统意图的运行，帮助系统工程师理解需要被捕获和定义的需求所在的背景环境。引出需求的技巧包括访谈、焦点群组、德尔菲法和软系统方法论，仅举几例。权衡分析和仿真工具亦可用于评价任务运行备选方案和选择所期望的任务备选方案。捕获和管理需求的工具是多种多样的。

执行该活动所捕获的源需求只是利益攸关者全部需求的一部分。就这一点而论，源需求由设计用于分解宽泛的需求说明书并揭示额外阐明的需要的若干活动而被扩展，这将导致书面源材料的修订或补加源文件，如会议记录。

4.2.2.3 初始化需求数据库 建立可追溯到源需要（后续可追溯到系统需求）的一个基线需求数据库，以用作 SE 流程中后续活动进行后期细化和/或修订的基础是必要的。首先需求数据库必须填充源文件，由此为支配其设计的整个系统需求集合提供基础。

4.2.2.4 开发生命周期概念 词汇"场景"通常用于描述行为的单个线程；在其他情况下，用于描述并行运行的多个单线程的扩展集合。场景和"如果"假设思维对于必须应对未来不确定性的计划者而言是最基本的工具。场景思维可追溯到早期哲学家如柏拉图和塞内卡（Heijden 等，2002）的著作。场景技术作为战略规划工具已经在历史上被军事战略家使用。构建场景用作在复杂和不确定环境中规划和决策的方法论。这种演练使人以创造性的方式思考，观察力的涌现减少忽略重要因素的机会，创建场景的行为增强组织内和组织间的沟通。场景构建是一项最基本的人类活动，该活动可涉及与现有/类似系统的运行者、潜在的最终用户而进行访谈以及接口工作组（IFWG）的会议。这种演练的结果可从使用建模工具和仿真以多种图解形式而被捕获。

Creation or upgrade of a system shares the same uncertainty regarding future use and emergent properties of the system. The stakeholder needs and requirements definition process captures the understanding of stakeholder needs in a series of life cycle concept documents, each focused on a specific life cycle stage: acquisition concept, deployment concept, OpsCon, support concept, and retirement concept. Each of these categories of life cycle concepts is discussed in Section 4.1.2.3. A primary goal of a concept document is to capture, early in the system life cycle, an implementation-free understanding of stakeholder needs by defining what is needed, without addressing how to satisfy the need. It captures behavioral characteristics required of the system in the context of other systems with which it interfaces, and captures the manner in which people will interact with the system for which the system must provide capabilities. Understanding these operational needs typically produces:

- *A source of specific and derived requirements that meet the needs and objectives of the customer and user*
- *Invaluable insight for SE and designers as they define design, develop, verify, and validate the system*
- *Diminished risk of latent system defects in the delivered operational systems*

If the system is for a military customer, there may be several required views of the system driven by architectural frameworks. These are defined, for example, in the US Department of Defense Architecture Framework (DODAF, 2010) and in the UK Ministry of Defense Architecture Framework (MODAF, n.d.) (OMG, 2013a).

The primary objective is to communicate with the end user of the system during the early specification stages to ensure that stakeholder needs (particularly the operational needs) are clearly understood and the rationale for performance requirements is incorporated into the decision mechanism for later inclusion in the system requirements and lower-level specifications. Interviews with operators of current/similar systems, potential users, interface meetings, IPO diagrams, functional flow block diagrams (FFBD), timeline charts, and N^2 charts provide valuable stakeholder input toward establishing a concept consistent with stakeholder needs. Other objectives are as follows:

1. To provide traceability between operational needs and the captured source requirements.
2. To establish a basis for requirements to support the system over its life, such as personnel requirements, support requirements, etc.

就系统未来的使用和涌现性而言,系统的创建或升级具有同样的不确定性。利益攸关者需要和需求定义流程在一系列生命周期概念文件中捕获对利益攸关者需要的理解,每个文件聚焦于某一特定的生命周期阶段:采办概念、部署概念、OpsCon、支持概念和退出概念。在 4.1.2.3 节中论述生命周期概念的每一类别。概念文件的主要目的是在系统生命周期的早期,通过定义需要什么但不涉及如何满足需要,捕获对利益攸关者需要(与实现无关)的理解。概念文件可在与系统交接的其他系统的背景环境中捕获系统所需求的行为特征,并捕获人员与(必须向系统提供能力的)系统之间进行交互的方式。理解这些运行需要通常会产生:

- 满足客户和用户的需要与目标的特定需求和衍生需求的来源

- SE 和设计者的宝贵的洞察力,伴随着他们定义、设计、开发、验证和确认系统

- 在已交付运行的系统中,降低系统潜在缺陷的风险

如果系统供军事客户使用,可能存在多个由架构框架驱动的所需系统视图。例如,美国国防部架构框架(DODAF,2010)和英国国防部架构框架(MODAF, n.d.)(OMG, 2013a)对此进行了定义。

主要目标是在早期的规范阶段与系统的最终用户沟通,以确保利益攸关者需要(特别是运行需要)被完全理解且性能需求的理由依据纳入决策机制,便于随后融入系统需求和更低层级的规范中。与现有/类似系统的运行者和潜在用户的访谈、接口会议、IPO 图、功能流块图(FFBD)、时间线图和 N^2 图为建立与利益攸关者需要一致的概念提供了有价值的利益攸关者的输入。其他目标如下:

1. 提供运行需要与捕获到的源需求之间的可追溯性。

2. 为需求建立一个基础,以在其整个生命期内支持系统,如人员需求、支持需求等。

TECHNICAL PROCESSES

3. To establish a basis for verification planning, system-level verification requirements, and any requirements for environmental simulators.

4. To generate operational analysis models to test the validity of external interfaces between the system and its environment, including interactions with external systems.

5. To provide the basis for computation of system capacity, behavior under-/overload, and mission effectiveness calculations.

6. To validate requirements at all levels and to discover implicit requirements overlooked from other sources.

Since the preliminary life cycle concepts have provided a broad description of system behavior, a starting point for further developing the concept is to begin by identifying outputs generated by external systems (modified as appropriate by passing through the natural system environment), which act as stimuli to the SOI and cause it to take specified actions and produce outputs, which in turn are absorbed by external systems. These single threads of behavior eventually cover every aspect of operational performance, including logistical modes of operation, operation under designated conditions, and behavior required when experiencing mutual interference with multi-object systems.

Aggregation of these single threads of behavior represents a dynamic statement of what the system is required to do and how it is to be acquired, deployed, operated, supported, and retired. No attempt is made at this stage to define a complete OpsCon or to allocate functions to hardware or software elements (this comes later during architectural design). The life cycle concepts are essentially definitions of the functional concepts and rationale from *the stakeholder perspective*. The life cycle concepts are further developed as follows:

1. Start with the source operational requirements; deduce a set of statements describing the higher level, mission-oriented needs.

2. Review the system needs with stakeholders and record the conflicts.

3. Define and model the operational boundaries.

4. For each model, generate a context diagram to represent the model boundary.

5. Identify all of the possible types of observable input and output events that can occur between the system and its interacting external systems.

3. 为验证计划、系统级验证需求和环境仿真器的任何需求建立一个基础。

4. 生成运行分析模型，以测试系统与其环境之间外部交联的有效性，包括与外部系统的相互作用。

5. 为系统能力、行为欠缺/过度和任务有效性的预测提供计算基础。

6. 确认所有层级中的需求，并查找被其他来源忽略的隐含需求。

由于初步生命周期概念已提供了系统行为的广泛描述，因此进一步开发概念的起点是首先识别由外部系统（通过自然系统环境适当修正）产生的输出，这些输出作为 SOI 的激励因素并导致 SOI 采取特定行动和产生输出，且转而被外部系统吸收。行为的这些单个线程最终涵盖各个方面的运行性能，包括后勤运行模式、指定条件下的运行以及经历与多个对象系统相互干扰时所要求的行为。

行为的这些单线程的集合表达系统被要求做的事情以及如何被采办、部署、运行、支持和退出的一个动态说明。在该阶段不需要尝试定义一个完整的 OpsCon 或向硬件或软件元素分配功能（将在后续架构设计阶段执行）。生命周期概念本质上是从利益攸关者的视角对功能概念的定义和理由依据。生命周期概念进一步开发如下：

1. 开始于源运行需求；推论用于描述更高层的面向任务的需要的一系列说明。

2. 与利益攸关者评审系统需要，并记录冲突。

3. 定义并建模运行的边界。

4. 为每个模型生成一个表示模型边界的背景环境图。

5. 识别在系统及与其接口的外部系统之间发生的所有可观察到的可能类型的输入和输出事件。

6. If the inputs/outputs are expected to be significantly affected by the environment between the system and the external systems, add concurrent functions to the IPO diagram to represent these transformations and add input and output events to the database to account for the differences in event timing between when an output is emitted and when an input is received.

7. Record the existence of a system interface between the system and the environment or external system.

8. For each class of interaction between a part of the system and an external system, create a functional flow diagram to model the sequence of interactions as triggered by the stimuli events generated by the external systems.

9. Add information to trace the function timing from performance requirements and simulate the timing of the functional flow diagrams to confirm operational correctness or to expose dynamic inconsistencies. Review results with users and operational personnel.

10. Develop timelines, approved by end users, to supplement the source requirements.

4.2.2.5 Generate the StRS A draft StRS should be generated to formally represent the stakeholder requirements. The StRS should be traceable to the stakeholder needs and to the BRS.

4.3 SYSTEM REQUIREMENTS DEFINITION PROCESS

4.3.1 Overview

4.3.1.1 Purpose As stated in ISO/IEC/IEEE 15288,

[6.4.3.1] The purpose of the System Requirements Definition process is to transform the stakeholder, user oriented view of desired capabilities into a technical view of a solution that meets the operational needs of the user.

4.3.1.2 Description System requirements are the foundation of the system definition and form the basis for the architecture, design, integration, and verification. Each requirement carries a cost. It is therefore essential that a complete but minimum set of requirements be established from defined stakeholder requirements early in the project life cycle. Changes in requirements later in the development cycle can have a significant cost impact on the project, possibly resulting in cancellation.

6. 如果预计输入/输出受系统与外部系统之间的环境的严重影响,则在 IPO 图上加上并发功能以表示这些转换,并在数据库上添加输入和输出事件,以解释发出输出与收到输入之间事件时间差的原因。

7. 记录系统和环境或外部系统之间存在的系统接口。

8. 对于系统的一个部分与一个外部系统之间的每一类相互作用,创建功能流程图,以便对外部系统产生的激励事件所触发的相互作用顺序进行建模。

9. 添加信息,以便从性能需求中追踪功能时序并仿真功能流程图的时序,以确认运行的正确性或揭示动态的不一致性。与用户和运行人员审查结果。

10. 开发时间基线,经最终用户批准,以补充源需求。

4.2.2.5 生成 StRS 应生成一个 StRS 草拟本来正式地表达利益攸关者需求。StRS 应可追溯到利益攸关者需要和 BRS。

4.3 系统需求定义流程

4.3.1 概览

4.3.1.1 目的 如 ISO/IEC/IEEE 15288 所述,

[6.4.3.1]系统需求定义流程的目的在于将面向利益攸关者和用户所期望的能力视角转换为满足用户运行需要的一个解决方案的一个技术视角。

4.3.1.2 描述 系统需求是系统定义的基础,且是构成架构、设计、综合和验证的基础。每个需求承载一份成本。因此,在项目生命周期的早期根据定义明确的利益攸关者需求建立一个完备但最小的需求集合是十分必要的。在开发周期的后期,需求变更会对项目成本造成巨大影响,也可导致项目的取消。

TECHNICAL PROCESSES

The system requirements definition process generates a set of system requirements from the supplier's perspective using the stakeholder requirements that reflect the user's perspective as the basis. The system requirements specify the system characteristics, attributes, functions, and performance that will meet the stakeholder requirements.

Requirements definition is both iterative and recursive (see Section 3.4.1). According to ISO/IEC/IEEE 29148, *Requirements engineering* (2011),

When the application of the same process or set of processes is repeated on the same level of the system, the application is referred to as iterative. Iteration is not only appropriate but also expected. New information is created by the application of a process or set of processes. Typically this information takes the form of questions with respect to requirements, analyzed risks or opportunities. Such questions should be resolved before completing the activities of a process or set of processes.

When the same set of processes or the same set of process activities are applied to successive levels of system elements within the system structure, the application form is referred to as recursive. The outcomes from one application are used as inputs to the next lower (or higher) system in the system structure to arrive at a more detailed or mature set of outcomes. Such an approach adds value to successive systems in the system structure.

Thus, iteration between this process and others is expected as more information is available and analysis is performed. This process continues to be recursively applied to define the requirements for each system element.

The output of the process must be compared for traceability to and consistency with the stakeholder requirements, without introducing implementation biases, before being used to drive the architecture definition process. The system requirements definition process adds the verification criteria to the defined system requirements. Figure 4.5 is the IPO diagram for the system requirements definition process.

4.3.1.3 Inputs/Outputs Inputs and outputs for the system requirements definition process are listed in Figure 4.5. Descriptions of each input and output are provided in Appendix E.

4.3.1.4 Process Activities The system requirements definition process includes the following activities:

- *Prepare for system requirements definition.*

系统需求定义流程使用作为基础反映用户视角的利益攸关者需求，从供应商的视角生成系统需求集合。系统需求明确规定满足利益攸关者需求的系统特征、属性、功能和性能。

需求分析既是迭代的又是递归的（见 3.4.1 节）。根据 ISO/IEC/IEEE 29148，需求工程（2011），

当在系统的同一层级上重复应用同一流程或流程集时，这种应用被称为迭代。迭代不仅是合适的而且是所期望的。通过应用一个流程或流程集就会创生新的信息。通常，该信息形式上表现为与需求、经分析的风险或机会相关的问题。这些问题应在完成一个流程或流程集的活动前得到解决。

当同一流程集合或同一流程活动集合被应用于系统结构内相继层级的系统元素时，这种应用形式被称为递归。一个应用的结果被用作系统结构中下一个更低（或更高）层级的系统的输入，以获得更详细或更成熟的结果集合。这种方法为系统结构中的相继系统增加价值。

因此，由于更多信息可供使用并进行分析，此流程与其他流程之间的迭代是期望的。此流程继续递归地应用于为每个系统元素的定义需求。

在用于驱动架构设计流程之前，在不引入实施偏见的情况下，必须比较该流程的输出与利益攸关者需求的可追溯性和一致性。系统需求定义流程为定义明确的系统需求增加验证准则。图 4.5 是系统需求定义流程的 IPO 图。

4.3.1.3 输入/输出　图 4.5 列出了系统需求定义流程的输入和输出。附录 E 提供每个输入和输出的描述。

4.3.1.4 流程活动　系统需求定义流程包括以下活动：

- 准备系统需求定义。

TECHNICAL PROCESSES

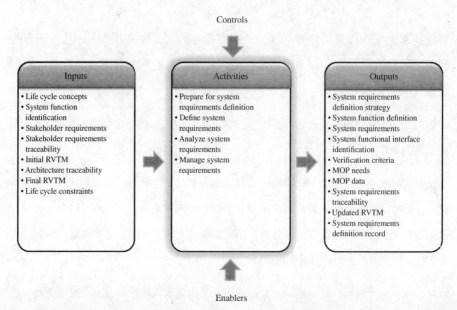

FIGURE 4.5 IPO diagram for the system requirements definition process. INCOSE SEH original figure created by Shortell and Walden. Usage per the INCOSE notices page. All other rights reserved.

- Establish the approach for defining the system requirements. This includes system requirements methods, tools, and the need for and requirements of any enabling systems, products, or services.

- In conjunction with the architecture definition process, determine the system boundary, including the interfaces, that reflects the operational scenarios and expected system behaviors. This task includes identification of expected interactions of the system with systems external to the system (control) boundary as defined in negotiated interface control documents (ICDs).

- *Define system requirements.*

  - Identify and define the required system functions. These functions should be kept implementation independent, not imposing additional design constraints. Define conditions or design factors that facilitate and foster efficient and cost-effective life cycle functions (e.g., acquisition, deployment, operation, support, and retirement). Also, include the system behavior characteristics.

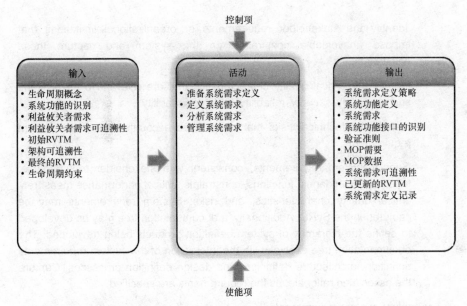

图 4.5 系统需求定义流程的 IPO 图。INCOSE SEH 原始图由 Shortell 和 Walden 创建。按照 INCOSE 通知页使用。版权所有。

- 建立定义系统需求的实施途径。这包括系统需求方法、工具以及对任何使能系统、产品或服务的需求的需要。

- 结合架构定义流程，确定系统边界，包括反映运行场景和期望的系统行为的接口。此任务包括按照商定的接口控制文件（ICD）的定义识别出系统与系统（控制）边界以外的系统之间的预期相互作用。

- *定义系统需求。*

  - 识别和定义所要求的系统功能。这些功能应保持与实施相独立，也不要施加额外的设计约束。定义出助促并推动有效率且具有成本效益的生命周期功能（如采办、部署、运行、支持和退出）的条件或设计因素，亦包括系统行为特征。

- Identify the stakeholder requirements or organizational limitations that impose unavoidable constraints on the system and capture those constraints.

- Identify the critical quality characteristics that are relevant to the system, such as safety, security, reliability, and supportability.

- Identify the technical risks that need to be accounted for in the system requirements.

- Specify system requirements, consistent with stakeholder requirements, functional boundaries, functions, constraints, critical performance measures, critical quality characteristics, and risks. System requirements may be captured in the SyRS. Additionally, a documentation tree may be developed to define the hierarchy of system definition products being developed. The documentation tree evolves with the interaction of the system requirements definition, architecture definition, and design definition processes. Capture the associated rationale, as the requirements are specified.

- *Analyze system requirements.*

    - Analyze the integrity of the system requirements to ensure that each requirement or set of requirements possess overall integrity. (See Section 4.3.2.2 for characteristics of a good requirement or set of requirements.)

    - Provide analysis results to applicable stake-holders to ensure that the specified system requirements adequately reflect the stakeholder requirements.

    - Negotiate modifications to resolve issues identified in the requirements.

    - Define verification criteria—critical performance measures that enable the assessment of technical achievement. System verification criteria include measures of performance (MOPs) and technical performance measures (TPMs), which are the "implementation" measures of success that should be traceable to the MOEs and MOSs (operational perspective) with the relationships defined.

- *Manage system requirements.*

    - Ensure agreement among key stakeholders that the requirements adequately reflect the stakeholder intentions.

    - Establish and maintain traceability between the system requirements and the relevant elements of the system definition (e.g., stakeholder requirements, architecture elements, interface definitions, analysis results, verification methods or techniques, and allocated, decomposed, and derived requirements.)

- 识别强加于系统的不可避免的约束的利益攸关者需求或组织的限制并捕获这些约束。

- 识别与系统相关的关键质量特征，如安全、安保、可靠性和可支持性。

- 识别出需要在系统需求中解释的技术风险。

- 明确规定与利益攸关者需求、功能边界、功能、约束、关键性能衡量指标、关键质量特征和风险一致的系统需求。系统需求可在 SyRS 中捕获。此外，可开发文档树，以定义正在开发的系统定义产品的层级结构。文档树与系统需求定义流程、架构定义流程和设计定义流程的交互而演进。随着需求被明确规定，捕获相关的理由依据。

- 分析系统需求。

  - 分析系统需求的完整性，以确保每个需求或需求集合都具有整体完整性。（关于良好需求或需求集合的特征，见 4.3.2.2 节。）

  - 为合适的利益攸关者提供分析结果以确保明确规定的系统需求充分地反映利益攸关者需求。

  - 协商修改以解决在需求中识别到的问题。

  - 定义验证准则——能够评估技术成果的关键性能测度。系统验证准则包括性能测度（MOP）和技术性能测度（TPM），它们是可追溯到具有已定义关系的 MOE 和 MOS（运行视角层面）成功"实施"的测度。

- 管理系统需求。

  - 确保关键利益攸关者之间就"需求充分地反映利益攸关者的意图"方面达成一致。

  - 建立并维持系统需求与系统定义的相关元素之间（如利益攸关者需求、架构元素、接口定义、分析结果、验证方法或技术以及分配的、分解的和导出的需求）的可追溯性。

- Maintain throughout the system life cycle the set of system requirements together with the associated rationale, decisions, and assumptions.

- Provide baseline information for configuration management.

4.3.2 Elaboration

This section elaborates and provides *"how-to"* information on the requirements analysis and management. Other key information on requirements can be found in ISO/ IEC TR 19760, *Systems engineering—A guide to the application of ISO/IEC 15288* (2003), and ISO/IEC/IEEE 29148, Requirements engineering (2011), and in EIA 632, Standard—*Processes for engineering a system* (ANSI/EIA, 2003), Requirements 14, 15, and 16, and Annex C3.1 a, b, and c.

4.3.2.1 Requirements Definition and Analysis

Concepts Requirements definition and analysis, like the set of SE processes, is an iterative activity in which new requirements are identified and constantly refined as the concept develops and additional details become known. The requirements are analyzed, and deficiencies and cost drivers are identified and reviewed with the customer to establish a requirements baseline for the project.

An objective of requirements analysis is to provide an understanding of the interactions between the various functions and to obtain a balanced set of requirements based on user objectives. Requirements are not developed in a vacuum. An essential part of the requirements development process is the OpsCon, the implicit design concept that accompanies it, and associated demands of relevant technology. Requirements come from a variety of sources, including the customer/users, regulations/codes, and corporate entities.

This complex process employs performance analysis, trade studies, constraint evaluation, and cost–benefit analysis. System requirements cannot be established without determining their impact (achievability) on lower-level elements. Therefore, requirements definition and analysis is an iteration and balancing process that works both *"top-down"* (called allocation and flow-down) and *"bottom-up."* Once the top-level set of system requirements has been established, it is necessary to allocate and flow them down to successively lower levels. As the allocation and flow-down process is repeated, it is essential that traceability be maintained to ensure that all system level requirements are satisfied in the resulting design. The resulting requirements database usually contains many attributes for each requirement and is also used in verification. Although it is an objective to avoid requirements that constrain or define aspects of the system implementation, it is not always possible. There are often necessary constraints that need to be reflected. This includes the following:

- 贯穿于系统生命周期内维持系统需求集合连同相关理由依据、决策和假设。

- 提供构型配置管理的基线信息。

4.3.2 详细阐述

本节详细阐述并提供关于需求分析和管理方面的"How-to"信息。关于需求方面的其他关键信息可在 ISO/IEC TR 19760、系统工程——ISO/IEC 15288（2003）应用指南、ISO/IEC/IEEE 29148 需求工程（2011）、EIA 632、标准——系统工程设计流程（ANSI/EIA, 2003）、第 14 条、第 15 条、第 16 条需求和附录 C3.1 的 a，b 和 c 中找到。

4.3.2.1 需求定义和分析

概念 需求定义和分析与 SE 流程集一样是一种迭代活动，在该活动中随着概念的开发以及增补的细节不断地得到了解，新需求亦不断地被识别和细化。对这些需求进行分析，并与客户一起识别和审查不足之处和成本动因，以建立项目的一个需求基线。

需求分析的一个目标是为各种不同功能之间相互作用提供一种理解并基于用户目标获取一个平衡的需求集合。需求不是凭空开发的。需求开发流程的一个最基本的部分是 OpsCon——与该流程共存的隐含设计概念和相关技术的有关要求。需求来源有很多，包括客户/用户、规则/准则和公司实体。

这个复杂流程采用性能分析、权衡研究、约束评价和成本/效益分析。只有确定系统需求对更低层级元素的影响（可实现性）才能建立系统需求。因此，需求的定义和分析是一种既可"自顶向下"（称作分配和向下流动），又可"自底向上"的迭代和进行平衡的流程。当顶层系统需求集合建立后，有必要将这些需求逐级向下分配、向下流动到更低的层级。由于分配和向下流动流程是重复的，为确保在产生的设计中满足所有系统层级的需求，有必要保持可追溯性。形成的需求数据库通常包含每个需求的多个属性，亦在验证中使用。尽管目的是避免产生约束或定义系统实施的各个方面的需求，但并非总能实现。通常存在需要反映的必要的约束。这包括以下内容：

- *Standards*—Identify standards required to meet quality or design considerations imposed as defined stakeholder requirements or derived to meet organization, industry, or domain requirements.

- *Utilization environments*—Identify the utilization environments and all environmental factors (natural or induced) that may affect system performance, impact human comfort or safety, or cause human error for each of the operational scenarios envisioned for system use.

- *Essential design considerations*—Identify design considerations including human systems integration (e.g., manpower, personnel, training, environment, safety, occupational health, survivability, habitability), system security requirements (e.g., information assurance, anti-tamper provisions), and potential environmental impact.

- *Design constraints*—Identify design constraints including physical limitations (e.g., weight, form/fit factors), manpower, personnel, and other resource constraints on operation of the system and defined interfaces with host platforms and interacting systems external to the system boundary, including supply, maintenance, and training infrastructures.

4.3.2.2 Characteristics and Attributes of Good Requirements In defining requirements, care should be exercised to ensure the requirement is appropriately crafted. The following characteristics should be considered for every requirement based on ISO/IEC/IEEE 29148, *Systems and software engineering—Life cycle processes—Requirements engineering* (2011), and the *INCOSE Guide for Writing Requirements* (INCOSE RWG, 2012):

- *Necessary*—Every requirement generates extra effort in the form of processing, maintenance, and verification. Only necessary requirements should therefore be included in specifications. Unnecessary requirements are of two varieties: (i) unnecessary specification of design, which should be left to the discretion of the designer, and (ii) a redundant requirement covered in some other combination of requirements.

- *Implementation independent*—Customer requirements may be imposed at any level they desire; however, when customer requirements specify design, it should be questioned. A proper requirement should deal with the entity being specified as a "black box" by describing what transformation is to be performed by the "box." The requirement should specify "what" is to be done at that level, not "how" it is to be done at that level.

- *标准*——识别为满足品质或设计因素所需的标准，这些因素作为已定义的利益攸关者需求而强加或为满足组织、行业或领域需求而导出。

- *使用环境*——识别可能影响系统性能，影响人员舒适度或安全，或对预想的系统使用的每个运行场景造成人为错误的使用环境及所有环境因素（自然的或诱发的）。

- *基本的设计考虑因素*——识别设计考虑因素，包括人员系统综合（如人力、人员、培训、环境、安全、职业健康、生存性和可居性）、系统安保需求（如信息保证、防篡改条款）以及潜在的环境影响。

- *设计约束*——识别设计约束，包括物理限制（如重量、形状/适配因素）、人力、人员，以及在系统运行、具有主平台和系统边界以外的交互系统的定义接口方面的其他资源约束，包括供应、维护和培训基础设施。

4.3.2.2 良好需求的特征和属性 定义需求时，应非常关注以确保需求被恰当精心地制定。系统和软件工程——生命周期流程——需求工程（2011）和《INCOSE 需求编写指南》（INCOSE RWG，2012），每个基于 ISO/IEC/IEEE 29148 的需求应考虑到下列特征：

- *必要的*——每个需求以处理、维护和验证的形式产生额外的努力。因此，应只将必要的需求包括在规范中。不必要的需求分为两种：① 不必要的设计规范应留给设计者慎重考虑；② 覆盖了在一些其他需求组合中的冗余的需求。

- *独立于实施的*——客户需求可强加在客户期望的任何层级上；然而，当客户需求明确规定设计时，应受到质疑。合适的需求应通过描述由"盒子"所实施的转换来处理作为"黑盒"而被指定的实体。需求应指定所在哪个层级做"什么"，而不是在该层级上"如何"做。

TECHNICAL PROCESSES

- *Unambiguous*—Requirements must convey what is to be done to the next level of development. Its key purpose is to communicate. Is the requirement clear and concise? Is it possible to interpret the requirement in multiple ways? Are the terms defined? Does the requirement conflict with or contradict another requirement? Each requirement statement should be written to address one and only one concept. Requirements with "and," "or," "commas," or other forms of redundancy can be difficult to verify and should be avoided as it can be difficult to ensure all personnel have a common understanding. Requirements must therefore be written with extreme care. The language used must be clear, exact, and in sufficient detail to meet all reasonable interpretations. A glossary should be used to precisely define often-used terms or terms, such as "process," that could have multiple interpretations.
- *Complete*—The stated requirement should be complete and measurable and not need further amplification. The stated requirement should provide sufficient capability or characteristics.
- *Singular*—The requirement statement should be of only one requirement and should not be a combination of requirements or more than one function or constraint.
- *Achievable*—The requirement must be technically achievable within constraints and requires advances in technology within acceptable risk. It is best if the implementing developer participate in requirements definition. The developer should have the expertise to assess the achievability of the requirements. In the case of items to be subcontracted, the expertise of potential subcontractors is very valuable in the generation of the requirements. Additionally, participation by manufacturing and customers/users can help ensure achievable requirements. When it is not possible to have the right mix of developer, subcontractor, and/or manufacturing expertise during the requirements generation, it is important to have their review at the first point possible to ensure achievability.
- *Verifiable*—Each requirement must be verified at some level by one of the four standard methods (inspection, analysis, demonstration, or test). A customer may specify, "The range shall be as long as possible." This is a valid but unverifiable requirement. This type of requirement is a signal that a trade study is needed to establish a verifiable maximum range requirement. Each verification requirement should be verifiable by a single method. A requirement requiring multiple methods to verify should be broken into multiple requirements. There is no problem with one method verifying multiple requirements; however, it indicates a potential for consolidating requirements. When the system hierarchy is properly designed, each level of specification has a corresponding level of verification during the verification stage. If element specifications are required to appropriately specify the system, element verification should be performed.

- *无歧义的*——需求必须传达在下一开发层级需要完成的工作。沟通是它的主要目的。需求清晰、简明吗？可以用多种方式解释需求吗？术语定义了吗？需求之间彼此冲突或矛盾吗？每个需求说明都应编写，为仅应对一个概念。具有"和"、"或"、"逗号"或其他形式冗余的需求难以验证，由于其难以确保被所有人共同理解，应予以避免。因此，编写需求时必须特别注意。使用的语言必须清晰、准确和足够详细以满足所有合理的解释。应使用术语表准确定义常用术语或名词，如"流程"，可能有多种解释。

- *完整的*——申明的需求应是完整、可度量且不需要进一步扩充的，亦应提供充分的能力或特征。

- *单一的*——需求说明应属于唯一一个需求，不应是需求的组合或一个以上的功能或约束。

- *可实现的*——需求在约束内必须是在技术上可实现的，且需要在可接受的风险内取得技术进展。如果负责实施的开发者参加需求定义工作是最好的。开发者应具备评估需求可实现性的专业知识。若项目被转包，在生成需求时，潜在承包商的专业知识是非常有价值的。另外，制造商和客户/用户的参与能够有助于确保需求的可实现性。当开发者、转包商和/或专业制造商的知识在需求生成期间不能正确融合时，为确保可实现性，重要的是在第一可能的时点对他们进行审查。

- *可验证的*——每个需求必须由四个标准方法（检验、分析、演示或试验）中的一种在某一层级上进行验证。一个客户也许指定"范围应尽可能地长"。这是一种有效但无法验证的需求。这种类型的需求预示权衡研究需要建立可验证的最大范围需求。每个验证需求应由单一方法进行验证。需要采用多种方法进行验证的需求应被分解成多个需求。采用一种方法验证多个需求没有问题；然而，这也表明了对需求进行合并的潜在可能。当系统层级结构被恰当设计时，每级规范在验证阶段都具有一个对应的验证层级。如果元素规范被要求适当地指定系统，则应进行元素验证。

TECHNICAL PROCESSES

- *Conforming*—In many instances, there are applicable government, industry, and product standards, specifications, and interfaces with which compliance is required. An example might be additional requirements placed on new software developments for possible reusability. Another might be standard test interface connectors for certain product classes. In addition, the individual requirements should conform to the organization's standard template and style for writing requirements—when all requirements within the same organization have the same look and feel, each requirement is easier to write, understand, and review.

Note: ISO/IEC/IEEE 29148 uses the term "consistent" instead of "conforming," stating that the requirement must be free of conflicts with other requirements. While that is correct, a requirement cannot, in and of itself, be consistent—consistency is more correctly a characteristic of the set of requirements (as described in the following paragraph).

In addition to the characteristics of individual requirements, the characteristics of a set of requirements as a whole should be addressed to ensure that the set of requirements collectively provides for a feasible solution that meets the stakeholder intentions and constraints. These include:

- *Complete*—The set of requirements contains everything pertinent to the definition of system or system element being specified.
- *Consistent*—The set of requirements is consistent in that the requirements are not contradictory nor duplicated. Terms and abbreviations are used consistently in all requirements, in accordance with the glossary.
- *Feasible/affordable*—The set of requirements can be satisfied by a solution that is obtainable within LCC, schedule, and technical constraints.

Note: ISO/IEC/IEEE 29148 uses the term "affordable" instead of "feasible," stating that the set of requirements are feasible within the life cycle constraints of cost, schedule, technical, and regulatory. The INCOSE Guide for Writing Requirements (INCOSE RWG, 2012) includes the word " feasible, " stating that " Feasible/Affordable is a more appropriate title for this characteristic of the requirement set."

- *Bounded*—The set of requirements define the required scope for the solution to meet the stake-holder needs. Consequently, all necessary requirements must be included; irrelevant requirements must be excluded.

In addition to the characteristics listed above, individual requirement statements may have a number of attributes attached to them (either as fields in a database or through relationships with other artifacts):

61

- *符合的*——许多情况下，需要与适用政府、行业和产品的标准、规范和接口相符。一个例子或许是为了可能的复用性对新软件开发施加的附加需求。另一个例子或许是用于某些产品类别的标准的测试接口连接器。此外，单独的需求应符合组织的需求编写的标准模板和风格——当同一组织内的所有需求具有相同的视觉和感觉时，每个需求都更容易编写、理解和审查。

注：ISO/IEC/IEEE 29148 使用术语"一致的"而不是"符合的"来说明该需求必须与其他需求不冲突。虽然这是正确的，但需求内部和本身无法一致——更加正确地说，一致性是需求集合的特征（正如下一段所描述）。

除了单个需求的特征外，为了确保需求集合共同提供满足利益攸关者的目的和约束的可行的解决方案，应当阐述作为整体的需求集合的特征。这些特征包括：

- *完整的*——需求集包含与特定系统或系统元素定义精确相关的所有内容。

- *一致的*——需求集合与不矛盾或不重复的需求一致。按照术语表，术语和缩写在所有需求中使用一致。

- *可行的/可承受的*——需求集合能够通过在 LCC（生命周期成本）、进度和技术约束之内获得的解决方案得以满足。

注：ISO/IEC/IEEE 29148 使用术语"可承受的"而不是"可行的"来说明需求集合在成本、进度、技术和管理的生命周期约束之内是可行的。INCOSE 需求编写指南（INCOSE RWG，2012）包括词语"可行的"，说明"可行的/可承受的是该需求集合特征的更恰当的标题。"

- *有界限的*——需求集合定义满足利益攸关者需要的解决方案的所需范围。因此，所有必要的需求必须包括在内；无关的需求必须被排除在外。

除了上述所列出的特征，单个的需求说明可能具有附属于这些特征的多个属性（作为数据库中的字段或通过与其他制品的关系）：

TECHNICAL PROCESSES

- *Trace to parent*—A child requirement is one that has been derived or decomposed from the parent— the achievement of all the children requirements will lead to the achievement of the parent requirement. Each of the children requirements must be able to be traced to its parent requirement (and thence to any antecedent requirement and ultimately to the system need/mission).

- *Trace to source*—Each requirement must be able to be traced to its source—this is different from tracing to a parent because it identifies where the requirement came from and/or how it was arrived at (rather than which other requirement is its parent).

- *Trace to interface definition*—The interactions between two systems are described in interface definitions that are often contained in a document that has a title such as an ICD. The interface requirements contained in each of the interacting systems will include reference to where the interaction is defined. This attribute provides a link trace between any of the interface requirements to where the interaction is defined.

- *Trace to peer requirements*—This attribute links requirements that are related to each other (other than parent/child) at the same level. Peer requirements may be related for a number of reasons, such as the following: they may be in conflict, or codependent, or bundled, or a complimentary interface requirement of another system to which the system has an interface.

- *Trace to verification method*—This could be a simple statement of the way in which the requirement is to be verified (inspection, demonstration, test, analysis, simulation), or it could be a more elaborate statement that effectively provides the outline of an appropriate test plan.

- *Trace to verification requirement(s)*—The verification method for each requirement simply states the planned method of verification (inspection, demonstration, test, analysis, simulation). In addition to stating the verification method, some organizations write a set of verification requirements in addition to the system requirements. This is beneficial as it forces the systems engineer to consider how each requirement is to be verified and in doing so helps identify and remove requirements that are not verifiable.

- *Trace to verification results*—The results of each verification will most often be contained in a separate document. This attribute traces each requirement to the associated verification results.

- *追溯到母需求*——子需求是从母需求中导出或分解而来的需求——所有子需求的实现将导致母需求的实现。每个子需求必须能够追溯到母需求（并由此追溯到任何先前的需求，最终追溯到系统需要/任务）。

- *追溯到源头*——每个需求必须能够追溯到其源头——这不同于追溯到母需求，因为它识别需求从哪里来和/或如何到此（而不是哪些是其母需求的其他需求）。

- *追溯到接口定义*——两个系统之间的交互在接口定义中描述，通常包含在具有以 ICD 命名的文件中。包含在每个接口系统中的接口需求将包括定义交互所用的参考。该属性提供定义交互所在的任意接口需求之间的连接追溯。

- *追溯到同级需求*——这个属性与同一层级上彼此相关的需求（母需求/子需求除外）相连接。同级需求可能因多种原因而相关联，如以下原因：它们可能冲突或相互依赖或捆绑的，或者是与该系统具有接口的另一个系统的一个互补性接口需求。

- *追溯到验证方法*——这可能是对需求要被验证（检验、演示、测试、分析、仿真）的方式的简单说明，或者它可能是有效地提供恰当测试计划概况的更详尽的说明。

- *追溯到验证需求*——每个需求的验证方法仅简单申明计划的验证（检验、演示、测试、分析、仿真）方法。除了申明验证方法外，一些组织还编写系统需求以外的一个验证需求集合。这是有益的，因为它迫使系统工程师考虑每个需求如何被验证，并且这么做有助于识别和去除那些不可验证的需求。

- *追溯到验证结果*——每个验证结果通常会包含在单独分开的文件中。该属性将每个需求追溯到相关的验证结果。

- *Requirements verification status*—It is useful to include an attribute that indicates whether the requirement has been verified or not.

- *Requirements validation status*—Requirements validation as the process of ensuring requirements and sets of requirements meet the rules and characteristics in the guide. Some organizations include an attribute field "*requirement validated*" to indicate whether the individual requirement has been validated.

- *Priority*—This is how important the requirement is to the stakeholder. It may not be a critical requirement (i.e., one the system must possess or it won't work at all), but simply something that the stakeholder(s) holds very dear. Priority may be characterized in terms of a level (1, 2, 3 or high, medium, low). Priority may be inherited from a parent requirement.

- *Criticality*—A critical requirement is one that the system must achieve or the system cannot function at all—perhaps can be viewed as one of the set of minimum essential requirements. Criticality may be characterized in terms of a level (1, 2, 3 or major, medium, minor). Criticality may be inherited from a parent requirement.

- *Risk*—This is the risk that the requirement cannot be achieved within technology, schedule, and budget. For example, the requirement may be possible technically (i.e., the requirement may be feasible) but have risk of achievement within available budget and schedule. Risk may be characterized in terms of a level (e.g., high, medium, low). Risk may be inherited from a parent requirement.

- *Key driving requirement (KDR)*—A KDR is a requirement that, to implement, can have a large impact on cost or schedule. A KDR can be of any priority or criticality—knowing the impact a KDR has on the design allows better management of requirements. When under schedule or budget pressure, a KDR that is low priority or low criticality may be a candidate for deletion.

- *Owner*—This is the person or element of the organization that has the right to say something about this requirement. The owner could be the source but they are two different attributes.

- *Rationale*—Rationale defines why the requirement is needed and other information relevant to better understand the reason for and intent of the requirement. Rationale can also define any assumptions that were made when writing the requirement, what design effort drove the requirement, the source of any numbers in the requirement, and note how and why a requirement is constrained (if indeed it is so).

- *需求验证状态*——包括指明需求是否已被验证过的属性在内是很有用的。

- *需求确认状态*——需求确认作为确保需求和需求集合的流程满足指南中的规则和特征。一些组织包括属性字段"需求已确认",用以表明单独的需求已经被确认。

- *优先级*——这是指需求对于利益攸关者而言有多重要。它可能是一个关键的需求(即,系统必须拥有的需求否则系统根本不工作),但也有仅仅是利益攸关者非常看重的事物。优先级可能按照层级(1、2、3 或高、中、低)来描述。优先级可能从母需求中继承。

- *关键性*——关键需求是系统必须实现的否则系统根本不能起作用的需求——或许被视为最低基本需求集合中的一个需求。关键性可能按照层级(1、2、3 或主要、中等、次要)来描述。关键性可能从母需求中继承。

- *风险*——风险是指在技术、进度和预算之内不能达成需求。例如,需求在技术上是可能的(即,需求或许是可行的),但在可用的预算和进度内具有达成风险。风险可能按照层级(如高、中、低)来描述。风险可能从母需求中得到。

- *关键驱动需求*(KDR)——KDR 是对成本或进度具有较大影响的待实施需求。KDR 可以具有任何优先级或关键性——知晓 KDR 对设计的影响使得对需求更好的管理。当处于进度或预算压力下时,处于低优先级或低关键性的 KDR 可能是需要删除的候选需求。

- *所有者*——它是对该需求有话语权的组织中的人员或元素。所有者可能是源头,但它们是两个不同的属性。

- *理由依据*——理由依据定义为什么需要该需求以及与更好地理解需求的原因和意图相关的其他信息。理由依据还可定义编写需求时做出的任何假设、哪些设计努力驱动了需求、需求中任何数字的源头、以及注意如何及为什么需求受到约束(若确实如此)。

- *Applicability*—This field may be used by an organization that has a family of similar product lines to identify the applicability of the requirement (e.g., to product line, region, or country).

- *Type*—It is often useful to attach to each requirement an attribute of type. While each organization will define types based on how they may wish to organize their requirements, examples of type include input, output, external interfaces, reliability, availability, maintainability, accessibility, environmental conditions, ergonomic, safety, security, facility, transportability, training, documentation, testing, quality provision, policy and regulatory, compatibility with existing systems, standards and technical policies, conversion, growth capacity, and installation. The type field is most useful because it allows the requirements database to be viewed by a large number of designers and stakeholders for a wide range of uses.

More detail on writing text-based requirements can be found in the *INCOSE Guide for Writing Requirements* (INCOSE RWG, 2012), which focuses on the writing of requirements and addresses the characteristics of individual requirement statements, the characteristics of sets of requirements, the attributes of individual requirement statements, and the rules for individual requirement statements.

4.3.2.3 Define, Derive, and Refine Functional/ Performance Requirements At the beginning of the project, SE is concerned primarily with user requirements analysis—leading to the translation of user needs into basic functions and a quantifiable set of performance requirements that can be translated into design requirements.

Defining, deriving, and refining functional and performance requirements apply to the total system over its life cycle, including its support requirements. These requirements need to be formally captured in a manner that defines the functions and interfaces and characterizes system performance such that they can be flowed down to hardware and software designers. This is a key SE activity and is the primary focus through System Requirements Review (SRR). During the requirements analysis, the support from most other disciplines (e.g., software, hardware, manufacturing, quality, verification, specialty) is necessary to ensure a complete, feasible, and accurate set of requirements that consider all necessary life cycle factors of the system definition. The customer is also a key stakeholder and validates the work as it progresses.

Establishing a total set of system requirements is a complex, time-consuming task involving nearly all project areas in an interactive effort. It must be done early, since it forms the basis for all design, manufacturing, verification, operations, maintenance, and retirement efforts and therefore determines the cost and schedule of the project. The activity is iterative for each stage, with continuous feedback as the level of design detail increases, and flows from the life cycle concepts, particularly the OpsCon.

- *适用性*——该领域由具有类似产品线系列的组织使用，用以识别需求的适用性（如对产品线、地区或国家的适用性）。

- *类型*——为每个需求赋予一个类型的属性通常是有用的。当每个组织将基于他们希望如何组织他们的需求来定义类型时，类型示例包括输入、输出、外部接口、可靠性、可用性、可维护性、可达性、环境条件、人体工程学、安全、安保、设施、可运输性、培训、文档、测试、质量条款、方针和法规、与现有系统的兼容性、标准和技术方针、转换、增长能力及安装。类型领域是最有用的，因为为了广泛的用途，它允许很多的设计者和利益攸关者查看需求数据库。

关于编写基于文本的需求的更多细节可在《INCOSE 需求编写指南》（INCOSE RWG，2012）中找到，该指南集中于需求的编写并提出单独需求说明的特征、需求集的特征、单独需求说明的属性，以及单独需求说明的规则。

4.3.2.3 定义、推导和细化功能/性能需求 在项目开始时，SE 主要关注用户需求分析——引导将用户需要转化为基本功能和可转化为设计需求的一系列量化的性能需求。

定义、推导和细化功能/性能需求在系统的全生命周期内应用于整体系统，也包括其支持需求。这些需求需要以定义功能和接口并描述系统性能的方式被正式捕获，从而使这些需求可向下流动到硬件和软件设计者。这是关键的 SE 活动，是系统需求评审（SRR）中的主要焦点。在需求分析期间，为确保完整、可行、准确的需求集合考虑到系统定义的所有必要的生命周期因素，来自众多其他学科（如软件、硬件、制造、质量、验证、专业）的支持是非常必要的。客户亦是关键的利益攸关者并随着工作的进展进行确认。

建立一个全面的系统需求集合是一项复杂的、耗时的任务，交互工作涉及几乎所有的项目域。由于这项工作构成所有设计、制造、验证、运行、维护和退出工作的基础，进而确定项目的成本和进度，所以必须尽早完成。该活动对于每个阶段都是迭代性的，随着设计细节层级的增加而持续反馈且跟随生命周期概念，特别是 OpsCon。

TECHNICAL PROCESSES

The result of the system requirements definition process should be a baseline set of complete, accurate, nonambiguous system requirements, recorded in the requirements database, accessible to all parties, and captured in an approved, released SyRS.

4.3.2.4 Define Other Nonfunctional Requirements

The life cycle concepts will also suggest requirements that are those dealing with operational conditions (e.g., safety; system security; reliability, availability, and maintainability; human systems integration, environ-mental engineering—see Chapter 10), as well as life cycle constraints (e.g., maintenance, disposal) that will strongly influence the definition of the solution elements.

4.3.2.5 Generate the SyRS The SyRS—which is often called the System Specification—is a baseline set of complete, accurate, nonambiguous system requirements, recorded in the requirements database and accessible to all parties. To be nonambiguous, requirements must be broken down into constituent parts in a traceable hierarchy such that each individual requirement statement is:

- Clear, unique, consistent, stand-alone (not grouped), and verifiable
- Traceable to an identified source requirement
- Not redundant, nor in conflict with, any other known requirement
- Not biased by any particular implementation

These objectives may not be achievable using source requirements. Often, requirements analysis is required to resolve potential conflicts and redundancies and to further decompose requirements so that each applies only to a single system function. Use of an automated requirements database will greatly facilitate this effort, but is not explicitly required.

During the system requirements definition process, it is often necessary to generate a "snapshot" report of clarified system requirements. To aid this process, it may be desirable to create a set of clarified requirement objects in the requirements database with information providing traceability from their corresponding originating requirement. Clarified requirements may be grouped as functional, performance, constraining, and nonfunctional.

4.4 ARCHITECTURE DEFINITION PROCESS

4.4.1 Overview

4.4.1.1 Purpose As stated in ISO/IEC/IEEE 15288,

系统需求定义流程的结果应是完整、准确、非歧义的系统需求集合的一个基线，被记录在需求数据库中，各方都能使用，并在经批准的、发布的 SyRS（系统规范）中捕获。

4.3.2.4　定义其他非功能性需求

生命周期概念还将提出处理运行条件（如，安全；系统安保；可靠性、可用性和可维护性；人员系统综合、环境工程——参见第 10 章）的需求、以及对解决方案元素的定义具有很大影响的生命周期约束（如，维护、处置）。

4.3.2.5　生成 SyRS　SyRS——通常被称为系统规范——是完整、准确、非歧义的系统需求集合的一个基线，被记录在需求数据库中，且各方都能使用。为了无歧义，必须将需求分解成可追溯层级结构中的组成部分，从而使每一单个的需求说明是：

- 清晰的、唯一的、一致的、独立的（不是成组的）和可验证的
- 可追溯到一个已识别的源需求
- 无冗余，与任何其他已知需求不冲突
- 不因任何特定的实现而带偏见

使用源需求可能无法达成这些目标。通常，需要用需求分析来解决潜在冲突和冗余并进一步分解需求，从而使每个需求只应用于一个系统功能。使用自动的需求数据库将极大地助促该工作，但这并非显性的要求。

在系统需求定义流程期间，通常有必要形成澄清后的系统需求的"快照"报告。为辅助该流程，在需求数据库内创建一个经过澄清后的需求对象集合也许是理想的，其中该需求数据库中的信息提供从对应原始需求开始的可追溯性。已澄清的需求可按功能、性能、约束和非功能方式分组。

4.4　架构定义流程

4.4.1　概览

4.4.1.1　目的　如 ISO/IEC/IEEE 15288 所述，

[6.4.4.1] The purpose of the Architecture Definition process is to generate system architecture alternatives, to select one or more alternative(s) that frame stakeholder concerns and meet system requirements, and to express this in a set of consistent views.

4.4.1.2 Description System architecture and design activities enable the creation of a global solution based on principles, concepts, and properties logically related and consistent with each other. The solution architecture and design have features, properties, and characteristics satisfying, as far as possible, the problem or opportunity expressed by a set of system requirements (traceable to mission/business and stakeholder requirements) and life cycle concepts (e.g., operational, support) and are implementable through technologies (e.g., mechanics, electronics, hydraulics, software, services, procedures). In this handbook, architecture and design activities are described as two separate processes to show that they are based on different and complementary notions. System architecture is more abstract, conceptualization oriented, global, focused to achieve the mission and OpsCon of the system, and focused on high-level structure in systems and system elements. It addresses the architectural principles, concepts, properties, and characteristics of the SOI. System design is more technology oriented through physical, structural, environmental, and operational properties forcing decisions for implementation by focusing on compatibility with technologies and other design elements and feasibility of construction and integration.

The architecture definition process aggregates and deals with incremental insights obtained about the specified requirements and emergent properties and behaviors of the system and system elements while managing suitability, viability, and affordability. The design definition process uses the artifacts of the architecture definition process (e.g., architecture description, feasibility analyses, system balance trades, vetted requirements). For information about the uses of architecture definition, refer to ISO/IEC/IEEE 42010 (2011).

The architecture definition process is used to create and establish alternative architectures through several views and models, to assess the properties of these alternatives (supported by the system analysis process), and to select appropriate technological or technical system elements that compose the system:

> An effective architecture is as design-agnostic as possible to allow for maximum flexibility in the design trade space. An effective architecture also highlights and supports trade-offs for the Design Definition process and possibly other processes such as Portfolio Management, Project Planning, System Requirements Definition, Verification, etc. (ISO/IEC/IEEE 15288:2015)

[6.4.4.1] 架构定义流程的目的是生成系统架构备选方案，选择出框定利益攸关者关注点且满足系统需求的一个或多个备选方案，并以一系列一致的视图对备选方案进行表达。

4.4.1.2 描述 系统架构和设计活动使得能够基于彼此相互逻辑地相关且协调一致的原则、概念和特性来创建一个全局解决方案。解决方案架构和设计具有尽可能满足系统需求集合（可追溯到任务/业务和利益攸关者需求）及生命周期概念（如运行、支持）所表达的问题或机会的特征、特性和特点，并且可通过技术（如机械、电子、液压、软件、服务、程序）来实施。本手册中，架构和设计活动作为两个独立的流程来描述，以表明它们基于不同且互补的观念。系统架构是更加抽象的、面向概念化的、全局的、聚焦于达成系统的任务和 OpsCon，并聚焦于系统和系统元素中的高层级结构。它阐述 SOI 的架构原则、概念、特性和特征。系统设计由聚焦于与技术及其他设计元素的兼容性以及构建和综合的可行性，通过物理、结构、环境和运行上的特性更加面向技术，促使决策得以实施。

架构定义流程汇集并应对所获得的关于明确规定的需求及系统和系统元素的涌现特性和行为的渐进的深入透视，同时管理适用性、生存性和可承受性。设计定义流程使用架构定义流程的制品（如架构描述、可行性分析、系统平衡的权衡、已审查的需求）。关于架构定义使用的信息，参见 ISO/IEC/IEEE 42010（2011）。

架构定义流程用于通过多种视图和模型来创建和建立备选的架构，评估这些备选方案的特性（由系统分析流程支持），并选择构成系统的恰当的技术元素或技术系统元素。

一个有效的架构尽可能独立于设计，以便使设计权衡空间具有最大的灵活性。有效的架构还强调并支持对设计定义流程和可能的其他流程（如项目群管理、项目规划、系统需求定义、验证等）的权衡。（ISO/IEC/IEEE 15288:2015）

TECHNICAL PROCESSES

This process is iterative and requires the participation of systems engineers or architects supported by relevant designers and specialists in the system domain. Also, iteration between this process and others is expected as more information is available and analysis is performed. This process also continues to be recursively applied to define the requirements for each system element. Iteration and recursion are further described in Section 4.3.1.2. Figure 4.6 is the IPO diagram for the architecture definition process.

4.4.1.3 Inputs/Outputs Inputs and outputs for the architecture definition process are listed in Figure 4.6. Descriptions of each input and output are provided in Appendix E.

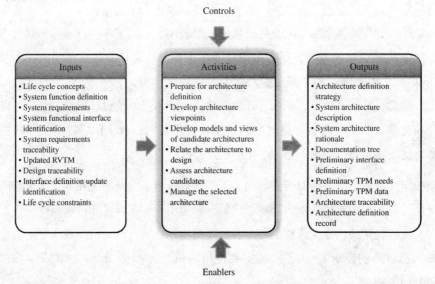

FIGURE 4.6 IPO diagram for the architecture definition process. INCOSE SEH original figure created by Shortell and Walden. Usage per the INCOSE notices page. All other rights reserved.

4.4.1.4 Process Activities The architecture definition process includes the following activities:

- *Prepare for architecture definition.*
    - Identify and analyze relevant market, industry, stakeholder, organizational, business, operations, mission, legal, and other information that will help to understand the perspectives that will guide the development of the architecture views and models. This information is intended to help build an understanding of the environment for which the solution is needed in order to establish better insight into the stakeholder concerns.

这个流程是迭代的，并要求系统工程师或架构师在系统领域内的相关设计者和专家的支持下参与。而且，此流程与其他流程之间的迭代是期望的，因为有更多信息可用并进行分析。此流程还继续递归地应用于定义每个系统元素的需求。4.3.1.2 节进一步描述了迭代和递归。图 4.6 是架构定义流程的 IPO 图。

4.4.1.3　输入/输出　图 4.6 列出了系统定义流程的输入和输出。附录 E 提供每个输入和输出的描述。

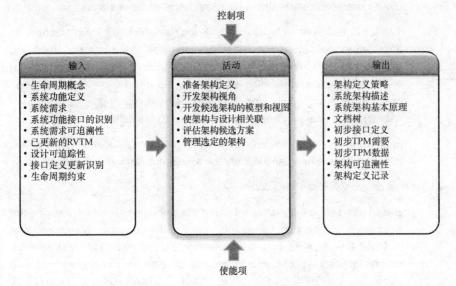

图 4.6　架构定义流程的 IPO 图。INCOSE SEH 原始图由 Shortell 和 Walden 创建。按照 INCOSE 通知页使用。版权所有。

4.4.1.4　流程活动　架构定义流程包括以下活动：

- *准备架构定义。*

    – 识别和分析相关市场、行业、利益攸关者、组织、业务、运行、任务、法律及其他信息，将有助于对指导架构视图和模型的开发视角的理解。此信息旨在帮助构建对所需要的解决方案的环境的理解，以便建立对利益攸关者的关切有更好的深入透视。

- In particular, analyze the system requirements and tag nonfunctional requirements, that is, those dealing with operational conditions (e.g., safety, security, dependability, human factors, simplicity of interfaces, environmental conditions), as well as life cycle constraints (e.g., maintenance, disposal, deployment) that will strongly influence the definition of the solution elements.

- Capture stakeholder concerns related to architecture. Usually, the stakeholder concerns focus on expectations or constraints that span one or more system life cycle stages. The concerns are often related to critical quality characteristics to the system that relate to those stages.

- Establish the approach for defining the architecture. This includes an architecture roadmap and strategy, as well as methods, modeling techniques, tools, and the need for any enabling systems, products, or services. The approach should also include the process requirements (e.g., measurement approach and methods), evaluation (e.g., reviews and criteria), and necessary coordination. Capture the evaluation criteria.

- Ensure the enabling elements or services will be available. As part of this task, plan for the need and identify the requirements for the enabling items.

- *Develop architecture viewpoints.*

    - Based on the identified stakeholder concerns, establish or identify the associated architecture viewpoints, the supporting kinds of models that facilitate the analysis and understanding of the viewpoint, and relevant architecture frameworks to support the development of the models and views.

- *Develop models and views of candidate architectures.*

    - Select or develop supporting modeling techniques and tools.

    - In conjunction with the system requirements definition process, determine the system context (i.e., how the SOI fits into the external environment) and boundary, including the interfaces, that reflect the operational scenarios and expected system behaviors. This task includes identification of expected interactions of the system with systems or other entities external to the system (control) boundary as defined in negotiated ICDs.

    - Determine which architectural entities (e.g., functions, input/output flows, system elements, physical interfaces, architectural characteristics, information/data elements, containers, nodes, links, communication resources, etc.) address the highest priority requirements (i.e., most important stakeholder concerns, critical quality characteristics, and other critical needs).

- 特别地，分析系统需求并标记非功能需求，即应对严重影响解决方案元素定义的运行条件（如安全、安保、可依赖性、人员因素、接口简化性、环境条件）以及生命周期约束（如维护、处置、部署）的那些需求。

- 捕获与架构相关的利益攸关者的关切。通常，利益攸关者的关切集中于跨一个或多个系统生命周期阶段的期望或约束。这些关切往往涉及与那些阶段相关的系统的关键品质特征。

- 建立定义架构的实施途径。这包括架构战略和路线图，以及方法、建模技术、工具和对任何使能系统、产品或服务的需要。该实施途径还应包括流程需求（如测量实施途径和方法）、评价（如评审和准则）以及必要的协调。捕获评价准则。

- 确保使能元素或服务将是可用的。作为此任务的一部分，对需要进行规划并为使能项识别需求。

- 开发架构视角。

  - 基于所识别的利益攸关者关切，建立或识别相关的架构视角、多种类型的助促视角分析和理解的支持性模型、以及支持模型和视图开发的相关架构框架。

- 开发候选架构的模型和视图。

  - 选择或开发支持建模的技术和工具。

  - 结合系统需求定义流程，确定系统背景环境（即 SOI 如何适配于外部环境）和边界，包括反映运行场景和期望的系统行为的接口。此任务包括识别系统与系统（控制）边界（按照商定的 ICD 的定义）外部的诸系统或其他诸实体的预期交互。

  - 确定哪些架构实体（如功能、输入/输出流、系统元素、物理接口、架构特征、信息/数据元素、容器、节点、链接、通信资源等）阐述最高优先权的需求（即最重要的利益攸关者关切点、关键品质特征及其他关键需要）。

TECHNICAL PROCESSES

- Allocate concepts, properties, characteristics, behaviors, functions, and/or constraints that are significant to architecture decisions of the system to architectural entities.

- Select, adapt, or develop models of the candidate architectures of the system, such as logical and physical models. It is sometimes neither necessary nor sufficient to use logical and physical models. The models to be used are those that best address key stakeholder concerns. Logical models may include functional, behavioral, or temporal models; physical models may include structural blocks, mass, layout, and other physical models (see Section 9.1 for more information about models).

- Determine need for derived system requirements induced by necessary added architectural entities (e.g., functions, interfaces) and by structural dispositions (e.g., constraints, operational conditions). Use the system requirements definition process to define and formalize them.

- Compose views from the models of the candidate architectures. The views are intended to ensure that the stakeholder concerns and critical requirements have been addressed.

- For each system element that composes the system, develop requirements corresponding to allocation, alignment, and partitioning of architectural entities and system requirements to system elements. To do this, invoke the stake-holder needs and requirements definition process and the system requirements definition process.

- Analyze the architecture models and views for consistency and resolve any issues identified. Correspondence rules from frameworks can be useful in this analysis (ISO/IEC/IEEE 42010, 2011).

- Verify and validate the models by execution or simulation, if modeling techniques and tools permit, and with traceability matrix of OpsCon. Where possible, use design tools to check their feasibility and validity. As needed, implement partial mock-ups or prototypes, or use executable architecture prototypes or simulators.

- *Relate the architecture to design.*

  - Determine the system elements that reflect the architectural entities. Since the architecture is intended to be design-agnostic, these system elements may be notional until the design evolves. To do this, partition, align, and allocate architectural entities and system requirements to system elements. Establish guiding principles for the system design and evolution. Sometimes, a " reference architecture " is created using these notional system elements as a means to convey architectural intent and to check for design feasibility.

- 分配概念、特性、特征、行为、功能和/或约束，这对于系统到架构实体的架构决策极为重要。

- 选择、适应或开发系统的候选架构的模型，如逻辑和物理模型。有时使用逻辑和物理模型既不必要也不充分。要使用的模型是最好地阐述关键利益攸关者关切的模型。逻辑模型可包括功能、行为或时间模型；物理模型可包括结构块、质量、布局及其他物理模型（关于模型的更多信息，见9.1节）。

- 确定由必须增加的架构实体（如功能、接口）和结构安排（如约束、运行条件）所诱发而导出的系统需求的需要。使用系统需求定义流程对它们进行定义和正式化。

- 从候选架构的诸模型构成视图。视图旨在确保利益攸关者关切点和关键需求已被阐述。

- 对于组成系统的每个系统元素，开发与架构实体的分配、对准和区划相对应的需求以及对系统元素的系统需求。为此，援引利益攸关者需要和需求定义流程以及系统需求定义流程。

- 分析架构模型和视图的一致性并解决所识别到的任何问题。在该分析中，来自于框架中的相应规则可能是有用的（ISO/IEC/IEEE 42010, 2011）。

- 如果建模技术和工具允许，通过执行或仿真并使用OpsCon的可追踪性矩阵来验证和确认模型。若可能，使用设计工具检查它们的可行性和有效性。若需要，实现部分初样或原型，或使用可执行的架构原型或仿真器。

- 使架构与设计相关联。

  - 确定反映架构实体的系统元素。由于架构旨在独立于设计，因此在设计演进之前这些系统元素也许是观念化的。为此，将系统需求和架构实体向系统元素区划、对准和分配。建立系统设计和演进的指导原则。有时，使用这些观念化的系统元素作为传达架构意图和检查设计可行性的手段来创建一个"参考架构"。

- Establish allocation matrices between architectural entities using their relationships.
- Perform interface definition for interfaces that are necessary for the level of detail and understanding of the architecture. The definition includes the internal interfaces between the system elements and the external interfaces with other systems.
- Determine the design characteristics that relate to the system elements and their architectural entities, such as by mapping (see Section 4.5).
- Determine need for derived system requirements induced by necessary added architectural entities (e.g., functions, interfaces) and by structural dispositions (e.g., constraints, operational conditions). Use the system requirements definition process to formalize them.
- For each system element that composes the parent system, develop requirements corresponding to allocation, alignment, and partitioning of architectural entities and system requirements to system elements. To do this, invoke the stake-holder needs and requirements definition process and the system requirements definition process.

- *Assess architecture candidates.*
  - Using the architecture evaluation criteria, assess the candidate architectures by applying the system analysis, measurement, and risk management processes.
  - Select the preferred architecture(s). This is done by applying the decision management process.

- *Manage the selected architecture.*
  - Capture and maintain the rationale for all selections among alternatives and decision for the architecture, architecture framework(s), viewpoints, kinds of models, and models of the architecture.
  - Manage the maintenance and evolution of the architecture, including the architectural entities, their characteristics (e.g., technical, legal, economical, organizational, and operational entities), models, and views. This includes concordance, completeness, and changes due to environment or context changes, technological, implementation, and operational experiences. Allocation and traceability matrices are used to analyze impacts onto the architecture. The present process is performed at any time evolutions of the system occur.
  - Establish a means for the governance of the architecture. Governance includes the roles, responsibilities, authorities, and other control functions.

— 使用架构实体的关系在架构实体之间建立分配矩阵。

— 为细节层级和架构理解所必需的接口进行接口定义。该定义包括系统元素之间的内部接口以及与其他系统的外部接口。

— 确定与系统元素及其架构实体相关的设计特征,例如通过映射(见 4.5 节)。

— 确定由必须增加的架构实体(如功能、接口)和结构安排(如约束、运行条件)所诱发的导出的系统需求的需要。使用系统需求定义流程将它们正式化。

— 对于构成母系统的每个系统元素,开发与架构实体以及系统需求到系统元素相对应的分配、对准和区划。为此,援引利益攸关者需要和需求定义流程以及系统需求定义流程。

- 评估架构候选方案。

— 使用架构评价准则,通过应用系统分析、测量和风险管理流程来评估候选架构。

— 选择优选架构。通过应用决策管理流程来完成。

- 管理选定的架构。

— 捕获并维持备选方案之间所有选择的理由依据以及对架构、架构框架、视角、模型种类和架构模型的决策。

— 管理架构的维护和演进,包括架构实体及其特征(如技术的、法律的、经济的、组织的和运行的)、模型和视图。这包括协调性、完整性以及由于环境或背景环境的变化、技术的、实施的和运行的经验。分配和可追踪性矩阵被用于分析对架构的影响。本流程在出现系统演进的任何时间可进行。

— 建立架构治理的手段。治理包括角色、职责、权限及其他控制功能。

TECHNICAL PROCESSES

– Coordinate review of the architecture to achieve stakeholder agreement. The stakeholder requirements and system requirements can serve as references.

Common approaches and tips:

- A function (e.g., to move) and its state of execution/ operational mode (e.g., moving) are similar but two complementary views. Consider a behavioral model of the system that transits from an operational mode to another one.

4.4.2 Elaboration

4.4.2.1 Architecture Representation The notion of system is abstract, but it is a practical means to create, design, or redesign products, services, or enterprises.

A system is one solution that could address/answer a problem or an opportunity; there may be several solutions to address the same problem or opportunity. The solution may be more or less complex, and the notion of system is useful to engineer complex solutions. A complex solution cannot be apprehended with a single view or model because of the number of characteristics or properties. These last are grouped reflecting typologies of data, and each type of data/characteristics is structured. The set of different types and interrelated "structures" can be understood as THE architecture of the system. The majority of interpretations of system architecture are based on the fairly intangible notion of structure.

Therefore, the system architecture is formally represented with sets of architectural entities such as functions, function flows, interfaces, resource flow items, information/data elements, physical elements, containers, nodes, links, communication resources, etc. These architectural entities may possess architectural characteristics such as dimensions, environmental resilience, availability, robustness, learnability, execution efficiency, mission effectiveness, etc. The entities are not independent but interrelated by the means of relationships.

4.4.2.2 Architecture Description of the System ISO/ IEC/IEEE 42010 specifies the normative features of architecture frameworks, viewpoints, and views as they pertain to architecture description. Viewpoints and views are sometimes specified in architecture frameworks such as Zachman (1987), DoDAF (2010), MoDAF (n.d.), The Open Group Architecture Framework (TOGAF), etc. Views are usually generated from models. Many SE practices use logical and physical models (or views) for modeling the system architecture. The architecture definition process includes also the possible usage of other view-points and views to represent how the system architecture addresses stakeholder concerns, for example, cost models, process models, rule models, ontological models, belief models, project models, capability models, data models, etc.Maier and Rechtin (2009) provide another view of the system architecture development process. Refer to Chapter 9 for a more detailed treatment of models.

– 协调架构评审以达成利益攸关者协议。利益攸关者需求和系统需求可作为参考。

常用方法和建议：

- 功能（如移动）及其执行/运行模式（如移动中）的状态相似但属于两个互补视图。考虑到从一种运行方式转移到另一种运行方式的系统的行为模型。

4.4.2 详细阐述

4.4.2.1 架构表达　系统的观念是抽象的，但它是创建、设计或再设计产品、服务或复杂组织体的实用手段。

系统是能够阐述/回答某一问题或机会的一种解决方案；阐述同一问题或机会可能有多种解决方案。解决方案可能复杂度有多有少，系统的观念对于工程实现复杂解决方案是有用的。由于特征或特性的数量，复杂解决方案无法使用单一视图或模型来理解。最后是对这些特征或特性按反映出数据的类型进行分组，并且对每种数据/特征的类型进行结构化。不同的类型和相互关联的"结构"集合可以被理解为系统的"特定"架构。系统架构的大多数解释都是基于相当隐性的结构观念。

因此，使用架构实体集合，如功能、功能流、接口、资源流动项、信息/数据元素、物理元素、容器、节点、连接、通信资源等，对系统架构进行正式地表达。这些架构实体可具有架构特征，如维度、环境恢复力、可用性、鲁棒性、可学习性、执行效率、任务有效性等。实体不是独立的，而是通过关系的方式相互关联的。

4.4.2.2 系统的架构描述　ISO/IEC/IEEE 42010 规定架构框架、视角和视图的规范性特征，因为它们都属于架构描述。视角和视图有时也在架构框架中明确规定，如 Zachman（1987）、DoDAF（2010）、MoDAF（n.d.）、The Open Group 架构框架（TOGAF）。视图通常从模型产生。许多 SE 实践使用逻辑和物理模型（或视图）对系统架构进行建模。架构定义流程还包括其他视角和视图的可能的使用，以表达系统架构如何阐述利益攸关者关切点，例如成本模型、流程模型、规则模型、本体论模型、信念模型、项目模型、能力模型、数据模型等。Maier 和 Rechtin（2009）提供系统架构开发流程的另一个视图。关于更详细的模型处理，参见第 9 章。

TECHNICAL PROCESSES

A viewpoint is intended to address a particular stakeholder concern (or set of closely related concerns).The viewpoint will specify the model kinds to be used in developing an architectural view that depicts how the architecture addresses that concern (or set of concerns). The viewpoint also specifies the ways in which the model(s) should be generated and how the models are used to compose the view.

An architecture framework contains standardized viewpoints, view templates, metamodels, model templates, etc. that facilitate the development of the views contained in an architecture description. ISO/IEC/IEEE42010 specifies the necessary features of an architecture framework.

4.4.2.3 Emergent Properties Emergence is the principle that whole entities exhibit properties, which are meaningful only when attributed to the whole, not to its parts. Every model of a human activity system exhibits properties as a whole entity that derive from its component activities and their structure, but cannot be reduced to them (Checkland, 1998).

System elements interact between themselves and can create desirable or undesirable phenomena called "emergent properties," such as inhibition, interference, resonance, or reinforcement of any property. Definition of the architecture of the system includes an analysis of interactions between system elements in order to prevent undesirable properties and reinforce desirable ones.

The notion of emergent property is used during architecture and design to highlight necessary derived functions and internal physical or environmental constraints. Corresponding derived requirements should be added to system requirements baseline when they impact the SOI.

4.4.2.4 Architecture in Product Lines The architecture plays a very important role in a product line. The architecture in a product line spans across several design variants, providing a cohesive basis for the product line designs by ensuring compatibility and interoperability across the product line.

4.4.2.5 Notion of Interface The notion of interface is one of the most important items to consider when defining the architecture of a system. The term "interface" comes from Latin words "inter" and "facere" and means "to do something between things." Therefore, the fundamental aspect of an interface is functional and is defined as inputs and outputs of functions.As functions are performed by physical elements, inputs/outputs of functions are also carried by physical elements, called physical interfaces. Consequentially, both functional and physical aspects are considered in the notion of interface. Examples of system elements and physical interfaces are shown in Table 4.1.

一个视角旨在阐述特定的利益攸关者的关切点（或密切相关的关切点集合）。视角将明确规定在开发架构视图中使用的模型种类，该架构视图图示架构如何阐述关切点（或关切点集合）。视角还明确规定模型应当被生成的方式以及模型如何用于构成视图的方式。

一个架构框架包含标准化的视角、视图模板、元模型、模型模板等，它们助促在架构描述中所包含的视图的开发。ISO/IEC/IEEE 42010规定架构框架的必要特征。

4.4.2.3 涌现性 涌现是全部实体呈现特性的原理，这些特性仅归属于整体时才有意义，而并非其部分。人类活动系统的每一个模型都是从其组件活动及其结构导出的全部实体来呈现特性，但并不能将其还原到这些组件活动和结构（Checkland, 1998）。

系统元素自身之间进行交互，并且可创生期望的或不期望的现象被称作"涌现性"，如任何特性的抑制、干扰、谐振或增强等。系统架构的定义包括对系统元素之间交互的分析，以便抑制不期望的特性并增强期望的特性。

在架构和设计期间使用涌现性的观念，以强调必需导出的功能及内部物理的或环境约束。当相应导出的需求影响 SOI 时，应将它们添加到系统需求基线中。

4.4.2.4 产品线中的架构 架构在产品线中发挥非常重要的作用。产品线中的架构跨诸多设计样种，通过确保跨产品线的兼容性和互操作性，为产品线设计提供紧密结合的基础。

4.4.2.5 接口的观念 当定义系统架构时，接口的观念是考虑的最重要项之一。术语"接口"来自于拉丁语"inter"（在……之间）和"facere"（做），是指"在事物之间做事情"。因此，接口的基本方面是功能性的，且被定义为功能的输入和输出。由于功能是通过物理元素实施的，因此功能的输入/输出也通过叫作物理接口的物理元素来实施。其结果是功能方面和物理方面都被考虑到接口的观念之中。表 4.1 指明系统元素和物理接口的示例。

TECHNICAL PROCESSES

TABLE 4.1 Examples of system elements and physical interfaces

Element	Product system	Service system	Enterprise system
System element	Hardware parts (mechanics, electronics, electrical, plastic, chemical, etc.) Operator roles Software pieces	Processes, databases, procedures, etc. Operator roles Software applications	Corporate, direction, division, department, project, technical team, leader, etc. IT components
Physical interface	Hardware parts, protocols, procedures, etc.	Protocols, documents, etc.	Protocols, procedures, documents, etc.

A representation of an interface in Figure 4.7 shows the function "send," located in one system element; the function "receive," located in the other one; and the physical interface that supports the input/output flow. In the context of complex exchanges between system elements in information technology (IT) systems, a protocol is seen as a physical interface that carries exchanges of data.

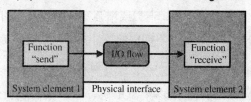

FIGURE 4.7 Interface representation. Reprinted with permission from Alain Faisandier. All other rights reserved.

4.4.2.6 Coupling Matrix Coupling matrices (also called N^2 diagrams) are a basic method to define the aggregates and the order of integration (Grady, 1994).They are used during architecture definition, with the goal of keeping the interfaces as simple as possible (see Fig. 4.8). Simplicity of interfaces can be a distinguishing characteristic and a selection criterion between alternate architectural candidates. The coupling matrices are also useful for optimizing the aggregate definition and the verification of interfaces (see Section 4.8.2.3).

4.4.2.7 Allocation and Partitioning of Logical Entities to Physical Entities
Defining a physical structural model of the architecture of a system consists of identifying system elements capable of performing the functions of logical models, identifying the physical interfaces capable to carry input/ output flows and control flows, and taking into account architectural characteristics that characterize the system in which they are included.

表 4.1　系统元素和物理接口的示例

元素	产品系统	服务系统	复杂组织体系统
系统元素	硬件部分（机械、电子、电气、塑料、化学等） 运行者角色 软件部分	流程、数据库、程序等 运行者角色 软件应用	公司、方向、分部、部门、项目、技术团队、领导者等 IT 组件
物理接口	硬件部件、协议、程序等	协议、文件等	协议、程序、文件等

图 4.7 中的接口表达指明位于一个系统元素中的功能"发送"；位于另一个系统元素中的功能"接收"；以及支持输入/输出流的物理接口。在信息技术（IT）系统中系统元素之间的复杂交换背景环境中，协议被视为进行数据交换的物理接口。

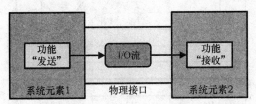

图 4.7 接口表征形式。经 Alain Faisandier 许可后转载。版权所有。

4.4.2.6 耦合矩阵　耦合矩阵（也称为 N^2 图）是定义综合的聚集总数与秩序的基本方法（Grady, 1994）。在架构定义期间使用耦合矩阵，目的是使接口保持尽可能简单（见图 4.8）。接口的简单性可以是备用的架构候选方案之间的区分特征和选择准则。耦合矩阵还用于优化聚集定义及接口的验证（见 4.8.2.3 节）。

4.4.2.7 逻辑实体到物理实体的分配和区划　定义系统架构的物理结构模型包括识别能够实施逻辑模型功能的系统元素，识别能够承载输入/输出流和控制流的物理接口，并将架构特征所在的系统的特征化考虑在内。

TECHNICAL PROCESSES

(a)

C1		X			X			
	C2	X	X				X	
X		C3			X			
	X		C4	X			X	
	X			C5		X		X
X		X	X	X	C6			
				X		C7		X
		X		X			C8	
X					X		X	C9

| A1 | A2 | A3 |

(b)

C1	X	X						
X	C3	X						
X	X	C6		X		X		
	X		C2	X	X			
			X	C4	X	X		
			X	X	C8			
			X			C5	X	X
						X	C7	X
X						X	X	C9

| B1 | B2 | B3 |

FIGURE 4.8 (a) Initial arrangement of aggregates; (b) final arrangement after reorganization. Reprinted with permission from Alain Faisandier. All other rights reserved.

The term allocation does not mean just to allocate logical entities to existing system elements. Partitioning and allocation means to separate, gather, or decompose logical entities into partitions and then to make the correspondence between these partitions and potential system elements. The system elements either exist (reusable, repurposed, or purchasable) can be developed and technically implemented.

Nonfunctional requirements and/or architectural characteristics are used as criteria to analyze, assess, and select candidate system elements and logical partitions. Examples of assessment criteria include similar transformations within the same technology, similar level of efficiency, exchange of same type of input/output flows (information, energy, materials), centralized or distributed controls, execution with close frequency level, dependability conditions, environment resistance level, and other enterprise constraints.

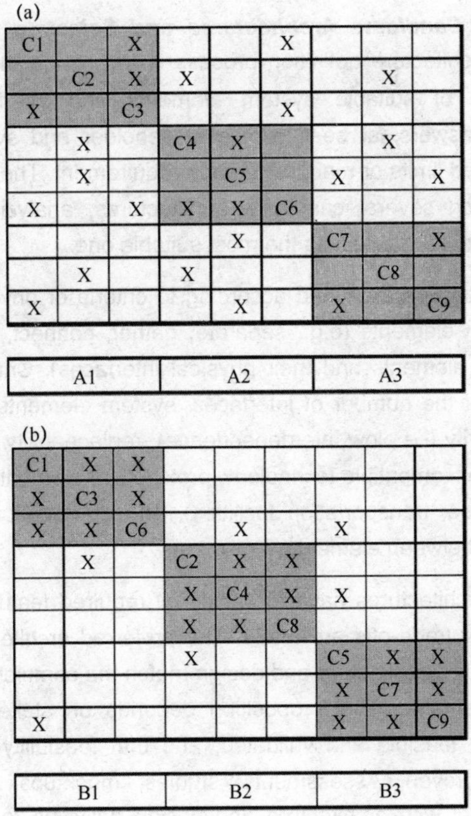

图 4.8（a）聚集的初始排列；（b）重组后的最终排列。经 Alain Faisandier 许可后转载。版权所有。

术语分配并不意味着仅向现有系统元素分配逻辑实体。区划和分配是指分离、收集或分解逻辑实体为若干区部，然后使这些区部与潜在系统元素之间相对应。系统元素或存在（可复用、重新赋予目的或可购买）可开发和在技术上实施。

非功能需求和/或架构特征用作分析、评估和选择候选系统元素及逻辑部分的准则。评估准则示例包括在相同技术内的相似转换、相似效率层级、同一类型输入/输出流（信息、能量、物质）的交换、集中式或分布式控制、具有相近频率层级的执行、可依赖性条件、环境抵制水平及其他复杂组织体约束。

TECHNICAL PROCESSES

4.4.2.8 *Defining Candidate Architectures and Selecting the Preferred One*
The goal of the architecture definition process is to provide the "best" possible architecture made of suitable system elements and interfaces, that is, the architecture that answers, at best, all the stakeholder and system requirements, depending on agreed limits or margins of each requirement. The preferred way to do this is by producing several candidate architectures; analyzing, assessing, and comparing them; and then selecting the most suitable one.

Candidate architectures are defined according to criteria or drivers in order to build up a set of system elements (e.g., separate, gather, connect, and disconnect the network of system elements and their physical interfaces). Criteria or drivers may include reduction of the number of interfaces, system elements that can be tested separately, modularity (i.e., low interdependence), replaceability of system elements during maintenance, compatible technology, proximity of elements in space, handling (e.g., weight, volume, transportation facilities), and optimization of resources and information shared between elements.

Viable candidate architectures have to satisfy all required features (e.g., functions, characteristics) after trade-offs are made. The preferred architecture represents an optimum such that the architecture and design match the complete set of stakeholder and system requirements. This proposition depends on stakeholder and system requirements being feasible and validated, and that feasibility and validation are demonstrated or proven. Assessments, studies, mock-ups, etc. are generally performed in parallel with architecture and design activities to obtain "proven" requirements.

Architecture definition activities include optimization to obtain a balance among architectural characteristics and acceptable risks. Certain analyses such as performance, efficiency, maintainability, and cost are required to get sufficient data that characterize the global or detailed behavior of the candidate architectures with respect to the stakeholder and system requirements. Those analyses are conducted with the system analysis process (see Section 4.6) and as specialty engineering activities (see Chapter 10).

4.4.2.9 *Methods and Modeling Techniques* Modeling, simulation, and prototyping used during architecture definition can significantly reduce the risk of failure in the finished system. Systems engineers use modeling techniques and simulation on large complex systems to manage the risk of failure to meet system mission and performance requirements. These are best performed by subject matter experts who develop and validate the models, conduct the simulations, and analyze the results. Refer to Chapter 9 for a more detailed treatment of models and simulations.

4.4.2.8 定义候选架构和选择优选架构 架构定义流程的目标是提供由适当系统元素和接口组成的"最好"可能的架构,即最好地回答所有利益攸关者和系统需求的架构,取决于每个需求的协定的限制或裕度。完成此任务的优选方式是通过产生多种候选架构;分析、评估和比较这些架构;然后选择一个最合适的架构。

按照准则或驱动因素来定义候选架构,以构建系统元素集合(如分离、收集、连接和断开系统元素及其物理接口的网络)。准则或驱动因素可包括接口及能够被分开单独测试的系统元素的数量的减少、模块化(即低的相互依赖性)、维护期间系统元素的可替换性、兼容技术、空间元素的近似度、搬移(如重量、体积、交通设施)以及元素之间共享的资源与信息的优化。

可行的候选架构必须在做出权衡后满足所有需求的特征(如功能、特征)。优选架构表达一种最佳情况即架构和设计与利益攸关者和系统需求的完整集合相吻合。这个主张取决于目前可行且经确认的利益攸关者和系统需求,而且可行性和确认需被演示验证或证实。为获得"经证实的"需求,评估、研究、初样等通常与架构和设计活动并行进行。

架构定义活动包括优化,以在架构特征和可接受风险之间取得平衡。需要进行一定的分析,如对性能、效率、可维护性和成本,以便获取相对于利益攸关者和系统需求特征化描述候选架构的全局或详细行为的充分数据。这些分析通过系统分析流程(见4.6节)来进行,且作为专业工程活动(见第10章)。

4.4.2.9 方法和建模技术 架构定义期间使用的建模、仿真和原型可大幅度地降低完成后的系统失败风险。系统工程师在大型复杂系统中使用建模技术和仿真来管理失败风险,以满足系统任务和性能需求。这些技术最好由主题专家实施,由他们开发并确认模型、进行仿真并分析结果。关于更详细的模型和仿真处理,参见第9章。

TECHNICAL PROCESSES

4.5 DESIGN DEFINITION PROCESS

4.5.1 Overview

4.5.1.1 Purpose   As stated in ISO/IEC/IEEE 15288,

[6.4.5.1] The purpose of the Design Definition process is to provide sufficient detailed data and information about the system and its elements to enable the implementation consistent with architectural entities as defined in models and views of the system architecture.

4.5.1.2 Description   System architecture deals with high-level principles, concepts, and characteristics represented by general views or models excluding details (see Section 4.4). System design supplements the system architecture providing information and data useful and necessary for implementation of the system elements. This information and data details the expected properties allocated to each system element and/or to enable the transition toward their implementation.

Design is the process of developing, expressing, documenting, and communicating the realization of the architecture of the system through a complete set of design characteristics described in a form suitable for implementation. Figure 4.9 is the IPO diagram for the design definition process.

Design concerns every system element (e.g., composed of implementation technologies such as mechanics, electronics, software, chemistry, human operations, and services) for which specific engineering processes are needed. The design definition process provides the detailed information and data that enable the implementation of a particular system element. This process provides feedback to the parent system architecture to consolidate or confirm the allocation and partitioning of architectural entities to system elements.

As a result, the design definition process provides the description of the design characteristics and design enablers necessary for implementation. Design characteristics include dimensions, shapes, materials, and data processing structures. Design enablers include formal expressions or equations, drawings, diagrams, tables of metrics with their values and margins, patterns, algorithms, and heuristics.

4.5.1.3 Inputs/Outputs   Inputs and outputs for the design definition process are listed in Figure 4.9. Descriptions of each input and output are provided in Appendix E.

4.5.1.4 Process Activities   The design definition process includes the following activities:

- *Prepare for design definition.*

4.5 设计定义流程

4.5.1 概览

4.5.1.1 目的 如 ISO/IEC/IEEE 15288 所述，

[6.4.5.1] 设计定义流程的目的是提供关于系统及其元素的充分详细的数据和信息，使实现能够与架构实体一致，正如系统架构的模型和视图中所定义的那样。

4.5.1.2 描述 系统架构应对由一般视图或模型（不包括细节）所表达的高层级原则、概念和特征（见 4.4 节）。系统设计对系统架构进行补充，提供对系统元素的实施有用且必要的信息和数据。此信息和数据详细描述分配给每个系统元素的和/或使能够向其实现转移的所期望的特性。

设计是通过以适用于实现的形式所描述的设计特征的完整集合来开发、表达、文件化和沟通系统架构的实现的过程。图 4.9 是设计定义流程的 IPO 图。

设计关注特定工程流程所需的每一个系统元素（例如所构成的实现技术构成，如机械、电子、软件、化学、人员操作和服务）。设计定义流程提供使特定系统元素能够实现的详细信息和数据。此流程为母系统架构提供反馈，以巩固或确认架构实体到系统元素的分配和分区。

因此，设计定义流程为实现提供必要的设计特征和设计使能项的描述。设计特征包括维度、形状、材料和数据处理结构。设计使能项包括正式表达式或方程、图样、图、具有其价值和欲度的指标表、特征模式、算法和启发法。

4.5.1.3 输入/输出 图 4.9 列出设计定义流程的输入和输出。附录 E 提供每个输入和输出的描述。

4.5.1.4 流程活动 设计定义流程包括以下活动：

- 准备设计定义。

TECHNICAL PROCESSES

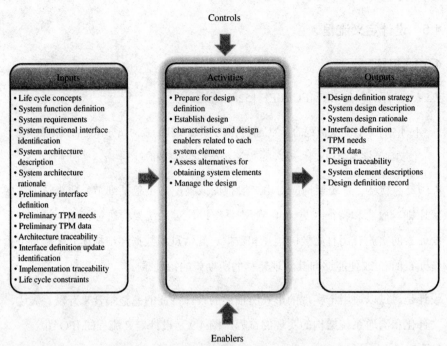

FIGURE 4.9 IPO diagram for the design definition process. INCOSE SEH original figure created by Shortell and Walden. Usage per the INCOSE notices page. All other rights reserved.

- Plan for technology management. Identify the technologies needed to achieve the design objectives for the system and its system elements. The technology management includes obsolescence management. Determine which technologies and system elements have a risk of becoming obsolete. Plan for their potential replacement, including identification of potential evolving technologies.

- Identify the applicable types of design characteristics for each system element considering the technologies that will be applied. Periodically assess the design characteristics and adjust as the system architecture evolves.

- Define and document the design definition strategy, including the need for and requirements of any enabling systems, products, or services.

- *Establish design characteristics and design enablers related to each system element.*

  - Perform requirements allocation to system elements for all requirements and system elements not fully addressed in the architecture definition process.

技术流程

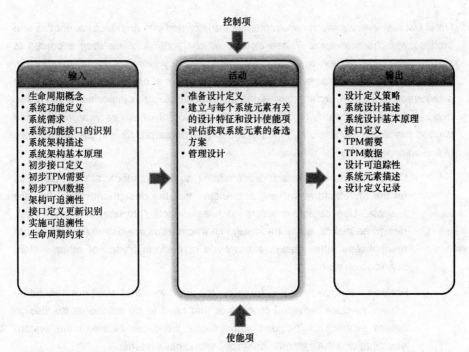

图 4.9 设计定义流程的 IPO 图。INCOSE SEH 原始图由 Shortell 和 Walden 创建。按照 INCOSE 通知页使用。版权所有。

- 规划技术管理。识别出达成系统及其系统元素的设计目标所需的技术。技术管理包括过时（技术）管理。确定哪些技术和系统元素具有变成过时的风险。为他们制定潜在更换规划，包括对潜在演进技术的识别。
- 识别每个系统元素的设计特征的适用类型，考虑到将要应用的技术。定期地评估设计特征并随系统架构的演进来调整。
- 定义和文件化设计定义策略，包括任何使能系统、产品或服务的需要和需求。

• 建立与每个系统元素相关的设计特征和设计使能项。

- 针对在架构定义流程中未全面阐述的所有需求和系统元素实施需求到系统元素的分配。

TECHNICAL PROCESSES

Note: Usually, every system requirement is transformed into architectural entities and architectural characteristics. Those entities or characteristics are then allocated to system elements, either by direct assignment or by some kind of partitioning. However, there will be some cases where it is impractical or impossible to transform a requirement into an architectural entity. In either case, it is important to do some degree of analysis or study to determine how best to flow down each requirement. It is good practice to try to do as much of the allocation as possible in the architecture definition process.

- Define the design characteristics relating to the architectural characteristics for the architectural entities, and ensure that the design characteristics are feasible. Use design enablers such as models (physical and analytical), design heuristics, etc. If the design characteristics are determined infeasible, then assess other design alternatives or perform trades of other system definition elements.

- Perform interface definition to define the interfaces that were not defined by the architecture definition process or that need to be refined as the design details evolve. This includes both internal interfaces between the system elements and the external interfaces with other systems.

- Capture the design characteristics of each system element. The resulting artifacts will be dependent on the design methods and techniques used.

- Provide rationale about selection of major implementation options and enablers.

- *Assess alternatives for obtaining system elements.*

  - Identify existing implemented elements. These include COTS, reused, or other nondeveloped system elements. Alternatives for new system elements to be developed may be studied.

  - Assess options for the system element, including the COTS system elements, the reused system elements, and the new system elements to be developed using selection criteria that is derived from the design characteristics.

  - Select the most appropriate alternatives.

  - If the decision is made to develop the system element, rest of the design definition process and the implementation process are used. If the decision is to buy or reuse a system element, the acquisition process may be used to obtain the system element.

注：通常，每个系统需求都被转换成架构实体和架构特征。然后通过直接分配或通过某种类型的分区将这些实体或特征分配到系统元素。但将会存在将需求转换成架构实体时不切实际或不可能的一些情况。不论哪种情况，进行某种程度的分析或研究以确定如何最好地向下细化每个需求是很重要的。在架构定义流程中试图进行尽可能多的分配是很好的实践。

- 为架构实体定义与架构特征相关的设计特征，并确保设计特征是可行的。使用设计使能项，如模型（物理的和解析的）、设计启发法等。如果设计特征被确定为不可行，那么评估其他设计备选方案或对其他系统定义元素进行权衡。
- 进行接口定义，以便对架构定义流程未定义的或需要随设计细节演进被细化的接口进行定义。该定义包括系统元素之间的内部接口以及与其他系统的外部接口。
- 捕获每个系统元素的设计特征。产生的制品将取决于所使用的设计方法和技术。
- 提供关于主要实现选项和使能项的选择的理由依据。

- 评估获取系统元素的备选方案。
  - 识别已实施的现有元素。这些元素包括 COTS、已复用元素或其他未开发的系统元素。可以研究待开发的新系统元素的备选方案。
  - 评估系统元素的选项，包括 COTS 系统元素、已复用的系统元素以及使用从设计特征导出的选择准则来开发的新系统元素。
  - 选择最恰当的备选方案。
  - 如果决定开发系统元素，则使用其余的设计定义流程和实现流程。如果决定购买或复用系统元素，则可使用采办流程来获取系统元素。

- *Manage the design.*
  - Capture and maintain the rationale for all selections among alternatives and decisions for the design, architecture characteristics, design enablers, and sources of system elements.
  - Manage the maintenance and evolution of the design, including the alignment with the architecture. Assess and control evolution of the design characteristics.
  - Establish and maintain bidirectional traceability between the architecture entities (including views, models, and viewpoints) to the stakeholder requirements and concerns; system requirements and constraints; system analysis, trades, and rationale; verification criteria and results; and design elements. The traceability between the design characteristics and the architectural entities also helps ensure architectural compliance.
  - Provide baseline information for configuration management.
  - Maintain the design baseline and the design definition strategy.

Common approaches and tips:
- Discipline engineers, or designers, perform the design definition of each concerned system element; they provide strong support (knowledge and competencies) to systems engineers, or architects, in the evaluation and selection of candidate system architectures and system elements. Inversely, systems engineers, or architects, must provide feedback to discipline engineers or designers to improve knowledge and know-how.

4.5.2 Elaboration

4.5.2.1 Architecture Definition versus Design Definition The architecture definition process focuses on the understanding and resolution of the stakeholder concerns. It develops insights into the relation between these concerns, the solution requirements, and the emergent properties and behaviors of the system. Architecture focuses on suitability, viability, and adaptability over the life cycle. An effective architecture is as design-agnostic as possible to allow for maximum flexibility in the design trade space. It focuses more on the "what" than the "how."

The design definition process, on the other hand, is driven by specified requirements, the architecture, and more detailed analysis of performance and feasibility. Design definition addresses the implementation technologies and their assimilation. Design provides the "how" or "implement-to" level of the definition.

- 管理设计。

  – 捕获和维护对设计、架构特征、设计使能项和系统元素来源的备选方案和决策之间的所有选择的理由依据。

  – 对设计的维护和演进进行管理,包括与架构对准。评估并控制设计特征的演进。

  – 建立并维持架构实体(包括视图、模型和视角)之间到利益攸关者需求和关切点,系统需求和约束,系统分析、权衡和理由依据,验证准则与结果,以及设计元素中之间的双向可追溯性。设计特征与架构实体之间的可追溯性还有助于确保架构的遵从。

  – 提供构型配置管理的基线信息。

  – 维持设计基线和设计定义策略。

常用方法和建议:

- 学科工程师或设计者实施每个受关切的系统元素的设计定义;他们在评价和选择候选系统架构和系统元素时为系统工程师或架构师提供强有力的支持(知识和才能)。反之,系统工程师或架构师必须为学科工程师或设计者提供反馈,以提高知识和"诀窍"。

4.5.2 详细阐述

4.5.2.1 架构定义与设计定义的对比 架构定义流程聚焦于利益攸关者关切点的理解和解析,并对这些关切点、解决方案需求及系统的涌现性和行为之间关系开发深入透视。架构聚焦于整个生命周期的适用性、生存力和适应性。有效的架构尽可能独立于设计,以便在设计权衡空间中有最大灵活性。它更聚焦于"做什么"而不是"如何做"。

另一方面,设计定义流程由特定的需求、架构以及性能和可行性更为详细地分析所驱动。设计定义阐述实现技术及其吸收过程。设计提供定义的"如何做"或"实施到的"层级。

TECHNICAL PROCESSES

4.5.2.2 Notions and Principles Used within Design

The purpose of system design is to make the link between the architecture of the SOI and the implementation of technological system elements that compose it. So, system design is understood as the complete set of detailed models, properties, or characteristics of each system element, described into a form suitable for implementation.

Every technological domain or discipline owns its peculiar laws, rules, theories, and enablers concerning transformational, structural, behavioral, and temporal properties of its composing parts of materials, energy, or information. These specific parts and/or their compositions are described with typical design characteristics and enablers. These allow achieving the implementation of the system element of interest through various transformations, linkages, and exchanges required by design characteristics (e.g., operability level, reliability rate, speed, safeguard level) that have been assigned during the architecture definition process.

- *Examples of generic design characteristics in mechanics of solids:* Shape, geometrical pattern, dimension, volume, surface, curves, resistance to forces, distribution of forces, weight, velocity of motion, temporal persistence
- *Examples of generic design characteristics in software*: Distribution of processing, data structures, data persistence, procedural abstraction, data abstraction, control abstraction, encapsulation, creational pat-terns (e.g., builder, factory, prototype, singleton), and structural patterns (e.g., adapter, bridge, composite, decorator, proxy)

4.5.2.3 Design Descriptors
Because it is sometimes difficult to define applicable requirements to a system element from the engineering data of the parent system (in particular from the expected architectural characteristics), it is possible to use the design descriptor technique as a complement. A design descriptor is the set of generic design characteristics and of their possible values. If similar, but not exact, system elements exist, it is possible to analyze these in order to identify their basic characteristics. Variations of the possible values of each characteristic determine potential candidate system elements.

4.5.2.4 Holistic Design
It is important to understand that design definition starts with the system as a whole consisting of system elements and ends with a definition (i.e., design) for each of these system elements (not just one of them) and how they are designed to work together as a complete system. System elements are identified in the architecture, although the architecture might only identify those elements that are architecturally significant.

4.5.2.2　设计内使用的观念和原则

系统设计的目的是在 SOI 的架构与构成它的技术系统元素的实现之间建立联系。因此系统设计被理解为每个系统元素的详细模型、特性或特征的完整集合，被描述成适合于实现的形式。

每个技术领域或学科都具有与其物质、能量或信息的组成部分的转换、结构、行为和时间特性有关的其自身的特殊定理、规则、理论和使能项。这些特定部分和/或其构成用典型的设计特征和使能项来描述。它们允许通过架构定义流程期间已经分配的设计特征（如可运行性水平、可靠性等级、速度、防护等级）所需的多种转换、联系和交换来达成所感兴趣之系统元素的实现。

- *固态机械学中的通用设计特征示例*：形状、几何特征模式、尺度、体积、表面、曲面、力的阻力、力的分布、重量、移动速度、时间持续性
- *软件通用设计特征示例*：处理的分布、数据结构、数据持续性、程序抽象、数据抽象、控制抽象、封装、创建的特征模式（如构建者、工厂、原型、单例）及结构特征模式（如适配器、桥接、复合、装饰、代理权）

4.5.2.3　设计描述符　由于有时难以从母系统的工程数据中（特别是从预期架构特征中）定义系统元素的适用需求，因此有可能使用设计描述符技术作为补充。设计描述符是通用设计特征及其可能值的集合。如果存在相似但不完全一样的系统元素，则有可能对它们进行分析，以便识别它们的基本特征。每个特征的可能值的变化确定潜在的候选系统元素。

4.5.2.4　整体设计　设计定义开始于由系统元素组成的作为一个整体的系统并结束于对每个系统元素（不仅是其中的一个）的定义（即设计）以及如何将这些系统元素设计成作为一个完整系统一起工作，理解这一点很重要。在架构中识别系统元素，尽管架构可能仅识别在架构上很重要的那些元素。

TECHNICAL PROCESSES

73 During the design definition process, it might be necessary to identify additional system elements to make the whole system work. This might entail embedding some enabling elements or services inside the system boundary. There is usually a trade-off between having an enabling item inside or outside the system. The architecture definition process could make this decision, but it might be better to allow design definition to handle this since it is often dependent on other design trade-offs and on design decisions that are made along the way.

Some of these additional system elements might be necessary to account for "missing" functions that were not identified in the architecture. For example, it might be determined that the various system elements should not all produce their own power backup but instead there should be a separate system element that performs this function for all the other elements. This would be the result of a design analysis to determine the best place to put this function. Or it could be the result of applying a design pattern to this particular problem.

It might be necessary to provide feedback to the architecture definition process regarding these design decisions and trade-offs to ensure there are no negative impacts on the architecture as a whole. The architecture might or might not be updated to reflect these design details, since this depends on whether it is important to capture these features as architecturally significant or not.

It is this holistic approach to design of a system that distinguishes this from design of an individual product or service. Holistic design is an approach to design that considers the system being designed as an interconnected whole, which is also part of something larger. Holistic concepts can be applied to the system as a whole along with the system in its context (e.g., the enterprise or mission in which the system participates), as well as the design of mechanical devices, the layout of spaces, and so forth. This approach to design often incorporates concerns about the environment, with holistic designers considering how their design will impact the environment and attempting to reduce environmental impact in their designs. Holistic design is about more than merely trying to meet the system requirements.

4.6 SYSTEM ANALYSIS PROCESS

4.6.1 Overview

4.6.1.1 Purpose As stated in ISO/IEC/IEEE 15288,

> [6.4.6.1] The purpose of the System Analysis process is to provide a rigorous basis of data and information for technical understanding to aid decision-making across the life cycle.

在设计定义流程期间，识别附加的系统元素以使整体系统工作可能是必要的。这可能需要在系统边界内部嵌入一些使能元素或服务。通常在系统内部或外部具有的使能项之间权衡。架构定义流程可做出这一决策，但让设计定义来处理可能会更好，因为设计定义往往取决于其他设计权衡以及沿此方式做出的设计决策。

一些附加的系统元素对于架构中未识别的"缺失"功能可能是必要的。例如，可以确定，各种不同的系统元素不应全部产生它们自己的功率备份，反而应该有为所有其他元素实施这种功能的分开单独的系统元素。这将是设计分析的结果，以便确定实施此功能的最佳场所。或者可以是将设计特征模式应用于这种特定问题的结果。

为架构定义流程提供关于这些设计决策和权衡的反馈可能是必要的，以便确保对作为整体的架构没有负面影响。要反映这些设计细节，架构可以更新或不更新，因为这取决于捕获这些特征对架构来说是否重要。

正是这种系统设计的整体方法将其与单个产品或服务的设计区分开来。整体设计是将所设计的系统看做互连的整体，并且该整体还构成某一更大的事物的部分的一种设计方法。整体概念被应用于系统与在其背景环境中的系统一起作为一个整体（如系统所参与的复杂组织体或任务），以及机械装置的设计、空间布局等。此设计方法往往纳入与环境有关的关切点，同时整体设计者考虑到他们的设计将如何影响环境并试图减少环境在他们的设计中的影响。整体设计比仅仅只试图满足系统需求要多得多。

4.6 系统分析流程

4.6.1 概览

4.6.1.1 目的 如 ISO/IEC/IEEE 15288 所述，

[6.4.6.1] 系统分析流程的目的是为技术理解提供严格的数据和信息基础，以辅助跨生命周期的决策。

TECHNICAL PROCESSES

4.6.1.2 Description This process performs quantitative assessments and estimations that are based on analyses such as cost analysis, affordability analysis, technical risk analysis, feasibility analysis, effectiveness analysis, and other critical quality characteristics. Those analyses use mainly quantitative modeling techniques, analytical models, and associated simulations, which are applied at varying levels of rigor and complexity depending on the level of fidelity needed. In some cases, it may be necessary to employ a variety of analytic functions or experimentation to obtain the necessary insight. The results serve as inputs into various technical decisions, providing confidence in the adequacy and integrity of the system definition toward achieving the appropriate system balance.

This process is used by (examples, but not limited to):

- Mission and business analysis process to analyze and estimate candidate OpsCon and/or candidate business models related to a potential SOI in terms of feasibility, costs, risks, and effectiveness

- Stakeholder requirements definition process and system requirements definition process to analyze issues relating to conflicts among the set of requirements, in particular those related to feasibility, costs, technical risks, and effectiveness (mainly performances, operational conditions, and constraints)

- Architectural definition process and design definition process to analyze and estimate architectural and design characteristics of candidate architectures and/or system elements, providing arguments in order to be able to select the most efficient ones in terms of costs, technical risks, effectiveness (e.g., performances, dependability, human factors), and other stakeholder concerns such as critical quality characteristics, affordability, maintenance, etc.

- Integration process, verification process, and validation process to estimate the related strategies

- Project assessment and control process to obtain estimates of the performance against established targets and thresholds, especially with respect to the technical measures (MOEs, MOSs, MOPs, and TPMs).

The results of analyses and estimations, as data, information, and arguments, are provided to the decision management process for selecting the most efficient alternative or candidate.

4.6.1.2 描述 此流程基于分析，例如成本分析、经济可承受性分析、技术风险分析、可行性分析、有效性分析及其他关键的品质特征，来进行定量的评估和估计。这些分析主要使用根据所需要的逼真度在各种严格性和复杂性层级上所应用的定量建模技术、分析模型及相关仿真。在某些情况中，使用各种分析函数或实验来获取必要的深入透视也许是必要的。结果用作各种不同的技术决策的输入，为实现恰当的系统平衡提供系统定义适当性和完整性的置信度。

使用此流程的是（示例，但不限于）：

- 任务和业务分析流程，用于在可行性、成本、风险和有效性方面分析和评估与潜在的 SOI 相关的候选 OpsCon 和/或候选业务模型。

- 利益攸关者需求定义流程和系统需求定义流程，用于分析与需求集合之间的冲突有关的问题，特别是与可行性、成本、技术风险和有效性（主要是性能、运行条件和约束）相关的那些问题。

- 架构定义流程和设计定义流程，用于分析和评估候选架构和/或系统元素的架构特征和设计特征，提供论据以便能够选择在成本、技术风险、有效性（如性能、可依赖性、人员因素）及其他利益攸关者关切点（如关键的品质特征、经济可承受性、维护等）方面最为高效的。

- 综合流程、验证流程和确认流程，用于评估相关的策略。

- 项目评估和控制流程，用以按照既定目标和阈值，尤其是相对于技术措施（MOE、MOS、MOP 和 TPM）获得性能评估。

为决策管理流程提供分析和评估的结果，如数据、信息和论据，以选择最高效的备选方案或候选方案。

TECHNICAL PROCESSES

74 In some cases, the results may be provided to the project assessment and control process, if the information is needed to monitor the progress of the system against its system objectives, performance thresholds, or growth targets, such as the projected or modeled reliability early in the development of the system as compared to its reliability growth curve.

Figure 4.10 is the IPO diagram for the system analysis process.

4.6.1.3 Inputs/Outputs Inputs and outputs for the system analysis process are listed in Figure 4.10. Descriptions of each input and output are provided in Appendix E.

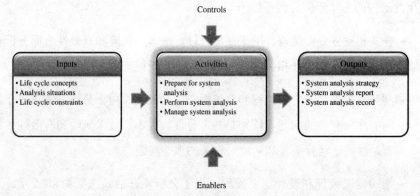

FIGURE 4.10 IPO diagram for the system analysis process. INCOSE SEH original figure created by Shortell and Walden. Usage per the INCOSE notices page. All other rights reserved.

4.6.1.4 Process Activities The system analysis process includes the following activities:

- *Prepare for system analysis.*

    - Define the scope, types, objectives, and level of accuracy of required analyses and their level of importance to the system stakeholders.

    - Define or select evaluation criteria (e.g., operational conditions, environmental conditions, performance, dependability, costs types, risks types). The criteria mainly come from stakeholder needs, nonfunctional requirements, and design characteristics. The criteria must be consistent with the decision management strategy (see Section 5.3).

    - Determine the candidate elements to be analyzed, the methods and procedures to be used, and the needed justification items.

在某些情况中，如果需要该信息来依照其系统目的、性能阈值或增长目标来监控系统的进展，该结果可提供给项目评估和控制流程，如在系统开发早期估计的或建模的可靠性与其可靠性增长曲线相比。

图 4.10 是系统分析流程的 IPO 图。

4.6.1.3 输入/输出 图 4.10 列出系统分析流程的输入和输出。附录 E 提供每个输入和输出的描述。

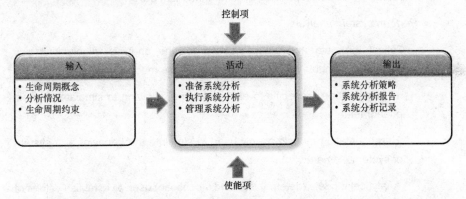

图 4.10 系统分析流程的 IPO 图。INCOSE SEH 原始图由 Shortell 和 Walden 创建。按照 INCOSE 通知页使用。版权所有。

4.6.1.4 流程活动 系统分析流程包括以下活动：

- *准备系统分析*。

  - 定义范围、类型、目标、所需分析的精确度等级及其对系统利益攸关者的重要性等级。

  - 定义或选择评价准则（如运行条件、环境条件、性能、可依赖性、成本类型、风险类型）。准则主要来自利益攸关者需要、非功能需求和设计特征。准则必须与决策管理策略（见 5.3 节）一致。

  - 确定待分析的候选元素、待使用的方法和程序及所需的合理性理由项。

TECHNICAL PROCESSES

- Determine the need and requirements for and obtain or acquire access to the enabling systems, products, or services necessary to perform analyses of the SOI.

- Schedule the analyses according to the availability of models, engineering data (e.g., OpsCon, business models, stakeholder requirements, system requirements, design characteristics, verification actions, validation actions), skilled personnel, and procedures.

- Document the corresponding system analysis strategy.

- *Perform system analysis.*

  - Collect the data and inputs needed for the analysis, highlighting any assumptions. Inputs can include models. Those models may be:

    ° Physical models, specific to each discipline, allowing to simulate physical phenomena

    ° Representation models mainly used to simulate the behavior of a system or system element

    ° Analytical models (deterministic and stochastic) used to establish values of estimates to approach the real operation of a system or system element

  - Carry out analyses as scheduled using defined methods and procedures for cost, risk, effectiveness, and validation of assumptions.

  - Conduct in-process peer reviews with appropriate subject matter experts to assess the validity, quality, and consistency of the evolving system with the stakeholder objectives and with previous analyses. Record and report in-process results.

- *Manage system analysis.*

  - Baseline the analysis results or reports using the configuration management process.

  - Maintain an engineering history of the system evolution from stakeholder needs definition to ultimate system retirement so that the project team can conduct bidirectional searches at any time during-or after-the system life cycle.

Common approaches and tips:

- 确定对进行 SOI 分析所必需的使能系统、产品或服务的需要和需求并获得或获取该使用权。
- 按照模型可用性、工程数据（如 OpsCon、业务模型、利益攸关者需求、系统需求、设计特征、验证措施、确认措施）、有技能的人员和程序对分析进度进行安排。
- 对相应的系统分析策略进行文件化。

• *执行系统分析。*
- 收集分析所需的数据和输入，强调任何假设。输入可以包括模型，这些模型可能是：
  ○ 每个学科特有的物理模型，允许对物理现象进行仿真
  ○ 主要用于对系统或系统元素的行为进行仿真的表征模型
  ○ 用于建立趋近系统或系统元素实际运行的估计值的分析模型（确定性和随机性）
- 使用成本、风险、有效性和假设确认的已定义的方法和程序按照进度安排执行分析。
- 与适当的主题专家进行流程中的同行评审，以便评估演进系统的有效性、品质以及与利益攸关者目标及之前分析的一致性。记录和报告流程中的结果。

• *管理系统分析。*
- 使用构型管理流程对分析结果或报告进行基线化。
- 维护系统从利益攸关者需要定义到最终的系统退役的演进的工程历史，从而使项目团队能够在系统生命周期期间或之后的任何时间进行双向搜索。

常用方法和建议：

TECHNICAL PROCESSES

- The methods are chosen based on time, cost, accuracy, technical drivers, and criticality of analysis. Due to cost and schedule, most systems only perform system analysis for critical characteristics.

- Models can never simulate all the behavior of a system: they operate only in one limited field with a restricted number of variables. When a model is used, it is always necessary to make sure that the parameters and data inputs are part of the operation field; if not, irregular outputs are likely.

- Models evolve during the project: by modification of parameters, by entering new data, and by the use of new tools.

- It is recommended to concurrently use several types of models in order to compare the results and/or to take into account another characteristic or property of the system.

- Results of a simulation shall always be given in their modeling context: tool used, selected assumptions, parameters and data introduced, and variance of the outputs.

4.6.2 Elaboration

During a system's life cycle, assessments should be performed when technical choices have to be made or justified, and not just to compare different solutions.

System analysis provides a rigorous approach to technical decision making. It is used to perform evaluations, including a set of analysis such as cost analysis, technical risk analysis, effectiveness analysis, and analysis of other properties.

4.6.2.1 Cost Analysis A cost analysis considers the full life cycle costs (LCC). The cost baseline can be adapted according to the project and the system. The full LCC may include labor and nonlabor cost items; it may include development, manufacturing, service realization, sales, customer utilization, supply chain, maintenance, and disposal costs (also see Section 10.1).

4.6.2.2 Technical Risk Analysis Technical risks should not be confused with project risks even if the method to manage them is the same. Technical risks address the system itself, not the project for its development. Of course, technical risks may interact with project risks. The system analysis process is often needed to perform the technical assessments that provide quantification and understanding of the probability or impact of a potential risk or opportunity (see Section 5.4 for risk management process details).

- 基于时间、成本、准确度、技术驱动因素和分析的关键性来选择方法。由于成本和进度,大多数系统仅执行关键特征的系统分析。

- 模型永远不能仿真系统的所有行为:它们通过有限数量的变量仅在一个有限专业内运行。在使用模型时,确保参数和数据输入是运行专业的一部分总是很必要的;否则,很可能存在不规则的输出。

- 模型在项目期间演进:通过修改参数、通过输入新数据及通过使用新工具。

- 建议并行使用多种类型的模型,以便对结果进行比较和/或将系统的其他特征或特性考虑进去。

- 仿真结果应总是在其建模背景环境中提供:所使用的工具、选定的假设、引入的参数和数据、以及输出的变化。

4.6.2 详细阐述

在系统的生命周期期间,应在必须做出或证明技术选择时执行评估,而并不仅仅是为了比较不同的解决方案。

系统分析提供严格的技术决策方法,用于进行评价,包括一系列分析,如成本分析、技术风险分析、有效性分析和其他特性分析。

4.6.2.1 成本分析 成本分析考虑到全生命周期成本(LCC)。成本基线可以按照项目和系统来调整。全 LCC 可能包括劳动成本项和非劳动成本项;可能包括开发成本、制造成本、服务实现成本、销售成本、客户使用成本、供应链成本、维护成本和处置成本(也可参见 10.1 节)。

4.6.2.2 技术风险分析 技术风险不应与项目风险相混淆,即使它们的管理方法相同。技术风险应对系统本身,而不是系统开发项目。当然,技术风险可与项目风险交互。系统分析流程往往需要执行技术评估,技术评估提供对潜在风险或机会的概率或影响的量化和理解(关于风险管理流程细节,参见 5.4 节)。

4.6.2.3 Effectiveness Analysis System effectiveness analysis is a term for a broad category of analyses that evaluate the degree or extent to which a system meets one or more criteria—the effectiveness of the system in meeting the criteria in its intended operational environment. The objective(s) and criteria may be derived from one or more desired system characteristics, such as MOEs, TPMs, or other attributes of the system, and influence the details of the analysis conducted. The analysis is more than simply determining if the criteria are met but also the degree to which they are met (or fall short or exceed), as this information is used to support trade-offs and evaluation of alternatives (such as which candidates to develop further, where improvements are needed or cost savings are possible). One of the challenges of effectiveness analysis is to prioritize and select the right set of effectiveness objectives and criteria; for example, if the product is made for a single use, the maintainability and the capability of evolution will not be relevant criteria.

4.6.2.4 Methods and Modeling Techniques Various types of models and modeling techniques can be used in the context of system analysis. These include, but are not limited to, physical models, structural models, behavior models, functional models, temporal models, mass models, cost models, probabilistic models, parametric models, layout models, network models, visualizations, simulations, mathematical models, and prototypes. For more information on models and modeling methods, see Section 9.1.

4.7 IMPLEMENTATION PROCESS

4.7.1 Overview

4.7.1.1 Purpose As stated in ISO/IEC/IEEE 15288,

> [6.4.7.1] The purpose of the Implementation process is to realize a specified system element.

The implementation process creates or fabricates a system element conforming to that element's detailed description (requirements, architecture, design, including interfaces). The element is constructed employing appropriate technology and industry practices.

4.7.1.2 Description During the implementation process, engineers follow the requirements allocated to the system element to fabricate, code, or build each individual element using specified materials, processes, physical or logical arrangements, standards, technologies, and/or information flows outlined in detailed drawings or other design documentation. System requirements are verified and stakeholder requirements are validated. If subsequent configuration audits reveal discrepancies, recursive interactions occur with predecessor activities or processes, as required, to correct them. Figure 4.11 is the IPO diagram for the implementation process.

4.6.2.3 有效性分析 系统有效性分析是广泛的分析类别中的一个术语，评价系统满足一个或多个准则的程度或范围——系统在其预计运行环境中满足准则的有效性。目标和准则可能来自一个或多个预期的系统特征，如 MOE、TPM 或系统的其他属性，并影响所进行的分析的细节。分析不仅仅是简单地确定是否满足准则，而是确定准则的满足程度（或不足或超过），因为该信息用于支持备选方案的权衡和评价（如哪些候选方案需进一步开发，哪里需要改进或成本节省是否有可能）。有效性分析的挑战之一是对有效性目标和准则的正确集进行优先级排序和选择；例如，如果产品制造是为单一的用途，则可维护性和演进能力将不是相关准则。

4.6.2.4 方法和建模技术 各种不同类型的模型和建模技术可在系统分析的背景环境中使用，包括但不限于物理模型、结构模型、行为模型、功能模型、时间模型、质量模型、成本模型、概率模型、参数模型、布局模型、网络模型、可视化、仿真、数学模型和产品原型。关于模型和建模方法的更多信息，参见 9.1 节。

4.7 实施流程

4.7.1 概览

4.7.1.1 目的 如 ISO/IEC/IEEE 15288 所述，

[6.4.7.1] 实施流程的目的是实现特定的系统元素。

实施流程创建或构造符合该元素的详细描述（需求、架构、设计，包括接口）的系统元素。使用适用技术和行业实践构造该元素。

4.7.1.2 描述 在实施流程期间，工程师遵循分配给系统元素的需求，以使用在详细图样或其他设计文件中概述的特定原料、流程、物理或逻辑布局、标准、技术和/或信息流来构造、编码或构建每个单独的元素。验证系统需求并确认利益攸关者需求。如果后续构型审计暴露出不符之处，按要求对之前的活动或流程进行递归性交互，以纠正它们。图 4.11 是实施流程的 IPO 图。

TECHNICAL PROCESSES

4.7.1.3 Inputs/Outputs Inputs and outputs for the implementation process are listed in Figure 4.11. Descriptions of each input and output are provided in Appendix E.

4.7.1.4 Process Activities Implementation process activities begin with detailed design and include the following:

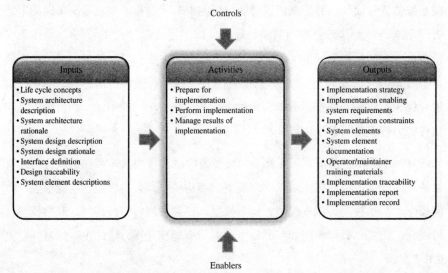

FIGURE 4.11 IPO diagram for the implementation process. INCOSE SEH original figure created by Shortell and Walden. Usage per the INCOSE notices page. All other rights reserved.

- *Prepare for implementation.*
    - Define fabrication/coding procedures, tools and equipment to be used, implementation tolerances, and the means and criteria for auditing configuration of resulting elements to the detailed design documentation. In the case of repeated system element implementations (such as for mass manufacturing or replacement elements), the implementation strategy is defined/refined to achieve consistent and repeatable element production and retained in the project decision database for future use.
    - Elicit from stakeholders, developers, and team-mates any constraints imposed by implementation technology, strategy, or implementation enabling systems.Record the constraints for consideration in the definition of the requirements, architecture, and design.
    - Document the plan for acquiring or gaining access to resources needed during implementation. The planning includes the identification of requirements and interfaces for the enabling system.

4.7.1.3 输入/输出　图 4.11 列出实施流程的输入和输出。附录 E 提供每个输入和输出的描述。

4.7.1.4 流程活动　实施流程活动从详细的设计开始，并包括下列内容：

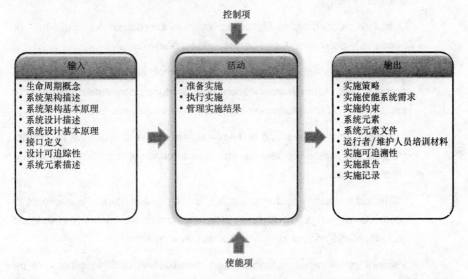

图 4.11　实施流程的 IPO 图。INCOSE SEH 原始图由 Shortell 和 Walden 创建。按照 INCOSE 通知页使用。版权所有。

- *准备实现*。

    - 定义构造/编码程序、要使用的工具和设备、实现允差以及对详细设计文件产生的元素的构型审计所用的手段和准则。至于重复的系统元素实现（如对于批量制造或可替换的元素），则定义/细化实现策略，以达到一致的和可重复的元素生产，并将实施策略保存在项目决策数据库中，以备将来使用。

    - 从利益攸关者、开发者和团队中引出由实现技术、策略或实现使能系统施加的任何约束。记录在需求、架构和设计的定义中需考虑的约束。

    - 对实施期间所需资源的使用权的获取或获得计划进行文件化。计划包括对使能系统的需求和接口的识别。

- *Perform implementation.*
  - Develop data for training users on correct and safe procedures for operating and maintaining that element, either as a stand-alone end item or as part of a larger system.
  - Complete detailed product, process, material specifications ("build-to" or "code-to" documents), and corresponding analyses.
  - Ensure the realization of the system elements per the detailed product, process, and material specifications and produce documented evidence of implementation compliance—specifically, these tasks are as follows:
    ° Conduct peer reviews and testing—Inspect and verify software for correct functionality, white box testing, etc. in accordance with software/ hardware best practices.
    ° Conduct hardware conformation audits— Compare hardware elements to detailed drawings to ensure that each element meets its detailed specifications prior to integration with other elements.
  - Prepare initial training capability and draft training documentation—To be used to provide the user community with the ability to operate, conduct failure detection and isolation, conduct contingency scenarios, and maintain the system as appropriate.
  - Prepare a hazardous materials log, if applicable.
  - Determine the packaging and storage requirements for the system element and ensure initiation of the packaging and/or storage, at the appropriate time.
- *Manage results of implementation.*
  - Identify and record implementation results. Maintain the records per organizational policy.
  - Record any anomalies encountered during the implementation process, and analyze and resolve the anomalies (corrective actions or improvements) using the quality assurance process (see Section 5.8).
  - Establish and maintain traceability of the implemented system elements with the system architecture, design, and system and interface requirements that are needed for implementation.

- 执行实施。
  - 开发用于培训用户有关运行和维护该元素的正确和安全程序方面的数据，作为单独的最终项，或作为更大的系统的部分。
  - 完成详细的产品、流程、资料规范（"按构建"或"按编码"文件化）以及相对应的分析。
  - 确保按照详细的产品、流程、资料规范实现系统元素并产生表明实施一致性的文件化的证据——特别是以下这些任务：
    - 进行同行评审和测试——按照软件/硬件的最佳实践，以"白盒"测试方式检查和验证软件的功能正确性。
    - 进行硬件符合性审计——将硬件元素与详细的图样进行比较，以确保每个元素在与其他元素综合前满足其详细规范。
  - 准备初始的培训能力并起草培训文件——用以向用户团体提供运行、进行失效检测和隔离、进行应急场景并适当地维护系统的能力。
  - 若适用，准备危险资料日志。
  - 确定系统元素的包装和存放需求，并确保在恰当的时间开始包装和/或存放。
- 管理实施结果。
  - 识别和记录实施结果。按照组织的方针维护记录。
  - 记录实施过程期间遭遇到的任何异常，并使用质量保证流程（见 5.8 节）来分析和化解该异常（纠正措施或改进）。
  - 建立并保持所实施的系统元素与系统架构、设计以及实施所需的系统和接口需求的可追溯性。

TECHNICAL PROCESSES

– Provide baseline information for configuration management.

Common approaches and tips:

- Keep the Integrated Product Development Team (IPDT) engaged to assist with configuration issues and redesign.
- Inspections are a proactive way to build in quality (Gilb and Graham, 1993).
- In anticipation of improving process control, reducing production inspections, and lowering maintenance activities, many manufacturing firms use Design for Six Sigma or lean manufacturing.
- Conduct hardware conformation audits or system element level hardware verification; ensure sufficient software unit verification prior to integration.
- Validate simulations; interface simulator drivers should be representative of tactical environments.

4.7.2 Elaboration

4.7.2.1 *Implementation Concepts* The implementation process typically focuses on the following four forms of system elements:

- *Hardware/physical*—Output is fabricated or adapted hardware or physical element. If the hardware element is being reused, it may require modification.
- *Software*—Output is software code and executable images
- *Operational resources*—Output includes procedures and training. These are verified to the system requirements and OpsCon.
- *Services*—Output includes specified services. These may be the result of one or more hardware, software, or operational elements resulting in the service.

The implementation process can support either the creation (fabrication or development) or adaptation of system elements. For system elements that are reused or acquired, such as COTS, the implementation process allows for adaption of the elements to satisfy the needs of the SOI. This is usually accomplished via configuration settings provided with the element (e.g., hardware configuration switches and software configuration tables). newly created products have more flexibility to be designed and developed to meet the needs of the SOI without modification.

– 提供构型配置管理的基线信息。

常用方法和建议：

- 保持综合产品开发团队（IPDT）的介入，以帮助解决构型配置问题和再设计。
- 检验是构建在内在质量的一种积极主动的方式（Gilb 和 Graham, 1993）。
- 为了改善过程控制，减少生产检验并降低维护活动，许多制造公司使用六西格玛设计或精益制造。
- 进行硬件符合性审计或系统元素级硬件验证；确保软件单元综合之前的验证充分。
- 确认仿真；接口仿真器的驱动因素应表征战术环境。

4.7.2 详细阐述

4.7.2.1 *实施概念* 实施流程通常集中于以下四种形式的系统元素：

- *硬件/物理的*——输出是构造的或适配的硬件或物理元素。如果硬件元素正在被复用，那么它可能需要修改。
- *软件*——输出是软件代码和可执行的映像文件。
- *运行资源*——输出包括程序和培训。按照系统需求和 OpsCon 对它们进行验证。
- *服务*——输出包括特定的服务。它们可能是导致产生服务的一个或多个硬件、软件或运行元素的结果。

实施流程可支持系统元素的创建（构造或开发）或适应性变化。对于复用的或采办的系统元素，如 COTS，实施流程允许元素的适应性变化，以满足 SOI 的需要。这通常通过为元素（如硬件构型配置转换和软件构型配置表）提供的构型配置的设置来完成。新创建的产品具有更多待设计和开发的灵活性，无须修改便满足 SOI 的需要。

TECHNICAL PROCESSES

4.8 INTEGRATION PROCESS

4.8.1 Overview

4.8.1.1 Purpose As stated in ISO/IEC/IEEE 15288,

> [6.4.8.1] The purpose of the Integration process is to synthesize a set of system elements into a realized system (product or service) that satisfies system requirements, architecture, and design.

4.8.1.2 Description Integration consists of progressively assembling the implemented system elements (hardware, software, and operational resources) that compose the SOI as defined and verifying the correctness of the static and dynamic aspects of interfaces between the implemented system elements. There is a strong focus on the interfaces to ensure that the intended operation of the system elements and interoperation with other systems is achieved. Any integration constraints are identified and considered during the definition of the requirements, architecture, and design. The interaction of the integration process with the system definition processes (i.e., system requirements definition, architecture definition, and design definition) early in the development is essential for avoiding integration issues during the system realization.

The integration process works closely with the verification and validation (V&V) processes. This process is iterated with the V&V processes, as appropriate. As the integration of system elements occurs, the verification process is invoked to check the correct implementation of architectural characteristics and design properties. The validation process may be invoked to check that the individual system elements provide the function intended. The process checks that all boundaries between system elements have been correctly identified and described, including physical, logical, and human– system interfaces and interactions (physical, sensory, and cognitive), and that all system element functional, performance, and design requirements and constraints are satisfied.

Figure 4.12 is the IPO diagram for the integration process.

4.8.1.3 Inputs/Outputs Inputs and outputs for the integration process are listed in Figure 4.12. Descriptions of each input and output are provided in Appendix E.

4.8.1.4 Process Activities The integration process includes the following activities:

- *Prepare for integration.*

4.8 综合流程

4.8.1 概览

4.8.1.1 目的 如 ISO/IEC/IEEE 15288 所述，

[6.4.8.1] 综合流程的目的是将系统元素的集合综合成为满足系统需求、架构和设计的实现的系统（产品或服务）中。

4.8.1.2 描述 综合包括按照定义逐步地聚集已实现的构成 SOI 的系统元素（硬件、软件和运行资源）以及验证已实现的系统元素之间的接口的静态和动态方面的正确性。对接口强烈关注，以便确保实现系统元素的预计运行以及与其他系统的交互运行。在需求、架构和设计的定义期间，任何综合约束都要被识别和考虑。在开发早期，综合流程与系统定义流程（即，系统需求定义、架构定义和设计定义）的交互对于避免在系统实现期间出现综合问题是根本性的。

综合流程与验证和确认（V&V）流程密切合作。该流程与 V&V 流程视情迭代。随着系统元素综合的发生，验证流程被调用以检查架构特征和设计特性的正确实现。确认流程也可被调用来检查单个的系统元素提供了意图的功能。该流程检查系统元素之间的所有边界已被正确地识别和描述，包括物理的、逻辑的和人员与系统的接口与交互（生理的、感觉的和认知的），以及检查满足所有系统元素的功能、性能和设计需求与约束。

图 4.12 是综合流程的 IPO 图。

4.8.1.3 输入/输出 图 4.12 列出综合流程的输入和输出。附录 E 提供每个输入和输出的描述。

4.8.1.4 流程活动 综合流程包括以下活动：

- 为综合的准备。

TECHNICAL PROCESSES

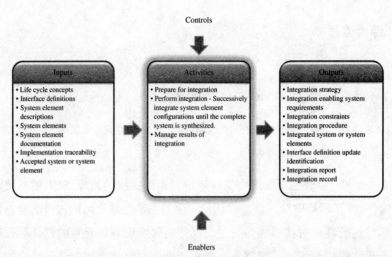

FIGURE 4.12 IPO diagram for the integration process. INCOSE SEH original figure created by Shortell and Walden. Usage per the INCOSE notices page. All other rights reserved.

- Define critical checkpoints to provide assurance of the correct behavior and operation of interfaces and functions of the system elements.

- Establish the integration strategy that minimizes integration time, costs, and risks:

    ° Define an optimized sequence order of assembly aggregates, composed of system elements, based on the system architecture definition, and on appropriate integration approaches and techniques.

    ° Define the configurations of the aggregates to be built and verified (depending on sets of parameters).

    ° Define the assembly procedures and related enablers.

- Identify integration constraints on the SOI, arising from the integration strategy, to be incorporated in the system requirements, architecture, and design (this includes requirements such as accessibility, safety for integrators, and required interconnections for sets of implemented system elements and for enablers).

- The acquisition of the enablers can be done through various ways such as rental, procurement, development, reuse, and subcontracting. An enabler may be a complete enabling system developed as a separate project from the project of the SOI.

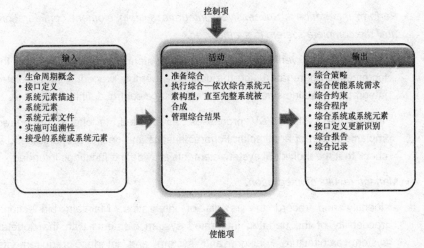

图 4.12 综合流程的 IPO 图。INCOSE SEH 原始图由 Shortell 和 Walden 创建。按照 INCOSE 通知页使用。版权所有。

- 定义关键检查点以为接口的正确行为和运行以及系统元素功能的提供保证。

- 建立使综合时间、成本和风险最小化的综合策略：

  ○ 基于系统架构定义和恰当的综合途径与技术，对由系统元素所构成的聚集定义一种优化的顺序。

  ○ 定义待构建和验证的聚合体的构型配置（取决于参数集合）。

  ○ 定义汇集的程序和相关使能项。

- 识别综合策略产生的将要纳入系统需求、架构和设计（这包括例如可达性、综合者的安全以及所实施的系统元素集与使能项的所需互连方面的要求）中的对 SOI 的综合约束。

- 使能项的采办可以通过各种不同的方式来完成，如租赁、采购、开发、复用和转包。使能项可能是作为与 SOI 项目独立项目而被开发的完整使能系统。

TECHNICAL PROCESSES

- *Perform integration - Successively integrate system element configurations until the complete system is synthesized.*

  – Assemble the verified and validated system elements to form the incremental aggregate using the defined assembly procedures, the related integration enabling systems, and the interface control definitions.

  – Invoke the system V&V processes, as needed, to check the correct implementation of architectural characteristics and design properties and to check that the individual system elements provide the functions intended.

- *Manage results of integration.*

  – Identify and record the results of integration. Maintain bidirectional traceability of the updated integrated system elements with the updated system architecture, design, and system and interface requirements. Maintain the records, including configuration updates, per organizational policy.

  – Record anomalies observed during the integration process (identifying corrective actions or improvements), and resolve them using the quality assurance process (see Section 5.8).

  – Update the integration strategy and schedule according to the progress of the project; in particular, the order of system elements assembly can be redefined or rescheduled because of unexpected events or unavailability of system elements as planned.

  – Coordinate integration activities with the project manager (e.g., for scheduling, acquisition of enablers, hiring of qualified personnel and resources), the architects or designers (e.g., for understanding of the architecture, errors, defects, nonconformance reports), and the configuration manager (e.g., for versions of submitted elements, architecture and design baselines, enablers, assembly procedures).

Common approaches and tips:

- Define an integration strategy that accounts for the schedule of availability of system elements (including the personnel that will use, operate, maintain, and sustain the system) and is consistent with defect/fault isolation and diagnosis practices.

- Development of integration enablers such as tools and facilities can take as long as the system itself. Development should be started as early as possible and as soon as the preliminary architecture definition is frozen.

- *实施综合 - 依次综合系统元素构型配置，直至完整的系统被集成。*

  – 使用定义的汇集程序、相关综合使能系统和接口控制定义来汇集经验证和确认的系统元素，以形成逐步的聚合。

  – 依需要调用系统 V&V 流程以检查架构特征和设计特性的正确实现，并检查单个系统元素提供意图的功能。

- *管理综合的结果。*

  – 识别并记录综合的结果。保持更新的综合系统元素与更新的系统架构、设计及系统与接口需求的双向可追溯性。按照组织的方针保持记录，包括构型配置的更新。

  – 记录综合流程中所观察到的异常（识别纠正措施或改进）并使用质量保证流程（见 5.8 节）化解异常。

  – 按照项目进展更新综合策略和进度；特别是，由于出现非预期事件或计划的系统元素不可用，则可以对系统元素汇集的顺序重新定义或重新安排。

  – 与项目经理（如时间安排、使能项的采办、合格人员的雇用和资源）、架构师或设计者（例如，对架构、误差、缺陷、不符合性报告的理解）及构型配置管理者（如已提交元素的版本、架构和设计基线、使能项、汇集程序）协调综合活动。

常用方法和建议：

- 定义综合策略，用于说明系统元素（包括使用、运行、维护和维持系统的人员）的可用性的进度并与缺陷/故障隔离和诊断的实际一致。
- 综合使能项（如工具和辅助设施）的开发会花费与系统本身同样长的时间。应尽可能早地在初步架构定义—冻结时就开始开发。

TECHNICAL PROCESSES

- The integration process of complex systems cannot be easily foreseen, and its progress may be difficult to control and observe. Therefore, it is recommended to plan integration with specific margins using flexible approaches and techniques, integrating sets by similar technologies, for example.

- Integrate aggregates in order to detect faults more easily. The use of the coupling matrix technique applies for all strategies and especially for the bottom-up integration strategy (see Section 4.8.2.3).

4.8.2 Elaboration

4.8.2.1 Concept of "Aggregate" The physical integration of a system is based on the notion of "aggregate." An aggregate is made up of several implemented system elements and their physical interfaces (system elements and connectors). Each aggregate is characterized by a configuration that specifies the implemented system elements to be physically assembled and their configuration status. A set of verification actions is applied on each aggregate. To perform these verification actions, a verification configuration that includes the aggregate plus verification tools is constituted. The verification tools are enabling elements and can be simulators (simulated system elements), stubs or caps, activators (launchers, drivers), harnesses, measuring devices, etc.

4.8.2.2 Integration by Level of System According to the Vee model, the system definition (top-down branch) is done by successive levels of decomposition; each level corresponds to physical architecture of systems and system elements. The integration (bottom-up branch) consists in following the opposite way of composition level by level.

On a given level, integration of implemented system elements is done on the basis of the physical architecture, using integration techniques or approaches as presented in the next section.

4.8.2.3 Integration Strategy and Approaches The integration of implemented system elements is performed according to a predefined strategy. The strategy relies on the way the architecture of the system has been defined. The strategy is described in an integration plan that defines the configuration of expected aggregates of implemented system elements, the order of assembly of these aggregates to carry out efficient verification actions and validation actions (e.g., inspections or tests). The integration strategy is thus elaborated in coordination with the selected verification strategy and validation strategy (see Sections 4.9 and 4.11).

- 复杂系统的综合流程无法容易地预见，且其进展可能难以控制和观察。因此，建议具有特定裕度的综合计划和采用灵活的途径和技术，例如采用相似技术的综合集合。

- 对聚合体进行综合是为了更容易地检测故障。耦合矩阵技术的使用适用于所有策略，尤其适用于自底向上的综合策略（见 4.8.2.3 节）。

4.8.2 详细阐述

4.8.2.1 "聚合体"的概念 系统的物理综合基于"聚合体"的观念。一个聚合体由多个已实现的系统元素及其物理接口（系统元素和连接器）组成。每个聚合体由一种构型配置所特征化，其规定了将要被物理汇集的已实现的系统元素及其构型配置状态。一个验证措施集合被应用到每个聚合体上。为实施这些验证措施，构建一个包括聚合体以及验证工具的验证构型。验证工具是使能元素，并且可以是仿真器（仿真的系统元素）、桩程序/数据包、执行器（发射器、驱动器）、控制机柜、测量装置等。

4.8.2.2 按系统的层级综合 按照 V 形模型，通过依次的分解层级来完成系统定义（自顶向下的分支）；每个层级对应于系统及系统元素的物理架构。综合（自底向上的分支）包括遵照与逐层级合成相反的方式。

在给定层级上，基于物理架构，使用下节提出的综合技术或途径来完成已实现的系统元素的综合。

4.8.2.3 综合策略和途径 按照预定义的策略实施已实现的系统元素的综合。该策略取决于系统架构定义的方式。综合计划中描述了该策略，对已实现的系统元素的预期聚合的构型配置以及这些集合体的汇集顺序进行定义，以便执行高效的验证措施和确认措施（如检验或测试）。因此，与所选择的验证策略和确认策略（见 4.9 节和 4.11 节）相协调对综合策略进行详细阐述。

TECHNICAL PROCESSES

To define an integration strategy, one can use one or several possible integration approaches and techniques. Any of these may be used individually or in combination. The selection of integration techniques depends on several factors, in particular the type of system element, delivery time, order of delivery of system elements, risks, constraints, etc. Each integration technique has strengths and weaknesses, which should be considered in the context of the SOI. Some integration techniques are summarized hereafter:

- *Global integration*—Also known as "big-bang integration"; all the delivered implemented system elements are assembled in only one step. This technique is simple and does not require simulating the system elements not being available at that time. But it is difficult to detect and localize faults; interface faults are detected late. It should be reserved for simple systems, with few interactions and few system elements without technological risks.

- *Integration "with the stream"* —The delivered system elements are assembled as they become available. This technique allows starting the integration quickly. It is complex to implement because of the necessity to simulate the system elements not yet available. It is impossible to control the end-to-end "functional chains"; so global tests are postponed very late in the schedule. It should be reserved for well-known and controlled systems without technological risks.

- *Incremental integration*—In a predefined order, one or a very few system elements are added to an already integrated increment of system elements. This technique allows a fast localization of faults: a new fault is usually localized in lately integrated system elements or dependent on a faulty interface. It requires simulators for absent system elements and many test cases: each system element addition requires the verification of the new configuration and regression testing. This technique is applicable to any type of architecture.

- *Subset integration*—System elements are assembled by subsets (a subset is an aggregate), and then subsets are assembled together; could be called "functional chains integration." This technique saves time due to parallel integration of subsets; delivery of partial products is possible. It requires less means (enablers) and fewer test cases than integration by increments. The subsets may be defined during the architecture definition; they may correspond to subsystems/ systems as defined in the architecture.

为定义综合策略，可以使用一个或多个可能的综合途径和技术。任何途径和技术都可单独地或以组合的方式使用。综合技术的选择取决于多种因素，特别是系统元素的类型、交付时间、系统元素交付的顺序、风险、约束等。每个综合技术都具有优势与劣势，应在 SOI 的背景环境中予以考虑。现将一些综合技术总结如下：

- *全局综合*——亦称为"大爆炸式的综合"；所有已交付的已实现的系统元素仅在一个步骤中汇集。该技术简单，且不需要仿真当时不可用的系统元素。但难以检测和定位故障；接口故障很晚才被检测到。应当在具有较少的交互和较少的系统元素且没有技术风险时为简单的系统预订该技术。

- *"随流"综合*——当交付的系统元素可用时进行汇集。该技术允许综合快速的开始。由于要仿真尚不可用的系统元素的必要性，因此实现复杂。控制端对端的"功能链"是不可能的；因此全局测试在进度中被推迟到很晚才进行。应当在没有技术风险时为非常清楚的受控的系统预订该技术。

- *渐进式综合*——按照预先定义的顺序，将一个或极少数的系统元素添加到已经综合的系统元素增量中。该技术允许快速的故障定位：新的故障通常定位在最近综合的系统元素中或依赖于一个错误接口。它要求对不存在的系统元素和许多测试用例使用仿真器；每个系统元素的添加都要求对新的构型配置的验证以及回归测试。该技术适用于任何类型的架构。

- *子集综合*——系统元素由子集（一个子集是一个集合体）汇集，然后将子集汇集在一起；可称为"功能链综合"。由于子集的平行综合，该技术节省时间；部分产品的交付是可能的。与渐进式综合相比，它需要较少的装置（使能项）和较少的测试用例。子集可在架构定义期间进行定义；它们对应于架构中定义的子系统/系统。

TECHNICAL PROCESSES

- *Top-down integration*—System elements or aggregates are integrated in their activation or utilization order. Availability of a skeleton of the system and early detection of architectural faults are possible; definition of test cases is close to reality; the reuse of test data sets is possible. But many stubs/caps need to be created; it is difficult to define test cases of the leaf system elements (lowest level). This technique is mainly used in intensive software systems. It starts from the system element of higher level; system elements of lower level are added until all leaf system elements are incorporated.

- *Bottom-up integration*—System elements or aggregates are integrated in the opposite order of their activation or utilization. The definition of test cases is easy; early detection of faults (usually localized in the leaf system element) is possible; the number of simulators to be used is reduced; an aggregate can be a subsystem. But the test cases shall be redefined for each step; drivers are difficult to define and realize; system elements of lower levels are "overtested"; this technique does not allow quick detection of the architectural faults. It is used for intensive software systems and for any kind of hardware system.

- *Criterion-driven integration*—The most critical system elements compared to the selected criterion are first integrated (e.g., dependability, complexity, technological innovation). The criteria are generally related to risks. This technique allows testing early and intensively critical system elements; early verification of architecture and design choices is possible. But the test cases and test data sets are difficult to define.

- *Reorganization of coupling matrices*—As noted in Section 4.4.2.6, coupling matrices are useful for highlighting interfaces during architecture definition as well as during integration. The integration strategy is defined and optimized by reorganizing the coupling matrix in order to group the system elements into aggregates and minimize the number of interfaces to be verified between aggregates. When verifying the interactions between aggregates, the matrix is an aid for fault detection. If by adding a system element to an aggregate, an error is detected, the fault can be either related to the system element, or to the aggregate, or to the interfaces. If the fault is related to the aggregate, it can relate to any system element or any interface between the system elements internal to the aggregate.

Usually, the integration strategy is defined as a combination of these approaches and techniques in order to optimize the integration work. The optimization takes into account the realization time of the system elements, their delivery scheduled order, their level of complexity, the technical risks, the availability of assembly tools, cost, deadlines, specific personnel capability, etc.

- *自顶向下综合*——系统元素或集合体按其激活或效用发生的顺序来综合。系统结构框架的可用性及架构错误的早期检测都是可能的；测试用例的定义接近于现实；测试数据集的复用是可能的。但很多桩程序/数据包需要创建；难于定义叶系元素（最低层级）的测试用例。该技术主要用在软件密集系统中。它从较高层级的系统元素开始；增加较低层级的系统元素，直至纳入所有叶系元素。

- *自底向上综合*——系统元素或聚合体按其激活或效用发生的相反顺序来综合。测试用例的定义很容易；错误（通常定位在叶系元素中）的早期检测是可能的；减少了待使用的仿真器数量；聚合体可以是子系统。但应当为每个步骤重新定义测试用例；驱动器难以定义和实现；较低层级的系统元素被"过度测试"；该技术不允许对架构错误快速检测。它用于软件密集系统和任何种类的硬件系统。

- *准则驱动的综合*——与所选定的准则相比，大多数关键的系统元素首先被综合（如可依赖性、复杂性、技术创新）。准则一般与风险相关。该技术允许对早期的特别关键的系统元素进行测试；架构和设计选择的早期验证是可能的。但测试用例和测试数据集难以定义。

- *耦合矩阵的重组*——如 4.4.2.6 节指出，耦合矩阵可用于在架构定义期间以及综合期间突出强调接口。通过重组耦合矩阵来定义和优化综合策略，以便将系统元素分组为聚合体并使聚合体之间待验证的接口数量最小化。当验证聚合体之间的交互时，矩阵是错误检测辅助工具。如果通过在聚合体中增加一个系统元素检测到一个误差，那么错误可以与系统元素相关，或者与聚合体相关，或者与接口相关。如果错误与聚合体相关，则它可以与任何系统元素或集合内部系统元素之间的任何接口相关。

通常，综合策略被定义为这些途径和技术的组合，以便优化综合工作。优化考虑到系统元素的实现时间、它们的交付的时间顺序、它们的复杂性层级、技术风险、汇集工具的可用性、成本、最后期限、特定的人员能力等。

TECHNICAL PROCESSES

4.9 VERIFICATION PROCESS

4.9.1 Overview

4.9.1.1 Purpose As stated in ISO/IEC/IEEE 15288,

> [6.4.9.1] The purpose of the Verification process is to provide objective evidence that a system or system element fulfils its specified requirements and characteristics.

4.9.1.2 Content/Description As described here, this process is an instance of a verification process applied to a SOI, or any system or system element that compose it, to establish that it has been "built right."

The verification process can be applied to any engineering element that has contributed to the definition and realization of the system itself (e.g., verification of a system requirement, a function, an input/output flow, a system element, an interface, a design property, a verification procedure). The purpose of the verification process is to provide evidence that no error/defect/fault has been introduced at the time of any transformation of inputs into outputs; it is used to confirm that this transformation has been made "right" according to the requirements and selected methods, techniques, standards, or rules. As is often stated, verification is intended to ensure that the "product is built right," while validation is intended to ensure that the "right product is built." Verification is a transverse activity to every life cycle stage of the system. In particular during the development of the system, verification applies onto any activity and product resulting from the activity.

Figure 4.13 is the IPO diagram for the verification process.

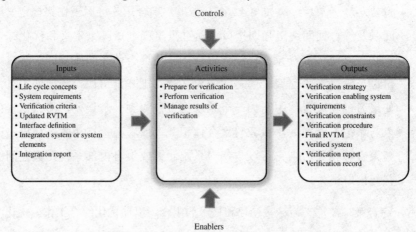

FIGURE 4.13 IPO diagram for the verification process. INCOSE SEH original figure created by Shortell and Walden. Usage per the INCOSE notices page. All other rights reserved.

4.9 验证流程

4.9.1 概览

4.9.1.1 目的 如 ISO/IEC/IEEE 15288 所述,

[6.4.9.1] 验证流程的目的是提供证明系统或系统元素满足其特定需求和特征的客观证据。

4.9.1.2 内容/描述 如此所描述,此流程是适用于 SOI 或任何系统或构成该系统的系统元素的一个验证流程的实例,用以确定系统或系统元素已经被"正确地构建"。

验证流程可应用于有助于系统本身的定义和实现(如系统需求、功能、输入/输出流、系统元素、接口、设计特性、验证程序的验证)的任何工程元素。验证流程的目的是提供证据以证明在输入到输出的任何转换的时刻没有引入误差/缺陷/错误;它用于根据需求及选定的方法、技术、标准或规则确认这种转换是"正确的"。正如通常所述的,验证旨在确保"产品被正确地构建",而确认旨在确保"正确的产品被构建"。验证是系统每一生命周期阶段的横向活动。特别是在系统开发期间,验证应用到任何活动以及活动产生的任何产品。

图 4.13 是验证流程的 IPO 图。

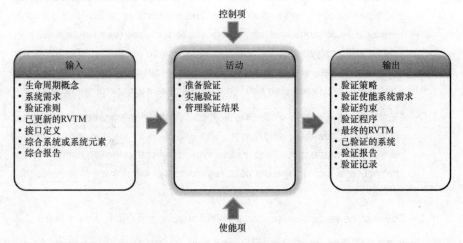

图 4.13 验证流程的 IPO 图。INCOSE SEH 原始图由 Shortell 和 Walden 创建。按照 INCOSE 通知页使用。版权所有。

TECHNICAL PROCESSES

4.9.1.3 Inputs/Outputs Inputs and outputs for the verification process are listed in Figure 4.13. Descriptions of each input and output are provided in Appendix E.

4.9.1.4 Process Activities The verification process includes the following activities:

- *Prepare for verification.*
  - Develop a strategy that prioritizes the verification actions to minimize costs and risks while maximizing operational coverage of system behaviors:
    ° Establish a list of the items for verification, including requirements, architectural characteristics, or design properties, and define the corresponding verification actions. The approach to verification should be identified and documented at the time that a requirement is first documented to ensure that the requirement as written is indeed verifiable. This may require restating a requirement or decomposing it into several verifiable statements. (For example, how do you satisfy the legitimate requirement of a system being "user friendly" ?)
    ° Establish a list of verification constraints that need to be considered. The constraints could impact the implementation of the verification actions and include contractual constraints, limitations due to regulatory requirements, cost, schedule, feasibility to exercise a function (such as in some ordinance), safety considerations, physical configurations, accessibility, etc.
    ° Considering the constraints, plan for the methods or techniques that will be applied for each verification action. The methods or techniques generally include inspection, analysis, demonstration, or test (see Section 4.9.2.2). Note that analysis often is considered to include modeling, simulation, and analogy when identifying the verification methods or techniques and success criteria are also defined that indicates successful verification.
    ° Establish the scope of the verification. Verification consumes resources: time, labor, facilities, and funds. The selection of what must be verified should be made according to the type of system, the objectives of the project, and the acceptable risks regarding the withdrawal of a verification action.
  - Develop the verification procedures that support the verification actions.
    ° Schedule the execution of verification actions in the project steps and define the configura-tion of submitted items to verification actions.

4.9.1.3 输入/输出 图 4.13 列出验证流程的输入和输出。附录 E 提供每个输入和输出的描述。

4.9.1.4 流程活动 验证流程包括以下活动：

- *准备的验证。*

    – 开发对验证措施进行优先级排序的一种策略，在使系统行为的运行覆盖范围最大化的同时使成本和风险最小化：

    ○ 建立验证项列表，包括需求、架构特征或设计特性，并定义相应的验证措施。在将一个需求首次文件化时就对验证方法进行识别和文件化，以确保书面的需求的确是可验证的。这可能要求对需求重述或将需求分解成多个可验证的申明。（例如，你如何满足"用户友好"的系统的合法需求？）

    ○ 建立需要考虑的验证约束列表。约束可能影响验证措施的实现，且包括合同约束、由于监管需求产生的限制、成本、进度、执行某种功能的可行性（如在某种条例中）、安全考虑因素、物理构型配置、可达性等。

    ○ 考虑到约束，要为应用于每个验证措施的方法或技术做计划。方法或技术一般包括检验、分析、演示或测试（见 4.9.2.2 节）。应注意，在识别验证方法或技术时往往认为分析包括建模、仿真和模拟，且还定义了表明成功验证的成功准则。

    ○ 建立验证的范围。验证消耗资源：时间、劳动、设施和资金。应当根据系统的类型、项目的目标以及考虑验证措施撤回的可接受风险来做出关于必须验证什么的选择。

    – 开发支持验证措施的验证程序。

    ○ 对项目步骤中的验证措施的执行进度安排并对验证措施提交项的构型配置进行定义。

- Identify verification constraints on the system or system elements, arising from the verification strategy, that relate to specific system requirements, architecture elements, or design elements. Typical constraints include performance characteristics, accessibility, and interface characteristics. Provide the constraint information for consideration in the system requirements definition, architecture definition, and design definition processes.

- Ensure that the necessary enabling systems, products, or services required for the verification actions are available, when needed. The planning includes the identification of requirements and interfaces for the enablers. The acquisition of the enablers can be done through various ways such as rental, procurement, development, reuse, and subcontracting. An enabler may be a complete enabling system developed as a separate project from the project of the SOI.

- *Perform verification.*

  - Implement the verification plan developed in the preceding subsection. That plan includes detailed descriptions for the selected verification actions:

    ° Item to be verified

    ° Expected results and success criteria

    ° Selected verification method or technique

    ° The data needed

    ° The corresponding enabling systems, products, or services

  - Using the verification procedures, execute the verification actions and record the results.

  - Analyze the verification results against any established expectations and success criteria to determine whether the element being verified indicates conformance.

- *Manage results of verification.*

  - Identify and record verification results and enter data in the Requirements Verification and Traceability Matrix (RVTM). Maintain the records per organizational policy.

  - Record anomalies observed during the verification process, and analyze and resolve the anomalies (corrective actions or improvements) using the quality assurance process (see Section 5.8).

- 识别出由验证策略产生的与特定系统需求、架构元素或设计元素相关的对系统或系统元素的验证约束。典型的约束包括性能特征、可达性及接口特征。在系统需求定义流程、架构定义流程和设计定义流程中提供需考虑的约束信息。
- 确保在需要时验证措施所需的必要的使能系统、产品或服务是可用的。规划包括使能项的需求和接口的识别。使能项的采办可以通过各种不同的方式来完成,如租赁、采购、开发、复用和转包。使能项可能是作为与 SOI 项目独立项目而被开发的完整的使能系统。

- *实施验证。*
  - 实现前一小节中开发的验证计划。该计划包括对选定的验证措施的详细描述:
    - 待验证项
    - 期望的结果和成功准则
    - 选定的验证方法或技术
    - 需要的数据
    - 相应的使能系统、产品或服务
  - 使用验证程序来执行验证措施并记录结果。
  - 对照已建立的任何预期和成功准则分析验证结果以确定所验证的元素是否表明符合性。

- *管理验证的结果。*
  - 识别和记录验证结果,并将数据输入到需求验证和可追溯性矩阵(RVTM)中。按照组织的方针维持记录。
  - 记录验证流程期间观察到的异常,并使用质量保证流程(见 5.8 节)来分析和化解该异常(纠正措施或改进)。

TECHNICAL PROCESSES

- Establish and maintain bidirectional traceability of the verified system elements with the system architecture, design, and system and interface requirements that are needed for verification.
- Provide baseline information for configuration management.
- Update the verification strategy and schedule according to the progress of the project; in particular, planned verification actions can be redefined or rescheduled as necessary.
- Coordinate verification activities with the project manager (e.g., for scheduling, acquisition of enablers, hiring of qualified personnel and resources), the architects or designers (e.g., for errors, defects, nonconformance reports), and the configuration manager (e.g., for versions of submitted items, requirements, architecture and design baselines, enablers, verification procedures).

Common approaches and tips:

- Beware the temptation to reduce the number of verification actions due to budget or schedule overruns. Remember that discrepancies and errors are more costly to correct later in the system life cycle.
- In the progress of the project, it is important to know, at any time, what has not been verified in order to estimate the risks about possibly dropping out some verification actions.
- Each system requirement should be quantitative, measurable, unambiguous, understandable, and testable. It is generally much easier and more cost effective to ensure that requirements meet these criteria while they are being written. Requirements adjustments made after implementation and/or integration are generally much more costly and may have wide-reaching redesign implications. There are several resources that provide guidance on creating appropriate requirements (see the System Requirements Definition Process, Section 4.3).
- Avoid conducting verification only late in the schedule when there is less time to handle discrepancies.
- Testing the actual system is expensive and is not the only verification technique. Other techniques such as simulation, analysis, review, etc. can be used on other engineering elements representing the SOI such as models, mock-ups, or partial prototypes.

- 建立和维持所验证的系统元素与系统架构、设计以及验证所需的系统和接口需求的双向可追溯性。

- 提供构型配置管理的基线信息。

- 按照项目进展更新验证策略和进度；特别是，可以根据需要对计划的验证措施重新定义或重新安排进度。

- 与项目经理（如时间安排、使能项的采办、合格人员的雇用和资源）、架构师或设计者（如误差、缺陷、不一致性报告）及构型配置管理者（如已提交项的版本、需求、架构和设计基线、使能项、验证程序）协调验证活动。

常用方法和建议：

- 谨防诱惑，不要因为预算超支或进度超期而减少验证措施的数量。应牢记，在系统生命周期的后期纠正差异和差错所需的费用会更高。

- 在项目的进展中，重要的是随时知晓什么尚未被验证，以便估计关于可能退出某些验证措施的风险。

- 每个系统需求应当是定量的、可测量的、无歧义的、可理解的且可测试的。通常，确保需求在编写的同时满足这些准则更容易且更具成本效益。在实现和/或综合之后做出的需求调整通常成本更为昂贵，并有可能导致重大的重新设计。有很多为创建适当需求提供指南的资源（见4.3 节，系统需求定义）。

- 当仅有很少时间去处理差异时，要避免仅在进度的后期进行验证

- 测试实际的系统是昂贵的，并且不是唯一的验证技术。其他技术，如仿真、分析、评审等，可用在表征 SOI 的其他工程元素上，如模型、样机或部分原型。

TECHNICAL PROCESSES

4.9.2 Elaboration

4.9.2.1 Notion of Verification Action A verification action describes what must be verified (e.g., a requirement, a characteristic, or a property as reference), on which item (e.g., requirement, function, interface, system element, system), the expected result (deduced from the reference), the verification technique to apply (e.g., inspection, analysis, demonstration, test), and on which level of decomposition of the system (e.g., SOI, intermediate level system element, leaf level system element).

The definition of a verification action applied to an engineering item (e.g., stakeholder requirement, system requirement, function, interface, system element, procedure, and document) includes the identification of the item on which the verification action will be performed, the reference used to define the expected result, and the appropriate verification technique.

The performance of a verification action onto the submitted item provides an obtained result which is compared with the expected result. The comparison enables the correctness of the item to be determined (see Fig. 4.14).

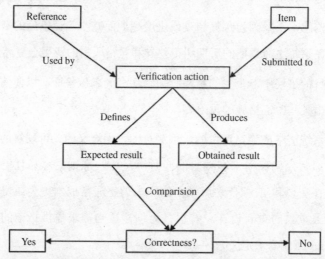

FIGURE 4.14 Definition and usage of a verification action. Reprinted with permission from Alain Faisandier. All other rights reserved.

Examples of verification actions:

- *Verification of a stakeholder requirement or a system requirement*—To check the application of syntactic and grammatical rules and characteristics defined in the stakeholder needs and requirements definition process and the system requirements definition process such as necessity, implementation free, unambiguous, consistent, complete, singular, feasible, traceable, and verifiable.

4.9.2 详细阐述

4.9.2.1 验证措施的观念 验证措施描述必须验证什么（如一个需求、一个特征或一个特性，作为参考）、关于什么项（如需求、功能，接口、系统元素、系统）、期望的结果（从参考中推断）、要应用的验证技术（如检验、分析、演示、测试）和关于系统分解的层级（如 SOI、中间层级系统元素、叶级系元素）。

应用于一个工程项（如利益攸关者需求、系统需求、功能、接口、系统元素、程序和文件）的验证措施的定义包括将要实施的验证措施项的识别、用于定义期望结果的参考以及恰当的验证技术。

验证措施在提交项上的实施提供获得的结果，将该结果与期望结果相比较。该比较能使待确定项的保持正确性（见图 4.14）。

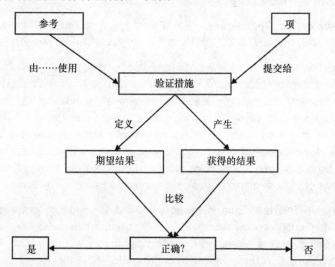

图 4.14 验证措施的定义和使用。经 Alain Faisandier 许可后转载。版权所有。

验证措施示例：

- *一个利益攸关者需求或一个系统需求的验证*——检查在利益攸关者需要和需求定义流程以及系统需求定义流程中所定义的句法和语法规则与特征的应用，如必要性、与实施无关的、无歧义的、一致的、完整的、唯一的、可行的、可追溯的和可验证的。

TECHNICAL PROCESSES

- *Verification of the architecture of a system*—To check the correct application of the appropriate patterns and heuristics used and the correct usage of modeling techniques or methods.

- *Verification of the design of a system element*—To check the correct usage of patterns, trade rules, or state of the art related to the concerned technology (e.g., software, mechanics, electronics, chemistry).

- *Verification of a system (product, service, or enterprise) or system element*—To check its realized characteristics or properties (e.g., as measured) against its specified requirements, expected architectural characteristics, and design properties (as described in the requirements, architecture, and design documents).

Considerations in selecting a verification approach include practical limitations of accuracy, uncertainty, repeatability that are imposed by the verification enablers, the associated measurement methods, and the availability, accessibility, and interconnection with the enablers.

4.9.2.2 Verification Techniques Basic verification techniques are as follows (IEEE 1012, 2012; ISO/IEC/ IEEE 29119, 2013; ISO/IEC/IEEE 29148, 2011):

- *Inspection*—This technique is based on visual or dimensional examination of an element; the verification relies on the human senses or uses simple methods of measurement and handling. Inspection is generally nondestructive and typically includes the use of sight, hearing, smell, touch, and taste; simple physical manipulation; mechanical and electrical gauging; and measurement. No stimuli (tests) are necessary. The technique is used to check properties best determined by observation (e.g., paint color, weight, documentation, listing of code). Peer reviews of process artifacts are also considered a type of inspection.

- *Analysis*—This technique is based on analytical evidence obtained without any intervention on the submitted element using mathematical or probabilistic calculation, logical reasoning (including the theory of predicates), modeling, and/or simulation under defined conditions to show theoretical compliance. Mainly used where testing to realistic conditions cannot be achieved or is not cost-effective.

- *Demonstration*—This technique is used to show correct operation of the submitted element against operational and observable characteristics without using physical measurements (no or minimal instrumentation or test equipment). It uses generally a set of actions selected to show that the element response to stimuli is suitable or to show that operators can perform their assigned tasks when using the element. Observations are made and compared with predetermined/expected responses.

- *一个系统架构的验证*——检查所使用的恰当特征模式和启发法的正确应用，以及建模技术或方法的正确使用。

- *一个系统元素设计的验证*——检查特征模式、权衡规则或与所考虑的相关技术（如软件、力学、电子、化学）有关的技术现状的正确使用。

- *一个系统（产品、服务或复杂组织体）或系统元素的验证*——对照其特定需求、预期的架构特征和设计特性（如需求、架构和设计文件中所述）检查其实现的特征或特性（如测量的）。

选择验证途径时的考虑因素包括验证使能项所施加的精度、不确定性、可重复性的实际限制、相关的测量方法、以及可用性、可达性及与使能项的相互连接。

4.9.2.2 验证技术 基本的验证技术如下（IEEE 1012, 2012；ISO/IEC/IEEE 29119, 2013；ISO/IEC/IEEE 29148, 2011）：

- *检验*——该技术基于元素的目视检查或尺寸检查；验证依赖于人的感觉或使用简单的测量和处理方法。检验一般是无损的，且通常包括视觉、听觉、嗅觉、触觉和味觉的使用，简单的物理操作；机械的和电气的计量；以及测量。激励（测试）是没有必要的。该技术用于检查由观察所确定的最好特性（如涂料颜色、重量、文件、代码列表）。流程制品的同行评审也被认为是一类检验。

- *分析*——该技术基于通过使用数学计算或概率计算、逻辑推理（包括预测理论）、建模和/或仿真在定义的条件下没有任何干预而获得的关于已提交元素的分析性证据，以便表明理论的合规性。主要用于现实条件的测试无法实现或不具成本效益的场合。

- *演示*——该技术用于对照运行特征和可观察的特征而不使用物理测量装置（没有仪器或测试设备或很少）就能展示已提交元素的正确运行。它一般使用一个选定的行动集合来展示元素对激励的响应是适合的或展示操作者在使用元素时能够实施分配给他们的任务。进行观察并与预定的/期望的响应相比较。

TECHNICAL PROCESSES

- *Test*—This technique is performed onto the submitted element by which functional, measurable characteristics, operability, supportability, or performance capability is quantitatively verified when subjected to controlled conditions that are real or simulated. Testing often uses special test equipment or instrumentation to obtain accurate quantitative data to be analyzed.

- *Analogy or similarity*—This technique (often considered as a type of analysis technique) is based on evidence of similar elements to the submitted element or on experience feedback. It is absolutely necessary to show by prediction that the context is invariant and that the outcomes are transposable (e.g., models, investigations, experience feedback). Analogy or similarity can only be used if the submitted element is similar in design, manufacture, and use; equivalent or more stringent verification actions were used for the similar element; and the intended operational environment is identical to or less rigorous than the similar element.

- *Simulation*—This technique (often considered as a type of analysis technique) is performed on models or mock-ups (not on the actual/physical elements) for verifying features and performance as designed.

- *Sampling*—This technique is based on verification of characteristics using samples. The number, tolerance, and other characteristics must be specified and be in agreement with the experience feedback.

Note: For techniques that do not include stimuli of the system element, no characteristics (exogenous attributes) can be observed only properties (endogenous attributes).

4.9.2.3 Integration, Verification, and Validation of the System There is sometimes a misconception that verification occurs after integration and before validation. In most of the cases, it is more appropriate to begin verification activities during development and to continue them into deployment and use.

Once system elements have been realized, their integration to form the whole system is performed. Integration assembles developed capabilities (via system elements) in preparation for verification actions as stated in the integration process (see Section 4.8).

4.9.2.4 Verification Level per Level Generally, the SOI has a number of layers of systems (made up of system elements at the next lower level). Thus, every system and system element is verified, and any findings possibly corrected before being integrated into the system of the higher level, as shown in Figure 4.15. In this figure, every time the term verify is used means that the corresponding verification process is invoked.

- *测试*——当服从于真实的或仿真的受控条件时，该技术可实施于所提交的元素，在定量验证功能、可测量特征、可操作性、可保障性或性能能力得到定量地验证。测试常常使用专用测试设备或仪器来获得待分析的精确的定量数据。

- *类比或相似性*——该技术（通常被视为一类分析技术）基于与已提交元素类似的元素的证据或基于经验反馈。它对于通过预测展示背景环境不变且结果可换位而言是绝对必要的（如模型、调查研究、经验反馈）。只有已提交元素在设计、制造和使用上类似，才可以使用类比或相似性；等效的或更严格的验证措施用于相似元素；意图的运行环境与相似元素相同严格或较低。

- *仿真*——该技术（通常被视为一类分析技术）在模型或原理原型上实施（不在实际的/物理元素上实施），用于按设计验证特征和性能。

- *抽样*——该技术基于使用样品对特征进行的验证。必须规定数量、公差及其他特征，并且必须与经验反馈一致。

注：对于不包括系统元素的激励的技术而言，观察不到特征（外生属性），仅观察到特性（内生属性）。

4.9.2.3　系统的综合、验证和确认　有时存在误解，即误认为验证发生在综合之后和确认之前。在大多数情况中，在开发期间开始验证活动并将验证活动持续投入到部署和使用更为恰当。

一旦实现了系统元素，则对它们进行综合以形成整体系统。在准备验证措施时，综合汇集已开发的能力（经由系统元素），如综合流程所述（见4.8节）。

4.9.2.4　逐层验证　通常，SOI 具有许多层系统（由下一较低层级上的系统元素组成）。因此，要验证每一系统和系统元素，并且在被综合到较高层级系统中之前对任何发现进行可能的纠正，如图 4.15 所示。在该图中，每次使用术语"验证"都意味着调用相应的验证流程。

TECHNICAL PROCESSES

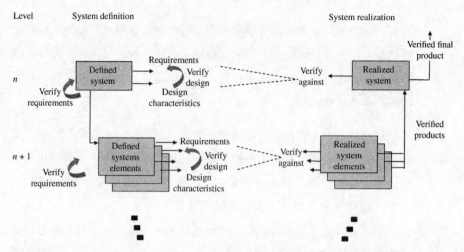

FIGURE 4.15 Verification level per level. Reprinted with permission from Alain Faisandier. All other rights reserved.

As necessary, systems and system elements are partially integrated in subsets (aggregates) in order to limit the number of properties to be verified within a single step. For each level, it is necessary to make sure by a set of verification actions that features stated at preceding level are not adversely affected. Moreover, a compliant result obtained in a given environment can turn noncompliant if the environment changes. So, as long as the system is not completely integrated and/or doesn't operate in the real operational environment, no result must be regarded as definitive.

4.10 TRANSITION PROCESS

4.10.1 Overview

4.10.1.1 Purpose As stated in ISO/IEC/IEEE 15288,

> [6.4.10.1] The purpose of the Transition process is to establish a capability for a system to provide services specified by stakeholder requirements in the operational environment.

Ultimately, the transition process enables the transfer of custody of the system and responsibility for system support from one organizational entity to another. This includes, but is not limited to, transfer of custody from the development team to the organizations that will subsequently operate and support the system. Successful conclusion of the transition process typically marks the beginning of the utilization stage of the SOI.

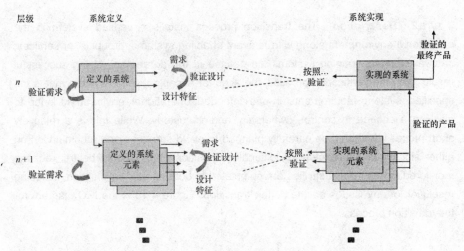

图 4.15 逐层验证。经 Alain Faisandier 许可后转载。版权所有。

必要时，将系统和系统元素部分地综合在子集（聚合体）中，以便限制单一步骤内待验证的特性数量。对于每一层级而言，有必要通过一个验证措施集合确保前一层级所说明的特征不会受到不利影响。此外，如果环境发生变化，在给定环境中所获得的符合性结果可转变为不符合。因此，只要系统没有被完全综合且/或不在真实的运行环境中运行，一定不能认为结果是决定性的。

4.10 转移流程

4.10.1 概览

4.10.1.1 目的　如 ISO/IEC/IEEE 15288 所述，

[6.4.10.1] 转移流程的目的在于为系统建立能够在运行环境中提供由利益攸关者需求所规定的服务的能力。

最终，转移流程使得能够将一个组织实体的系统监护和系统支持职责转移给另一组织实体。这包括但不限于将开发团队的监护转移给将随后运行和保障该系统的组织。转移流程的成功结束通常标志着 SOI 的使用阶段的开始。

TECHNICAL PROCESSES

4.10.1.2 Description The transition process installs a verified system in the operational environment along with relevant enabling systems, products, or services, such as operator training systems, as defined in the agreement. Using successful results from the verification process, the acquirer accepts that the system meets the specified system requirements in the intended operational environment prior to allowing a change in control, ownership, and/or custody. While this is a relatively short process, it should be carefully planned to avoid surprises and recrimination on either side of the agreement. Additionally, transition plans should be tracked and monitored to ensure all activities are completed to both parties' satisfaction, including resolution of any issues arising during transition. Figure 4.16 is the IPO diagram for the transition process.

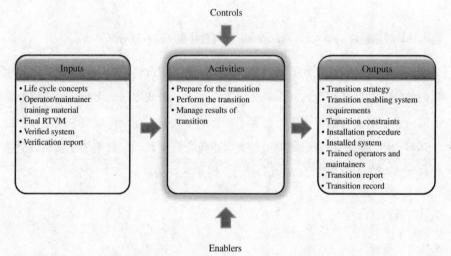

FIGURE 4.16 IPO diagram for the transition process. INCOSE SEH original figure created by Shortell and Walden. Usage per the INCOSE notices page. All other rights reserved.

4.10.1.3 Inputs/Outputs Inputs and outputs for the transition process are listed in Figure 4.16. Descriptions of each input and output are provided in Appendix E.

4.10.1.4 Process Activities The transition process includes the following activities:

- *Prepare for the transition.*
  - Plan for the transition of the system. The strategy should include operator training, logistics support, delivery strategy, and problem rectification/resolution strategy.

4.10.1.2 描述 转移流程在运行环境中安装一个经验证的系统，以及相关的使能系统、产品或服务，如协议中定义的操作者培训系统。通过使用验证流程中的成功结果，采办方在允许改变控制、所有权和/或监护之前，接受系统在意图的运行环境中满足规定的系统需求。尽管该工作是一个相对较短的流程，为避免协议任一方出现意外和相互指责，应予以仔细计划。此外，为确保所有活动完成后双方都满意，应跟踪并监控转移计划，包括对转移期间出现的任何问题的解决。图 4.16 是转移流程的 IPO 图。

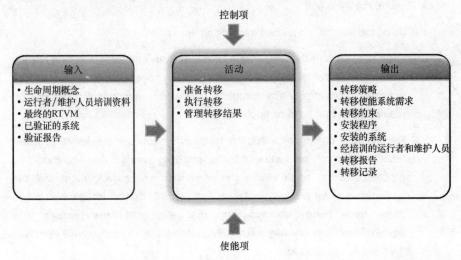

图 4.16 转移流程的 IPO 图。INCOSE SEH 原始图由 Shortell 和 Walden 创建。按照 INCOSE 通知页使用。版权所有。

4.10.1.3 输入/输出 图 4.16 列出转移流程的输入和输出。附录 E 提供每个输入和输出的描述。

4.10.1.4 流程活动 转移流程包括以下活动：

- *准备转移。*

    - 计划系统的转移。策略应包括运行者培训、后勤保障、交付策略和问题纠正/解决策略。

- Develop installation procedures.

- Ensure that the necessary enabling systems, products, or services required for transition are available, when needed. The planning includes the identification of requirements and interfaces for the enablers. The acquisition of the enablers can be done through various ways such as rental, procurement, development, reuse, and subcontracting. An enabler may be a complete enabling system developed as a separate project from the project of the SOI.

- *Perform the transition.*

  - Using the installation procedures, install the system.

  - Train the users in the proper use of the system and affirm users have the knowledge and skill levels necessary to perform operation and maintenance activities. This includes a complete review and handoff of operator and maintenance manuals, as applicable.

  - Receive final confirmation that the installed system can provide its required functions and can be sustained by the enabling systems and services. This process typically ends with a formal, written acknowledgement that the system has been properly installed and verified, that all issues and action items have been resolved, and that all agreements pertaining to development and delivery of a fully supportable system have been fully satisfied or adjudicated.

  - After the demonstration of functionality in the operational site and any review for operational readiness, the system can be placed into service.

- *Manage results of transition.*

  - Postimplementation incidents and problems are captured and may lead to corrective actions or changes to the requirements. The quality assurance process is used for the incident and problem resolution that is reported during performance of the transition process.

  - Record anomalies observed during the transition process. These provide awareness of the results, information needed to address anomalies, and a historical record. Anomalies may stem from the transition strategy, supporting enabling systems, interfaces, etc. The project assessment and control process is used to analyze the anomalies and determine what, if any, action is needed.

- 开发安装程序。

- 确保用于转移所必要的使能系统、产品或服务是可用的。规划包括使能项的需求和接口的识别。使能项的采办可以通过各种不同的方式来完成，如租赁、采购、开发、复用和转包。使能项可能是被开发为与 SOI 项目相独立项目的完整使能系统。

- *实施转移。*

  - 使用安装程序安装系统。

  - 培训用户以正确地使用系统，确定用户具有实施运行和维护活动所必需的知识和技能水平。若适用，这包括对操作者和维护手册的全面评审和转换。

  - 接收最终确认，即安装的系统能够提供它所需的功能并且能够由使能系统和服务来维持。该流程通常以正式的书面确认结束，即系统已恰当地安装和验证，所有问题和行动项已解决，以及完全可支持的系统开发和交付有关的所有协议已完全被满足或裁定。

  - 在运行场地中对功能性进行演示以及对运行准备度进行评审之后，可将系统投入服务。

- *管理转移的结果。*

  - 捕获实现后的事件和问题，而且可能引起需求的纠正措施或变更。质量保证流程用于对执行转移流程期间报告的事件和问题的解决。

  - 记录转移流程期间观察到的异常。这些异常提供对结果、处理异常所需的信息以及历史记录的意识。异常可能源于转移策略、支持性使能系统、接口等。项目评估和控制流程用于分析异常并确定需要什么行动（若存在）。

TECHNICAL PROCESSES

- Maintain bidirectional traceability of the transitioned system elements with the transition strategy, system architecture, design, and system requirements.
- Provide baseline information for configuration management.

Common approaches and tips:

- *When acceptance activities cannot be conducted within the operational environment, a representative locale is selected*
- *This process relies heavily on quality assurance and configuration management documentation.*

4.11 VALIDATION PROCESS

4.11.1 Overview

4.11.1.1 Purpose As stated in ISO/IEC/IEEE 15288,

[6.4.11.1] The purpose of the Validation process is to provide objective evidence that the system, when in use, fulfills its business or mission objectives and stakeholder requirements, achieving its intended use in its intended operational environment.

4.11.1.2 Description The validation process is applied to a SOI, or any system or system element that composes it, at the appropriate points in the life cycle stages to provide confidence that the right system (or system element) has been built.

The validation process can be applied to any system element or engineering item of the system or its definition that has been defined or realized (e.g., validation of a stakeholder requirement, a system requirement, a function, an input/output flow, a system element, an interface, a design property, an integration aggregate, a validation procedure). Thus, the validation process is performed to help ensure that the system or any system element meets the need of its stakeholder in the life cycle (i.e., the engineering processes and the transformation of inputs produced what was intended— the "right" result).

Validation is a transverse activity to every life cycle stage of the system. In particular, during the development of the system, validation applies to any process/activity and product resulting from the process/activity.

The validation process works closely with other life cycle processes. For example, the business or mission analysis process establishes a targeted operational capability. The operational capability (e.g., mission or business profile and operational scenarios) is transformed by the stakeholder needs and requirements definition process into stakeholder needs and requirements. The validation process works concurrently with these processes to define the applicable validation actions and validation procedures through the life cycle to ensure the system evolves in a way that there is a high level of confidence that the operational capability will be provided, as refined in the stakeholder needs and requirements.

— 维护转移的系统元素与转移策略、系统架构、设计及系统需求的双向可追溯性。

— 提供构型管理的基线信息。

常用方法和建议：

- 当验收活动无法在运行环境内实施时，则选择一个有代表性的场所。
- 该流程很大程度上依赖于质量保证和构型配置管理文件。

4.11 确认流程

4.11.1 概览

4.11.1.1 目的 如 ISO/IEC/IEEE 15288 所述，

[6.4.11.1] 确认流程的目的在于提供客观证据，证明系统在使用时符合其业务或使命任务目标及利益攸关者的需求，从而在其意向的运行环境中实现其意向的使用。

4.11.1.2 描述 确认流程在生命周期阶段的恰当时刻应用于 SOI 或任何系统或构成该系统的系统元素，以便为已构建了正确的系统（或系统元素）提供信心。

确认流程可应用于已定义或已实现的任何系统元素或系统的工程项或其定义（如一个利益攸关者需求、一个系统需求、一个功能、一个输入/输出流、一个系统元素、一个接口、一个设计特性、一个综合聚合体、一个确认程序的确认）。因此，执行确认流程以帮助确保系统或任何系统元素在生命周期内满足利益攸关者的需要（即，工程流程和输入转换产生意图的事物——"正确的"结果）。

确认是系统每一生命周期阶段的横向活动。特别是在系统开发期间，确认应用到任何流程/活动以及流程/活动产生的任何产品。

确认流程与其他生命周期流程密切配合。例如，业务或任务分析流程建立一个目标的运行能力。运行能力（如任务或业务概况及运行场景）被利益攸关者需要和需求定义流程转换为利益攸关者需要和需求。确认流程与这些流程并行工作以便在生命周期内定义适用的确认措施和确认程序，从而确保对系统将会提供运行能力具有高层级置信度的方式演进，如利益攸关者需要和需求所细化的一样。

TECHNICAL PROCESSES

Figure 4.17 is the IPO diagram for the validation process.

4.11.1.3 Inputs/Outputs Inputs and outputs for the validation process are listed in Figure 4.17. Descriptions of each input and output are provided in Appendix E.

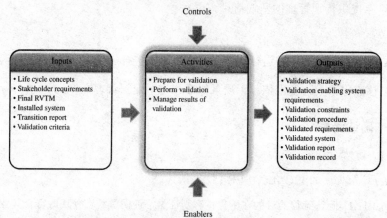

FIGURE 4.17 IPO diagram for the validation process. INCOSE SEH original figure created by Shortell and Walden. Usage per the INCOSE notices page. All other rights reserved.

4.11.1.4 Process Activities The validation process includes the following activities:

- *Prepare for validation.*

    - Establish the validation strategy, which is often part of a validation plan, that optimizes the number and type of validation actions while minimizing costs and risks:

        ○ Identify the stakeholders who will be involved in the validation activities and define their roles and responsibilities. This can include the acquirer, the supplier, and a third-party representative.

        ○ The scope of the validation plan is dependent on the life cycle stage and the progress within it. Validation may be appropriate for the full system, a system element, or an artifact, such as the ConOps or a prototype, in addition to the delivered system.

        ○ Establish a list of validation constraints that need to be considered. The constraints could impact the implementation of the validation actions and include contractual constraints, limitations due to regulatory requirements, cost, schedule, feasibility to exercise a function (such as in some ordinance), safety considerations, physical configurations, accessibility, etc.

图 4.17 是确认流程的 IPO 图。

4.11.1.3 输入/输出 图 4.17 列出确认流程的输入和输出。附录 E 提供每个输入和输出的描述。

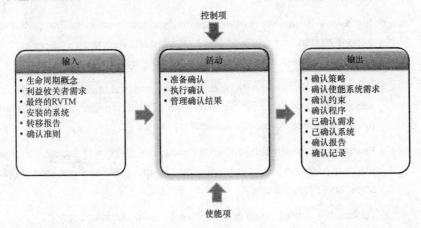

图 4.17 确认流程的 IPO 图。INCOSE SEH 原始图由 Shortell 和 Walden 创建。按照 INCOSE 通知页使用。版权所有。

4.11.1.4 流程活动 确认流程包括以下活动：

- *准备确认。*

    – 建立确认策略，这通常是确认计划的一部分，在优化确认措施的数量与类型的同时使成本和风险最小化：

        ○ 识别将要参与确认活动的利益攸关者并定义他们的角色与职责。这可包括采办方、供应商和第三方代表。

        ○ 确认计划的范围取决于生命周期阶段及生命周期阶段内的进展。除了所交付的系统外，确认还可适于全系统、系统元素或制品，如 ConOps 或一个原型。

        ○ 建立需要考虑的确认约束列表。这些约束可能影响确认措施的实施，且包括合同约束、由于监管需求产生的限制、成本、进度、行使某种功能的可行性（如在某种条例中）、安全考虑因素、物理构型配置、可达性等。

TECHNICAL PROCESSES

- With appropriate consideration to the constraints, select suitable validation approach to be applied, such as inspection, analysis, demonstration, or test, depending on the life cycle stage. Identify any enablers needed.
- It may be necessary to prioritize the validation actions. The validation actions should be prioritized and evaluated against constraints, risks, type of system, project objectives, and other relevant criteria.
- Determine if there are any validation gaps and that the resulting validation actions will provide an acceptable level of confidence that the system or system element will meet the identified needs.
- Ensure appropriate scheduling via the project planning process to meet the requirements for the execution of the validation actions in the applicable project steps.
- Define the configuration of submitted items to validation actions.

– Identify validation constraints on the system, arising from the validation strategy, to be incorporated in the stakeholder requirements. This includes practical limitations of accuracy, uncertainty, repeatability that are imposed by the validation enablers, the associated measurement methods, and the availability, accessibility, and interconnection with enablers.

– Ensure that the necessary enabling systems, products, or services required for the validation actions are available, when needed. The planning includes the identification of requirements and interfaces for the enablers. An enabler may be a complete enabling system developed as a separate project from the project of the SOI.

- *Perform validation.*

– Develop the validation procedures that support the validation actions.

– Ensure readiness to conduct validation: availability and configuration status of the system/ item, the availability of the validation enablers, qualified personnel or operators, resources, etc.

– Conduct validation actions in accordance with the procedures. This should include performing the actions in the operational environment or one as close to it as possible. During the conduct of the validation actions, record the results of the actions, as they are performed.

○ 通过适当考虑约束，选择要应用的适当的确认方法，如检验、分析、演示或测试，取决于生命周期阶段。识别所需的任何使能项。

○ 对确认措施进行优先级排序可能是必要的。应对照约束、风险、系统类型、项目目标及其他相关准则对确认措施进行优先级排序和评价。

○ 确定是否具有任何确认差距以及产生的确认措施是否将提供可接受的置信度等级，即系统或系统元素将满足所识别的需要。

○ 为在适用的项目步骤中满足确认措施的执行要求，通过项目规划流程进行恰当的进度安排。

○ 定义确认措施的提交项的构型配置。

— 识别出由确认策略产生的被纳入利益攸关者需求中的系统的确认约束。包括确认使能项所施加的准确度、不确定性、可重复性的实际限制、相关的测量方法，以及可用性、可达性和与使能项的相互连接。

— 确保在需要时确认措施所需的必要的使能系统、产品或服务是可用的。规划包括使能项的需求和接口的识别。使能项可能是被开发为 SOI 项目中的独立项目的完整使能系统。

• *实施确认。*

— 开发支持确认措施的确认程序。

— 确保实施确认的准备就绪：系统/项的可用性和构型配置状态、确认使能项的可用性、合格的人员或操作者、资源等。

— 按照程序进行确认措施。应包括在运行环境或尽可能接近运行环境的某一环境中实施行动。在进行确认措施期间，随着行动的实施，记录其结果。

TECHNICAL PROCESSES

- *Manage results of validation.*
    - Identify and record validation results and enter data in the validation report (including any necessary updates to the RVTM). Maintain the records per organizational policy.
    - Record anomalies observed during the validation process, and analyze and resolve the anomalies (corrective actions or improvements) using the quality assurance process (see Section 5.8).
        o Ensure the results and anomalies and/or non-conformances are analyzed using the project assessment and control process.
        o Compare the obtained results with the expected results; deduce the degree of conformance of the submitted item (i.e., provides the services expected by the stakeholder); decide whether the degree of conformance is acceptable.
        o Problem resolution is handled through the quality assurance and project assessment and control processes. Changes to the system or system element definition (i.e., requirements, architecture, design, or interfaces) and associated engineering artifacts are performed within other technical processes.
    - Obtain acquirer (or other authorized stakeholders) acceptance of validation results.
    - Maintain bidirectional traceability of the validated system elements with the validation strategy, business/mission analysis, stakeholder requirements, system architecture, design, and system requirements.
    - Provide baseline information for configuration management.
    - Update the validation strategy and schedule according to the progress of the project; in particular, planned validation actions can be redefined or rescheduled as necessary.

Common approaches and tips:

- Validation methods during the business and mission analysis process include assessment of OpsCon through operational scenarios that exercise all system operational modes and demonstrating system-level performance over the entire operating regime. The architects and designers use the results of this activity to forecast success in meeting the expectations of users and the acquirer, as well as to provide feedback to identify and correct performance deficiencies before implementation (Engel, 2010).

- *管理确认的结果。*
  - 识别和记录确认结果,并将数据输入确认报告中(包括必要的 RVTM 更新)。按照组织的方针维护记录。
  - 记录确认流程期间所观察到的异常,并使用质量保证流程(见 5.8 节)来分析和化解该异常(纠正措施或改进)。
    - 确保使用项目评估和控制流程对结果和异常和/或不符合性进行分析。
    - 将所获得的结果与期望的结果进行比较;推演出提交项的符合性程度(即,提供利益攸关者所期望的服务);决定该符合性的程度是否可接受。
    - 问题的解决是通过质量保证流程和项目评估和控制流程来处理的。在其他技术流程内执行系统或系统元素定义(即,需求、架构、设计或接口)及相关工程制品的变更。
  - 获取采办方(或其他经授权的利益攸关者)对确认结果的验收。
  - 维护确认的系统元素与确认策略、业务/任务分析、利益攸关者需求、系统架构、设计及系统需求的双向可追溯性。
  - 提供构型配置管理的基线信息。
  - 按照项目进展更新确认策略和进度;特别是,可以根据需要对已计划的确认措施重新定义或重新安排。

常用方法和建议:

- 业务和任务分析流程期间的确认方法包括通过运用所有系统运行模式的运行场景的对 OpsCon 的评估,以及演示在整个运行域之上的系统级性能。架构师和设计者使用该活动的结果来预测是否成功满足用户和采办方的期望,并在实现之前为识别和纠正性能缺陷提供反馈(Engel, 2010)。

TECHNICAL PROCESSES

- It is recommended to start the drafting of the validation plan as soon as the first OpsCon and scenarios and stakeholder requirements are known. Early consideration of potential validation methods or techniques allows the project to anticipate constraints, estimate cost, and start the acquisition of validation enablers such as test facilities, simulators, etc., avoiding cost overruns and schedule slippages.

- Validation is applied to the operational system, but is most effective if it is also applied earlier by analysis, simulation, emulation, etc. of anticipated operational characteristics.

- A key output of validation is the assurance provided of the loci of system dynamic and integrity limits. These envelopes provide actionable knowledge for users to determine system suitability, anticipated effectiveness, survivability, and refurbishment.

- Validation also reveals the effects the SOI may have on collateral, enabling, or interoperating systems. Validation actions and analysis should include these system interactions in the scope.

- Involve the broadest range of stakeholders with validation. Often, the end users and other relevant stakeholders are involved in the validation activities.

- If possible, involve users/operators with validation. Validation will often involve going back directly to the users to have them perform some sort of acceptance test under their own local conditions.

4.11.2 Elaboration

4.11.2.1 Notion of Validation Action A validation action describes what must be validated (e.g., an operational scenario, a requirement, or a set of requirements as reference), on which item (e.g., requirement, function, interface, system element, system), the expected result (deduced from the reference), the validation technique to apply (e.g., inspection, analysis, demonstration, test), and on which level of the system hierarchy (e.g., SOI, intermediate level system element, leaf level system element).

The definition of a validation action applied to an engineering item (e.g., stakeholder requirement, system requirement, function, interface, system element, procedure, document) includes the identification of the item on which the validation action will be performed, the reference used to define the expected result, and the appropriate validation technique.

- 建议在最初 OpsCon 和场景及利益攸关者需求已知后就立刻开始起草确认计划。尽早考虑潜在的确认方法或技术可使项目预期约束，估计成本并开始采办确认使能项，如测试设施、仿真器等，从而避免成本超支和进度拖延。

- 确认应用于运行系统，但如果还通过对预期的运行特征的分析、仿真、模仿等更早地应用是最有效的。

- 确认的一个关键输出是保证提供系统动态和完整性限制的所在。这些范围为用户提供可行动的知识，以确定系统适用性、期望的有效性、生存性和翻新性。

- 确认还展现出 SOI 对附属的、使能的或互操作系统可能具有的影响。确认措施和分析应包括这些系统在该范围内的交互。

- 使最广泛范围的利益攸关者参与确认。通常，最终用户及其他相关的利益攸关者参与确认活动。

- 若可能，使用户/操作者参与确认。确认往往会涉及直接回溯到用户，使他们在自己的本地条件下实施某种验收试验。

4.11.2 详细阐述

4.11.2.1 确认措施的观念 确认措施描述必须确认什么（如一个运行场景、一个需求或一个需求集合，作为参考）、关于什么项（如需求、功能、接口、系统元素、系统）、期望的结果（从参考中推断）、要应用的确认技术（如检验、分析、演示、测试）和关于系统层级结构的哪一层级（如 SOI、中间层级系统元素、叶层级系统元素）。

应用于工程项（如利益攸关者需求、系统需求、功能、接口、系统元素、程序和文件）的确认措施的定义包括将要实施的确认措施项的识别、用于定义期望结果的参考以及恰当的确认技术。

TECHNICAL PROCESSES

The performance of a validation action onto the submitted item provides an obtained result which is compared with the expected result. The comparison enables the project to judge the element's acceptability regarding the relevance in the context of use (see Fig. 4.18).

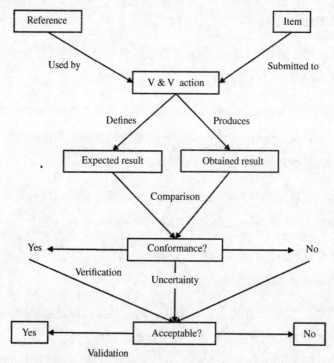

FIGURE 4.18 Definition and usage of a validation action. Reprinted with permission from Alain Faisandier. All other rights reserved.

Examples of validation actions:

- *Validation of a requirement*—To make sure its content is justified and relevant to stakeholder needs or expectations.

- *Validation of an engineering artifact (architecture, design, etc.)*—To make sure its content is justified and relevant to stakeholder needs or expectations and contributes to achieving the mission or business profile and operational scenarios.

- *Validation of a system (product, service, or enterprise)*—To demonstrate that the product, service, or enterprise satisfies its stakeholder requirements, mission or business profile, and operational scenarios.

确认措施在提交项上的实施提供了一个获得的结果，将该结果与期望的结果相比较。该比较使项目能够考虑在使用的背景环境中相关因素下判断元素的可接受性（见图 4.18）。

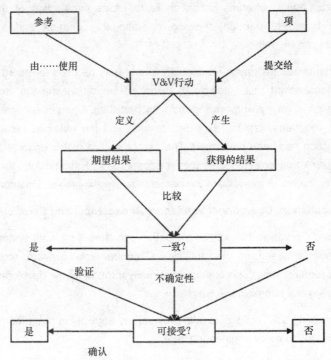

图 4.18 确认措施的定义和使用。经 Alain Faisandier 许可后转载。版权所有。

确认措施示例：

- 一个需求的确认——要确保其内容被证明合理且与利益攸关者需要或期望相关。

- 一种工程制品（架构、设计等）的确认——要确保其内容被证明合理，与利益攸关者需要或期望相关，且有助于实现任务或业务概况及运行场景。

- 系统（产品、服务或复杂组织体）的确认——要演示产品、服务或复杂组织体满足其利益攸关者需求、任务或业务概况及运行场景。

TECHNICAL PROCESSES

4.11.2.2 Validation Techniques The validation techniques are the same as those used for verification (see Section 4.9.2.2), but the purposes are different; verification is used to show compliance with the specified system requirements and to detect errors/defects/ faults, whereas validation is to prove satisfaction of the desired operational capability through showing operational scenarios and stakeholder requirements can be met.

A Requirements and Validation Traceability Matrix may be used to record data such as validation action list, selected validation method/technique to validate implementation of every engineering item (in particular, operational scenarios and stakeholder requirements), the expected results, and the obtained results when a validation action has been performed. The use of such a matrix enables the project team to ensure that selected operational scenarios and stakeholder requirements have been validated, or to evaluate the percentage of validation actions completed.

4.11.2.3 Validation, Operational Validation, Acceptance, and Certification

Validation and Operational Validation Validation concerns the global system seen as a whole and is based on the totality of requirements (system requirements, stakeholder requirements). It is obtained gradually throughout the development stage by pursuing several nonexclusive ways:

- Cumulating V&V actions' results provided by application of the corresponding processes to every engineering item
- Performing final validation actions onto the complete integrated system in an industrial environment (as close as possible to the future operational environment)
- Performing operational validation actions onto the complete system in its operational environment (context of use)

The goal is to completely validate the system capability to meet all requirements prior to the production and utilization stages. Problems uncovered in these stages are very costly to correct. As such, early discovery of deviations from requirements reduces overall project risk and helps the project deliver a successful, low-cost system. Validation results are an important element of decision gate reviews.

Acceptance Acceptance is an activity conducted prior to transition such that the acquirer can decide that the system is ready to change ownership from supplier to acquirer. A set of operational validation actions is often exercised, or a review of validation results is systematically performed.

4.11.2.2 确认技术 确认技术与验证所用的技术（见 4.9.2.2 节）相同，但目的不同；验证用于展示对特定的系统需求的符合性并检测误差/缺陷/故障，而确认则用于通过展示能够满足运行场景和利益攸关者需求来证明满足期望的运行能力。

一个需求和确认可追溯性矩阵可用于记录数据，如确认措施列表、确认每一工程项（特别是运行场景和利益攸关者需求）的实现所用的选定的确认方法/技术、期望的结果以及当一个确认措施已经被实施时所获得的结果。这种矩阵的使用使项目团队能够确保已选定的运行场景和利益攸关者需求已经被确认或评价已完成的确认措施的百分比。

4.11.2.3 确认、运行确认、验收和认证

确认和运行确认 确认关注被看做整体的全局系统，且基于需求（系统需求、利益攸关者需求）的全体。该系统通过追求多种非排它的方式贯穿于开发阶段逐渐地获得：

- 累积 V&V 行动的结果，这些结果是通过在每一工程项上应用相应流程而提供的
- 在工业环境（尽可能接近未来的运行环境）中的完整综合的系统上实施最终的确认措施
- 在完整系统的运行环境（使用的背景环境）中对完整系统实施运行确认措施

目标在于全面确认系统在生产和使用阶段之前满足所有需求的能力。在这些阶段中纠正暴露的问题需要非常高的成本。正因如此，尽早发现需求的偏离可降低整体项目的风险，并有助于项目交付一个成功的低成本的系统。确认结果是决策门评审的一个重要元素。

验收 验收是在转移之前进行的活动，从而使采办方可决定系统已准备好将所有权从供应商移交给采办方。通常运用一个运行的确认措施集合，或系统地实施确认结果的一种评审。

Certification Certification is a written assurance that the product or article has been developed, and can perform its assigned functions, in accordance with legal or industrial standards (e.g., for aircraft). The development reviews, verification results, and validation results form the basis for certification. However, certification is typically performed by outside authorities, without direction as to how the requirements are to be verified. For example, this method is used for electronics devices via Conformité Européenne (CE) certification in Europe and Underwriters laboratories (UL) certification in the United States and Canada.

Readiness for Use As part of the analysis of the validation results, the project team needs to make a readiness for use assessment. This may occur many times in the life cycle, including the first article delivery, completion of production (if more than a single system is produced), and following maintenance actions. In the field, particularly after maintenance, it is necessary to establish whether the system is ready for use.

Qualification The system qualification requires that all V&V actions have been successfully performed. These V&V actions cover not only the SOI itself but also all the interfaces with its environment (e.g., for a space system, the validation of the interface between space segment and ground segment). The qualification process has to demonstrate that the characteristics or properties of the realized system, including margins, meet the applicable system requirements and/or stakeholder requirements. The qualification is concluded by an acceptance review and/or an operational readiness review.

As illustration of this, for a space system, the last step of the qualification is covered by the first launch or the first flight. This first flight needs to be milestoned by a flight readiness review that will verify that the flight and ground segments including all supporting systems such as tracking systems, communication systems, and safety systems are ready for launch. A complementary review can be held just before launch (launch readiness review) to authorize the launch. A successful launch participates to the qualification process, but the final system qualification is achieved only after in-orbit tests for a spacecraft or even several flights for a launcher in order to cover the different missions for which the system has been developed.

4.11.2.4 Validation Level per Level Generally, the SOI has been decomposed during architecture definition in a set of layers of systems; thus, every system and system element is validated and possibly corrected before being integrated into the parent system of the higher level, as shown in Figure 4.19. In this figure, every time the term validate is used means that the corresponding validation process is invoked.

认证 认证是产品或物品按照法定标准或工业标准（如飞机）已被开发且可实施其指定功能的书面保证。开发评审、验证结果和确认结果形成认证的基础。然而，认证通常由外部权威机构实施，且没有关于需求如何被验证的引导。例如，该方法用于具备欧洲的欧洲统一（CE）认证及美国和加拿大的保险商试验所（UL）认证的电子设备。

使用准备度 作为确认结果的分析的一部分，项目团队需要准备就绪对使用进行评估。这可能在生命周期内多次出现，包括首件交付、生产完成（如果生产一个以上的单一系统）以及后续维护行动。在该领域内，特别是在维护后，有必要确定系统是否已准备好投入使用。

鉴定 系统鉴定要求所有 V&V 行动已成功实施。这些 V&V 行动不仅涵盖 SOI 本身，而且还涵盖与其环境的所有接口（如，对于一个太空系统而言，太空部分和地面部分之间的接口的确认）。鉴定流程必须演示已实现的系统的特征或特性（包括裕度）满足适用的系统需求和/或利益攸关者需求。鉴定以验收评审和/或运行准备度评审而得出结论。

如此所阐述，对于一个太空系统而言，鉴定的最后一个步骤涵盖在首次发射或首次飞行中。首次飞行需要通过一个飞行准备度评审来确立为里程碑，该评审将验证飞行部分和地面部分（包括所有支持系统，如跟踪系统、通信系统和安全系统）已准备好发射。可在发射前举行补充评审（发射准备度评审）以授权发射。一个成功的发射参与到鉴定流程中，但最终的系统鉴定仅在航天器在轨测试乃至一个发射装置多次飞行之后完成，以便涵盖系统已经开发的不同任务。

4.11.2.4 逐层确认 一般地，在架构定义期间 SOI 已按照系统的一系列层级被分解；因此，每一系统和系统元素，在被综合到较高层级母系统中之前就被确认进行可能的纠正，如图 4.19 所示。在该图中，每次使用术语"确认"都意味着调用相应的确认流程。

TECHNICAL PROCESSES

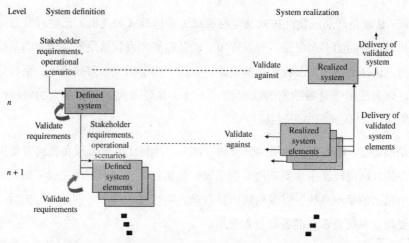

FIGURE 4.19 Validation level per level. Note: System requirements are validated against the stakeholder requirements. Reprinted with permission from Alain Faisandier. All other rights reserved.

As necessary, systems and system elements are partially integrated in subsets (aggregates) in order to limit the number of properties to be validated within a single step. For each level, it is necessary to make sure by a set of final validation actions that features stated at preceding level are not adversely affected. Moreover, a compliant result obtained in a given environment can turn noncompliant if the environment changes. So, as long as the system is not completely integrated and/or doesn't operate in the real operational environment, no result must be regarded as definitive.

4.12 OPERATION PROCESS

4.12.1 Overview

4.12.1.1 Purpose As stated in ISO/IEC/IEEE 15288,

[6.4.12.1] The purpose of the Operation process is to use the system to deliver its services.

This process is often executed concurrent with the maintenance process.

4.12.1.2 Description The operation process sustains system services by preparing for the operation of the system, supplying personnel to operate the system, monitoring operator–system performance, and monitoring the system performance. When the system replaces an existing system, it may be necessary to manage the migration between systems such that persistent stakeholders do not experience a breakdown in services.

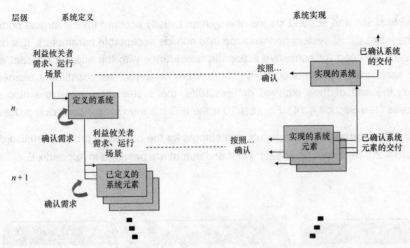

图 4.19 逐层确认。注：按照利益攸关者需求确认系统需求。经 Alain Faisandier 许可后转载。版权所有。

必要时，将系统和系统元素部分地综合在子集（聚合体）中，以便限制单一步骤内要确认的特性数量。对于每一层级而言，有必要通过最终的确认措施集合确保前面层级所说明的特征不会受到不利影响。并且，如果环境发生变化，给定环境中获得的符合性结果可转变为不符合。因此，只要系统没有被完全综合且/或不在真实的运行环境中运行，一定不能认为结果是决定性的。

4.12 运行流程

4.12.1 概览

4.12.1.1 目的　如 ISO/IEC/IEEE 15288 所述，

[6.4.12.1] 运行流程的目的是使用系统来交付其服务。

该流程经常与维护流程并行执行。

4.12.1.2 描述　运行流程通过准备系统的运行、提供运行系统的人员、监控操作者——系统性能，并监控系统性能来维持系统服务。当该系统替换一个现有系统时，可能有必要管理系统之间的迁移，从而使持续存在的利益攸关者免遭服务的中断。

TECHNICAL PROCESSES

The utilization and support stages of a system usually account for the largest portion of the total LCC. If system performance falls outside acceptable parameters, this may indicate the need for corrective actions in accordance with the support concept and any associated agreements. When the system or any of its constituent elements reach the end of their planned or useful life, the system may enter the disposal process (see Section 4.14). Figure 4.20 is the IPO diagram for the operation process.

4.12.1.3 Inputs/Outputs Inputs and outputs for the operation process are listed in Figure 4.20. Descriptions of each input and output are provided in Appendix E.

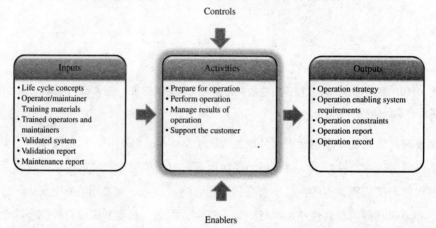

FIGURE 4.20 IPO diagram for the operation process. INCOSE SEH original figure created by Shortell and Walden. Usage per the INCOSE notices page. All other rights reserved.

4.12.1.4 Process Activities The operation process includes the following activities:

- *Prepare for operation.*

  – Plan for operation, including the development of the strategy:

    o Establish availability of equipment, services, and personnel and performance tracking system.

    o Verify schedules for personnel and facilities (on a multi-shift basis if appropriate).

    o Define business rules related to modifications that sustain existing or enhanced services.

    o Implement the OpsCon and environmental strategies.

一个系统的使用和保障阶段通常占总 LCC 的最大部分。若系统性能超出可接受的参数范围，这可能表明需要按照保障方案及相关协议采取纠正措施。当系统或其组成元素达到其计划的或使用寿命的终点时，系统可进入退出流程（见 4.14 节）。图 4.20 是运行流程的 IPO 图。

4.12.1.3 输入/输出 图 4.20 列出运行流程的输入和输出。附录 E 提供每个输入和输出的描述。

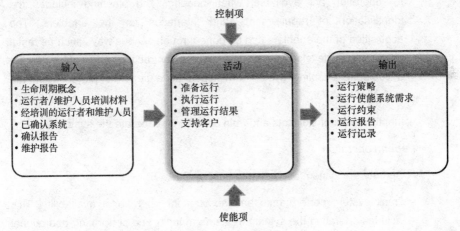

图 4.20 运行流程的 IPO 图。INCOSE SEH 原始图由 Shortell 和 Walden 创建。按照 INCOSE 通知页使用。版权所有。

4.12.1.4 流程活动 运行流程包括以下活动：

- *运行的准备。*

  – 为运行做计划，包括策略的开发：

    o 建立设备、服务及人员和性能跟踪系统的可用性。

    o 验证人员和设施的时间安排（若合适，以轮班制为基础）。

    o 定义与维持现有或增强服务的修改相关的业务规则。

    o 实现 OpsCon 和环境策略。

TECHNICAL PROCESSES

- ○ Review operational performance measures, thresholds, and criteria.
- ○ Verify all personnel have had the applicable system safety training.
- Feed back any operation constraints on the system or system elements to the system requirements definition, architecture definition, or design definition processes.
- Ensure that the necessary enabling systems, products, or services required for operation are available, when needed. The planning includes the identification of requirements and interfaces for the enablers. The acquisition of the enablers can be done through various ways such as rental, procurement, development, reuse, and subcontracting. An enabler may be a complete enabling system developed as a separate project from the project of the SOI.
- Identify operator skill sets and train operators to operate the system.

- *Perform operation.*

  - Operate the system according to the OpsCon.
  - Track system performance and account for operational availability. This includes operating the system in a safe manner and performing operational analysis to determine system noncompliance.
  - When abnormal operational conditions warrant, conduct planned contingency actions. Perform system contingency operations, if necessary.

- *Manage results of operation.*

  - Document the results of operations.
  - Record anomalies observed during the operation process, and analyze and resolve the anomalies (corrective actions or improvements) using the quality assurance process (see Section 5.8). Implementation of procedures for restoring operations to a safe state should be followed. Maintain bidirectional traceability of the operations elements with the operation strategy, business/mission analysis, ConOps, OpsCon, and stake-holder requirements.

- *Support the customer.*

  - Perform tasks needed to address customer requests.

- ○ 评审运行性能测度、阈值和准则。

- ○ 验证所有人员已接受适用的系统安全培训。

– 向系统需求定义流程、架构定义流程或设计定义流程反馈对系统或系统元素的任何运行约束。

– 确保在需要时运行所需的必要的使能系统、产品或服务是可用的。规划包括使能项的需求和接口的识别。使能项的采办可以通过各种不同的方式来完成，如租赁、采购、开发、复用和转包。使能项可能是与被开发为 SOI 项目相独立的项目的完整使能系统。

– 识别操作者技能集合并培训操作者以运行该系统。

- 实施运行。

– 按照 OpsCon 运行系统。

– 跟踪系统性能并以运行可用性解释原因。这包括以安全的方式运行系统以及实施运行分析以确定系统的不符合性。

– 当异常运行条件合理地出现时，实施计划的应急行动。若必要，实施系统应急运行。

- 管理运行的结果。

– 将运行的结果文件化。

– 记录运行流程期间观察到的异常，并使用质量保证流程（见 5.8 节）来分析和化解该异常（纠正措施或改进）。应遵循程序的实施将运行恢复到安全状态。维护运行元素与运行策略、业务/任务分析、ConOps、OpsCon 和利益攸关者需求的双向可追溯性。

- 支持客户。

– 执行应对客户请求所需的任务。

TECHNICAL PROCESSES

4.12.2 Elaboration

4.12.2.1 Operation Enabling Systems The operation process includes other enabling systems that need to be considered that may be different from other processes. The following amplifies on these:

- *Operational environment*—The circumstances surrounding and potentially affecting something that is operating. For example, electronic or mechanical equipment may be affected by high temperatures, vibration, dust, and other parameters that constitute the operating environment.

- *Training systems*—Provides operators with knowledge and skills required for proper system operations.

- *Technical data*—Procedures, guidelines, and check-lists needed for proper operation of the system, including prerequisites for operation, procedures for activation and checkout of the system, operating procedures, procedures for monitoring system performance, procedures for problem resolution, and procedures for shutting the system down.

- *Facilities and infrastructure*—Facilities (e.g., buildings, airfields, ports, roadways) and infrastructure (e.g., IT services, fuel, water, electrical service) required for system operation.

- *Sustaining engineering*—Monitors system performance, conducts failure analysis, and proposes corrective actions to sustain required operational capabilities.

- *Maintenance planning and management*—With the goal of minimizing system downtime, planned and/or preventive maintenance is performed to sustain system operations. The maintenance system responds to operator trouble/problem reports, conducts corrective maintenance, and restores system for operations.

4.13 MAINTENANCE PROCESS

4.13.1 Overview

4.13.1.1 Purpose As stated in ISO/IEC/IEEE 15288,

[6.4.13] The purpose of the Maintenance process is to sustain the capability of the system to provide a service.

4.12.2 详细阐述

4.12.2.1 运行使能系统   运行流程包括需要考虑的可能与其他流程不同的其他使能系统。下面详述这些系统：

- *运行环境*——围绕并潜在地影响正在运行的事物的环境。例如，电子或机械设备可能受高温、振动、灰尘及构成运行环境的其他参数的影响。

- *培训系统*——为操作者提供恰当的系统运行所需的知识和技能。

- *技术数据*——恰当的系统运行所需的程序、指南和检查单，包括运行的前提、系统激活和检查的程序、运行程序、监控系统性能的程序、问题解决程序及关闭系统的程序。

- *设施及基础建设*——系统运行所需的设施（如，建筑、机场、港口、车道）及基础建设（如，IT 服务、燃料、水、电气服务）。

- *维持工程*——监控系统性能，进行失效分析并提出纠正措施以维持所需的运行能力。

- *维护规划和管理*——以最小化系统停机时间为目的，执行有计划的维护和/或预防性维护以维持系统运行。维护系统响应操作者的故障/问题报告，进行纠正性维护并使系统恢复运行。

4.13 维护流程

4.13.1 概览

4.13.1.1 目的   如 ISO/IEC/IEEE 15288 所述，

[6.4.13] 维护流程的目的是保持系统提供某种服务的能力。

TECHNICAL PROCESSES

4.13.1.2 Description Maintenance includes the activities to provide operations support, logistics, and material management. Based on feedback from ongoing monitoring of the operational environment, problems are identified, and corrective, remedial, or preventive actions are taken to restore full system capability. This process contributes to the system requirements definition process, architecture definition process, and design definition process when considerations of constraints imposed in later life cycle stages are used to influence the system requirements, architecture, and design. Figure 4.21 is the IPO diagram for the maintenance process.

4.13.1.3 Inputs/Outputs Inputs and outputs for the maintenance process are listed in Figure 4.21. Descriptions of each input and output are provided in Appendix E.

4.13.1.4 Process Activities The maintenance process includes the following activities:

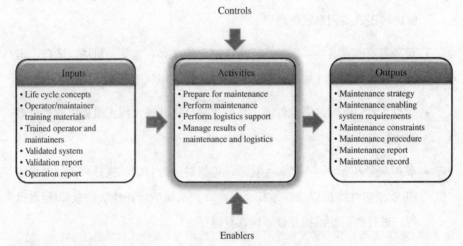

FIGURE 4.21 IPO diagram for the maintenance process. INCOSE SEH original figure created by Shortell and Walden. Usage per the INCOSE notices page. All other rights reserved.

- *Prepare for maintenance.*
  - Plan for maintenance, including the development of the strategy, that includes the following:
    o Sustained service across the life cycle in order to meet customer requirements and achieve customer satisfaction. The strategy should define the types of maintenance actions (preventive, corrective, modification) and levels of maintenance (operator, in situ, factory).

技术流程

4.13.1.2 描述 维护包括提供运行支持、后勤和资料管理的活动。基于对运行环境持续监测的反馈，识别问题并采取纠正措施、补救措施或预防措施来恢复完整的系统能力。当生命周期阶段后期施加的约束因素被用于影响系统需求、架构和设计时，该流程有助于系统需求定义流程、架构定义流程和设计定义流程。图 4.21 是维护流程的 IPO 图。

4.13.1.4 流程活动 维护流程包括以下活动：

4.13.1.3 输入/输出 图 4.21 列出维护流程的输入和输出。附录 E 提供每个输入和输出的描述。

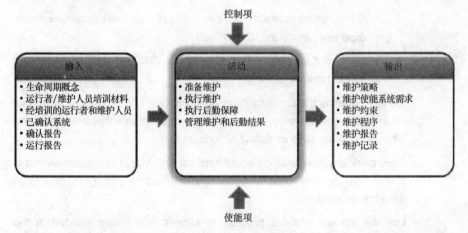

图 4.21 维护流程的 IPO 图。INCOSE SEH 原始图由 Shortell 和 Walden 创建。按照 INCOSE 通知页使用。版权所有。

- *维护的准备。*

  – 维护的计划，包括策略的开发，包括以下内容：

    ○ 跨越生命周期维持的服务，以便满足客户需求并达到客户满意。该策略应定义维护行动的类型（预防、纠正、修改）以及维护层级（操作者、现场、工厂）。

TECHNICAL PROCESSES

- o The various types of maintenance actions, including corrective maintenance (addressing failures or problems), adaptive maintenance (addressing changes needed to accommodate system evolution), perfective maintenance (addressing enhancements), and scheduled preventive maintenance (addressing routine servicing to prevent failures) while minimizing operational downtime.
- o The approach to address logistics needs across the life cycle, considering the following:
    - △ Approach to manage spare parts or system elements (e.g., number, type, storage, location, shelf life, conditions, renewal frequency). Analysis includes requirements and plans for packaging, handling, storage, and transportation (PHS&T).
    - △ Anticounterfeit approach (i.e., prevention of counterfeit system elements, especially parts, in the supply chain).
    - △ Identify or define sources of technical data and training needed to support maintenance.
- – Identify or define the skills, training, qualifications, and number of personnel for maintenance. Consider any regulatory requirements that drive the need for specific skills, such as safety or security.
- – Feed back any maintenance constraints on the system or system elements to the system requirements definition, architecture definition, or design definition processes.
- – Use the system analysis process to support any trades needed of the maintenance strategy and approach to ensure affordability, feasibility, supportability, and sustainability of the system maintenance.
- o Ensure that the necessary enabling systems, products, or services required for mainte-nance are available, when needed.The planning includes the identification of requirements and interfaces for the enablers. The acquisition of the enablers can be done through various ways such as rental, procurement, development, reuse, and subcontracting. An enabler may be a complete enabling system developed as a separate project from the project of the SOI.
- o Assign trained, qualified personnel to be maintainers.
- • *Perform maintenance.*
    - – Develop maintenance procedures for preventive and corrective maintenance.

- 各种不同类型的维护行动，包括并同时使运行停机时间最小化。纠正性维护（应对失效或问题）、适应性维护（应对调节系统演进所需的变更）、完善性维护（应对增强）及计划的预防性维护（应对防止故障的常规服务）。
- 应对后勤的方法需要跨越生命周期考虑因素如下：
  △ 管理备件或系统元素的方法（如，数量、类型、存储、位置、保存期限、条件、更新频率）。分析包括包装、处理、存储和运输（PHS&T）的需求及计划。
  △ 防伪方法（即，防止伪造的系统元素，尤其是供应链中的零件）。
  △ 识别或定义技术数据的来源以及支持维护所需的培训。
  △ 识别或定义技能、培训、资格及维护人员的数目。考虑驱动特定技能（如保安或安全）所需要的任何管理需求。

– 将系统或系统元素的任何维护约束反馈到系统需求定义流程、架构定义流程或设计定义流程。

– 使用系统分析流程来支持维护策略和方法所需的权衡，以确保系统维护的经济可承受性、可行性、可支持性和可保持性。

- 确保在需要时维护所要求的必要的使能系统、产品或服务是可用的。规划包括使能项的需求和接口的识别。使能项的采办可以通过各种不同的方式来完成，如租赁、采购、开发、复用和转包。使能项可能是作为与 SOI 项目相独立的一个项目的完整使能系统而开发的。
- 将经过培训的合格人员指定为维护人员。

- *实施维护*。

  – 开发预防性维护和纠正性维护的维护程序。

TECHNICAL PROCESSES

- Identify, record, and resolve system anomalies.
- Restore system operation after a failure.
- Plan future maintenance by reviewing and analyzing anomaly reports.
- Initiate analysis and corrective action to remedy previously undetected design errors.
- Per the identified schedule, perform preventive maintenance actions using defined maintenance procedures.
- Determine the need for adaptive or perfective.

• *Perform logistics support.*

- Conduct acquisition logistics actions—Includes performing trade studies and analysis to determine the most cost-effective means to support the system across the life cycle. Design influence considers features that impact the inherent reliability and maintainability of system services contrasted with the affordability of other support options including the use of spare parts. Design considerations are often constrained by availability requirements, the impact of supply chain management, manpower restrictions, and system afford-ability.Acquisition logistics plans for and develops strategies for system life cycle supportability (maintenance, supply support, support equipment, staffing, and enabling systems) concurrently with the definition of the system requirements.

- Conduct operational logistics actions— Operational logistics is the concurrent tuning of both the SOI and enabling systems throughout the operational life to ensure effective and efficient delivery of system capabilities. Operational logistics actions enable the system to achieve required operational readiness. The actions include staffing, maintenance management, supply support, support equipment, technical data needs (e.g., manuals, instructions, lists), training support, sustaining engineering, computing resources, and facilities. Operational logistics provides a rich source of data concerning the operational performance of the system. This data should be used to support trend analysis and provides a direct feedback loop on system effectiveness and efficiency and insight into customer satisfaction.

• *Manage results of maintenance and logistics.*

- Document the results of the maintenance process.

- 识别、记录和化解系统异常。

- 失效后使系统恢复运行。

- 通过评审和分析异常报告来规划未来的维护。

- 启动分析和纠正措施以补救之前未检测到的设计误差。

- 按照识别的进度，使用定义的维护程序实施预防性维护措施。

- 确定对适应性或完善性的需要。

- *实施后勤保障。*

  - 实施采办后勤措施——包括执行权衡研究和分析以确定跨生命周期保障系统的最具成本效益的方式。设计影响系统服务的固有可靠性和可维护性的影响考虑特征，与其他保障选项（包括备件的使用）的经济可承受性形成对比。设计考虑因素通常受到可用性需求、供应链管理的影响、人力限制和系统经济可承受性的约束。采办后勤与系统需求定义并行地计划和开发系统生命周期可保障性的策略（维护、供应支持、支持设备、人员配备和使能系统）。

  - 实施运行后勤措施——运行后勤贯穿于运行生命内并行调整 SOI 和使能系统，以确保系统能力的有效和高效的交付。运行后勤措施使系统能够实现所需的运行准备度。这些措施包括人员配备、维护管理、供应保障、保障设备、技术数据需要（如，手册、操作指南、清单）、培训支持、维持工程、计算资源和设施。运行后勤提供与系统的运行性能有关的丰富的数据来源。该数据应当用于支持趋势分析并提供对系统有效性和效率的直接反馈回路和对客户满意度的深入理解。

- *管理维护和后勤的结果。*

  - 将维护流程结果文件化。

TECHNICAL PROCESSES

- Record anomalies observed during the maintenance process, and analyze and resolve the anomalies (corrective actions or improvements) using the quality assurance process (see Section 5.8). Identify and record trends of main-tenance and logistics actions.

- Maintain bidirectional traceability of the maintenance actions and the applicable system elements and system definition artifacts.

- Obtain feedback from customers to understand the level of satisfaction with the maintenance and logistics support.

4.13.2 Elaboration

Maintenance is a term often used to refer to a phase during which the system is already operational and the primary activities are related to sustaining the capability of the system, system enhancement, upgrades, and modernization (Patanakul and Shenhar, 2010). In this context, the objective is to design the system from the start to allow for the possibility of maintaining value over the entire system life cycle, thereby promoting value sustainment (Ross et al., 2008). Since operations and maintenance costs generally comprise a significant percent of total LCC, maintenance is an important part of the system definition, often linked to logistics engineering, disposal, and environmental impact analysis.

Maintenance helps ensure that a system continues to satisfy its objective over its intended lifetime. In that timeframe, system expectations will expand, environments in which the system is operated will change, technology will evolve, and elements of the system may become unsupportable and need to be replaced. The COTS desktop computing environment is a case in point. Since the introduction of USB printers, it is difficult to find cables to support parallel port printers.

Maintenance is an integrated effort designed to address aging systems and a need to maintain those systems in operation. A maintenance program may include reengineering electronic and mechanical elements to cope with obsolescence, developing automated test equipment, and extending the life of aging systems through technology insertion enhancements and proactive maintenance (Herald et al., 2009).There are two main mitigation strategies that help users cope with the performance capability gap: a performance-based logistics (PBL) approach known as performance-based life cycle product support and technology refreshment programs (TRPs) (Sols et al., 2012).

- 记录维护流程期间观察到的异常，并使用质量保证流程（见 5.8 节）来分析和化解该异常（纠正措施或改进）。识别和记录维护及后勤行动的趋势。

- 对维护行动与适用的系统元素和系统定义制品的双向可追溯性进行维持。

- 从客户中获得反馈以理解对维护和后勤保障的满意度水平。

4.13.2 详细阐述

维护是通常用于指系统已运行的一个术语，主要活动涉及维持系统的能力、系统增强、升级和现代化（Patanakul and Shenhar，2010）。在该背景环境中，目的是从设计系统一开始，以考虑到在整个系统生命周期中维持价值的可能性，从而促进价值的保持（Ross 等，2008）。由于运行和维护成本一般包括总 LCC 中很大的百分比，因此维护是系统定义的一个重要部分，通常与后勤工程、退出和环境影响分析相关联。

维护有助于确保系统在其预期生命中持续的满足其目标。在这个时间段中，系统的期望会扩展，系统运行的环境会变化，技术会演进，并且系统元素也会变得无法支持并需要替换。COTS 桌面计算环境就是一个很好的例子。由于通用串行总线（USB）打印机的引入，就很难找到支持并行端口打印机的电缆。

维护是一种设计用于应对正在老化的系统以及使这些系统保持运行所需要的综合性工作。为了应对报废一个维护项目可包括"再造"电的和机械的元素，开发自动测试设备，以及通过技术嵌入增强和主动维护延长老化系统的寿命（Herald 等，2009）。有两种主要缓解策略帮助用户应对性能能力的差距：基于性能的后勤（PBL）方法，被称为基于性能的生命周期产品支持，以及技术更新计划（TRP）（Sols 等，2012）。

TECHNICAL PROCESSES

An open system architecture approach has been found to improve the ability to insert new capabilities (threat evolution), reduce development time, reduce maintenance cost, generate affordable COTS obsolescence management, and increase human systems integration. Open architectures enable a more capable, reliable, adaptable, and resilient system that accommodates continuous organizational and technology changes (Jackson and Ferris, 2013).

4.13.2.1 Maintenance Concept The maintenance concept is an important life cycle concept document. The following are recommended details that should be considered:

- *Types of maintenance*
  - Corrective maintenance—Processes and procedures for restoring system services to normal operations (e.g., remove and replace hardware, reload software, apply a software patch). Includes post-maintenance test procedures to verify that the system is ready for operations. Time spent troubleshooting can be greatly reduced by well engineered diagnostic capabilities such as built in test (BIT).
  - Preventive maintenance—Processes and procedures for scheduled/routine maintenance actions (e.g., cleaning, filter changes, visual inspections) needed to sustain optimal system operational performance.
  - System modifications—Processes and procedures intended to extend (sustain) the useful life of the system or to provide new (upgrade) system capabilities.
- *Levels of maintenance/repair*
  - User/operator maintenance—Some routine and simple maintenance tasks are able to be per-formed by the system operators or users. Operator maintenance includes routine (e.g., filter changes, recording data) and corrective (e.g., "software" resets, install a ready spare, tire change) tasks.
  - *In situ* maintenance and repair (sometimes referred to as "field" maintenance)—Maintenance tasks performed by a trained maintainer at, or near, the operational location.
  - Factory (maintenance, repair, and overhaul) (sometimes referred to as "depot" maintenance)— Major maintenance tasks that require advanced maintenance skills and special tooling/equipment not available at the operational location.

已发现开放式系统架构方法可提高嵌入新能力（威胁演进）的能力，缩短开发时间，减少维护成本，产生可负担得起的 COTS 报废管理，并提高人员系统的综合。开放式架构使得适应持续的组织和技术变化的系统更具有能力、更可靠、更能适应且更有弹性（Jackson 和 Ferris，2013）。

4.13.2.1 维护方案 维护方案是一份重要的生命周期概念文件。以下是应当考虑的建议细节：

- *维护的类型*
  - 纠正性维护——使系统服务恢复到正常运行的流程和程序（如，移除并更换硬件，重新加载软件，应用软件补丁）。包括维护后的测试程序，以验证系统已准备好运行。可通过良好设计的诊断能力，如内置测试（BIT），大幅度缩短花费在排除故障上的时间。

  - 预防性维护——维持最佳系统运行性能所需的计划安排的/常规维护行动（如，清理、过滤器更换、目视检查）的流程和程序。

  - 系统修改——旨在延长（维持）系统的使用寿命或提供新的（升级）系统能力的流程和程序。

- *维护/维修的层级*
  - 用户/操作者维护——一些常规的简单维护任务能够由系统操作者或用户来执行。操作者维护包括常规的（如，过滤器更换、记录数据）和纠正性的（如，"软件"重置，安装现成的备件，轮胎更换）任务。

  - 现场维护和维修（有时称为"实地"维护）——由运行场所或附近的经过培训的维护人员所执行的维护任务。

  - 工厂（维护、维修和翻修）（有时称为"后方"维护）——运行场所不可用的需要高级维护技能和专用工装/设备的主要维护任务。

TECHNICAL PROCESSES

4.13.2.2 Maintenance Enabling Systems The maintenance process includes other enabling systems that need to be considered that can be different from other processes. The following amplifies on these:

- *Operational environment*—The circumstances surrounding and potentially affecting something that is operating. For example, electronic or mechanical equipment may be affected by high temperatures, vibration, dust, and other parameters that constitute the operating environment.

- *Supply support/PHS&T*—Consists of all actions, necessary to determine requirements and to acquire, catalog, receive, store, transfer, package, transport, issue, and dispose of spares, repair parts, and supplies needed to sustain required level of operations (e.g., system availability). The reliability of the system is impacted when a maintenance action is delayed by the supply system.

- *Training systems*—Provides maintainers with knowledge and skills required for proper system maintenance.

- *Technical data*—Procedures, guidelines, and checklists needed for proper maintenance of the system, including preventive actions (clean and adjust), analysis and diagnostics, fault isolation, fault localization, parts lists, corrective maintenance (remove and replace), calibration, modification and upgrade instructions, post-maintenance validation, etc. For systems with condition based maintenance (CBM) capabilities, the technical documentation will include information on how those capabilities are used to support maintenance.

- *Facilities and infrastructure*—Facilities (e.g., buildings, warehouses, hangars, waterways) and infrastructure (e.g., IT services, fuel, water, electrical service, machine shops, dry docks, test ranges) required for system maintenance.

- *Tools and support equipment*—Common and special purpose tools (e.g., hand tools, meters) and support equipment (e.g., test sets, cranes) used to support system maintenance.

4.13.2.2 *维护使能系统* 维护流程包括需要考虑的可能与其他流程不同的其他使能系统。下面详述这些系统：

- *运行环境*——围绕并潜在地影响正在运行的事物的环境。例如，电子或机械设备可能受高温、振动、灰尘及构成运行环境的其他参数的影响。

- *供应支持/PHS&T* ——由确定需求以及采办、编目、接收、存储、传递、包装、运输、发布和处置维持所需运行层级（如，系统可用性）所需的备件、维修零件和供应所必需的所有措施组成。当由于供应系统而延迟了维护行动时，系统的可靠性将受到影响。

- *培训系统*——为维护人员提供恰当的系统维护所需的知识和技能。

- *技术数据*——恰当的系统维护所需的程序、指南和检查清单，包括预防性措施（清理和调整）、分析和诊断、故障隔离、故障定位、零件清单、纠正性维护（移除并更换）、校准、修改和升级指令、维护后的确认等。对于具有基于状态的维护（CBM）能力的系统而言，技术文件将包括关于这些能力如何被用于保障维护的信息。

- *设施及基础建设*——系统维护所需要的设施（如，建筑、仓库、机库、水道）及基础建设（如，IT 服务、燃料、水、电气服务、加工车间、干船坞、试验场）。

- *工具和支持设备*——用于支持系统维护的常用和特殊用途工具（如，手工工具、仪表）和支持设备（如测试装置、起重机）。

TECHNICAL PROCESSES

- *Design interface/sustaining engineering*—Design interface attempts to "design in " supportability features of the system. Supportability considerations minimize the logistics footprint, maximize reliability, ensure effective maintainability, and address the long-term issues related to obsolescence management, technology refreshment, modifications and upgrades, and overall usage under all operating conditions. Sustaining engineering ensures the continued operation and maintenance of a system with managed (i.e., known) risk. Activities include collection and analysis of use and maintenance data; analysis of safety hazards, failure causes and effects, reliability and maintainability trends, and operational usage profile changes; root cause analysis of in-service problems (including operational hazards, problem reports, parts obsolescence, corrosion effects, and reliability degradation); development of required design changes to resolve operational issues; and other activities necessary to ensure cost-effective support to achieve required levels of readiness and performance over the system's life cycle.

- *Maintenance planning and management*—The focus of the maintenance planning process is to define accessibility, diagnostics, repair, and sparing requirements; identify factors that impact the system's designed utilization rates (e.g., maintenance man hours per maintenance action, maintenance ratios); identify life cycle supportability design, installation, maintenance, and operating constraints and guide-lines; and provide level of Repair Analysis (LORA).

4.13.2.3 Maintenance Techniques The following are some maintenance techniques that can be used for the SOI. This is by no means an exhaustive list:

- *Condition-Based Maintenance (CBM)* is a strategy to improve system reliability (by reducing the amount of time the system is unavailable while conducting routine or corrective maintenance). Well-designed systems include the cost effective use of sensors (e.g., airflow, vibration, thermal, viscosity) and integrated, analysis-based decision support capabilities to determine/predict the need for maintenance actions required to prevent a failure. System technical documents will include a description of any embedded condition based maintenance capabilities and how resultant maintenance actions will be performed.

- *设计接口/维持工程*——设计接口试图将系统的可支持性特征融入设计之中。可支持性因素使后勤占用空间最小化，使可靠性最大化，确保有效的可维护性并在所有运行条件下应对与报废管理、技术更新、修改和升级以及总体使用相关的长期问题。维持工程确保一个系统持续运行和维护的风险受管理（即，已知的）。活动包括收集和分析使用数据和维护数据；分析安全危害、失效原因和影响、可靠性和可维护性趋势及运行上的使用概况变化；对运行中的问题（包括运行危害、问题报告、零件报废、腐蚀效应和可靠性退化）的根本原因分析；开发所要求的设计变更以化解运行问题；以及确保具成本效益的保障所必需的其他活动，以便在整个系统生命周期中实现所需的准备度和性能水平。

- *维护规划和管理*——维护规划流程的关注要点是定义可达性、诊断、维修和备件需求；识别影响系统的设计利用率的因素[每一维护行动的维护人员的维护时间（小时）、维护比率]；识别生命周期可支持性设计、安装、维护和运行约束及指南；以及提供维修层级分析（LORA）。

4.13.2.3 维护技术 以下是可以用于 SOI 的一些维护技术。这并不是一份详尽列表：

- *基于状态的维护（CBM）*是提高系统可靠性（通过在实施常规或纠正性维护的同时缩短系统不可用的时间量）的策略。设计良好的系统包括有成本效益地使用传感器（如，气流、振动、热量、粘度）以及综合的基于分析的决策支持能力，以确定/预测对防止失效所需的维护行动的需要。系统技术文件将包括对于嵌入的任何基于状态的维护能力以及产生的维护行动将如何被执行方面的描述。

TECHNICAL PROCESSES

- *Reliability-centered maintenance (RCM)* is a cost effective maintenance strategy to address dominant causes of equipment failures [supported by failure modes, effects, and criticality analysis (FMECA) and fault tree analysis (FTA)]. It provides a systematic approach to defining a routine maintenance program composed of cost-effective tasks that preserve important functions. SAE International (SAE) JA1011:2009 provides detailed information on the RCM process. RCM is a strategy to improve system reliability (by reducing the amount of time the system is unavailable while conducting routine or preventive maintenance).

- *Performance-based life cycle product support* is a strategy for cost-effective system support. Rather than contracting for maintenance (goods and services) needed to sustain operations, the customer and the service provider(s) agree on the delivery of performance outcomes (defined by performance metric(s) for a system or product). When properly implemented, the provider is incentivized (increased profits or contract extensions) to develop sustainment strategies (e.g., system reliability investment, inventory management practices, logistics arrangements) necessary to meet the required performance outcomes at reduced costs. PBL approaches usually result in sustained performance outcomes at a lower cost than those achieved under a non-PBL or transactional product support arrangements for system material and maintenance.

4.14 DISPOSAL PROCESS

4.14.1 Overview

4.14.1.1 Purpose As stated in ISO/IEC/IEEE 15288,

[6.4.14.1] The purpose of the Disposal process is to end the existence of a system element or system for a specified intended use, appropriately handle replaced or retired elements, and to properly attend to identified critical disposal needs.

The disposal process is conducted in accordance with applicable guidance, policy, regulations, and statutes throughout the system life cycle.

- *以可靠性为中心的维护（RCM）*是具有成本效益的维护策略，用以应对[由失效模式、影响与危害性分析（FMECA）和故障树分析（FTA）支持的]设备失效的主导原因。它提供定义常规维护项目的系统化方法，该常规维护项目由保存重要功能的具成本效益的任务组成。国际汽车工程师学会（SAE）JA1011:2009 提供关于 RCM 流程的详细信息。RCM 是提高系统可靠性的策略（通过在实施常规或预防性维护的同时缩短系统不可用的时间量）。

- *基于性能的生命周期产品支持*是具成本效益的系统支持的策略。客户和服务提供商对性能结果（由系统或产品的性能测度所定义）的交付达成一致，而不是将维持运行所需的维护（货物与服务）给分包商。当恰当地实施时，鼓励（增加利润或续约合同）提供商开发以降低的成本满足所需性能结果所必需的维持策略（如，系统可靠性投资、库存管理实践、后勤安排）。PBL 方法通常导致以比按照非 PBL 方法或系统材料与维护的事务产品支持安排所实现的性能结果更低的成本维持性能结果。

4.14 处置流程

4.14.1 概览

4.14.1.1 目的 如 ISO/IEC/IEEE 15288 所述，

[6.4.14.1] 处置流程的目的是结束系统元素或系统的存在以用于特定的预期使用，恰当地处理被替代或退出的元素，以及恰当地处理所识别的关键处置需要。

处置流程按照适用的指南、方针、法规和条例贯穿于系统生命周期内实施。

TECHNICAL PROCESSES

4.14.1.2 Description Disposal is a life cycle support process because concurrent consideration of disposal during the development stage generates requirements and constraints that must be balanced with defined stakeholders' requirements and other design considerations.Further, environmental concerns drive the designer to consider reclaiming the materials or recycling them into new systems. This process can be applied for the incremental disposal requirements at any point in the life cycle, for example, prototypes that are not to be reused or evolved, waste materials during manufacturing, or parts that are replaced during maintenance. Figure 4.22 is the IPO diagram for the disposal process.

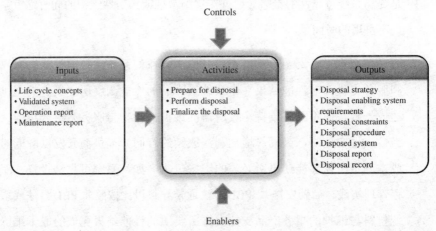

FIGURE 4.22 IPO diagram for the disposal process. INCOSE SEH original figure created by Shortell and Walden. Usage per the INCOSE notices page. All other rights reserved.

4.14.1.3 Inputs/Outputs Inputs and outputs for the disposal process are listed in Figure 4.22. Descriptions of each input and output are provided in Appendix E.

4.14.1.4 Process Activities The disposal process includes any steps necessary to return the environment to an acceptable condition; handle all system elements and waste products in an environmentally sound manner in accordance with applicable legislation, organizational constraints, and stakeholder agreements; and document and retain records of disposal activities, as required for monitoring by external oversight or regulatory agencies. In general, disposal process includes the following activities:

4.14.1.2 描述 由于在开发阶段中并行考虑处置所产生的需求和约束，必须与定义明确的利益攸关者需求及其他设计因素相平衡，因此处置是一个生命周期支持流程。此外，环境问题驱使设计者考虑回收物料或在新系统中循环再利用。该流程可在生命周期的任一时刻应用于渐进处置需求，例如不被复用或演进的原型、制造期间的废物或维护期间被取代的零件。图 4.22 是处置流程的 IPO 图。

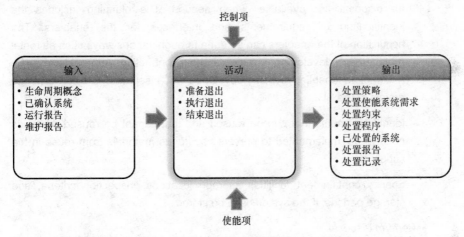

图 4.22 处置流程的 IPO 图。INCOSE SEH 原始图由 Shortell 和 Walden 创建。

4.14.1.3 输入/输出 图 4.22 列出处置流程的输入和输出。附录 E 提供每个输入和输出的描述。

4.14.1.4 流程活动 处置流程包括将环境返回到可接受状态所必需的任何步骤；按照适用的法规、组织的约束和利益攸关者协议，以对环境无害的方式处理所有系统元素和废弃产品；按照外部监督机构或管理机构的监测要求，文件化并保留处置活动记录。总之，处置流程包括下列活动：

- *Prepare for disposal.*
  - Review the retirement concept (may be called concept of disposal), including any hazardous materials and other environmental impacts to be encountered during disposal.
  - Plan for disposal, including the development of the strategy.
  - Impose associated constraints on the system requirements.
  - Ensure that the necessary enabling systems, products, or services required for disposal are available, when needed. The planning includes the identification of requirements and interfaces for the enablers. The acquisition of the enablers can be done through various ways such as rental, procurement, development, reuse, and subcontracting. An enabler may be a complete enabling system developed as a separate project from the project of the SOI.
  - Identify elements that can be reused and that cannot be reused. Methods need to be implemented to prevent hazardous materials from reuse in the supply chain.
  - Specify containment facilities, storage locations, inspection criteria, and storage periods, if the system is to be stored.
- *Perform disposal.*
  - Decommission the system elements to be terminated.
  - Disassemble the elements for ease of handling. Include identification and processing of reusable elements.
  - Extract all elements and waste materials that are no longer needed—this includes removing materials from storage sites, consigning the elements and waste products for destruction or permanent storage, and ensuring that the waste products cannot get back into the supply chain.
  - Dispose of deactivated system element per disposal procedure.
  - Remove affected staff and capture the tacit knowledge for future needs.
- *Finalize the disposal.*
  - Confirm no adverse effects from the disposal activities and return the environment to its original state.

- *准备退出*。
    - 评审退出概念（可能被称为处置概念），包括任何危险物料及在退出期间遇到的其他环境影响。
    - 计划处置，包括策略的开发。
    - 对系统需求施加相关约束。
    - 确保在需要时处置所需的必要的使能系统、产品或服务是可用的。规划包括使能项的需求和接口的识别。使能项的采办可以通过各种不同的方式来完成，如租赁、采购、开发、复用和转包。使能项可能是被开发为 SOI 项目中的独立项目的完整使能系统。
    - 识别能够被复用以及不能被复用的元素。需要对方法进行实施，以防止在供应链中复用危险物料。
    - 若要存储系统，则规定防范设施、存储位置、检验准则和存储期。
- *实施处置*。
    - 停用将要被终结的系统元素。
    - 分解元素以便于处理。包括可复用元素的识别和处理。
    - 提取不再需要的所有元素和废料——包括从储存场地移除物料，将元素和废弃产品销毁或永久性储存，以及确保废品不会返回到供应链中。
    - 按照处置程序退出已停用的系统元素。
    - 使受影响的人员离开并获取隐性知识供未来之需。
- *终结处置*。
    - 确认处置活动中没有不利影响并使环境返回为其初始状态。

TECHNICAL PROCESSES

— Maintain documentation of all disposal activities and residual hazards.

Common approaches and tips:

- The project team conducts analyses to develop solutions for ultimate disposition of the system, constituent elements, and waste products based on evaluation of alternative disposal methods available.Methods addressed should include storing, dismantling, reusing, recycling, reprocessing, and destroying end products, enabling systems, system elements, and materials.

- Disposal analyses include consideration of costs, disposal sites, environmental impacts, health and safety issues, responsible agencies, handling and shipping, supporting items, and applicable federal, state, local, and host-nation regulations.

- Disposal analyses support selection of system elements and materials that will be used in the system design and should be readdressed to consider design and project impacts from changing laws and regulations throughout the project life cycle.

- Disposal strategy and design considerations are updated throughout the system life cycle in response to changes in applicable laws, regulations, and policy.

- Consider donating an obsolete system—Many items, both systems and information, of cultural and historical value have been lost to posterity because museums and conservatories were not considered as an option during the retirement stage.

- Concepts such as zero footprint and zero emissions drive current trends toward corporate social responsibility that influence decision making regarding cleaner production and operational environments and eventual disposal of depleted materials and systems.

- The ISO 14000 series includes standards for environmental management systems and life cycle assessment (ISO 14001, 2004).

- Instead of designing cradle-to-grave products, dumped in landfills at the end of their "life," a new concept is transforming industry by creating products for cradle-to-cradle cycles, whose materials are perpetually circulated in closed loops. Maintaining materials in closed loops maximizes material value without damaging ecosystems (McDonough, 2013).

— 保存所有处置活动和残留危险物的文件。

常用方法和建议：

- 项目团队基于对可用的备选处置方法的评估来实施分析以开发系统、组成元素及废弃产品最终处置的解决方案。所涉及的方法应包括储存、分解、复用、回收、重新处理并销毁最终产品、使能系统、系统元素和物料。

- 处置分析包括成本因素、处置场地、环境影响、健康和安全性问题、责任机构、处理和装运、支持项及适用的联邦、国家、当地与所在国的规定。

- 处置分析对将用于系统设计的系统元素和物料的选择提供支持，并应被重新审视以考虑贯穿于项目生命周期内不断变化的法律和法规对设计和项目的影响。

- 针对适用法律、法规和方针的变化，贯穿于系统生命周期内更新处置策略和设计因素。

- 考虑捐赠废弃的系统——由于展览馆和艺术博物馆在退出阶段期间并没有被当作一个选项，许多具有文化价值和历史价值的项目、系统和信息没有留给后代。

- 诸如零占用空间和零排放等概念驱使当前企业社会责任发展的趋势，这影响与更清洁的生产和运行环境以及废弃的物料和系统的最终处置相关的决策。

- ISO 14000 系列包括环境管理系统和生命周期评估的标准（ISO 14001, 2004）。

- 为了取代设计在其"生命"终结时被废弃于垃圾堆中的从摇篮到坟墓的产品，新的概念正在通过创建"从摇篮到摇篮"循环的产品来转变行业，所述产品的物料在闭环中永久性地循环。将材料保持在闭环中，在不损坏生态系统的情况下，使物料价值最大化（McDonough, 2013）。

5 TECHNICAL MANAGEMENT PROCESSES

Within the system life cycle, the creation or upgrade of products and services is managed by the conduct of projects. For this reason, it is important to understand the contribution of systems engineering (SE) to the management of the project. Technical management processes are defined in ISO/IEC/IEEE 15288 as follows:

> [6.3] The Technical Management Processes are used to establish and evolve plans, to execute the plans, to assess actual achievement and progress against the plans and to control execution through to fulfillment. Individual Technical Management Processes may be invoked at any time in the life cycle and at any level …

Systems engineers continually interact with project management. Systems engineers and project managers bring unique skills and experiences to the program on which they work. A life cycle from project manager's point of view (project start–project end) is defined differently than from the systems engineer's point of view (product idea–product disposal). But there is a "shared space" where systems engineers and project managers have to collaborate to drive the team's performance and success (Langley et al., 2011).

Technical management processes include project planning, project assessment and control, decision management, risk management, configuration management, information management, measurement, and quality assurance (QA). These processes are found throughout an organization as they are essential to generic management practices and apply both inside and outside the project context. This chapter of the handbook focuses on processes relevant to the technical coordination of a project.

5.1 PROJECT PLANNING PROCESS

5.1.1 Overview

5.1.1.1 Purpose As stated in ISO/IEC/IEEE 15288,

> [6.3.1.1] The purpose of the Project Planning process is to produce and coordinate effective and workable plans.

INCOSE Systems Engineering Handbook: A Guide for System Life Cycle Processes and Activities, Fourth Edition. Edited by David D. Walden, Garry J. Roedler, Kevin J. Forsberg, R. Douglas Hamelin and Thomas M. Shortell.

© 2015 John Wiley & Sons, Inc. Published 2015 by John Wiley & Sons, Inc.

5 技术管理流程

在系统生命周期内，产品和服务的创建或升级是通过项目的实施来管理的。因此，理解系统工程（SE）对项目管理的贡献是非常重要的。技术管理流程在 ISO/IEC/IEEE 15288 中定义如下：

> [6.3] 技术管理流程用于建立和演进计划、执行计划、按计划评估实际的成果和进度，以及对计划的执行进行控制，直至完成。单独的技术管理流程可在生命周期的任何时候和任何层级被引用……

系统工程师与项目管理人员不断地交互。系统工程师和项目经理为他们所工作的项目群带来独特的技能和经验。从项目经理的视角（项目开始—项目结束）对生命周期的定义不同于从系统工程师的视角（产品构想—产品退出）的定义。但是存在一个"共享空间"，在该空间系统工程师和项目经理必须协同以驱动团队的绩效和成功（Langley 等，2011）。

技术管理流程包括项目规划、项目评估和控制、决策管理、风险管理、构型管理、信息管理、测量和质量保证（QA）。由于这些流程对于通用管理实践而言是必要的且在项目背景环境的内部和外部均适用，因此会发现它们贯穿于整个组织中。本手册的这一章集中于与项目的技术协调相关的流程。

5.1 项目规划流程

5.1.1 概览

5.1.1.1 目的 如 ISO/IEC/IEEE 15288 所述，

> [6.3.1.1] 项目规划流程的目的在于产生并协调有效且切实可行的计划。

INCOSE 系统工程手册：系统生命周期流程和活动指南，第 4 版。编撰：David D. Walden, Garry J. Roedler, Kevin J. Forsberg, R. Douglas Hamelin 和 Thomas M. Shortell。
John Wiley & Sons 公司版权所有©2015。由 John Wiley & Sons 公司于 2015 年出版。

TECHNICAL MANAGEMENT PROCESSES

5.1.1.2 Description Project planning starts with the identification of a new potential project and continues after the authorization and activation of the project until its termination. The project planning process is performed in the context of the organization. The life cycle model management process (see Section 7.1) establishes and identifies relevant policies and procedures for managing and executing a technical effort; identifying the technical tasks, their interdependencies, risks, and opportunities; and providing estimates of needed resources and budgets. The planning includes the determination of the need for specialized equipment, facilities, and specialists during the project to improve efficiency and effectiveness and decrease cost overruns. This requires coordination across the set of processes. For example, different disciplines work together in the performance of system requirements definition, architecture definition, and design definition processes to evaluate the parameters such as manufacturability, testability, operability, maintainability, and sustainability against product performance. Project tasking may be concurrent to achieve the best results.

Project planning establishes the direction and infrastructure necessary to enable the assessment and control of the project progress and identifies the details of the work and the right set of personnel, skills, and facilities with a schedule for needed resources from within and outside the organization. Project planning establishes the direction and infrastructure necessary to enable the assessment and control of the project progress and identifies the details of the work and the right set of personnel, skills, and facilities with a schedule for needed resources from within and outside the organization. Figure 5.1 is the IPO diagram for the project planning process.

5.1.1.3 Inputs/Outputs Inputs and outputs for the project planning process are listed in Figure 5.1. Descriptions of each input and output are provided in Appendix E.

5.1.1.4 Process Activities The project planning process includes the following activities:

- *Define the project.*
    - Analyze the project proposal and related agreements to define the project objectives, scope, and constraints.
    - Establish tailoring of organization procedures and practices to carry out planned effort. Chapter 8 contains a detailed discussion on tailoring.

5.1.1.2　描述　项目规划开始于对新的潜在项目的识别，并在项目的授权和激活后继续，直至项目终止。项目规划流程在组织的背景环境中被实施。生命周期模型管理流程（见 7.1 节）建立并识别关于管理和执行技术工作的相关方针和程序；识别技术任务及相互依赖关系、风险和机会；并提供对所需资源和预算的估计。规划包括在项目期间确定对专业化设备、设施及专家的需要以提高效率和有效性并减少成本超支。这要求跨流程集的协调。例如，在系统需求定义流程、架构定义流程和设计定义流程的实施中不同的学科的共同工作，以按照产品性能来评估参数，如可制造性、可测试性、可运行性、可维护性和可持续性。为达成最佳结果，项目的任务可以是并行的。

项目规划建立能够评估和控制项目进度所必需的指导和基础设施，并使用组织内外所需资源的进度表来识别工作细节以及人员、技能和设施的正确集合。图 5.1 是项目规划流程的 IPO 图。

5.1.1.3　输入/输出　图 5.1 列出项目规划流程的输入和输出。附录 E 提供每个输入和输出的描述。

5.1.1.4　流程活动　项目规划流程包括以下活动：

- 定义项目。

    – 分析项目建议书和相关协议书以定义项目的目的、范围和约束。

    – 建立组织程序和实践的剪裁，以开展计划的工作。第 8 章包含对剪裁的详细论述。

TECHNICAL MANAGEMENT PROCESSES

– Establish a work breakdown structure (WBS) based on the evolving system architecture.

– Define and maintain a life cycle model that is tailored from the defined life cycle models of the organization. This includes the identification of major milestones, decision gates, and project reviews.

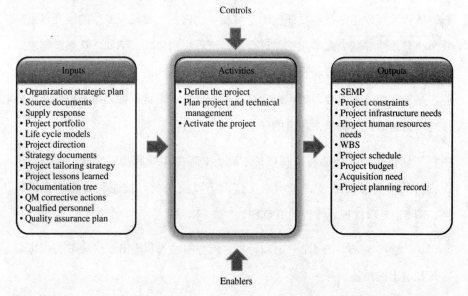

FIGURE 5.1 IPO diagram for the project planning process. INCOSE SEH original figure created by Shortell and Walden. Usage per the INCOSE notices page. All other rights reserved.

- *Plan project and technical management.*

  – Establish the roles and responsibilities for project authority.

  – Define top-level work packages for each task and activity identified. Each work package should be tied to required resources including procurement strategies.

  – Develop a project schedule based on objectives and work estimates.

  – Define the infrastructure and services required.

  – Define costs and estimate project budget.

  – Plan the acquisition of materials, goods, and enabling system services.

- 基于演进的系统架构建立工作分解结构（WBS）。

- 定义和维护从组织已定义的生命周期模型中经过剪裁的一个生命周期模型。这包括对主要里程碑、决策门和项目评审的识别。

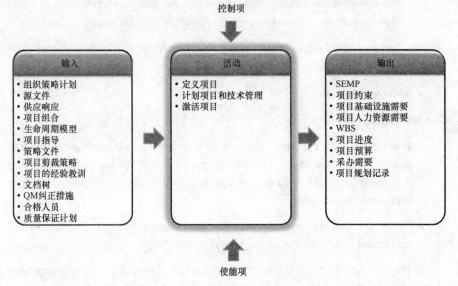

图 5.1 项目规划流程的 IPO 图。INCOSE SEH 原始图由 Shortell 和 Walden 创建。按照 INCOSE 通知页使用。版权所有。

- 规划项目和技术管理。

  - 为项目权限建立角色和职责。

  - 为已识别的每个任务和活动定义顶层工作包。每个工作包应受所需资源（包括采购策略）的限制。

  - 基于目标和工作预估开发项一个目进度安排表。

  - 定义所要求的基础设施和服务。

  - 定义成本并估计项目预算。

  - 计划资料、物品和使能系统服务的采办。

TECHNICAL MANAGEMENT PROCESSES

- Prepare a systems engineering management plan (SEMP) or systems engineering plan (SEP), including the reviews that will be performed across the life cycle.

- Generate or tailor quality management, configuration management, risk management, information management, and measurement plans to meet the needs of the project (may be the SEMP or SEP for smaller projects).

- Establish the criteria to be used for major milestones, decision gates, and internal reviews.

- *Activate the project.*

Common approaches and tips:

- The SEMP is an important outcome of planning that identifies activities, key events, work packages, and resources. It references other planning documents that are tailored for use on the project as discussed in later sections of this chapter.

- IPDTs are frequently used to break down communications and knowledge stovepipes within organizations (Martin, 1996).

- The creation of the WBS is an activity where SE and project management intersect (Forsberg et al., 2005). Sometimes, software engineers presume that these tools apply only to large hardware projects, and they avoid the project management tools needed for success. Dr. Richard Fairley has documented an excellent approach to dispel this notion (Fairley, 2009).

- Skipping or taking shortcuts in the planning process reduces the effectiveness of other technical management processes.

- Agile project management methods also include planning—the cycles may be shorter and more frequent, but planning is an essential process.

- Defining project objectives and the criteria for success are critical to successful projects. The project value to the stakeholders should be clearly defined to guide project decision making. The project value should be expressed in technical performance measures (TPMs) (Roedler and Jones, 2006).

- 准备系统工程管理计划（SEMP）或系统工程计划（SEP），包括将要跨生命周期实施的评审。

- 生成或剪裁质量管理计划、构型配置管理计划、风险管理计划、信息管理计划和测度计划以满足项目的需要（对于较小项目而言，可能是 SEMP 或 SEP）。

- 建立用于主要里程碑、决策门和内部评审的准则。

- 激活项目。

常用方法和建议：

- SEMP 是识别活动、关键事件、工作包和资源规则的一个重要成果。它引用其他规划文件，这些文件被剪裁以便在项目上使用，本章后面几节有论述。

- IPDT 经常被用于打破组织内的沟通障碍和知识壁垒（Martin，1996）。

- 创建 WBS 是 SE 与项目管理人员交叉的一个活动（Forsberg 等，2005）。有时，软件工程师假定这些工具仅适用于大型硬件项目，他们忽略成功所需的项目管理工具。Richard Fairley 博士已翔实记录了一种优异的方法以消除这种观念（Fairley，2009）。

- 跳过或在规划流程中采取捷径会降低其他技术管理流程的有效性。

- 敏捷项目管理方法也包括规划——周期可能较短且较频繁，但规划是一个基本的流程。

- 定义项目的目标和成功的准则对于成功的项目而言是极为关键的。应当明确地定义项目对于利益攸关者的价值，以便引导项目决策。应当以技术性能测度（TPM）来表达项目价值（Roedler 和 Jones，2006）。

TECHNICAL MANAGEMENT PROCESSES

- Incorporate risk assessment early in the planning process to identify areas that need special attention or contingencies. Always attend to the technical risks (PMI, 2013).

- The Project Management Institute is a source of guidelines for project planning.

- The ISO/IEC/IEEE 16326 standard for project management also provides additional guidance on this subject (ISO/IEC/IEEE 16326, 2009).

5.1.2 Elaboration

5.1.2.1 Project Planning Concepts Project planning estimates the project budget and schedule against which project progress will be assessed and controlled. Systems engineers and project managers must collaborate in project planning. Systems engineers perform technical management activities consistent with project objectives. Technical management activities include planning, scheduling, reviewing, and auditing the SE process as defined in the SEMP and the SE Master Schedule (SEMS).

5.1.2.2 SEMP The SEMP is the top-level plan for managing the SE effort. It defines how the project will be organized, structured, and conducted and how the total engineering process will be controlled to provide a product that satisfies stakeholder requirements. A well written SEMP provides guidance to a project and helps the organization avoid unnecessary discussions about how to perform SE. Organizations generally maintain a template of the SEMP suitable for tailoring and reuse. Effective project control requires that there be a SEMP, which the systems engineer keeps current and uses on a daily basis to manage the team's actions.

The SEMS is an essential part of the SEMP and a tool for project control because it identifies the critical path of technical activities in the project. Verification activities may also receive special attention in the SEMS. In addition, the schedule of tasks and dependencies helps justify requests for personnel and resources needed throughout the development life cycle.

The SEMP and SEMS are supported by a project or contract WBS that defines a project task hierarchy. Work authorization is the process by which the project is baselined and financially controlled. A description of the organization procedures for starting work on a part of the WBS may be defined in the SEMP.

- 尽早地将风险评估纳入规划流程中，以识别需要特殊关注的领域或意外事件。要始终关注技术风险（PMI，2013）。
- 项目管理协会是项目规划指南的来源。
- 针对项目管理的 ISO/IEC/IEEE 16326 标准亦提供有关该主题的附加指南（ISO/IEC/IEEE 16326，2009）。

5.1.2 详细阐述

5.1.2.1 项目规划方案 项目规划对评估和控制项目进展所依据的项目预算和进度进行估计。系统工程师和项目经理必须在项目规划中密切协同。系统工程师执行与项目目标相一致的技术管理活动。如在 SEP 和 SE 主进度计划（SEMS）中所定义的，技术管理活动包括规划、进度安排、评审和审计 SE 流程。

5.1.2.2 SEMP SEMP 是管理 SE 工作的顶层规划。其定义如何组织、构成和实施项目以及总体工程流程将如何被控制以提供满足利益攸关者需求的产品。一份完好写明的 SEMP 为项目提供指南，并帮助组织省去有关如何实施 SE 的不必要的讨论。组织通常保留适于剪裁和复用的 SEMP 模板。有效的项目控制要求有一个 SEMP，系统工程师保持现行有效且日常使用的 SEMP，以管理团队的行动。

由于 SEMS 识别项目中的技术活动的关键路径，因此 SEMS 是 SEMP 的一个基本组成部分，也是一个项目的控制工具。验证活动在 SEMS 中也可能受到特别关注。此外，任务和依赖性的进度安排有助于证明贯穿于开发生命周期内所需人员和资源请求的合理性。

通过定义项目任务层级结构的项目或合同 WBS 来支持 SEMP 和 SEMS。工作权限是项目被基线化并受到财务控制所依据的流程。针对启动 WBS 部分工作的组织程序的描述可在 SEMP 中定义。

TECHNICAL MANAGEMENT PROCESSES

TPMs (see Section 5.7.2.8) are a tool used for project control, and the extent to which TPMs will be employed should be defined in the SEMP (Roedler and Jones, 2006).

A SEMP should be prepared early in the project, submitted to the customer (or to management for in-house projects), and used in technical management for the concept and development stages of the project or the equivalent in commercial practice. The creation of the SEMP involves defining the SE processes, functional analysis approaches, what trade studies will be included in the project, schedule, and organizational roles and responsibilities, to name a few of the more important aspects of the plan. The SEMP also reports the results of the effort undertaken to form a project team and outlines the major deliverables of the project, including a decision database, specifications, and baselines. Participants in the creation of the SEMP should include senior systems engineers, representative subject matter experts, project management, and often the customer.

The format of the SEMP can be tailored to fit project, customer, or company standards. To maximize reuse of the SEMP for multiple projects, project-specific appendices are often used to capture detailed and dynamic information, such as the decision database, a schedule of milestones and decision gate reviews, and the methodology to be used in resolving problems uncovered in reviews.

The process inputs portion of the SEMP identifies the applicable source documents (e.g., customer specifications from the RFP, SOW, standards, etc.) to be used in the performance of the project and in developing associated deliverables (e.g., the system specification and technical requirements document). It also may include previously developed specifications for similar systems and company procedures affecting performance specifications. A technical objectives document should be developed and may be one of the source documents for the decision database. The document may also be part of the ConOps for the system (see Section 4.2.2.4).

The SEMP should include information about the project organization, technical management, and technical activities. A complete outline of a SEMP is available in ISO/IEC/IEEE 24748-4 (2014), which is aligned with ISO/IEC/IEEE 15288 and this handbook. As a high-level overview, the SEMP should include the following:

- Organization of the project and how SE interfaces with the other parts of the organization

- Responsibilities and authority of the key positions

TPM（见 5.7.2.8 节）是一种用于项目控制的工具，同时 TPM 的使用程度应在 SEMP 中定义（Roedler 和 Jones，2006）。

应在项目早期准备 SEMP，提交给客户（或内部项目的管理人员），并且用于项目的概念阶段和开发阶段的技术管理中，或同等地用于商业实践中。SEMP 的创建涉及定义 SE 流程、功能分析方法、哪些权衡研究将被包含在项目中、进度以及组织的角色和职责，以列举该计划的一些较为重要的方面。SEMP 亦报告为了形成一个项目团队所承担工作的结果，并概述项目的主要可交付物，包括决策数据库、规范和基线。创建 SEMP 的参与者应包括高级系统工程师、主题专家代表、项目管理人员并常常包含客户。

SEMP 的格式可被剪裁，以适配项目、客户或公司的标准。为了针对多个项目最大化复用 SEMP，项目特定的附录通常用于捕获详细的和动态的信息，如决策数据库、里程碑和决策门的评审的进度安排、以及在解决评审中所暴露问题时所要使用的方法论。

SEMP 的流程输入部分识别在实施项目中和开发相关可交付物（如系统规范和技术需求文件）中所要使用的适用源文件（如 RFP 中的客户规范、SOW、标准等），亦可包括为类似系统预先开发的规范和影响性能规范的公司程序。应开发一个技术目标文件，且该文件可作为决策数据库的源文件之一。该文件亦可作为系统的 ConOps 的一部分（见 4.2.2.4 节）。

SEMP 应包括有关项目组织、技术管理和技术活动的信息。可在与 ISO/IEC/IEEE 15288 和本手册一致的 ISO/IEC/IEEE 24748-4（2014）中得到 SEMP 的完整概述。作为高层级概览，SEMP 应包括以下内容：

- 项目的组织以及 SE 如何与组织的其他部分进行交互

- 关键职位的职责和权限

TECHNICAL MANAGEMENT PROCESSES

- Clear system boundaries and scope of the project
- Project assumptions and constraints
- Key technical objectives
- Infrastructure support and resource management (i.e., facilities, tools, IT, personnel, etc.)
- Approach and methods used for planning and executing the technical processes described in this handbook (see Chapter 4)
- Approach and methods used for planning and executing the technical management processes described in this handbook (see Chapter 5)
- Approach and methods used for planning and executing the applicable specialty engineering processes described in this handbook (see Chapter 10)

The SEMP may sometimes address affordability/cost-effectiveness/life cycle cost (LCC) analysis (see Section 10.1) and value engineering (see Section 10.14) practices to provide insight into system/cost-effectiveness. For example, can the project be engineered to have significantly more value with minimal additional cost? If so, does the customer have the resources for even the modest cost increase for the improvement? Can the solutions be achieved within their budgets and schedules? This assures the customer that obvious cost-effective alternatives have been considered (ISO/IEC/IEEE 24748-4, 2014).

Technical reviews are essential to ensure that the system will meet the requirements and that the requirements are understood by the development team. Formal reviews are essential to determine readiness to proceed to the next stage of the system life cycle. The number and frequency of these reviews and their associated decision gates must be tailored for specific projects. The SEMP should list what technical reviews will be conducted and the methodology to be used in solving problems uncovered during those reviews. The SEMP should list what technical reviews will be conducted and the methodology to be used in solving problems uncovered during those reviews.

The system life cycles shown in Figure 3.3 illustrate the appropriate time for reviews and decision gates. They may not fit all projects, and some projects may need more or fewer reviews. Additionally, formal, documented decision gates, with the customer in attendance, can impose significant cost on the project. Projects should plan to use more frequent, informal, in-house reviews to resolve most issues and strive to exit decision gates with no major customer-imposed action items.

- 清晰的项目边界和项目范围

- 项目的假设和约束

- 关键技术目标

- 基础设施的保障和资源管理（即设施、工具、IT、人员等）

- 用于规划和执行本手册所描述的技术管理流程（见第 4 章）的途径和方法

- 用于规划和执行本手册所描述的技术流程（见第 5 章）的途径和方法

- 用于规划和执行本手册所描述的适用的专业工程流程（见第 10 章）的途径和方法

SEMP 有时可能要应对可支付性/成本效益/生命周期成本（LCC）分析（见 10.1 节）和价值工程（见 10.14 节）实践，以提供对系统/成本效益的深入理解。例如，项目能否被设计成以最低的附加成本实现更多的价值？若是，即使成本有适量的增长客户是否有资源来进行改进？解决方案能否在其预算和进度安排内完成？这确保客户已经考虑到明显的成本效益备选方案（ISO/IEC/IEEE 24748-4, 2014）。

技术评审对于确保系统将满足需求以及开发团队理解需求而言是根本性的。正式评审对于确定继续进行系统生命周期下一阶段的准备度而言是根本性的。这些评审及其相关决策门的数量和频度必须针对特定项目而进行剪裁。SEMP 应列出将要实施的技术评审以及在解决评审期间暴露出的问题所采用的方法论。

图 3.3 所示的系统生命周期阐明评审和决策门的合适时间。它们可能不适配所有项目，有一些项目可能需要更多或更少的评审。此外，客户参与的文件化的正式决策门可使项目产生大量成本。项目应计划使用更频繁的且非正式的内部评审，以化解大多数问题；并极力退出那些不含主要客户施加的行动项的决策门。

TECHNICAL MANAGEMENT PROCESSES

Transitioning critical technologies should be done as a part of risk management (see Section 5.4) but is called out separately here for special emphasis. Critical technologies should be identified, and the steps outlined for risk management should be followed. Additionally, completed and planned risk management work should be explicitly referenced in the SEMP.

The system being proposed may be complex enough that the customer will require training to use it. During the project, it may be necessary to train those who will develop, manufacture, verify, deploy, operate, support, conduct training, or dispose of the system. A plan for this training is required in the SEMP and should include the following:

1. Analysis of performance
2. Behavior deficiencies or shortfalls
3. Required training to remedy deficiencies or shortfalls
4. Schedules to achieve required proficiencies

Verification is usually planned using a verification matrix that lists all the requirements and anticipated verification methods. The possible methods of verification include inspection, analysis, demonstration, and test. The SEMP should state, at least in preliminary general terms, that a verification plan will be written to define the items to be verified and which methods will be used to verify performance. The plan should also define who is to perform and witness the verification of each item. This should also relate to the SEMS for time phasing of the verification process. Detailed procedures are usually not written for inspection, analysis, and demonstration methods. Simulations may be used for testing, when quantifiable results are needed, or for demonstration, when qualitative results are satisfactory.

A well-written SEMP provides guidance to a project and helps the organization avoid unnecessary discussions about how to perform SE. In addition, a schedule and organization are defined that help the project procure the personnel needed throughout the development life cycle and assess progress.

转移关键技术应作为风险管理的一部分（见 5.4 节）来执行，但为特殊强调，则在本节被单独引用。应识别关键技术，并遵守针对风险管理所概括的步骤。此外，已完成的和已规划的风险管理工作应在 SEMP 中被明确地引用。

正在被建议的系统可能非常复杂以至于客户需要培训才能使用。在项目期间，可能有必要对那些将从事开发、制造、验证、部署、运行、维持、实施培训或退出系统的人员进行培训。该培训的计划需要在 SEMP 中并应包括下列内容：

1. 性能的分析

2. 行为的缺陷或不足

3. 补救缺陷或不足所要求的培训

4. 达到所要求的娴熟程度的进度安排

通常使用列出所有需求和预期验证方法的验证矩阵来对验证进行规划。可能的验证方法包括检测、分析、演示和测试。SEMP 应至少在初步的一般条款中阐明：将写明验证计划以定义待验证项并定义使用哪些方法来验证性能。计划亦应定义每项验证由谁执行和见证。该计划亦应涉及 SEMS，用以对验证流程进行分段时间安排。检验、分析和演示方法中通常不写明详细的程序。当需要量化的结果时，仿真可被用于测试；或当定性结果符合要求时，仿真则可用于演示。

完好写明的 SEMP 为项目提供指南，并帮助组织省去有关如何执行 SE 的不必要的论述。此外，对进度和组织进行定义，帮助项目获得贯穿于开发生命周期内所需的人员并帮助评估进展。

TECHNICAL MANAGEMENT PROCESSES

5.2 PROJECT ASSESSMENT AND CONTROL PROCESS

5.2.1 Overview

5.2.1.1 Purpose   As stated in ISO/IEC/IEEE 15288,

> [6.3.2.1] The purpose of the Project Assessment and Control process is to assess if the plans are aligned and feasible; determine the status of the project, technical and process performance; and direct execution to ensure that the performance is according to plans and schedules, within projected budgets, to satisfy technical objectives.

Assessments are scheduled periodically and for all milestones and decision gates. The intention is to maintain good communications within the project team and with the stakeholders, especially when deviations are encountered.

The process uses these assessments to direct the efforts of the project, including redirecting the project when the project does not reflect the anticipated maturity.

5.2.1.2 Description   The project planning process (see Section 5.1) identified details of the work effort and expected results. The project assessment and control process collects data to evaluate the adequacy of the project infrastructure, the availability of necessary resources, and the compliance with project performance measures. Assessments also monitor the technical progress of the project and may identify new risks or areas that require additional investigation. A discussion of the creation and assessment of TPMs is found in Section 5.7.2.8.

The rigor of the project assessment and control process is directly dependent on the complexity of the system of interest (SOI). Project control involves both corrective and preventive actions taken to ensure that the project is performing according to plans and schedules and within projected budgets. The project assessment and control process may trigger activities within the other process areas in this chapter. Figure 5.2 is the IPO diagram for the project assessment and control process.

5.2 项目评估和控制流程

5.2.1 概览

5.2.1.1 目的　如 ISO/IEC/IEEE 15288 所述，

> [6.3.2.1] 项目评估和控制流程的目的是评估计划是否协调一致且切实可行；确定项目、技术和流程性能的状态；以及指导执行以确保在预计的预算内性能符合计划和进度安排，以便满足技术目标。

定期地安排评估并用于所有里程碑和决策门。其意图在于，保持项目团队内以及与利益攸关者的良好沟通，尤其是在遇到偏离时。

该流程使用这些评估来指导项目的工作，包括在项目不能反映预期的成熟度时，对项目进行重新定向。

5.2.1.2 描述　项目规划流程（见 5.1 节）识别工作投入的细节和预期的结果。项目评估和控制流程收集数据，以评价项目基础设施的合适性、必要资源的可用性以及与项目性能测度的符合性。评估亦监控项目的技术进度并可识别新的风险或需要额外调查研究的区域。关于 TPM 的创建和评估的论述可在 5.7.2.8 节中找到。

项目评估和控制流程的严格性直接依赖于所感兴趣之系统（SOI）的复杂性。项目控制涉及所采取的纠正措施和预防措施，以确保项目按照计划和进度在预计的预算内执行。项目评估和控制流程可能触发本章中的其他流程域的活动。图 5.2 是项目评估和控制流程的 IPO 图。

TECHNICAL MANAGEMENT PROCESSES

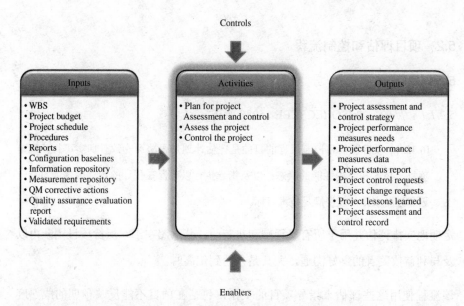

FIGURE 5.2   IPO diagram for the project assessment and control process. INCOSE SEH original figure created by Shortell and Walden. Usage per the INCOSE notices page. All other rights reserved.

5.2.1.3 Inputs/Outputs   Inputs and outputs for the project assessment and control process are listed in Figure 5.2. Descriptions of each input and output are provided in Appendix E.

5.2.1.4 Process Activities   The project assessment and control process includes the following activities:

- *Plan for project assessment and control.*
  - *Develop a strategy for project assessment and control for the system.*
- *Assess the project.*
  - Review measurement results associated with the project.
  - Determine actual and projected cost against budget, actual and projected time against schedule, and deviations in project quality.
  - Evaluate the effectiveness and efficiency of the performance of project activities.
  - Evaluate the adequacy and the availability of the project infrastructure and resources.

技术管理流程

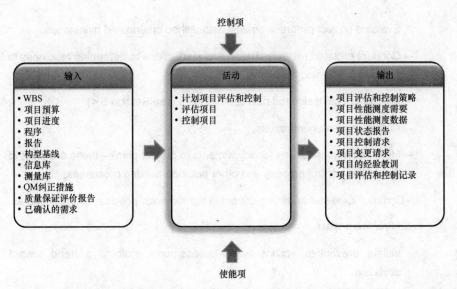

图 5.2 项目评估和控制流程的 IPO 图。INCOSE SEH 原始图由 Shortell 和 Walden 创建。按照 INCOSE 通知页使用。版权所有。

5.2.1.3 输入/输出 图 5.2 列出项目评估和控制流程的输入和输出。附录 E 提供每个输入和输出的描述。

5.2.1.4 流程活动 项目评估和控制流程包括以下活动：

- *项目评估和控制的计划。*

    – 为系统开发项目评估和控制的策略。

- *评估项目。*

    – 评审与项目相关联的测度结果。

    – 对照预算确定实际的成本和预计的成本、对照进度确定实际的时间和预计的时间并确定项目质量的偏离。

    – 评价项目活动绩效的有效率和效果。

    – 评价项目基础设施和资源的合适性和可用性。

TECHNICAL MANAGEMENT PROCESSES

- Evaluate project progress against established criteria and milestones.
- Conduct required reviews, audits, and inspections to determine readiness to proceed to the next milestone.
- Monitor critical tasks and new technologies (see Section 5.4).
- Analyze assessment results.
- Make recommendations for adjustments to project plans—these are input to the project control process and other decision-making processes.
- Communicate status as designated in agreements, policies, and procedures.

- *Control the project.*
    - Initiate preventive actions when assessments indicate a trend toward deviation.
    - Initiate problem resolution when assessments indicate nonconformance with performance success criteria.
    - Initiate corrective actions when assessments indicate deviation from approved plans.
    - Establish work items and changes to schedule to reflect the actions taken.
    - Negotiate with suppliers for any goods or services acquired from outside the organization.
    - Make the decision to proceed, or not to proceed, when assessments support a decision gate or milestone event.

Common approaches and tips:

- One way for project management to remain updated on project status is to conduct regular team meetings. Short stand-up meetings on a daily or weekly schedule are effective for smaller groups.
- Prevailing wisdom suggests that "what gets measured gets done," but projects should avoid the collection of measures that are not used in decision making.
- The Project Management Institute provides industry wide guidelines for project assessment, including Earned Value Management techniques.

- 对照建立的准则和里程碑评估项目进展。
- 实施要求的评审、审计和检验以确定继续进行下一里程碑的准备度。
- 监控关键任务和新技术（见 5.4 节）。
- 分析评估结果。
- 提出项目计划的调整建议——将这些建议输入到项目控制流程和其他决策流程。
- 按照协议、方针和程序中的指定来沟通状态。

- *控制项目*。
  - 当评估表明有偏离趋势时启动预防性措施。
  - 当评估表明与性能成功准则不一致时启动问题求解措施。
  - 当评估表明与已批准的计划存在偏离时启动纠正措施。
  - 建立工作项和进度变更以反映所采取的行动。
  - 与供应商谈判从组织外部采办的任何商品或服务。
  - 当评估支持决策门或里程碑事件时，做出继续进行或不继续进行的决策。

常用方法和建议：

- 开展定期的团队会议是使项目状态保持更新的一种项目管理的方式。每天或每周安排的简短的站立会议对于较小的群体是有效的。
- 普遍看法认为"所测即所做"，但项目应避免收集在决策中未使用的测度。
- 项目管理协会为项目评估提供全行业的指南，包括挣值管理技术。

TECHNICAL MANAGEMENT PROCESSES

- Project teams need to identify critical areas and control them through monitoring, risk management, or configuration management.
- An effective feedback control process is an essential element to enable the improvement of project performance.
- Agile project management techniques schedule frequent assessments and make project control adjustments on tighter feedback cycles than other plan-driven development models.
- Tailoring of organization processes and procedures (see Chapter 8) should not jeopardize any certifications. Processes must be established with effective review, assessment, audit, and upgrade.

5.3 DECISION MANAGEMENT PROCESS

5.3.1 Overview

5.3.1.1 Purpose As defined by ISO/IEC/IEEE 15288,

[6.3.3.1] The purpose of the Decision Management process is to provide a structured, analytical framework for objectively identifying, characterizing and evaluating a set of alternatives for a decision at any point in the life cycle and select the most beneficial course of action.

Table 5.1 provides a partial list of decision situations (opportunities) that are commonly encountered throughout a system's life cycle. Buede (2009) provides a much larger list.

TABLE 5.1   Partial list of decision situations (opportunities) throughout the life cycle

Life cycle stage	Decision situation (opportunity)
Concept	Assess technology opportunity/initial business case
	Craft a technology development strategy
	Inform, generate, and refine an initial capability document
	Inform, generate, and refine a capability development document
	Conduct analysis of alternatives supporting program initiation decision
Development	Select system architecture
	Select system element
	Select lower-level elements
	Select test and evaluation methods
Production	Perform make-or-buy decision
Utilization, support	Select production process and location
Retirement	Select maintenance approach
	Select disposal approach

- 项目团队需要识别关键域并通过监控、风险管理或构型管理来控制它们。
- 有效的反馈控制流程是使项目绩效能够改进的一个根本性元素。
- 敏捷项目管理技术在时间上安排频繁的评估,并以比其他计划驱动型开发模型更为严格的反馈周期进行项目控制调整。
- 组织流程和程序的剪裁(见第 8 章)不应危及任何认证。流程的建立必须具有有效的评审、评估、审计和升级。

5.3 决策管理流程

5.3.1 概览

5.3.1.1 目的 如 ISO/IEC/IEEE 15288 所述,

[6.3.3.1] 决策管理流程的目的是提供结构化的分析框架,以便在生命周期的任意时刻客观地识别、描述和评价决策的备选方案集合,并选择最有益的行动路线。

表 5.1 提供贯穿于系统的生命周期通常遇到的决策情景(机会)的部分列表。Buede(2009)提供一个范围更大的列表。

表 5.1 贯穿于生命周期内的决策情景(机会)的部分列表

生命周期阶段	决策情景(机会)
概念/方案	评估技术机会/最初的业务案例
	制定一种技术开发策略
	提供依据、生成并细化最初的能力文件
	提供依据、生成并细化能力开发文件
	对支持计划启动决策的备选方案的进行分析
开发	选择系统架构
	选择系统元素
	选择较低层级的元素
	选择测试和评估方法
生产	制定自制或外购决策
使用、保障 退役	选择生产流程和场所
	选择维护方法
	选择处置方法

Consider the number of decisions involved in crafting a technology development strategy, generating an initial capability document, selecting a system architecture, converging on a detailed design, constructing test and evaluation plans, making production make-or-buy decisions, creating production ramp-up plans, crafting maintenance plans, and defining disposal approaches. New product developments entail an array of interrelated decisions that require the holistic perspective of the SE discipline. In fact, it can be argued that all SE activities should be conducted within the context of supporting good decision making. If an SE activity cannot point to at least one of the many decisions embedded in a system life cycle, one must wonder why the activity is being conducted at all. Positioning decision management as a critical SE activity will ensure the efforts are rightfully interpreted as relevant and meaningful and thus maximize the discipline's value proposition to new product developers and their leadership.

5.3.1.2 Description A formal decision management process is the transformation of a broadly stated decision situation into a recommended course of action and associated implementation plan. The process can be executed by a resourced decision team that consists of a decision maker with full responsibility, authority, and accountability for the decision at hand, a decision analyst with a suite of reasoning tools, subject matter experts with performance models, and a representative set of end users and other stakeholders (Parnell et al., 2013). The decision process is executed within the policy and guidelines established by the sponsoring agent. The decision process realizes this transformation through a structured process. It must be recognized that this process, as with most any SE process, contains subjective elements, and two equally qualified teams can come to different conclusions and recommendations. A well structured trade study process, however, will be able to capture and communicate the impact that different value judgments have on the overall decision and even facilitate the search for alternatives that remain attractive across a wide range of value schemes. Figure 5.3 is the IPO diagram for the decision management process.

5.3.1.3 Inputs/Outputs Inputs and outputs for the decision management process are listed in Figure 5.3. Descriptions of each input and output are provided in Appendix E.

5.3.1.4 Process Activities The decision management process includes the following activities:

考虑到在制定技术开发策略、生成最初的能力文件、选择系统架构、集中于详细设计、构建测试和评价计划、做出生产自制或外购决策、创建生产提升计划、制定维护计划以及定义退出方法中所涉及的决策的数目。新产品的开发需要一系列相互关联的决策，这些决策需要 SE 学科的整体视角。实际上，可以认为所有 SE 活动应在支持良好决策的背景环境中进行。如果 SE 活动不能指出在系统生命周期中嵌入的至少一种决策，那么必须知道究竟为什么进行该活动。定位决策管理作为一种关键的 SE 活动将确保工作被正当地解释为相关的且有意义的，并因此使新产品的开发者及其领导力的学科价值主张最大化。

5.3.1.2 描述　正式的决策管理流程是将大致说明的决策情景转换为建议的行动路线及相关的实施计划。该流程可由一个资源丰富的决策团队来执行，该团队包括对即将进行的决策具有全部职责、权限和责任的决策者、具有一套推理工具的决策分析者、具有性能模型的主题专家以及最终用户及其他利益攸关者的代表集合（Parnell 等，2013）。在发起代理人建立的方针和指南内执行决策流程。决策流程通过结构化的流程实现这种转换。必须认识到，该流程与大多数的任何 SE 流程一样都包含主观要素，并且两个同等合格的团队可得出不同的结论和建议。然而，结构良好的权衡研究流程将能够捕获和沟通不同的价值判断对总体决策的影响，甚至便于寻找跨广范围价值方案内保持有吸引力的备选方案。图 5.3 是决策管理流程的 IPO 图。

5.3.1.3 输入/输出　图 5.3 列出决策管理流程的输入和输出。附录 E 提供每个输入和输出的描述。

5.3.1.4 流程活动　决策管理流程包括以下活动：

TECHNICAL MANAGEMENT PROCESSES

- *Prepare for decisions.*
  - Define the decision management strategy for the system.
  - Establish and challenge the decision statement, and clarify the decision to be made. This is one of the most important steps since an incomplete or incorrect decision statement will inappropriately constrain the options considered or even lead the team down the wrong path. Consider the differences that would result from the following decision statements:
    o What car should I buy?
    o What vehicle should I buy?
    o What vehicle should I acquire (buy or lease)?
    o Do I need to travel?
- *Analyze the decision information.*
  - Frame, tailor, and structure decision.
  - Develop objectives and measures.
  - Generate creative alternatives.
  - Assess alternatives via deterministic analysis.
  - Synthesize results.

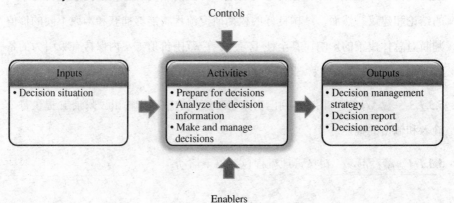

FIGURE 5.3 IPO diagram for the decision management process. INCOSE SEH original figure created by Shortell and Walden. Usage per the INCOSE notices page. All other rights reserved.

- *决策准备。*
    - 为系统定义决策管理策略。
    - 建立和挑战决策说明,并阐明将要做出的决策。这是最重要的步骤之一,因为一个不完整的或不正确的决策说明将不适当地约束所考虑的选项,或甚至导致团队走向错误的路径向下发展。考虑由以下决策说明导致的差异:
        - 我应该买什么车?
        - 我应该买什么运载器?
        - 我应该采办(买或租)什么运载器?
        - 我需要旅行吗?
- *分析决策信息。*
    - 框定、剪裁和构建决策。
    - 开发目标和测度。
    - 生成创造性的备选方案。
    - 经由确定性分析评估备选方案。
    - 综合结果。

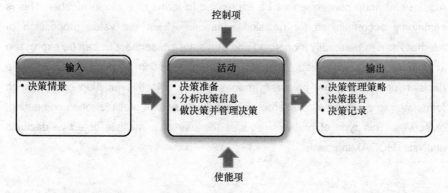

图 5.3 决策管理流程的 IPO 图。INCOSE SEH 原始图由 Shortell 和 Walden 创建。按照 INCOSE 通知页使用。版权所有。

- Identify uncertainty and conduct probabilistic analysis, as appropriate.
- Assess impact of uncertainty.
- Improve alternatives.
- Communicate trade-offs.
- Present recommendation and implementation plan.
- *Make and manage decisions.*
    - Record the decision, with the relevant data and supporting documentation.– record the decision, with the relevant data and supporting documentation.
    - Communicate new directions from the decision.

5.3.2 Elaboration

Systems engineers will likely face many types of decision situations throughout the life cycle of a project. Systems engineers must choose the analytical approach that best fits the frame and structure of the decision problem at hand. For instance, when there are "... clear, important, and discrete events that stand between the implementation of the alternatives and the eventual consequences..." (Edwards et al., 2007), a decision tree is often a well suited analytical approach, especially when the decision structure has only a few decision nodes and chance nodes. As the number of decision nodes and chance nodes grows, the decision tree quickly becomes unwieldy and loses some of its communicative power. Furthermore, decision trees require end node consequences be expressed in terms of a single number. This is commonly accomplished for decision situations where the value proposition of alternatives can be readily monetized and end state consequences can be expressed in dollars, euros, yen, etc. When the value proposition of alternatives within a decision problem cannot be easily monetized, an objective function can often be formulated to synthesize an alternative's response across multiple, often competing, objectives. This type of problem requires the use of a multiple objective decision analysis (MODA) approach.

- 识别不确定性并酌情进行概率分析。

- 评估不确定性的影响。

- 改进备选方案。

- 沟通权衡。

- 提出建议和实施计划。

- *制定并管理决策。*

  - 记录决策,以及相关数据和支持性文件。

  - 沟通决策中的新方向。

5.3.2 详细阐述

系统工程师将可能贯穿于项目的生命周期面对多种类型的决策情景。系统工程师必须选择最适配于手头上的决策问题的框架和结构的分析方法。例如,当存在"……处于备选方案的实施和最终的结果的实施之间的清晰、重要、独立的事件"时(Edwards 等,2007),决策树往往是非常适合的分析方法,尤其是当决策结构仅有几个决策节点和机会节点时。随着决策节点和机会节点的数量的增长,决策树很快地变得难以控制,并失去它的一些沟通力。而且,决策树需要以一个数字来表达最终节点的结果。这通常针对决策情景来完成,其中备选方案的价值主张可以被立即货币化且最终状态的结果可以用美元、欧元、日元等来表示。当决策问题之内备选方案的价值主张无法容易地被货币化时,通常可制定一个目标功能,以便跨多个目标(通常是竞争目标)综合备选方案的响应。这种类型的问题需要使用多目标决策分析(MODA)方法。

The decision management method most commonly employed by systems engineers is the trade study and more often than not employs some form of MODA approach. The aim is to define, measure, and assess shareholder and stakeholder value and then synthesize this information to facilitate the decision maker's search for an alternative that represents the optimally balanced response to often competing objectives. MODA approaches generally differ in the degree to which an alternative's response to objectives (and subobjectives) is aggregated, the mathematics used to aggregate such responses, the techniques used to elicit value statements from stakeholders, the treatment of uncertainty, the robustness of sensitivity analysis, the use of screening techniques, and the versatility and quality of trade space visualization outputs. If time and funding allow, systems engineers may want to conduct trade studies using several techniques, compare and contrast results, and reconcile any differences to ensure findings are robust. Although there are many possible ways to specifically implement MODA, the discussion contained in the balance of this section represents a short summary of best practices regardless of the specific implementation technique employed.

5.3.2.1 Framing and Tailoring Decisions To help ensure the decision makers and stakeholders fully understand the decision context and to enhance the overall traceability of the decision, the decision analyst should capture a description of the system baseline as well as a notion for how the envisioned system will be used along with some indication of system boundaries and anticipated interfaces. Decision context includes such details as the time frame allotted for the decisions, an explicit list of decision makers and stakeholders, a discussion regarding available resources, and expectations regarding the type of action to be taken as a result of the decision at hand as well as decisions anticipated in the future (Edwards et al., 2007).

5.3.2.2 Developing Objectives and Measures Defining how a decision will be made may seem like a straightforward assignment but often becomes an arduous task of seeking clarity amidst a large number of ambiguous stakeholder need statements, engaging in uncomfortable discussions regarding the relative priority of each requirement, and establishing walkaway points and stretch goals. As Keeney puts it,

> Most important decisions involve multiple objectives, and usually with multiple-objective decisions, you can't have it all. You will have to accept less achievement in terms of some objectives in order to achieve more on other objectives. But how much less would you accept to achieve how much more? (Keeney, 2002)

系统工程师最常使用的决策管理方法是权衡研究，且多半使用某种形式的 MODA 方法。目的是定义、测量和评估股东和利益攸关者的价值，然后综合该信息以促进决策者寻找表征对通常的竞争性目标的最佳平衡响应的备选方案。MODA 方法的不同之处一般在于备选方案对目标（和子目标）的响应的汇集程度、用于汇集这些响应的数学运算、用于从利益攸关者中引出价值说明的技术、不确定性的处理、灵敏度分析的鲁棒性、筛选技术的使用以及权衡空间可视化输出的通用性和质量。如果时间和资金允许，系统工程师可能想使用多种技术开展权衡研究，比较和对比结果，并调和差异以确保发现结果的鲁棒性。尽管存在具体实施 MODA 的许多可能的方式，不论使用了哪种特定实施技术，本节其余部分所包含的论述表征最佳实践的简短摘要。

5.3.2.1　框定和剪裁决策　为有助于确保决策者和利益攸关者全面地理解决策背景环境以及为增强决策的总体可追溯性，决策分析者应捕获系统基线的描述以及关于预想的系统将如何与系统边界和预期界面的某种表示一起使用方面的观点。决策背景环境包括像为决策分配的时间框架、决策者和利益攸关者的明确列表、关于可用资源的论述以及关于行动类型（被认为是手头上的决策以及在将来预期的决策的结果）的预期之类的细节（Edwards 等，2007）。

5.3.2.2　开发目标和测度　定义如何做决策似乎像是一种简单的分配，但通常变成一项在许多模糊不清的利益攸关者需要说明之中寻求清晰的艰巨任务，从而对每个需求的相对优先权要介入不太舒服的讨论中并建立易取胜点和挑战性目标。正如 Keeney 所说，

> 最重要的决策涉及多个目标，通常涉及多个目标决策，你无法拥有全部。你将必须针对某些目标接受较少的成果，以便在其他目标上实现更多成果。但你会接受以多么少的结果实现多么多的？（Keeney, 2002）

The first step is to use the information obtained from the stakeholder requirements definition process, requirements analysis process, and requirements management processes to develop objectives and measures.

For each fundamental objective, a measure must be established so that alternatives that more fully satisfy the objective receive a better score on the measure than those alternatives that satisfy the objective to a lesser degree. A measure (also known as attribute, criterion, and metric) must be unambiguous, comprehensive, direct, operational, and understandable (Keeney and Gregory, 2005).

5.3.2.3 Generating Creative Alternatives For many trade studies, the alternatives will be systems composed of many interrelated system elements. It is important to establish a meaningful product structure for the SOI and to apply this product structure consistently throughout the decision analysis effort in order to aid effectiveness and efficiency of communications about alternatives. The product structure should be a useful decomposition of the physical elements of the SOI. Each alternative is composed of specific design choices for each physical element. The ability to quickly communicate the differentiating design features of given alternatives is a core element of the decision-making exercise.

5.3.2.4 Assessing Alternatives via Deterministic Analysis With objectives and measures established and alternatives identified and defined, the decision team should engage subject matter experts, ideally equipped with operational data, test data, models, simulation, and expert knowledge. The decision team can best prepare for subject matter expert engagement by creating structured scoring sheets. Assessments of each concept against each criterion are best captured on separate structured scoring sheets for each alternative/objective combination. Each score sheet contains a summary description of the alternative under examination and a summary of the scoring criteria to which it is being measured.

5.3.2.5 Synthesizing Results At this point in the process, the decision team has generated a large amount of data as summarized in the objective measure consequence table created as the final task of the process step. Now, it is time to explore the data, make sense of the data, and display results in a way that facilitates understanding.

第一步是使用从利益攸关者需求定义流程、需求分析流程和需求管理流程中获得的信息来开发目标和测度。

必须针对每个基本目标建立测度,从而使更全面地满足目标的备选方案比较低程度满足目标的备选方案得到更好的测度分数。一个测度(亦称为属性、准则和度量指标)必须无歧义、全面、直接、可运行且可理解(Keeney 和 Gregory,2005)。

5.3.2.3 *生成创造性的备选方案* 对于许多权衡研究而言,备选方案将是由许多相互关联的系统元素组成的系统。重要的是为 SOI 建立有意义的产品结构并贯穿于决策分析工作内一致地应用该产品结构,以便辅助有关备选方案沟通的效果和效率。产品结构应当是对 SOI 物理元素的一种有用分解。每个备选方案由每个物理元素的特定设计选择组成。对给定备选方案的有差别的设计特征进行快速沟通的能力是决策运用的核心元素。

5.3.2.4 *经由确定性分析评估备选方案* 由于目标和测度已建立且备选方案已识别和定义,决策团队应有具备运行数据、测试数据、模型、仿真和专家知识的理想的主题专家介入。决策团队可通过创建结构化的记分表为主题专家的介入做最好的准备。最好在每个备选方案/目标组合的分开的结构化记分表上获取按照每个准则对每个概念的评估。每个记分表包含对正在检查的备选方案的概要描述以及测量所依据的记分准则的概要。

5.3.2.5 *综合结果* 在该流程的这个点上,决策团队已生成了大量数据,正如在流程步骤的最终任务所创建的目标测度结果表中所总结。现在是探究该数据、赋予该数据意义并以有助于理解的方式显示结果的时候。

5.3.2.6 Identifying Uncertainty and Conducting Probabilistic Analysis As part of the assessment, it is important for the subject matter expert to explicitly discuss potential uncertainty surrounding the assessed score and variables that could impact one or more scores. One source of uncertainty that is common within SE trade-off analysis exploring various system architectures is that system concepts are generally described as a collection of system element design choices but lack discussion of the component-level design decisions that are normally made downstream during detailed design. Many times, the subject matter expert can assess an upper, nominal, and lower bound measure response by making three separate assessments (i) assuming a low performance, (ii) assuming moderate performance, and (iii) assuming high performance. Once all scores and their associated uncertainty are appropriately captured, Monte Carlo simulations can be executed to identify uncertainty that impacts the decision findings and identify areas of uncertainty that are inconsequential to decision findings.

5.3.2.7 Accessing Impact of Uncertainty: Analyzing Risk and Sensitivity Decision analysis uses many forms of sensitivity analysis including line diagrams, tornado diagrams, waterfall diagrams, and several uncertainty analyses including Monte Carlo simulation, decision trees, and influence diagrams (Parnell et al., 2013).

5.3.2.8 Improving Alternatives One could be tempted to end the decision analysis here, highlight the alternative that has the highest total value, and claim success. Such a premature ending however would not be considered best practice. Mining the data generated for the first set of alternatives will likely reveal opportunities to modify some system element design choices to claim untapped value and reduce risk.

5.3.2.9 Communicating Trade-Offs This is the point in the process where the decision team identifies key observations regarding what stakeholders seem to want and what they must be willing to give up in order to achieve it. It is here where the decision team can highlight the design decisions that are least significant and/or most influential and provide the best stakeholder value. In addition, the important uncertainties and risks should also be identified. Observations regarding combinatorial effects of various design decisions are also important products of this process step. Finally, competing objectives that are driving the trade should be explicitly highlighted as well.

5.3.2.6 识别不确定性和进行概率分析 作为评估的一部分，对于主题专家而言重要的是围绕已评估的分数和可能影响一个或多个分数的变量明确地讨论潜在的不确定性。在探究各种不同的系统架构的 SE 权衡分析之内一个公共的不确定性来源是系统概念一般被描述为系统元素设计选择的一个集合，但在详细设计期间缺乏对通常在下游的组件级设计决策的论述。很多时候，主题专家可通过做出三个单独的评估（i）假设低性能，（ii）假设中等性能和（iii）假设高性能来评估上限、额定界限和下限的测度响应。一旦所有分数及其相关不确定性被适当地获取，可执行蒙特卡罗仿真以识别影响决策调查结果的不确定性并识别对决策调查结果不重要的不确定性的区域。

5.3.2.7 评估不确定性的影响：分析风险和灵敏度 决策分析使用多种形式的灵敏度分析，包括线型图、龙卷风图和瀑布图，以及多种不确定性分析，包括蒙特卡罗模拟、决策树和影响图（Parnell 等，2013）。

5.3.2.8 改进备选方案 可试图在这里结束决策分析，强调具有最高总值的备选方案，并宣布成功。然而这种不成熟的结尾不会被视为最佳实践。挖掘第一个备选方案集合所产生的数据将有可能揭示修改一些系统元素设计选择的机会，以索取未开发的价值并减少风险。

5.3.2.9 沟通权衡 这是决策团队识别关键观察结果（关于利益攸关者似乎想要什么以及为了实现它必须愿意放弃什么）的过程中的要点。在这里决策团队可强调最不重要的和/或最有影响力的设计决策并提供最佳的利益攸关者价值。此外，还应识别重要的不确定性和风险。关于各种不同的设计决策的组合影响的观察结果也是该流程步骤的重要产物。最后，还应明确地强调那些驱动权衡的竞争性目标。

5.3.2.10 Presenting Recommendations and Implementing Action Plan It is often helpful to describe the recommendation in the form of clearly worded, actionable task list to increase the likelihood of the decision analysis leading to some form of action, thus delivering some tangible value to the sponsor. Reports are important for historical traceability and future decisions. Take the time and effort to create a comprehensive, high quality report detailing study findings and supporting rationale. Consider static paper reports augmented with dynamic hyperlinked electronic reports.

5.4 RISK MANAGEMENT PROCESS

5.4.1 Overview

5.4.1.1 Purpose As stated in ISO/IEC/IEEE 15288,

> [6.3.4.1] The purpose of the risk Management process is to identify, analyze, treat and monitor the risks continually.

5.4.1.2 Description Numerous standards, guidelines, and informational publications address the subjects of risk and risk management. In some cases, the application of specific standards may be mandated by industry regulations or customer contractual agreement.

Because there is significant variation, and even contradiction, in the risk and risk management concepts and practices presented in these publications, it is important that process owners, implementers, and users of risk management processes ensure that their understanding of, and approach to, risk management is sufficient, consistent, and appropriate for their specific context, scope, and objectives.

Definitions of Risk According to E. H. Conrow, "Traditionally, risk has been defined as the likelihood of an event occurring coupled with a negative consequence of the event occurring. In other words, a risk is a potential problem—something to be avoided if possible, or its likelihood and/or consequences reduced if not" (2003).

Below are several definitions consistent with this traditional concept of risk:

- ISO/IEC/IEEE 16085:2006, *Systems and software engineering—Life cycle processes—Risk management*, defines risk as the "combination of the probability of an event and its consequence," with the following three notes:

5.3.2.10 提出建议和实施行动计划 以措辞清晰的、可行动的任务列表的形式描述建议通常很有帮助,以便增加导致产生某种形式的行动决策分析的可能性,从而向发起人交付某种有形的价值。报告对于历史可追溯性和未来决策而言很重要。花费时间和精力创建一个全面的高品质报告,详述研究发现和支持性理由依据。考虑带有动态超链接电子报告增强的静态纸质报告。

5.4 风险管理流程

5.4.1 概览

5.4.1.1 目的 如 ISO/IEC/IEEE 15288 所述,

[6.3.4.1] 风险管理流程的目的是连续地识别、分析、处理和监控风险。

5.4.1.2 描述 许多标准、指南和信息出版物涉及风险和风险管理的学科。在某些情况下,特定标准的应用可能由行业规定或客户合同协议来授权。

由于这些出版物中提出的风险和风险管理的概念和实践中具有显著差异,甚至矛盾,因此重要的是风险管理流程的流程所有者、实施者和使用者确保他们对风险管理的理解和风险管理方法是充分的、一致的、适于他们的特定背景环境、范围和目标。

风险的定义 根据 E. H. Conrow,"传统上,风险被定义为一个事件的发生伴随有该事件发生的负面后果的可能性。换言之,风险是一种潜在的问题——若可能应予以避免的事情,或者若不能避免则其可能性和/或后果应予以减轻"(2003)。

下面是与这种传统的风险概念一致的几个定义:

- ISO/IEC/IEEE 16085:2006,《系统和软件工程——生命周期流程——风险管理》将风险定义为"事件的可能性及其后果的组合",具有以下三个注解:

TECHNICAL MANAGEMENT PROCESSES

- The term risk is generally used only when there is at least the possibility of negative consequences.

- In some situations, risk arises from the possibility of deviation from the expected outcome or event.

- See ISO/IEC Guide 51 (1999) for issues related to safety.

Comments: (i) This ISO/IEC/IEEE 16085 definition is taken from ISO Guide 73: 2002, definition 3.1.1 (since replaced with the revised definition in ISO Guide 73:2009). (ii) This is the definition referenced in ISO/IEC/IEEE 15288.

As a corollary to risk, Conrow defines opportunity as "the potential for the realization of wanted, positive consequences of an event" (Conrow, 2003). The idea of considering opportunities and positive outcomes (in addition to negative outcomes) as an integral part of a risk management process has gained favor with some experts and practitioners. New risk and risk management concepts intended to support this broadened scope for risk management are evolving. Notable definitions of risk that reflect this broaden scope are:

- ISO guide 73:2009, *Risk management—Vocabulary*, defines risk as the "effect of uncertainty on objectives" with the following five notes:

  - An effect is a deviation from the expected—positive and/or negative.

  - Objectives can have different aspects (such as financial, health and safety, and environmental goals) and can apply at different levels (such as strategic, organization-wide, project, product, and process).

  - Risk is often characterized by reference to potential events and consequences or a combination of these.

  - Risk is often expressed in terms of a combination of the consequences of an event (including changes in circumstances) and the associated likelihood of occurrence.

  - Uncertainty is the state, even partial, of deficiency of information related to, understanding or knowledge of, an event, its consequence, or likelihood.

According to the *Practice Standard for Project Risk Management* (PMI, 2009), when assessing the importance of a project risk, consider the two key dimensions of risk: uncertainty and the effect on the project's objectives. The uncertainty dimension may be described using the term "probability," and the effect dimension may be called "impact" (though other descriptors are possible, such as "likelihood" and "consequence").

- 术语风险一般仅在至少具有负面后果可能性时使用。

- 在某些情景下，风险产生于偏离预期结果或事件的可能性。

- 与安全相关的问题参见 ISO/IEC 指南 51（1999）。

注释：（i）该 ISO/IEC/IEEE 16085 定义取自 ISO 指南 73:2002，定义 3.1.1（此后在 ISO 指南 73:2009 中被替换为修正定义）。（ii）这是 ISO/IEC/IEEE 15288 中引用的定义。

作为风险的结果，Conrow 将机会定义为"实现某一事件的想要的、正面的结果的可能性"（Conrow，2003）。将机会和正面结果（除负面结果之外）视为风险管理流程的组成部分的理念得到了一些专家和实践者的支持。预计用于支持这种扩大的风险管理范围的新风险和风险管理概念正在演进。反映这种扩大范围的著名的风险定义是：

- ISO 指南 73:2009，《风险管理——词汇表》将风险定义为"不确定性对目标的影响"，具有以下五个注解：

    - 一种效应是对预期的—正面的和/或负面的结果的偏离。

    - 目标可具有不同的方面（如财务目标、健康和安全目标及环境目标）并且可在不同层级（如策略上的、全组织的、项目、产品和流程）上应用。

    - 通过参考潜在事件和后果或它们的组合来描述风险特征。

    - 风险通常根据事件的后果（包括环境变化）及相关的事件发生可能性的组合来表示。

    - 不确定性是与事件、事件的后果或可能性的理解或知识相关的信息缺乏或甚至部分缺乏的状态。

根据《项目风险管理的实践标准》（PMI，2009），在评估项目风险的重要性时，考虑风险的两个关键因素：不确定性和对项目目标的影响。不确定性因素可使用术语"可能性"来描述，效应维度可被称为"影响"（尽管其他描述符也有可能，如"可能性"和"后果"）。

Risk includes both distinct events, which are uncertain but can be clearly described, and general conditions, which are less specific but may give rise to uncertainty. The two types of project risk, those with negative and those with positive effects, are called, respectively, "threats" and "opportunities."

Process Enablers   It has been found that an organization's structure and culture can have a significant effect on the performance of the risk management process. ISO 31000, *Risk management—Principles and Guidelines* (2009), outlines a model that advocates the establishment of principles for managing risk and a framework for managing risk that work in concert with the process for managing risk.

Risk Management Process   Risk management *is a disciplined approach to dealing with the uncertainty that is present throughout the entire system life cycle.* A primary objective of risk management is to identify and manage (take proactive steps) to handle the uncertainties that threaten or reduce the value provided by a business enterprise or organization. Since risk cannot be reduced to zero, another objective is to achieve a proper balance between risk and opportunity.

This process is used to understand and avoid the potential cost, schedule, and performance (i.e., technical) risks to a system and to take a proactive and structured approach to anticipate negative outcomes and respond to them before they occur. Organizations manage many forms of risk, and the risk associated with system development is managed in a manner that is consistent with the organization strategy.

Every new system or modification of an existing system is based on the pursuit of an opportunity. Risk is always present in the life cycle of systems, and the risk management actions are assessed in terms of the opportunity being pursued.

External risks are often neglected in project management. External risks are risks caused by or originating from the surrounding environment of the project (Fossnes, 2005). Project participants often have no control or influence over external risk factors, but they can learn to observe the external environment and eventually take proactive steps to minimize the impact of external risks on the project. The typical issues are time-dependent processes, rigid sequence of activities, one dominant path for success, and little slack.

风险同时包括不确定的但能够被明确描述的以及不太特定的但可能引发不确定性的一般情况的截然不同的事件。在这两种项目风险中那些具有负面影响和正面影响的项目风险分别被称为"威胁"和"机会"。

流程使能项　已经发现组织的结构和文化可对风险管理流程的性能具有显著影响。ISO 31000,《风险管理——原则和指南》(2009)概述了一种模型,该模型提倡建立与风险管理流程协同工作的风险管理原则和风险管理框架。

风险管理流程　风险管理是处理贯穿于整个系统生命周期内存在的不确定性的学科方法。风险管理的主要目标是识别和管理(采取积极措施)以便对威胁或降低业务复杂组织体或组织所提供价值的不确定性进行处理。由于无法将风险降为零,因此另一个目标是在风险和机会之间实现恰当的平衡。

该流程用于理解和避免系统的潜在成本、进度和性能(即技术)风险;采取积极的结构化方法预见负面结果,并在产生负面结果之前对其做出响应。组织管理多种形式的风险,与系统开发有关的风险采用与组织策略一致的方式来管理。

每个新系统或现有系统的完善都是基于对机会的追求。风险始终存在于系统的生命周期内,并且风险管理措施根据所追求的机会来评估。

外部风险在项目管理中经常被忽略。外部风险是由项目的周围环境所引起或产生的风险(Fossnes, 2005)。项目参与者通常无法控制或影响外部风险因素,但他们可以学习观察外部环境并最终采取主动的措施,以使外部风险对项目的影响最小化。典型的问题是依赖时间的流程、严格的活动顺序、取得成功的一个主导路径和稍事放松。

TECHNICAL MANAGEMENT PROCESSES

Typical strategies for coping with risk include transference, avoidance, acceptance, or taking action to reduce the anticipated negative effects of the situation. Most risk management processes include a prioritization scheme whereby risks with the greatest negative effect and the highest likelihood are treated before those deemed to have lower negative consequences and lower likelihood. The objective of risk management is to balance the allocation of resources such that the minimum amount of resources achieves the greatest risk mitigation (or opportunity realization) benefits. Figure 5.4 is the IPO diagram for the risk management process.

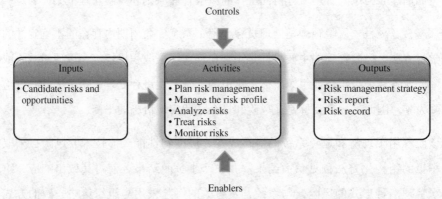

FIGURE 5.4 IPO diagram for the risk management process. INCOSE SEH original figure created by Shortell and Walden. Usage per the INCOSE notices page. All other rights reserved.

5.4.1.3 Inputs/Outputs Inputs and outputs for the risk management process are listed in Figure 5.4. Descriptions of each input and output are provided in Appendix E.

5.4.1.4 Process Activities The risk management process includes the following activities:

- *Plan risk management.*
  - Define and document the risk strategy.
- *Manage the risk profile.*
  - Establish and maintain a risk profile to include context of the risk and its probability, consequences, risk thresholds, and priority and the risk action requests along with the status of their treatment.
  - Define and document risk thresholds and acceptable and unacceptable risk conditions.
  - Periodically communicate the risks with the appropriate stakeholders.

应对风险的典型策略包括转移、规避、接受或采取行动以降低该情景下预期的负面影响。大多数风险管理流程包括一个优先级排序方案,据此负面影响最大且发生概率最高的风险能够在负面结果较小且发生概率较低的风险前被处理。风险管理的目标在于平衡资源配置,从而使最少量的资源实现最大的风险缓解(或机会实现)效益。图 5.4 是决策管理流程的 IPO 图。

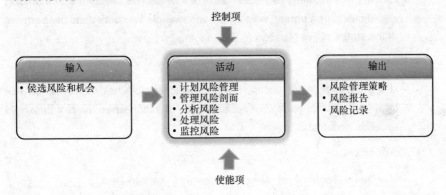

图 5.4 风险管理流程的 IPO 图。INCOSE SEH 原始图由 Shortell 和 Walden 创建。按照 INCOSE 通知页使用。版权所有。

5.4.1.3 输入/输出 图 5.4 列出风险管理流程的输入和输出。附录 E 提供每个输入和输出的描述。

5.4.1.4 流程活动 风险管理流程包括以下活动:

- *计划风险管理。*

  – 定义和文件化风险策略。

- *管理风险剖面。*

  – 建立和维护风险剖面,以包括风险背景环境及其概率、后果、风险阈值、优先级、风险行动请求及其处理状态。

  – 定义和文件化风险阈值以及可接受的和不可接受的风险条件。

  – 与适当的利益攸关者定期地沟通风险。

TECHNICAL MANAGEMENT PROCESSES

- *Analyze risks.*
  - Define risk situations and identify the risks.
  - Analyze risks for likelihood and consequence to determine the magnitude of the risk and its priority for treatment.
  - Define a treatment scheme and resources for each risk, including identification of a person who will be responsible for continuous assessment of the status of the situation.
- *Treat risks.*
  - Using the criteria for acceptable and unacceptable risk, consider the risk treatment alternative, and generate a plan of action when the risk threshold exceeds acceptable levels.
- *Monitor risks.*
  - Maintain a record of risk items and how they were treated.
  - Maintain transparent risk management communications.

Common approaches and tips:

- In the project planning process (see Section 5.1), a risk management plan (RMP) is tailored to satisfy the individual project procedures for risk management.
- A risk management process establishes documentation, maintained as the risk profile, that includes a description, priority, treatment, responsible person, and status of each risk item.
- One rule of thumb for identifying risks is to pose each risk candidate in an "if <situation>, then <consequence>, for <stakeholder >" format. This form helps to determine the validity of a risk and assess its magnitude or importance. If the statement does not make sense or cannot be put in this format, then the candidate is probably not a valid risk. For example, a statement that describes a situation but not a consequence impacting a specific stakeholder implies that the potential event will not affect the project. Similarly, a statement of potential consequence to a stakeholder without a clear situation or event chain scenario description is probably not adequately understood and requires more analysis. Similarly, a statement of potential consequence to a stakeholder without a clear situation or event chain scenario description is probably not adequately understood and requires more analysis.

- 分析风险。
  - 定义风险情景并识别风险。
  - 分析发生风险的可能性和后果以确定风险的程度及其处理的优先级。
  - 为每个风险定义一个处理方案和一些资源，包括识别负责连续评估风险情景状态的人员。
- 处理风险。
  - 使用可接受的和不可接受的风险的准则，在风险阈值超过可接受的水平时考虑风险处理备选方案并生成一个行动计划。
- 监控风险。
  - 维持风险项和如何处理风险的记录。
  - 维持透明的风险管理沟通。

常用方法和建议：

- 在项目计划流程（见 5.1 节）中，对风险管理计划（RMP）进行剪裁以满足风险管理的单个项目程序。
- 风险管理流程建立文件，并作为风险剖面予以保存，其包括每个风险项的描述、优先级、处理、负责人和状态。
- 识别风险的一个经验法则是使每个候选风险置于"对于<利益攸关者>，如果<情景>，则<后果>"的格式之中。这种形式有助于确定风险的有效性并评估其程度或重要性。如果陈述没有意义或无法置于这种格式中，则候选风险可能不是有效的风险。例如，描述的情景对特定利益攸关者将不产生影响的陈述意味着该潜在事件不会影响该项目。同样，只描述对利益攸关者的潜在后果而没有清晰的情景或事件链场景描述的陈述，可能得不到充分地理解且需要更多的分析。

TECHNICAL MANAGEMENT PROCESSES

- All personnel are responsible for using the procedure to identify risks.

- Risks can be identified based on prior experience, brainstorming, lessons learned from similar programs, and checklists.

- Risk identification activity should be applied early and continuously throughout the life cycle of the project.

- Document everything so if unforeseen issues and challenges arise during execution, the project can recreate the environment within which the planning decisions were made and know where to update the information to correct the problem.

- Negative feedback toward personnel who identify a potential problem will discourage the full cooperation of engaged stakeholders and could result in failure to address serious risk-laden situations. Conduct a transparent risk management process to encourage suppliers and other stakeholders to assist in risk mitigation efforts. Some situations can be difficult to categorize in terms of probability and consequences; involve all relevant stakeholders in this evaluation to capture the maximum variety in viewpoints.

- Many analyses completed throughout the technical processes, such as FMECA, may identify candidate risk elements.

- The measures for risk management vary by organization and by project. As with any measure, use measurement analysis or statistics that help manage the risk.

- Experience has shown that terms such as "positive risk" and concept models that define opportunity as a subset of risks serve only to confuse. Projects that are subject to regulatory standards or customer requirements regarding risk management process and output must use extreme care when integrating opportunity management with risk management. The Project Management Institute is a good source for more information for project risk management (PMI, 2013).

- 所有人员对识别风险所采用的程序负责。

- 可基于以前的经验、头脑风暴、类似项目计划中的经验教训和检查清单来识别风险。

- 风险识别活动应在早期应用并持续地贯穿于项目的生命周期。

- 将每件事情都文件化,以便如果在执行期间出现未预见的问题和挑战,项目可重新创建环境,在该环境内做出计划决策并知道在何处更新信息以纠正问题。

- 向识别潜在问题的人员做出负面反馈会阻碍与参与的利益攸关者的全面合作,并可能导致在应对充满严重风险的情景时失败。实施透明的风险管理流程,以鼓励供应商及其他利益攸关者协助风险缓解工作。有些情景很难按照概率和后果分类;在此评价中涉及所有相关的利益攸关者,以获取观点的最多种类。

- 贯穿于技术流程完成的许多分析,例如 FMECA,可识别候选风险元素。

- 风险管理的测度因组织和项目而异。对于任何测度,使用有助于管理风险的测量分析或统计。

- 经验已表明,诸如"正面风险"的术语以及将机会定义为风险子集的概念模型只会制造混淆。关于风险管理流程和输出的受规定标准或客户需求支配的项目必须在综合机会管理与风险管理时特别谨慎。项目管理协会是更多项目风险管理信息的良好来源(PMI, 2013)。

TECHNICAL MANAGEMENT PROCESSES

5.4.2 Elaboration

5.4.2.1 *Risk Management Concepts* Most projects are executed in an environment of uncertainty. Risks (also called threats) are events that if they occur can influence the ability of the project team to achieve project objectives and jeopardize the successful completion of the project (Wideman, 2002). Well-established techniques exist for managing threats, but there is some debate over whether the same techniques are applicable to recognizing opportunities. In an optimal situation, opportunities are maximized at the same time as threats are minimized, resulting in the best chance to meet project objectives (PMI, 2000). The Øresund Bridge case (see Section 3.6.2) illustrates this in that the man-made Peberholm Island was created from the materials dredged from the Strait to meet environmental requirements and is now a sanctuary for a rare species of tern.

The measurement of risk has two components (see Figure 5.5):

- The likelihood that an event will occur
- The undesirable consequence of the event if it does occur

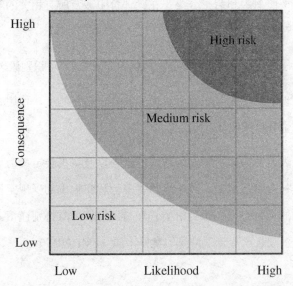

FIGURE 5.5 Level of risk depends upon both likelihood and consequences. INCOSE SEH v1 Figures 4.5 and 4.6. Usage per the INCOSE notices page. All other rights reserved.

5.4.2 详细阐述

5.4.2.1 风险管理概念 大多数项目都是在不确定性环境中执行的。风险（亦称为威胁）是一旦发生即会影响项目团队达成项目目标的能力并危及项目成功完成的那些事件（Wideman, 2002）。对管理威胁存在有良好建立的技术，但同样的技术是否可用于正在识别的机会，还存在一些争论。在最佳情景下，将威胁最小化的同时使机会最大化，从而产生满足项目目标的最佳机会（PMI, 2000）。厄勒海峡大桥案例（见 3.6.2 节）详细阐明这一情况，人造 Peberholm 岛是采用从海峡中挖掘出来的材料建造的用以满足环境要求，并且现在是燕鸥稀有物种的保护区。

风险的衡量有两个部分（见图 5.5）：

- 事件将会发生的可能性
- 事件的发生不期望的后果

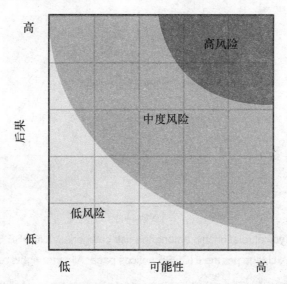

图 5.5 风险等级取决于可能性和后果两者。INCOSE SEH 第 1 版中图 4.5 和 4.6。按照 INCOSE 通知页使用。版权所有。

The likelihood that an undesirable event will occur is often expressed as a probability. The consequence of the event is expressed in terms that depend on the nature of the event (e.g., lost investment, inadequate performance, etc.). The combination of low likelihood and low undesirable consequences gives low risk, while high risk is produced by high likelihood and highly undesirable consequences.

By changing the adjective from undesirable to desirable, the noun changes from risk to opportunity, but the diagram remains the same. As suggested by the shading, most projects experience a comparatively small number of high-risk or high-opportunity events.

There is no alternative to the presence of risk in system development. The only way to remove risk is to set technical goals very low, to stretch the schedule, and to supply unlimited funds. None of these events happen in the real world, and no realistic project can be planned without risk. The challenge is to define the system and the project that best meet overall requirements, allow for risk, and achieve the highest chances of project success. Figure 5.6 illustrates the major interactions between the four risk categories: technical, cost, schedule, and programmatic. The arrow labels indicate typical risk relationships, though others certainly are possible.

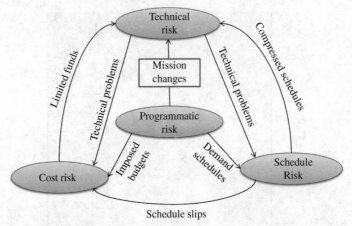

FIGURE 5.6 Typical relationship among the risk categories. INCOSE SEH v1 Figures 4.5, 4.6, and 4.7. Usage per the INCOSE notices page. All other rights reserved.

- *Technical risk*—The possibility that a technical requirement of the system may not be achieved in the system life cycle. Technical risk exists if the system may fail to achieve performance requirements; to meet operability, producibility, testability, or integration requirements; or to meet environmental protection requirements. A potential failure to meet any requirement that can be expressed in technical terms is a source of technical risk.

不期望事件将会发生的可能性通常以概率来表达。事件的后果以依赖于事件的本质属性（如损失的投资、不良的性能等）的术语来表达。低的可能性与低的不期望后果的组合产生低的风险，而高的可能性与高的不期望后果的组合产生高风险。

通过将形容词"不期望的"改为"期望的"，可将名词"风险"改为"机会"，但图保持不变。如阴影部分所示，大多数项目都经历了相对少量的高风险或高机会的事件。

系统开发中存在的风险是不可替代的。消除风险的唯一方式是将技术目标设置得非常低，延长进度，并且提供无限的资金。现实世界中这些事件都不会发生，并且任何现实项目都无法在脱离风险的情况下而计划。挑战在于定义最能满足整体需求的系统和项目，考虑到风险，并获得使项目成功的最大机会。图5.6 阐明技术、成本、进度和项目计划性这四种风险类别之间的主要相互作用。箭头符号指出典型的风险关系，尽管其他情况也可能必然存在。

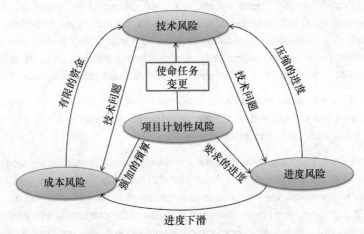

图 5.6　风险类别之间的典型关系。INCOSE SEH 第 1 版中图 4.5、4.6 和 4.7。按照 INCOSE 通知页使用。版权所有。

- *技术风险*——系统的技术需求在系统生命周期内无法实现的可能性。如果系统未能实现性能需求；未能满足可运行性、可生产性、可测试性或综合需求；或未能满足环境保护需求，则存在技术风险。未能满足能够以技术术语表达的任何需求的可能性是技术风险的来源。

TECHNICAL MANAGEMENT PROCESSES

- *Cost risk*—The possibility that available budget will be exceeded. Cost risk exists if the project must devote more resources than planned to achieve technical requirements, if the project must add resources to support slipped schedules due to any reason, if changes must be made to the number of items to be produced, or if changes occur in the organization or national economy. Cost risk can be predicted at the total project level or for a system element. The collective effects of element-level cost risks can produce cost risk for the total project.

- *Schedule risk*—The possibility that the project will fail to meet scheduled milestones. Schedule risk exists if there is inadequate allowance for acquisition delays. Schedule risk exists if difficulty is experienced in achieving scheduled technical accomplishments, such as the development of software. Schedule risk can be incurred at the total project level for milestones such as deployment of the first system element. The cascading effects of element level schedule risks can produce schedule risk for the total project.

- *Programmatic risk*—Produced by events that are beyond the control of the project manager. These events are often produced by decisions made by personnel at higher levels of authority, such as reductions in project priority, delays in receiving authorization to proceed with a project, reduced or delayed funding, changes in organizational or national objectives, etc. Programmatic risk can be a source of risk in any of the other three risk categories.

5.4.2.2 Risk Management Approach Once a risk management strategy and risk profile have been established, the three key risk management process activities are analyze risks, treat risks, and monitor risks.

Analyze Risks Analyzing risks involves identifying risks and evaluating their relative likelihood and consequence. The basis for this evaluation may be qualitative or quantitative; regardless, the objective is to set priorities and focus attention on areas of risk with the greatest consequences to the success of the project. All stakeholders and project personnel should feel welcome to contribute to identifying and analyzing risks.

If a project is unprecedented, brainstorming using strength–weakness–opportunity–threat (SWOT) or Delphi techniques may be appropriate. However, most projects represent a new combination of existing systems or system elements or represent the insertion of incremental advances in technology. This means that key insights can be gained concerning a current project's risk by examining the successes, failures, problems, and solutions of similar prior projects. The experience and knowledge gained, or lessons learned, can be applied to identify potential risk in a new project and to develop risk-specific management strategies.

- *成本风险*——将要超过可用预算的可能性。如果项目必须比原计划投入更多的资源来实现技术需求，如果项目必须增加资源以支持由任何原因导致的进度延期，如果必须改变待生产项的数量，或如果组织或国民经济发生变化，则存在成本风险。可在整个项目层级上或针对系统元素预测成本风险。元素层级上的成本风险的集合效应可产生整个项目的成本风险。

- *进度风险*——项目未能满足预定里程碑的可能性。如果用于采办延期的补贴不足，则存在进度风险。如果在实现诸如软件开发的预定技术成果时遇到困难，则存在进度风险。整个项目层级的里程碑，如第一个系统元素的部署，都可能招致进度风险。元素层级的进度风险的级联效应可产生整个项目的进度风险。

- *项目计划性风险*——由项目经理无法控制的事件所产生。这些事件常常是由更高级权限上的人员做出的决策所产生的，如项目优先级降低，接受继续项目的授权的延迟，资金减少或延误，组织或国家目标的变化等。项目计划性风险可以是任何其他三种风险类别中风险源。

5.4.2.2 *风险管理方法* 风险管理策略和风险剖面一旦建立，则三个关键的风险管理流程活动就是：分析风险、处理风险和监控风险。

分析风险 分析风险涉及识别风险和评价风险的相关可能性和后果。这种评估的基准可能是定性的或定量的；无论采用哪种方式，其目的是设置优先级并将关注点集中在对项目的成功具有最大后果的风险域。所有利益攸关者和项目人员都应乐于对识别和分析风险做出贡献。

如果项目是前无先例的，则使用优势—劣势—机会—威胁（SWOT）分析或德尔菲技术的头脑风暴也许是适当的。然而，大多数项目表征现有系统或系统元素的新组合，或表征渐进式技术进步的引入。这意味着可通过检查之前类似项目的成功、失败、问题和解决方案来得到有关当前项目风险的关键见解。所获得的经验和知识或吸取的教训可应用于识别新项目中的潜在风险并且开发风险特定的管理策略。

The first step is to determine the information needs. This could vary from assessing the risk in development of a custom computer chip to identifying the risks associated with a major system development. Next, systems engineers define the basic characteristics of the new system as a basis for identifying past projects that are similar in technology, function, design, etc. Based on the availability of data, analogous systems or system elements are selected and data gathered. Often, the data collection process and initial assessment lead to a further definition of the system for the purposes of comparison. Comparisons to prior systems may not be exact or the data may need to be adjusted to be used as a basis for estimating the future. The desired output is insight into cost, schedule, and technical risks of a project based on observations of similar past projects.

Uncertainty is characterized by a distribution of outcomes based on the likelihood of occurrence and severity of consequences. As noted previously, risk involves both the likelihood and consequences of the possible outcomes. In its most general form, risk analysis should capture the spectrum of outcomes relative to the desired project technical performance, cost, and schedule requirements. Risk generally needs to be analyzed subjectively because adequate statistical data are rarely available. Expert interviews and models are common techniques for conducting risk analysis.

Risk Perception   It is important to recognize that the severity of consequences (or impact) associated with a risk is an attribute of the person or group potentially affected by the risk, rather than an attribute of the "risk event" per se. In other words, the occurrence of a risk event will have different effects on different people depending on (i) their specific situation and perspective at the time of the occurrence and (ii) their unique personal values, perceptions, and sensitivities. It is possible, for example, for one person (or group) to be negatively impacted by an event or situation, while another person (or group) is positively impacted by the same event or situation. This is to be expected in win–lose and competitive event scenarios. Likewise, when two or more people (or groups) are affected in a uniformly negative or positive manner, at least some variation in their assessment of the level of negative or positive impact can be expected. In general, a risk event having significant variation in effect for different individuals or groups should be separated into an appropriate number of uniquely identified individual risk statements and detailed further to include reference to the effected person or group and the specific effects that are unique to them.

第一步是确定信息需要。这一步骤可从评估定制的计算机芯片的开发风险转变为识别与一个重大系统开发有关的风险。下一步，系统工程师定义新系统的基本特性，作为识别在技术、功能和设计等方面相似的以往项目的基础。基于数据的可用性，选择类似系统或子系统并收集数据。为了进行比较，数据收集流程和初步评估常常导致系统的进一步定义。与以往系统的比较可能不准确或数据可能需要被调整以便用作预测未来的基础。期望的输出是基于对以往类似项目的观察结果，对项目的成本、进度和技术风险的理解。

不确定性是通过基于发生的可能性和后果的严重性结果的分布而被特征化的。如上所述，风险涉及可能结果的可能性和后果。从最广义的形式来看，风险分析应获取与所期望的项目技术性能、成本和进度需求相关结果的谱系。因为很少能得到充分的统计数据，所以风险一般需要进行主观的分析。专家访谈和模型是进行风险分析的常用技术。

风险感知 与风险相关的后果（或影响）的严重性是可能受风险影响的个人或群体的属性，而不是"风险事件"本身的属性，这一认识是很重要的。换言之，风险事件的发生将对不同人员具有不同影响，取决于（i）事件发生时人员特有的情况和视角以及（ii）人员的独特个人价值、感知和灵敏度。例如，一个人（或群体）可能受到事件或情景的负面影响，而另一个人（或群体）受到同一事件或情景的正面影响。这可在输—赢和竞争性事件场景中预料到。同样，当两个或更多个人（或群体）以统一的负面或正面方式受影响时，至少可以预期到他们对负面或正面影响水平的评估存在部分变化。一般而言，对不同的个人或群体具有显著差异的风险事件应当被分成适当数量的唯一识别的单独风险说明，并被进一步详细说明，以便提及受影响人员或群体以及他们所特有的特定影响。

TECHNICAL MANAGEMENT PROCESSES

To achieve accuracy of risk estimation and overall effectiveness of the risk management effort, it is very important that risk estimates are based on clear, unambiguous risk descriptions and an adequate understanding of the values and situations of those potentially affected by the occurrence of the risk event. When possible, direct communication with the affected individuals and groups is preferred. Variation in perceived risk levels can often be reduced through clarification of risk event scenarios and better definition of risk assessment criteria and rating scales.

ISO Guide 73:2009, *Risk management—Vocabulary*, defines risk perception as a "stakeholder's view on a risk" noting that "risk perception reflects the stakeholder's needs, issues, knowledge, belief and values." A stakeholder is defined as a "person or organization that can affect, be affected by, or perceive themselves to be affected by a decision or activity."

Generally, the stakeholders consulted with to define system needs, expectations, and requirements (see Section 4.2) should also be consulted with to identify and assess risks. It is good practice to establish and maintain traceability between stakeholders and risks, as well as stakeholders and requirements. Risk statements and estimates that do not reference the stakeholder(s) affected by the risk event or situation should be viewed as vague and incomplete or at least potentially inaccurate.

Expert Interviews   Efficient acquisition of expert judgments is extremely important to the overall accuracy of the risk management effort. The expert interview technique consists of identifying the appropriate experts, questioning them about the risks in their area of expertise, and quantifying these subjective judgments. One result is the formulation of a range of uncertainty or a probability density (with respect to cost, schedule, or performance) for use in any of several risk analysis tools.

Since expert interviews result in a collection of subjective judgments, the only real "error" can be in the methodology for collecting the data. If it can be shown that the techniques for collecting the data are not adequate, then the entire risk assessment can become questionable. For this reason, the methodology used to collect the data must be thoroughly documented and defensible. Experience and skill are required to encourage the expert to divulge information in the right format. Typical problems encountered include identification of the wrong expert, obtaining poor quality information, unwillingness of the expert to share information, changing opinions, getting biased viewpoints, obtaining only one perspective, and conflicting judgments. When conducted properly, expert interviews provide reliable qualitative information. However, the transformation of that qualitative information into quantitative distributions or other measures depends on the skill of the analyst.

为实现风险管理工作的风险评估准确性和总体效率，基于清晰、无歧义的风险描述以及对可能受风险事件的发生影响的个人或群体的价值和情景的充分理解来进行风险评估是非常重要的。若可能，与受影响的个人和群体的直接沟通是首选的。通常可通过澄清风险事件场景以及更好地定义风险评估准则和评定量来减少感知的风险等级的差异。

ISO 指南 73:2009，《风险管理——词汇表》将风险感知定义为"利益攸关者对风险的视角"，指出"风险感知反映利益攸关者的需要、问题、知识、信念和价值"。利益攸关者被定义为"能够影响决策或活动、受决策或活动影响或感知自己将要受决策或活动影响的个人或组织"。

通常，在定义系统需要、预期和需求方面进行商议的利益攸关者（见 4.2 节）还应在识别和评估风险方面进行商议。建立并维护利益攸关者与风险之间可追溯性是一种良好的做法，如同利益攸关者与需求之间的一样。未提及受风险事件或情景影响的利益攸关者的风险说明和估计应当被视为含糊不清的、不完整的或至少有可能不准确的。

专家访谈　　高效地获取专家的判断对于风险管理工作的总体准确性极其重要。专家访谈技术包括识别合适的专家，询问他们所在专业领域的风险，并且量化这些主观判断。一个结果是一系列不确定性或概率密度（关于成本、进度或性能）的公式化表达，在许多风险分析工具中使用。

由于专家访谈产生主观判断的集合，唯一真正的"误差"可出现在收集数据的方法论中。如果能够表明收集数据的技术不充分，则整体风险评估变得令人存疑。为此，收集数据所用的方法论必须被完整文件化且正当合理。需要采用经验和技巧鼓励专家以正确的格式披露信息。所遇到的典型问题包括识别不合适的专家、获得劣质信息、专家不愿意共享信息、改变意见、得到带偏见的观点、固守单一视角以及相互冲突的判断。若实施得当，专家访谈提供可靠的定性信息。然而，定性信息转换为定量分布或其他测度则取决于分析者的技能。

TECHNICAL MANAGEMENT PROCESSES

Risk Assessment Techniques ISO 31010, *Risk management—Risk assessment techniques* (2009), provides detailed descriptions and application guidance for approximately 30 assessment techniques ranging from brainstorming and checklists to Failure Mode and Effects Analysis (FMEA), fault tree analysis (FTA), Monte Carlo simulation, and Bayesian statistics and Bayes nets.

Treat Risks Risk treatment approaches (also referred to as risk handling approaches) need to be established for the moderate and high-risk items identified in the risk analysis effort. These activities are formalized in the RMP. There are four basic approaches to treat risk:

1. Avoid the risk through change of requirements or redesign.
2. Accept the risk and do no more.
3. Control the risk by expending budget and other resources to reduce likelihood and/or consequence.
4. Transfer the risk by agreement with another party that it is in their scope to mitigate. Look for a partner that has experience in the dedicated risk area.

The following steps can be taken to avoid or control unnecessary risks:

- *Requirements scrubbing*—Requirements that significantly complicate the system can be scrutinized to ensure that they deliver value equivalent to their investment. Find alternative solutions that deliver the same or comparable capability.

- *Selection of most promising options*—In most situations, several options are available. A trade study can include project risk as a criterion when selecting the most promising alternative.

- *Staffing and team building*—Projects accomplish work through people. Attention to training, team-work, and employee morale can help avoid risks introduced by human errors.

For high-risk technical tasks, risk avoidance is insufficient and can be supplemented by the following approaches:

- Early procurement
- Initiation of parallel developments
- Implementation of extensive analysis and testing

风险评估技术 ISO 31010，《风险管理——风险评估技术》（2009）提供大约 30 种评估技术的详细描述和应用指南，范围从头脑风暴和检查单到失效模式和影响分析（FMEA）、故障树分析（FTA）、蒙特卡罗仿真以及贝叶斯统计与贝叶斯网。

处理风险　需要建立风险处理方法（亦称为风险处置方法），用于在风险分析工作中识别出的中度风险项和高风险项。这些活动在 RMP 中被正式化。存在四种处理风险的基本方法：

1. 通过需求的变更或再设计来规避风险。
2. 接受风险且不采取更多的行动。
3. 通过支出预算及其他资源以降低可能性和/或减轻后果来控制风险。
4. 通过与另一方就风险处于其缓解范围内达成协议而转移风险。寻找在指定的风险域具有经验的合作者。

可采取下列步骤规避或控制不必要的风险：

- *需求清理*——可以对使系统严重复杂化的需求进行仔细检查，以确保其交付价值等同于其投资。找出交付相同的或能力可比的备选解决方案。

- *选择最有期望的选项*——大多数情景下，可得到若干选项。当选择最有期望的备选方案时，权衡研究可包括项目风险作为一个准则。

- *人员编制和团队建设*——项目由人员完成工作。关注培训、团队合作和员工士气可有助于规避人为差错导致的风险。

对于高风险的技术任务，风险规避是不充分的，可通过下列方法补充：

- 提早采购
- 启动并行开发
- 实施广泛的分析和试验

TECHNICAL MANAGEMENT PROCESSES

- Contingency planning

The high-risk technical tasks generally imply high schedule and cost risks. Cost and schedule are impacted adversely if technical difficulties arise and the tasks are not achieved as planned. Schedule risk is controlled by early procurement of long-lead items and provisions for parallel-path developments. However, these activities also result in increased early costs. Testing and analysis can provide useful data in support of key decision points. Finally, contingency planning involves weighing alternative risk mitigation options.

For each risk that is determined credible after analysis, a risk Treatment Plan (also referred to as a risk Mitigation Action Plan) should be created that identifies the risk treatment strategy, the trigger points for action, and any other information to ensuring the treatment is effectively executed. The risk Treatment Plan can be part of the risk record on the risk profile. For risks that have significant consequences, a contingency plan should be created in case the risk treatment is not successful. It should include the triggers for enacting a contingency plan.

In China, the authorities built the short maglev train line in Shanghai (see Section 3.6.3) as a proof of concept. In spite of the high investment, this represented lower risk to the project than attempting a longer line with an unproven technology. The results collected from this project are inspiring others to consider maglev alternatives for greater distances.

Monitor Risks   Project management uses measures to simplify and illuminate the risk management process.

Each risk category has certain indicators that may be used to monitor project status for signs of risk. Tracking the progress of key system technical parameters can be used as an indicator of technical risk.

The typical format in tracking technical performance is a graph of a planned value of a key parameter plotted against calendar time. A second contour showing the actual value achieved is included in the same graph for comparative purposes. Cost and schedule risk are monitored using the products of the cost/schedule control system or some equivalent technique. Normally, cost and schedule variances are used along with a comparison of tasks planned to tasks accomplished. A number of additional references exist on the topic of risk management (AT&T, 1993; Barton et al., 2002; Michel and Galai, 2001; Shaw and Lake, 1993; Wideman, 2004).

- 应急计划

高风险的技术任务通常意味着高进度风险和高成本风险。如果出现技术上的困难且任务未按计划实现，成本和进度都会受到不利影响。进度风险是由长期项的提前采购和并行路径开发的预防措施来控制的。然而，这些活动亦导致早期成本增加。试验和分析可提供支持关键决策点的有用数据。最后，应急计划涉及权衡备选风险缓解选项。

对于每种经分析后被确定为可信的风险，应创建一个识别风险处理策略、行动的触发点和任何其他信息的风险处理计划（亦称为风险缓解行动计划），以确保有效地进行处理。风险处理计划可以是风险剖面中风险记录的一部分。对于具有严重后果的风险，如果风险处理不成功，则应创建一个应急计划。该应急计划应包括制订应急计划的触发器。

中国当局在上海建造了短途磁悬浮列车线（见 3.6.3 节）用来进行概念验证。尽管投资很高，但这说明该项目的风险比使用未被证实的技术尝试长途磁悬浮列车线的风险更低。从该项目中收集到的结果正启发他人考虑用于更长距离的磁悬浮备选方案。

监控风险 项目管理使用一些测度来简化和阐明风险管理流程。

每个风险类别都具有某些用于监控项目的风险信号状态的指标。对关键系统技术参数的进展的跟踪可用作技术风险的一个指标。

跟踪技术性能的典型格式是按照日历时间绘制的关键参数计划值的图表。出于比较的目的，在这一相同图表中包含第二条曲线用于表明所实现的实际值。使用成本/进度控制系统产品或一些等效技术来监控成本风险和进度风险。正常情况下，使用成本偏差和进度偏差，并同时比较计划中的任务和已完成的任务。关于风险管理主题存在着大量的附加参考文献（AT&T，1993；Barton 等，2002；Michel 和 Galai，2001；Shaw 和 Lake，1993；Wideman，2004）。

Risk management process scope, context, and objectives The risk management process described in this section is generic and may be applied at any stage of the SE life cycle (see Section 3.3), at any level in a system hierarchy (see Section 2.3), or to an SoS (see Section 2.4). In addition, an organization may decide to include opportunity management (i.e., risk and opportunity management) or management of positive consequences (in addition to negative consequences) as part of one or more risk management processes. As a foundation for efficiency and effectiveness, the scope and context of the risk management process should be clearly defined and consistent with requirements and expectations for the process.

Defining the System and Its Boundaries ISO 31000, *Risk management—Principles and guidelines on implementation* (2009), provides guidance and rationale for establishing the external and internal context of a risk management process.

System models (see Section 9.1) describing the system to which the risk management process applies (whether enterprise, product, or service) can facilitate the risk management process by providing about what the system "is" and "does," how it behaves in different scenarios, the location of its boundaries, and the definition of internal and external interfaces. System models can greatly enhance communication and can help ensure the comprehensiveness needed for full risk identification.

The scope of a risk management process also includes a time dimension. It is rare that a single risk management process is used throughout the lifetime of a system for all risks in all life cycle stages. For example, a product development organization might utilize a project risk management process during the development stage, while separate risk management processes performed by different organizations may be used years later for the utilization and support stages. By defining the calendar time and life cycle stage scopes of each risk management process, the likelihood of gaps and overlap is reduced.

Risk Management and the System Life Cycle Once the scope and context of a system have been established from a hierarchical standpoint, it is possible to define and model the system (and its associated risks) in relation to its life cycle. Risks at the early stages of the life cycle (exploratory research and concept definition) are quite different than the risks at the final stage (retirement). It is often necessary to consider risks in other stages while performing activities in the current stage. For example, risks associated with the retirement stage (e.g., human exposure to hazard waste during disposal) should be considered as part of risk management performed to evaluate concept options for disposability.

风险管理流程的范围、背景环境和目标　本节所描述的风险管理流程是通用的，且可在 SE 生命周期的任何阶段（见 3.3 节）、系统层级结构的任何层级（见 2.3 节）或 SoS（见 2.4 节）中应用。此外，组织可决定包括机会管理（即风险和机会管理）或正面结果（以及负面后果）的管理作为一个或多个风险管理流程的一部分。作为效率和有效性的基础，风险管理流程的范围和背景环境应当被明确地定义，并且与流程的需求和预期一致。

定义系统及其边界　ISO 31000，《风险管理——原则与实施指南》（2009）提供建立风险管理流程的外部和内部背景环境的指南和基本原理。

系统模型（见 9.1 节）描述风险管理流程所适用的系统（无论是复杂组织体、产品还是服务），可通过提供关于系统"是"什么、"做"什么、系统如何在不同的场景中运行、系统边界的位置以及内外接口的定义来促进风险管理流程。系统模型能够极大地加强沟通，并且能够有助于确保完整的风险识别所需的全面性。

风险管理流程的范围还包括时间维度。对于所有生命周期阶段中的所有风险而言，贯穿于系统的使用寿命使用一个风险管理流程是很罕见的。例如，产品开发组织可能在开发阶段期间使用项目风险管理流程，而不同的组织所执行的单独的风险管理流程可能在后期的使用阶段和维持阶段使用多年。通过定义每个风险管理流程的日历时间和生命周期阶段范围，降低差距和重叠的可能性。

风险管理和系统生命周期　一旦从层级的观点建立系统的范围和背景环境，就有可能针对系统的生命周期对系统（及其相关风险）进行定义和建模。生命周期早期阶段（探索性研究和概念定义）的风险与最后阶段（退役）的风险截然不同。通常有必要在执行当前阶段的活动的同时考虑其他阶段的风险。例如，与退役阶段有关的风险（如处置期间人体暴露于危害废物）应当被视为用于评估处置性的概念选项的风险管理的一部分。

TECHNICAL MANAGEMENT PROCESSES

The performance and output of risk management activities related to safety risks encountered during the development, production, utilization, support, and retirement stages may be governed by regulations and standards (e.g., ISO 14971, *Medical devices— Application of risk management to medical devices*) or by customer contractual agreement. As necessary, specialty engineering activities such as safety analysis (see Section 10.10), usability analysis/human systems integration (Section 10.13), and system security engineering (see Section 10.11) should be utilized and coordinated through a risk management process that complies with regulatory and/or customer requirements.

5.4.2.3 Opportunity Management Concepts SE and project management are all about pursuing an opportunity to solve a problem or fulfill a need. Opportunities enable creativity in resolving concepts, architectures, designs, and strategic and tactical approaches, as well as the many administrative issues within the project. It is the selection and pursuit of these strategic and tactical opportunities that determine just how successful the project and system will be. Of course, opportunities usually carry risks, and each opportunity will have its own set of risks that must be intelligently judged and properly managed to achieve the full value (Forsberg et al., 2005).

Opportunities represent the potential for improving the value of the project results. The project champions (e.g., the creators, designers, integrators, and implementers) apply their "best-in-class" practices in the pursuit of opportunities. After all, the fun of working on projects is doing something new and innovative. It is these opportunities that create the project's value. *Risks* are defined as chances of injury, damage, or loss. Risks are the chances of not achieving the results as planned. Each of the strategic and tactical opportunities pursued has associated risks that undermine and detract from the opportunity's value. These are the risks that must be managed to enhance the opportunity value and the overall value of the project (see Figure 5.7). Opportunity management and risk management are, therefore, essential to—and performed concurrently with—the planning process but require the application of separate and unique techniques that justify this distinct technical management element.

There are two levels of opportunities and risks. Because a project is the pursuit of an opportunity, the *macro* level is the project opportunity itself. The approach to achieving the macro opportunity and the mitigation of associated project-level risks are structured into the strategy and tactics of the project cycle, the selected decision gates, the teaming arrangements, key personnel selected, and so on. The *element* level encompasses the tactical opportunities and risks within the project that become apparent at lower levels of decomposition and as project life cycle stages are planned and executed. This can include emerging, unproven technology; incremental and evolutionary methods that promise high returns; and the temptation to circumvent proven practices to deliver products better, faster, and cheaper.

在开发阶段、生产阶段、使用阶段、维持阶段和退役阶段期间经历的与安全风险有关的风险管理活动的运行情况和输出可能受法规和标准（如 ISO 14971，《医疗设备——风险管理在医疗设备中的应用》）或客户合同协议的管控。必要时，诸如安全性分析（见 10.10 节）、可用性分析/人员系统综合（见 10.13 节）及系统安保工程（见 10.11 节）的专业工程活动应当遵循法规和/或客户需求的风险管理流程来使用和协调。

5.4.2.3　机会管理概念　SE 和项目管理都是关于追求解决问题或实现需要的机会。机会能够使概念/方案、架构、设计、战略和战术方法，以及项目内许多行政问题的解决具有创造性。正是对这些战略和战术机会的选择和追求决定了项目和系统的成功程度。当然，机会常常伴随着风险，并且每个机会都将有自己的风险集合，为实现全面价值，这些风险必须被明智地判断和恰当地管理（Forsberg 等，2005）。

*机*会表征提高项目结果价值的潜力。项目捍卫者（如创造者、设计者、综合者和实施者）在追求机会时应用他们的"最佳"实践。毕竟，项目工作中的乐趣在于做一些新鲜的、创新的事。正是这些机会创造了项目的价值。风险被定义为伤害、损害或损失的可能性。风险是未按计划达到结果的可能性。所追求的每个战略和战术机会都具有破坏和减损机会价值的相关风险。为提高项目的机会价值和整体价值（见图 5.7），必须管理这些风险。因此，机会管理和风险管理对于计划流程是根本性的，并与该流程并行实施，但需要应用能够证明这种截然不同的技术管理要素合理性的分开的且独特的技术。

机会和风险存在两个层级。由于项目就是追求机会，因此宏观层级是项目机会本身。达到宏观机会的方法和相关项目层级风险的缓解构成项目周期的战略和战术、所选择的决策门、团队筹划和所选择的关键人员等。元素层级包括项目内的战术机会和风险，在较低分解层级上并且在项目生命周期阶段计划和执行时变得明显。这可包括新兴的、未经证实的技术；承诺高回报的渐进方法和演进方法；以及冒着规避已被证实的常规做法的风险，以更好、更快和更低的成本交付产品。

TECHNICAL MANAGEMENT PROCESSES

Overall project value can be expressed as benefit divided by cost. Opportunities and their risks should be managed jointly to enhance project value. This is based on the relative merits of exploiting each opportunity and mitigating each risk. In the context of the opportunity and the resultant value, we carry a spare tire to mitigate the risk of a flat tire by reducing the probability and impact of having a delayed trip. The high value we place on getting where we want to go far exceeds the small expense of a spare. When deciding to pursue the opportunity of a long automobile trip, we may take extra risk management precautions, such as preventive maintenance and spares for hard-to-find parts.

The assessment of opportunity and risk balance is situational. For example, few today have a car with more than one spare tire (multiple spares were common practice in the early 1900s). However, a few years ago, an individual decided to spend a full month driving across the Australian Outback in late spring. He was looking for solitude in the wilderness (his opportunity). On advice from experienced friends, he took four spare tires and wheels. They also advised him that the risk of mechanical breakdown was very high on a 30-day trip, and the consequence would almost certainly be fatal. The risk of two vehicles breaking down at the same time was acceptably low. So, he adjusted the opportunity for absolute solitude by joining two other adventurers. They set out in three cars. Everyone survived in good health, but only two cars returned, and two of his "spare" tires were shredded by the rough terrain. The "balanced" mitigation approach proved effective.

5.5 CONFIGURATION MANAGEMENT PROCESS

5.5.1 Overview

5.5.1.1 Purpose As stated in ISO/IEC/IEEE 15288,

> [6.3.5.1] The purpose of the Configuration Management process is to manage and control system elements and configurations over the life cycle. CM also manages consistency between a product and its associated configuration definition.

整个项目的价值可被表达为效益除以成本。机会及其风险应一并被管理，以提高项目价值。这基于发挥每个机会并缓解每个风险的相对优点。在机会和由此产生的价值的背景环境中，我们携带一个备用轮胎以缓解轮胎漏气的风险，降低耽误旅行的概率和影响。我们想要到达的更远地方所带来的高价值远远超出一个备件的小开销。当决定追求长途汽车旅行的机会时，我们可能采取额外的风险管理措施，如预防性维护和备用难以找到的零件。

机会和风险平衡的评估视情而论。例如，当今很少有人在车上备有多于一个的备胎（备有多个备胎在 20 世纪早期曾是一种常见做法）。然而，在几年前的春末，有人决定花费整整一个月的时间驾车穿越澳大利亚内陆。他在荒野中寻求独处（他的机会）。根据有经验的朋友的忠告，他携带了四个备用轮胎和车轮。他们还忠告他 30 天的旅行具有很高的机械故障风险并且几乎可以肯定的是，其后果会是致命的。可以接受的是，两辆车同时故障的风险是低的。因此，他通过加入两位其他冒险者，调整了绝对独处的机会。他们驾驶三辆车出发。每个人都得以幸存并且健康状况良好，但只有两辆车返回，"备用"轮胎中的两个被崎岖地形撕裂。"平衡的"缓解方法被证明是有效的。

5.5 构型配置管理流程

5.5.1 概览

5.5.1.1 目的 如 ISO/IEC/IEEE 15288 所述，

[6.3.5.1] 构型配置管理流程的目的是在生命周期内管理和控制系统元素及构型配置。CM 还管理产品及其相关构型配置定义之间的一致性。

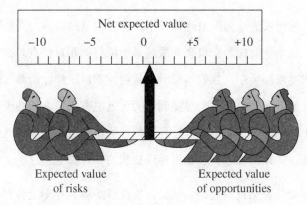

FIGURE 5.7 Intelligent management of risks and opportunities. Derived from (Forsberg et al., 2005) page 224. Reprinted with permission from Kevin Forsberg. All other rights reserved.

This is accomplished by ensuring the effective management of the evolving configuration of a system, both hardware and software, during its life cycle. Fundamental to this objective is the establishment, control, and maintenance of software and hardware baselines. Baselines are business, budget, functional, performance, and physical reference points for maintaining development and control. These baselines, or reference points, are established by review and acceptance of requirements, design, and product specification documents. The creation of a baseline may coincide with a project milestone or decision gate. As the system matures and moves through the life cycle stages, the software or hardware baseline is maintained under configuration control.

5.5.1.2 Description Evolving the concept definition and system definition is a reality that must be addressed over the life of a system development effort and throughout the utilization and support stages of the system. Configuration management ensures that product functional, performance, and physical characteristics are properly identified, documented, validated, and verified to establish product integrity; that changes to these product characteristics are properly identified, reviewed, approved, documented, and implemented; and that the products produced against a given set of documentation are known. Figure 5.8 is the IPO diagram for the configuration management process.

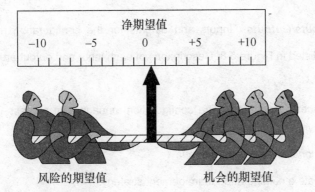

图 5.7 风险和机会的智能管理。来自 Forsberg 等（2005），第 224 页。经 Kevin Forsberg 许可后转载。版权所有。

通过确保在其生命周期内有效管理一个系统(硬件和软件)的持续演进的构型配置来实现这一目的。该目标的基础是软件和硬件基线的建立、控制和维护。基线是保持开发和控制的业务、预算、功能、性能和物理参考点。这些基线或参考点是通过评审和验收需求、设计和产品规范文件而建立的。基线的创建可与项目里程碑或决策门同时发生。随着系统趋于成熟及在生命周期阶段的推进，软件或硬件基线要保持在构型配置控制之下。

5.5.1.2 描述 持续演进的概念定义和系统定义是在系统开发工作的整个生命期内以及系统的使用和维持阶段内必须应对的实际问题。构型配置管理确保：产品功能、性能和物理特性被恰当地识别、文件化、确认和验证，以建立产品完整性；这些产品特性的变更被适当地识别、评审、审批、文件化和实施；以及按照一个给定的已知文档的集合生产的产品。图 5.8 是构型管理流程的 IPO 图。

TECHNICAL MANAGEMENT PROCESSES

5.5.1.3 Inputs/Outputs Inputs and outputs for the configuration management process are listed in Figure 5.8. Descriptions of each input and output are provided in Appendix E.

5.5.1.4 Process Activities The configuration management process includes the following activities:

- *Plan configuration management.*
  - Create a configuration management strategy.
  - Implement a configuration control cycle that incorporates evaluation, approval, validation, and verification of engineering change requests (ECRs).
- *Perform configuration identification.*
  - Identify system elements and information items to be maintained under configuration control as configuration items (CIs).
  - Establish unique identifiers for the CIs.
  - Establish baselines for the CIs at appropriate points through the life cycle, including agreement of the baselines by the acquirer and supplier.

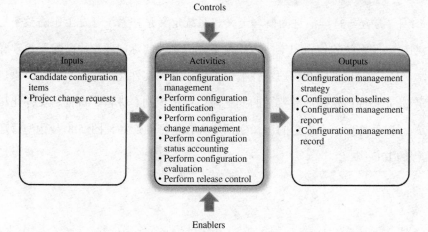

FIGURE 5.8 IPO diagram for the configuration management process. INCOSE SEH original figure created by Shortell and Walden. Usage per the INCOSE notices page. All other rights reserved.

5.5.1.3 输入/输出 图 5.8 列出构型管理流程的输入和输出。附录 E 提供每个输入和输出的描述。

5.5.1.4 流程活动 构型管理流程包括以下活动：

- *规划构型管理。*

  – 创建一个构型管理策略。

  – 实施一种将工程变更请求（ECR）的评估、批准、确认和验证纳之在内的构型控制周期。

- *实施构型识别。*

  – 识别要在构型配置控制之下作为配置项（CI）予以维护的系统元素和信息项。

  – 为 CI 建立唯一标识符。

  – 在生命周期内的适当时刻为各 CI 建立基线，包括采办方和供应商达成的基线协议。

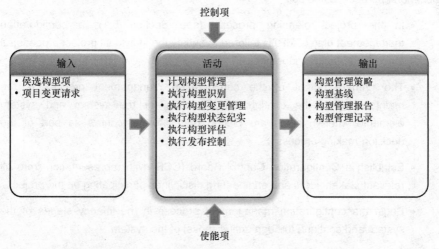

图 5.8 构型配置管理流程的 IPO 图。INCOSE SEH 原始图由 Shortell 和 Walden 创建。按照 INCOSE 通知页使用。版权所有。

TECHNICAL MANAGEMENT PROCESSES

- *Perform configuration change management.*
  - Control baseline changes throughout the system life cycle. This includes the identification, recording, review, approval, tracking, and processing of requests for change (RFCs) and requests for variance (RFVs) (also known as deviations).
- *Perform configuration status accounting.*
  - Develop and maintain configuration control documentation and configuration management data, and communicate the status of controlled items to the project team.
- *Perform configuration evaluation.*
  - Perform configuration audits and configuration management surveillance reviews associated with milestones and decision gates to validate the baselines.
- *Perform release control.*
  - Perform prioritization, tracking, scheduling, and closing changes, including relevant supporting documentation.

Common approaches and tips:

- In the project planning process (see Section 5.1), a configuration management plan (CMP) is tailored to satisfy the individual project procedures for configuration management.
- The primary output of the configuration management process is the maintenance of the configuration baseline for the system and system elements wherein items are placed under formal control as part of the decision-making process.
- Establish a Configuration Control Board (CCB) with representation from all relevant stakeholders and engineering disciplines participating on the project.
- Begin the configuration management process in the infancy stages of the system and continue through until disposal of the system.
- Configuration management documentation is maintained throughout the life of the system.

- *实施构型配置变更管理。*
  - 贯穿于系统生命周期控制基线变更。这包括变更请求（RFC）和偏差请求（RFV）（亦称为偏离）的识别、记录、评审、审批、跟踪和处理。
- *实施构型状态报告。*
  - 开发并维护构型控制文档和构型管理数据，并与项目团队交流受控项的状态。
- *实施构型评估。*
  - 实施有关里程碑和决策门的构型审计和构型管理监督评审以确认基线。
- *实施发布控制。*
  - 执行优先级排序、跟踪、时间安排和关闭变更，包括相关的支持文档。

常用方法和建议：

- 在项目计划流程（见 5.1 节）中，对构型管理计划（CMP）进行剪裁以满足构型管理的各个项目程序。
- 构型管理流程的主要输出是对系统和系统元素的构型基线的维护，其中这些项作为决策流程的一部分被置于正式控制之下。
- 建立由参与项目的所有相关的利益攸关者和工程专业学科的代表组成的构型控制委员会（CCB）。
- 在系统初期阶段开始构型管理流程并一直持续到系统的处置。
- 贯穿于系统生命期维护构型管理文件。

TECHNICAL MANAGEMENT PROCESSES

- Additional guidance regarding configuration management activities can be found in ISO 10007 (2003), IEEE Std 828 (2012), and ANSI/EIA 649B (2011).

- Domain-specific practices, such as SAE Aerospace recommended Practice (ARP) 4754A, *Guidelines for Development of Civil Aircraft and Systems (2010)*, provide additional application detail for the domain.

5.5.2   Elaboration

5.5.2.1   Configuration Management Concepts   The purpose of configuration management is to establish and maintain the control of requirements, documentation, and artifacts produced throughout the system's life cycle and to manage the impact of change on a project. Baselines consolidate the evolving configuration states of system elements to form documented baselines at designated times or under defined circumstances. Baselines form the basis for the next change. Selected baselines typically become formalized between acquirer and supplier, depending on the practices of the industry and the contractual involvement of the acquirer in the configuration change process. There are generally three major types of baselines at the system level: functional baseline, allocated baseline, and product baseline. These may vary by domain or local strategy (ISO/IEC/IEEE 15288:2015).

Change is in evitable, as indicated in Figure 5.9. Systems engineers ensure that the change is necessary and that the most cost-effective recommendation has been proposed.

Initial planning efforts for configuration management are performed at the onset of the project and defined in the CMP, which establishes the scope of items that are covered by the plan; identifies the resources and personnel skill level required; defines the tasks to be performed; describes organizational roles and responsibilities; and identifies configuration management tools and processes, as well as methodologies, standards, and procedures that will be used on the project. Configuration control maintains integrity by facilitating approved changes and preventing the incorporation of unapproved changes into the items under configuration control. Such activities as check-in and checkout of source code, versions of system elements, and deviations of manufactured items are part of configuration management. Independent configuration audits assess the evolution of a product to ensure compliance to specifications, policies, and contractual agreements. Formal audits may be performed in support of decision gate review.

- 关于构型管理活动的附加指导可在 ISO 10007（2003）、IEEE Std 828（2012）和 ANSI/EIA 649B（2011）中找到。

- 领域特定的实践，如 SAE 航空航天所推荐的实践（ARP）4754A、《民用飞机和系统的开发指南》（2010）为领域提供附加的应用细节。

5.5.2 详细阐述

5.5.2.1 构型管理概念 构型管理的目的是建立并维护对贯穿于系统的生命周期所产生的需求、文档及制品的控制，以及管理变更对项目的影响。基线加强系统元素的演进构型状态的管理，以便在指定时间或在定义的环境下形成文件化的基线。基线构成下一变更的基础。选择的基线通常在采办方和供应商之间达成正式化，取决于在构型变更流程中的行业实践和采办方合同所涉及的事宜。在系统层级通常具有三种主要类型的基线：功能基线、分配基线和产品基线。这些基线可随领域或当地策略（ISO/IEC/IEEE 15288:2015）而变化。

变更是不可避免的，如图 5.9 所示。系统工程师要确保变更是必要的，并已提出最具成本效益的建议。

初始的构型管理计划工作在项目一开始便执行并在 CMP 中被定义，这样就建立了计划所涵盖的构型配置项的范围；识别所要求的资源和人员技能水平；定义需要实施的任务；描述组织的角色和职责；并识别构型管理工具和流程，以及用于项目的方法论、标准和程序。构型控制通过促进已批准的变更和阻止将未经批准的变更纳入构型控制之下的构型项来维护完整性。这种活动如源代码的检入和检出、系统元素的版本和制造项的偏离是构型管理的一部分。独立的构型审计评估产品的演进以确保符合规范、方针和合同协议。正式的审计可在决策门评审的支持下执行。

TECHNICAL MANAGEMENT PROCESSES

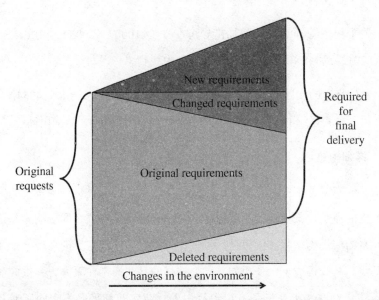

FIGURE 5.9 Requirements changes are inevitable. Figure 5.9 requirements changes are inevitable. Derived from (Forsberg et al., 2005) Figure 9.3. Reprinted with permission from Kevin Forsberg. All other rights reserved.

A request to change the current configuration of a system is typically made using an Engineering Change Proposal (ECP). An ECP may originate in a number of ways. The customer may request an ECP to address a change in requirements or a change in scope; an unexpected breakthrough in technology may result in the supplier of a system element proposing an ECP; or a supplier may identify a need for changes in the system under development. Circumstances like these that will potentially change the scope or the requirements are appropriate reasons to propose an ECP and to conduct an analysis to understand the effect of the change on existing plans, costs, and schedules. The ECP must be approved before the change is put into effect. It is never appropriate to propose an ECP to correct cost or schedule variances absent of change in scope. A minor change that falls within the current project scope usually does not require an ECP but should be approved and result in the generation of an engineering notice (En). It is also important to ask, "What is the impact of not making the change?" especially as the system matures, since changes made later in the life cycle have an increasing risk of hidden impacts, which can adversely affect system cost, schedule, and technical performance.

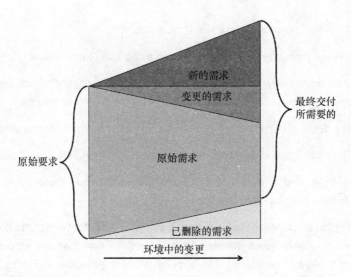

图 5.9 需求变更是不可避免。来自 Forsberg 等（2005），图 9.3。经 Kevin Forsberg 许可后转载。版权所有。

通常使用工程变更建议书（ECP）提出更改系统当前构型的要求。ECP 可能以多种方式生成。客户可要求 ECP 以应对需求的变更或范围的变更；技术上意想不到的突破可导致系统元素供应商提出 ECP；或供应商可识别到正在开发中的系统的变更需要。像这样可能改变范围或需求的情况，是提出 ECP 和实施分析的恰当理由，以理解变更对现有计划、成本和进度的影响。ECP 必须在变更生效前得到批准。在范围不变时提出 ECP 以修正成本或进度偏离是不合适的。当前项目范围的微小变更通常不需要 ECP，但应进行审批并得以产生工程通知单（EN）。特别是随着系统逐渐成熟，由于在生命周期后期做出的变更增加隐藏影响的风险，可对系统成本、进度和技术性能产生不利影响，因此，询问"不做出变更的影响是什么？"亦是非常重要的。

The most desirable outcomes of an ECP cycle are:

1. System functionality is altered to meet a changing requirement.
2. New technology or a new product extends the capabilities of the system beyond those initially required in ways that the customer desires.
3. The costs of development, or of utilization, or of support are reduced.
4. The reliability and availability of the system are improved.

Outcomes three and four reduce LCC and potentially save more money than is invested to fund the proposed change.

ECPs and ENs help ensure that a system evolves in ways that allow it to continue to satisfy its operational requirements and its objectives and that any modification is known to all relevant personnel. The airplane system illustrated in Figure 2.2 is an example of a product family that depends on accurate identification of system elements and characteristics to support the mix and match consumer market.

5.5.2.2 Configuration Management Approach Configuration management establishes and maintains control over requirements, specifications, configuration definition documentation, and design changes. Configuration identification, configuration control, configuration status accounting, and configuration audits of the functional and physical configuration (i.e., validation and distribution) are the primary focus of configuration management.

There will always be a need to make changes; however, SE must ensure (i) that the change is necessary and (ii) that the most cost-effective solution has been proposed. Configuration management must, therefore, apply technical and administrative direction, surveillance, and services to do the following:

- Identify and document the functional and physical characteristics of individual CIs such that they are unique and accessible in some form.
- Assign a unique identifier to each version of each CI.
- Establish controls to allow changes in those characteristics.
- Concur in product release and ensure consistent products via the creation of baseline products.
- Record, track, and report change processing and implementation status and collect measures pertaining to change requests or problems with the product baseline.

ECP 周期最理想的结果是：

1. 改变系统功能性以满足变更需求。

2. 新技术或新产品拓展以客户期望的方式超出在初期所要求的系统能力。

3. 降低开发成本、使用成本或维持成本。

4. 提高系统的可靠性和可用性。

第三、四个结果降低 LCC 并且可能较将资金投入到资助建议变更中更节省费用。

ECP 和 EN 有助于确保系统以连续满足其运行需求及其目标的方式演进，并且所有相关人员都知晓任何修改。图 2.2 阐述的飞机系统是依赖于系统元素和特性的精准识别来支持消费者市场混搭和匹配的产品系列的示例。

5.5.2.2 *构型管理方法* 构型管理建立并维护对需求、规范、构型定义文件和设计变更的控制。功能和物理构型（即确认和分布）的构型识别、构型控制、构型状态报告和构型审计是构型管理的主要焦点。

总是需要做出变更；然而，SE 必须确保（i）变更是必要的；及（ii）提出最具成本效益的解决方案。因此，构型管理必须应用技术和行政指导、监视和服务完成下列内容：

- 识别和文件化各个 CI 的功能和物理特性，使这些特性在某种形式上是唯一且可实现的。

- 为每个版本的每个 CI 分配一个唯一识别符。

- 建立控制以允许改变上述特性。

- 同意产品发布并通过创建基线产品确保产品一致。

- 记录、跟踪和报告变更处理和实施状态，并收集与产品基线变更请求或问题有关的测度。

- Maintain comprehensive traceability of all transactions.

Configuration Identification Configuration identification uniquely identifies the elements within a baseline configuration. This unique identification promotes the ability to create and maintain master inventory lists of baselines. As part of the SE effort, the system is decomposed into CIs, which serve as the critical elements subjected to rigorous formal control. The compilation of all the CIs is called the CI list. This list may reflect items that are developed, vendor produced, or provided by the customer for integration into the final system. These items may be deliverable items under the contract or used to produce the deliverable items.

Change Management Configuration change management, or change control, manages the collection of the items to be base-lined and maintains the integrity of the CIs identified by facilitating approved changes (e.g., via ECRs) and preventing the incorporation of unapproved changes into the baseline. Change control should be in effect beginning at project initiation.

Change Classification Effective configuration control requires that the extent of analysis and approval action for a proposed engineering change be in concert with the nature of the change. The problem statement includes a description of the proposed change, the reason for the proposed change, the impacts of the change on cost and schedule, and all affected documentation. Change classification is a primary basis of configuration control. All changes to baselined items are classified as outside of the scope of the requirements or within the scope of the requirements. A change outside the scope of project requirements is a change to a project baseline item that affects the form, fit, specification, function, reliability, or safety. The coordinating review board determines if this proposed change requires a change notice for review and approval.

Changes are sometimes categorized into two main classes: Class I and Class II. A Class I change is a major or significant change that affects cost, schedule, or technical performance. Normally, Class I changes require customer approval prior to being implemented. A Class II change is a minor change that often affects documentation errors or internal design details. Generally, Class II changes do not require customer approval.

- 保持所有事务的全面可追溯性。

构型识别　构型识别唯一地识别基线构型内的元素。这种唯一识别提升创建和维护基线的主目录清单的能力。作为 SE 工作的一部分，系统被分解为用于受严格的正式控制的若干关键元素的 CI。所有 CI 的汇总被称为 CI 列表。该列表可反映所开发的、卖方生产的或客户提供的用于综合到最终系统内的项。这些项可以是依据合同的可交付项，或用于生产的可交付项。

变更管理　构型变更管理或变更控制通过促进已批准的变更（如借助 ECR）和防止将未经批准的变更纳入基线，来管理待基线化项的收集并维护已识别的 CI 的完整性。变更控制应在项目开始时生效。

变更分类　有效的构型控制要求建议的工程变更的分析和审批行动的范围与变更的本质属性协调一致。问题说明包括对建议变更的描述、建议变更的理由、变更对成本和进度的影响及所有受影响的文件。变更分类是构型控制的主要基础。对基线化项的所有变更被分类为需求范围以外或需求范围以内。项目需求范围以外的变更是影响形式、适配、规格、功能、可靠性或安全性的项目基线项的变更。协调评审委员会决定这种建议性变更是否需要针对评审和审批的变更通知。

变更有时被分成两大类：I 类和 II 类。I 类变更是影响成本、进度或技术性能的主要或重大变更。正常情况下，I 类变更需要在实施前由客户审批。II 类变更是经常影响文件错误或内部设计细节的微小变更。通常情况下，II 类变更不需要由客户审批。

TECHNICAL MANAGEMENT PROCESSES

CCB An overall CCB is implemented at project initiation to provide a central point to coordinate, review, evaluate, and approve all proposed changes to base-lined documentation and configurations, including hardware, software, and firmware. The review board is composed of members from the various disciplines, including SE, software and hardware engineering, project management, product assurance, and configuration management. The chairperson is delegated the necessary authority to act on behalf of the project manager in all matters falling within the review board responsibilities. The configuration management organization is delegated responsibility for maintaining status of all proposed changes. Satellite or subordinate boards may be established for reviewing software or hardware proposed changes below the CI level. If those changes require a higher approval review, they are forwarded to the overall review board for adjudication.

Changes that fall within review board jurisdiction should be evaluated for technical necessity, compliance with project requirements, compatibility with associated documents, and project impact. As changes are written while the hardware and/or software products are in various stages of manufacture or verification, the review board should require specific instructions for identifying the effectivity or impact of the proposed software or hardware change and disposition of the in-process or completed hardware and/or software product. The types of impacts the review board should assess typically include the following:

- All parts, materials, and processes are specifically approved for use on the project.
- The design depicted can be fabricated using the methods indicated.
- Project quality and reliability assurance requirements are met.
- The design is consistent with interfacing designs.

Methods and Techniques Change control forms provide a standard method of reporting problems and enhancements that lead to changes in formal baselines and internally controlled items. The following forms provide an organized approach to changing hardware, software, or documentation:

- *Problem/change reports*—Can be used for documenting problems and recommending enhancements to hardware/software or its complementary documentation. These forms can be used to identify problems during design, development, integration, verification, and validation.

CCB 全体 CCB 从项目启动就开始实施，以为协调、评审、评价和审批对基线化文件和构型的所有建议变更（包括硬件、软件和固件）提供一个中心点。评审委员会由各种不同专业学科的成员组成，包括 SE、软件和硬件工程、项目管理、产品保证和构型管理。主席被授予代表项目管理者处理属于评审委员会职责内的所有事项的必要权限。构型管理组织被授权维护所有建议变更的状态的职责。可建立附属或下级委员会，用于评审在 CI 层级以下的软件或硬件的建议变更。如果上述变更需要较高的审批评审，则将变更送至全体评审委员会，以备裁决。

属于评审委员会裁决的变更应评价其技术必要性、与项目需求的一致性、与相关文件的兼容性以及对项目的影响。由于变更被记录，同时硬件和/或软件产品处于制造或验证的各种不同阶段，评审委员会应需要特定的说明，用于识别流程中的或已完成的硬件和/或软件产品的建议软件或硬件变更以及处置的有效性或影响。评审委员会应评估的影响的类型，通常包括下述内容：

- 被专门批准用于该项目的所有零件、资料和流程。

- 使用所指示方法可制造所描述的设计。

- 项目质量和可靠性保证需求被满足。

- 设计与接口设计一致。

方法和技术 变更控制形式提供报告问题和增强措施的一种标准方法，促使正式基线和内部受控项发生变更。下列形式提供变更硬件、软件或文件的有组织的方法：

- *问题/变更报告*——可用于将问题文件化并建议提高硬件/软件或其补充文件。这些形式可用于在设计、开发、综合、验证和确认期间识别问题。

- *Specification change notice (SCN)*—Used to propose, transmit, and record changes to baselined specifications.

- *ECPs*—Used to propose Class I changes to the customer. These proposals describe the advantages of the proposed change and available alternatives and identify funding needed to proceed.

- *ECRs*—Used to propose Class II changes.

- *Request for deviation/waiver*—Used to request and document temporary deviations from configuration identification requirements when permanent changes to provide conformity to an established baseline are not acceptable.

Configuration Status Accounting Status accounting provides the data on the status of controlled products needed to make decisions regarding system elements throughout the product life cycle. Configuration management maintains a status of approved documentation that identifies and defines the functional and physical characteristics, status of proposed changes, and status of approved changes. This subprocess synthesizes the output of the identification and control subprocesses. All changes authorized by the configuration review boards (both overall and subordinate) culminate in a comprehensive traceability of all transactions. Such activities as check-in and checkout of source code, builds of CIs, deviations of manufactured items, and waiver status are part of the status tracking. By statusing and tracking project changes, a gradual change from the *build-to* to the *as-built* configuration is captured. Suggested measures for consideration include the following:

- Number of changes processed, adopted, rejected, and open

- Status of open change requests

- Classification of change requests summary

- Number of deviations or waivers by CI

- Number of problem reports open, closed, and in-process

- Complexity of problem reports and root cause

- Labor associated with problem resolution and verification stage when problem was identified

- Processing times and effort for deviations, waivers, ECPs, SCNs, ECRs, and problem reports

- *规范变更通知单（SCN）*——用于建议、传送和记录对基线化规范的变更。

- *ECP*——用于向客户建议 I 类变更。这些建议描述所建议变更和可用备选方案的优势并识别继续进行所需的资金。

- *ECR*——用于建议 II 类变更。

- *偏离/让步的请求*——当提供与已建立的基线的一致性而做出的永久性变更是不可接受时，用于对与构型识别需求的临时偏离提出请求和文件化。

构型状态报告　状态报告提供贯穿于产品生命周期关于系统元素的决策所需要的受控产品的状态方面的数据。构型管理维护已批准的文件的状态，该状态识别和定义功能和物理特性、建议变更的状态和已批准的变更的状态。这一子流程综合了识别与控制子流程的输出。构型评审委员会（全体和下级）授权的所有变更均以所有事务的全面可追溯性而结束。这种活动如同源代码的检入和检出、CI 的构建、制造项的偏离和让步状态一样是状态跟踪的一部分。通过记录项目变更的状态并跟踪项目变更，获取从"要构建的"（build-to）构型到实际构建的（as-built）构型的逐渐变更。建议考虑的测度包括下列内容：

- 被处理的、采用的、拒绝的和未完的变更的数量

- 未关闭的变更请求的状态

- 变更请求概要的分类

- CI 偏离或让步的数量

- 未关闭的、已关闭的和流程中的问题报告的数量

- 问题报告和根本原因的复杂性

- 当问题曾被识别时与问题解决和验证阶段相关的工作

- 偏离、让步、ECP、SCN、ECR 和问题报告的处理时间和工作

TECHNICAL MANAGEMENT PROCESSES

- Activities causing a significant number of change requests and rate of baseline changes

Configuration Evaluations Configuration evaluations, or audits, are performed independently by configuration management and product assurance to evaluate the evolution of a product and ensure compliance to specifications, policies, and contractual agreements. Formal audits, or functional and physical configuration audits, are performed at the completion of a product development cycle.

The functional configuration audit is intended to validate that the development of a CI has been completed and has achieved the performance and functional characteristics specified in the system specification (functional baseline). The physical configuration audit is a technical review of the CI to verify the as-built maps to the technical documentation (physical baseline). Finally, configuration management performs periodic in-process audits to ensure that the configuration management process is followed.

5.6 INFORMATION MANAGEMENT PROCESS

5.6.1 Overview

5.6.1.1 Purpose As stated in ISO/IEC/IEEE 15288,

[6.3.6.1] The purpose of the Information Management process is to generate, obtain, confirm, transform, retain, retrieve, disseminate and dispose of information, to designated stakeholders.

Information management plans, executes, and controls the provision of information to designated stakeholders that is unambiguous, complete, verifiable, consistent, modifiable, traceable, and presentable. Information includes technical, project, organizational, agreement, and user information. Information is often derived from data records of the organization, system, process, or project.

Information management needs to provide relevant, timely, complete, valid, and, if required, confidential information to designed parties during and, as appropriate, after the system life cycle. It manages designed information, including technical, project, organizational, agreement, and user information.

Information management ensures that information is properly stored, maintained, secured, and accessible to those who need it, thereby establishing/maintaining integrity of relevant system life cycle artifacts.

- 引起大量变更请求的活动以及基线变更速率

构型评价　构型评价或审计通过构型管理和产品保证被独立地实施以评价产品的演进并确保符合规范、方针和合同协议。在产品开发周期结束时实施正式的审计或功能的和物理的构型审计。

功能构型审计旨在确认 CI 开发已完成且达成系统规范（功能基线）指定的性能和功能特性。物理构型审计是 CI 的技术评审，以验证实际构建的和技术文件（物理基线）的对应关系。最后，构型管理实施周期性的流程中审计以确保构型管理流程得到遵循。

5.6 信息管理流程

5.6.1 概览

5.6.1.1 目的　如 ISO/IEC/IEEE 15288 所述，

> [6.3.6.1] 信息管理流程的目的是生成、获得、确认、转换、保留、检索、传播和处置提供给指定利益攸关者的信息。
>
> 信息管理计划、执行和控制向指定利益攸关者提供信息，使其是无歧义的、完整的、可验证的、一致的、可修改的、可追溯的和可呈现的。信息包括技术信息、项目信息、组织信息、协议信息和用户信息。信息通常来源于组织、系统、流程或项目的数据记录。

信息管理需要在系统生命周期期间以及之后（视情况而定）提供相关的、及时的、完整的、有效的且（若要求）保密的信息给指定的相关方，并管理指定的信息，包括技术信息、项目信息、组织信息、协议信息和用户信息。

信息管理确保信息被适当地存储、维护和保密，能被需要信息的人员获得，由此建立/维护相关的系统生命周期制品的完整性。

TECHNICAL MANAGEMENT PROCESSES

5.6.1.2 Description Information exists in many forms, and different types of information have different values within an organization. Information assets, whether tangible or intangible, have become so pervasive in contemporary organizations that they are indispensable. The impact of threats to secure access, confidentiality, integrity, and availability of information can cripple the ability to get work done. As information systems become increasingly interconnected, the opportunities for compromise increase (Brykczynski and Small, 2003). The following are important terms in information management:

- Information is what an organization has compiled or its employees know. It can be stored and communicated, and it might include customer information, proprietary information, and/or protected (e.g., by copyright, trademark, or patent) and unprotected (e.g., business intelligence) intellectual property.

- Information assets are intangible information and any tangible form of its representation, including drawings, memos, email, computer files, and databases.

- Information security generally refers to the confidentiality, integrity, and availability of the information assets (ISO 17799, 2005).

- Information security management includes the controls used to achieve information security and is accomplished by implementing a suitable set of controls, which could be policies, practices, procedures, organizational structures, and software.

- Information Security Management System is the life cycle approach to implementing, maintaining, and improving the interrelated set of policies, controls, and procedures that ensure the security of an organization's information assets in a manner appropriate for its strategic objectives.

Information management provides the basis for the management of and access to information throughout the system life cycle, including after disposal if required. Designated information may include organizational, project, agreement, technical, and user information. The mechanisms for maintaining historical knowledge in the prior processes—decision making, risk, and configuration management—are under the responsibility of information management. Figure 5.10 is the IPO diagram for the information management process.

5.6.1.3 Inputs/Outputs Inputs and outputs for the information management process are listed in Figure 5.10. Descriptions of each input and output are provided in Appendix E.

5.6.1.2 描述 信息以多种形式存在，不同类型的信息在组织内具有不同的价值。无论是有形的还是无形的信息资产，在当代组织中都变得非常普遍，以至于它们是不可或缺的。信息的安全访问、机密性、完整性和可用性的威胁所产生的影响会摧毁完成工作的能力。随着信息系统相互关联的日益增加，为了折中的机会也会随之增加（Brykczynski 和 Small, 2003）。以下是信息管理中的重要术语：

- 信息是组织汇编的或其员工知晓的内容。它可以被存储和交流，并且可包括客户信息、专用的信息和/或受保护的（如受到版权、商标或专利的保护）和不受保护的（如商业情报）知识产权。

- 信息资产是隐性信息及任何显性形式的表征形式，包括图样、备忘录、电子邮件、计算机文件和数据库。

- 信息安保通常涉及信息资产的机密性、完整性和可用性（ISO 17799, 2005）。

- 信息安保管理包括用于实现信息安保的控制，并通过实施一系列适当控制（可包括策略、实践、程序、组织结构和软件）来完成。

- 信息安保管理系统是以适合其战略目标的方式实施、维护和改善一系列确保组织信息资产的安保性的相互关联的策略、控制和程序的生命周期方法。

信息管理贯穿于系统生命周期内（如需要，还包括处置之后）为信息的管理和访问提供基础。指定的信息可包括组织信息、项目信息、协议信息、技术信息和用户信息。前期流程中（决策、风险和构型管理）用于维护历史知识的机制属于信息管理职责范围。图 5.10 是信息管理流程的 IPO 图。

5.6.1.3 输入/输出 图 5.10 列出信息管理流程的输入和输出。附录 E 提供每个输入和输出的描述。

TECHNICAL MANAGEMENT PROCESSES

5.6.1.4 Process Activities The information management process includes the following activities:

- Prepare for information management.
  - Support establishing and maintaining a system data dictionary—see project planning outputs.
  - Define system-relevant information, storage requirements, access privileges, and the duration of maintenance.
  - Define formats and media for capture, retention, transmission, and retrieval of information.
  - Identify valid sources of information and designate authorities and responsibilities regarding the origination, generation, capture, archival, and disposal of information in accordance with the configuration management process.

Controls

Inputs	Activities	Outputs
• Candidate information items • Project change requests	• Prepare for information management • Perform information management	• Information management strategy • Information repository • Information management report • Information management record

Enablers

FIGURE 5.10 IPO diagram for the information management process. INCOSE SEH original figure created by Shortell and Walden. Usage per the INCOSE notices page. All other rights reserved.

- Perform information management.
  - Periodically obtain or transform artifacts of information.
  - Maintain information according to integrity, security, and privacy requirements.

5.6.1.4 流程活动 信息管理流程包括以下活动：

- 准备信息管理。

  - 支持建立和维护系统数据字典——参见项目规划输出。

  - 定义系统相关的信息、存储需求、访问权限和维护的持续期。

  - 定义信息的获取、保留、传递和检索的格式和媒介。

  - 识别有效的信息源并按照构型管理流程指定关于信息的起源、产生、获取、归档和处置的权限与职责。

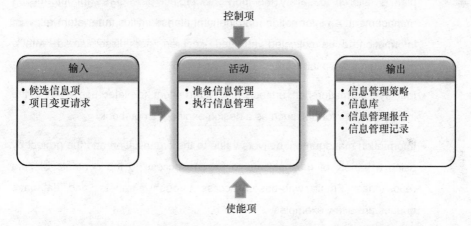

图 5.10 信息管理流程的 IPO 图。INCOSE SEH 原始图由 Shortell 和 Walden 创建。按照 INCOSE 通知页使用。版权所有。

- 实施信息管理。

  - 定期获得或转换信息的制品。

  - 根据完整性需求、安保需求和隐私需求维护信息。

- Retrieve and distribute information in an appropriate form to designated parties, as required by agreed schedules or defined circumstances.

- Archive designated information for compliance with legal, audit, knowledge retention, and project closure requirements.

- Dispose of unwanted, invalid, or unverifiable information according to organizational policy, security, and privacy requirements.

Common approaches and tips:

- In the project planning process (see Section 5.1), an information management plan is tailored to satisfy the individual project procedures for information management. An information management plan identifies the system-relevant information to be collected, retained, secured, and disseminated, with a schedule for disposal.

- Identify information-rich artifacts and store them for later use even if the information is informal, such as a design engineer's notebook.

- Information management delivers value to the organization and the project by using a variety of mechanisms to provide access to the contents of data repositories. Email, web-based access through intranets, and database queries are a few examples.

- ISO 17799, *Code of Practice for Information Security Management*, is an international standard that provides a best practice framework for implementing security controls.

- ISO 10303, *Automation Systems and Integration— Product Data Representation and Exchange—* informally referred to as "Standard for the Exchange of Product Model Data" *(STEP)*. It includes Application Protocol (AP) 239, *Product Life Cycle Support* (PLCS), which addresses information requirements for complex systems.

- 按照协定的进度或定义的环境要求以适当形式检索信息并分发给指定方。

- 对指定信息归档以符合法律、审计、知识保留和项目收尾的需求。

- 根据组织策略、安保和隐私需求处置不需要的、无效的或无法验证的信息。

常用方法和建议：

- 在项目规划流程（见 5.1 节）中，对信息管理计划进行剪裁，以满足信息管理的单独项目程序。信息管理计划识别需要收集、保留、安保和传播的系统相关的信息以及处置的进度。

- 识别信息丰富的制品并将之存储以备日后使用，即使是不正式的信息，例如设计工程师的笔记本。

- 信息管理使用各种机制来提供对数据仓库内容的访问，为组织和项目提供价值。电子邮件、通过企业内联网基于 Web 的访问以及数据库查询就是一些例子。

- ISO 17799，《信息安保管理实施条例》是一个国际标准，提供一种实施安保控制的最佳实践框架。

- ISO 10303，自动化系统和综合——产品数据的表征和交换——非正式地称为"产品模型数据交换标准"（STEP）。它包括应用协议（AP）239，产品生命周期支持（PLCS），其应对复杂系统的信息需求。

5.6.2 Elaboration

5.6.2.1 Information Management Concepts The purpose of information management is to maintain an archive of information produced throughout the system's life cycle. The initial planning efforts for information management are defined in the information management plan, which establishes the scope of project information that is maintained; identifies the resources and personnel skill level required; defines the tasks to be performed; defines the rights, obligations, and commitments of parties for generation, management, and access; and identifies information management tools and processes, as well as methodologies, standards, and procedures that will be used on the project. Typical information includes source documents from stakeholders, contracts, project planning documents, verification documentation, engineering analysis reports, and the files maintained by configuration management. Today, information management is increasingly concerned with the integration of information via databases, such as the decision database, and the ability to access the results from decision gate reviews and other decisions taken on the project; requirements management tools and databases; computer-based training and electronic interactive user manuals; websites; and shared information spaces over the Internet, such as INCOSE Connect. The STEP—ISO 10303 standard provides a neutral computer-interpretable representation of product data throughout the life cycle. ISO 10303-239 (AP 239), *PLCS*, is an international standard that specifies an information model that defines what information can be exchanged and represented to support a product through life (PLCS, 2013). INCOSE is a cosponsor of ISO 10303-233, *Application Protocol: Systems Engineering* (2012). AP 233 is used to exchange data between a SysML™ and other SE application and then to applications in the larger life cycle of systems potentially using related ISO STEP data exchange capabilities.

With effective information management, information is readily accessible to authorized project and organization personnel. Challenges related to maintaining databases, security of data, sharing data across multiple platforms and organizations, and transitioning when technology is updated are all handled by information management. With all the emphasis on knowledge management, organizational learning, and information as competitive advantage, these activities are gaining increased attention.

5.6.2 详细阐述

5.6.2.1 信息管理概念 信息管理的目的是维护贯穿于系统生命周期内所产生信息的存档。信息管理计划中定义信息管理的初始计划工作，建立所维护的项目信息的范围；识别所需的资源和人员技能水平；定义需要执行的任务；针对产生、管理和访问定义各方的权利、义务和投入；识别信息管理的工具和流程，以及将用于项目的方法论、标准和程序。典型信息包括利益攸关者的源文件、合同、项目规划文件、验证文档、工程分析报告和由构型管理维护的文件。如今，信息管理越来越关注经由数据库对信息的综合，如决策数据库，以及访问来自项目的决策门评审及其他项目决策中结果的能力；需求管理工具和数据库；基于计算机的培训和电子交互式用户手册；网站；以及互联网上的共享信息空间，如 INCOSE Connect。STEP-ISO 10303 标准提供一种可对贯穿于生命周期内的产品数据进行中性计算机可解释的表征。ISO 10303-239（AP239）PLCS 是一项规定信息模型的国际标准，定义可进行交换和表征的信息，以便在生命周期内支持产品（PLCS，2013）。INCOSE 是 ISO 10303-233，应用协议：系统工程（2012）的联合发起者。AP 233 被用于 SysML ™ 和其他 SE 应用之间的数据交换，然后通过潜在地使用相关的 ISO STEP 数据交换能力将数据交换到更大的系统生命周期的应用中。

通过有效的信息管理，经授权的项目和组织人员可很容易地获取信息。当技术更新时，与维护数据库、数据安保、跨多个平台和组织共享数据以及转移有关的多个挑战全部通过信息管理来处理。随着把全部重点放在作为竞争优势的知识管理、组织学习和信息上，这些活动正得到越来越多的关注。

TECHNICAL MANAGEMENT PROCESSES

5.7 MEASUREMENT PROCESS

5.7.1 Overview

5.7.1.1 Purpose As stated in ISO/IEC/IEEE 15288,

[6.3.7.1] The purpose of the Measurement process is to collect, analyze, and report objective data and information to support effective management and *demonstrate* the quality of the products, services, and processes.

5.7.1.2 Description The SE measurement process will help define the types of information needed to support program management decisions and implement SE best practices to improve performance. The key SE measurement objective is to measure the SE process and work products with respect to program/project and organization needs, including timeliness, meeting performance requirements and quality attributes, product conformance to standards, effective use of resources, and continuous process improvement in reducing cost and cycle time.

The *Practical Software and Systems Measurement Guide* (DoD and US Army, 2003), Section 1.1, states:

Measurement provides objective information to help the project manager:

- *Communicate effectively throughout the project organization*
- *Identify and correct problems early*
- *Make key trade-offs*
- *Track specific project objectives*
- *Defend and justify decisions*

Specific measures are based on information needs and how that information will be used to make decisions and take action. Measurement thus exists as part of a larger management process and includes not just the project manager but also systems engineers, analysts, designers, developers, integrators, logisticians, etc. The decisions to be made motivate the kinds of information to be generated and, therefore, the measurements to be made.

Another concept of successful measurement is the communication of meaningful information to the decision makers. It is important that the people who use the measurement information understand what is being measured and how it is to be interpreted. Figure 5.11 is the IPO diagram for the measurement process.

5.7 测度流程

5.7.1 概览

5.7.1.1 目的 如 ISO/IEC/IEEE 15288 所述，

[6.3.7.1] 测度流程的目的是收集、分析和报告客观数据和信息，以支持有效管理并证实产品、服务和流程的质量。

5.7.1.2 描述 SE 测度流程将有助于定义支持项目群管理决策以及实施 SE 的最佳实践所需的信息类型，以改进绩效。关键的 SE 测度的目的是测度与项目群/项目和组织需要有关的 SE 流程和工作成果，包括时效性、满足性能需求和质量属性、产品与标准符合性、资源的有效利用以及减少成本和缩短周期中持续的流程改进。

《实用软件和系统测度指南》（美国国防部和美国陆军，2003），1.1 节阐明：

测度提供客观信息，以帮助项目经理：

- *贯穿于项目组织内有效地沟通*

- *及早识别并纠正问题*

- *做出关键权衡*

- *跟踪特定项目目标*

- *捍卫和证明决策其正当性*

特定的测度基于信息需要以及信息将如何被用于做出决策和采取行动。因此，测度作为较大管理流程的一部分而存在，并且不仅包括项目经理，还包括系统工程师、分析者、设计者、开发者、综合者和后勤人员等。要做出的决策会促使产生多种类型的信息，并由此进行测度。

成功测度的另一个概念是与决策者进行有意义的信息的沟通。重要的是使用测度信息的人员理解测度什么和如何解释。图 5.11 是测度流程的 IPO 图。

TECHNICAL MANAGEMENT PROCESSES

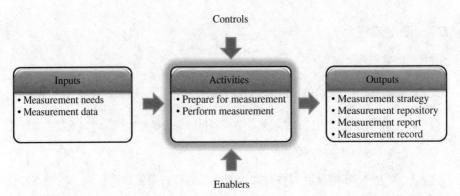

FIGURE 5.11 IPO diagram for the measurement process. INCOSE SEH original figure created by Shortell and Walden. Usage per the INCOSE notices page. All other rights reserved.

5.7.1.3 Inputs/Outputs Inputs and outputs for the measurement process are listed in Figure 5.11. Descriptions of each input and output are provided in Appendix E.

5.7.1.4 Process Activities The measurement process includes the following activities:

- *Prepare for measurement.*

  - Identify the measurement stakeholders and their measurement needs, and develop a strategy to meet them.

  - Identify and select relevant prioritized measures that aid with the management and technical performance of the program.

  - Define the base measures, derived measures, indicators, data collection, measurement frequency, measurement repository, reporting method and frequency, trigger points or thresholds, and review authority.

- *Perform measurement.*

  - Gather, process, store, verify, and analyze the data to obtain measurement results (information products).

  - Document and review the measurement information products with the measurement stakeholders and recommend action, as warranted by the results.

技术管理流程

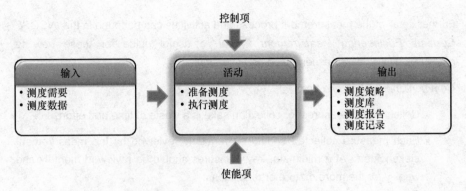

图 5.11 测度流程的 IPO 图。INCOSE SEH 原始图由 Shortell 和 Walden 创建。按照 INCOSE 通知页使用。版权所有。

5.7.1.3 *输入/输出* 图 5.11 列出测量流程的输入和输出。附录 E 提供每个输入和输出的描述。

5.7.1.4 *流程活动* 测度流程包括以下活动:

- *测度的准备。*

  – 识别测度利益攸关者及其测度需要,并开发策略以满足需要。

  – 识别并选择有助于项目群的管理和技术性能的相关优先测度。

  – 定义基础测度、推演测度、指标、数据采集、测量频率、测量库、报告方法和频度、触发点或阈值以及评审权限。

- *实施测度。*

  – 为获得测度结果(信息产物),收集、处理、存储、验证和分析数据。

  – 按照结果来文件化和评审与测度利益攸关者和建议行动相关的测度信息产物。

TECHNICAL MANAGEMENT PROCESSES

Further detail of the measurement process and activities can be found in the *INCOSE Systems Engineering Measurement Primer*, a useful guide for those new to measures as well as for experienced practitioners.

Common approaches and tips:

- Collection of measures for collection sake is a waste of time and effort.
- Each measure collected should be regularly reviewed by the measurement stakeholders. At a minimum, key measures should be reviewed monthly and weekly for the more mature organizations.
- Some contracts identify measures of effectiveness (MOEs) that must be met. The derived measures of performance (MOPs) and TPMs that provide the necessary insight into meeting the MOEs are automatic measures to be included within the measurement plan. Other measures to consider should provide insight into technical and programmatic execution of the program (Roedler and Jones, 2006).
- The best measures require minimal effort to collect and are repeatable, straightforward to understand, and presented in a format on a regular (weekly or monthly) basis with trend data.
- Many methods are available to present the data to the measurement stakeholders. Line graphs and control charts are two of the more frequently used. Tools are available to help with measurement.
- If a need for corrective action is perceived, further investigation into the measures may be necessary to identify the root cause of the issue to ensure that corrective actions address the cause instead of a symptom.
- Measurement by itself does not control or improve process performance. Measurement results must be provided to decision makers in a manner that provides the needed insight for the right decisions to be made.

5.7.2 Elaboration

5.7.2.1 Measurement Concepts Measurement concepts have been expanded upon in the previous works that the SE measurement practitioner should reference for further insights:

- *INCOSE Systems Engineering Measurement Primer, v2.0* (INCOSE, 2010b)
- *Technical Measurement Guide*, Version 1.0 (INCOSE and PSM, 2005)

测度流程和活动的进一步细节可在 INCOSE 系统工程测量法入门中找到，对于测量新手和有经验的实践者而言是一本有用的指南。

常用方法和建议：

- 为了收集而收集测度是一种时间和工作的浪费。

- 所收集的每个测度均应由测度利益攸关者定期评审。至少关键的测度应由更成熟的组织按周和按月评审。

- 有些合同识别那些必须要被满足的有效性测度（MOE）。提供对满足 MOE 的必要理解的推演性能测度（MOP）和 TPM 是包括在测量计划内的自动测度。需考虑的其他测度应提供对项目群的技术和按计划执行的深入见解（Roedler 和 Jones, 2006）。

- 最佳的测度需要最少的收集工作且是可重复的、易于理解的以及定期（按周或按月）以某种格式给出趋势数据。

- 许多方法可用于向测度利益攸关者提供数据。线形图和控制图是更常用图中的两种。工具可用于帮助测度。

- 如果认识到纠正措施的需要，为识别问题的根本原因，以确保纠正措施应对的是原因而不是现象，对测度的进一步研究也许是必要的。

- 测度本身并不控制或改进流程性能。测度结果必须按照为做出正确决策提供所需要的深入理解的方式提供给决策者。

5.7.2 详细阐述

5.7.2.1 *测度概念* 测度概念在已开展的 SE 测量实践者应参考的一些工作上扩展，以便进一步深入理解：

- 《INCOSE 系统工程测量入门》2.0 版本（INCOSE，2010b）

- 《技术测量指南》1.0 版本（INCOSE 和 PSM，2005）

TECHNICAL MANAGEMENT PROCESSES

- *Systems Engineering Leading Indicators Guide, Version 2.0* (LAI, INCOSE, PSM, and SEARI,2010)

- ISO/IEC/IEEE 15939, Systems and Software Engineering—Measurement Process (2007)

- PSM Guide V4.0c, Practical Software and Systems Measurement (DoD and US Army, 2003).

- *CMMI® (Measurement and Quantitative Management Process Areas)*, Version 1.3 (Software Engineering Institute, 2010)

- *Practical Software Measurement: Objective Information for Decision Makers* (Mcgarry et al., 2001)

- *System Development Performance Measurement Report* (NDIA, 2011)

- SEBoK Part 3: SE and Management/Systems Engineering Management/ Measurement (SEBoK, 2014)

5.7.2.2   Measurement Approach   As discussed in the *Systems Engineering Measurement Primer* referenced earlier, measurement may be thought of as a feedback control system. The graphic below shows that measurements may be taken at three primary points in the system. However, the value in measurement comes not from the act of measurement but rather from the eventual analysis of the data and the implementation of action to either correct a variance from a target value or to improve current performance to a more desirable level. The decision as to when to act comes from comparison of the actual measured value against a target value for that measure. The target value and the allowable difference between the target and actual before action is taken are based upon evaluation of risk to the project or product performance meeting their required goals.

Figure 5.12 further indicates that there might be advantage to separating the planning of the measurement process into measures related to the development process (staffing, requirements approved, etc.) and measures related to the product performance (product weight, product speed, etc.). The advantage in this separation would be the recognition that the stakeholders and data collectors for process-related measures will likely be a different population than stakeholders and data collectors for product-related measures.

- 《系统工程领先指标指南》2.0 版本（LAI、INCOSE、PSM 和 SEARI，2010）

- ISO/IEC/IEEE 15939，系统和软件工程——测度流程（2007）

- PSM 指南，4.0c 版本，实用软件和系统测度（美国国防部和美国陆军，2003）

- CMMI®（测量和定量管理流程域），1.3 版本（软件工程研究所，2010）

- 实用软件测度：决策者的客观信息（Mcgarry 等，2001）

- 系统开发性能测量报告（NDIA，2011）

- SEBoK 第 3 部分：SE 和管理/系统工程管理/测度（SEBoK，2014）

5.7.2.2 测度方法 正如前面提到的《系统工程测量入门》中所论述，测量可被认为是一种反馈控制系统。下图表明可在系统中的三个主要部分进行测量。然而，测量中的价值并不是来自测量行为，而是来自数据的最终分析和行动的实施，以便纠正与目标值的偏差或将当前性能提高到更理想的水平。关于何时行动的决策来自测度的实际测量值与目标值的比较。目标值以及采取行动前目标值与实际值之间允许的差异，基于对项目风险或产品性能的评估，从而满足其所要求的目的。

图 5.12 进一步表明，将测度流程的规划分成与开发流程（人员配备、需求审批等）有关的测度和与产品性能（产品重量、产品速度等）有关的测度可能有好处。这种划分的好处将是认识到：流程相关测度的利益攸关者和数据收集者将有可能是与产品相关测度的利益攸关者和数据收集者不同的群体。

TECHNICAL MANAGEMENT PROCESSES

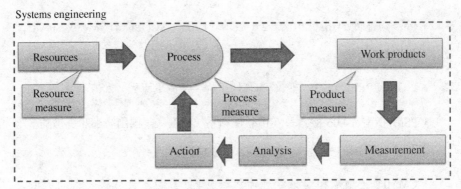

FIGURE 5.12 Measurement as a feedback control system. From INCOSE-TP-2010-005-02 Figure 2.1. Usage per the INCOSE notices page. All other rights reserved.

5.7.2.3 Process-Oriented Measures A reasonable approach in organizing process-oriented measures is to categorize them based upon the organizational goal supported. Organizational goals may be broken into the following categories:

- Cost (development)
- Schedule (development)
- Quality (process)

The specific measures selected under each goal would be those necessary to track process activities thought to produce the greatest risk to meeting the goals of the product or organization.

5.7.2.4 Leading Indicators An important subgroup of process-related measures is leading indicators. A leading indicator is a measure for evaluating the effectiveness of how a specific activity is applied on a program in a manner that provides information about impacts that are likely to affect the system performance or SE effectiveness objectives.

A leading indicator may be an individual measure, or collection of measures, that is predictive of future system performance before the performance is realized. Leading indicators aid leadership in delivering value to customers and end users while assisting in taking interventions and actions to avoid rework and wasted effort.

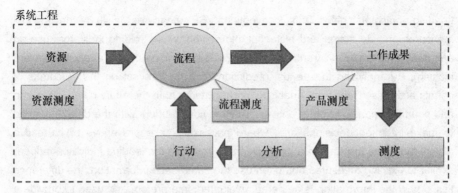

图 5.12 作为反馈控制系统的测量。来自 INCOSE-TP-2010-005-02，图 2.1。按照 INCOSE 通知页使用。版权所有。

5.7.2.3 *面向流程的测度* 面向流程的测度的合理组织方法是基于被支持的组织目标对它们进行分类。组织目标可被分解成以下类别：

- 成本（开发）
- 进度（开发）
- 质量（流程）

按每个目的选择的特定测度对于跟踪认为产生最大风险的流程活动而言是必要的，以便满足产品或组织的目的。

5.7.2.4 *领先指示器* 与流程相关测度的一个重要子群是领先指标。领先指标是一种以提供关于可能影响系统性能或 SE 的信息的方式来对如何将特定活动应用于项目群的有效性进行评价的测度。

领先指示器可以是一个独立测度或测度集合，能够在性能被实现之前预测未来的系统性能。领先指示器帮助领导层向客户和最终用户交付价值，同时有助于采取干预和行动来避免返工和浪费的工作。

Leading indicators differ from conventional SE measures in that conventional measures provide status and historical information, while leading indicators use an approach that draws on trend information to allow for predictive analysis (forward looking). By analyzing the trends, predictions can be forecast on the outcomes of certain activities. Trends are analyzed for insight into both the entity being measured and potential impacts to other entities. This provides leaders with the data they need to make informed decisions and, where necessary, take preventive or corrective action during the program in a proactive manner. While the leading indicators appear similar to existing measures and often use the same base information, the difference lies in how the information is gathered, evaluated, interpreted, and used to provide a forward looking perspective. Examples of leading indicator measures include the following:

- *Requirements trends*—Rate of maturity of the system definition against the plan. Requirements trends characterize the stability and completeness of the system requirements that could potentially impact design and production.
- *Interface trends*—Interface specification closure against plan. Lack of timely closure could pose adverse impact to system architecture, design, implementation, and/or verification and validation (V&V), any of which could pose technical, cost, and schedule impact.
- *Requirements validation trends*—Progress against the plan in ensuring that the customer requirements are valid and properly understood. Adverse trends would pose impacts to system design activity with corresponding impacts to technical, cost, and schedule baselines and customer satisfaction.

For a more detailed treatment of this topic, including measurement examples, please consult the *Systems Engineering Leading Indicators Guide* (Roedler et al., 2010).

5.7.2.5 Product-Oriented Measures As shown in referenced document Technical Measurement (Roedler and Jones, 2006), product measures can be thought of as an interdependent hierarchy (see Figure 5.13).

5.7.2.6 MOEs and MOPs MOEs and MOPs are two concepts that represent types of measures typically collected. MOEs are defined in the *INCOSE-TP-2003-020-01, Technical Measurement Guide* (Roedler and Jones, 2006), as follows:

The "operational" measures of success that are closely related to the achievement of the mission or operational objective being evaluated, in the intended operational environment under a specified set of conditions; i.e., how well the solution achieves the intended purpose. (Adapted from DoD 5000.2, DAU, and INCOSE)

领先指示器与常规 SE 测度的不同之处在于：常规测度提供状态和历史信息，而领先指示器使用一种借助趋势信息的方法来进行预测分析（前瞻）。通过分析诸多趋势，可预测某些活动的结果。分析趋势以便理解所测度的实体以及对其他实体的潜在影响。这为领导者提供他们需要的数据以做出明智的决策，并且必要时在计划期间以积极方式采取预防性措施或纠正措施。尽管领先指示器与现有测度看似相似并且通常使用相同的基础信息，但区别在于信息如何被收集、评价、解释并被用于提供前瞻视角。领先指示器测度的示例包括以下内容：

- *需求趋势*——按计划的系统定义成熟度的等级。需求趋势描述那些潜在地影响设计和生产的系统需求的稳定性和完整性。

- *接口趋势*——按计划关闭接口规范。缺少及时的关闭可能对系统架构、设计、实施和/或检验和确认（V&V）造成不利影响，这些事项中的任何一个均可能造成技术、成本和进度上的影响。

- *需求确认趋势*——以确保客户需求有效且被恰当理解的计划而进展。不利趋势可能对系统设计活动造成影响，并同时对技术、成本和进度基线与客户满意度产生相应影响。

关于该主题的更详细的处理，包括测量示例，请查阅《系统工程领先指标指南》（Roedler 等，2010）。

5.7.2.5 面向产品的测度 正如参考文件技术测量（Roedler 和 Jones，2006）中所示，产品测度可被认为是相互依赖的层级结构（见图 5.13）。

5.7.2.6 MOE 和 MOP MOE 和 MOP 是表征通常所收集的测度类型的两个概念。MOE 在 INCOSE-TP-2003-020-01，《技术测量指南》（Roedler 和 Jones，2006）中定义如下：

在一系列特定条件下的预期运行环境中，与所评价的任务或运行目标的达成紧密相关的成功的"运行"测度；即解决方案达成预期目的的程度。（改编自 DoD 5000.2，DAU 和 INCOSE）

TECHNICAL MANAGEMENT PROCESSES

MOPs are defined as follows:

The measures that characterize physical or functional attributes relating to the system operation, measured or estimated under specified testing and/or operational environment conditions. (Adapted from DoD 5000.2, DAU, INCOSE, and EPI 280-04, LM Integrated Measurement guidebook)

133

FIGURE 5.13 Relationship of technical measures. From INCOSETP200302001 Figure 1.1. Usage per the INCOSE notices page. All other rights reserved.

MOEs, which are stated from the acquirer (customer/user) viewpoint, are the acquirer's key indicators of achieving the mission needs for performance, suitability, and affordability across the life cycle.

Although they are independent of any particular solution, MOEs are the overall operational success criteria (e.g., mission performance, safety, operability, operational availability, etc.) to be used by the acquirer for the delivered system, services, and/or processes. INCOSE-TP-2003-020-01, *Technical Measurement Guide* (Roedler and Jones, 2006), Section 3.2.1, provides further discussion on this topic.

MOPs measure attributes considered as important to ensure that the system has the capability to achieve operational objectives. MOPs are used to assess whether the system meets design or performance requirements that are necessary to satisfy the MOEs. MOPs should be derived from or provide insight for MOEs or other user needs. INCOSE-TP-2003-020-01, *Technical Measurement Guide*, Section 3.2.2, provides further discussion on this topic.

MOP 定义如下：

在特定试验和/或运行环境条件下，测量或估计的用于描述与系统运行有关的物理或功能属性的测度。（改编自 DoD 5000.2，DAU，INCOSE 和 EPI 280-04，LM 综合测量指导手册）

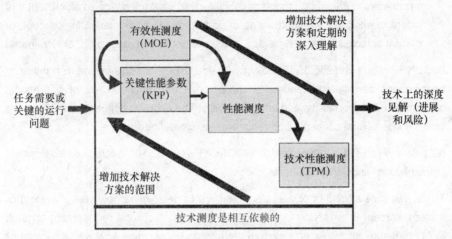

图 5.13 技术测度的关系。来自 INCOSE-TP-2003-020-01，图 1.1。按照 INCOSE 通知页使用。版权所有。

根据采办方（客户/用户）观点所述，MOE 是跨生命周期实现性能、适用性和可承受性的任务需要的采办方的关键指标。

尽管它们独立于任何特定的解决方案，但 MOE 是采办方用于已交付系统、服务和/或流程的整体运行成功准则（如任务性能、安全性、运行性、运行有效性等）。INCOSE-TP-2003-020-01，《技术测量指南》（Roedler 和 Jones，2006），3.2.1 节提供关于该主题的进一步论述。

MOP 测量那些被认为很重要的属性，以确保系统具有实现运行目标的能力。MOP 用于评估系统是否符合那些满足 MOE 所必需的设计或性能需求。MOP 应从 MOE 或其他用户需要中推演或提供对它们的深入理解。INCOSE-TP-2003-020-01《技术测量指南》3.2.2 节提供关于该主题的进一步论述。

TECHNICAL MANAGEMENT PROCESSES

5.7.2.7 Key Performance Parameters Key Performance Parameters (KPPs) are defined in the INCOSE-TP-2003-020-01, *Technical Measurement Guide* (Roedler and Jones, 2006), as follows:

> Key Performance Parameters are a critical subset of the performance parameters representing those capabilities and characteristics so significant that failure to meet the threshold value of performance can be cause for concept, or system selected to be reevaluated or the project to be reassessed, or terminated.

Each KPP has a threshold and objective value. KPPs are the minimum number of performance parameters needed to characterize the major drivers of operational performance, supportability, and interoperability. The acquirer defines the KPPs at the time the operational concepts and requirements are defined.

5.7.2.8 TPMs TPMs are defined in INCOSE-TP-2003-020-01, *Technical Measurement Guide* (Roedler and Jones, 2006), as follows:

TPMs measure attributes of a system element to deter-mine how well a system or system element is satisfying or expected to satisfy a technical requirement or goal. TPMs measure attributes of a system element to deter- mine how well a system or system element is satisfying or expected to satisfy a technical requirement or goal.

TPMs are used to assess design progress, compliance to performance requirements, or technical risks and provide visibility into the status of important project technical parameters to enable effective management, thus enhancing the likelihood of achieving the technical objectives of the project. TPMs are derived from or pro-vide insight for the MOPs focusing on the critical technical parameters of specific architectural elements of the system as it is designed and implemented. Selection of TPMs should be limited to critical technical thresholds or parameters that, if not met, put the project at cost, schedule, or performance risk. The TPMs are not a full listing of the requirements of the system or system element. The SEMP should define the approach to TPMs (Roedler and Jones, 2006).The SEMP should define the approach to TPMs (Roedler and jones, 2006).

Without TPMs, a project manager could fall into the trap of relying on cost and schedule status alone, with perhaps the verbal assurances of technical staff to assess project progress. This can lead to a product developed on schedule and within cost that does not meet all key requirements. Values are established to provide limits that give early indications if a TPM is out of tolerance, as illustrated in Figure 5.14.

5.7.2.7 关键性能参数 关键性能参数（KPP）在 INCOSE-TP-2003-020-01，《技术测量指南》（Roedler 和 Jones，2006）中定义如下：

> 关键性能参数是性能参数的关键子集，表征那些非常重要的能力和特性以至于如果不满足性能阈值便可导致所选择的概念或系统被重新评价或项目被重新评估或终止。

每个 KPP 具有一个阈值和客观值。KPP 是描述运行性能、保障性和互操作性的主要驱动因素所需的最少数量的性能参数。采办方在定义运行概念和需求时定义 KPP。

5.7.2.8 TPM TPM 在 INCOSE-TP-2003-020-01，《技术测量指南》（Roedler 和 Jones，2006）中定义如下：

TPM 测度系统元素的属性以确定系统或系统元素满足或期望满足技术需求或目标的程度。

TPM 用于评估设计进展、性能需求符合性或技术风险，并提供重要项目技术参数的状态的清晰度，使其能进行有效管理，从而提高达成项目技术目标的可能性。TPM 来自或提供对 MOP 的深入理解，专注于所设计和实施的系统的特定架构元素的关键技术参数。TPM 的选择应限于关键技术阈值或参数，若不能满足，项目则会遭受成本、进度或性能风险。TPM 不是系统或系统元素的全部需求列表。SEMP 应定义 TPM 的方法（Roedler 和 Jones，2006）。

若没有 TPM，项目经理就可能会落入只依赖于成本和进度状态的陷阱之中，也许会为评估项目进展向技术人员做出口头保证。这可能导致在进度和成本范围内开发的产品不满足所有关键需求。如果 TPM 超出容差范围，如图 5.14 所示，就会为早期给出指示所提供的限制建立价值。

TECHNICAL MANAGEMENT PROCESSES

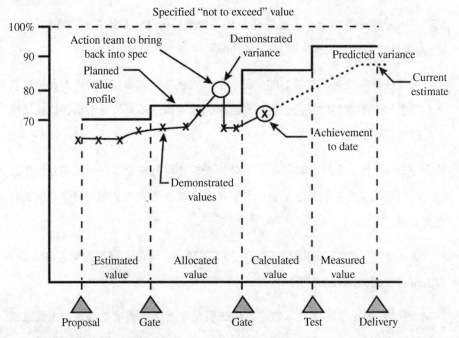

FIGURE 5.14 TPM monitoring. Reprinted with permission from Kevin Forsberg. All other rights reserved.

Periodic recording of the status of each TPM provides the continuing verification of the degree of anticipated and actual achievement of technical parameters. Measured values that fall outside an established tolerance band alert management to take corrective action. INCOSE-TP-2003-020-01, *Technical Measurement Guide*, Section 3.2.3, provides further discussion on this topic.

5.8 QUALITY ASSURANCE PROCESS

5.8.1 Overview

5.8.1.1 Purpose As stated in ISO/IEC/IEEE 15288,

[6.3.8.1] The purpose of the Quality Assurance process is to help ensure the effective application of the organization's Quality Management process to the project.

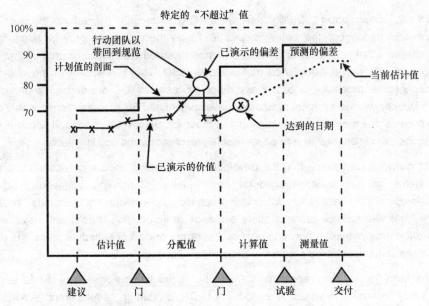

图 5.14 TPM 监控。经 Kevin Forsberg 许可后转载。版权所有。

每个 TPM 的状态的周期性记录提供技术参数的预期和实际实现程度的持续验证。超出已确定的误差带范围的测量值提醒管理人员采取纠正措施。INCOSE-TP-2003-020-01《技术测量指南》3.2.3 节提供关于该主题的进一步论述。

5.8 质量保证流程

5.8.1 概览

5.8.1.1 目的 如 ISO/IEC/IEEE 15288 所述，

> [6.3.8.1] 质量保证流程的目的是帮助确保组织的质量管理流程在项目中的有效应用。

TECHNICAL MANAGEMENT PROCESSES

5.8.1.2 Description Quality assurance (QA) is broadly defined as the set of activities throughout the entire project life cycle necessary to provide adequate confidence that a product or service conforms to stakeholder requirements or that a process adheres to established methodology (ASQ, 2007). A subset of the quality management process, the quality assurance process activities are defined to provide an independent assessment of whether development and SE processes are capable of outcomes that meet requirements and that those processes are performed accurately, precisely, and consistently with all applicable prescriptions and documentation.

QA provides confidence that the developing organization, including subcontractors, adheres to established procedural requirements. Controlling variation in the development process is key to reducing variation in development outcomes. Thus, the quality assurance process offers a means of introducing checks and balances into the development process to ensure that errors or cost or schedule pressures do not result in uncontrolled process or procedural changes.

The term "quality assurance" (or QA) is often used interchangeably with the term "quality control." However, the focus of QA is during development activities (proactive), while "quality control" is typically associated with "inspection" after development activities (reactive).

QA is implemented through procedures for monitoring development and production processes and verifying that QA activities are effective in reducing defects in product or service outcomes. Additionally, QA is responsible for identification, analysis, and control of anomalies or errors discovered during life cycle activities. The level of rigor in QA must be appropriate to product or service requirements for the system under development.

Figure 5.15 is the IPO diagram for the quality assurance process.

5.8.1.3 Inputs/Outputs Inputs and outputs for the quality assurance process are listed in Figure 5.15. Descriptions of each input and output are provided in Appendix E.

5.8.1.4 Process Activities The quality assurance process includes the following activities:

- *Prepare for quality assurance.*
    - Establish and maintain the QA strategy (often captured in a QA plan).
    - Establish and maintain QA guidelines—policies, standards, and procedures.
    - Define responsibilities and authorities.

5.8.1.2 描述 质量保证（QA）被广泛地定义为产品或服务符合利益攸关者需求或流程遵循已建立的方法论提供足够的信心所必需的贯穿于整个项目生命周期内的活动集（ASQ，2007）。定义了质量管理流程和质量保证流程的活动子集，以提供开发流程和 SE 流程的结果是否能够满足需求且那些流程是否被准确地、精确地和与所有适用规定和文件一致地执行方面的独立评估。

QA 提供开发组织（包括转包商）遵守已建立的程序需求的信心。在开发流程中控制差异对于减少开发结果中的变化而言很关键。因而，质量保证流程提供将检查和平衡引入开发流程中的手段，以确保误差或成本或进度压力不会导致流程不受控或程序上的变化。

术语"质量保证"（或 QA）通常与术语"质量控制"可互换使用。然而，QA 的关注点是在开发活动（主动式的）期间，而"质量控制"通常与开发活动之后的"检验"相关联（反应式的）。

通过程序来实施 QA，以监控开发流程和生产流程并验证 QA 活动对于减少产品或服务结果中的缺陷是有效的。此外，QA 负责识别、分析和控制生命周期活动期间发现的异常或误差。QA 的严格层级必须适合开发中的系统的产品或服务需求。

图 5.15 是质量保证流程的 IPO 图。

5.8.1.3 输入/输出 图 5.15 列出质量保证流程的输入和输出。附录 E 提供每个输入和输出的描述。

5.8.1.4 流程活动 质量保证流程包括以下活动：

- 质量保证的准备。

    – 建立并维护 QA 策略（通常在 QA 计划中获取）。

    – 建立并维护 QA 指南——方针、标准和程序。

    – 定义职责和权限。

- *Perform product or service evaluations.*
  - Perform the evaluations at appropriate times in the life cycle as defined by the QA plan, ensuring V&V of the outputs of the life cycle processes. Ensure that QA perspectives are appropriately represented during design, development, verification, validation, and production activities.
  - Evaluate product verification results as evidence of QA effectiveness.

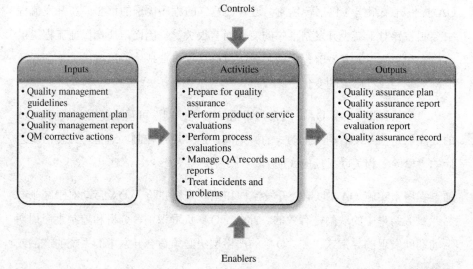

FIGURE 5.15 IPO diagram for the quality assurance process. INCOSE SEH original figure created by Shortell and Walden. Usage per the INCOSE notices page. All other rights reserved.

- *Perform process evaluations.*
  - Implement prescribed surveillance on processes to provide an independent evaluation of whether the developing organization is in compliance with established procedures.
  - Evaluate enabling tools and environments for conformance and effectiveness.
  - Flow applicable procedural and surveillance requirements throughout the project supply chain and evaluate subcontractor processes for conformance to allocated requirements.

- *实施产品或服务评估。*

    - 按照 QA 计划的定义在生命周期的适当时间进行评价,确保生命周期流程的输出的 V&V。确保在设计、开发、验证、确认和生产活动期间恰当地表征 QA 视角。

    - 评估产品验证结果作为 QA 有效性的证据。

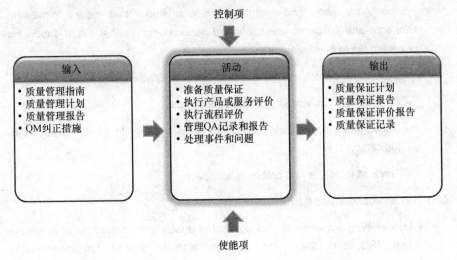

图 5.15 质量保证流程的 IPO 图。INCOSE SEH 原始图由 Shortell 和 Walden 创建。按照 INCOSE 通知页使用。版权所有。

- *实施流程评价。*

    - 对流程实施规定的监视,以提供开发组织是否符合已建立程序方面的独立评估。

    - 针对符合性和有效性对使能工具和环境进行评估。

    - 使适用的程序需求和监视需求贯穿于项目供应链内流动,并评估转包商流程与分配的需求的符合性。

TECHNICAL MANAGEMENT PROCESSES

- *Manage QA records and reports.*
  - Create, maintain, and store records and reports in accordance with applicable requirements.
  - Identify incidents and problems associated with product and process evaluations.
- *Treat incidents and problems.*

Note: Incidents are short-term anomalies or observations that require immediate attention, and problems are confirmed nonconformities that would cause the project to fail to meet requirements.

  - Document, classify, report, and analyze all anomalies.
  - Perform root cause analysis and note trends.
  - Recommend appropriate actions to resolve anomalies and errors, when indicated.
  - Track all incidents and problems to closure.

Common approaches and tips:

- Use existing agreements and applicable quality certifications or registrations (e.g., ISO 9001, CMMI, etc.), together with the organization's overarching quality management policy to provide essential guidance for QA approaches.
- Analyze statistics from process audits, verification results, product discrepancy reports, customer satisfaction monitoring, and accident and incident reporting to verify whether QA activities are effective.
- Continually demonstrate uncompromising integrity in monitoring development and SE processes. There must not be a perception that the development organization or project management is inappropriately influencing the judgment of QA personnel.
- Establish organizational independence and consistent support from senior leadership. The QA team should not be completely beholden to the project manager. Implement an escalation process so that QA issues not addressed by the project can be escalated to organizational leadership, as appropriate.

- *管理 QA 记录和报告。*

    – 按照适用需求创建、维护并存储记录和报告。

    – 识别与产品和流程评价相关联的事件和问题。

- *处理事件和问题。*

注：意外事件是需要立即关注的短期异常或观察结果，问题是已经确认的不符合性将会导致项目不能满足需求。

    – 文件化、分类、报告并分析所有异常。

    – 进行根因分析并注意趋势。

    – 建议适当的措施以便在有指征时解决异常和误差。

    – 跟踪所有事件和问题，直至关闭。

常用方法和建议：

- 使用现有协议和适用质量认证或注册（如 ISO 9001、CMMI 等）连同组织的首要的质量管理方针，以便为 QA 方法提供基本指导。

- 分析来自流程审计、验证结果、产品差异报告、客户满意度监控、以及事故和意外事件报告的统计数据以验证 QA 活动是否有效。

- 在监控开发流程和 SE 流程时不断地展示毫不妥协的完整性。一定不能认为开发组织或项目管理人员不恰当地影响 QA 人员的判断。

- 建立组织独立性和高层领导的一致支持。QA 团队不应完全受制于项目经理。实施升级流程，使得项目没有应对的 QA 问题能够被适当地升级到组织的领导层。

TECHNICAL MANAGEMENT PROCESSES

5.8.2 Elaboration

5.8.2.1 QA Concepts

Quality cannot be "inspected into" products or services after they are developed. QA performs an important role in ensuring that all elements of the developing organization execute activities in accordance with approved plans and procedures as a means of building quality into products or services. Through this process control role, QA enables systematic process improvement. Deming described this relationship between quality and process improvement: "Quality comes not from inspection, but from improvement of the production process" (Deming, 1986).

QA applies the policies and standards that govern development activities and procurement of raw materials in support of quality goals and objectives of the project. For example, the NASA has adopted the SAE Aerospace Standard (AS) 9100, Quality Systems—Aerospace—Model for Quality Assurance in Design, Development, Production, Installation and Servicing, as a means of building quality into its systems and controlling the statistical variation in all system elements by requiring adherence to a common quality standard (SAE Aerospace Quality Standard AS9100:C, 2009). Similarly, an acquiring organization may mandate that suppliers achieve a given CMMI level as a means of assuring that a given supplier is capable of delivering a consistent level of quality in its development processes.

QA also plays a key role during the verification activities themselves. The presence of independent QA personnel during verification activities provides an unbiased perspective on the integrity of verification procedures and the appropriate calibration of verification equipment and facilities. QA personnel also provide an independent assessment that verification results are accurately recorded. For example, it is not uncommon to require a QA signature on verification results reports to attest that the verification procedures were followed and the results were accurately reported.

5.8.2.2 QA Methods Common QA techniques include:

- *Checklist*—A tool for ensuring all important steps or actions in an operation have been taken (ASQ, 2007).

- *Quality audit*—An independent review to deter-mine whether project activities comply with established policies, processes, or procedures.

- *Root cause analysis*—A method of problem solving using specific techniques designed to address the underlying (root) causes of defects or anomalies. Popular root cause analysis techniques include Ishikawa (fishbone) diagrams, FTA, failure mode effects and criticality analysis, and the five why's technique.

5.8.2 详细阐述

5.8.2.1 QA 概念

产品或服务开发之后不能进行"内在的"质量检验。QA 在确保开发组织的所有元素按照经批准的计划和程序执行活动作为建造高质量产品或服务的手段中执行重要角色。尽管该流程控制角色，但 QA 能够使系统的流程改进。Deming 将质量和流程改进之间的关系描述为："质量不是来自检验，而是来自生产流程的改进"（Deming，1986）。

QA 应用管控对开发活动和原材料采购的方针和标准，以便支持项目的质量目的和目标。例如，NASA 采用 SAE 航空航天标准（AS）9100，设计、开发、生产、安装和服务中质量保证的质量体系—航空航天—模型，作为通过要求遵循公共质量标准（SAE 航空航天质量标准 AS9100:C, 2009）来将质量建造在系统之中并控制所有系统元素的统计差异的手段。同样，采办组织可授权供应商实现给定的 CMMI 等级，作为确保指定供应商在其开发流程中能够交付一致的质量水平的手段。

QA 在其本身的验证活动期间还扮演关键角色。验证活动期间独立的 QA 人员的存在可对验证程序的完整性和验证设备及设施的恰当校准方面提供无偏见的视角。QA 人员还提供独立评估：验证结果是否被准确地记录。例如，验证结果报告上需要 QA 签名以证明遵循了验证程序且结果被准确地报告，这并不少见。

5.8.2.2 QA 方法　公共 QA 技术包括：

- *检查清单*——确保在运行中已采取所有重要步骤或措施的一种工具（ASQ, 2007）。

- *质量审计*——确定项目活动是否符合已建立的方针、流程或程序的一种独立评审。

- *根因分析*——指定一种采用特定技术解决问题的方法以应对缺陷和异常的根因。流行的根因分析技术包括石川（鱼骨）图、FTA、失效模式影响与危害性分析及五个为什么技术。

6 AGREEMENT PROCESSES

The initiation of a project begins with user need. Once a need is perceived and resources are committed to establish a project, it is possible to define the parameters of an acquisition and supply relationship. One instance for which this relationship exists is whenever an organization with a need does not have the ability to satisfy that need without assistance. Agreement processes are defined in ISO/IEC/IEEE 15288 as follows:

> [6.1(b)] [Agreement] processes define the activities necessary to establish an agreement between two organizations.

Acquisition is also an alternative for optimizing investment when a supplier can meet the need in a more economical or timely manner. The acquisition and supply processes are the subject of Sections 6.1 and 6.2, respectively.

Virtually all organizations interface with one or more organizations from industry, academia, government, customers, partners, etc. An overall objective of agreement processes is to identify these external interfaces and establish the parameters of these relationships, including identifying the inputs required from the external entities and the outputs that will be provided to them. This network of relationships provides the context of the business environment of the organization and access to future trends and research. Some relationships are defined by the exchange of products or services.

The acquisition and supply processes are two sides of the same coin. Each process establishes the context and constraints of the agreement under which the other system life cycle processes are performed, regardless of whether the agreement is formal (as in a contract) or informal. The unique activities for the agreement processes are related to contracts and managing business relationships. An important contribution of ISO/IEC/ IEEE 15288 is the recognition that systems engineers are relevant contributors in this domain (Arnold and Lawson, 2003). The maglev train case (see Section 3.6.3) is an example where the government representatives of China and Germany participated in the relationship.

INCOSE Systems Engineering Handbook: A Guide for System Life Cycle Processes and Activities, Fourth Edition. Edited by David D. Walden, Garry J. Roedler, Kevin J. Forsberg, R. Douglas Hamelin and Thomas M. Shortell.

© 2015 John Wiley & Sons, Inc. Published 2015 by John Wiley & Sons, Inc.

6 协议流程

项目的启动从用户的需要开始。一旦一种需要被感知且建立一个项目的资源得到承诺,就有可能定义采办和供应关系的范围。存在这种关系的一个实例是当具有某一需要的组织在无援助的情况下没有满足该需要的能力。协议流程在 ISO/IEC/IEEE 15288 中定义如下:

[6.1(b)][协议]流程定义两个组织间建立协议所必需的活动。

当供应商能够以更经济或更及时的方式满足需要时,采办亦是优化投资的一个备选方案。采办流程和供应流程分别是 6.1 节和 6.2 节的主题。

实际上,所有组织与来自工业界、学术界、政府、客户和合作者等的一个或多个组织都建立联系。协议流程的总体目标是识别这些外部接口并建立这些关系的范围,包括识别外部实体所需的输入和将要提供给外部实体的输出。这种关系网络提供组织业务环境的背景环境并有机会获取未来的趋势与研究。一些关系由产品或服务的交换来定义。

采办流程和供应流程就像是同一硬币的两个面。每个流程均建立执行其他系统生命周期流程所依据的协议的背景环境和约束,不论协议是正式的(诸如以合同形式)还是非正式的。协议流程的独特活动与合同和管理业务关系有关。ISO/IEC/IEEE 15288 的重要贡献是认识到系统工程师是该领域的相关贡献者(Arnold 和 Lawson, 2003)。磁悬浮列车案例(见 3.6.3 节)是中国和德国政府代表参与这种关系的实例。

INCOSE 系统工程手册:系统生命周期流程和活动指南,第 4 版。编撰:David D. Walden、Garry J. Roedler、Kevin J. Forsberg、R. Douglas Hamelin 和 Thomas M. Shortell。
John Wiley & Sons 公司版权所有©2015。由 John Wiley & Sons 公司于 2015 年出版。

AGREEMENT PROCESSES

Agreement negotiations are handled in various ways depending on the specific organizations and the formality of the agreement. For example, in formal contract situations with government agencies, there is usually a contract negotiation activity that may include multiple roles for both the acquirer and supplier to refine the contract terms and conditions. The systems engineer is usually in a supporting role to the project manager during negotiations and is responsible for impact assessments for changes, trade studies on alternatives, risk assessments, and other technical input needed for decisions.

A critical element to each party is the definition of acceptance criteria, such as:

1. Percent completion of the SyRS
2. Requirements stability and growth measures, such as the number of requirements added, modified, or deleted during the preceding time interval (e.g., month, quarter, etc.)
3. Percent completion of each contract requirements document: SOW, RFP, etc.

These criteria protect both sides of the business relationship—the acquirer from being coerced into accepting a product with poor quality and the supplier from the unpredictable actions of a fickle or indecisive buyer.

It is important to note that the previous criteria are negotiated for use during the agreement. During negotiations, it is also critical that both parties are able to track progress toward an agreement. Identifying where there are agreements or disagreements for documents and clauses is vital.

Note also that the agreement processes can be used for coordinating within an organization between different business units or functions. In this case, the agreement will usually be more informal, not requiring a contract or other legally binding set of documentation.

Two agreement processes are identified by ISO/IEC/ IEEE 15288: the acquisition process and the supply process. They are included in this handbook because they conduct the essential business of the organization related to the system of interest (SOI) and establish the relationships between organizations relevant to the acquisition and supply (i.e., buying and selling) of products and services.

协议的协商依特定的组织和协议的正式程度,来以各种不同的方式来处理。例如,在与政府机构的正式合同情况下,通常具有合同协商活动,可包括采办方和供应商的多重角色,以便细化合同条款和条件。系统工程师通常在协商期间对项目经理起着支持的作用并负责变化的影响评估、备选方案的权衡研究、风险评估以及决策所需的其他技术输入。

各方的一个关键元素是验收准则的定义,如:

1. SyRS 的完成百分比
2. 需求稳定性和增长测度,如在先前的时间区间(如月、季度等)增加、修改或删除的需求的数量
3. 每个合同需求文件(SOW 和 RFP 等)的完成百分比

这些准则保护业务关系中的双方——采办方不被迫接受某一劣质产品;供应商免遭善变或优柔寡断的买方不可预知的行动。

注意在协议期间协商使用先前的准则是非常重要的。协商期间,双方能够对协议跟踪进展也是很关键的。识别对文件和条款有一致同意或不同意的地方是至关重要的。

还要注意到,协议流程可用于在组织内部不同的业务单元或功能之间进行协调。在这种情况下,协议将通常更加非正式的,不需要合同或有法律约束力的其他文档集。

两个协议流程与 ISO/IEC/ IEEE 15288:采办流程和供应流程密切相关。因为这两个协议流程开展与所感兴趣之系统(SOI)相关的组织的基本业务,并且建立与产品和服务的采办和供应(即购买和出售)相关的组织之间的关系,所以它们都被包括在本手册中。

AGREEMENT PROCESSES

6.1 ACQUISITION PROCESS

6.1.1 Overview

6.1.1.1 Purpose As stated in ISO/IEC/IEEE 15288,

> [6.1.1.1] The purpose of the Acquisition process is to obtain a product or service in accordance with the acquirer's requirements.

The acquisition process is invoked to establish an agreement between two organizations under which one party acquires products or services from the other. The acquirer experiences a need for an operational system, for services in support of an operational system, for elements of a system being developed by a project, or for services in support of project activities. Often, our experience with the acquisition process is typified by the purchase of commodities or commercial products, such as telephones or automobiles. SE is often required to facilitate the procurement of more complex services and products. The start of an acquisition/supply process begins with the determination of, and agreement on, user needs. The goal is to find a supplier that can meet those needs.

6.1.1.2 Description The role of the acquirer demands familiarity with the technical, technical management, and organizational project-enabling processes as it is through them that the supplier will execute the agreement. An acquirer organization applies due diligence in the selection of a supplier to avoid costly failures and impacts to the organization budgets and schedules. This section is written from the perspective of the acquirer organization. Figure 6.1 is the IPO diagram for the acquisition process.

6.1.1.3 Inputs/Outputs Inputs and outputs for the acquisition process are listed in Figure 6.1. Descriptions of each input and output are provided in Appendix E.

6.1.1.4 Process Activities The acquisition process includes the following activities:

- *Prepare for the acquisition.*
    - Develop and maintain acquisition plans, policies, and procedures to meet the organization strategies, goals, and objectives as well as the needs of the project management and technical SE organizations.

6.1 采办流程

6.1.1 概览

6.1.1.1 目的 如 ISO/IEC/IEEE 15288 所述，

[6.1.1.1] 采办流程的目的是按照采办方的需求获得产品或服务。

调用采办流程，以建立两个组织之间一方从另一方获取产品或服务所依据的协议。采办方对运行系统、支持运行系统的服务、项目正在开发的系统的元素或支持项目活动的服务体验一种需要。通常，我们在采办流程方面的经验主要体现在采购商品或商用产品，如电话或汽车。常常要求 SE 来辅助较为复杂的服务和产品的采办。采办/供应流程的启动从用户需要的确定和达成协议开始。目标是找到一个能够满足这些需要的供应商。

6.1.1.2 描述 采办方的角色要求熟悉技术流程、技术管理流程和组织的项目使能流程，因为供应商将通过它们执行该协议。采办方组织在选择供应商时应用尽职调查，以避免高代价的失败以及对组织预算和进度的影响。本节是从采办方组织的视角来编写的。图 6.1 是采办流程的 IPO 图。

6.1.1.3 输入/输出 图 6.1 列出采办流程的输入和输出。附录 E 提供每个输入和输出的描述。

6.1.1.4 流程活动 采办流程包括以下活动：

- *为采办准备*。
    - 开发和维护采办计划、方针和程序，以满足组织的战略、目的和目标以及项目管理组织和技术 SE 组织的需要。

- Identify needs in a request for supply—such as an RFP or request for quotation (RFQ) or some other mechanism to obtain the supply of the system, service, or product. Through the use of the technical processes, including system requirements definition, the acquiring organization produces a set of requirements that will form the basis for the technical information of the agreement.
- Identify a list of potential suppliers—suppliers may be internal or external to the acquirer organization.

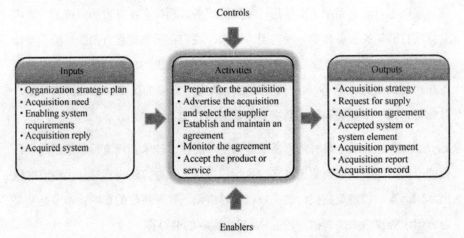

FIGURE 6.1 IPO diagram for the acquisition process. INCOSE SEH original figure created by Shortell and Walden. Usage per the INCOSE notices page. All other rights reserved.

- *Advertise the acquisition and select the supplier.*
  - Distribute the RFP, RFQ, or other documented request for supply and select appropriate suppliers—using selection criteria, rank suppliers by their suitability to meet the overall need and establish supplier preferences and corresponding justifications. Viable suppliers should be willing to conduct ethical negotiations, able to meet technical obligations, and willing to maintain open communications throughout the acquisition process.

- 识别请求供应的需要——如邀标书（RFP）或报价申请书（RFQ）或获得系统、服务或产品供应的一些其他机制。通过使用技术流程，包括系统需求定义，采办组织生成将要形成协议的技术信息基础的一个需求集合。

- 识别潜在的供应商的列表——供应商可能是采办方组织内部的或外部的。

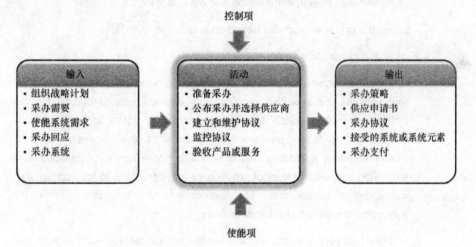

图 6.1 采办流程的 IPO 图。INCOSE SEH 原始图由 Shortell 和 Walden 创建。按照 INCOSE 通知页使用。版权所有。

- 公布采办并选择供应商。

  - 分发 RFP、RFQ 或其他文件化的供应申请书并选择合适的供应商——使用选择准则，按照供应商满足整体需要的适用性对供应商排序并建立供应商优选及相应的正当理由。切实可行的供应商应愿意开展符合道德规范的协商，能够满足技术义务，并愿意贯穿于采办流程保持公开沟通。

AGREEMENT PROCESSES

- Evaluate supplier responses to the RFP or supply request—ensure the (SOI) meets acquirer needs and complies with industry and other standards. Assessments from the project portfolio management and quality management processes and recommendations from the requesting organization are necessary to determine the suitability of each response and the ability of the supplier to meet the stated commitments. Record recommendations from evaluation of responses to the RFP. This can range from formal documentation to less formal interorganizational interactions (e.g., between design engineering and marketing).
- Select the preferred supplier based on acquisition criteria.

• *Establish and maintain an agreement.*

- Negotiate agreement—the supplier commits to provide a product or service that satisfies the specified requirements and acceptance criteria for the system of interest. Both supplier and acquirer agree to participate in verification, validation, and acceptance activities; the acquirer agrees to render payment according to the schedule. Both agree to participate in exception and change control procedures and contribute to transparent risk management procedures. The agreement will establish criteria for assessing progress toward final delivery.
- Establish delivery acceptance criteria—the procurement specification, in the context of the overall agreement, should clearly state the criteria by which the acquirer will accept delivery from the supplier. A verification matrix can be used to clarify these criteria.

• *Monitor the agreement.*

- Manage acquisition process activities, including decision making for agreements, relationship building and maintenance, interaction with organization management, responsibility for the development of plans and schedules, and final approval authority for deliveries accepted from the supplier.
- Maintain communications with supplier, stakeholders, and other organizations regarding the project.

- 评价供应商对 RFP 或供应申请书的响应——确保（SOI）满足采办方需要且符合行业标准和其他标准。项目群管理和质量管理流程中的评估以及请求组织的建议，对于确定每个响应的适用性及供应商满足所述承诺的能力是十分必要的。对响应 RFP 的评估中的建议进行记录，可覆盖从正式文件到并非正式的组织间互动（如设计工程和市场之间）。

- 基于采办准则选择首选供应商。

- *建立和维护协议。*

 - 协商协议——供应商承诺提供满足所感兴趣之系统的特定需求和验收准则的产品或服务。供应商和采办方都同意参加验证、确认和验收活动；采办方同意根据进度支付款项。供应商和采办方都同意参加例外和变更控制程序，并促成透明的风险管理程序。协议应建立评估最终交付进度的准则。

 - 建立交付验收准则——整体协议背景环境中的采购规范应清晰地阐明采办方将验收供应商交付的准则。验证矩阵可用于阐明这些准则。

- *监控协议。*

 - 管理采办流程活动，包括协议的决策、关系的建立和维护、与组织管理的互动、开发计划和进度的职责以及从供应商接收的交付品的最终审批权限。

 - 与项目有关的供应商、利益攸关者和其他组织保持沟通。

AGREEMENT PROCESSES

- Status progress against the agreed-to schedule to identify risks and issues, to measure progress toward mitigation of risks and adequacy of progress toward delivery and cost and schedule performance, and to determine potential undesirable outcomes for the organization. The project assessment and control process provides necessary evaluation information regarding cost, schedule, and performance.

- Amend agreements when impacts on schedule, budget, or performance are identified.

- *Accept the product or service.*

    - Accept delivery of products and services—in accordance with all agreements and relevant laws and regulations.

    - Render payment—or other agreed consideration in accordance with agreed payment schedules.

    - Accept responsibility in accordance with all agreements and relevant laws and regulations.

    - When an acquisition process cycle concludes, a final review of performance is conducted to extract lessons learned for continued process performance.

Note: The agreement is closed through the portfolio management process, which manages the full set of systems and projects of the organization.

Common approaches and tips:

- Establish acquisition guidance and procedures that inform acquisition planning, including recommended milestones, standards, assessment criteria, and decision gates. Include approaches for identifying, evaluating, choosing, negotiating, managing, and terminating suppliers.

- Establish a technical point of responsibility within the organization for monitoring and controlling individual agreements. This person maintains communication with the supplier and is part of the decision-making team to assess progress in the execution of the agreement. The possibility of late delivery or cost overruns should be identified and communicated to the organization as early as noted.

Note: There can be multiple points of responsibility for an agreement that focus on technical, programmatic, marketing, etc.

- 按照商定的进度跟踪进展，以识别风险和问题，测量风险缓解的进展、交付进展的充分性以及成本和进度绩效，并确定组织的潜在的不合需要的结果。项目评估和控制流程提供关于成本、进度和绩效的必要的评价信息。

- 当识别出对进度、预算或执行的影响时，修订协议。

- 验收产品或服务。

  - 验收产品和服务的交付——按照所有协议和相关法律法规。

  - 支付款项——或符合协定的支付进度的其他协定的考虑因素。

  - 按照所有协议和相关法律法规承担职责。

  - 当采办流程周期结束时，进行执行的最终评审，以吸取经验教训用于持续的流程执行。

注：协议通过项目组合管理流程结束，该流程管理组织的系统和项目的完整集合。

常用方法和建议：

- 建立采办指南和程序，通告采办计划，包括建议的里程碑、标准、评估准则和决策门。包括识别、评价、选择、协商、管理和终止供应商的方法。

- 建立组织内监控和控制单独协议的技术职责项。该责任人保持与供应商的沟通，并且是评估协议执行进展的决策团队的一部分。应尽早注意识别延期交付或成本超支的可能性，并传达给组织。

注：可具有多个集中于技术、计划、市场等方面的协议职责项。

AGREEMENT PROCESSES

- Define and track measures that indicate progress on agreements. Appropriate measures require the development of tailored measures that do not drive unnecessary and costly efforts but do provide the information needed to ensure the progress is satisfactory and that key issues and problems are identified early to allow time for resolution with minimal impact to the delivery and quality of the product and service.

- Include technical representation in the selection of the suppliers to critically assess the capability of the supplier to perform the required task. This helps reduce the risk of contract failure and its associated costs, delivery delays, and increased resource commitment needs. Past performance is highly important, but changes to key personnel should be identified and evaluated.

- Communicate clearly with the supplier about the real needs and avoid conflicting statements or making frequent changes in the statement of need that introduce risk into the process.

- Maintain traceability between the supplier's responses to the acquirer's solicitation. This can reduce the risk of contract modifications, cancellations, or follow-on contracts to fix the product or service.

- Institute for supply management has useful guidance for purchasing and marketing (ISM, n.d.).

6.2 SUPPLY PROCESS

6.2.1 Overview

6.2.1.1 Purpose As stated in ISO/IEC/IEEE 15288,

> [6.1.2.1] The purpose of the Supply process is to provide an acquirer with a product or service that meets agreed requirements.

The supply process is invoked to establish an agreement between two organizations under which one party supplies products or services to the other. Within the supplier organization, a project is conducted according to the recommendations of this handbook with the objective of providing a product or service to the acquirer that meets the contracted requirements. In the case of a mass produced commercial product or service, a marketing function may represent the acquirer and establish customer expectations.

- 定义并跟踪指示协议进展的测度。适当的测度要求开发经剪裁的测度，其不驱动不必要且高成本的工作，但确实提供确保进度符合要求且提早识别关键事项和问题所需的信息，以允许有时间采用对产品和服务的交付及质量影响最小的方式来解决。

- 在选择供应商时，包括技术代表，要苛刻地评估供应商执行所要求任务的能力。这会帮助降低合同失败及其相关成本、延期交付和增加资源承诺需要的风险。以往的表现非常重要，但应识别并评价关键人员的变更。

- 与供应商就实际需要进行清晰地沟通，在说明将风险引入到流程的需要时，避免产生冲突的说明或做出频繁更改。

- 维护从供应商响应到采办方请求之间的可追溯性。这会降低合同的修改、取消或后续的风险，或按合同对产品和服务进行修补。

- 供应管理协会具有采购和市场管理的有用指南（ISM, n.d.）。

6.2 供应流程

6.2.1 概览

6.2.1.1 目的 如 ISO/IEC/IEEE 15288 所述，

[6.1.2.1] 供应流程的目的是向采办方提供满足协定需求的产品或服务。

调用供应流程，以建立两个组织之间一方向另一方供应产品或服务所依据的协议。在供应商组织内，以向采办方提供满足合同需求的产品或服务为目的，按照本手册的建议实施项目。至于大规模生产的商用产品或服务，市场功能可代表采办方并建立客户期望。

AGREEMENT PROCESSES

6.2.1.2 Description The supply process is highly dependent upon the technical, technical management, and organizational project-enabling processes as it is through them that the work of executing the agreement is accomplished. This means that the supply process is the larger context in which the other processes are applied under the agreement. This section is written from the perspective of the supplier organization. Figure 6.2 is the IPO diagram for the supply process.

6.2.1.3 Inputs/Outputs Inputs and outputs for the supply process are listed in Figure 6.2. Descriptions of each input and output are provided in Appendix E.

6.2.1.4 Process Activities The supply process includes the following activities:

- *Prepare for the supply.*
  - Develop and maintain strategic plans, policies, and procedures to meet the needs of potential acquirer organizations, as well as internal organization goals and objectives including the needs of the project management and technical SE organizations.
  - Identify opportunities.
- *Respond to a tender.*
  - Select appropriate acquirers willing to conduct ethical negotiations, able to meet financial obligations, and willing to maintain open communications throughout the supply process.
  - Evaluate the acquirer requests and propose a SOI that meets acquirer needs and complies with industry and other standards. Assessments from the project portfolio management, human resource management, quality management, and business or mission analysis processes are necessary to determine the suitability of this response and the ability of the organization to meet these commitments.
- *Establish and maintain an agreement.*
  - Supplier commits to meet the negotiated requirements for the SOI; meet delivery milestones, verification, validation, and acceptance conditions; accept the payment schedule; execute exception and change control procedures; and maintain transparent risk management procedures. The agreement will establish criteria for assessing progress toward final delivery.

6.2.1.2 描述 供应流程很大程度上取决于技术流程、技术管理流程和组织的项目使能流程，因为协议的执行工作正是通过它们完成的。这意味着供应流程是在协议中采用其他流程所对应的更大的背景环境。本节是从供应商组织的视角来编写的。图 6.2 是供应流程的 IPO 图。

6.2.1.3 输入/输出 图 6.2 列出供应流程的输入和输出。附录 E 提供每个输入和输出的描述。

6.2.1.4 流程活动 供应流程包括以下活动：

- 为供应准备。

    - 开发和维护策略计划、方针和程序，以满足潜在的采办方组织的需要以及内部组织的目的和目标，包括项目管理组织和技术 SE 组织的需要。

    - 识别机会。

- 响应投标。

    - 选择合适采办方，其愿意开展符合道德规范的协商，能够满足财务义务，并愿意贯穿于供应流程保持公开沟通。

    - 评价采办方的请求并提出满足采办方需要且符合行业标准和其他标准的 SOI。项目组合管理流程、人力资源管理流程、质量管理流程、业务或任务分析流程中的评估，对于确定该响应的适用性及组织满足这些承诺的能力是十分必要的。

- 建立和维护协议。

    - 供应商承诺满足 SOI 的协商需求；满足交付里程碑、验证、确认和验收条件；验收支付进度；执行例外和变更控制程序；以及维护透明的风险管理程序。协议应建立评估最终交付进度的准则。

AGREEMENT PROCESSES

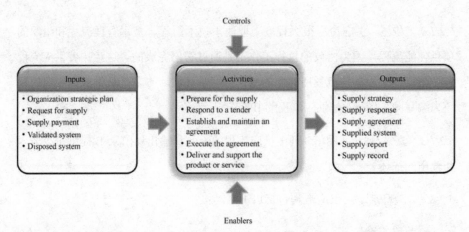

FIGURE 6.2 IPO diagram for the supply process. INCOSE SEH original figure created by Shortell and Walden. Usage per the INCOSE notices page. All other rights reserved.

- *Execute the agreement.*

  – Start the project and invoke the other processes defined in this handbook.

  – Manage supply process activities, including decision making for agreements, relationship building and maintenance, interaction with organization management, responsibility for the development of plans and schedules, and final approval authority for deliveries made to acquirer.

  – Maintain communications with acquirers, sub-suppliers, stakeholders, and other organizations regarding the project.

  – Evaluate carefully the terms of the agreement to identify risks and issues, progress toward mitigation of risks, and adequacy of progress toward delivery; evaluate cost and schedule performance; and determine potential undesirable outcomes for the organization.

- *Deliver and support the product or service.*

  – After acceptance and transfer of the final products and services, the acquirer will provide payment or other consideration in accordance with all agreements, schedules, and relevant laws and regulations.

协议流程

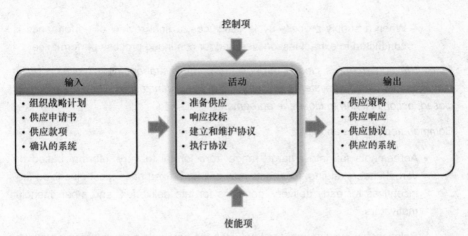

图 6.2 供应流程的 IPO 图。INCOSE SEH 原始图由 Shortell 和 Walden 创建。按照 INCOSE 通知页使用。版权所有。

- *执行协议*。
  - 启动项目并调用本手册定义的其他流程。
  - 管理供应流程活动，包括协议的决策、关系的建立和维护、与组织管理的互动、开发计划和进度的职责及提供给采办方的交付物的最终审批权限。
  - 与项目有关的采办方、下级供应商、利益攸关者和其他组织保持沟通。
  - 仔细地评估协议的条款，以识别风险和问题、风险缓解的进展以及交付进展的充分性；评价成本和进度的执行；并确定组织的潜在非期望结果。
- *交付和支持产品或服务*。
  - 在验收和转移最终产品和服务后，采办方将按照所有协议、进度和相关法律法规提供款项或其他考虑因素。

AGREEMENT PROCESSES

- When a supply process cycle concludes, a final review of performance is conducted to extract lessons learned for continued process performance.

Note: The agreement is closed through the portfolio management process, which manages the full set of systems and projects of the organization. When the project is closed, action is taken to close the agreement.

Common approaches and tips:

- Agreements fall into a large range from formal to very informal based on verbal understanding. Contracts may call for a fixed price, cost plus fixed fee, incentives for early delivery, penalties for late deliveries, and other financial motivators.

- Relationship building and trust between the parties is a nonquantifiable quality that, while not a substitute for good processes, makes the human interactions agreeable.

- Develop technology white papers or similar documents to demonstrate and describe to the (potential) acquirer the range of capabilities in areas of interest. Use traditional marketing approaches to encourage acquisition of mass-produced products.

- Maintain an up-to-date Internet presence, even if the organization does not engage in electronic commerce.

- When expertise is not available within the organization (e.g., legal and other governmental regulations, laws, etc.), retain subject matter experts to provide information and specify requirements related to agreements.

- Invest sufficient time and effort into understanding acquirer needs before the agreement. This can improve the estimations for cost and schedule and positively affect agreement execution. Evaluate any technical specifications for the product or service for clarity, completeness, and consistency.

- Involve personnel who will be responsible for agreement execution to participate in the evaluation of and response to the acquirer's request. This reduces the start-up time once the project is initiated, which in turn is one way to recapture the cost of writing the response.

- Make a critical assessment of the ability of the organization to execute the agreement; otherwise, the high risk of failure and its associated costs, delivery delays, and increased resource commitment needs will reflect negatively on the reputation of the entire organization.

- 当供应流程周期结束时，开展绩效的最终评审以吸取经验教训用于持续的流程执行。

注：协议通过项目群管理流程结束，该流程管理组织的系统和项目的完整集。当项目结束时，采取行动结束协议。

常用方法和建议：

- 基于口头的理解，协议落入从正式到非常不正式的一个大的范围。合同可要求固定价格、成本加固定费用、提前交付奖励、延迟交付惩罚及其他财务激励。

- 各方之间关系的建立和信任是一种不可估量的品质，虽然其不是良好流程的替代物，但可以使人员的互动令人愉快。

- 开发技术白皮书或类似文件，以便向（潜在）采办方演示并描述所感兴趣之领域的能力范围。使用传统的市场方法鼓励采办大规模生产的产品。

- 维护一个现代化的互联网平台，即使组织未从事电子商务。

- 当在组织内专业知识不可用时（如法律及其他政府法规、法律等），聘用领域专家以提供信息并规定有关协议的需求。

- 达成协议前投入充分的时间和精力理解采办方需要。这可以改进对成本和进度的估计并积极地影响协议的执行。评价产品或服务的清晰度、完整性和一致性的任何技术规范。

- 使将要负责协议执行的人员参加对采办方请求的评估和响应。一旦项目启动，即可减少启动时间，这转而是挽回编写该响应的成本的一种方式。

- 对组织执行协议的能力进行批判性评估；否则，合同失败及其相关成本、延期交付和增加资源承诺需要的风险将对整体组织的声誉产生消极影响。

7 ORGANIZATIONAL PROJECT-ENABLING PROCESSES

Organizational project-enabling processes are the purview of the organization (also known as enterprise) and are used to direct, enable, control, and support the system life cycle. Organizational project-enabling processes are defined in ISO/IEC/IEEE 15288 as follows:

> [6.2] The Organizational Project-Enabling Processes help ensure the organization's capability to acquire and supply products or services through the initiation, support and control of projects. They provide resources and infrastructure necessary to support projects ....

This chapter focuses on the capabilities of an organization relevant to the realization of a system; as stated above, they are not intended to address general business management objectives, although sometimes the two overlap.

Organizational units cooperate to develop, produce, deploy, utilize, support, and retire (including dispose of) the system of interest (SOI). Enabling systems may also need to be modified to meet the needs of new systems, developed, or acquired if they do not exist. Examples include development, manufacturing, training, verification, transport, logistics, maintenance, and disposal systems that support the SOI.

Six organizational project-enabling processes are identified by ISO/IEC/IEEE 15288. They are life cycle model management, infrastructure management, portfolio management, human resource management, quality management (QM), and knowledge management (KM). The organization will tailor these processes and their interfaces to meet specific strategic and communications objectives in support of the system projects.

INCOSE Systems Engineering Handbook: A Guide for System Life Cycle Processes and Activities, Fourth Edition. Edited by David D. Walden, Garry J. Roedler, Kevin J. Forsberg, R. Douglas Hamelin and Thomas M. Shortell.

© 2015 John Wiley & Sons, Inc. Published 2015 by John Wiley & Sons, Inc.

7 组织的项目使能流程

组织的项目使能流程是组织（亦称作复杂组织体）的范围并用来指导、使能、控制和支持系统生命周期。组织的项目使能流程在 ISO/IEC/IEEE 15288 中定义如下：

> [6.2] 组织的项目使能流程有助于确保组织通过启动、支持和控制项目来采办并供应产品或服务的能力。这些流程提供支持项目所必需的资源和基础设施。

本章集中于与组织实现一个系统有关的能力；如上所述，这些流程并不旨在应对一般的业务管理目标，尽管两者有时会重叠。

组织的多个单元合作以开发、生产、部署、使用、维持和退役（包括处置）所感兴趣之系统（SOI）。使能系统可能还需要修改，以满足新系统的需要，如果不存在，还需要开发或采办。示例包括支持 SOI 的开发、制造、培训、验证、运输、后勤、维护和处置系统。

ISO/IEC/IEEE 15288 识别六个组织级项目使能流程，分别是生命周期模型管理、基础设施管理、项目群管理、人力资源管理、质量管理（QM）和知识管理（KM）。组织将对这些流程及其接口进行剪裁，以满足对系统项目的支持而特定战略和沟通目标。

INCOSE 系统工程手册：系统生命周期流程和活动指南，第 4 版。编撰：David D. Walden、Garry J. Roedler、Kevin J. Forsberg、R. Douglas Hamelin 和 Thomas M. Shortell。

John Wiley & Sons 公司版权所有©2015。由 John Wiley & Sons 公司于 2015 年出版。

ORGANIZATIONAL PROJECT-ENABLING PROCESSES

7.1 LIFE CYCLE MODEL MANAGEMENT PROCESS

7.1.1 Overview

7.1.1.1 Purpose As stated in ISO/IEC/IEEE 15288,

[6.2.1.1] The purpose of the Life Cycle Model Management process is to define, maintain, and assure availability of policies, life cycle processes, life cycle models, and procedures for use by the organization with respect to the scope of [ISO/IEC/IEEE 15288].

The value propositions to be achieved by instituting organization-wide processes for use by projects are as follows:

- Provide repeatable/predictable performance across the projects in the organization (this helps the organization in planning and estimating future projects and in demonstrating reliability to customers)

- Leverage practices that have been proven successful by certain projects and instill those in other projects across the organization (where applicable)

- Enable process improvement across the organization

- Improve ability to efficiently transfer staff across projects as roles are defined and performed consistently

- Enable leveraging lessons that are learned from one project for future projects to improve performance and avoid issues

- Improve startup of new projects (less reinventing the wheel)

In addition, the standardization across projects may enable cost savings through economies of scale for support activities (tool support, process documentation, etc.).

7.1.1.2 Description This process (i) establishes and maintains a set of policies and procedures at the organization level that support the organization's ability to acquire and supply products and services and (ii) provides integrated system life cycle models necessary to meet the organization's strategic plans, policies, goals, and objectives for all projects and all system life cycle stages. The processes are defined, adapted, and maintained to support the requirements of the organization, SE organizational units, individual projects, and personnel. Life cycle model management processes are supplemented by recommended methods and tools. The resulting guidelines in the form of organization policies and procedures are still subject to tailoring by projects, as discussed in Chapter 8. Figure 7.1 is the IPO diagram for the life cycle model management process.

7.1 生命周期模型管理流程

7.1.1 概览

7.1.1.1 目的 如 ISO/IEC/IEEE 15288 所述，

[6.2.1.1] 生命周期模型管理流程的目的是定义、维护和确保组织在[ISO/IEC/IEEE 15288]的范围内所使用的方针、生命周期流程、生命周期模型和程序的可用性。通过制定项目所使用的组织范围内的流程来达成的价值主张，如下：

- 提供组织内跨多个项目的可重复/可预测的绩效（这可帮助组织计划和评估未来项目以及向客户展示可靠性）
- 更强有力地发挥已由某些项目证实是成功的实践的作用，并将其渗透到跨组织的其他项目中（如果适用）
- 使能跨组织的流程改进
- 随着角色被一致地定义和执行，提高跨项目高效地调动人员的能力
- 从一个项目中学到的可推而广之的经验教训能够用于未来的项目，以便提高绩效和避免出现问题
- 改善新项目的启动（不重复已有的）

此外，跨项目的标准化可通过支持活动（工具支持、流程文件等）的规模效益来促使成本节约。

7.1.1.2 描述 该流程（i）建立和维护可支持组织采办和供应产品与服务能力的一系列组织层级的方针和程序，并且（ii）提供满足所有项目和所有系统生命周期阶段的组织策略计划、方针、目的和目标所必需的综合的系统生命周期模型。对上述流程进行定义、调整以及维护，以支持组织、SE 组织单元、单独项目和人员的需求。生命周期模型管理流程由推荐的方法和工具补充。由此得出的组织方针和程序形式的指南仍是按项目进行剪裁的内容，如第 8 章所述。图 7.1 是生命周期模型管理流程的 IPO 图。

ORGANIZATIONAL PROJECT-ENABLING PROCESSES

7.1.1.3 Inputs/Outputs Inputs and outputs for the life cycle model management process are listed in Figure 7.1. Descriptions of each input and output are provided in Appendix E.

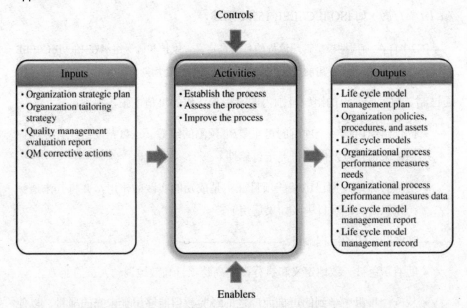

FIGURE 7.1 IPO diagram for the life cycle model management process. INCOSE SEH original figure created by Shortell and Walden. Usage per the INCOSE notices page. All other rights reserved.

7.1.1.4 Process Activities The life cycle model management process includes the following activities:

- *Establish the process.*
  - Identify sources (organization, corporate, industry, academia, stakeholders, and customers) of life cycle model management process information.
  - Distill the information from multiple sources into an appropriate set of life cycle models that are aligned with the organization and business area plans and infrastructure.
  - Establish life cycle model management guidelines in the form of plans, policies, procedures, tailoring guidance, models, and methods and tools for controlling and directing the life cycle models.

7.1.1.3 输入/输出 图 7.1 列出生命周期模型管理流程的输入和输出。附录 E 提供每个输入和输出的描述。

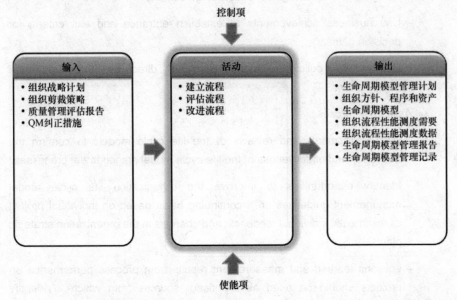

图 7.1 生命周期模型管理流程的 IPO 图。INCOSE SEH 原始图由 Shortell 和 Walden 创建。按照 INCOSE 通知页使用。版权所有。

7.1.1.4 流程活动 生命周期模型管理流程包括以下活动：

- 建立流程。

    - 识别生命周期模型管理流程信息的来源（组织、公司、工业界、学术界、利益攸关者和客户）。

    - 从多个来源中提取信息，形成一套适用的、与组织和业务域规划和基础设施协调一致的生命周期模型。

    - 建立计划、方针、程序、剪裁指导、模型和方法及工具形式的生命周期模型管理指南，用以控制和指导生命周期模型。

- Define, integrate, and communicate life cycle model roles, responsibilities, authorities, requirements, measures, and performance criteria based on the life cycle model management process guidelines.

- Use business achievements to establish entrance and exit criteria for decision gates.

- Disseminate policies, procedures, and directives throughout the organization.

- *Assess the process.*

    - Use assessments and reviews of the life cycle models to confirm the adequacy and effectiveness of the life cycle model management processes.

    - Identify opportunities to improve the organization life cycle model management guidelines on a continuing basis based on individual project assessments, individual feedback, and changes in the organization strategic plan.

    - Lessons learned and measurement results from process performance on projects should be used as significant sources from which to identify improvements.

- *Improve the process.*

    - Prioritize and implement the identified improvement opportunities.

    - Communicate with all relevant organizations regarding the creation of and changes in the life cycle model management guideline.

Common approaches and tips:

- Base the policies and procedures on an organization level strategic and business area plan that provides a comprehensive understanding of the organization's goals, objectives, stakeholders, competitors, future business, and technology trends.

- Ensure that policy and procedure compliance review is included as part of the business decision gate criteria.

— 基于生命周期模型管理流程指南，定义、综合并沟通生命周期模型的角色、职责、权限、需求、测度和绩效准则。

— 使用业务成果为决策门建立进入和退出准则。

— 贯穿于组织传播方针、程序和指令。

- 评估流程。

    — 使用生命周期模型的评估和评审来确认生命周期模型管理流程的充分性和有效性。

    — 基于单独项目评估、单独反馈和组织战略规划的变更来识别持续改进组织生命周期模型管理指南的机会。

    — 从项目的流程绩效中得到的经验教训和测量结果，应用于识别改进的重要来源。

- 改进流程。

    — 对识别的改进机会进行优先级排序并实施。

    — 针对生命周期模型管理指南的创建和变更与所有相关组织进行沟通。

常用方法和建议：

- 将策略和程序建立在提供对组织目的、目标、利益攸关者、竞争者、未来业务和技术趋势的全面了解的组织级的战略和业务域规划的基础上。

- 确保方针和程序的符合性评审作为业务决策门准则的一部分而被纳入。

- Develop a life cycle model management process information database with essential information that provides an effective mechanism for disseminating consistent guidelines and providing announcements about organization-related topics, as well as industry trends, research findings, and other relevant information. This provides a single point of contact for continuous communication regarding the life cycle model management guidelines and encourages the collection of valuable feedback and the identification of organization trends.

- Establish an organization center of excellence for life cycle model management processes. This organization can become the focal point for the collection of relevant information, dissemination of guidelines, and analysis of assessments and feedback. They can also develop checklists and other templates to support project assessments to ensure that the predefined measures and criteria are used for evaluation.

- Manage the network of external relationships by assigning personnel to identify standards, industry and academia research, and other sources of organization management information and concepts needed by the organization.

The network of relationships includes government, industry, and academia. Each of these external interfaces provides unique and essential information for the organization to succeed in business and meet the continued need and demand for improved and effective systems and products for its customers. It is up to the life cycle model management process to fully define and utilize these external entities and interfaces (i.e., their value, importance, and capabilities that are required by the organization):

- Legislative, regulatory, and other government requirements

- Industry SE and management-related standards, training, and capability maturity models

- Academic education, research results, future concepts and perspectives, and requests for financial support

- Establish an organization communication plan for the policies and procedures. Most of the processes in this handbook include dissemination activities. An effective set of communication methods is needed to ensure that all stakeholders are well informed.

组织的项目使能流程

- 开发具有基本信息的生命周期模型管理流程信息数据库，为传播一致的指南和提供与组织相关主题以及行业趋势、研究成果和其他相关信息有关的发布提供一个有效机制。这为与生命周期模型管理指南相关的持续沟通提供一个接触点，并鼓励收集有价值的反馈和识别组织趋势。

- 为生命周期模型管理流程建立一个卓越中心组织。该组织可成为收集相关信息、传播指南以及分析评估和反馈的聚焦点。上述组织还可以开发检查清单和其他模板以支持项目评估，从而确保采用预定的测度和准则进行评估。

- 通过指派人员识别组织所需要的标准、工业界和学术界的研究，以及组织管理信息和概念的其他来源，以管理外部关系网络。

关系网络包括政府、工业界和学术界。这些外部接口中的每个将会为组织提供独特的、基本的信息，以便在商业上取得成功，并满足其客户对改进的、有效的系统和产品的持续的需要和要求。根据生命周期模型管理流程来全面地定义和利用这些外部实体和接口（即：其价值、重要性和组织所需的能力）：

    – 法律、法规及其他政府需求

    – 行业 SE 和管理相关的标准、培训及能力成熟度模型

    – 学术教育、研究成果、未来概念和视角以及对财务支持的请求

- 为方针和程序建立组织沟通计划。本手册中大部分流程均包括传播活动。需要一套有效的沟通方法，以确保所有利益攸关者都能很好地得到信息。

- Ensure the methods and tools for enabling the application of life cycle model management processes are effective and tailored to the implementation approach of the organization and its projects. A responsible organization can be created or designated to coordinate the identification and development of partnerships and/or relationships with tool vendors and working groups. They can recommend the use of methods and tools that are intended to help personnel avoid confusion, frustration, and wasting valuable time and money. These experts may also establish an integrated tool environment between interacting tools to avoid cumbersome (and inaccurate) data transfer.

- Include stakeholders, such as engineering and project management organizations, as participants in developing the life cycle model management guidelines. This increases their commitment to the recommendations and incorporates a valuable source of organization experience.

- Develop alternative life cycle models based on the type, scope, complexity, and risk of a project. This decreases the need for tailoring by engineering and project organizations.

- Provide clear guidelines for tailoring and adaptation.

- Work continually to improve the life cycle models and processes.

7.1.2 Elaboration

7.1.2.1 Standard SE Processes An organization engaged in SE provides the requirements for establishing, maintaining, and improving the standard SE process and the policies, practices, and supporting functional processes (see Fig. 7.2) necessary to meet customer needs throughout the organization. Further, it defines the process for tailoring the standard SE process for use on projects and for making improvements to the project-tailored SE processes.

Organizational management must review and approve the standard SE process and changes to it. Organizations should consider establishing an SE Process group (SYSPG) to oversee SE process definition and implementation.

- 确保使生命周期模型管理流程应用的方法和工具有效且根据组织及其项目的实施方法进行剪裁。可创建或指定一个责任组织来协调与工具卖方和工作组的合作和/或关系的识别和发展。他们能够推荐使用旨在帮助人员避免困惑、沮丧和浪费宝贵的时间和资金的方法和工具。这些专家还可以在交互工具之间建立综合工具环境，以避免繁琐（和不准确）的数据传输。

- 在生命周期模型管理指南的开发过程中，纳入诸如工程和项目管理组织的利益攸关者作为参与者。这可加强其建议的承诺并纳入宝贵的组织经验来源。

- 基于项目的类型、范围、复杂性和风险来开发替代的生命周期模型。这可降低工程和项目组织剪裁的需要。

- 提供清晰的剪裁和调整指南。

- 持续地工作以改进生命周期模型和流程。

7.1.2 详细阐述

7.1.2.1 标准 SE 流程 一个介入 SE 的组织提供建立、维护和改进标准 SE 流程以及贯穿于组织内满足客户需要所必需的方针、实践和支持性功能流程（见图 7.2）的需求。此外，该组织还定义对在项目中使用的标准 SE 流程进行剪裁的流程以及对项目剪裁的 SE 流程进行改进的流程。

组织管理人员必须对标准 SE 流程及其变更进行评审和审批。组织应考虑建立一个 SE 流程组（SYSPG），以监督 SE 流程的定义和实施。

ORGANIZATIONAL PROJECT-ENABLING PROCESSES

An organization defines or selects a standard set of SE processes for use as a reference SE process model, which is tailored by projects to meet specific customer and stakeholder needs. The reference model should tailor industry, government, or other agency "best practices" based on multiple government, industry, and organization reference SE process documents. The reference SE model includes an SE process improvement requirements, activity, or process to ensure ongoing evaluation and improvement actions that take advantage of lessons learned. Projects are expected to follow this process, as tailored to meet project-specific SE process needs. The standard process must be tailorable, extensible, and scalable to meet a diverse range of projects, from small study projects to large projects requiring thousands of participants.

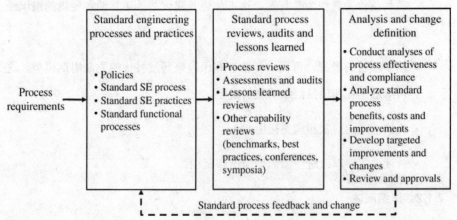

FIGURE 7.2 Standard SE process flow. INCOSE SEH v2 Figure 5.3. Usage per the INCOSE notices page. All other rights reserved.

The standard SE process model can be established by selection of specific processes and practices from this handbook, industry SE process references (such as ISO/ IEC/IEEE 15288 and ANSI/EIA-632), and government SE process references, as appropriate, and is applicable to every engineering capability maturity focus area or process area in the CMMI® approach (CMMI Product Team, 2010).

A high-performing organization also reviews the process (as well as work products), conducts assessments and audits (e.g., CMMI assessments and ISO audits), retains corporate memory through the understanding of lessons learned, and establishes how benchmarked processes and practices of related organizations can affect the organization. Successful organizations should analyze their process performance, its effectiveness and compliance to organizational and higher directed standards, and the associated benefits and costs and then develop targeted improvements.

组织定义或选择一个 SE 流程的标准集合,用作参考 SE 流程模型,在项目中对该模型进行剪裁以满足特定客户和利益攸关者的需要。参考模型应基于政府、行业和组织的多个参考 SE 流程文件来剪裁行业、政府或其他机构的"最佳实践"。参考 SE 模型包括 SE 流程改进需求、活动或流程以确保进行中的评估和改进行动利用了经验教训。期望项目遵循该流程,并进行剪裁以满足项目特有的 SE 流程需要。标准流程必须是可剪裁、可延伸和可扩展的,以满足各种各样的项目,从小型研究项目到需要成千上万个参与者的大型项目。

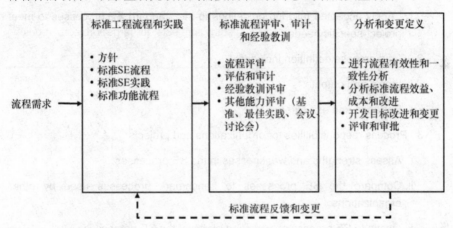

图 7.2 标准 SE 流程图。INCOSE SEH 第 2 版中图 5.3。按照 INCOSE 通知页使用。版权所有。

标准 SE 流程模型通过从本手册、行业 SE 流程参考(例如 ISO/IEC/IEEE 15288 和 ANSI/EIA-632)和政府 SE 流程参考中适当地选择特定流程和实践的方式来建立,并适用于 CMMI®方法中的每个工程能力成熟度关注域或流程域(CMMI 产品团队,2010)。

高绩效组织还评审该流程(以及工作产物),进行评估和审计(例如 CMMI 评估和 ISO 审计),通过理解经验教训来保留组织记忆,并建立能影响该组织的相关组织的基准流程和实践。成功的组织应分析其流程绩效、有效性及其与组织的更高指导标准的一致性以及相关效益和成本,然后进行有针对性的改进。

ORGANIZATIONAL PROJECT-ENABLING PROCESSES

The basic requirements for standard and project-tailored SE process control, based on CMMI and other resources, are as follows:

1. Process responsibilities for projects:

    i. Identify SE processes.

    ii. Document the implementation and maintenance of SE processes.

    iii. Use a defined set of standard methods and techniques to support the SE processes.

    iv. Apply accepted tailoring guidelines to the standard SE processes to meet project-specific needs.

2. Good process definition includes:

    i. Inputs and outputs.

    ii. Entrance and exit criteria.

3. Process responsibilities for organizations and projects:

    i. Assess strengths and weaknesses in the SE processes.

    ii. Compare the SE processes to benchmark processes used by other organizations.

    iii. Institute SE process reviews and audits of the SE processes.

    iv. Institute a means to capture and act on lessons learned from SE process implementation on projects.

    v. Institute a means to analyze potential changes for improvement to the SE processes.

    vi. Institute measures that provide insight into the performance and effectiveness of the SE processes.

    vii. Analyze the process measures and other information to determine the effectiveness of the SE processes.

Although it should be encouraged to identify and capture lessons learned throughout the performance of every project, the SE organization must plan and follow through to collect lessons learned at predefined milestones in the system life cycle. The SE organization should periodically review lessons learned together with the measures and other information to analyze and improve SE processes and practices. The results need to be communicated and incorporated into training. It should also establish best practices and capture them in an easy-to-retrieve form.

基于 CMMI 及其他资源，对标准和项目剪裁的 SE 流程控制的基本需求如下：

1. 项目流程职责：

    i. 识别 SE 流程。

    ii. 将 SE 流程的实施和维护文件化。

    iii. 使用定义的标准方法和技术集合来支持 SE 流程。

    iv. 将已接受的剪裁指南应用于标准 SE 流程以满足项目特有的需要。

2. 良好的流程定义包括：

    i. 输入和输出。

    ii. 进入和退出准则。

3. 组织和项目流程职责：

    i. 评估 SE 流程中的优势与劣势。

    ii. 应将 SE 流程与其他组织使用的基准流程进行比较。

    iii. 实行 SE 流程评审和 SE 流程审计。

    iv. 实行一种用于从项目的 SE 流程实施中获取经验教训并基于该经验教训行动的手段。

    v. 实行一种用于分析潜在的变更以改进 SE 流程的手段。

    vi. 实行提供对 SE 流程的绩效和有效性的深入理解的测度。

    vii. 分析流程测度和其他信息以确定 SE 流程的有效性。

尽管应当鼓励在每一项目的执行期间识别和获取经验教训，然而 SE 组织必须制定计划并持续跟进以便在系统生命周期内预定的里程碑上收集经验教训。SE 组织应定期评审经验教训连同测度和其他信息，以分析和改进 SE 流程和实践。需要对结果进行沟通并纳入培训。还应建立最佳实践并以易于检索的形式获取最佳实践。

For more information on process definition, assessment, and improvement, see the resources in the bibliography, including the CMMI.

7.2 INFRASTRUCTURE MANAGEMENT PROCESS

7.2.1 Overview

7.2.1.1 Purpose As stated in ISO/IEC/IEEE 15288,

> [6.2.2.1] The purpose of the Infrastructure Management process is to provide the infrastructure and services to projects to support organization and project objectives throughout the life cycle.

7.2.1.2 Description The work of the organization is accomplished through projects, which are conducted within the context of the infrastructure environment. This infrastructure needs to be defined and understood within the organization and the project to ensure alignment of the working units and achievement of overall organization strategic objectives. This process exists to establish, communicate, and continuously improve the system life cycle process environment. Figure 7.3 is the IPO diagram for the infrastructure management process.

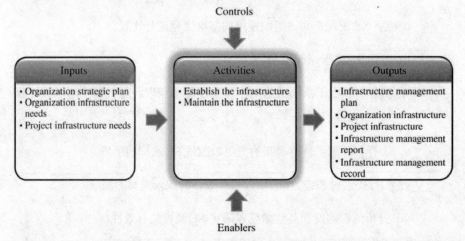

FIGURE 7.3 IPO diagram for the infrastructure management process. INCOSE SEH original figure created by Shortell and Walden. Usage per the INCOSE notices page. All other rights reserved.

关于流程定义、评估和改进的更多信息，参见参考书目中的资源，包括 CMMI。

7.2 基础设施管理流程

7.2.1 概览

7.2.1.1 目的 如 ISO/IEC/IEEE 15288 所述，

[6.2.2.1] 贯穿于生命周期中，基础设施管理流程的目的是为项目提供基础设施和服务，以支持组织和项目目标。

7.2.1.2 描述 通过在基础设施环境的背景环境内进行的项目来完成组织的工作。该基础设施需要在组织和项目内被定义和理解，以确保工作单元与整体组织策略目标的达成协调一致。该流程的存在是用于建立、沟通和持续改进系统生命周期流程环境。图 7.3 是基础设施管理流程的 IPO 图。

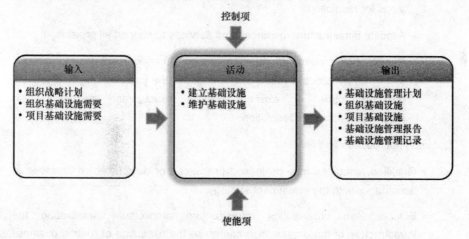

图 7.3 基础设施管理流程的 IPO 图。INCOSE SEH 原始图由 Shortell 和 Walden 创建。按照 INCOSE 通知页使用。版权所有。

ORGANIZATIONAL PROJECT-ENABLING PROCESSES

7.2.1.3 Inputs/Outputs Inputs and outputs for the infrastructure management process are listed in Figure 7.3. Descriptions of each input and output are provided in Appendix E.

7.2.1.4 Process Activities The infrastructure management process includes the following activities:

- *Establish the infrastructure.*
  - Gather and negotiate infrastructure resource needs with organization and projects.
  - Establish the infrastructure resources and services to ensure organization goals and objectives are met.
  - Manage resource and service conflicts and shortfalls with steps for resolution.

- *Maintain the infrastructure.*
  - Manage infrastructure resource availability to ensure organization goals and objectives are met. Conflicts and resource shortfalls are managed with steps for resolution.
  - Allocate infrastructure resources and services to support all projects.
  - Control multi-project infrastructure resource management communications to effectively allocate resources throughout the organization, and identify potential future or existing conflict issues and problems with recommendations for resolution.

Common approaches and tips:

- Qualified resources may be leased (insourced or outsourced) or licensed in accordance with the investment strategy.
- Establish an organization infrastructure architecture. Integrating the infrastructure of the organization can make the execution of routine business activities more efficient.
- Establish a resource management information system with enabling support systems and services to maintain, track, allocate, and improve the resources for present and future organization needs. Computer-based equipment tracking, facilities allocation, and other systems are recommended for organizations with over 50 people.

7.2.1.3 输入/输出 图 7.3 列出基础设施管理流程的输入和输出。附录 E 提供每个输入和输出的描述。

7.2.1.4 流程活动 基础设施管理流程包括以下活动：

- *建立基础设施*。

    – 汇集并协商与组织和项目相关的基础设施资源需要。

    – 建立基础设施资源和服务，以确保满足组织目的和目标。

    – 为解决问题分步骤地管理资源和服务的冲突与不足。

- *维护基础设施*。

    – 管理基础设施资源可用性，以确保满足组织目的和目标。为解决问题分步骤地管理冲突和资源不足。

    – 分配基础设施资源和服务，以支持所有项目。

    – 控制多项目基础设施资源管理沟通，以便贯穿于组织内有效地分配资源并识别潜在的未来或现有冲突问题和难题，同时在解决时提供建议。

常用方法和建议：

- 合格的资源可依据投资策略的租借（内包或外包）或授权使用。

- 建立组织基础设施架构。综合组织基础设施可使日常业务活动的执行更加高效。

- 建立具有使能支持系统和服务的一个资源管理信息系统，以维护、跟踪、分配和改善资源以便满足当前和未来的组织需要。建议将基于计算机的设备跟踪、设施分配和其他系统用于超过 50 人的组织。

ORGANIZATIONAL PROJECT-ENABLING PROCESSES

- Attend to physical factors, including facilities and human factors, such as ambient noise level and computer access to specific tools and applications.
- Begin planning in early life cycle stages of all system development efforts to address utilization and support resource requirements for system transition, facilities, infrastructure, information/ data storage, and management. Enabling resources should also be identified and integrated into the organization's infrastructure.

7.2.2 Elaboration

7.2.2.1 Infrastructure Management Concepts Projects all need resources to meet their objectives. Project planners determine the resources needed by the project and attempt to anticipate both current and future needs. The infrastructure management process provides the mechanisms whereby the organization infrastructure is made aware of project needs and the resources are scheduled to be in place when requested. While this can be simply stated, it is less simply executed. Conflicts must be negotiated and resolved, equipment must be obtained and sometimes repaired, buildings need to be refurbished, and information technology services are in a state of constant change. The infrastructure management organization collects the needs, negotiates to remove conflicts, and is responsible for providing the enabling organization infrastructure without which nothing else can be accomplished. Since resources are not free, their costs are also factored into investment decisions. Financial resources are addressed under the portfolio management process, but all other resources, except for human resources (see Section 7.4), are addressed under this process.

Infrastructure management is complicated by the number of sources for requests, the need to balance the skills of the labor pool against the other infrastructure elements (e.g., computer-based tools), the need to maintain a balance between the budgets of individual projects and the cost of resources, the need to keep apprised of new or modified policies and procedures that might influence the skills inventory, and myriad unknowns.

Resources are allocated based on requests. Infrastructure management collects the needs of all the projects in the active portfolio and schedules or acquires nonhuman assets, as needed. Additionally, the infrastructure management process maintains and manages the facilities, hardware, and support tools required by the portfolio of organization projects. Infrastructure management is the efficient and effective deployment of an organization's resources when and where they are needed. Such resources may include inventory, production resources, or information technology. The goal is to provide materials and services to a project when they are needed to keep the project on target and on budget. A balance should be found between efficiency and robustness. Infrastructure management relies heavily on forecasts into the future of the demand and supply of various resources.

- 注意物理因素，包括设施和人为因素，如环境噪声级及特定工具和应用的计算机接入。
- 在所有系统开发工作的生命周期早期阶段开始计划，以应对系统转移、设备、基础设施、信息/数据存储和管理的使用和支持资源需求。使能资源还应被识别并综合到组织的基础设施中。

7.2.2 详细阐述

7.2.2.1 基础设施管理概念 所有项目均需要资源以满足其目标。项目计划人员确定项目所需的资源并试图预测当前和未来需要。基础设施管理流程提供使组织基础设施知悉项目需要，并在申请时将资源安排到位的机制。此机制虽然可以简单地陈述，但执行却不简单。必须协商并解决冲突，必须获得设备并偶尔会修理，建筑需要翻新，以及信息技术服务应处于不断变化的状态。基础设施管理组织收集需要，进行协商以消除冲突，并负责提供使能组织基础设施，没有该基础设施，任何事情均无法实现。由于资源并不是免费的，因此其成本也要纳入到投资决策的考虑因素中。在项目组合管理流程中应对财务资源，但该流程也应对除人力资源（见 7.4 节）外的所有其他资源。

由于请求来源的数量、对劳动力群体的技能与其他基础设施元素（如基于计算机的工具）进行平衡的需要，使单独项目预算与资源成本之间保持平衡的需要、时刻了解可能影响技能储备的新的或改进的方针和程序的需要以及无数未知因素，使基础设施管理复杂化。

基于请求来分配资源。需要时，基础设施管理收集进行中项目组合所有项目的需要并调度或采办非人力资产。此外，基础设施管理流程维护并管理组织项目的项目组合所需的设施、硬件和支持工具。基础设施管理是在需要的时间和地点对组织资源的高效且有效的部署。这种资源可包括储备、生产资源或信息技术。目的是在需要时为项目提供材料和服务，以便将项目保持在目标和预算内。应在效率和鲁棒性之间找到一种平衡。基础设施管理在很大程度上依赖于未来不同资源的需求和供应的预测。

ORGANIZATIONAL PROJECT-ENABLING PROCESSES

The organization environment and subsequent investment decisions are built on the existing organization infrastructure, including facilities, equipment, personnel, and knowledge. Efficient use of these resources is achieved by exploiting opportunities to share enabling systems or to use a common system element on more than one project. These opportunities are enabled by good communications within the organization. Integration and interoperability of supporting systems, such as financial, human resources (see Section 7.4), and training, is critically important to executing organization strategic objectives. Feedback from active projects is used to refine and continuously improve the infrastructure.

Further, trends in the market may suggest changes in the supporting environment. Assessment of the availability and suitability of the organization infrastructure and associated resources provides feedback for improvement and reward mechanisms. All organization processes require mandatory compliance with government and corporate laws and regulations. Decision making is governed by the organization strategic plan.

7.3 PORTFOLIO MANAGEMENT PROCESS

7.3.1 Overview

7.3.1.1 Purpose As stated in ISO/IEC/IEEE 15288,

> [6.2.3.1] The purpose of the Portfolio Management process is to initiate and sustain necessary, sufficient and suitable projects in order to meet the strategic objectives of the organization.

Portfolio management also provides organizational output regarding the set of projects, systems, and technical investments of the organization to external stakeholders, such as parent organizations, investors/funding sources, and governance bodies.

7.3.1.2 Description Projects create the products or services that generate income for an organization. Thus, the conduct of successful projects requires an adequate allocation of funding and resources and the authority to deploy them to meet project objectives. Most business entities manage the commitment of financial resources using well-defined and closely monitored processes.

The portfolio management process also performs ongoing evaluation of the projects and systems in its portfolio. Based on periodic assessments, projects are determined to justify continued investment if they have the following characteristics:

组织环境及随后的投资决策建立在现有组织基础设施上，包括设施、设备、人员和知识。通过寻求机会来分享使能系统或在一个以上项目中使用公共系统元素的方式来实现这些资源的高效使用。这些机会通过组织内的良好沟通来获得。支持系统的集成和互操作，如财务资源、人力资源（见 7.4 节）和培训，对于执行组织策略目标至关重要。用进行中的项目的反馈来完善并持续改进基础设施。

此外，市场趋势可能意味着支持性环境的变化。对组织基础设施和相关资源的可用性和适用性的评估为改进和激励机制提供反馈。所有组织流程都要求必须符合政府和企业的法律和法规。决策取决于组织战略计划。

7.3 项目群管理流程

7.3.1 概览

7.3.1.1 目的 如 ISO/IEC/IEEE 15288 所述，

> [6.2.3.1] 项目群管理流程的目的是启动并维持必要的、充分的且合适的项目，以便满足组织的战略目标。

项目群管理还向外部利益攸关者，如母体组织、投资者/资金来源和治理机构，提供与组织的项目、系统和技术投资的集合有关的组织输出。

7.3.1.2 描述 项目创造为组织产生收益的产品或服务。因此，成功项目的实施需要合理地分配资金和资源以及权限以便对它们进行部署，从而满足项目目标。大部分业务实体均利用明确定义并严密监控的流程来管理财务资源的投入。

项目群管理流程还对其项目群中的项目和系统进行持续评价。如果项目具有以下特征，基于定期评估对其进行确定，以证明持续投资的合理性：

- Contribute to the organization strategy
- Progress toward achieving established goals
- Comply with project directives from the organization
- Are conducted according to an approved plan
- Provide a service or product that is still needed and providing acceptable investment returns

Otherwise, projects may be redirected or, in extreme instances, cancelled. Figure 7.4 is the IPO diagram for the portfolio management process.

7.3.1.3 Inputs/Outputs Inputs and outputs for the portfolio management process are listed in Figure 7.4. Descriptions of each input and output are provided in Appendix E.

7.3.1.4 Process Activities The portfolio management process includes the following activities:

- *Define and authorize projects.*
  - Identify, assess, and prioritize investment opportunities consistent with the organization strategic plan.
  - Establish business area plans—use the strategic objectives to identify candidate projects to fulfill them.
  - Establish project scope, define project management accountabilities and authorities, and identify expected project outcomes.
  - Establish the domain area of the product line defined by its main features and their suitable variability.
  - Allocate adequate funding and other resources.
  - Identify interfaces and opportunities for multi-project synergies.

- 有助于组织战略

- 向着实现既定目标进展

- 符合组织的项目指南

- 按照批准的计划实施

- 提供依然需要的且提供可接受的投资回报的产品或服务

否则，项目可重新定向或在极端的情况下被取消。图 7.4 是项目群管理流程的 IPO 图。

7.3.1.3　*输入/输出* 　图 7.4 列出项目群管理流程的输入和输出。附录 E 提供每个输入和输出的描述。

7.3.1.4　*流程活动* 　项目群管理流程包括以下活动：

- 定义和授权项目。

    - 识别并评估与组织战略计划一致的投资机会并进行优先级排序。

    - 建立业务域计划——使用战略目标来识别候选项目以实施这些项目。

    - 建立项目范围，定义项目管理责任和权限，并识别预期项目结果。

    - 建立根据产品线的主要特征及其合适的可变性所定义的产品线的领域范围。

    - 分配充足的资金和其他资源。

    - 识别多项目协同的界面和机会。

ORGANIZATIONAL PROJECT-ENABLING PROCESSES

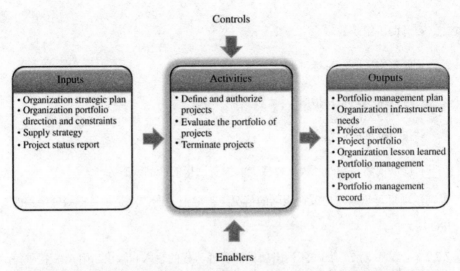

FIGURE 7.4 IPO diagram for the portfolio management process. INCOSE SEH original figure created by Shortell and Walden. Usage per the INCOSE notices page. All other rights reserved.

- Specify the project governance process including organizational status reporting and reviews.
- Authorize project execution.
- *Evaluate the portfolio of projects.*
  - Evaluate ongoing projects to provide rationale for continuation, redirection, or termination.
- *Terminate projects.*
  - Close, cancel, or suspend projects that are completed or designated for termination.

Common approaches and tips:

- The process of developing the business area plans helps the organization assess where it needs to focus resources to meet present and future strategic objectives. Include representatives from relevant stakeholders in the organization community.
- When investment opportunities present themselves, prioritize them based on measurable criteria such that projects can be objectively evaluated against a threshold of acceptable performance.

组织的项目使能流程

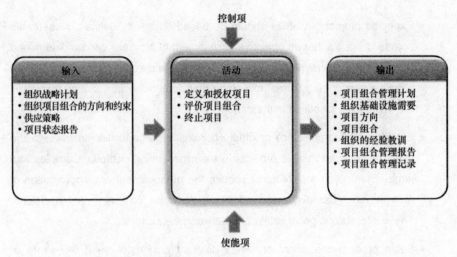

图 7.4 项目群管理流程的 IPO 图。INCOSE SEH 原始图由 Shortell 和 Walden 创建。按照 INCOSE 通知页使用。版权所有。

- 规定项目治理流程，包括组织状态的报告和评审。

- 授权项目执行。

- *评价项目群*。

- 评价进行中的项目，以便为持续、重新定向或终止提供依据。

- *终止项目*。

- 结束、取消或暂停已完成或指定终止的项目。

常用方法和建议：

- 开发业务域计划的流程帮助组织评估何处需要集中资源，以满足当前和未来的战略目标。包括组织团体中相关的利益攸关者的代表。

- 当投资机会出现时，基于可衡量准则排列优先级，以便依据可接受的性能阈值对项目进行客观评价。

- Expected project outcomes should be based on clearly defined, measurable criteria to ensure that an objective assessment of progress can be determined. Specify the investment information that will be assessed for each milestone. Initiation should be a formal milestone that does not occur until all resources are in place as identified in the project plan.

- Establish a program office or other coordination organization to manage the synergies between active projects in the organization portfolio. Complex and large organization architectures require the management and coordination of multiple interfaces and make additional demands on investment decisions. These interactions occur within and between the projects.

- Set a product line approach when different customers need the same or similar systems (i.e., common features), with some customizations (i.e., variants).

- Include risk assessments (see Section 5.4) in the evaluation of ongoing projects. Projects that contain risks that may pose a challenge in the future might require redirection. Cancel or suspend projects whose disadvantages or risks to the organization outweigh the investment.

- Include opportunity assessments in the evaluation of ongoing projects. Addressing project challenges may represent a positive investment opportunity for the organization. Avoid pursuing opportunities that are inconsistent with the capabilities of the organization and its strategic goals and objectives or contain unacceptably high technical risk, resource demands, or uncertainty.

- Allocate resources based on the requirements of the projects; otherwise, the risk of cost and schedule overruns may have a negative impact on quality and performance of the project.

- Establish effective governance processes that directly support investment decision making and communications with project management.

- 预期项目结果应基于明确定义的可衡量的准则,以确保可以确定对进展的客观评估。规定每个里程碑将要评估的投资信息。启动应是一个在项目计划中所识别的所有资源均到位前才会出现的正式里程碑。

- 建立一个项目计划办公室或其他协调组织,以管理组织项目群中当前项目之间的协同。复杂的大型组织架构需要多个界面的管理和协调并对投资决策提出附加要求。这些交互作用发生在项目内和项目之间。

- 当不同的客户需要相同的或类似的系统(即公共特征)以及一些定制(即差异)时设置产品线方法。

- 将风险评估包含到正在进行中的项目的评价中(见 5.4 节)。含有可能在未来面临挑战风险的项目可能需要重新定向。取消或暂停对于组织而言是劣势的或风险超出投资的项目。

- 将机会评估包含到正在进行中的项目的评价中。对于组织来说,应对项目挑战可能代表积极的投资机会。避免寻求与组织能力及其战略目的和目标不一致或包含高得无法接受的技术风险、资源需求或不确定性的机会。

- 基于项目的需求来分配资源;否则,成本超支和进度延期的风险可能会对项目的质量和绩效产生负面影响。

- 建立直接支持投资决策和与项目管理人员沟通的有效治理流程。

7.3.2 Elaboration

7.3.2.1 Define the Business Case and Develop Business Area Plans Portfolio management balances the use of financial assets within the organization. Organization management generally demands that there be some beneficial return for the effort expended in pursuing a project. The business case and associated business area plans establish the scope of required resources (e.g., people and money) and schedule and set reasonable expectations. An important element of each control design gate is a realistic review of the business case as the project matures. The result is reverification or perhaps restatement of the business case. The Iridium case, described in Section 3.2, illustrates the dangers of failing to keep a realistic perspective. Similarly, despite the technological triumph of implementing the world's first maglev train line (Section 3.6.3), the exorbitant initial cost and slow return on investment are causing authorities to question plans to build another line.

The business case may be validated in a variety of ways. For large projects, sophisticated engineering models, or even prototypes of key system elements, help prove that the objectives of the business case can be met and that the system will work as envisioned prior to committing large amounts of resources to full-scale engineering and manufacturing development. For very complex systems, such a demonstration can be conducted at perhaps 20% of development cost. For smaller projects, when the total investment is modest, proof-of-concept models may be constructed during the concept stage to prove the validity of business case assumptions.

Investment opportunities are not all equal, and organizations are limited in the number of projects that can be conducted concurrently. Further, some investments are not well aligned with the overall strategic plan of the organization. For these reasons, opportunities are evaluated against the portfolio of existing agreements and ongoing projects, taking into consideration the attainability of the stakeholders' requirements.

7.4 HUMAN RESOURCE MANAGEMENT PROCESS

7.4.1 Overview

7.4.1.1 Purpose As stated in ISO/IEC/IEEE 15288,

> [6.2.4.1] The purpose of the Human resource Management process is to provide the organization with necessary human resources and to maintain their competencies, consistent with business needs.

7.3.2 详细阐述

7.3.2.1 定义商业案例并开发业务区域计划 项目群管理对组织内财务资产的使用进行平衡。组织管理通常要求为追求项目所做的努力得到利益回报。业务案例和相关业务域计划建立所需资源（如人员和资金）和进度的范围，并设定合理的期望。随着项目的成熟，每个控制设计门的重要元素是业务案例的实际评审事项。结果是对业务案例进行重新验证或可能进行重新描述。3.2 节中所述的铱星案例阐明未能保持现实视角的危险。同样，尽管在实施世界第一条磁悬浮列车线上取得了技术成就（见 3.6.3 节），但过高的初始成本和缓慢的投资回报使得有关部门对建设另一条列车线的计划产生了怀疑。

业务案例可能以各种方式得到确认。对于大型项目，复杂的工程模型，或甚至是关键系统元素的原型有助于证明可满足商业案例的目标，并证明系统将会在将大量资源投入到大规模工程和制造开发之前按预想结果运行。对于非常复杂的系统，此类演示将以大概百分之二十的开发成本来实施。对于较小项目，如果总投资适中，可在概念阶段构建概念验证模型，以证明业务案例假设的有效性。

投资机会并非完全相等，且组织可并行实施的项目数量有限。此外，一些投资与组织的总体战略计划并不完全一致。由于这些原因，在考虑到利益攸关者需求的可实现性的情况下，针对现有协议和进展中的项目的项目群对机会进行评价。

7.4 人力资源管理流程

7.4.1 概览

7.4.1.1 目的 如 ISO/IEC/IEEE 15288 所述，

[6.2.4.1] 人力资源管理流程的目的是为组织提供必要的人力资源并维持其能力与业务需要相一致。

ORGANIZATIONAL PROJECT-ENABLING PROCESSES

7.4.1.2 Description Projects all need resources to meet their objectives. This process deals with human resources. Nonhuman resources, including tools, data-bases, communication systems, financial systems, and information technology, are addressed using the infrastructure management process (see Section 7.2).

Project planners determine the resources needed for the project by anticipating both current and future needs. The human resource management process provides the mechanisms whereby the organization management is made aware of project needs and personnel are scheduled to be in place when requested. While this can be simply stated, it is less simply executed. Conflicts must be resolved, personnel must be trained, and employees are entitled to vacations and time away from the job.

The human resource management organization collects the needs, negotiates to remove conflicts, and is responsible for providing the personnel without which nothing else can be accomplished. Since qualified personnel are not free, their costs are also factored into investment decisions. Figure 7.5 is the IPO diagram for the human resource management process.

7.4.1.3 Inputs/Outputs Inputs and outputs for the human resource management process are listed in Figure 7.5. Descriptions of each input and output are provided in Appendix E.

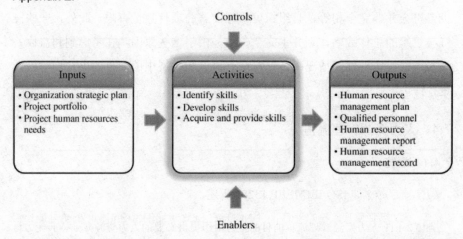

FIGURE 7.5 IPO diagram for the human resource management process. INCOSE SEH original figure created by Shortell and Walden. Usage per the INCOSE notices page. All other rights reserved.

7.4.1.2 描述 所有项目均需要资源以实现其目的。该流程处理人力资源。非人力资源,包括工具、数据库、通信系统、财务系统和信息技术是使用基础设施管理流程来处理的(见 7.2 节)。

项目计划者通过预计当前和未来的需要来确定项目所需的资源。人力资源管理流程提供使组织管理层知悉项目需要并在请求时将人员调度到位的机制。此机制虽然可以简单地被陈述,但执行却不简单。必须解决冲突,必须对人员进行培训,且员工有权享有假期和离岗时间。

人力资源管理组织收集需要,进行协商以消除冲突,并负责提供人员,没有这类人员,任何事情均无法完成。由于合格人员并不是免费的,因此其成本也要纳入到投资决策的考虑因素中。图 7.5 是人力资源管理流程的 IPO 图。

7.4.1.3 输入/输出 图 7.5 列出人力资源管理流程的输入和输出。附录 E 提供每个输入和输出的描述。

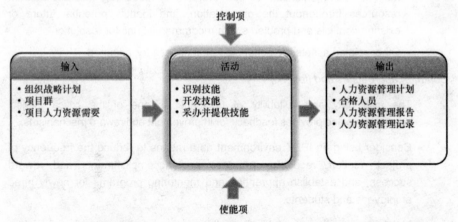

图 7.5 人力资源管理流程的 IPO 图。INCOSE SEH 原始图由 Shortell 和 Walden 创建。按照 INCOSE 通知页使用。版权所有。

497

ORGANIZATIONAL PROJECT-ENABLING PROCESSES

7.4.1.4 Processes Activities The human resource management process includes the following activities:

- *Identify skills.*
  - The skills of the existing personnel are identified to establish a "skills inventory."
  - Review current and anticipated projects to determine the skills needed across the portfolio of projects.
  - Skill needs are evaluated against available people with the prerequisite skills to determine if training or hiring activities are indicated.

- *Develop skills.*
  - Obtain (or develop) and deliver training to close identified gaps of project personnel.
  - Identify assignments that lead toward career progression.

- *Acquire and provide skills.*
  - Provide human resources to support all projects.
  - Train or hire qualified personnel when gaps indicate that skill needs cannot be met with existing personnel.
  - Maintain communication across projects to effectively allocate human resources throughout the organization, and identify potential future or existing conflicts and problems with recommendations for resolution.
  - Other related assets are scheduled or, if necessary, acquired.

Common approaches and tips:

- The availability and suitability of personnel is one of the critical project assessments and provides feedback for improvement and reward mechanisms.

- Consider using an IPDT environment as a means to reduce the frequency of project rotation, recognize progress and accomplishments and reward success, and establish apprentice and mentoring programs for newly hired employees and students.

- Maintain a pipeline of qualified candidates that are interested in joining the organization as employees or temporary staff. Focus recruitment, training, and retention efforts on personnel with experience levels, skills, and subject matter expertise demanded by the projects. Personnel assessments should review proficiency, motivation, and ability to work in a team environment, as well as the need to be retrained, reassigned, or relocated.

7.4.1.4 *流程活动* 人力资源管理流程包括以下活动：

- *识别技能。*
  - 识别现有人员的技能以建立"技能储备"。
  - 评审当前的和预期的项目以确定跨项目群所需的技能。
  - 针对具有必备技能的可用人员来评价技能需要，以确定培训或招聘活动是否明确。
- *开发技能。*
  - 获得（或开发）并提供培训，以弥补已识别到的项目人员缺口。
  - 识别出导向职业发展的分配任务。
- *采办并提供技能。*
  - 提供人力资源以支持所有项目。
  - 当缺口表明现有人员无法满足技能需要时，培训或雇用合格的人员。
  - 跨项目保持沟通以便贯穿于组织有效地分配人力资源，并识别潜在的未来或现有的冲突和问题，为解决方案提出建议。
  - 将其他相关资产列入计划，或在必要时采办所述资产。

常用方法和建议：

- 人员的可用性和适用性是关键的项目评估之一，并为改进和激励机制提供反馈。
- 考虑将 IPDT 环境用作降低项目轮换频率的手段，认同进步和成绩并奖励成功，并为新雇用的员工和学生建立实习和指导方案。
- 对有兴趣加入到组织作为员工或临时员工的合格候选人员维护一个通道。注重对具有项目所要求的经验水平、技能和专业知识的人员的招聘、培训和留用工作。人员评估应评审熟练水平、动机和在团队环境中的工作能力，以及再培训、再分配或再安置的需要。

ORGANIZATIONAL PROJECT-ENABLING PROCESSES

- Personnel are allocated based on requests and conflicts are negotiated. The goal is to provide personnel to a project when they are needed to keep the project on target and on budget.

- A key concern is keeping project personnel from becoming overcommitted, especially persons with specialized skills.

- Skills inventory and career development plans are important documentation that can be validated by engineering and project management.

- Qualified personnel and other resources may be hired temporarily—insourced or outsourced in accordance with the organizational strategy.

- Encourage personnel to engage in external networks as a means of keeping abreast of new ideas and attracting new talent to the organization.

- Maintain an organization career development program that is not sidetracked by project demands. Develop a policy that all personnel receive training or educational benefits on a regular cycle. This includes both undergraduate and graduate studies, in-house training courses, certifications, tutorials, workshops, and conferences.

- Remember to provide training on organization policies and procedures and system life cycle processes.

- Establish a resource management information infrastructure with enabling support systems and services to maintain, track, allocate, and improve the resources for present and future organization needs. Computer-based human resource allocation and other systems are recommended for organizations over 50 people.

- Use the slack time in the beginning of a project to obtain and train the necessary people to avoid a shortfall of skilled engineers, technologists, managers, and operations experts.

- Trends in the market may suggest changes in the composition of project teams and the supporting IT environment.

- All organization processes require mandatory compliance with government and corporate laws and regulations.

- Employee performance reviews should be conducted regularly, and career development plans should be managed and aligned to the objectives of both the employee and the organization. Career development plans should be reviewed, tracked, and refined to provide a mechanism to help manage the employee's career within the organization.

- 基于请求来分配人员并协商冲突。目的是在需要时为项目提供人员，使项目保持在目标和预算内。

- 一个关键问题是避免项目人员尤其是具有专门技能的人员过量使用。

- 技能储备和职业发展计划是可由工程和项目管理人员确认的重要文件。

- 合格人员和其他资源可根据组织策略以内包或外包的形式临时雇用。

- 鼓励人员将外部网络用作一种时刻了解新理念并向组织吸引新人才的手段。

- 维持不受项目需求牵引的组织职业发展计划。开发所有人员定期接受培训或教育收益的方针。这包括本科课程和研究生课程、内部培训课程、认证、指导教程、研讨会和会议。

- 记住提供关于组织方针和程序以及系统生命周期流程的培训。

- 建立具有使能支持系统和服务的资源管理信息基础设施，以维持、跟踪、分配和改进资源以便满足当前的和未来的组织需要。建议超过 50 人的组织使用基于计算机的人力资源分配和其他系统。

- 利用项目初期的空闲时间来获得并培训必需人员，以避免有技能的工程师、技术人员、管理人员和运营专家的短缺。

- 市场趋势可能意味着项目团队和 IT 支持环境构成的变化。

- 所有组织流程都要求必须符合政府和企业的法律和法规。

- 员工绩效评审应定期实施，且职业发展计划应得以管理并与员工和组织的目标对准。应对职业发展计划进行评审、跟踪和改进，以提供有助于在组织内管理员工职业的机制。

7.4.2 Elaboration

7.4.2.1 Human Resource Management Concepts

The human resource management process maintains and manages the people required by the portfolio of organization projects. Human resource management is the efficient and effective deployment of qualified personnel when and where they are needed. A balance should be found between efficiency and robustness. Human resource management relies heavily on forecasts into the future of the demand and supply of various resources.

The primary objective of this process is to provide a pool of qualified personnel to the organization. This is complicated by the number of sources for requests, the need to balance the skills of the labor pool against the other infrastructure elements (e.g., computer-based tools), the need to maintain a balance between the budgets of individual projects and the cost of resources, the need to keep apprised of new or modified policies and procedures that might influence the skills inventory, and myriad unknowns.

Project managers face their resource challenges competing for scarce talent in the larger organization pool. They must balance access to the experts they need for special studies with stability in the project team with its tacit knowledge and project memory. Today's projects depend on teamwork and optimally multidisciplinary teams. Such teams are able to resolve project issues quickly through direct communication between team members. Such intrateam communication shortens the decision-making cycle and is more likely to result in improved decisions because the multidisciplinary perspectives are captured early in the process.

7.5 QUALITY MANAGEMENT PROCESS

7.5.1 Overview

7.5.1.1 Purpose   As stated in ISO/IEC/IEEE 15288,

> [6.2.5.1] The purpose of the Quality Management process is to assure that products, services and implementations of the quality management process meet organizational and project quality objectives and achieve customer satisfaction.

7.4.2 详细阐述

7.4.2.1 人力资源管理概念

人力资源管理流程维持并管理组织的项目群所需的人员。人力资源管理是在需要的时间和地点对合格的人员进行高效且有效的部署。应在效率和鲁棒性之间找到一种平衡。人力资源管理在很大程度上依赖于对各种不同资源的需求与供应的前景预测。

该流程的主要目的是向组织提供合格的人员储备。由于请求来源的数量、对劳动力储备的技能与其他基础设施元素（如基于计算机的工具）进行平衡的需要，维持单独项目预算与资源成本之间平衡的需要、时刻了解可能影响技能储备的新的或改进的策略和程序的需要以及无数未知因素，使该流程复杂化。

项目经理面临着在较大的组织储备中争夺稀缺人才的资源挑战。他们必须对特定研究中所需专家的使用与具有其隐性知识和项目经验的项目团队的稳定性进行平衡。现今的项目依靠团队合作和最佳的多学科团队。这种团队能够通过团队成员之间的直接沟通快速地解决项目问题。由于在流程初期就获得了多学科的视角，这种团队内的沟通将缩短决策周期并更可能形成改善的决策。

7.5 质量管理流程

7.5.1 概览

7.5.1.1 目的 如 ISO/IEC/IEEE 15288 所述，

[6.2.5.1] 质量管理流程的目的是确保产品、服务和质量管理流程的实施满足组织的和项目的质量目标并达到客户满意。

ORGANIZATIONAL PROJECT-ENABLING PROCESSES

7.5.1.2 Description The quality management process makes visible the goals of the organization toward customer satisfaction. Since primary drivers in any project are time, cost, and quality, inclusion of a quality management process is essential to every organization. Many of the system life cycle processes are concerned with quality issues, and this forms some of the justification for exerting time, money, and energy into establishing these processes in the organization. Application of this handbook is one approach toward inserting a quality discipline into an organization.

Quality management (QM) process establishes, implements, and continuously improves the focus on customer satisfaction and organization goals and objectives. There is a cost to managing quality as well as a benefit. The effort and time required to manage quality should not exceed the overall value gained from the process. Figure 7.6 is the IPO diagram for the quality management process.

7.5.1.3 Inputs/Outputs Inputs and outputs for the quality management process are listed in Figure 7.6. Descriptions of each input and output are provided in Appendix E.

7.5.1.4 Process Activities The quality management process includes the following activities:

- *Plan quality management.*
  - Identify, assess, and prioritize quality guidelines consistent with the organization strategic plan. Establish QM guidelines—policies, standards, and procedures.
  - Establish organization and project QM goals and objectives.
  - Establish organization and project QM responsibilities and authorities.
- *Assess quality management.*
  - Evaluate project assessments.
  - Assess customer satisfaction against compliance with requirements and objectives.
  - Continuously improve the QM guidelines.
- *Perform quality management corrective action and preventive action.*
  - Recommend appropriate action, when indicated.
  - Maintain open communications—within the organization and with stakeholders.

7.5.1.2 描述 质量管理流程使追求客户满意度的组织的目标清晰可见。由于任何项目中的主要驱动因素都是时间、成本和质量，因此包含质量管理流程对于每个组织而言都是必要的。许多系统生命周期流程都涉及质量问题，这样就形成了在组织内建立这些流程中投入时间、资金和精力的一些正当理由。本手册的应用是向组织内引入质量规程的一种途径。

质量管理（QM）流程建立、实施并持续改善对顾客满意度及组织目的和目标的关注度。管理质量既有代价也有收益。管理质量所需的工作和时间应不超过从该流程中获得的整体价值。图 7.6 是质量管理流程的 IPO 图。

7.5.1.3 输入/输出 图 7.6 列出质量管理流程的输入和输出。附录 E 提供每个输入和输出的描述。

7.5.1.4 流程活动 质量管理流程包括以下活动：

- *计划质量管理。*
    - 识别并评估与组织战略计划一致的质量指南并进行优先级排序。建立 QM 指南——策略、标准和程序。
    - 建立组织和项目的 QM 目的和目标。
    - 建立组织和项目的 QM 职责和权限。
- *评估质量管理。*
    - 对项目评估进行评价。
    - 依据与需求和目标的符合性评估客户满意度。
    - 持续改善 QM 指南。
- *实施质量管理纠正措施和预防措施。*
    - 当被指明时，推荐适当的措施。
    - 维持开放沟通——在组织内以及与利益攸关者。

Common approaches and tips:

- Quality is a daily *focus*—not an afterthought!

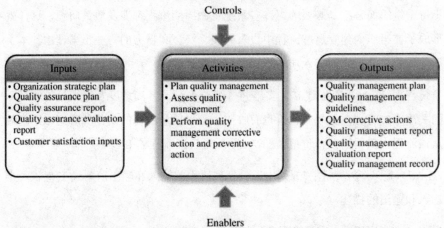

FIGURE 7.6 IPO diagram for the quality management process. INCOSE SEH original figure created by Shortell and Walden. Usage per the INCOSE notices page. All other rights reserved.

- Strategic documentation, including quality policy, mission, strategies, goals, and objectives, provides essential inputs for analysis and synthesis of quality impacts, requirements, and solutions. Existing agreements also provide direction regarding the appropriate level of attention given to quality within the organization.

- Management commitment to quality is reflected in the strategic planning of the organization—the rest of the organization will follow. Everyone in the organization should know the quality policy.

- Development of a QM intranet and information database with essential information provides an effective mechanism for disseminating consistent guidelines and providing announcements about related topics, as well as industry trends, research findings, and other relevant information. This provides a single point of contact for continuous communication regarding the QM guidelines and encourages the collection of valuable feedback and the identification of organization trends.

组织的项目使能流程

常用方法和建议:

- 质量是一种日常焦点——并非事后考虑的事情!

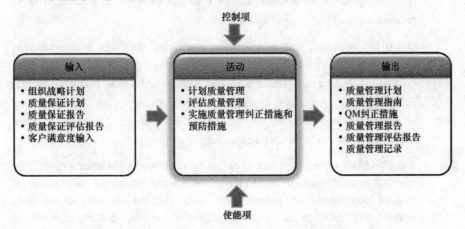

图 7.6 质量管理流程的 IPO 图。INCOSE SEH 原始图由 Shortell 和 Walden 创建。按照 INCOSE 通知页使用。版权所有。

- 策略文件(包括质量方针、任务、策略、目的和目标)为质量影响、需求和解决方案的分析和综合提供基本输入。现有协议还提供对组织内合适的质量关注层级的指导。

- 管理在质量中的投入反映在组织的战略规划中——接着是组织的其余部分。组织中的每个成员均应知悉质量方针。

- QM 内联网和具有基本信息的信息数据库的开发为传播一致的指南并发布相关主题及行业趋势、研究成果和其他相关信息提供一种有效的机制。这为 QM 指南有关的持续沟通提供一个唯一联络点并鼓励收集有价值的反馈以及对组织趋势的识别。

ORGANIZATIONAL PROJECT-ENABLING PROCESSES

- Analyze statistics from process audits, tests and evaluations, product discrepancy reports, customer satisfaction monitoring, accident and incident reporting, and the implementation of changes to items of a product (e.g., recalled product and/or production lines).

- QM is big business, and a plethora of standards, methods, and techniques exist to help an organization. A short list includes the ISO 9000 series, total quality management (TQM), and Six Sigma (statistical process control) (Brenner, 2006). According to ISO 9001 (2008), quality is the "Ability of a set of inherent characteristics of a product, system, or process to fulfill requirements of customers and other interested parties."

- A successful strategy is to aim at achieving customer satisfaction primarily by preventing nonfulfillment of requirements. Ideally, customer satisfaction is linked to compliance with requirements. Two indicators that the process is not working are situations where (i) the project is compliant but the customer is unhappy or (ii) the project is not compliant but the customer is happy.

- The consistent involvement and commitment of top management with timely decision making is mandatory for the quality program. This is reflected in staffing and training of project auditors.

- Project assessments include measurements that can be evaluated to determine the performance of a project team and the progress toward a quality outcome.

- Trends in tailoring of project-specific quality plans provide clear indications of potential improvements in the overall organization guidelines.

- The team of people working in this process will also find a wealth of material in ISO standards and other sources.

- The quality program should have an escalation mechanism so that findings/issues not addressed by the project can be raised to senior management. Projects are under schedule/resource constraints and may not always be responsive to quality findings. In these situations, there should be a mechanism to raise this to senior management to alert them to potential impacts and help them to deter-mine how to proceed.

7.5.2 Elaboration

7.5.2.1 QM Concepts The purpose of QM is to outline the policies and procedures necessary to improve and control the various processes within the organization that ultimately lead to improved business performance.

- 分析来自流程审计、测试和评估、产品差异报告、客户满意度监测、事故和事件报告以及产品项变更实施（如召回的产品和/或生产线）的统计数据。

- QM 是大事，且存在过多的标准、方法和技术可帮助组织。一份简短的列表包括 ISO 9000 系列、全面质量管理（TQM）和六西格玛（统计流程控制）（Brenner, 2006）。依据 ISO 9001（2008），质量是"产品、系统或流程的一系列固有特性实现客户和其他有关方的需求的能力"。

- 成功策略的目的主要在于通过避免无法实现需求来达到客户满意。理想情况下，客户满意度与是否符合需求有关。以下两种情况表明流程不可行：（i）项目符合但客户不乐意；（ii）项目不符合但客户乐意。

- 最高管理层及时决策的一致参与和承诺，对于质量计划而言是强制性的。这反映在项目审计人员的配备和培训中。

- 项目评估包括可以被评估的用于确定项目团队的绩效和质量结果进展的诸多测量。

- 项目特有的质量计划的剪裁趋势清晰地表明一般组织指南中的潜在改善。

- 在该流程中工作的团队也会在 ISO 标准和其他资源中发现大量资料。

- 质量计划应具有升级机制，以便向高级管理层上报项目未应对的调查结果/问题。项目受到进度/资源约束，可能不会总是响应质量调查结果。在这些情况下，应存在一种机制，向高级管理层上报这种问题，从而警示他们存在潜在影响并帮助他们决定如何继续。

7.5.2 详细阐述

7.5.2.1 QM 概念 QM 的目的是概括必要的策略和程序，以改进并控制组织内最终使得业务绩效改善的各种不同的流程。

The primary objective of QM is to produce an end result that meets or exceeds stakeholder expectations. For example, using a quality system program, manufacturers establish requirements for each type or family of product to achieve products that are safe and effective. To meet this objective, they establish methods and procedures to design, produce, distribute, service, and document devices that meet the quality system requirements. QM is closely related to the V&V processes.

Quality assurance (QA) is generally associated with activities such as failure testing, statistical control, and total quality control. Many organizations use statistical process control as a means to achieve Six Sigma levels of quality. Traditional statistical process controls use random sampling to test a fraction of the output for variances within critical tolerances. When these are found, the manufacturing processes are corrected before more bad parts can be produced.

Quality experts (Crosby, 1979; Juan, 1974) have determined that if quality cannot be measured, it cannot be systematically improved. Assessment provides the feedback needed to monitor performance, make midcourse corrections, diagnose difficulties, and pinpoint improvement opportunities. A widely used paradigm for QA management is the plan–do–check–act approach, also known as the Shewhart cycle (Shewhart, 1939).

Quality pioneer W. Edwards Deming stressed that meeting user needs represents the defining criterion for quality and that all members of an organization need to participate actively in "constant and continuous" quality improvement—to commit to the idea that "good enough isn't" (Deming, 1986). His advice marked a shift from inspecting for quality after production to building concern for quality into organization processes. As an example, in 1981, Ford launched a "Quality is job 1" campaign that went beyond getting good workers and supporting them with high-quality training, facilities, equipment, and raw materials. By characterizing quality as a "job," everyone in the organization was motivated to concern themselves with quality and its improvement— for every product and customer (Scholtes, 1988).

Total quality control deals with understanding what the stakeholder/customer really wants. If the original need statement does not reflect the relevant quality requirements, then quality can be neither inspected nor manufactured into the product. For instance, the Øresund bridge consortium included not only the bridge material and dimensions but also operating, environmental, safety, reliability, and maintainability requirements.

QM 的主要目标是产生满足或超出利益攸关者期望的最终结果。例如，制造商利用质量体系计划建立每个类型或系列产品的需求，以获得安全有效的产品。为达到此目标，制造商建立了满足质量体系需求的设计、生产、分布、服务和文件化设施的方法和程序。QM 与 V&V 流程密切相关。

质量保证（QA）通常与诸如故障测试、统计控制和全面质量控制等活动相关联。许多组织将统计流程控制用作实现六西格玛质量水平的一种手段。对于在临界公差之内的偏差，传统统计流程控制使用随机抽样来测试一部分输出。当发现偏差时，在更多的劣质零件生产之前就纠正制造流程。

质量专家（Crosby, 1979；Juan, 1974）已经确定如果质量无法测量，就无法系统地改进。评估为监测性能、进行中间纠正、诊断难题和指出改进机会提供所需的反馈。QA 管理广泛使用的范式是"计划–实施–检查–行动"的方法，亦称为休哈特循环（Shewhart, 1939）。

质量先锋爱德华兹·戴明强调满足用户需要表达定义质量准则，还强调组织的所有成员都需要积极参与"持续不断"的质量改进，从而遵循"足够好还不行"理念（Deming, 1986）。他的建议标志着从"在生产后检验质量"转向"把对质量的关切构建到组织的流程中"。例如，1981 年，福特发起了一次"质量是第一工作"的运动，这次运动远不止获得优秀员工，还用高质量培训、设施、设备和原材料来支持他们。通过将质量描述成"工作"，激励组织中的每个成员将自己与质量及质量改进紧密联系起来——为了每个产品和客户（Scholtes, 1988）。

全面质量控制涉及理解利益攸关者/客户的真正愿望。如果原始需要说明没有反映相关的质量需求，那么质量既不能检验，也不能制造到产品之中。例如，厄勒海峡大桥联合企业不仅包括大桥材料和尺寸，而且包括运行、环境、安全、可靠性和可维护性需求。

ORGANIZATIONAL PROJECT-ENABLING PROCESSES

Product certification is the process of certifying that a certain product has passed performance or QA tests or qualification requirements stipulated in regulations, such as a building code or nationally accredited test standards, or that it complies with a set of regulations governing quality or minimum performance requirements. Today, medical device manufacturers are advised to use good judgment when developing their quality system and apply those sections of the Food and Drug Administration Quality System regulation that are applicable to their specific products and operations. The regulation 21-CFR-820.5 is continuously updated since its release in 1996. As such, it ought not to be possible to repeat the errors of the Therac-25 project (see Section 3.6.1).

7.6 KNOWLEDGE MANAGEMENT PROCESS

7.6.1 Overview

7.6.1.1 Purpose As stated in ISO/IEC/IEEE 15288,

The purpose of the Knowledge Management process is to create the capability and assets that enable the organization to exploit opportunities to re-apply existing knowledge.

7.6.1.2 Description KM is a broad area that transcends the bounds of SE and project management, and there are a number of professional societies that focus on it. KM includes the identification, capture, creation, representation, dissemination, and exchange of knowledge across targeted groups of stakeholders. It draws from the insights and experiences of individuals and/or organizational groups. The knowledge includes both explicit knowledge (conscious realization of the knowledge, often documented and easily communicated) and tacit knowledge (internalized in an individual without conscious realization) and can come from either individuals (through experience) or organizations (through processes, practices, and lessons learned) (Alavi and Leidner, 1999; Roedler, 2010).

Within an organization, explicit knowledge is usually captured in its training, processes, practices, methods, policies, and procedures. In contrast, tacit knowledge is embodied in the individuals of the organization and requires specialized techniques to identify and capture the knowledge, if it is to be passed along within the organization.

KM efforts typically focus on organizational objectives such as improved performance, competitive advantage, innovation, the sharing of lessons learned, integration, and continuous improvement of the organization (Gupta and Sharma, 2004). So it is generally advantageous for an organization to adopt a KM approach that includes building the framework, assets, and infrastructure to support the KM.

产品认证是认证某一产品已通过诸如"建筑细则"或"国家认可测试标准"规章中规定的性能或 QA 测试或资格要求,或者认证该产品符合一系列控制质量或最低性能需求的规章的过程。如今,医疗器械制造商被建议在开发质量体系时要运用良好的判定,并应用食品与药品管理局质量体系规章中那些适用于制造商的特定产品和运行的条款。规章 21-CFR-820.5 自 1996 年发布后不断更新。正因如此,不应该有重复 Therac-25 项目(见 3.6.1 节)错误的可能。

7.6 知识管理流程

7.6.1 概览

7.6.1.1 目的 如 ISO/IEC/IEEE 15288 所述,

知识管理流程的目的是创建出使组织能够利用机会再次应用现有知识的能力和资产。

7.6.1.2 描述 KM 是超越 SE 和项目管理范围的一个广泛区域,并且有很多专业学会专注于 KM。KM 包括跨利益攸关者目标群对知识的识别、获取、创建、表达、传播和交换。它从对个人和/或组织群组的深入理解和经验中提取。知识包括显性知识(知识的有意识的认识,通常被文件化且容易传达)和隐性知识(内化于个体的无意识的认识)并且可来自个人(通过经验)或组织(通过流程、实践和经验教训)(Alavi 和 Leidner, 1999;Roedler, 2010)。

在组织内,显性知识通常在组织的培训、流程、实践、方法、策略和程序中获取。相反,隐性知识体现在组织的个人中,并且如果要在组织内传递则需要专业技术来识别和获取该知识。

KM 工作通常集中于组织的目标,如改进绩效、竞争优势、创新、经验教训的共享、综合及组织的持续改进(Gupta 和 Sharma, 2004)。因此对于组织而言采用一种包括构建框架、资产和基础设施的 KM 方法来支持 KM 是有利的。

ORGANIZATIONAL PROJECT-ENABLING PROCESSES

The motivation for putting KM in place includes:

- Information sharing across the organization
- Reducing redundant work due to not having the information needed at the right time
- Avoiding "reinventing the wheel"
- Facilitating training, focusing on best practices
- Capturing knowledge that would "go out the door" with retirements and attrition

The last item in this list is a major concern as we see a negative slope in the supply of systems engineers. As the percentage of experienced systems engineers retiring is increasing, it becomes even more important to capture the tacit knowledge that otherwise could be lost and then make that knowledge available to the developing systems engineers.

In this handbook, KM is viewed from an organizational project-enabling perspective, that is, how the organization supports the program or project environment with the resources in its KM system. The support provided to the project can come in several ways, including:

- Knowledge captured from technical experts
- Lessons learned captured from previous similar projects
- Domain engineering information that is applicable for reuse on the project, such as part of a product line or family of systems
- Architecture or design patterns that are commonly encountered
- Other reusable assets that may be applicable to the SOI

Figure 7.7 is the IPO diagram for the knowledge management process.

7.6.1.3 Inputs/Outputs Inputs and outputs for the knowledge management process are listed in Figure 7.7. Descriptions of each input and output are provided in Appendix E.

7.6.1.4 Process Activities The knowledge management process includes the following activities:

组织的项目使能流程

使 KM 落实到位的动机包括：

- 跨组织的信息共享

- 减少由于在正确的时间没有信息而造成的冗余工作

- 避免"重复已有的工作"

- 促进培训，集中于最佳实践

- 获取随退休和消耗而"溜走的"的知识

该列表中的最后一项是主要关注点，因为我们看到系统工程师的供给出现负增长。随着有经验的系统工程师退休的比率逐渐增加，获取隐性知识（否则隐性知识可能会丢失），然后使开发中的系统工程师获得该知识变得更加重要。

在本手册中，从组织的项目使能视角来看待 KM，即组织如何用在 KM 系统中的资源支持计划或项目环境中。为项目提供的支持可采取多种方式，包括：

- 从技术专家中获取的知识

- 从之前的类似项目中获取的经验教训

- 适用于在项目上复用的领域工程信息，如产品线或系统系列的一部分

- 普遍遇到的架构或设计特征模式

- 可能适用于 SOI 的其他可复用资产

图 7.7 是知识管理流程的 IPO 图。

7.6.1.3 输入/输出 图 7.7 列出知识管理流程的输入和输出。附录 E 提供每个输入和输出的描述。

7.6.1.4 流程活动 知识管理流程包括以下活动：

- *Plan knowledge management.*
  - Establish a KM strategy that defines how the organization and projects within the organization will interact to ensure the right level of knowledge is captured to provide useful knowledge assets. This needs to be done in a cost-effective manner, so there is a need to prioritize the efforts.
  - Establish the scope of the KM strategy—the organization and projects need to identify the specific knowledge information to capture and manage.
  - Establish which projects will be subject to the knowledge management process. If the knowledge assets are not used, then the effort has been wasted. If there is no identified project that will benefit from the knowledge asset, then it probably should not be considered.
- *Share knowledge and skills throughout the organization.*
  - Capture, maintain, and share knowledge and skills per the strategy.

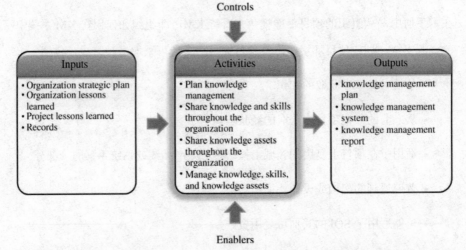

FIGURE 7.7 IPO diagram for the knowledge management process. INCOSE SEH original figure created by Shortell and Walden. Usage per the INCOSE notices page. All other rights reserved.

  - The infrastructure should be established to include mechanisms to easily identify and access the assets and to determine the level of applicability for the project considering its use.

- *计划知识管理。*

    - 建立 KM 策略，该策略定义组织和该组织内的项目将如何相互作用，以确保获取适当水平的知识来提供有用的知识资产。这需要以具有成本效益的方式进行，因此需要对工作进行优先级排序。

    - 建立 KM 策略的范围——组织和项目需要识别特定的知识信息以便获取和管理。

    - 确定哪些项目将采用知识管理流程。如果没有使用知识资产，那么努力就是徒劳无功的。如果没有识别出受益于知识资产的项目，那么可能不应该考虑这些项目。

- *贯穿于组织共享知识和技能。*

    - 按照策略获取、维护和共享知识及技能。

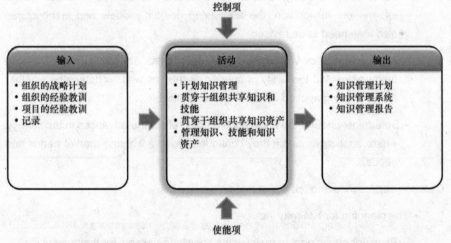

图 7.7 知识管理流程的 IPO 图。INCOSE SEH 原始图由 Shortell 和 Walden 创建。按照 INCOSE 通知页使用。版权所有。

    - 应建立基础设施，以包括易于识别和访问资产的机制以及确定考虑使用该资产的项目的适用性水平的机制。

ORGANIZATIONAL PROJECT-ENABLING PROCESSES

- *Share knowledge assets throughout the organization.*
  - Establish a taxonomy for the reapplication of knowledge.
  - Establish a representation for domain models and domain architectures. The intent of this is to ensure an understanding of the domain to help identify and manage opportunities for common system elements and their representations, such as architecture or design patterns, reference architectures, common requirements, etc.
  - Define or acquire the knowledge assets applicable to the domain, including system and software elements, and share them across the organization. As the system and system elements are defined in the technical processes, the information items that represent those definitions should be captured and included as knowledge assets for the domain.
- *Manage knowledge, skills, and knowledge assets.*
  - As the domain, family of systems, or product line changes, ensure the associated knowledge assets are revised or replaced to reflect the latest information. In addition, the associated domain models and architectures also may need to be revised.
  - Assess and track where the knowledge assets are being used. This can help understand the utility of specific assets, as well as determine whether they are being applied where they are applicable.
  - Determine whether the knowledge assets reflect the advances in technology, where applicable, and if they continue to evolve with the market trends and needs.

Common approaches and tips:

- The planning for KM may include:
  - Plans for obtaining and maintaining knowledge assets for their useful life
  - Characterization of the types of assets to be collected and maintained along with a scheme to classify them for the convenience of users
  - Criteria for accepting, qualifying, and retiring knowledge assets
  - Procedures for controlling changes to the knowledge assets
  - A mechanism for knowledge asset storage and retrieval

- *贯穿于组织共享知识资产。*
  - 建立知识再运用的结构分类法。
  - 建立领域模型和领域架构的表达形式。其目的是确保理解领域以便有助于识别和管理公共系统元素及其表达形式的机会,如架构或设计特征模式、参考架构、公共需求等。
  - 定义或采办适用于该领域的知识资产,包括系统和软件元素,并跨组织共享知识资产。由于技术流程中定义了系统及系统元素,因此应获取表达那些定义的信息项,并作为知识资产包括在该领域内。

- *管理知识、技能和知识资产。*
  - 由于领域、系统系列或产品线发生变化,确保相关知识资产被修正或取代以反映最新信息。此外,还可能需要修改相关领域模型和架构。
  - 评估并跟踪正在使用知识资产的地方。这可有助于理解特定资产的实用性以及确定它们是否应用于合适的地方。
  - 确定知识资产是否反映技术进步(若适用),以及是否继续随市场趋势和需要而演进。

常用方法和建议:

- KM 的计划可包括:
  - 在其使用寿命内获得和保留知识资产的计划
  - 与方案一同收集和保留的资产类型的描述,将资产分类方便用户使用
  - 知识资产的接受、认定和退出准则
  - 控制知识资产变更的程序
  - 知识资产的存储与检索机制

- In developing an understanding of the domain, it is important to identify and manage both the commonalities (such as features, capabilities, or functions) and the differences or variations of the system elements (including where a common system element has variations in parameters depending on the system instance). The domain representations should include:
  - Definition of the boundaries
  - Relationships of the domains to other domains
  - Domain models that incorporate the commonalities and differences allowing for sensitivity analysis across the range of variation
  - An architecture for a family of systems or product line within the domain, including their commonalities and variations

7.6.2 Elaboration

7.6.2.1 General KM Implementation Since KM focuses on capturing the organizational, project, and individual knowledge for use throughout the organization in the future, it is important to capture end-of-project lessons learned prior to the project personnel moving on to new assignment. An effective knowledge management process has the knowledge capture mechanisms in place to capture the relevant information throughout the life of the project rather than trying to piece it together at the end. This includes identification of systems that are part of a product line or family of systems and system elements that are designed for reuse. For the first instance of these systems and system elements, the KM system needs to capture the domain engineering artifacts in a way to facilitate their use in the future. For subsequent instances, the KM system needs to provide the domain engineering information and capture any variations, updates of technology, and lessons learned. Issues important to the organization:

- Definition and planning of KM activities for domain engineering and asset preservation, including tasks dedicated to domain engineering of product lines or families of systems and to the preservation of reusable assets.
- Integration of architecture management into the KM system including frameworks, architecture reuse, architecture reference models, architecture patterns, platform-based engineering, and product line architecture.

- 在开发对领域的理解时，重要的是识别和管理系统元素的共性（如特征、能力或功能）和差异或变化（包括取决于系统实例的公共系统元素的参数中哪里发生了变化）。领域表达形式应包括：

  – 边界的定义

  – 领域与其他领域的关系

  – 领域模型将共性和差异纳入其中，从而允许跨变化范围进行灵敏度分析

  – 领域内系统系列或产品线的架构，包括它们的共性和变化

7.6.2 详细阐述

7.6.2.1 一般 KM 实施　由于 KM 集中于获取组织的知识、项目知识和个人知识为将来贯穿于组织内使用，在项目人员开展新的任务之前获取结束项目的经验教训是很重要的。有效的知识管理流程具有恰当的知识获取机制，以便贯穿于项目生命获取相关信息，而不是试图在最后将信息拼合在一起。这包括识别作为产品线或系统系列一部分的系统以及被设计用于复用的系统元素。在这些系统和系统元素的第一个实例中，KM 系统需要以促进将来使用的方式获取领域工程制品。在后续实例中，KM 系统需要提供领域工程信息并获取任何变化、技术更新和经验教训。对组织重要的问题：

- 用于领域工程和资产保存的 KM 活动的定义和规划，包括致力于产品线或系统系列的领域工程的任务以及致力于可复用资产保存的任务。

- 架构管理在 KM 系统中的综合，包括框架、架构复用、架构参考模型、架构特征模式、基于平台的工程和产品线架构。

ORGANIZATIONAL PROJECT-ENABLING PROCESSES

- Characterization of the types of assets to be collected and maintained including an effective means for users to find the applicable assets
- Determination of the quality and validity of the assets

7.6.2.2 Potential Reuse Issues There are serious traps in reuse, especially with respect to COTS and non-developmental item (NDI) elements:

- Are the new system or system element requirements and operational characteristics exactly like the prior one? Trap: the prior solution was intended for a different use, environment, or performance level, or it was a concept only that was never built.

- Did the prior system or system element work correctly? Trap: it worked perfectly, but the new application is outside the qualified range (e.g., using a standard car for a high-speed track race).

- Is the new system or system element to operate in the same environment as the prior one? Trap: it is not certain, but there is no time to study it. One NASA Mars probe was lost because the development team used a radiator design exactly as was used on a successful satellite in Earth orbit. When the Mars mission failed, the team then realized that Earth orbiting environment, while in space, is different from a deep space mission.

- Is the system/system element definition defined and understood (i.e., requirements, constraints, operating scenarios, etc.)? Trap: too often, the development team assumes that if a reuse solution will be applied (especially for COTS), there is no need for well-defined system definition. The issues may not show up until systems integration, causing major cost and schedule perturbations.

- Is the solution likely to have emergent requirements/behaviors where the reuse is being considered? Trap: a solution that worked in the past was used without consideration for the evolution of the solution. If COTS is used, there may be no way to adapt or modify it for the emergent requirements.

A properly functioning KM system paired with well-defined processes and engineering discipline can help avoid these problems.

161

522

- 待收集和保留的资产类型的描述，包括供用户发现适用资产的有效手段
- 资产的质量和有效性的确定

7.6.2.2 潜在的复用问题　在复用中，尤其是 COTS 和非开发项（NDI）元素中存在严重的陷阱：

- 新的系统或系统元素需求及运行特性与之前的恰巧相似吗？陷阱：之前的解决方案预计用于不同的使用、环境或性能水平，或者仅是一种从未构建的概念。

- 之前的系统或系统元素工作正确吗？陷阱：之前的系统或系统元素工作很完美，但新的应用处于合格范围之外（如，在高速场地赛中使用标准车辆）。

- 新的系统或系统元素与之前的在同一环境中运行吗？陷阱：不确定，但没有时间来研究它。一个 NASA 火星探测器由于开发团队使用了恰巧成功地应用在地球轨道卫星上的散热器的设计而损失了。在火星任务失败时，团队才意识到空间中的地球轨道环境不同于深空任务。

- 系统/系统元素定义被定义和理解了吗（即，需求、约束、运行场景等）？陷阱：太频繁，开发团队假设如果复用解决方案被应用（尤其是用于 COTS），则不需要定义明确的系统定义。问题在系统综合之前可能不会被展现，造成主要成本和进度严重混乱。

- 解决方案在被考虑复用时是否有可能具有涌现的需求/行为？陷阱：一个在过去有效的解决方案，在没有考虑到解决方案的演进的情况下被使用。若使用 COTS，可能没有办法顾及涌现的需求来调整或修改它。

与明确定义的流程和工程学科相对应的恰当运行的 KM 系统，可有助于避免这些问题。

8 TAILORING PROCESS AND APPLICATION OF SYSTEMS ENGINEERING

Standards and handbooks address life cycle models and systems engineering (SE) processes that may or may not fully apply to a given organization and/or project. Most are accompanied by a recommendation to adapt them to the situation at hand.

The principle behind tailoring is to ensure that the process meets the needs of the project while being scaled to the level of rigor that allows the system life cycle activities to be performed with an acceptable level of risk. Tailoring scales the rigorous application to an appropriate level based on need. The life cycle model may be tailored as described in Chapter 3. Processes may be tailored as described in this section. All processes apply to all stages, tailoring determines the process level that applies to each stage, and that level is never zero. There is always some activity in each process in each stage.

Figure 8.1 is a notional graph for balancing formal process against the risk of cost and schedule overruns (Salter, 2003). Insufficient SE effort is generally accompanied by high risk. However, as illustrated in Figure 8.1, too much formal process also introduces high risk. If too much rigor or unnecessary process activities or tasks are performed, increased cost and schedule impacts will occur with little or no added value. Tailoring occurs dynamically over the system life cycle depending on risk and the situational environment and should be continually monitored and adjusted as needed.

For ISO/IEC/IEEE 15288, process tailoring is the deleting or adapting the processes to satisfy particular circumstances or factors of the organization or project using the process. While ISO/IEC/IEEE 15288 tailoring focuses on the deletion of unnecessary or unwarranted process elements, it does allow for additions and modifications as well.[1]

This chapter describes the process of tailoring the life cycle models and SE processes to meet organization and project needs. Refer to ISO/IEC TR 24748-1 (2010) and ISO/IEC TR 24748-2 (2010) for more information on adaptation and tailoring. This chapter also describes the application of the SE processes in various product sectors and domains, for product lines, for services, for enterprises, and in very small and micro enterprises (VSMEs).

INCOSE Systems Engineering Handbook: A Guide for System Life Cycle Processes and Activities, Fourth Edition. Edited by David D. Walden, Garry J. Roedler, Kevin J. Forsberg, R. Douglas Hamelin and Thomas M. Shortell.

© 2015 John Wiley & Sons, Inc. Published 2015 by John Wiley & Sons, Inc.

8 系统工程的剪裁流程和应用

标准和手册涉及可能完全适用于或不完全适用于特定组织和/或项目的生命周期模型和系统工程（SE）流程。大多数标准和手册都附一个使其适应当前情况的建议。

剪裁背后的原则是确保流程在被缩减到允许按照可承受风险等级执行系统生命周期活动的严格程度的同时满足项目的需要。剪裁基于需要将严格应用缩减至一个合适的等级。可如第 3 章所述剪裁生命周期模型。可如本节所述剪裁流程。所有流程均适用于所有阶段，剪裁确定适用于每个阶段的流程层级，并确定该层级从不为空。在每个阶段的每个流程中总是存在一些活动。

图 8.1 是平衡正式流程与成本超支和进度延期风险的一个概念图（Salter, 2003）。不充分的 SE 工作通常伴随有高风险。然而，如图 8.1 所示，太多正式流程亦会引入高风险。如果执行太多严格的或不必要的流程活动或任务，将使成本和进度的影响增加，几乎没有增值。剪裁依据风险和具体环境动态地出现在系统生命周期中，且应按需要被持续地监测和调整。

对于 ISO/IEC/IEEE 15288，流程剪裁是删除或调整流程，以满足使用该流程的组织或项目的特殊情况或因素。虽然 ISO/IEC/IEEE 15288 剪裁聚焦于删除不必要的或不合理的流程要素，但其亦允许添加和修改。

本章描述剪裁生命周期模型和 SE 流程的流程，以满足组织需要和项目需要。更多关于调整和剪裁的信息，参见 ISO/IEC TR 24748-1（2010）和 ISO/IEC TR 24748-2（2010）。本章亦描述 SE 流程在各种不同的产品行业和领域、产品线、服务、复杂组织体以及极小型和微型复杂组织体（VSME）中的应用。

INCOSE 系统工程手册：《系统生命周期流程和活动指南》，第 4 版。编撰：David D.、Walden、GarryJ.、Roedler、Kevin J.、Forsberg、R.Douglas Hamelin 和 Thomas M.Shortell。
John Wiley & Sons 公司版权所有©2015。由 John Wiley & Sons 公司于 2015 年出版。

TAILORING PROCESS AND APPLICATION OF SYSTEMS ENGINEERING

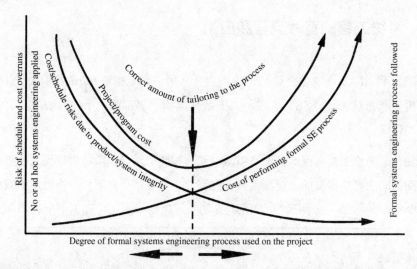

FIGURE 8.1 Tailoring requires balance between risk and process. INCOSE SEH original figure created by Michael Krueger, adapted from Ken Salter. Usage per the INCOSE notices page. All other rights reserved.

8.1 TAILORING PROCESS

8.1.1 Overview

8.1.1.1 Purpose As stated in ISO/IEC/IEEE 15288,

> [A.2.1] The purpose of the Tailoring process is to adapt the processes of [ISO/IEC/IEEE 15288] to satisfy particular circumstances or factors.

8.1.1.2 Description At the organization level, the tailoring process adapts external standards in the context of the organizational processes to meet the needs of the organization. At the project level, the tailoring process adapts organizational processes for the unique needs of the project. Figure 8.2 is the IPO diagram for the tailoring process.

8.1.1.3 Inputs/Outputs Inputs and outputs for the tailoring process are listed in Figure 8.2. Descriptions of each input and output are provided in Appendix E.

8.1.1.4 Process Activities The tailoring process includes the following activities:

- *Identify and record the circumstances that influence tailoring.*
  - Identify tailoring criteria for each stage— Establish the criteria to determine the process level that applies to each stage.

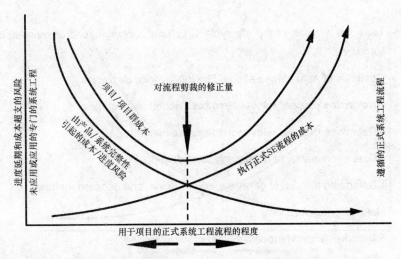

图 8.1 剪裁要求风险和流程之间平衡。INCOSE SEH 原始图由 Michael Krueger 创建，借鉴于 Ken Salter。按照 INCOSE 通知页使用。版权所有。

8.1 剪裁流程

8.1.1 概览

8.1.1.1 目的 如 ISO/IEC/IEEE 15288 所述，

[A.2.1] 剪裁流程的目的是调整[ISO/IEC/IEEE 15288]的流程以满足特殊的情况或因素。

8.1.1.2 描述 在组织层级上，剪裁流程调整组织流程背景环境中的外部标准以满足组织的需要。在项目层级上，剪裁流程调整组织流程以满足项目的独特需要。图 8.2 是剪裁流程的 IPO 图。

8.1.1.3 输入/输出 图 8.2 列出了剪裁流程的输入和输出。附录 E 提供每个输入和输出的描述。

8.1.1.4 流程活动 剪裁流程包括以下活动：

- *识别和记录影响剪裁的外部环境。*
  - 为每个阶段识别剪裁准则——建立用来确定适用于每个阶段的流程等级的准则。

- *Take due account of the life cycle structures recommended or mandated by standards.*
- *Obtain input from parties affected by the tailoring decisions.*
  - Determine process relevance to cost, schedule, and risks.
  - Determine process relevance to system integrity.
  - Determine quality of documentation needed.
  - Determine the extent of review, coordination, and decision methods.
- *Make tailoring decisions.*
- *Select the life cycle processes that require tailoring.*
  - Determine other changes needed for the process to meet organizational or project needs beyond the tailoring (e.g., additional outcomes, activities, or tasks).

Controls

Inputs	Activities	Outputs
• Organization strategic plan • Life cycle models	• Identity and record the circumstances that influence tailoring • Take due account of the life cycle structures recommended or mandated by standards • Obtain input from parties affected by the tailoring decisions • Make tailoring decisions • Select life cycle processes that require tailoring	• Organization tailoring strategy • Project tailoring strategy

Enablers

FIGURE 8.2 IPO diagram for the tailoring process. INCOSE SEH original figure created by Shortell and Walden. Usage per the INCOSE notices page. All other rights reserved.

Common approaches and tips:

- Eliminate unnecessary outcomes, activities, and tasks, and add additional ones.

系统工程的剪裁流程和应用

- *适当考虑标准建议或强制的生命周期结构。*
- *从受剪裁决策影响的各方获取输入。*
    - *确定与成本、进度和风险相关的流程。*
    - *确定与系统整体性相关的流程。*
    - *确定所需文件的质量。*
    - *确定评审、协调和决策方法的程度。*
- *做出剪裁决策。*
- *选择要求剪裁的生命周期流程。*
    - *确定流程为满足超出剪裁的组织或项目需要所需的其他变更（例如，其他成果、活动或任务）。*

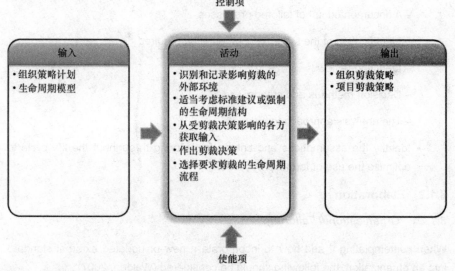

图 8.2 剪裁流程的 IPO 图。INCOSE SEH 原始图由 Shortell 和 Walden 创建。按照 INCOSE 通知页使用。版权所有。

常用方法和建议：

- 消除不必要的成果、活动和任务并增加另外的成果、活动和任务。

- Base decisions on facts and obtain approval from an independent authority.
- Use the decision management process to assist in tailoring decisions.
- Conduct tailoring at least once for each stage.
- Drive the tailoring based on the environment of the system life cycle stages.
- Constrain the tailoring based on agreements between organizations.
- Control the extent of tailoring based on issues of compliance to stakeholder, customer, and organization policies, objectives, and legal requirements.
- Influence the extent of tailoring of the agreement process activities based on the methods of procurement or intellectual property.
- Remove extra activities as the level of trust builds between parties.
- Identify a set of formal processes and outcomes/ activities/tasks at the end of the tailoring process. This includes but is not limited to:
    - A documented set of tailored processes
    - Identification of the system documentation required
    - Identified reviews
    - Decision methods and criteria
    - The analysis approach to be used
- Identify the assumptions and criteria for tailoring throughout the life cycle to optimize the use of formal processes.

8.1.2 Elaboration

8.1.2.1 Organizational Tailoring

When contemplating if and how to incorporate a new or updated external standard into an organization, the following should be considered (Walden, 2007):

- Understand the organization.
- Understand the new standard.
- Adapt the standard to the organization (not vice versa).

- 基于事实决策并获取独立的权威机构的批准。
- 使用决策管理流程以协助剪裁决策。
- 每个阶段进行至少一次剪裁。
- 基于系统生命周期阶段的环境驱动剪裁。
- 基于组织间的协议约束剪裁。
- 基于有关符合利益攸关者、客户和组织的策略、目标和法律要求的议题控制剪裁的程度。
- 基于采购方法或知识产权影响协议流程活动的剪裁程度。
- 额外活动随着双方之间建立信任水平而被消除。
- 在剪裁流程结束时识别一系列正式的流程和成果/活动/任务。这包括但不限于：

    – 一系列文件化的已剪裁的流程

    – 识别所需的系统文件

    – 识别的评审

    – 决策方法和准则

    – 待使用的分析方法

- 识别贯穿于生命周期期的有关剪裁的假设和准则，以便优化正式流程的使用。

8.1.2 详细阐述

8.1.2.1 组织的剪裁

当考虑是否以及如何将新的或已更新的外部标准纳入组织时，应考虑以下内容（Walden，2007）：

- 理解组织。
- 理解新的标准。
- 使标准适应组织（反之则不然）。

- Institutionalize standards compliance at the "right" level.
- Allow for tailoring.

8.1.2.2 Project Tailoring Project tailoring applies specifically to the work executed through programs and projects. Factors that influence tailoring at the project level include:

- Stakeholders and customers (e.g., number of stakeholders, quality of working relationships, etc.)
- Project budget, schedule, and requirements
- Risk tolerance
- Complexity and precedence of the system

Today's systems are more often jointly developed by many different organizations. Cooperation must transcend the boundaries of any one organization. Harmony between multiple suppliers is often best maintained by agreeing to follow a set of consistent processes and standards. Consensus on a set of practices is helpful but adds complexity to the tailoring process.

8.1.2.3 Traps in Tailoring Common traps in the tailoring process include, but are not limited to, the following:

1. Reuse of a tailored baseline from another system without repeating the tailoring process
2. Using all processes and activities "just to be safe"
3. Using a pre-established tailored baseline
4. Failure to include relevant stakeholders

8.2 TAILORING FOR SPECIFIC PRODUCT SECTOR OR DOMAIN APPLICATION

The discipline of SE can be applied on any size and type of system. However, that does not mean that it should be blindly applied in the same fashion on every system. While the same SE fundamentals apply, different domains require different points of emphasis in order to be successful.

This section provides a starting point for people looking to apply SE in different domains. An objective of this section is to provide guidance on how to introduce the concepts of SE to a new field. As an example, the audience for this guidance might be SE champions that already work in the field and want to promote SE.

- 使"正确"层级上的标准符合性制度化。
- 允许剪裁。

8.1.2.2 项目剪裁 项目剪裁特别适用于在项目群和项目中执行的工作。影响项目层级剪裁的因素包括：

- 利益攸关者和客户（如利益攸关者的数量、工作关系的质量等）
- 项目预算、进度和需求
- 风险容限
- 系统的复杂性和优先级

如今的系统更多的时候是由许多不同组织联合开发的。合作必须超越任何一个组织的边界。通过同意遵循一系列一致的流程和标准，多个供应商之间的和谐通常会得到最好的保持。在一系列实践上达成共识是有益的，但会增加剪裁流程的复杂性。

8.1.2.3 剪裁中的陷阱 剪裁流程中的常见陷阱包括但不限于以下内容：

1. 复用来自于另一个系统的剪裁基线而不重复剪裁流程
2. "为了安全起见"，使用所有流程和活动
3. 使用预先建立的剪裁基线
4. 未能包括相关的利益攸关者

8.2 特定产品行业或领域应用的剪裁

SE 的学科能够应用于任何一种规模和类型的系统。然而，那并不意味着应以相同的方式将 SE 学科盲目地应用于每个系统。虽然应用相同的 SE 基本原理，但是为了取得成功不同的领域要求着重点不同。

本节为希望在不同领域应用 SE 的人员提供一个起点。本节的目的是为如何将 SE 概念引入到一个新的领域提供指南导。例如，本指导的受众可能是已经从事于该领域工作并想要促进 SE 的 SE 先锋。

The domains are listed in alphabetical order and should not be considered to be exhaustive. In addition, one needs to recognize the evolving maturity of SE in these applications. The following descriptions capture the current state of the practice and may not necessarily be representative of the state of practice at the time this handbook is being read/used.

8.2.1 Automotive Systems

SE has been traditionally applied with success by industries that develop and sell large complex systems, with a relatively long life cycle and small production volumes. In contrast, the automotive industry has manufactured high volumes of consumer products with a great diversity for more than 120 years, with quite a success. A highly cost-driven industry, the automotive sector is characterized by a permanent quest for cost-efficiency and massive reuse of parts and engineering artifacts.

Modern commercial automobiles have become complex, high-technology products in a relatively short time span. The complexity drivers usually pointed out by the actors of the industry are:

- The increasing number of vehicle functionalities supported by software, electronics, and mechatronic technologies

- Increasing safety constraints

- Increasing environmental constraints, such as air pollution and decreasing fossil fuel stocks, leading to a partial or total electrification of the vehicle power train

- Globalization and the car market growth in emerging countries, which induce greater variations in the product definition and a different repartition of worldwide production of vehicles

- New trends such as "advanced mobility services," "smart cities and smart transportation systems," and autonomous vehicles, which induce changes in the purpose and scope of the traditional automobile product

Most engineering specialties and domain specific practices involved in the automotive industry follow international standards. Some important automotive-related standards, associations, and SE standards are shown in Table 8.1. These standards usually define characteristics, measurement procedures, or testing procedures for specific vehicle systems or specific aspects of the product and are often adapted to establish regulations by local authorities.

这些领域被按照字母顺序列出并且不应被视为是详尽的。此外，需要认识到 SE 在这些应用中正在演进的成熟度。以下描述选取当时的实际状态并且可能不一定表示阅读/使用本手册时的实际状态。

8.2.1 汽车系统

传统上，SE 已经成功应用在开发和销售具有相当长生命周期和低生产量的大型复杂系统的行业。相反，在超过 120 年的时间里，汽车行业已经制造出了大量和多种多样的消费产品，并且相当成功。汽车业作为一个高成本驱动的行业，其特点是永远追求成本效益并大量复用零件和工程制品。

现代商用汽车已经在相对较短的时间跨度内已经变成为复杂的高科技产品。通常由行业参与者指出的复杂性驱动因素是：

- 由软件技术、电子技术和机电技术所支持的车辆功能数量的持续增加
- 安全约束的持续增加
- 环境约束的持续增加，例如空气污染和化石燃料储备的减少，导致车辆动力传动系的部分或全部电气化
- 全球化和新兴国家汽车市场的增长，引起产品定义的更大变化以及车辆在全世界范围生产的不同划分
- 诸如"先进移动服务"、"智能城市和智能交通系统"以及自主车辆等新趋势，引发传统汽车产品的用途和范围的变化

汽车行业中涉及的大部分工程专业和领域特定实践遵循国际标准。与汽车有关的一些重要标准、协会和 SE 标准如表 8.1 所示。这些标准通常定义特定车辆系统或产品特定方面的特征、测量程序或测试程序，并且往往被调整以适应地方当局建立法规。

TABLE 8.1 Standardization-related associations and automotive standards

Organization/standard	Description
SAE International, formerly the Society of Automotive Engineers	One of the main organizations that coordinate the development of technical standards for the automotive industry. Currently, SAE International is a globally active professional association and standards organization for engineering professionals in various industries, whose principal emphasis is placed on transport industries such as automotive, aerospace, and commercial vehicles
Japan Society of Automotive Engineers (JSAE)	An organization that sets automotive standards in Japan, analogous to the SAE International
Association for Standardization of Automation and Measuring Systems (ASAM)	An incorporated association under German law whose members are primarily international car manufacturers, suppliers, and engineering service providers from the automotive industry. The ASAM standards define protocols, data models, file formats, and application programming interfaces (APIs) for the use in the development and testing of automotive electronic control units
AUTomotive Open System ARchitecture (AUTOSAR)	An open and standardized automotive software architecture, jointly developed by automobile manufacturers, suppliers, and tool developers. Some of its key goals include the standardization of basic system functions, scalability to different vehicle and platform variants, transferability throughout the network, integration from multiple suppliers, maintainability
The GENIVI Alliance	A nonprofit consortium whose goal is to establish a globally competitive, Linux-based operating system, middleware, and platform for the automotive in-vehicle infotainment (IVI) industry. GENIVI specifications cover the entire product life cycle and software updates and upgrades over the vehicle's lifetime
ISO/TS 16949	An international standard for particular requirements for the application of ISO 9001 quality management systems for automotive production and relevant service part organizations
IEC 62196	An international standard for set of electrical connectors and charging modes for electric vehicles maintained by the International Electrotechnical Commission (IEC)

表 8.1 与标准化有关的协会和汽车标准

组织/标准	描述
国际汽车工程师学会，之前称为汽车工程师学会	协调汽车行业技术标准开发的主要组织之一。目前，国际汽车工程师学会是一个各种不同行业的工程专业人员组成的全球性活跃的专业协会和标准化组织，其主要的着重点是诸如汽车、航空航天和商业运输工具等的运输行业
日本汽车工程师学会（JSAE）	制定日本汽车标准的组织，类似国际汽车工程师学会
自动化及测量系统标准协会（ASAM）	遵守德国法律的法人团体，其成员主要是来自汽车行业的国际汽车制造商、供应商和工程服务提供商。ASAM 标准定义供开发和测试汽车电子控制装置使用的协议、数据模型、文件格式以及应用编程接口（API）
汽车开放系统架构（AUTOSAR）	由汽车制造商、供应商和工具开发商联合开发的开放的标准化汽车软件架构。AUTOSAR 的一些关键目标包括基本系统功能、不同车辆和平台变化的可伸缩性、贯穿整个网络的可转移性、来自多个供应商的综合以及可维护性的标准化
GENIVI 联盟	一个非盈利联合体，其目标是建立一个用于汽车车载信息娱乐（IVI）行业的具有全球竞争力的基于 Linux 的操作系统、中间件和平台。GENIVI 规范涵盖整个产品生命周期以及车辆使用寿命期间的软件更新和升级
ISO/TS 16949	关于对 ISO 9001 质量管理系统应用于汽车生产和相关维修零件组织的特殊需求的国际标准
IEC 62196	国际电工委员会（IEC）维护的、用于一系列电动车辆电连接器和充电模式的国际标准

Automotive standards are normally used for product qualification or "certification" purposes, although it should be noted that certification procedures in the auto-motive industry differ greatly from those applied in other sectors, like aerospace. Usually, the scope of these standards is either the product or the manufacturing process, with specific regulations ruling each vehicle system. One remarkable exception to this type of standards is the functional safety standard ISO 26262 (2011), which defines activities and work products covering a full "safety life cycle" and allowing a systematic and traceable mastery of safety risks.

SE as a discipline started to develop rather recently in some automotive organizations, around the mid-1990s or early 2000s, mainly as a response to the aforementioned challenges. While the organizations that have shown interested in SE agree that most of these challenges could be leveraged by applying a systems approach, SE is not yet a standard practice in the industry.

The application of SE is not mandatory in the automotive industry to develop and sell products. The cultural and organizational changes that are required for an effective global industry-wide implementation of SE are beginning to take place.

The automotive industry is characterized by an extensive reuse of legacy engineering artifacts (requirements, architectures, validation plans) and system constituents (systems, system elements, parts) and by the management of huge product lines (at vehicle, system, and part levels), with high stakes associated to the efficient management of those product lines.

When applying SE in the automotive industry, one should take into consideration these specificities. An adapted SE process for the automotive domain should then be an ideal combination of product-driven and stakeholder-/requirement-driven approaches. The weight of the former type of approach will be more important for "classical" vehicle systems and based on typical variant management techniques (product derivation or configuring), while the relative weight of the second type of approach will be more important for innovative systems and "untraditional" vehicle scopes or usages.

In particular, when implementing the SE technical processes for the first time, the importance of upstream SE activities (e.g., operational analysis and requirement elaboration) must be underlined. Ideally, these activities should take place *off-line*, that is, "disconnected" from the development of the product, so that the produced SE artifacts baselines are properly defined and verified.

汽车标准通常用于产品鉴定或"认证"目的，虽然应注意到汽车行业的认证程序与其他行业（如航空航天）应用的认证程序有很大的不同。通常，这些标准的范围或是产品或是制造流程，并对每种车辆系统具有特定的条规。这种类型标准的一个值得注意的例外情况是功能安全标准 ISO 26262（2011），其定义诸多覆盖一个全部"安全生命周期"以及使安全风险的系统的可追溯掌控的活动和工作产物。

最近，大约在 20 世纪 90 年代中期或 21 世纪初，在一些汽车组织中开始将 SE 作为一个学科进行开发，主要作为对上述挑战的响应。虽然对 SE 感兴趣的组织一致认为大多数挑战可通过应用系统法而得到高效解决，但是 SE 仍然不是该行业中的一种标准实践。

在汽车行业中，SE 在开发和销售产品中的应用不是强制性的。全球性行业范围 SE 有效实施所需的文化和组织变革正在开始。

汽车行业的特点是广泛复用以往遗留的工程制品（需求、架构、确认计划）和系统的组成部分（系统、系统元素、零件）以及管理庞大的产品线（车辆、系统和零件级），具有与有效管理这些产品线相关联的高风险。

当在汽车行业中应用 SE 时，均应考虑到这些特殊性。然后，适应于汽车领域的 SE 流程应是产品驱动方法与利益攸关者驱动方法/需求驱动法的一种理想的组合。前一类方法的权重对于"经典"车辆系统更为重要的并基于典型的变异管理技术（产品衍生或构型配置），同时，第二类方法的相对权重对于创新系统和"非传统"车辆范围或用途更为重要。

特别是，当第一次实施 SE 技术流程时，必须强调 SE 上游活动（例如运行分析和需求详述）的重要性。理想情况下，这些活动应离线进行，即，与产品的开发"分离"，以便恰当的定义和验证所产生的 SE 制品基线。

Since the involvement of carmakers in the development the different vehicle systems differs from one car maker to the other and from one vehicle domain to the other, care should also be taken in the tailoring of agreement processes, which will depend on the particular project context. The formalization of the exchanges between the Original Equipment Manufacturer (OEM) and their providers during the lifetime of a project should help tailoring these processes and could set the groundwork for their future standardization.

8.2.2 Biomedical and Healthcare Systems

SE has come to be recognized as a major contributor to biomedical and healthcare product and development, particularly due to the need for architectural soundness, requirements traceability, and subdisciplines such as safety, reliability, and human factors engineering. Biomedical and healthcare systems range from low complexity to high complexity. They also range from low risk to high risk. Many of these systems must work in harsh environments, including inside the human body. SE within the biomedical and healthcare environment may not be as mature as the defense and aerospace domain, for example, but the growing complexity of medical devices, their connectivity, and their use environment is quickly increasing the breadth and depth of the practice. Standards such as ISO 14971 (application of risk management to medical devices) and IEC 60601 (medical device safety) are driving organizations to take a deeper look into safety and the engineering practices behind it. Other government standards and regulations such as the US Code of Federal Regulations (CFR) Title 21 Part 820 (medical devices/quality system regulations) shape the development process for medical devices sold therein. Thus, systems engineers are increasingly being brought on board to leverage their life cycle management skills and validate that the final product does indeed meet the needs of its stakeholders.

In the biomedical and healthcare environment, it is important to understand that risk management is generally centered around user safety rather than technical or business risks and that traceability is often a key factor in regulatory audits. Organizations that have strong SE practices are therefore in a better position to avoid development pitfalls and to effectively defend their design decisions if a regulatory audit does occur. Going forward, biomedical and healthcare organizations that effectively leverage SE standards such as ISO/IEC/IEEE15288 and ISO 29148 will be in a much better position to develop safe and effective products. In general, such standards do not need to be excessively tailored, although firms with new or maturing practices may want to focus on lean implementations in order to obtain early and effective adoption.

因为汽车制造商参与开发，不同的车辆系统在一个汽车制造商与另一个汽车制造商或一个车辆领域与另一个车辆领域之间是不同的，因此，在剪裁协议流程中应予谨慎，这将取决于特定项目背景环境。项目生命周期期间的原始设备制造商（OEM）及其提供商之间的交换的正规化应有助于剪裁这些流程并能够为未来的标准化奠定基础。

8.2.2 生物医学和健康医疗系统

SE 已被认为是生物医学和健康医疗产品和开发的一个主要贡献者，特别是由于架构坚实、需求追溯性以及诸如安全性、可靠性与人因工程学等分支学科的需要。生物医学和健康医疗系统的范围是从低复杂度系统到高复杂度，亦从低风险到高风险。大多数生物医学和健康医疗系统必须在严酷的环境中工作，包括人体内部。例如，生物医学和健康医疗环境内的 SE 可能不像国防和航空航天领域的 SE 那么成熟，但医疗器械及其连接性和使用环境的复杂度的提高正在快速增加实践的广度和深度。诸如 ISO 14971（医疗器械风险管理的应用）和 IEC 60601（医疗器械安全性）等标准正在驱动组织对安全性及其背后的工程实践进行更加深入的调查。其他政府标准和法规为诸如《美国联邦法规汇编（CFR）》第 21 册，第 820 部分（医疗器械/质量体系法规）所推荐的医疗器械塑形开发流程。因此，正在越来越多地引入系统工程师以充分发挥他们的生命周期管理技能并确认最终产品确实满足其利益攸关者的需要。

在生物医学和健康医疗环境中，重要的是理解风险管理通常以用户安全而不是技术风险或商业风险为中心，并且理解可追溯性往往是监管审计中的一个关键因素。因此，如果进行监管审计，那么具有很强的 SE 实践的组织在避开开发陷阱和有效保护其设计决策方面占有优势。今后，有效且强有力地发挥诸如 ISO/IEC/IEEE 15288 和 ISO 29148 的 SE 标准的作用的生物医学和健康医疗组织将会在开发安全且有效的产品方面占具更多优势。通常，尽管具备新的或成熟的实践的公司可能想要聚焦于精益实施以便获得早期和有效的采用，但这样的标准不需要过度的剪裁。

8.2.3 Defense and Aerospace Systems

While SE has been practiced in some form from antiquity, what has now become known as the modern definition of SE has its roots in defense and aerospace systems of the twentieth century. It has been recognized as a distinct activity in the late 1950s and early 1960s as a result of the technological advances taking place that led to increasing levels of system complexity and systems integration challenges. The need for SE increased with the large-scale introduction of digital computers and software. Defense and aerospace evolved to address systemic approaches to issues such as the widespread adaptation of COTS technologies and the use of system-of-system (SoS) approaches.

Defense and aerospace systems are characterized as complex technical systems with many stakeholders and compressed development timelines. The systems must also be highly available and work in extreme conditions all over the world—from deserts to rain forests and to arctic outposts. Defense and aerospace systems are also characterized by long system life cycles, so logistics is of prime importance. Most defense and aerospace systems have a strong human interaction, so usability/human systems integration is critical for successful operations.

Since SE has a strong heritage in defense and aerospace, much of the SE processes in this handbook can be used as is in a straightforward manner, with just the normal project tailoring for the unique aspects of the project. It is important to note that as ISO/IEC/IEEE 15288 has evolved into a more domain- and country-neutral SE standard, some care must be taken to ensure that the defense and aerospace focus is reasserted upon application. The defense organizations of many countries have specific policies, standards, and guidebooks to guide the application of SE in their environment.

8.2.4 Infrastructure Systems

Infrastructure systems are significant physical structures used for commodity transfer (e.g., roads, bridges, railroads, mass transit, and electricity, water, wastewater, and oil and gas distribution) or are major industrial plants (e.g., oil and gas platforms, refineries, mines, smelters, nuclear reprocessing, water and wastewater treatment, and steelworks). The domain is multidisciplinary, vast and diverse, and characterized by large-scale projects often with loosely defined boundaries, evolving system architectures, long implementation (which can exceed a few decades) and asset life periods, and multiphase life cycles. Infrastructure systems are distinguished from manufacturing and production by most often being one-off, large objects and where construction takes place on a site rather than in a factory.

8.2.3 防务和航空航天系统

虽然自古以来已经以某种形式实践过 SE，但是如今为人所知的 SE 为现代定义之根，则源于 20 世纪的防务和航空航天系统。由于技术进步的出现，促使系统复杂性和系统综合挑战等级的持续增加，已经在 50 年代末期和 60 年代早期将 SE 视为一个截然不同的活动。对 SE 的需要随着数字计算机和软件的大规模引入而增加。防务和航空航天的持续演进将系统方法用于针对诸如 COTS 技术的广泛适应和系统之系统（SoS）方法的使用等议题。

防务和航空航天系统的特征在于具有诸多利益攸关者和紧缩的开发时间表的复杂技术系统。该系统还必须在全球极端条件（从沙漠到雨林再到北极前哨）下高度可用和工作。防务和航空航天系统的特征亦在于长的系统寿命周期，因此，后勤是极其重要的。大多数防务和航空航天系统具有很强的人员交互作用，因此，可用性/人员系统综合对于成功运行十分关键。

因为 SE 具有强烈的防务和航空航天传承，所以本手册中的大多数 SE 流程能够以直接的方式使用，仅对项目的独特方面进行正常项目剪裁即可。重要的是要注意到当 ISO/IEC/IEEE 15288 已演进为一个更加领域中立且国家中立的 SE 标准时，必须关注确保依据应用而重申防务和航空航天的聚焦点。诸多国家的防务组织已拥有用于指导 SE 在其环境中应用的特定策略、标准和指导手册。

8.2.4 基础设施系统

基础设施系统是用于公共品转移的重要物理结构（例如公路、桥梁、铁路、公共交通工具以及电、水、废水和石油天然气的分配系统）或是主要的工业设施（例如石油天然气平台、精炼厂、矿山、冶炼厂、核燃料再处理、水和废水处理以及炼钢厂）。这个领域是多学科的、广大的和多种多样的，且其特征在于通常只具有松散定义的边界、持续演进的系统架构、漫长的实施过程（可超过几十年）和资产生命时期和多阶段生命周期的大型项目。基础设施系统与制造和生产的区别在于其很多时候是一次性的、大型物体并且是在现场建造而不是在工厂中。

Infrastructure projects tend to exhibit a degree of uniqueness, complexity, and cost uncertainty. A significant proportion of time and cost is spent during the construction stage. These difficulties are exacerbated by changes in project environments (economical, political, legislative, technological, etc.) and hence to the stakeholders' expectations and design solutions that take place over an extended period of time.

Unlike other SE domains, most infrastructure projects cannot be standardized and do not involve a prototype. In addition, significant external interfaces exist, and failures in one element may cascade to other elements. For example, falling debris from the World Trade Center attack damaged the water system and flooded the New York Stock Exchange, resulting in severe impacts to an interconnected system of infrastructure systems.

Within infrastructure, many of the engineering disciplines (e.g., civil, structural, mechanical, chemical, process, and electrical) have well-established, traditional practices and are guided by industry codes and standards. SE practices are more developed in the high-technology subsystems that involve software development (e.g., modern communications-based train control systems, intelligent transportation systems, telemetry and process control) and particularly in industries where system safety is paramount.

The benefit of SE to the infrastructure domain lies in the structured approach to delivering and operating a multidisciplinary, integrated, and configurable system and needs to align with the associated project management and asset management practices. Many of the processes described in this handbook are being employed to handle that complexity but using a different terminology. The "Guide for the Application of Systems Engineering in Large Infrastructure Projects" (INCOSE-TP-2010-007-01, 2012) emphasizes the relationship between system, work, and organizational breakdown structures and the importance of good configuration management, not only through the project life cycle but during the whole asset lifetime. These items and the interactions between them are determined by a multitude of constraints, with the physical structure of the object and access being the most obvious but also many others, such as availability of suitable contractors and of manpower and machines, fabrication and shipping durations, and ones arising from the environment in which the construction takes place.

Infrastructure projects tend to quickly define the high-level design solution (e.g., a high-speed railway or nuclear power plant), and therefore, the greatest benefit of applying SE principles is gained in the systems integration and construction stage. Careful analysis to identify all the potentially affected organizations, structures, systems, people, and processes is required to ensure proposed changes will not adversely impact other areas and will lead to the required outcome.

基础设施项目倾向于呈现独特性、复杂性和成本不确定性的一种程度。相当比例的时间和成本是在建造阶段花费的。这些困难由于在项目环境（经济环境、政治环境、法律环境、技术环境等）中的变化而被加剧因此利益攸关者期望和设计解决方案的改变也会在延长的时间内发生。

与其他 SE 领域不同，大多数基础设施项目不能被标准化并且不涉及原型机。此外，存在重要的外部接口，并且一个元素的故障可级联到其他元素。例如，世界贸易中心遇袭时坠落的碎片破坏了供水系统，并淹没了纽约证券交易所，对基础设施系统的互连系统产生了严重的影响。

在基础设施内部，大多数工程学科（例如，土木、结构、机械、化学、过程以及电气）具有业已建立的传统实践并且受行业规范和标准的指导。在涉及软件开发（例如，基于现代通信的列车控制系统、智能交通系统、遥测和过程控制）的高技术子系统中以及特别是在系统安全最为重要的行业中，对 SE 实践进行了更多开发。

SE 对基础设施领域的效益在于交付和运行多学科的、综合的和可配置的系统的结构化方法，并且需要与相关的项目管理实践和资产管理实践一致。本手册中描述的大多数流程正是用于处理复杂性的，但使用不同的术语。《大型基础设施项目中系统工程的应用指南》（INCOSE-TP-2010-007-01，2012）强调不仅是项目生命周期中的而且是整个资产使用寿命期间的系统、工作和组织分解结构间的关系以及良好构型管理的重要性。这些项目及其之间的交互由多个约束确定，且对象和接近的物理结构是最明显的，但还有诸多其他约束，例如，合适的承包商以及人力和机器的可用性、制造和装运的持续时间以及由建造所处的环境引起的约束。

基础设施项目倾向于快速定义高层级设计解决方案（例如，高速铁路或核发电厂），因此，获得应用 SE 原理的最大效益是在系统综合和建造阶段获得应用 SE 原则的最大效益。需要进行仔细分析来识别所有可能受到影响的组织、结构、系统、人员和流程，以确保建议的变更不会对其他领域造成不利影响并且将产生所要求的成果。

The domain could benefit from better organization and integration of activities leading into the construction stage of the project life cycle. This would help manage the risks and uncertainty associated with cost estimating and changing scope and could improve construction productivity, hence making the industry more cost-effective. SE efforts tailored to meet the uniqueness characterized by infrastructure projects control the project's internal and external dynamics (including uncertainty caused by natural events) and consider the project and postproject conditions will allow the domain to respond to the unique challenges it faces due to its broad, diverse, unique, and unpredictable makeup.

The documentation of infrastructure processes is largely company and project specific, although based on general standards, such as the ISO 9000/10000 series (quality management), ISO 10845 (construction procurement), and ISO 12006 (organization of information about construction works), as well as various national and international standards for the contractual (legal) framework in which a construction project is embedded. The choice of contractual framework and the allocation of liability and commercial risk is a major factor in designing the construction process.

8.2.5 Space Systems

Space systems are those systems that leave the Earth's atmosphere or are closely associated with their support and deployment. Due to the extremely high costs and physical exertion of deploying assets into Earth orbit or beyond, space systems typically require high reliability with no maintenance other than software changes. This makes it necessary for all system elements to work the first time or be compensated by complex operational workarounds.

SE was developed in large part due to the demands of the space race and associated defense technologies such as ballistic missiles. The discipline is very mature in this domain and does not require adaptation.

Key emphases of SE in the space domain are validation and verification, testing, and integration of highly reliable, well-characterized systems. Risk management is also key in determining when to incorporate new technologies and how to react to changing requirements through multiyear developments and programmatic challenges. The traditional SE Vee approach has its basis in space systems, as they are typically relatively new designs conceived, built, and deployed by agencies or prime contractors. Declining budgets are making the unity of vision less common as consortia and partnerships are made to pool resources.

该领域可受益于对引入项目生命周期施工阶段的活动进行更好地组织和综合。这将有助于管理风险以及与成本估计和变化范围相关的风险和不确定性并可提高建造的生产力，因此使得行业更具成本效益。以基础设施项目控制项目的内部和外部动态（包括自然事件造成的不确定性）为特征的，经过对 SE 努力的剪裁而得以满足的 SE 努力，并认为项目和项目后条件将允许该领域对由广泛的、多种多样的、独特的且不可预见因素所组成的独特挑战做出相应。

尽管基于诸如 ISO 9000/10000 系列（质量管理）、ISO 10845（施工采购）以及 ISO 12006（关于施工的信息的组织）的通用标准以及用于施工项目所嵌入的合同（法律）框架的各种不同的国家和国际标准，但基础设施流程的文件在很大程度上是于公司和项目所特定的。合同框架的选择以及法律责任义务与商业风险的分配是设计建造流程中的主要因素。

8.2.5 空间系统

空间系统是离开地球大气层或与其维持和部署密切相关的那些系统。由于将资产部署到地球轨道或地球轨道以外需要极其高昂的代价和物力，空间系统通常要求高可靠性，且除了软件更改以外不需进行维护。这使得所有系统元素必需一次性工作或由复杂的运行应急手段进行补偿。

SE 在很大程度上是因为空间竞赛和诸如弹道导弹相关的防务技术的要求而发展的。该学科在这个领域是非常成熟的并且不需要适应性改变。

SE 在空间领域的着重点是确认与验证、测试和高可靠性的综合、具有良好特征化的系统。风险管理亦是确定在多年开发和规划性挑战中何时纳入新技术以及如何对变更需求做出反应的关键。因为空间系统通常是由专门机构或主承包商构想相对新颖的设计、构建和部署，所以传统 SE 的 V 形方法的基础在于空间系统。随着联合和伙伴关系被要求作为资源池（筹资），所以持续降低的预算正在使得愿景的统一不太常见了。

A large number of standards are used in the space domain. Telecommunications is a major source, since spectra and noninterference must be negotiated on a global basis. Electrical and data standards are also used in many parts of both in-space and ground support systems. National space agencies and militaries often set standards as well (e.g., the European Cooperation for Space Standardization, US "MIL" Standards). An excellent example of a space-based SE handbook is freely available from the NASA (2007b). As space systems are deployed by more countries, standardization via ISO and IEEE is becoming more common for an increasing set of interoperability issues.

8.2.6 (Ground) Transportation Systems

The use of SE has emerged as the standard approach for complex capital programs in the fields of aerospace engineering, defense, and information technology. In the transportation industry, however, the assimilation of SE principles and methods has grown more slowly. The historical orientation of most transportation agencies has been to structure capital projects by engineering disciple and adopt low-bid procurement methods.

However, in the past few decades, the use of SE in transportation has been growing, with several progressive transit agencies (including London Underground and Network Rail in the United Kingdom, ProRail in the Netherlands, New York City Transit in the United States, and the Land Transport Authority in Singapore) having established SE departments and many others beginning to embed SE principles into their capital programs.

Emerging experience from these agencies suggests the following aspects are most relevant to consider when applying and tailoring SE principles to the transportation domain:

- *Recognize the single-discipline tendency*—the historical orientation in transportation agencies is to organize projects by engineering discipline. This can generate disciplinary silos that flow from early engineering through procurement and into system design, build, and test. SE must work across these silos, which can introduce tension to organizations used to working in the traditional way.
- *Working with in-service SoS*—many transportation systems are large, distributed, in-service SoS. This means that transportation SE work is often focused on systems upgrade and must work with and around the existing operation while still maintaining appropriate levels of public service.

大量的标准被用于空间领域。因为频谱和非干扰性必须在全球基础上进行协商，所以电信是一个主要来源。电气和数据标准亦被用于空中和地面支持系统的诸多部分。国家航天局和军方也经常设定标准（例如，欧洲空间标准化组织、美国"MIL"标准）。基于空间的 SE 手册的卓越示例可从 NASA（2007b）免费获取。随着更多国家部署空间系统，对于不断增加的一系列互操作性问题，借助 ISO 和 IEEE 的标准化正变得更加常见。

8.2.6 （地面）运输系统

SE 的使用已经成为航空航天工程、防务和信息技术领域中用于复杂资本规划的标准方法。然而，在运输业中，对 SE 原则和方法的消化吸收却一直缓慢。大多数运输机构的传统导向依然是被工程信徒们结构资本项目并采用最低中标价的采购方法。

然而，在过去的几十年中，SE 在运输业中的使用正在增加，若干先进的运输机构（包括英国伦敦地铁和铁路网公司、荷兰铁路设施管理局、美国纽约公共交通部以及新加坡陆路交通管理局）已经建立了 SE 部门并且诸多其他运输机构开始在其资本规划中嵌入 SE 原则。

从这些机构显现出的经验表明下述方面是与按运输领域应用并剪裁 SE 原则时所考虑的内容最为相关的：

- *意识到单学科倾向*——运输机构的历史定位是按工程学科组织项目。这可形成从早期工程经由采购流向系统设计、构建和试验的学科"壁垒"。SE 必须穿过这些"壁垒"起作用，才可向以传统方式工作的组织引入张力。

- *与服役中的系统之系统一起工作*——许多运输系统是大型的、分布式的、服役中的系统之系统。这意味着运输 SE 工作在依然保持适当的公共服务水平的同时往往聚焦于系统升级并且必须与现有运行一起工作并围绕其工作。

- *Delivering socioeconomic benefits are often the key driver, yet the operator has an important role*—transportation systems primarily existing to provide socioeconomic benefits to the public. The key drivers for many transportation agencies are therefore often focused on improving customer experience and public safety. However, care should be taken to also focus on operational benefits, and operability and maintainability should be considered throughout.

- *Demonstrating the value of SE*—many transportation agencies have a historic divide between capital delivery and operations that persists into today's organizations. Project managers face cost and schedule pressure to deliver quickly, and this can sometimes generate opposition from those who perceive that the emphasis SE places on early-phase analysis injects delay into the project manager's schedule, while the benefits accrued from the analysis are felt most strongly by the operational stakeholders who are outside of the project manager's immediate organization.

- *Considering the commercial and public pressures*—there is increasing dependence on private investment to fund the larger transportation infrastructure projects, which increases the demand for a faster return on investment (ROI), which in turn places greater restrictions on time available to explore the problem space. Furthermore, while the desire to select to a jump to technology solution is endemic in many sectors, many transportation agencies face acute pressure from the public, the media, and politicians to demonstrate early signs of progress in terms that internal and external stakeholders understand— typically in specifics of technology. These pressures can sometimes negatively influence an agency's ability to apply SE to its projects.

8.3 APPLICATION OF SYSTEMS ENGINEERING FOR PRODUCT LINE MANAGEMENT

Product line management (PLM) is a combination of product, process, management, and organization to migrate from single-system engineering to a product line approach. PLM can support the goal of improved organizational competitiveness as it decreases the development cost, increases the quality, and enlarges the product catalog.

- *尽管交付社会经济效益往往是关键驱动因素，但操作者仍具有重要的作用*——运输系统主要是为公众提供社会经济效益而存在。因此，诸多运输机构的关键驱动因素往往聚焦于改善客户体验和公共安全。然而，亦应注意聚焦于运行效益，并且应始终考虑到可操作性和可维护性。

- *展示 SE 的价值*——许多运输机构在资本交付和运行之间具有一个历史性的分工，并在今天的组织中延续：项目经理面临迅速交付的成本和进度压力，并且这有时能够引起一些人的反对，这些人认为 SE 所强调的早期分析使项目经理的进度延迟，而由该分析所产生的效益却被项目经理当前组织以外的运营利益攸关者最强烈地感受到了。

- *考虑商业和公众压力*——对用于为大型运输基础设施项目提供资金的私人投资的依赖性持续在增长，这就增加了对更快的投资回报（ROI）的需要，进而对于探索问题空间的可用时间带来了更大的限制。此外，虽然选择跳跃性的技术解决方案的愿望在诸多行业中是常见的，但许多运输机构都面临来自公众、媒体和政治家对以内部和外部利益攸关者理解的方式说明早期进展迹象（通常是技术细节）的巨大压力。这些压力有时对机构在其项目中应用 SE 的能力产生负面影响。

8.3 用于产品线管理的系统工程应用

产品线管理（PLM）是从单系统工程迁移至产品线方法的产品、流程、管理和组织的组合。PLM 可通过降低开发成本、提高质量并扩充产品目录而支持提升组织竞争力的目标。

For the customer, a product line approach allows an organization to offer a family of products, which are adapted to the customer. This can lead to optimization of the product or service, the cost and time of acquisition, the quality of the system, and the cost of ownership. For the organization, PLM leads to optimization of its commercial position and its industrial loads, usually highlighting the identical functionality leading to standardization. These relationships are shown in Figure 8.3.

FIGURE 8.3 Product line viewpoints. Reprinted with permission from Alain le Put. All other rights reserved.

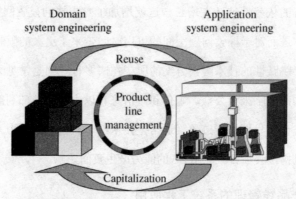

FIGURE 8.4 Capitalization and reuse in a product line. Reprinted with permission from Alain le Put. All other rights reserved.

When implementing PLM, both the domain and the application SE processes must change as shown in Figure 8.4. The domain products (or generic products) address the product line requirements and are the results of domain SE. The generic products produced by the domain SE activities are capitalized from the application artifacts (e.g., generic requirements, generic architecture, and generic tests). These artifacts may be common or variable. The application products are the results of application SE. These products are the result of the instantiation of the generic products (i.e., reuse).

对于客户，产品线方法允许组织提供一族产品以适应于客户。这能够促使产品或服务、采办的成本和时间、系统质量以及拥有成本的优化。对于组织，PLM 促使其商业地位及其工业负担的优化，通常强调引向标准化的相同功能性。这些关系如图 8.3 所示。

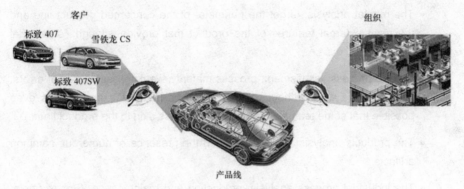

图 8.3 产品线视角。经 Alain le Put 许可后转载。版权所有。

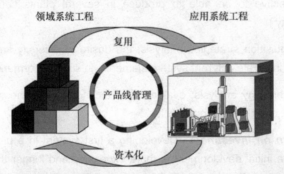

图 8.4 产品线中的资本化和复用。经 Alain le Put 许可后转载。版权所有。

当实施 PLM 时，领域 SE 流程和应用 SE 流程必须按图 8.4 所示进行变化。领域产品（或通用产品）满足产品线需求并且是领域 SE 的结果。该领域 SE 活动产生的通用产品依据应用制品（例如通用需求、通用架构以及通用试验）进行资本化。这些制品可能是共同的或可变的。应用产品是应用 SE 的结果。这些产品是通用产品（即，复用）的实例化的结果。

8.3.1 Product Line Scoping

Product line scoping defines the main features of the product line, which provide sufficient reuse potential to justify the setting up or the evolution of a product line organization. Product line scoping is mainly based on:

- The market analysis: target the perimeter of the concerned product line and the main external features of the product that provide enough commercial potential.

- The SE process assessment: process maturity, weaknesses source of errors, and repetitive work without greater added value. (For example, it is even possible that some teams already have a process akin to the product lines.)

- The products analysis: product tree and the presence of numerous common artifacts.

- The industrial process analysis: production and maintenance. (For example, the objective to be able to produce in several plants can be source of variability.)

- The acquisition strategies analysis: the desire to diversify suppliers and the need to integrate different system elements (in size, performance, etc.).

- The technology analysis: evaluation of technology readiness levels, design maturity, and domain applicability.

8.3.1.1 Return on Investment Developing a first system in a product line is an investment. The initial development is more expensive and longer than for a single system development. But the following systems developed in the same product line are less expensive and may be delivered earlier (i.e., reduced time to market). In either case, the ROI must be evaluated, measured and managed.

As illustrated in Figure 8.5 three projects are developed either in a single-system engineering classical approach (top part) or in a product line SE approach (middle part). The ROI (bottom part) is initially negative (e.g., investment phase) and then positive (benefits of the product line approach). In this example, the break-even point is reached for the third project.

8.3.1 产品线范围界定

产品线范围界定定义产品线的主要特征，其提供足够的复用潜力以证明产品线组织的建立或演变的正确性。产品线范围界定主要基于：

- 市场分析：将相关产品线的周围以及提供足够商业潜力的产品的主要外部特征作为目标。

- SE 流程评估：流程成熟度、错误的劣势来源以及无较大附加值的重复性工作。（例如，甚至有可能的是一些团队已经具有一个类似产品线的流程。）

- 产品分析：产品树和许多共用制品的存在。

- 行业流程分析：生产和维护。（例如，能够在多个工厂中产生的目标可以是可变性之源。）

- 采办策略分析：使供应商多样化的愿望以及综合（尺寸、性能等）不同的系统元素的需要。

- 技术分析：技术成熟度、设计成熟度以及领域适用性的评估。

8.3.1.1 投资回报　在产品线中开发第一个系统是一种投资。最初的开发比一个单一系统的开发更加昂贵且时间更长。但是，之后在相同产品线中开发的诸多系统是比较便宜的并且可能更早交付（即，缩短进入市场的时间）。无论是哪种情况，均应对 ROI 进行评估、衡量和管理。

如图 8.5 所示，三个项目均是以单系统工程经典方法（上部）或产品线 SE 方法（中部）进行开发的。ROI（底部）最初是负数（例如，投资阶段），然后是正数（产品线方法的效益）。在本示例中，第三个项目达到收支平衡点。

8.4 APPLICATION OF SYSTEMS ENGINEERING FOR SERVICES

This section introduces the concept of service SE, where SE methodologies are adapted to include a disciplined, systemic, and service-oriented, customer-centric approach among different stakeholders and resources for near-real-time value cocreation and service delivery. The twenty-first-century technology-intensive global services economy can be characterized as *"information driven, customer-centric, e-oriented, and productivity focused."* It requires transdisciplinary collaborations between society, science, enterprises, and engineering (Chang, 2010). Several researchers and businesses are utilizing a socioeconomic and technological perspective to investigate end-user (customer) interactions with enterprises by developing formal methodologies for value cocreation and productivity improvements. These methodologies have evolved into service SE, which mandates a disciplined and systemic approach (service oriented, customer-centric) among different stakeholders and resources in the design and delivery of the service to help customize and personalize service transactions to meet particular customer needs (Hipel et al., 2007; Maglio and Spohrer, 2008; Pineda et al., 2014; Tien and Berg, 2003; Vargo and Akaka, 2009).

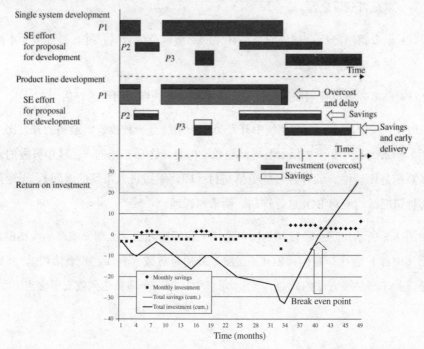

FIGURE 8.5 Product line return on investment. Reprinted with permission from Alain le Put. All other rights reserved.

8.4 用于服务的系统工程应用

本节引入服务 SE 的概念,其中调整 SE 方法论以包括不同利益攸关者与资源间用于近实时价值共创和服务交付的科律性的、系统的、面向服务的、以客户为中心的方法。21 世纪技术密集型的全球服务经济的特征在于"信息驱动、以客户为中心、面向 e 以及聚焦于生产率。"这要求社会、科学、复杂组织体和工程之间的跨学科密切协作(Chang,2010)。多个研究和业务正在利用社会经济学和技术视角以通过开发用于价值共创和生产率改进的正式方法论的方式调查研究最终用户(客户)与复杂组织体的交互作用。这些方法论已经演进为服务 SE,其在设计和交付服务过程中强制执行不同利益攸关者和资源之间的科律性的且系统的方法(面向服务的、以客户为中心)以有助于用户化和个性化服务交易,从而满足特定客户的需要(Hipel 等,2007;Maglio 和 Spohrer,2008;Pineda 等,2014;Tien 和 Berg,2003;Vargo 和 Akaka,2009)。

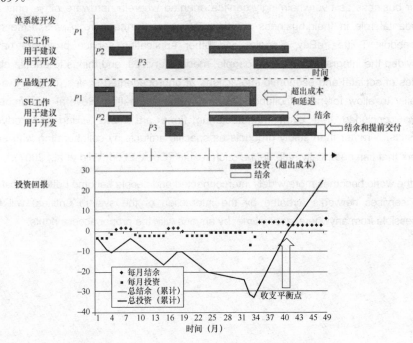

图 8.5 产品线投资回报。经 Alain le Put 许可后转载。版权所有。

Service systems can be viewed as an SoS, where individual, heterogeneous, functional systems are linked together to realize new features/functionalities of a meta- system and to improve robustness, lower cost, and increase reliability. For service systems, understanding the integration needs among loosely coupled systems and system entities along with the information flows required for both governance and operations, administration, maintenance, and provisioning (OAM&P) of the service presents major challenges in the definition, design, and implementation of services (Domingue et al., 2009; Maier, 1998). Cloutier et al. (2009) presented the importance of Network-Centric Systems (NCS) for dynamically binding different system entities in engineered systems rapidly to realize adaptive SoS that, in the case of service systems, are capable of knowledge emergence and real-time behavior emergence for service discovery and delivery.

The typical industry example given of this progression toward services is the International Business Machines (IBM), which still produces some hardware but view their business as overwhelmingly service oriented wherein hardware plays only an incidental role in their business solutions services. Furthermore, Apple, Amazon, Facebook, Twitter, eBay, Google, and other application service providers have provided the integrated access to people, media, services, and things to enable new styles of societal and economic interactions at unprecedented scales, flexibility, and quality to allow for mass collaboration and value cocreation. Several researchers have concluded that even manufacturing firms are more willing to produce "results," rather than solely products as specific artifacts by collaborating with end users that can potentially help define such results (Cook, 2004; Wild et al., 2007).

As the world becomes more widely interconnected and people become better educated, the services networks (created by the interaction of the system entities) will be accessible from anywhere, at any time, by anyone with the proper access rights.

服务系统可被视为一个系统之系统，其中单个的、不同种类的功能系统被链接在一起以实现元系统的新特征/功能性并提高鲁棒性、降低成本和增加可靠性。对于服务系统，理解松散耦合系统与系统实体之间的综合需要以及服务的治理和运行、管理、维护与供应（OAM&P）所需的信息流在服务的定义、设计以及实施方面展现主要挑战（Domingue 等，2009；Maier，1998）。Cloutier 等（2009）提出了以网络为中心的系统（NCS）对于在设计的系统中迅速地动态结合不同的系统实体以实现适应性系统之系统的重要性，就服务系统而言，以网络为中心的系统能够进行用于服务发现和交付的知识涌现以及实时行为涌现。

所给出的向着服务进展的典型行业示例是 IBM，该公司仍生产某些硬件，但将其业务视为是完全面向服务，其中硬件仅在其业务解决方案服务中起到附属作用。此外，苹果公司、亚马逊、脸谱网、推特网、易趣网、谷歌以及其他的应用服务提供商已经提供了对人、媒体、服务以及事物的综合访问，以便能够以前所未有的规模、灵活性和质量实现新型的社会与经济交互，以允许大规模协作和价值共创。许多研究人员已得出的结论是：即使是制造工厂也更愿意通过与能够潜在地帮助定义结果的最终用户密切合作来生产"结果"，只作为特定制品的产品（Cook，2004 年；Wild 等，2007）。

随着世界各国更加广泛地相互联系以及人们受到更好的教育，（系统实体相互作用所创建的）服务网络将在任何地方、任何时间、由具有适当访问权的任何人使用。

8.4.1 Fundamentals of Service

Services are activities that cause a transformation of the state of an entity (a person, product, business, region, or nation) by mutually agreed terms between the service provider and the customer. Individual services are relatively simple, although they may require customization and significant backstage support (e.g., knowledge management, logistics, decision analysis, forecasting) to assure quality and timely delivery. A *service system* enables a service and/or set of services to be accessible to the customer (individual or enterprise) where stakeholders interact to create a particular service value chain to be developed and delivered with a specific objective (Spohrer and Maglio, 2010). In service systems practice, the *service value chain* is described in terms of links among the system entities connected via the NCS. A value proposition can be viewed as a request from one service system to another to run an algorithm (the value proposition) from the perspectives of multiple stakeholders according to culturally determined value principles (Spohrer, 2011). Spath and Fahnrich (2007) defined a service metamodel comprising nine types of system entities and their corresponding attributes as shown in Table 8.2.

Thus, a service or service offering is created by the relationships among service system entities (including information flows) through business processes into strategic capabilities that consistently provide superior value to the customer. From an SE perspective, participating system entities dynamically configure four types of resources: people, technology/environment infrastructure, organizations/institutions, and shared information/symbolic knowledge in the service delivery process.

Systems are part of other systems that are often expressed by *system hierarchies* (Skyttner, 2006) to create a multilevel hierarchy. Thus, the service system is composed of service system entities that interact through processes defined by governance and management rules to create different types of outcomes in the context of stakeholders with the purpose of providing improved customer interaction and value cocreation. This concept can be extended to create the service system hierarchy shown in Figure 8.6.

8.4.1 服务的基本原则

服务是按服务提供商和客户共同商定的条款引起实体（个人、产品、业务、区域或国家）状态转换的活动。尽管可能要求客户定制化以及大量的后台支持（例如，知识管理、后勤、决策分析、预测）以确保质量和按时交付，但个人服务是相对简单的。服务系统使某一服务和/或系列服务能够为客户（个人或复杂组织体）易达，其中利益攸关者相互作用以创建一个按照特定目标开发和交付的特定服务价值链（Spohrer 和 Maglio，2010）。在服务系统实践中，依照经由 NCS 连接的系统实体间的联系描述服务价值链。价值主张可被视为从一个服务系统到另一个服务系统的请求，以按照文化确定价值的原则从多个利益攸关者的视角来运行一个算法（价值主张）（Spohrer，2011）。Spath 和 Fahnrich（2007）定义了包含如表 8.2 所示的九类系统实体及其相应属性的服务元模型。

因此，通过从业务流程到战略能力的服务系统实体（包括信息流）之间的关系创建服务和服务提供物，始终如一地向客户提供卓越价值。从 SE 的视角来看，参与的系统实体动态地配置四类资源：服务交付流程中的人员、技术/环境基础设施、组织/机构以及共享的信息/符号知识。

系统是通常表示为系统层级结构（Skyttner，2006）的其他系统的一部分以创建多层级的层级结构。因此，服务系统是由在治理和管理规则所定义的流程中相互作用的诸多服务系统实体组成的，以在利益攸关者的背景环境中创建不同类型的成果，目的是提供改进的客户相互作用和价值共创。可扩展本概念以创建如图 8.6 所示的服务系统层级结构。

The *fundamental attributes* of a service system include togetherness, structure, behavior, and emergence. As mentioned earlier, today's global economy is very competitive, and the service system's trajectory should be well controlled as time goes by (Qiu, 2009) since services are "real time in nature and are consumed at the time they are co-produced" (Tien and Berg, 2003), that is, during service transactions. The service system should evolve and adapt to the conditions within the business space in a manner to ensure that the customized service behaves as expected. This adaptive behavior of service system implies that its design must be truly transdisciplinary and must include methodologies from social science (e.g., sociology, psychology, and philosophy) to natural science (mathematics, biology, etc.) to management (e.g., organization, economics, and entrepreneurship) (Hipel et al., 2007).

TABLE 8.2 Attributes of System Entities

Entity type	Attributes
Customer	Features, attitudes, preferences, requirements
Goals	Business, service, customer
Inputs	Physical, information, knowledge, constraints
Outputs	Physical, information, knowledge, waste, customer satisfaction
Processes	Service provision, service delivery, service operations, service support, customer relationships, planning and control
Human enablers	Service providers, support providers, management, owner organization, customer
Physical enablers	Enterprise, organizations, buildings, equipment, enabling technologies at customer premises (e.g., desktop 3D printers), furnishings, location, etc.
Informatics enablers	Information; knowledge; methods, processes, and tools (MPTs); decision support; skill acquisition
Environment	Political, economic, social, technological, environmental factors

服务系统的基本属性包括融合、结构、行为和涌现性。如前所述,如今的全球经济是非常具有竞争性的,并且服务系统的轨迹应随着时间的流逝得到很好的控制(Qiu,2009),因为服务"在本质属性上是实时的并且是在共同产生时被消费的"(Tien 和 Berg,2003),即服务交易期间。服务系统应以确保客户化服务按期望行为的方式演进并适应于业务空间范围内的条件。服务系统的这种适应行为意味着其设计必须是真正跨学科的并且必须包括从社会科学(例如,社会学、心理学和哲学)到自然科学(数学、生物学等)再到管理学(例如,组织、经济和企业家精神)的方法论(Hipel 等,2007)。

表 8.2　系统实体的属性

实体类型	属性
客户	特征、态度、偏好、需求
目标	业务、服务、客户
输入	物理、信息、知识、约束
输出	物理、信息、知识、废物、客户满意度
流程	服务条款、服务交付、服务操作、服务支持、客户关系、规划和控制
人员使能项	服务提供商、支持提供商、管理、所有者组织、客户
物理使能项	复杂组织体、组织、建筑、设备、客户处的使能技术(例如,桌面3D打印机)、装备、位置等
信息使能项	信息;知识;方法、流程、工具(MPT);决策支持;技能采办
环境	政治、经济、社会、技术、环境因素

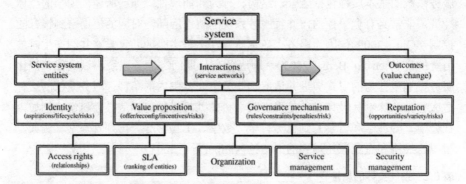

FIGURE 8.6 Service system conceptual framework. Reprinted with permission from Dr. James C. Spohrer. All other rights reserved.

8.4.2 Properties of Services

Services not only involve the interaction between the service provider and the consumer to produce value but have other intangible attributes like quality of service (e.g., ambulance service availability and response time to an emergency request). The demand for service may have loads dependent on time of day, day of week, season, or unexpected needs (e.g., natural disasters, product promotion campaigns, holiday season, etc.), and services are rendered at the time they are requested. Thus, the design and operations of service systems "is all about finding the appropriate balance between the resources devoted to the systems and the demands placed on the system, so that the quality of service to the customer is as good as possible" (Daskin, 2010).

A *service-level agreement* (SLA) represents the negotiated service-level requirements of the customer and establishes valid and reliable service performance measures to ensure that service providers meet and maintain the prescribed quality of service, user-perceived performance, and degree of satisfaction of the user. The service system's SLAs are then the composition of these categories evaluated on a systemic level to ensure consistency, equity, and sustainability of the service (Spohrer, 2011; Theilmann and Baresi, 2009; Tien and Berg, 2003).

The twenty-first century is witnessing accelerated technology development and global mass collaboration as an established mode of operation. Value cocreation is achieved as loosely entangled actors or entities come together in unprecedented ways to meet mutual and broader market requirements.

系统工程的剪裁流程和应用

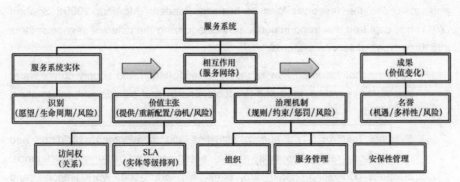

图 8.6 服务系统概念框架。经 Dr.James C. Spohrer 许可后转载。版权所有。

8.4.2 服务的特性

服务不仅涉及服务提供商和消费者之间产生价值的相互作用还具有类似服务质量的其他无形属性（例如，救护车服务可用性以及对紧急请求的响应时间）。服务需求可能具有依赖于当日时间、星期几、季节或意料之外的需要（例如，自然灾害、产品促销活动、度假旺季等）的工作量，并在要求时提供服务。因此，服务系统的设计和运行"全是有关发现专用于系统的资源与对系统的要求之间的适当平衡，以便提供给客户的服务质量尽可能地好"（Daskin，2010）。

服务等级协议（SLA）表示经协商的客户服务等级需求并建立有效且可靠的服务性能测度，以确保服务提供商满足和保持规定的服务质量、用户感知的性能以及用户的满意度。那么服务系统的 SLA 是以确保服务的一致性、公平和持续性的系统等级评估的这些类别的构成（Spohrer，2011；Theilmann 和 Baresi，2009; Tien 和 Berg，2003）。

21 世纪正见证加速的技术开发和全球大规模协作作为一种业已建立的运作模式。随着松散卷入的参与者或实体以前所未有的方式聚集来满足相互和更广泛的市场需求时，价值共创得以实现。

This transformation will radically reshape the nature of work, the boundaries of the enterprise, and the responsibilities of business leaders (McAfee, 2009). Spohrer (2011) has captured this trend in service by categorizing the different service sectors into three types of service systems:

- *Systems that focus on flow of things*: transportation and supply chain, water and waste recycling, food and products, energy and electric grid, information, and cloud.

- *Systems that focus on human activities and development*: buildings and construction, retail, hospitality/media, entertainment, banking and finance, business consulting, healthcare, family life, education, work life/jobs, and entrepreneurship.

- *Systems that focus on governing* (city, state, nation): the classification helps in identifying different objectives and constraints for the design and operations of the service system (e.g., strategic policies under limited budget, education, strategic readiness for quick response, national defense, maximizing profit while minimizing cost) and determining the overlap and synergies required among different service entities and required science disciplines.

8.4.3 Scope of Service SE

Current enterprises must plan, develop, and manage the enhancements of their infrastructure, products, and services, including marketing strategies to include product and service offerings based on new, unexplored, or unforeseen customer needs with clearly differentiated value propositions. Taking the service SE approach is imperative for the service-oriented, customer-centric, holistic view to select and combine service system entities to define and discover relationships among service system entities to plan, design, adapt, or self-adapt to cocreate value. Major challenges faced by service SE include the dynamic nature of service systems evolving and adapting to constantly changing operations and/or business environments and the need to overcome silos of knowledge. Interoperability of service system entities through interface agreements must be at the forefront of the service SE design process for the harmonization of OAM&P procedures of the individual service system entities (Pineda,2010).

此转换将彻底地重塑工作的本质属性、复杂组织体的边界以及业务领导者的职责（McAfee，2009）。Spohrer（2011）已经通过将不同服务行业分类为三种类型的服务系统捕捉到服务的这种趋势：

- 聚焦于事物流动的系统：运输和供应链、水和废物再循环、食品和产品、能量和电网、信息和云。

- 聚焦于人类活动和开发的系统：建筑和建造、零售、食宿/媒体、娱乐、银行和金融、业务咨询、健康医疗、家庭生活、教育、工作生活/职业以及企业家精神。

- 聚焦于治理（城市、州、国家）的系统：该分类有助于识别用于设计和运行服务系统的不同目标和约束（例如，限定预算下的战略方针、教育、快速响应的战略准备、国防、最小化成本的同时最大化利润）和确定不同服务实体和所需的科学学科之间需要的重叠和协同效应。

8.4.3 服务 SE 的范围

当前的复杂组织体必须计划、开发和管理其基础设施、产品和服务的增强措施（包括市场策略）以包括基于新的、未探讨的或未预见到的、具有明显不同的价值主张的客户需要的产品和服务提供物。采用服务 SE 方法对于面向服务的、以客户为中心的整体视角是绝对必要的，以便选择和组合服务系统实体来定义和发现服务系统实体间的关系，从而计划、设计、适应或自适应以共创价值。服务 SE 面对的主要挑战包括演进和适应于不断变化的运行和/或业务环境的服务系统的动态本质属性以及克服知识隔阂的需要。为了单个服务系统实体的 OAM&P 程序的协调一致，通过接口协议的服务系统实体的互操作性必须位于服务 SE 设计流程的最前端（Pineda，2010）。

In addition, service systems require open collaboration among all stakeholders, but recent research on mental models of multidisciplinary teams shows integration and collaboration into cohesive teams has proven to be a major challenge (Carpenter et al., 2010).

Interoperability among the different service system entities becomes highly relevant in service SE since the constituent entities are designed according to stakeholder needs; the entity is usually managed and operated to satisfy its own objectives independently of other system entities. The objectives of individual service system entities may not necessarily converge with the overall objectives of the service system. Thus, the *service system design process* (SSDP) adapts traditional SE life cycle best practices, as illustrated by Luzeaux and Ruault (2010), Lin and Hsieh (2011), Lefever (2005), and the Office of Government Commerce (2009).

Another important role of service SE is the *management of the service design process* whose main focus is to provide the planning, organizational structure, collaboration environment, and program controls required to ensure that stakeholder's needs are met from an end-to-end customer perspective. The service design management process aligns business objectives and business operational plans with end-to-end service objectives including customer management plans, service management and operations plans, and operations technical plans (Hipel et al., 2007; Pineda et al., 2014).

8.4.4 Value of Service SE

Service SE brings a customer focus to promote service excellence and to facilitate service innovation through the use of emerging technologies to propose creation of new service systems and value cocreation. Service SE uses disciplined approaches to minimize risk by coordinating/orchestrating social aspects, governance (including security), environmental, human behavior, business, customer care, service management, operations, and technology development processes. Service systems engineers must play the role of an integrator, considering the interface requirements for the interoperability of service system entities—not only for technical integration but also for the processes and organization required for optimal customer experience during service operations. The service design definition process includes the definition of methods, processes, and procedures necessary to monitor and track service requirements verification and validation as they relate to the OAM&P procedures of the whole service system and its entities. These procedures ensure that failures by any entity are detected and do not propagate and disturb the operations of the service (Luzeaux and Ruault, 2010).

此外，服务系统要求在所有利益攸关者之间进行开放式协作，但是最近关于多学科团队心智模型的研究表明整合和协作为富有凝聚力的团队已被证明是一个主要挑战（Carpenter 等，2010）。

不同服务系统实体之间的互操作性在服务 SE 中变得高度相关，因为组成实体均是根据利益攸关者的需要设计的；实体的管理和运行是为了满足其自身的目的而独立于系统的其他实体。单个服务系统实体的目的可能不必趋近于服务系统总的目标。因此，如 Luzeaux 和 Ruault（2010）、Lin 和 Hsieh（2011）、Lefever（2005）以及政府商务办公室（2009）所述，服务系统设计流程（SSDP）要适应传统 SE 生命周期最佳实践。

服务 SE 的另一重要作用是对服务设计流程的管理，该流程的主要聚焦点是提供从端对端的客户视角确保利益攸关者的需要得到满足所需要的规划、组织结构、协作环境以及项目控制。服务设计管理流程使业务目标和业务运行计划与端对端的服务目标一致，包括客户管理计划、服务管理和运行计划以及运行技术计划（Hipel 等，2007；Pineda 等，2014）。

8.4.4 服务 SE 的价值

服务 SE 将客户聚焦点带到以通过使用新兴技术来促进服务卓越并助推服务创新，以便提出新服务系统和价值共创的创建。服务 SE 采用科律化的方法以通过协调/精心安排社会各方面、治理（包括安保性）、环境、人员行为、业务、客户关注、服务管理、运行和技术开发流程来使风险最小化。服务系统工程师必须扮演综合者的角色，考虑服务系统实体互操作性的接口需求——不仅用于技术综合也用于服务运行期间最佳客户体验所需的流程和组织。服务设计定义流程包括监控和跟踪服务需求验证与确认所必需的方法、流程和程序的定义，因为它们与整体服务系统及其实体的 OAM&P 程序有关。这些程序确保任何实体所导致的失效均被检测到并且不传播和不干扰服务的运行（Luzeaux 和 Ruault，2010）。

The world's economies continue to move toward the creation and delivery of more innovative services. To best prepare tomorrow's leaders, new disciplines are needed that include and ingrain different skills and create the knowledge to support such global services. Service systems engineers fit the T-shaped model of professionals (Maglio and Spohrer, 2008) who must have a deeply developed specialty area as well as a broad set of skills and capabilities in addition to the service system management and engineering skills, as summarized by Chang (2010).

8.5 APPLICATION OF SYSTEMS ENGINEERING FOR ENTERPRISES

This section illustrates the applications of SE principles in enterprise SE for the planning, design, improvement, and operation of an enterprise in order to transform and continuously improve the enterprise to survive in a globally competitive environment. Enterprise SE is the application of SE principles, concepts, and methods to the planning, design, improvement, and operation of an enterprise. Enterprise SE is an emerging discipline that focuses on frameworks, tools, and problem-solving approaches for dealing with the inherent complexities of the enterprise. Furthermore, enterprise SE addresses more than just solving problems; it also deals with the exploitation of opportunities for better ways to achieve the enterprise goals. A good overall description of enterprise SE is provided in the book by Rebovich and White (2011). For more detailed information on this topic, please see the enterprise SE articles in Part 4 of SEBoK (2014).

8.5.1 Enterprise

An enterprise consists of a purposeful combination (e.g., a network) of interdependent resources (e.g., people, processes, organizations, supporting technologies, and funding) that interact with each other to coordinate functions, share information, allocate funding, create workflows, and make decisions and their environment(s) to achieve business and operational goals through a complex web of interactions distributed across geography and time (Rebovich and White, 2011).

An enterprise must do two things: (1) develop things within the enterprise to serve as either external offerings or as internal mechanisms to enable achievement of enterprise operations and (2) transform the enterprise itself so that it can most effectively and efficiently perform its operations and survive in its competitive and constrained environment.

世界经济持续走向更多创新服务的创建和交付。为了最好地培养未来的领导者，需要诸多包容多彩的不同技能的新学科并创建支持所述全球服务的知识。如 Chang（2010）所述，服务系统工程师适配职业人员的 T 形模型（Maglio 和 Spohrer，2008），除了服务系统管理和工程技能外，其还必须具有深入开发的专业领域与一系列广泛的技能和能力。

8.5 用于复杂组织体的系统工程应用

本节阐明 SE 原理在复杂组织体 SE 中的应用，用于复杂组织体的规划、设计、改进和运行，以便复杂组织体的转型和持续改进，以求在全球竞争环境中的生存。复杂组织体 SE 是 SE 原理、概念和方法，对复杂组织体规划、设计、改进和运行的应用。复杂组织体 SE 是为应对复杂组织体固有复杂性而聚焦于框架、工具和问题求解途径的新兴学科。而且，复杂组织体 SE 涉及的不仅仅是求解问题；其亦涉及对实现复杂组织体目标的更好方式的机会的探索。Rebovich 和 White（2011）所著的书中对复杂组织体 SE 提供了足够全面的描述。关于此主题的更多详细信息，请参见 SEBoK（2014）第 4 部分中的复杂组织体 SE 条款。

8.5.1 复杂组织体

复杂组织体是由相互依赖的资源（例如，人员、流程、组织、支持技术以及资金）的有目的的组合（例如，网络）组成的，这些资源通过跨越地域和时间分布的相互作用的复杂网络彼此相互作用以协调功能、共享信息、分配资金、创建工作流、做出决策并建立环境以实现业务和运行目标（Rebovich 和 White，2011）。

复杂组织体必须做的两件事：（1）在复杂组织体内部开发事物以服务于外部提供物或作为内部机制使复杂组织体运行达成；（2）转变复杂组织体本身以使其能够最高效率和最佳效果地实施其运行并在具有竞争性和受约束的环境中生存。

It is worth noting that an enterprise is not equivalent to an "organization" according to the definition earlier. This is a frequent misuse of the term enterprise. Figure 8.7 shows that an enterprise includes not only the organizations that participate in it but also people, knowledge, and other assets such as processes, principles, policies, practices, doctrine, theories, beliefs, facilities, land, intellectual property, and so on.

Giachetti (2010) distinguishes between enterprise and organization by saying that an organization is a view of the enterprise. The organization view defines the structure and relationships of the organizational units, people, and other actors in an enterprise. Using this definition, we would say that all enterprises have some type of organization, whether formal, informal, hierarchical, or self-organizing network.

To enable more efficient and effective enterprise transformation, the enterprise needs to be looked at "as a system," rather than merely as a collection of functions connected solely by information systems and shared facilities (Rouse, 2009). While a system perspective is required for dealing with the enterprise, this is rarely the task or responsibility of people who call themselves systems engineers.

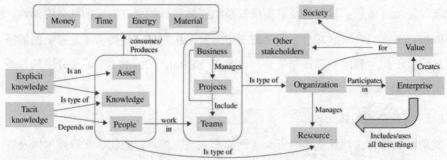

Notes:
1. All entities shown are decomposable, except people. For example, a business can have sub businesses, a project can have subprojects, a resource can have sub resources, an enterprise can have sub enterprises.
2. All entities have ether names. For example, a program can be a project comprising server all subprojects. (Often called merely projects). Business can be an agency, team can be group, value can be utility, etc.
3. There is no attempt to be prescriptive in the names chosen for this diagram. The main goal of this is to show how this chapter uses these terms and how they are related to each other in a conceptual manner.

FIGURE 8.7 Organizations manage resources to create enterprise value. From SEBoK (2014). Reprinted with permission from BKCASE Editorial Board. All other rights reserved.

值得注意的是，根据之前的定义，复杂组织体与"组织"是不相等的。这是常见的对复杂组织体术语的误用。图 8.7 表明复杂组织体不仅包括加入其中的组织还包括人员、知识、以及诸如流程、原则、策略、实践、学说、理论、信仰、设施、土地、知识产权等的其他资产。

Giachetti（2010）对复杂组织体和组织进行了区分，他表示组织是复杂组织体的一种视角。组织视角定义复杂组织体中的组织单元、人员以及其他角色者的结构和关系。使用此定义，我们将会说所有的复杂组织体均具有某种组织形式，无论是正式的、非正式的、层级的还是自组织网络。

为了使复杂组织体的转型能够更高效率且更有效果，复杂组织体需要被视为"一个系统"而不只是通过信息系统和共享设施完全连接的功能集合（Rouse，2009）。虽然需要用系统的视角应对复杂组织体，但这却是那些称自己为系统工程师的人员的罕有的任务或责任。

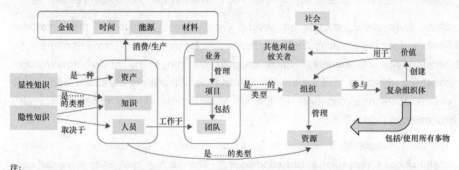

注：
1. 除了人员，所有显示的实体均是可分解的。例如，业务可具有子业务，项目可具有子项目，资源可具有子资源，复杂组织体可具有子复杂组织体。
2. 所有实体都有别名。例如，项目群计划可以是一个包括服务于所有子项目的项目（通常仅被称为项目）。业务可称为机构，团队可称为群组，价值可称为效用等。
3. 这里并未试图对选定用于此图的名称进行规定。本图的主要目的是表明本章如何使用这些术语以及这些术语如何以概念的方式彼此相关。

图 8.7　组织管理资源以创建复杂组织体价值。来自 SEBoK（2014）。经 BKCASE 编辑委员会许可后转载。版权所有。

8.5.2 Creating value

The primary purpose of an enterprise is to create value for society, for other stakeholders, and for the organizations that participate in that enterprise. This is illustrated in Figure 8.7, which shows all the key elements that contribute to this value creation process.

There are three types of organizations of interest to an enterprise: businesses, projects, and teams. A typical business participates in multiple enterprises through its portfolio of projects. Large SE projects can be enterprises in their own right, with participation by many different businesses, and may be organized as a number of subprojects.

8.5.3 Capabilities in the Enterprise

The enterprise acquires or develops systems or individual elements of a system. The enterprise can also create, supply, use, and operate systems or system elements.

Since there could possibly be several organizations involved in this enterprise venture, each organization could be responsible for particular systems or perhaps for certain kinds of elements. Each organization brings their own organizational capability with them, and the unique combination of these organizations leads to the overall operational capability of the whole enterprise. These concepts are also illustrated in Figure 8.7.

The word "capability" is used in SE in the sense of "the ability to do something useful under a particular set of conditions." This section discusses three different kinds of capabilities: organizational capability, system capability, and operational capability. It uses the word "competence" to refer to the ability of people relative to the SE task. Individual competence (sometimes called "competency") contributes to, but is not the sole determinant of, organizational capability. This competence is translated to organizational capabilities through the work practices that are adopted by the organizations. New systems (with new or enhanced system capabilities) are developed to enhance enterprise operational capability in response to stakeholder's concerns about a problem situation.

As shown in Figure 8.8, operational capabilities provide operational services that are enabled by system capabilities. These system capabilities are inherent in the system that is conceived, developed, created, and/or operated by an enterprise. Enterprise SE concentrates its efforts on maximizing operational value for various stakeholders, some of whom may be interested in the improvement of some problem situation.

8.5.2 创造价值

复杂组织体的初始目的是为社会、其他利益攸关者以及参加复杂组织体的组织创造价值。如图 8.7 所示，展示了有助于此价值创造流程的所有关键元素。

复杂组织体感兴趣的组织类型有三种：业务、项目和团队。一个典型的业务通过其项目群参与多个复杂组织体。大型 SE 项目可以是独立的复杂组织体，有许多不同的业务参与，并且可被组织为多个子项目。

8.5.3 复杂组织体的能力

复杂组织体采办或开发系统或系统的单个元素。复杂组织体亦可创建、供给、使用和运行系统或系统元素。

因为可能有多个组织参与此复杂组织体的风险投资，每个组织可能负责特定的系统或可能负责一定类型的元素。每个组织均会随之带来他们自身的组织能力，并且这些组织的独特组合会导向整个复杂组织体的总体运营能力。图 8.7 也阐明了这些概念。

SE 中使用的词语"能力"在一定意义上是"在一个特定条件集合下做有用的事情的能力"。本节论述三种不同类型的能力：组织能力、系统能力以及运行能力。其使用的词语"才能"指的是与 SE 任务相关的人员的才能。个人才能（有时称为"才能"）有助于组织能力，但不是唯一的决定因素。这种才能通过组织所选用的工作实践转化为组织能力。开发新系统（具有新的或增强的系统能力）以提高复杂组织体响应利益攸关者对问题情境的关注的运行能力。

如图 8.8 所示，运行能力提供由系统能力使能的运行服务。这些系统能力是由复杂组织体构想、开发、创建和/或运行的系统所固有的。复杂组织体 SE 将其努力集中在最大化各种不同利益攸关者的运行价值，其中一些利益攸关者可能对某些问题情境的改善感兴趣。

TAILORING PROCESS AND APPLICATION OF SYSTEMS ENGINEERING

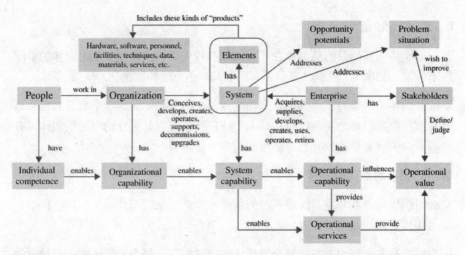

FIGURE 8.8 Individual competence leads to organizational, system, and operational capability. From SEBoK (2014). Reprinted with permission from BKCASE Editorial Board. All other rights reserved.

Enterprise SE, however, addresses more than just solving problems; it also deals with the exploitation of opportunities for better ways to achieve the enterprise goals. These opportunities might involve lowering of operating costs, increasing market share, decreasing deployment risk, reducing time to market, and any number of other enterprise goals. The importance of addressing opportunity potentials should not be underestimated in the execution of enterprise SE practices.

The operational capabilities of an enterprise will have a contribution to operational value (as perceived by the stakeholders). Notice that the organization or enterprise can deal with either the system as a whole or with only one (or a few) of its elements. These elements are not necessarily hard items, like hardware and software, but can also include "soft" items, like people, processes, principles, policies, practices, organizations, doctrines, theories, beliefs, and so on.

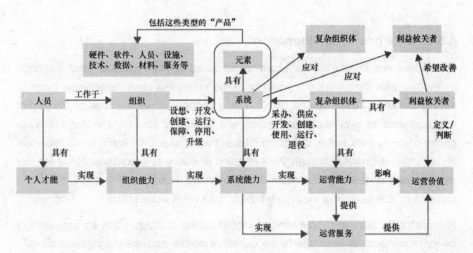

图 8.8 个人胜任力产生组织能力、系统能力以及运营能力。来自于 SEBoK (2014)。经 BKCASE 编辑委员会许可后转载。版权所有。

然而，复杂组织体 SE 涉及的不仅仅是求解问题；其亦涉及对实现复杂组织体目标的更好方式的机会的探索。这些机会可能涉及降低运行成本、增加市场份额、减少部署风险、缩短上市时间以及许多其他的复杂组织体目标。在执行复杂组织体 SE 实践过程中不应低估应对机会潜力的重要性。

复杂组织体的运行能力将对运行价值做出贡献（正如利益攸关者所知悉的）。要注意，组织或复杂组织体能够应对作为一个整体的系统抑或只是其中一个（或几个）系统元素。这些元素不一定是类似硬件和软件的"硬"项目，而且可包括"软"项目诸如人员、流程、原则、方针、实践、组织、学说、理论、信仰等。

8.5.4 Enterprise Transformation

Enterprises are constantly transforming, whether at the individual level (wherein individuals alter their work practices) or at the enterprise level (large-scale planned strategic changes) (Srinivansan, 2010). These changes are a response on the part of the enterprise to evolving opportunities and emerging threats. It is not merely a matter of doing work better, but doing different work, which is often a more important result. Value is created through the execution of business processes. However, not all processes necessarily contribute to overall value (Rouse, 2005). It is important to focus on process and how they contribute to the overall value stream.

Rebovich says there are "new and emerging modes of thought that are increasingly being recognized as essential to successful systems engineering in enterprises" (2006). For example, in addition to the traditional SE process areas, MITRE has included the following process areas in their enterprise SE process to close the gap between enterprise SE and traditional SE (DeRosa, 2005):

- Strategic technical planning
- Enterprise architecture
- Capabilities-based planning analysis
- Technology planning
- Enterprise analysis and assessment

These enterprise SE processes are shown in the context of the entire enterprise in Figure 8.9 (DeRosa, 2006). The enterprise SE processes are shown in the middle with business processes on the left and traditional SE processes on the right. Further information on using SE practices to transform an enterprise is provided in Transforming the Enterprise Using a Systems Approach (Martin, 2011).

8.5.4 复杂组织体转型

无论是在个人层级（在其中的个人改变其工作实践）还是在复杂组织体层级（大规模规划的战略变化），复杂组织体均持续地转型（Srinivansan, 2010）。这些变化是复杂组织体的部分对演进机会和显现威胁能做出的响应。其不仅事关更好的工作，还关系到做不同的工作，而这往往是更重要的结果。价值是通过执行业务流程创造的。然而，不是所有的流程一定有助于总体价值（Rouse, 2005）。重要的是聚焦于流程以及这些流程如何有助于总体价值流。

Rebovich 表示存在"诸多新的和新兴的思考模式，并且越来越多的人意识到：这些模式对于在复杂组织体中成功地进行系统工程是根本性的"（2006）。例如，除了传统的 SE 流程领域外，MITRE 已经在其复杂组织体 SE 流程中包括了下述流程领域以弥补复杂组织体 SE 与传统 SE 之间的差距（DeRosa, 2005）：

- 战略技术规划
- 复杂组织体架构
- 基于能力的规划分析
- 技术规划
- 复杂组织体分析和评估

这些复杂组织体 SE 流程如图 8.9 示于整个复杂组织体的背景环境之中（DeRosa, 2006）。复杂组织体 SE 流程显示在图中部，左边是业务流程，右边是传统 SE 流程。《使用系统方法转型复杂组织体》（Martin, 2011）中提供了更多关于使用 SE 实践转型复杂组织体的信息。

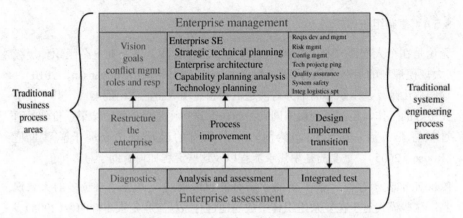

FIGURE 8.9 Enterprise SE process areas in the context of the entire enterprise. Reprinted with permission from the MITRE Corporation. All other rights reserved.

8.6 APPLICATION OF SYSTEMS ENGINEERING FOR VERY SMALL AND MICRO ENTERPRISES

VSMEs are defined as organizations that have a small number of employees, many times less than 50 and as few as 1. The contributions of VSMEs to the global economy are well documented. By some estimates, over 98% of economic value is generated globally by enterprises with less than 25 people. In addition, VSMEs contribute to large enterprise systems and SoS and are important and essential to system success. VSME guidance is generic and applicable to SE functions in any domain.

Tailoring of SE processes in any life cycle stage or domain is typical for all projects but is crucial for small enterprises. Of course, as with any project tailoring for processes, the tailoring needs to be risk driven, accounting for things such as the criticality attributes of the project. Due to their small size, VSMEs often find it difficult to apply international standards to their business needs and to justify the application of standards to their business practices. The typical VSME does not have a comprehensive infrastructure, and the few personnel usually are performing multiple roles.

系统工程的剪裁流程和应用

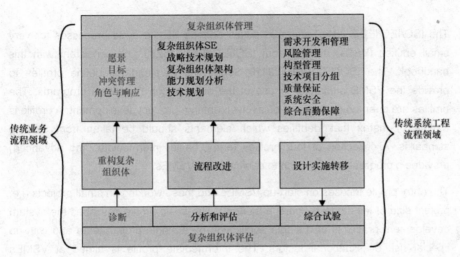

图 8.9 整个复杂组织体背景环境中的复杂组织体 SE 流程领域。经 MITRE 公司许可后转载。版权所有。

8.6 极小型和微型组织体的系统工程的应用

VSME 被定义为具有少量雇员的组织，雇员人数很多时候小于 50 或为 1。VSME 对全球经济的贡献得到了较好的文件记载。据估计，超过 98%的经济价值是由世界上雇员人数少于 25 人的组织体创造的。此外，VSME 有助于大型复杂组织体系统和系统之系统并且对系统成功是非常重要和根本性的。VSME 指导对任何领域中的 SE 功能均是通用和适用的。

在任何生命周期阶段或领域中剪裁 SE 流程对于所有项目而言是都典型的方式，但对小型组织体则是至关重要的。当然，与任何项目的流程剪裁一样，剪裁需要风险驱动，考虑诸如项目关键属性等事物。由于规模小，VSME 往往发现难以将国际标准应用到其业务需要并证明将标准应用到其业务实践的正确性。典型的 VSME 不具有全面的基础设施，并且少数员工通常担当多种角色。

TAILORING PROCESS AND APPLICATION OF SYSTEMS ENGINEERING

The ISO/IEC/IEEE 29110 standard series defines SE life cycle processes for very small entities (VSEs) derived from ISO/IEC/IEEE 15288 and consistent with this handbook. The ISO/IEC/IEEE 29110 Series (2014) standards define profiles to provide the VSME guidance for use of the standard on noncritical programs. The profiles are oriented on the level of involvement in a product development. A profile is a type of matrix that identifies which elements should be taken from existing standards. A collection of four profiles (entry, basic, intermediate, and advanced) provides a progressive approach to serving most VSMEs.

The entry profile focuses on start-up VSMEs and those working on small projects (i.e., project size of less than six person-months). The basic profile describes the system development practices of a single application by a single project team and with no special risk or situational factors. The intermediate profile is aimed at VSMEs developing multiple projects, while the advanced profile applies to VSMEs that want to grow as independent system development businesses.

For critical programs, such as mission critical or safety critical, this guidance does not apply, since the criticality of the programs would dictate a much greater level of rigor and comprehensive SE.

ISO/IEC/IEEE 29110 标准系列定义源自 ISO/IEC/IEEE 15288 并符合本手册要求的极小型实体（VSE）的 SE 生命周期流程。ISO/IEC/IEEE 29110 系列（2014）标准定义配置文件以提供关于将该标准用于非关键性项目群的 VSME 指导。基于参与产品开发的程度适应配置文件。配置文件是一种识别哪些元素应取自现有标准的矩阵。四种配置文件（入门、基本、中间和高级）的集合提供一种服务大多数 VSME 的一种开明的途径。

入门配置文件聚焦于开始阶段的 VSME 以及那些从事于小项目的 VSME（即，项目规模小于 6 个人一月的工作量）。基本配置文件描述单一项目团队的单一应用的系统开发实践，并且无特别风险或情境因素。中间配置文件以开发多个项目的 VSME 为目标，并且高级配置文件应用于想要随着独立系统开发业务而增长的 VSME。

对于关键计划，诸如任务关键的与安全关键的，本指导不适用，因为项目群的关键性将规定严格程度更高且更加全面的 SE。

9 CROSS-CUTTING SYSTEMS ENGINEERING METHODS

The previous chapters provided a serial description of the systems engineering (SE) processes used across the system life cycle. This chapter provides insight into methods that cut across the SE processes, reflecting various aspects of the iterative and recursive nature of SE.

9.1 MODELING AND SIMULATION

Stakeholders of the SE life cycle have used models and simulations for some time both to check their own thinking and to communicate their concepts to others. The benefit is twofold: (i) models and simulations confirm the need for the systems and the anticipated system behaviors before proceeding with the development of an actual system, and (ii) models and simulations present a clear, coherent design to those who will develop, test, deploy, and evolve the system, thereby maximizing productivity and minimizing error. The ability to detect limitations and incompatibilities via system models and simulations early in a project helps avoid higher project cost and schedule overruns later in a project, especially during system operation. The value of modeling and simulation increases with the size, be it physical or complexity, of the system or system of systems (SoS) under development.

Early in the SE life cycle, the objective of modeling and simulation is to obtain information about the system before significant resources are committed to its design, development, construction, verification, or operation. To that end, modeling and simulation helps generate data in the domain of the analyst or reviewer, not available from existing sources, in a manner that is affordable and timely to support decision making. An adequate, accurate, and timely model and simulation informs stakeholders of the implications of their preferences, provides perspective for evaluating alternatives, and builds confidence in the capabilities the system will provide. They also help the development, deployment, and operational staffs comprehend the design requirements, appreciate imposed limits from technology and management, and ensure an adequate degree of sustainability. Finally, adequate, accurate, and timely models and simulations help the organization and its suppliers provide the necessary and sufficient personnel, methods, tools, and infrastructure for system realization.[1]

INCOSE Systems Engineering Handbook: A Guide for System Life Cycle Processes and Activities, Fourth Edition. Edited by David D. Walden, Garry J. Roedler, Kevin J. Forsberg, R. Douglas Hamelin and Thomas M. Shortell.

© 2015 John Wiley & Sons, Inc. Published 2015 by John Wiley & Sons, Inc.

9 跨领域/学科系统工程方法

前面几章提供了对跨系统生命周期使用的系统工程（SE）流程的一系列描述。本章提供对跨领域/学科 SE 流程的方法的深入理解，反映 SE 的迭代和递归本质属性的各个不同方面。

9.1 建模和仿真

SE 生命周期的利益攸关者一段时间内一直使用模型和仿真来检查他们自己的思考并与他人沟通他们的概念。其好处是双重的：（i）在开始开发实际的系统之前，模型和仿真确认对系统及预期的系统行为的需要；以及（ii）模型和仿真向那些将要开发、测试、部署和演进系统的人员呈现一个清晰的、连贯一致的设计，从而使生产率最大化，使错误最小化。这种在项目早期通过系统模型和仿真来检测限制性和不兼容性的能力，有助于在项目后期尤其是系统运行期间避免较高的项目成本和进度延期。无论是物理方面还是复杂性方面，建模和仿真的价值随着正在开发中的系统或系统之系统（SoS）的规模而增长。

在 SE 生命周期早期，建模和仿真的目的是在重要资源投入到系统的设计、开发、建造、验证或运行前获取有关该系统的信息。为此，在分析者或评审者的领域，建模和仿真以可承受且及时的支持决策的方式帮助生成无法从现有来源中得到的数据。充分的、精确的且及时的模型和仿真告知利益攸关者他们的偏好可能引发的后果，为评价备选方案提供视角，并对系统将要提供的能力建立信心。模型和仿真亦帮助开发人员、部署人员和运行人员理解设计需求，领会来自于技术和管理所施加的限制，并确保充分的可持续度。最后，充分的、精确的且及时的模型和仿真有助于组织及其供应商提供系统实现所需的必要且充足的人员、方法、工具和基础设施。[2]

INCOSE 系统工程手册：系统生命周期流程和活动指南，第 4 版。编撰：David D.Walden、Garry J.Roedler、Kevin J.Forsberg、R.Douglas Hamelin 和 Thomas M.Shortell。
John Wiley & Sons 公司版权所有©2015。由 John Wiley & Sons 公司于 2015 年出版。

The long-term benefits of modeling and simulation are commensurate with the gap between the extent, variety, and ambiguity of the problem and the competencies of downstream staffing. A relatively simple model of an intended system may be sufficient for a highly competent staff, whereas a much more elaborate simulation may be necessary for a less competent staff, especially one faced with producing a novel, large-scale system that is capable of autonomously coping with unpredictable mission situations. Ultimately, the benefit of modeling and simulation is proportional to the stakeholders' perception of the timeliness, trustworthiness, and ease of use and maintenance of the model or simulation. Consequently, the planned resources anticipated to be spent in development, verification, validation, accreditation, operation, and maintenance of the model must be consistent with the expected value of the information obtained through use of the model.

9.1.1 Models Versus Simulations

The terms model and simulation are often mistakenly interchanged in discussions. Each term has its own specific meaning. The term "model" has many definitions but generally refers to an abstraction or representation of a system, entity, phenomenon, or process of interest (DoD 5000.59, 2007). The many other definitions of model generally refer to a model as a representation of some entity in the physical world. The representations are intended to describe selected aspects of the entity, such as its geometry, functions, or performance. In the context of SE, a model that represents a system and its environment is of particular importance to the systems engineer who must analyze, specify, design, and verify systems, as well as share information with other stakeholders. Different types of models are used to represent systems for different modeling purposes.

The term "simulation" is the implementation of a model (or models) in a specific environment that allows the model's execution (or use) over time. In general, simulations provide a means for analyzing complex dynamic behavior of systems, software, hardware, people, and physical phenomena.

建模和仿真的长期好处与问题的程度、多样性及模糊性和下游人员的能力之间的差距相称。预期系统的相对简单的模型对于能力很强的人员而言可能足已，而更详尽的仿真对于能力较差的人员而言可能非常有必要，尤其是当面临产生一个新颖的、大规模的能够自主应对不可预见的任务态势的系统时更是如此。最后，建模和仿真的好处与利益攸关者对及时性、可信赖度以及使用和维护模型或仿真便利性的洞察力成比例。因此，预期花费在模型开发、验证、确认、鉴定、运行和维护方面的计划资源必须与通过使用该模型所获取的信息的预期价值相一致。

9.1.1 模型与仿真的对比

术语"模型"和"仿真"在论述中往往会被错误地互换。每个术语都有自己的特定含义。术语"模型"有很多种定义，但一般是指对所感兴趣的系统、实体、现象或流程的抽象或表达（美国国防部 5000.59，2007）。"模型"的很多其他定义一般是指一种模型作为一些实体在物质世界中的表达。这些表达旨在描述实体所选定的方面，诸如实体的几何、功能或性能。在 SE 的背景环境中，表达系统及其环境的模型对于必须分析、指定、设计和验证系统以及与其他利益攸关者分享信息的系统工程师而言特别重要。不同类型的模型用于表示系统的不同建模目的。

术语"仿真"是某一模型（或多个模型）在特定环境中允许随时间推移时的模型执行（或使用）。一般来说，仿真提供一种分析系统、软件、硬件、人员和物理现象的复杂动态行为的方式。

A computer simulation includes the analytical model that is represented in executable code, the input conditions and other input data, and the computing infrastructure. The computing infrastructure includes the computational engine needed to execute the model, as well as input and output devices. The great variety of approaches to computer simulation is apparent from the choices that the designer of computer simulation must make.

In addition to representing the system and its environment, the simulation must provide efficient computational methods for solving the equations. Simulations may be required to operate in real time, particularly if there is an operator in the loop. Other simulations may be required to operate much faster than real time and perform thousands of simulation runs to provide statistically valid simulation results.

9.1.2 Purpose of Modeling

System models can be used for many purposes. One of the first principles of modeling is to clearly define the purpose of the model. Some of the purposes that models can serve throughout the system life cycle include:

- *Characterizing an existing system*: Many existing systems are poorly documented, and modeling the system can provide a concise way to capture the existing system architecture and design. This information can then be used to facilitate maintaining the system or to assess the system with the goal of improving it. This is analogous to creating an architectural model of an old building with overlays for electrical, plumbing, and structure before proceeding to upgrade it to new standards to withstand earthquakes.

- *Mission and system concept formulation and evaluation*: Models can be applied early in the system life cycle to synthesize and evaluate alternative mission and system concepts. This includes clearly and unambiguously defining the system's mission and the value it is expected to deliver to its beneficiaries. Models can be used to explore a trade space by modeling alternative system designs and assessing the impact of critical system parameters such as weight, speed, accuracy, reliability, and cost on the overall measures of merit. In addition to bounding the system design parameters, models can also be used to validate that the system requirements meet stakeholder needs before proceeding with later life cycle activities such as architecting and design.

计算机仿真包括以可执行代码、输入条件和其他输入数据表示的解析模型以及计算基础设施。计算基础设施包括执行该模型所需的计算引擎以及输入和输出装置。从计算机仿真的设计者必须做出的选择中可明显地看到多种多样的计算机仿真方法。

除了表达系统及其环境之外，仿真还必须提供对方程求解的高效的计算方法。特别是当操作者在回路中时，仿真可能被要求实时运行。其他仿真可能被要求比真实时间更快地运行，并且完成成千上万次的仿真运行，以提供统计上有效的仿真结果。

9.1.2 建模目的

系统模型可以用于多种目的。建模的首要原则之一是明确地定义模型的目的。模型贯穿于系统生命周期内可服务的一些目的包括：

- *特征化一个现有系统*：很多现有系统的文件化很差，而且对系统建模可以提供一种简洁的方式来获取现有系统的架构和设计。后续，可以利用该信息来辅助系统的维护或者以改善系统为目标对系统进行评估。这类似于在按照新标准升级旧建筑抵御地震之前，创建一个层叠的供电、管道和结构的旧建筑的架构模型。

- *使命任务及系统概念的正规形成和评估*：模型可以在系统生命周期早期用于综合并评价备选的使命任务及系统概念。这包括明确清晰地定义系统的使命任务以及期望交付给其受益者的价值。通过对备选的系统设计建模并评估关键系统参数（例如，与整体测度优化有关的重量、速度、精度、可靠性和成本）的方式，模型可以用于探索一个权衡空间。除了束缚系统设计参数之外，模型还可以用于在后续开始的诸如架构开发和设计等生命周期活动之前，确认系统需求是否满足利益攸关者的需要。

- *System architecture design and requirements flowdown*: Models can be used to support architecting system solutions, as well as flow mission and system requirements down to system elements. Different models may be required to address different aspects of the system design and respond to the broad range of system requirements. This may include models that specify functional, interface, performance, and physical requirements, as well as other nonfunctional requirements such as reliability, maintainability, safety, and security.

- *Support for systems integration and verification*: Models can be used to support integration of the hardware and software elements into a system, as well as to support verification that the system satisfies its requirements. This often involves integrating lower-level hardware and software design models with system-level design models, which verify that system requirements are satisfied. Systems integration and verification may also include replacing selected hardware and design models with actual hardware and software products in order to incrementally verify that system requirements are satisfied. This is referred to as hardware-in-the-loop testing and software-in-the-loop testing. Models can also be used to define the test cases and other aspects of the test program to assist in test planning and execution.

- *Support for training*: Models can be used to simulate various aspects of the system to help train users to interact with the system. Users may be operators, maintainers, or other stakeholders. Models may be a basis for developing a simulator of the system with varying degrees of fidelity to represent user interaction in different usage scenarios.

- *Knowledge capture and system design evolution*: Models can provide an effective means for capturing knowledge about the system and retaining it as part of organizational knowledge. This knowledge, which can be reused and evolved, provides a basis for supporting the evolution of the system, such as changing system requirements in the face of emerging, relevant technologies, new applications, and new customers. Models can also enable the capture of families of products.

- *系统架构设计和需求向下流动*：模型可以被用来支持构架系统解决方案，并将使命任务和系统需求向下流动到系统元素。可能需要不同的模型以应对系统设计的不同方面并响应系统需求的宽广范围，可以包括指定功能、接口、性能和物理以及其他诸如可靠性、可维护性、安全性和安保性等非功能需求的模型。

- *支持系统综合和验证*：模型可以用于支持将硬件和软件元素综合到某一系统中以及支持系统满足其需求的验证。这通常涉及较低层级硬件和软件设计模型与系统级设计模型的综合，它验证系统需求是否得到满足。系统综合和验证还可以包括用真实的硬件和软件产品取代所选定的硬件和设计模型，以便渐进地验证系统需求是否得以满足。这被称为硬件在回路的测试和软件在回路的测试。模型还可以被用来定义测试用例和测试计划的其他方面，以便为测试规划和执行提供帮助。

- *支持培训*：模型可以用于仿真系统的多个不同侧面，以便为培训用户与系统交互提供帮助。用户可以是操作人员、维护人员或其他利益攸关者。模型可以是开发不同保真度的系统模拟器的基础，以呈现用户在不同使用场景中的交互。

- *知识捕获和系统设计演进*：模型可以提供一种捕获系统相关知识并将这些知识作为组织知识的一部分而保存的有效手段。这种可被复用和演进的知识为支持系统的演进提供一个基础，例如，在面对新兴的有关技术、新应用和新客户时而变更系统需求。模型还可以使产品族得以捕获。

Models represent the essential characteristics of the system of interest (SOI), the environment in which the system operates, and the interactions with enabling and interfacing systems and operators. Models and simulations can be used within most system life cycle processes, for example:

- *Business or mission analysis*—descriptive model of the problematic situation ensures the right problem(s) is being addressed.
- *Requirements (stakeholder and system) definition*—Enables justification of requirements and avoids over-/under-specification.
- *Architecture definition*—Evaluate candidate options against selection criteria and enable active agents to discover the best architecture, including the integration to other systems.
- *Design definition*—obtain needed design data, adjust parameters for optimization, and update system model fidelity as actual data for system elements become available.
- *Verification and validation*—Simulate the system's environment, evaluate verification and validation data (simulation uses observable data as inputs for computation of critical parameters that are not directly observable), and validate the fidelity of the simulation (false-positives/false-negatives).
- *Operations*—Simulations that reflect actual behavior and simulate operations in advance of execution for planning, validation, and operator training.

9.1.3 Model Scope

The model must be scoped to address its intended purpose. In particular, the types of models and associated modeling languages selected must support the specific needs to be met. For example, suppose models are constructed to support an aircraft's development. A system architecture model may describe the interconnection among the airplane parts, a trajectory analysis model may analyze the airplane trajectory, and a fault tree analysis model may assess potential causes of airplane failure. For each type of model, the appropriate breadth, depth, and fidelity should be determined to address the model's intended purpose.

The model breadth reflects the system requirements coverage in terms of the degree to which the model must address the functional, interface, performance, and physical requirements, as well as other nonfunctional requirements. For an airplane functional model, the model breadth may be required to address some or all of the functional requirements to power up, take off, fly, land, power down, and maintain the aircraft's environment.

模型表达所感兴趣之系统（SOI）系统运行的环境和使能系统及交接系统与操作人员的交互的最基本的特征。模型和仿真可以在大多数系统生命周期流程内使用，例如：

- *业务或任务分析*——问题情境的描述性模型确保应对的是正确的问题。

- *需求（利益攸关者和系统）定义*——使能需求的合理性证明并避免过度/不足的规范。

- *架构定义*——按照选择准则评价候选选项，并使能主动的代理人发现最佳架构，包括综合到其他系统。

- *设计定义*——获得所需的设计数据，为优化而调整参数，并且在系统元素的实际数据可用时更新系统模型的保真度。

- *验证和确认*——仿真系统的环境，评价验证和确认数据（仿真使用可观测的数据作为输入，以计算非直接可观测的关键参数），并且确认仿真的保真度（假阳性/假阴性）。

- *运行*——在执行计划、进行确认和培训操作人员之前反映实际行为并模拟运行的仿真。

9.1.3 模型范围

为应对意图目的，模型必须界定其范围。尤其是，所选定的模型及相关建模语言的类型必须支持将要满足的特定需要。例如，假设构建模型来支持飞机的开发。系统架构模型可描述飞机部件之间的互连，轨迹分析模型可分析飞机的轨迹，故障树分析模型可评估飞机故障的潜在原因。对于每种类型的模型而言，为应对模型的意图目的，应确定合适的广度、深度和保真度。

模型广度在模型必须应对功能、接口、性能和物理需求以及其他非功能需求的程度方面反映系统需求的覆盖范围。对于一个飞机功能模型而言，可能要求模型广度应对加电、起飞、飞行、着陆、断电和维护飞机环境的一些或全部功能需求。

The model's depth indicates the coverage of system decomposition from the system context down to the system elements. For the air transport SoS example shown in Figure 2.2, a model's scope may require it to define the system context, ranging from the aircraft, the control tower, and the physical environment down to the navigation system and its system elements, such as the inertial measurement unit, and perhaps down to the lower-level parts of the inertial measurement unit.

The model's fidelity indicates the level of detail the model must represent for any given part of the model. For example, a model that specifies the system interfaces may be fairly abstract and represent only the logical information content, such as aircraft status data; or it may be much more detailed to support higher-fidelity information, such as the encoding of a message in terms of bits, bytes, and signal characteristics. Fidelity can also refer to the precision of a computational model, such as the time step required for a simulation.

9.1.4 Types of Models and Simulations

There are many different types of models and simulations to address different aspects of a system and different types of systems. The specific type of model or simulation selected for a phase of the system's life cycle depends on the intended use and particular characteristics of the system that are of interest and the level of model accuracy required, in other words its *"fitness for purpose."* Generally, a specific model or simulation focuses on some subset of the total system characteristics, such as timing, process behavior, or various performance measures.

9.1.4.1 Types of Models "The image of the world around us, which we carry in our head, is just a model" (Forrester, 1961). Most systems start as a mental model that is elaborated and translated in several stages to form a final model or simulation product. A model maybe a mental image of selected concepts, and relationships between them, that can be translated to sketches, textual specifications, graphics/images, mock-ups, scale models, prototypes, or emulations. Often, separate models are prepared for distinct viewpoints, such as functional, performance, reliability, survivability, operational availability, and cost.

It is useful to classify models to assist in selecting the right kind of model for the intended purpose and scope. Models can be classified in many different ways. The model taxonomy shown in Figure 9.1 is one such taxonomy and provides a useful classification for one instance as an illustration and not necessarily providing an exhaustive set of model classes; other classes may exist:

模型深度表明系统从系统背景环境分解到系统元素的覆盖范围。对于图 2.2 所示的空中运输 SoS 示例,模型的范围可能要求其定义系统的背景环境,范围从飞机、控制塔台和物理环境一直到导航系统及其系统元素,例如惯性测量单元,并且或许会到惯性测量单元的较低层级部分。

模型保真度表明模型必须针对模型的给定部分表示的细节层级。例如,具体指出系统接口的模型可能相当抽象并且仅代表诸如飞机状态数据等逻辑信息内容;或者它可能更加详细,以支持保真度更高的信息,例如按照比特、字节和信号特征对信息进行编码。保真度还可以指计算模型的精度,例如仿真需要的时间步长。

9.1.4 模型和仿真的类型

有很多不同类型的模型和仿真来应对某一系统的不同方面以及不同类型的系统。选用于系统生命周期某一阶段的特定类型的模型或仿真取决于所感兴趣的系统的预期用途和特殊特征以及所需要的模型精度层级,换句话说就是其"目的适配性"。一般情况下,特定模型或仿真聚焦于全部系统特征的某一子集,例如时序安排、流程行为或多种性能测度。

9.1.4.1 模型的类型 "我们头脑中所承载的对于我们周围世界的图像就是一个模型"(Forrester, 1961)。大多数系统始于一个心智模型,以若干阶段被详细阐述和转化而形成最终模型或仿真产物。模型也许是所选定的概念以及它们之间的关系的心智图像,可以被转化为草图、文本规范、图形/图像、概念原型、缩比模型、原型或模仿。往往是为诸如功能、性能、可靠性、生存性、运行可用性和成本等迥然不同的视角准备单独的模型。

为了有助于选择正确类型的模型用于预期目的和范围,对模型进行分类是有用的。模型可以按很多不同的方式进行分类。图 9.1 所示模型分类法就是这样一种分类法,并且作为图解为一个实例提供有用的分类,但不一定提供模型类别的详尽集合;也许还存在其他类别:

- *Physical mock-ups*—A model that represents an actual system, such as a model airplane or wind tunnel model, or a more abstract representation, such as a model that is often represented using a computer.
- *Abstract models*—An abstract model can have many different expressions to represent a system, entity, phenomena, or process, which vary in degrees of formalism. Therefore, the initial classification of models that distinguishes between informal models and formal models is noted, with this guidance focusing on formal models.

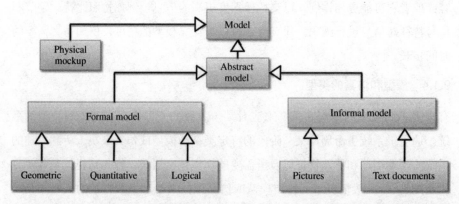

FIGURE 9.1 Sample model taxonomy. Reprinted with permission from Sandy Friedenthal. All other rights reserved.

- *Informal models*—One can represent a system using a simple drawing tool or with words. However, unless there is clear agreement on the meaning of the terms in the less formal representations, there is a potential lack of precision and the possibility of ambiguity in the representation. While such informal representations can be useful, a model must meet certain expectations for it to be considered within the scope of modeling and simulation for SE.
- *Formal models*—Formal models can be further classified as geometric, quantitative (i.e., mathematical), and/or logical models. Geometric model represents the geometric or spatial relationships of the system or entity. Quantitative models represent quantitative relationships (e.g., mathematical equations) about the system or entity that yield numerical results. Logical models, also referred to as conceptual models, represent logical relationships about the system such as a whole–part relationship, an inter- connection relationship between parts, or a precedence relationship between activities, to name a few. The logical models are often depicted in graphs (nodes and arcs) or tables.

- *物理实体模型*——代表诸如模型飞机或风洞模型等实际系统的模型，或者一种更抽象的表达，这种模型往往利用计算机表示。

- *抽象模型*——抽象模型可具有很多不同表达方式来表示正式程度有差异的系统、实体、现象或流程。因此，在关注于正规模型的这一指导下，指明区分非正规模型和正规模型的初始模型分类。

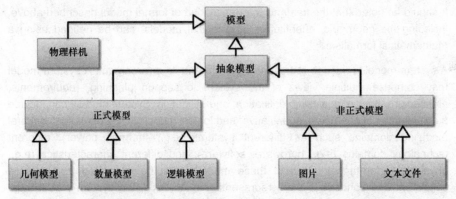

图 9.1 模型分类法示例。经 Sandy Friedenthal 许可后转载。版权所有。

- *非正规模型*——非正规模型可以利用简单的绘图工具或文字来表示系统。但是，除非存在关于以不太正规的表征的术语含义的明确协定，否则这种表征可能缺乏精确度并且有歧义的可能性。当这种非正规的表征可能有用时，模型必须满足将在 SE 建模和仿真的范围内考虑到的对它的某些期望。

- *正规模型*——正式模型可以被进一步地分类为几何的、定量的（即，数学的）和/或逻辑的模型。几何模型表达系统或实体的几何或空间关系。定量模型表达关于产生数值结果的系统或实体的数量关系（例如，数学方程）。逻辑模型也称为概念模型，表达关于系统的逻辑关系，例如，整体与部分的关系、两个部分之间的互连关系或两个活动之间的优先关系，仅仅举这几个例子。逻辑模型往往以图形（节点和弧线）或表格形式描述。

The following example illustrates the above taxonomy. An aircraft may be represented by a three-dimensional geometric model that specifies the detailed geometry of the aircraft. The aircraft may also be represented by a quantitative model that represents its possible flight trajectories in terms of its acceleration, speed, position, and orientation. The aircraft may be further represented by a logical model that describes the source and destination of signals across the aircraft or potential causes of airplane failure, such as how an engine failure can result in a loss of power and cause the airplane to lose altitude. It is apparent that many different models may be used to represent a SOI.

It should be noted that the semantics for each kind of formal model described above, including the geometric, quantitative, and logical models, can be defined using a mathematical formalism.

A system model is used to represent a system and its environment. A system model may comprise multiple views of the system to support planning, requirements, architecture, design, analysis, verification, and validation. System models can include a combination of geometric, quantitative, and logical models. They often span several modeling domains such as different systems (e.g., thermal, power), different technology domains (e.g., hardware, software), and different characteristics (e.g., physical, performance). Each of these models must be integrated to ensure a consistent and cohesive system representation. As such, the system model must enable representation of general-purpose system modeling concepts such as behavior and structure that can be shared across modeling domains.

A. Wayne Wymore is credited with one of the early efforts to formally define a system model using a mathematical framework in *A Mathematical Theory of Systems Engineering: The Elements* (1967). Wymore established a rigorous mathematical framework for designing systems in a model-based context.

Some examples of system models may include the following (from ISO/IEC/IEEE 15288):

- *A functional model* that captures the system functions and their functional interfaces
- *A behavioral model* that captures the overall behavior of the system functions
- *A temporal model* that captures the timing-related aspects of the architecture.
- *A structural model* that captures the system elements and their physical interfaces
- *A mass model* that captures the mass-related aspects of the system
- *A layout model* that captures the absolute and relational spatial placements of the system elements
- *A network model* that captures the flow of resources among the applicable system functions or elements

以下示例阐明上述分类法。飞机可能通过规定飞机的详细几何结构的三维几何模型来表示。飞机还可能通过在其加速度、速度、位置和方位方面表明其可能的飞行轨迹的定量模型来表示。此外，飞机可能通过描述贯穿飞机的信号的来源和目的地或飞机故障的潜在原因的逻辑模型来表示，例如，发动机故障如何造成功力损失并致使飞机失去高度。显然，许多不同模型可用于表示 SOI。

应注意的是，上述每种类型的正规模型，的语义可以利用数学形式来定义，包括几何模型、数量模型和逻辑模型。

系统模型被用来表示系统及其环境。系统模型可能包括系统的多重视图，以支持规划、需求、架构、设计、分析、验证和确认。系统模型可以包括几何模型、定量模型和逻辑模型的组合。系统模型往往跨越若干建模领域，例如不同的系统（如，热力、电力）、不同的技术域（如，硬件、软件）和不同的特征（如，物理的、性能的）。这些模型中的每一个都必须被综合，以确保一致的且紧密结合的系统表达。正因如此，系统模型必须能够表达诸如行为和结构可以跨建模域共享的通用的系统建模概念。

A. Wayne Wymore 被赞誉为利用《系统工程数学理论：元素》（1967）中的一个数学框架正规地定义系统模型的早期贡献者之一。Wymore 在基于模型的背景环境中建立了用于设计系统的严格数学框架。

系统模型的一些示例可能包括以下模型（来自 ISO/IEC/IEEE 15288）：

- 捕获系统功能及其功能界面的*功能模型*
- 捕获系统功能的全部行为的*行为模型*
- 捕获架构的时间安排相关方面的*时间模型*
- 捕获系统元素及其物理界面的*结构模型*
- 捕获系统的质量相关方面的*质量模型*
- 捕获系统元素的绝对的和相关的空间布置的*布局模型*
- 捕获可用的系统功能或元素之间的资源流的*网络模型*

9.1.4.2 Types of Simulations Simulations can be described under one or more of the following types:

- *Physical simulations* utilize physical models and aim to replicate a relatively small number of system attributes to a high degree of accuracy (fidelity). Often, such simulations require physical models of specific environmental attributes with similar levels of fidelity. Such simulations are often costly to construct and the limited number of system and environment attributes restricts the range of questions that can be answered. This kind of simulation is used when cheaper computer-based simulations cannot be constructed to answer questions. Examples of physical simulations are wind tunnels tests, environmental tests, and mock-ups that elucidate manufacturing processes.

- *Computer-based* simulations can be divided into subtypes based on models of computation (MoC), for example, discrete event, continuous time solving, or finite element. Each requires the mathematical model to conform to a certain structure, and some can be combined to create a hybrid MoC. Where stochastic processes are being modeled or there is uncertainty in system inputs, Monte Carlo simulations can be performed in order to perform a statistical analysis on output of many simulation runs. There is a trend in simulation environments to separate the execution architecture (that in effect implements the MoC algorithm) and the models of the systems of interest, with the latter being implemented in a modular form. Doing this helps deal with complex models and improves the possibility of reusing models in different simulations. Computer based simulations can be made to cover a broad scope of system attributes and indeed can become quite complex in their own right by including models of many types of systems interacting in many different ways. When the level of complexity becomes such that the expertise necessary to create fit-for-purpose models for different parts of the overall simulation is distributed between many subject matter experts, the construction of such simulations becomes in itself an exercise in SE.

- *Hardware and/or human-in-the-loop simulations* execute in real time and use computer-based models to close the loop on inputs and outputs with a hardware and or human element of the system. Such simulations have a high level of fidelity but can be costly, especially if physical stimulation is required, for example, motion or visual scene generation.

Within the US defense community, it is common to refer to simulations as live, virtual, or constructive, where:

- *Live simulation* refers to live operators operating real systems

9.1.4.2 *仿真的类型* 仿真可根据下述其中一种或多种类型描述：

- *物理仿真*利用物理模型并且旨在以非常高的精确度（保真度）复制相对较少的系统属性。这类仿真往往要求具有相似保真度水平的特定环境属性的物理模型。构建这类仿真往往成本很高，并且有限数量的系统和环境属性限定可回答的问题的范围。当低成本的基于计算机的仿真无法被构建回答问题时，使用这类仿真。物理仿真的示例是风洞试验、环境试验和解释制造工艺的实体模型。

- *基于计算机的仿真*可以基于计算模型（MoC）分为多个子类型，例如，离散事件、连续时间解算或有限元。每个子类型都要求数学模型符合某一结构，并且一些子类型可以组合在一起以创建混合型 MoC。用于随机流程的建模或系统输入存在不确定性时，可以进行蒙特卡洛仿真，以便对多次仿真运行的输出进行统计分析。仿真环境的一个趋势是将执行架构（事实上该架构实施 MoC 算法）与所感兴趣之系统的模型分开，后者以模块化形式实施。这种做法有助于应对复杂模型并改善以不同仿真复用模型的可能性。为涵盖广泛的系统属性，可以进行基于计算机的仿真，并且通过将以许多不同方式交互的多种形式的系统的模型包括在内，其本身的确可以变得相当复杂。当复杂性层次致使为总体仿真的不同部分建立适用性模型，所需的专业知识分布于许多领域专家之间时，这类仿真的构建本身成为 SE 中的一项实践工作。

- *硬件和/或人在回路中的仿真实时执行*，并使用基于计算机的模型闭环具有系统的硬件和/或人员要素的输入和输出的回路。这类仿真具有很高的保真度水平但成本很高，特别是如果需要物理激励，例如，运动或视觉景象的生成。

在美国防务行业内，这通常指的是现场仿真、虚拟仿真或构造仿真，其中：

- *现场仿真*指现场的操作人员操作真实的系统

CROSS-CUTTING SYSTEMS ENGINEERING METHODS

- *Virtual simulation* refers to live operators operating simulated systems
- *Constructive simulation* refers to simulated operators operating with simulated systems

The virtual and constructive simulations may also include actual system hardware and software in the loop as well as stimulus from a real system environment.

9.1.5 Developing Models and Simulations

The completed model or simulation can be considered a system or a product in its own right. Therefore, the general steps in the development and application of a model or simulation are closely aligned to the SE processes described within this handbook. Models need to be planned and tracked, just like any other developmental effort.

Executed in parallel and critical to any model and simulation development effort is the verification, validation, and accreditation (VV&A) process that certifies that a model or simulation is acceptable for use for a specific purpose. Given the consequences of using the knowledge gained from the model or simulation, the user of that knowledge must be confident that the knowledge is of sufficient credibility (i.e., it is "fit for purpose"). This implies that the associated risks of employing the model or simulation in the decision process are minimized such that it is perceived that the model and simulation is of more use than not (i.e., the risks of not using the model or simulation are greater than the risks of using the model or simulation). The US DOD Modeling and Simulation Coordination office has committed significant resources to produce comprehensive guidance on VV&A (M&SCo, 2013).

9.1.6 Model and Simulation Integration

Many different types of models and simulations may be used as part of a model-based approach. A key activity is to facilitate the integration of models and simulations across multiple domains and disciplines. As an example, system models can be used to specify the elements of the system. The logical model of the system architecture may be used to identify and partition the elements of the system and define their interconnection or other relationships among the elements. Quantitative models for performance, physical, and other quality characteristics, such as reliability, may be employed to determine the required values for specific element properties to satisfy the system requirements. An executable system model that represents the interaction of the system elements may be used to validate that the element requirements can satisfy the system behavioral requirements. The aforementioned models each represent different facets of the same system. The different engineering disciplines for electrical, mechanical, and software each create their own models representing different facets of the same system. The different models must be sufficiently integrated to ensure a cohesive system solution.

- *虚拟仿真*指现场的操作人员操作仿真的系统
- *构造仿真*指仿真的操作人员操作仿真的系统

虚拟仿真和构造仿真还可能包括实际系统硬件和软件在回路中以及来源于真实系统环境的激励。

9.1.5 开发模型和仿真

完整的模型或仿真本身可被视为一个系统或产品。因此，开发和应用模型或仿真的一般步骤与本手册描述的 SE 流程相当一致。正如任何其他开发工作，模型需要被规划和追溯。

验证、确认和鉴定（VV&A）流程并行执行并对模型和仿真开发工作是至关重要的，其证明模型或仿真适用于特定目的。考虑到后续将使用从模型或仿真中获取的知识，这些知识的用户必须确信这些知识有足够的可信度（即，它"适配于目的"）。这意味着，在决策流程中采用模型或仿真的相关风险被最小化，以致意识到要尽可能多地使用模型和仿真（即，不使用模型或仿真的风险大于使用模型或仿真的风险）。美国国防部建模与仿真协调办公室已投入重要的资源来产生关于 VV&A（M&SCo, 2013）的全面指导。

9.1.6 模型和仿真综合

许多不同类型的模型和仿真可用于基于模型的方法的一部分。一个关键活动是促进模型和仿真跨多个领域和学科的综合。例如，系统模型可用来特定系统的元素。系统架构的逻辑模型可用来识别和划分系统的元素并定义元素的互连或元素之间的其他关系。用于诸如性能、物理和其他品质特征如可靠性等的定量模型可用来确定特定元素特性的所需值，以满足系统需求。表达系统元素的交互的一个可执行系统模型可用来确认元素需求是否可以满足系统行为需求。上述模型各自表现同一系统的不同方面。电气、机械和软件的不同工程学科各自创建其表现同一系统的不同方面的模型。不同的模型必须被充分综合，以确保协调一致的系统解决方案。

To support the integration, the models and simulations must establish semantic interoperability to ensure that a construct in one model has the same meaning as a corresponding construct in another model. A simple example is the name of a particular element that may appear in a higher-level system element model, a reliability model, and electrical design model. This modeling information must be exchanged between modeling tools and consistently represented in the different models.

One approach to semantic interoperability is to use model transformations between different models. Transformations are defined which establish correspondence between the concepts in one model and the concepts in another. In addition to establishing correspondence, the tools must have a means to exchange the model data and share the information. There are multiple means for exchanging data between tools, including file exchange, use of API, and a shared repository.

The use of modeling standards for modeling languages, model transformations, and data exchange is an important enabler of integration across modeling domains.

9.1.7 Model Management

Since the system models and simulations are primary artifacts of the SE effort, their management is of particular importance. The management of models and simulations throughout the system life cycle includes configuration management concerns related to versioning and change control. These are complex processes in their own right, particularly when distributed teams may update different aspects of different pieces. Change management techniques such as branch and merge may be employed along with other integration approaches. Another important aspect of model and simulation management is the ongoing validation. As changes are introduced to the models and simulations, the team needs to ensure they remain a sufficient representation of the system for their intended purpose.

9.1.8 Modeling Standards

Different types of models are needed to support the analysis, specification, design, and verification of systems. Modeling standards play an important role in defining agreed-upon system modeling concepts that can be represented for a particular domain of interest and enable the integration of different types of models across domains of interest.

Standards for system modeling languages can enable cross-discipline, cross-project, and cross-organization communication. This communication offers the potential to reduce the training requirements for practitioners when transitioning from one project to another and enables the reuse of system artifacts within and across projects and organizations. Standard modeling languages also provide a common foundation for advancing the practice of SE, as do other SE standards.

为支持综合，模型和仿真必须建立语义的互操作性，以确保一个模型中的构建与另一个模型中的相应构建具有相同含义。举一个简单的例子，某一特殊子类元素的名称可能出现在一个较高层级系统元素模型、一个可靠性模型和电气设计模型中的特别元素的名称。该建模信息必须在建模工具之间交换并在不同的模型中得到一致地表达。

达到语义互操作性的一种途径是在不同的模型之间使用模型转换。对在一个模型中的概念与另一个模型中的概念之间建立对应关系的转换进行定义。除了建立对应关系之外，该工具还必须有一种交换模型数据并共享信息的手段。工具之间交换数据的手段有多种，包括文件交换、使用API以及一个共享库。

将建模标准用于建模语言、模型转换和数据交换是一个跨建模领域综合的重要使能器。

9.1.7 模型管理

因为系统模型和仿真是 SE 工作的主要制品，所以对它们的管理特别重要。贯穿系统生存周期内的模型和仿真管理包括与版本控制和更改控制相关的构型管理的关注点。这些流程自身就是复杂的，特别是当分布的团队可能更新不同部分的不同方面时。诸如分支和合并等变更管理技术可能与其他综合方法一同被采用。模型和仿真管理的另一个重要方面是持续确认。当变更被引入到模型和仿真中时，团队需要确保模型和仿真保持系统意图的目的的充分表达。

9.1.8 建模标准

需要不同类型的模型来支持系统的分析、规范、设计和验证。建模标准在定义协定一致的系统建模概念中起着重要作用，可以由所感兴趣之特定领域表示并且使能不同类型的模型跨越所感兴趣的领域来综合。

系统建模语言的标准可以使能跨学科、跨项目和跨组织沟通。这种沟通提供了在从一个项目转换为另一个项目时减少对实践者的培训要求的可能性，并且使得系统制品能在项目和组织内部和之间复用。标准建模语言还为推进 SE 实践提供一个共同基础，其他 SE 标准也是如此。

Modeling standards include standards for modeling languages, data exchange between models, and the transformation of one model to another to achieve semantic interoperability. A partial list of representative modeling standards can be found under the modeling standards section of SEBoK (2014).

9.1.9 Modeling Languages

Modeling languages are generally intended to be both human interpretable and computer interpretable and are specified in terms of both syntax and semantics.

The abstract syntax specifies the model constructs and the rules for constructing the model from its constructs. In the case of a natural language like English, the constructs may include types of words such as verbs, nouns, adjectives, and prepositions, and the rules specify how these words can be used together to form proper sentences. The abstract syntax for a mathematical model may specify constructs to define mathematical functions, variables, and their relationship. The abstract syntax for a logical model may also specify constructs to define logical entities and their relationships such as the interconnection relationship between parts or the precedence relationship between actions. A well-formed model conforms to its rules, just as a well-formed sentence must conform to the grammatical rules of the natural language.

The concrete syntax specifies the symbols used to express the model constructs. The natural language, such as English or German, can be expressed in text or Morse code. A modeling language may be expressed using graphical symbols and/or text statements. For example, a functional flow model may be expressed using graphical symbols consisting of a combination of graphical nodes and arcs annotated with text, while a simulation modeling language may be expressed using a text syntax in a programming language such as FORTRAN or C.

The semantics of a language define the meaning of the constructs. For example, an English word does not have explicit meaning until the word is defined. A sentence can be grammatically correct, but can still be gibberish if the words are not defined, or be misunderstood if the meaning of the words is ambiguous in the context of their use. The language must give meaning both to the concept of a verb or noun and to the meaning of a specific word that is a verb or noun. Similarly, a modeling construct that is expressed as a symbol, such as a box or arrow on a flowchart, does not have meaning until it is defined. The box and arrow each represent different concepts. These concepts must be defined, and the specific boxes and arrows should also be defined. The definitions can be expressed in natural language or other formalisms. For example, the symbols $\sin(x)$ and $\cos(x)$ represent the sine and cosine function, which are defined precisely in mathematics. If the position of a pendulum is defined in terms of $\sin(\theta)$ and $\cos(\theta)$, the meaning of the pendulum position is understood in terms of these formalisms.

建模标准包括建模语言、模型之间数据交换以及一个模型到另一个模型的转换的标准以达成语义互操作性。在 SEBoK（2014）的建模标准一节可找到具有代表性的建模标准的部分列表。

9.1.9 建模语言

建模语言一般的意图是既要人类可解释的又要计算机可解释，并且在句法及语义方面做出规定。

抽象的句法规定模型的构建以及根据其构造而构建模型的规则。就像英语等自然语言，构造可能包括诸如动词、名词、形容词和介词等单词类型，而规则规定这些单词如何一同使用来构成正确的句子。数学模型的抽象句法能够规定定义数学函数、变量及其关系的构造。逻辑模型的抽象句法还能规定定义逻辑实体及其关系的构造，例如两个部分之间的互连关系或两个活动之间的前后置关系。构建良好的模型符合其规则，正如构建良好的句子必须符合自然语言的语法规则。

具体句法规定用于表达模型构造的符号。诸如英语或德语等自然语言可以用文字或莫尔斯码来表达。建模语言可使用图形符号和/或文字说明来表达。例如，功能流模型可使用由带文本注释的图形节点和弧线组合构成的图形符号来表达，而仿真建模语言可使用诸如 FORTRAN 或 C 等编程语言中的一种文本句法来表达。

一种语言的语义定义构造的含义。例如，英语单词在被定义之前没有明确的含义。一个句子在句法上可能是正确的，但如果单词没有被定义则仍然可能令人费解，或者如果单词的含义在其使用的上下文中有歧义则可能会被曲解。语言必须向动词或名词的概念以及某一特定词动词或名词词性的意思赋予含义。同样，以诸如流程图上的方框或箭头等符号表达的建模构造在被定义之前没有含义。方框和箭头各自代表不同的概念。这些概念必须被定义，并且特定的方框和箭头也应被定义。这些定义可以用自然语言或其他形式来表达。例如，$\sin(x)$ 和 $\cos(x)$ 符号表示在数学中准确定义的正弦函数和余弦函数。如果摆锤的位置依照 $\sin(\theta)$ 和 $\cos(\theta)$ 来定义，那么摆锤位置的含义依照这些正规形式来理解。

CROSS-CUTTING SYSTEMS ENGINEERING METHODS

The SysML™ from the OMG has emerged as an important modeling language for systems (OMG, 2013b). Summary descriptions of the SysML diagram types shown in Figure 9.2 are as follows:

- A *package diagram* (pkg) is used to organize the model into packages that contain other model elements. This facilitates model navigation and reuse, as well as access and change control.
- A *requirements diagram* (req) captures text-based requirements. Having requirements within the model enables fine-grained traceability from requirement to requirement and between requirements and design, analysis, and verification elements in the model.
- System structure is represented using block diagrams:
  - A *block definition diagram* (bdd) describes the system hierarchy and classifications of system elements.
  - An *internal block diagram* (ibd) depicts the internal structure of a system in terms of how its parts are interconnected using ports and connectors, describing how the parts within the system are interconnected.
- Behavior is captured in use case, activity, sequence, and state machine diagrams:
  - A *use case diagram* (uc) provides a high-level description of the system functionality in terms of how users and external systems use the system to achieve their goals.

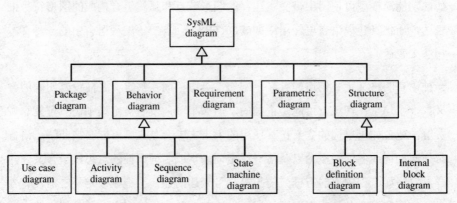

FIGURE 9.2   SysML diagram types. From Friedenthal et al. (2012), Figure 3.2. Reprinted with permission from Sandy Friedenthal. All other rights reserved.

来自于 OMG 的 SysML™ 已经作为系统的重要建模语言而出现（OMG，2013b）。如图 9.2 所示的 SysML 图类型的概括描述如下：

- *包图*（pkg）用于将模型组织成包含其他模型元素的包。这有助于模型导航和复用，以及访问及变更控制。

- *需求图*（req）捕获基于文本的需求。模型内具有需求使得从需求到需求以及需求与模型中的设计、分析和验证元素之间能够进行精细粒度的追溯。

- 系统结构使用块图来表示：
  - *块定义图*（bdd）描述系统层级结构层次以及系统元素的分类。
  - *内部块图*（ibd）在系统各个部分如何利用端口和连接器互连方面描述系统的内部结构，即描述系统内的各个部分如何互连。

- 行为是在用例图、活动图、顺序图和状态机图中捕获的。
  - *用例图*（uc）在用户和外部系统如何使用该系统达成其目标方面，提供对系统功能性的高层级描述。

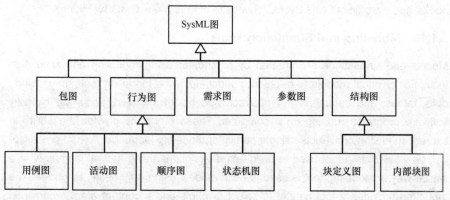

图 9.2 SysML 图类型。来自于 Friedenthal 等（2012），图 3.2。经 Sandy Friedenthal 许可后转载。版权所有。

- An *activity diagrams* (act) represents the transformation of inputs to outputs through a controlled sequence of actions.
- A *sequence diagram* (sd) represents interaction in terms of the time-ordered exchange of messages between collaborating parts of a system.
- A *state machine diagram* (stm) describes the states of a system or its parts; the transitions between the states; the actions that occur within states or upon transition, entry, or exit; and the events that trigger transitions.

- A *parametric diagram* (par) represents constraints on system property values as necessary to support detailed engineering analysis. These constraints may include performance, reliability, and mass properties, among others. SysML™ can be integrated with other engineering analysis models and tools to execute the analysis.

SysML includes an allocation relationship to represent allocation of functions to elements, allocation of logical to physical elements, and other types of allocations. SysML™ is a general-purpose modeling language that is intended to support many different model-based methods, such as structured analysis methods and object-oriented methods. A particular method may require only a subset of the diagrams. For example, a simplified functional analysis method may only require activity diagrams augmented by bdds, ibds, and perhaps requirements diagrams.

For general information on SysML™, along with links to tool vendors, articles, and books, see the official OMG SysML™ website at http://www.OMGSysML.org.

9.1.10 Modeling and Simulation Tools

Models and simulations are created by a modeler using modeling and simulation tools. For physical models (e.g. physical mock-ups), the modeling tools may include drills, lathes, and hammers. For abstract models, the modeling tools are typically software programs running on a computer. These programs provide the ability to express modeling constructs using a particular modeling language. A word processor can be viewed as a tool used to build text descriptions using natural language. In a similar way, a modeling tool is used to build models using a modeling language. The tool often provides a tool palette to select symbols and a content area to construct the model from the graphical symbols or other concrete syntax. A modeling tool typically checks the model to evaluate whether it conforms to the rules of the language and enforces such rules to help the modeler create a well-formed model. This is similar to the way a word processor checks the text to see that it conforms to the grammar rules for the natural language.

- *活动图*（act）表达通过行动的一种受控顺序从输入到输出的转换。
- *顺序图*（sd）依照消息在一个系统的各个合作部分之间按时序交换来表达交互。
- *状态机图*（stm）描述系统及其各个部分的状态；状态之间的转移；在状态内出现的或在转移、进入或退出时出现的行动；以及触发转移的事件。

• *参数图*（par）表达关于支持详细工程分析所需的系统特性量值的约束。这些约束可能包括性能、可靠性、质量特性和其他因素等。SysML™可与执行分析的其他工程分析模型和工具综合在一起。

SysML 包括一种表示功能到元素的分配、逻辑到物理元素的分配关系及其他分配类型。SysML™是一种将用于支持诸如结构分析方法和面向对象方法等许多不同的基于模型的方法的通用建模语言。某一特殊的方法可能仅需要上述图的子集。例如，简化的功能分析方法可能仅需要由 bdd、ibd 或许还有需求图所扩充的活动图。

关于 SysML™的一般信息以及对工具供应商、文献和手册的链接，参见 OMG SysML™官方网站，网址：http://www.OMGSysML.org。

9.1.10 建模和仿真工具

模型和仿真由建模者利用建模和仿真工具创建。对于物理模型（如，物理实体原型）而言，建模工具可能包括钻头、车床和锤子。对于抽象模型而言，建模工具通常是在计算机上运行的软件程序。这些程序提供利用特定的建模语言表达建模构造的能力。文字处理可被视为一种用来利用自然语言构建文本描述的工具。同理，建模工具用于利用建模语言构建模型。工具往往提供选择符号的工具板以及根据图形符号或其他具体句法构建模型的内容域。建模工具通常对模型进行检查，以评价模型是否符合语言的规则以及是否强制执行这些规则以帮助建模者创建良好的模型。这类似于文字处理对文本进行检查以查看其是否符合自然语言的语法规则。

Some modeling and simulation tools are commercially available products, while others may be created or customized to provide unique modeling solutions. Modeling and simulation tools are often used as part of a broader set of engineering tools, which constitute the system development environment. There is increased emphasis on tool support for standard modeling languages that enable models and modeling information to be interchanged among different tools.

9.1.11 Indicators of Model Quality

The quality of a model should not be confused with the quality of the design that the model represents. For example, one may have a high-quality, computer-aided design model of a chair that accurately represents the design of the chair, yet the design itself may be flawed such that when one sits in the chair, it falls apart. A high quality model should provide a representation sufficient to assist the design team in assessing the quality of the design and uncovering design issues.

Model quality is often assessed in terms of the adherence of the model to modeling guidelines and the degree to which the model addresses its intended purpose. Typical examples of modeling guidelines include naming conventions, application of appropriate model annotations, proper use of modeling constructs, and applying model reuse considerations. Specific guidelines are different for different types of models. For example, the guidelines for developing a geometric model using a computer-aided design tool may include conventions for defining coordinate systems, dimensioning, and tolerances.

9.1.12 Model and Simulation-Based Metrics

Models and simulations can provide a wealth of information that can be used for both technical and management metrics to assess the modeling and simulation efforts and, in many cases, the overall SE effort. Different types of models and simulations provide different types of information. In general, models and simulations provide information that enables one to:

- Assess progress
- Estimate effort and cost
- Assess technical quality and risk
- Assess model quality

有些建模和仿真工具是商业化的产品，而有些工具则可被创建或定制以提供独特的建模解决方案。建模和仿真工具往往用于构成系统开发环境的广泛工程工具集合的一部分。支持模型和建模信息能在不同工具间互换的标准建模语言倍受关注。

9.1.11 模型质量的指标

模型的质量不应与模型所表达的设计的质量相混淆。例如，一个人可能具有一个准确地表达椅子的设计的、高质量的计算机辅助椅子设计模型，但该设计本身可能是有缺陷的，如此一来，当一个人坐在椅子上时，椅子就散架。高质量的模型应提供足以协助设计团队评估设计的质量并发现设计问题的表达。

模型质量往往依照模型对建模指南的遵守以及模型应对其意图目的的程度来评估。建模指南的典型示例包括命名约定、恰当模型注释的应用、建模构造的恰当使用以及对模型复用考虑的应用。特定的指南对于不同类型的模型而言是不同的。例如，使用计算机辅助设计工具开发几何模型的指南可能包括对坐标系、尺寸标注和公差进行定义的约定。

9.1.12 基于模型和仿真的衡量标准

模型和仿真可以提供大量的既可用于技术衡量标准又可用于管理衡量标准的信息，以评估建模和仿真工作，并且在很多情况下，评估总体 SE 工作。不同类型的模型和仿真提供不同类型的信息。一般来说，模型和仿真提供的信息使人们能够：

- 评估进展

- 估算工作和成本

- 评估技术质量和风险

- 评估模型质量

A model's progress can be assessed in terms of the completeness of the modeling effort relative to the defined scope of the model. Models may also be used to assess progress in terms of the extent to which the requirements have been satisfied by the design or verified through testing. When augmented with productivity metrics, the model can be used to estimate the cost of performing the required SE effort to deliver the system.

Models and simulations can be used to identify critical system parameters and assess technical risks in terms of any uncertainty that lies in those parameters. The models and simulations can also be used to provide additional metrics that are associated with its purpose. For example, when the model's purpose is to support mission and system concept formulation and evaluation, then a key metric may be the number of alternative concepts that are explored over a specified period of time.

9.2 MODEL-BASED SYSTEMS ENGINEERING

This section provides an overview of the model-based SE (MBSE) methodology including a summary of benefits of a model-based approach over a more traditional document-based approach, the purpose and scope of an MBSE approach, references to a survey of MBSE methods used to perform MBSE, and a brief discussion on model management.

9.2.1 MBSE Overview

A number of model and simulation practices have been formalized into SE processes. These processes are the foundation of MBSE. The *INCOSE Systems Engineering Vision 2020* (2007) defines MBSE as *"the formalized application of modeling to support system requirements, design, analysis, verification, and validation activities beginning in the conceptual design phase and continuing throughout development and later life cycle phases."*

MBSE enhances the ability to capture, analyze, share, and manage the information associated with the specification of a product, resulting in the following benefits:

- *Improved communications* among the development stakeholders (e.g., the customer, program management, systems engineers, hardware and software developers, testers, and specialty engineering disciplines)

- *Increased ability to manage system complexity* by enabling a system model to be viewed from multiple perspectives and to analyze the impact of changes

- *Improved product quality* by providing an unambiguous and precise model of the system that can be evaluated for consistency, correctness, and completeness

模型的进展可以依照建模工作相对于模型的定义范围而言的完整性来评估。模型还可能用于依照设计满足需求或通过测试验证需求的范围来评估进展。当增添了生产率衡量标准时，模型可以用于估算为交付系统进行所需 SE 工作的成本。

模型和仿真可用于识别关键系统参数并依照那些参数中存在的不确定性来评估技术风险。模型和仿真还可用于来提供与其目的相关联的附加衡量标准。例如，当模型的目的是支持任务及系统概念的形成和评价时，关键的衡量标准则可能是在规定时间内探索到的备选的方案的数目。

9.2 基于模型的系统工程

本节提供基于模型的 SE（MBSE）方法论的概览，包括基于模型的方法超越较为传统的基于文件的方法的益处的概括、MBSE 方法的目的和范围、参考 MBSE 方法的综述来执行 MBSE 以及模型管理的简要论述。

9.2.1 MBSE 概览

许多模型和仿真实践已经被正规化为 SE 流程。这些流程是 MBSE 的基础。《INCOSE 系统工程愿景 2020》（2007）中将 MBSE 定义为"支持以概念设计阶段开始并持续贯穿于开发和后期的生命周期阶段的系统需求、设计、分析、验证和确认活动的正规化建模应用"。

MBSE 增强获取、分析、共享和管理与产品规范相关的信息的能力，从而带来以下好处：

- 改善开发利益攸关者（如，客户、项目管理人员、系统工程师、硬件及软件开发者、测试人员和专业工程学科）之间的沟通

- 提高通过使能从多个角度观察系统模型来管理系统复杂性的能力，以及分析变更影响的能力

- 通过提供可评估一致性、正确性和完整性的无歧义的且精确的系统模型，提升产品质量

- Enhanced knowledge capture and reuse of the information by capturing information in more standardized ways and leveraging built-in abstraction mechanisms inherent in model-driven approaches. This in turn can result in reduced cycle time and lower maintenance costs to modify the design

- *Improved ability to teach and learn SE fundamentals* by providing a clear and unambiguous representation of the concepts

MBSE is often contrasted with a traditional document based approach to SE. In a document-based SE approach, there is often considerable information generated about the system that is contained in documents and other artifacts such as specifications, interface control documents, system description documents, trade studies, analysis reports, and verification plans, procedures, and reports. The information contained within these documents is often difficult to maintain and synchronize, and difficult to assess in terms of its quality (correctness, completeness, and consistency).

In an MBSE approach, much of this information is captured in a system model or set of models. The system model is a primary artifact of the SE process. MBSE formalizes the application of SE through the use of models. The degree to which this information is captured in models and maintained throughout the life cycle depends on the scope of the MBSE effort. Leveraging an MBSE approach to SE is intended to result in significant improvements in system requirements, architecture, and design quality; lower the risk and cost of system development by surfacing issues early in the system definition; enhance productivity through reuse of system artifacts; and improve communications among the system development team.

9.2.1.1 Survey of MBSE Methodologies In general, a methodology can be defined as the collection of related processes, methods, and tools used to support a specific discipline (Martin, 1996). That more general notion of methodology can be specialized to MBSE methodology, which we characterize as the collection of related processes, methods, and tools used to support the discipline of SE in a "model-based" or "model-driven" context (Estefan, 2008).

In 2008, a survey of candidate MBSE methodologies was published under the auspices of an INCOSE technical publication (Estefan, 2008). Six (6) candidate MBSE methodologies were surveyed: INCOSE Object-Oriented Systems Engineering Method (OOSEM), IBM Rational Telelogic Harmony-SE, IBM Rational Unified Process for Systems Engineering (RUP-SE), Vitech MBSE Methodology, JPL State Analysis (SA), and Dori Object-Process Methodology (OPM). Additional information on these methodologies is available on the INCOSE MBSE initiative Wiki (INCOSE, 2010a).

- 通过以更加标准化的方式捕获信息并高效地利用模型驱动方法固有的内置抽象机制，增强知识捕获及信息的复用；这反过来会导致缩短的周期时间和更低的维护成本，以改进设计

- 通过提供概念的清晰且无歧义的表达，提升教授与学习 SE 基本原理的能力

MBSE 常常与传统的基于文件的 SE 方法形成对比。在基于文件的 SE 方法中，常常生成的关于系统的大量信息，被包含在文件以及其他制品中，例如规范、接口控制文件、系统描述文件、权衡研究、分析报告以及验证计划、程序和报告。这些文件内包含的信息往往很难维护和同步，并且很难评估其质量（正确性、完整性和一致性）。

在 MBSE 方法中，在某一系统模型或模型集合中捕获大量的这类信息。系统模型是 SE 流程的首要制品。MBSE 通过使用模型使 SE 的应用正规化。在模型中捕获这些信息并贯穿于生命周期维护这些信息的程度取决于 MBSE 工作的范围。强有力地发挥 SE 的 MBSE 方法的作用意在使系统需求、架构和设计质量得到显著改善；通过早期暴露系统定义的问题，降低系统开发的风险和成本；通过复用系统制品，提高生产率；并改善系统开发团队之间的沟通。

9.2.1.1 MBSE 方法论综述 一般来说，方法论可以被定义为用于支持特定学科的有关流程、方法和工具的集合（Martin，1996）。更为一般的方法论概念将会专用于 MBSE 方法论，我们将其描述为在"基于模型的"或"模型驱动的"背景环境中用于支持 SE 学科的有关流程、方法和工具的集合（Estefan，2008）。

2008 年，多种备选的 MBSE 方法论综述在 INCOSE 技术出版物（Estefan，2008）的支持下发表。概述了六种备选的 MBSE方法论：INCOSE 面向对象的系统工程方法（OOSEM）、IBM RationalTelelogic harmony-SE、IBM Rational 统一的系统工程流程（RUP-SE）、Vitech MBSE 方法论、JPL 状态分析（SA）和 Dori 对象—流程—方法论（OPM）。关于这些方法论的其他信息在 INCOSE MBSEinitiative Wiki 上可以找到（INCOSE，2010a）。

Two example methods that are included in the SE handbook are the functions-based SE (FBSE) method in Section 9.3 and OOSEM in Section 9.4. Although the functions-based method is not referred to as model based, there are other functions-based methods such as the Vitech MBSE Methodology that are explicitly model based. OOSEM is defined as an end-to-end MBSE method, where the artifacts of the method are modeling artifacts that are managed and controlled throughout the SE process.

9.3 FUNCTIONS-BASED SYSTEMS ENGINEERING METHOD

9.3.1 Introduction

FBSE is an approach to SE that focuses on the functional architecture of the system. A *function* is a characteristic task, action, or activity that must be performed to achieve a desired outcome. A function may be accomplished by one or more system elements comprising equipment (hardware), software, firmware, facilities, personnel, and procedural data.

The objective of FBSE is to create a functional architecture for which system products and processes can be designed and to provide the foundation for defining the system architecture through the allocation of functions and subfunctions to hardware/software, databases, facilities, and operations (e.g., personnel).

FBSE describes what the system will do, not how it will do it. Ideally, this process begins only after all of the system requirements have been fully identified. Often, this will not be possible, and these tasks will have to be done iteratively, with the functional architecture being further defined as the system requirements evolve.

9.3.1.1 Method Overview The FBSE process is iterative, even within a single stage in the system life cycle. The functional architecture begins at the top level as a set of functions that are defined in the applicable requirements document or specification, each with functional, performance, and limiting requirements allocated to it (in the extreme, top-level case, the only function is the system, and all requirements are allocated to it). As shown in Figure 9.3, the next lower level of the functional architecture is developed and evaluated to determine whether further decomposition is required. If it is, then the process is iterated through a series of levels until a functional architecture is complete.

SE 手册中包括的两个示例方法是 9.3 节中的基于功能的 SE（FBSE）方法和 9.4 节中的 OOSEM。尽管基于功能的方法并不称为基于模型，但诸如 Vitech MBSE 方法论等其他基于功能的方法显然是基于模型的。OOSEM 被定义为端到端的 MBSE 方法，其中，该方法的制品是贯穿于 SE 流程受管理和控制的建模制品。

9.3 基于功能的系统工程方法

9.3.1 简介

FBSE 是一种聚焦于系统的功能架构的 SE 方法。功能是为达成期望结果而必须实施的特有任务、行动或活动。一项功能可能通过由设备（硬件）、软件、固件、设施、人员和程序数据构成的一个或多个系统元素来实现。

FBSE 的目的在于，创建一个系统产品和流程能够被设计的功能架构，并通过给硬件/软件、数据库、设施和运行（如人员）分配功能和子功能来为定义系统架构提供基础。

FBSE 描述系统要做什么，而不是它要如何做。理想情况下，该流程只有在所有系统需求都已被完全识别后才开始。这往往是不可能的，并且这些任务将必须迭代地进行，功能架构随着系统需求的演进而被进一步定义。

9.3.1.1 方法概览 FBSE 流程是迭代的，即使在系统生命周期的单个阶段内（也要迭代）。功能架构从顶层开始，作为适用的需求文件或规范中定义的功能集合，每层具有向其分配的功能、性能和限制需求（在极端的顶层情况下，只有功能是系统，且所有需求都被分配给该系统）。如图 9.3 所示，开发并评价功能架构的下一个更低层级，以确定是否需要进一步分解。若是，则通过一系列层级来迭代该流程，直到功能架构完成。

CROSS-CUTTING SYSTEMS ENGINEERING METHODS

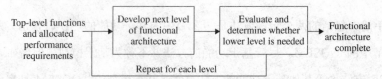

FIGURE 9.3 Functional analysis/allocation process. INCOSE SEH v1 Figure 4.3-1. Usage per the INCOSE notices page. All other rights reserved.

FBSE should be conducted iteratively:

- To define successively lower-level functions required to satisfy higher-level functional requirements and to define alternative sets of functional requirements
- With requirements definition, to define mission and environment-driven performance and to determine that higher-level requirements are satisfied
- To flow down performance requirements and design constraints
- With architecture and design, to refine the definition of product and process solutions

At each level of the process, alternative decompositions and allocations may be considered and evaluated for each function and a single version selected. After all of the functions have been identified, then all the internal and external interfaces to the decomposed sub-functions are established. These steps are shown in Figure 9.4.

FBSE examines a defined function to identify all the subfunctions necessary to accomplish that function; all usage modes must be included in the analysis. This activity is conducted to the level of depth needed to support required architecture and design efforts. Identified functional requirements are analyzed to determine the lower-level functions required to accomplish the parent requirement. Every function that must be performed by the system to meet the operational requirements is identified and defined in terms of allocated functional, performance, and other limiting requirements. Each function is then decomposed into subfunctions, and the requirements allocated to the function are each decomposed with it. This process is iterated until the system has been completely decomposed into basic sub-functions, and each sub-function at the lowest level is completely, simply, and uniquely defined by its requirements. In the process, the interfaces between each of the functions and sub-functions are fully defined, as are the interfaces to the external world.

Identified subfunctions are arranged in a functional architecture to show their relationships and interfaces (internal and external). Functional requirements should be arranged in their logical sequence so that lower-level functional requirements are recognized as part of higher-level requirements. Functions should have their input, output, and functional interface requirements (both internal and external) defined and be traceable from beginning to end conditions. Time critical requirements must also be analyzed.

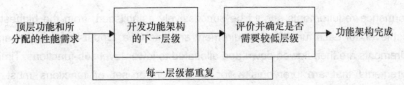

图 9.3 功能分析/分配流程。INCOSE SEH 第 1 版中图 4.3-1。按照 INCOSE 通知页使用。版权所有。

FBSE 应迭代地开展：

- 为满足更高层级的功能需求，要定义所需的相继的更低层级的功能，并且定义功能需求的备选集合

- 用需求定义去定义任务和环境驱动的性能并确定更高层级需求得到满足

- 让性能需求和设计约束向下游流动

- 用架构和设计去细化产品和流程解决方案的定义

在流程的每个层级，可能要考虑并评价每个功能的备选分解和分配，并选择一个单一版本。所有功能已被识别后，则要建立被分解的子功能的所有内部的和外部的接口。这些步骤如图 9.4 所示。

FBSE 考查一个定义的功能，以识别实现该功能必需的所有子功能；在此分析中必须包括所有的使用模式。该活动在支持所要求的架构和设计工作所需要的深度层级上进行。分析已识别的功能需求，以确定实现母体需求所需的较低层级的功能。根据已分配的功能、性能及其他限制需求，识别并定义必须由系统执行以满足运行需求的每个功能。然后，每个功能被分解到子功能，且分配到该功能的多个需求各自都使用该功能进行分解。对该流程进行迭代，直到系统已被完全分解为基本的子功能且每个处于最低层级的子功能由其需求完全地、简单地、唯一地定义为止。在该流程中，全面定义每个功能和子功能之间的接口以及与外界的接口。

已识别的子功能被置于一个功能架构中，以表明子功能之间的关系和接口（内部的和外部的）。功能需求应按其逻辑顺序设置，以使更低层级的功能需求被公认为更高层级需求的一部分。功能应使其输入、输出和功能接口需求（内部的和外部的）被定义并可从开始追溯到结束状况。时间关键的需求也必须得到分析。

Performance requirements should be successively established, from the highest to lowest level, for each functional requirement and interface. Upper-level performance requirements are then flowed down and allocated to lower-level sub-functions. Timing requirements that are prerequisite for a function or set of functions must be determined and allocated. The resulting set of requirements should be defined in measurable terms and in sufficient detail for use as design criteria. Performance requirements should be traceable from the lowest level of the current functional architecture, through the analysis by which they were allocated, to the higher-level requirement they are intended to support. All of these types of product requirements must also be verified.

Note that while performance requirements may be decomposed and allocated at each level of the functional decomposition, it is sometimes necessary to proceed through multiple levels before allocating the performance requirements. Also, sometimes, it is necessary to develop alternative functional architectures and conduct a trade study to determine a preferred one. With each iteration of FBSE, alternative decompositions are evaluated and all interfaces are defined.

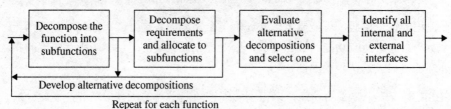

FIGURE 9.4 Alternative functional decomposition evaluation and definition. INCOSE SEH v1 Figure 4.3-2. Usage per the INCOSE notices page. All other rights reserved.

The products of FBSE can take various formats depending on the specific stage of the project and on the specific technique used to develop the functional architecture. The following are some key outputs generated from FBSE:

1. *Input–process–output (IPO) diagrams*—Top-level diagram of a data flow that is related to a specific level of system decomposition. This diagram portrays all inputs and outputs of a system but shows no decomposition.

2. *Behavior diagrams*—describe behavior that specifies system-level stimulus responses using constructs that specify time sequences, concurrencies, conditions, synchronization points, state information, and performance.

为每个功能需求和接口从最高层级到最低层级依次建立性能需求。然后，更高层级性能需求向下流动并分配给更低层级的子功能。必须确定和分配作为某一功能或功能集合的前提的时序需求。形成的需求集合应以可测度的术语、足够详细地被定义，以用作设计准则。性能需求应通过分配所依据的分析，可从当前功能架构的最低层级追溯到性能需求意图支持的更高层级需求。所有这些类型的需求亦必须被验证。

需注意，当性能需求可在功能分解的每个层级被分解和分配时，有时有必要在分配性能需求之前在多个层级中进行分解与分配。此外，为确定一个首选分解，有时有必要开发备选功能架构并进行权衡研究。利用 FBSE 的每个迭代，评价备选分解并定义所有接口。

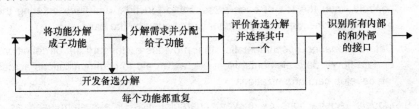

图 9.4 备选功能分解的评价和定义。INCOSE SEH 第 1 版中图 4.3-2。按照 INCOSE 通知页使用。版权所有。

FBSE 的产物可根据项目的特定阶段和用于开发功能架构的特定技术而采取各种不同的格式。以下是 FBSE 产生的一些关键输出：

1. *输入—流程—输出*（*IPO*）*图*——与系统分解的特定层级有关的数据流的顶层图。该图描绘系统的所有输入和输出，但未示分解。

2. *行为图*——利用规定时序、并发、条件、同步点、状态信息和性能的构造来描述规定系统级激励响应的行为。

3. *Control flow diagrams*—depict the set of all possible sequences in which operations may be performed by a system or a software program. There are several types of control flow diagrams, including box diagrams, flowcharts, and state transition diagrams.
4. *Data flow diagrams (DFDs)*—Provide an inter- connection of each of the behaviors that the system must perform. All inputs to the behavior designator and all outputs that must be generated are identified along with each of the data stores that each must access. Each of the DFDs must be checked to verify consistency with the IPO diagram or higher-level DFD.
5. *Entity relationship (ER) diagrams*—depict a set of entities (e.g., functions or architecture elements) and the logical relationships between them.
6. *Functional flow block diagrams (FFBDs)*—relate the inputs and outputs and provide some insight into flow between the system functions.
7. *Integrated definition for functional modeling (IDEF) diagrams*—Show the relationship between functions by sequential input and output flows. Process controls enter the top of each represented function, and lines entering the bottom show the supporting mechanism needed by the function.
8. *Data dictionaries*—documentation that provides a standard set of definitions of data flows, data elements, files, etc. as an aid to communications across the development organizations.
9. *Models*—Abstractions of relevant characteristics of a system used as a means to understand, communicate, design, and evaluate a system. They are used before the system is built and while it is being verified or in service.
10. *Simulation results*—output from a simulation of the system that behaves or operates like the SOI when provided a set of controlled inputs..

The objective of the functional decomposition activity is to develop a hierarchy of FFBDs that meet all the functional requirements of the system. Note, however, that this hierarchy is only a portion of the functional architecture. The architecture is not complete until all of the performance and limiting requirements have been appropriately decomposed and allocated to the elements of the hierarchy, as described earlier.

A description of each function in the hierarchy should be developed to include the following:

1. Its place in a network (e.g., FFBD or IDEF0/1 diagrams) characterizing its interrelationship with the other functions at its level

3. *控制流图*——描述可能由系统或软件程序所执行的运行中所有可能序列的集合。有多种类型的控制流图，包括框图、流程图和状态转换图。

4. *数据流图*（*DFD*）——提供系统必须实施的每个行为的互连。行为指定者的所有输入和必须产生的所有输出连同必须分别访问的每个数据存储区一起识别。为了验证与 IPO 图或更高层 DFD 的一致性，必须检查每个 DFD。

5. *实体关系*（*ER*）*图*——描述一个实体集合（如功能或架构元素）及它们之间的逻辑关系。

6. *功能流方块图*（*FFBD*）——关联输入与输出，并且提供对系统功能之间的流的理解。

7. *综合的功能建模定义*（*IDEF*）*图*——通过顺序的输入及输出流表明功能之间的关系。流程控制进入每个所表达功能的顶部，进入底部的线表明功能所需的支持机制。

8. *数据字典*——提供数据流、数据元素和文件等的标准定义集合的文档，用于辅助跨越开发组织沟通。

9. *模型*——系统相关特性的抽象表示，作为一种理解、沟通、设计和评估系统的手段。在系统被建造之前以及正在被验证或在使用中，使用模型。

10. *仿真结果*——当提供一个受控的输入集合时，具有像 SOI 行为和运行方式的系统仿真的输出。

功能分解活动的目的在于开发满足系统的所有功能需求的 FFBD 一个层级结构。然而，需注意，这种层级结构只是功能架构的一部分。如上所述，所有性能需求和限制需求已被适当地分解并分配给层级结构的元素后，架构才完成。

应开发对层级结构中每个功能的描述，以包括如下：

1. 其在网络（如 FFBD 或 IDEF0/1 图）中的位置描述与该层级上其他功能的相互关系的特征

CROSS-CUTTING SYSTEMS ENGINEERING METHODS

2. The set of functional requirements that have been allocated to it and define what it does

3. Its inputs and outputs, both internal and external

These various outputs characterize the functional architecture. There is no one preferred output that will support this analysis. In many cases, several of these are necessary to understand the functional architecture and the risks that may be inherent in the system architecture.in many cases, several of these are necessary to understand the functional architecture and the risks that may be inherent in the system architecture. Using more than one of these formats allows for a "check and balance" of the analysis process and will aid in communication across the system design team.

9.3.2 FBSE Tools

Tools that can be used to perform FBSE include:

- Analysis tools
- Modeling and simulation tools
- Prototyping tools
- Requirements traceability tools

9.3.3 FBSE Measures

The following measures can be used to measure the overall process and products of FBSE:

- Number of allocation-related trade studies completed as a percent of the number identified
- Percent of analyses completed
- Number of functions without a requirements allocation
- Number of functions not decomposed
- Number of alternative decompositions
- Number of internal and external interfaces not completely defined
- Depth of the functional hierarchy as a percentage versus the target depth
- Percent of performance requirements that have been allocated at the lowest level of the functional hierarchy

2. 已被分配到层级结构上且定义它做什么的功能需求集合

3. 其内部和外部的输入及输出

这些多种输出描述功能架构的特征。不存在支持该分析的首选输出。在许多情况下,其中的几种输出对于理解功能架构以及可能在系统架构中固有的风险而言是必要的。使用一个以上这类格式允许分析流程的"检查和平衡",并会有助于跨系统设计团队的沟通。

9.3.2 FBSE 工具

可以用来执行 FBSE 的工具包括:

- 分析工具
- 建模和仿真工具
- 原型构建工具
- 需求可追溯性工具

9.3.3 FBSE 测度

下列测度可用于衡量 FBSE 的总体流程和产物:

- 已完成的与分配有关的权衡研究的数量在已识别的数量中所占的百分比
- 已完成的分析的百分比
- 无需求分配的功能的数量
- 未分解的功能的数量
- 备选分解的数量
- 未完全定义的内部和外部接口的数量
- 功能层级结构的深度与目标深度的对比
- 已分配在功能层级结构的最低层级上的性能需求所占百分比

9.4 OBJECT-ORIENTED SYSTEMS ENGINEERING METHOD

9.4.1 Introduction

OOSEM (Estefan, 2008) integrates object-oriented concepts with model-based and traditional SE methods to help architect flexible and extensible systems that accommodate evolving technologies and changing requirements. OOSEM supports the specification, analysis, design, and verification of systems. OOSEM can also facilitate integration with object-oriented software development, hardware development, and verification and validation methods.

Object-oriented SE evolved from work in the mid-1990s at the Software Productivity Consortium in collaboration with lockheed Martin Corporation. The methodology was applied in part to a large, distributed information system development at lockheed Martin that included hardware, software, database, and manual procedure elements.

The INCOSE Chesapeake Chapter established the OOSEM Working group in November 2000 to help evolve the methodology further. OOSEM is described in INCOSE and industry papers (Friedenthal, 1998; lykins et al., 2000) and in A Practical Guide to SysML: The Systems Modeling Language, by Friedenthal et al. (2012).

The OOSEM objectives are as follows:

- Capture information throughout the life cycle sufficient to specify, analyze, design, verify, and validate systems
- Integrate MBSE methods with object-oriented software, hardware, and other engineering methods
- Support system-level reuse and design evolution

Figure 9.5 depicts the techniques and concepts that constitute OOSEM. OOSEM incorporates foundational SE practices, object-oriented concepts, and other unique techniques to deal with system complexity. Practices recognized as essential to SE are core tenets of OOSEM. These include requirements analysis, trade studies, and Integrated Product and Process Development (IPPD). See Section 9.7 for more about IPPD, which emphasizes multidisciplinary teamwork in the development process.

Object-oriented concepts that are leveraged in OOSEM include blocks (i.e., classes in UML) and objects, along with the concepts of encapsulation and inheritance. These concepts are supported directly by SysML. Techniques that are unique to OOSEM include parametric flowdown, system/logical decomposition, requirements variation analysis, and several others.

9.4 面向对象的系统工程方法

9.4.1 简介

OOSEM（Estefan，2008）综合面向对象的概念和基于模型的、传统的 SE 方法，以帮助构架适应技术不断演进和需求不断变化的灵活的、可扩展的系统。OOSEM 支持系统的规范、分析、设计和验证。OOSEM 也可以促进与面向对象的软件开发、硬件开发和验证及确认方法的综合。

面向对象的 SE 于 20 世纪 90 年代中期从与洛克希德—马丁公司合作的软件生产率联合体的工作中演进而成。该方法论部分地应用于洛克希德—马丁公司的大型分布式信息系统开发，包括硬件、软件、数据库和手册程序元素。为了帮助该方法论进一步演进，INCOSE Chesapeake 分会在 2000 年 11 月建立了 OOSEM 工作组。由 Friedenthal 等人在 INCOSE 和行业论文（Friedenthal,1998；lykins 等，2000）以及《SysML 实用指南：系统建模语言》中描述 OOSEM。

OOSEM 的目的如下：

- 贯穿于生命周期充分地规范、分析、设计、验证和确认系统而捕获信息
- 综合 MBSE 方法和面向对象的软件、硬件及其他工程方法
- 支持系统级复用和设计演进

图 9.5 描述构成 OOSEM 的技术和概念。OOSEM 纳入基础的 SE 实践、面向对象的概念及其他独特技术来应对系统复杂性。被认为对 SE 至关重要的实践是 OOSEM 的核心宗旨，包括需求分析、权衡研究以及综合产品和流程开发（IPPD）。关于 IPPD 的更多信息，参见 9.7 节，其中强调了开发流程中的多学科团队合作。

在 OOSEM 中更强有力地发挥作用的面向对象的概念包括"块"（即，UML 中的"类"）和对象，以及"封装"和继承的概念。这些概念由 SysML 直接支持。OOSEM 特有的技术包括参数向下流动、系统/逻辑分解、需求差异分析及若干其他方面。

CROSS-CUTTING SYSTEMS ENGINEERING METHODS

9.4.2 Method Overview

The OOSEM supports a development process as illustrated in Figure 9.6.

This development process includes sub-processes to:

- *Manage system development*—To plan and control the technical effort, including planning, risk management, configuration management, and measurement
- *Define system requirements and design*, including specifying the system requirements, developing the system architecture, and allocating the system requirements to system elements
- *Develop system elements*—To design, implement and test the element, which satisfies the allocated requirements
- *Integrate and test the system*—To integrate the system elements and verify that they satisfy the system requirements, individually and together

This process is consistent with a typical "Vee" process as described in Chapter 3. It can be applied recursively and iteratively at each level of the system hierarchy. For example, if the system hierarchy includes multiple system element levels, the process may be applied at the system level to specify the first level of system element requirements. Then the process can be applied again for each system element at the first level to specify requirements for the system elements at the second level, and so forth.

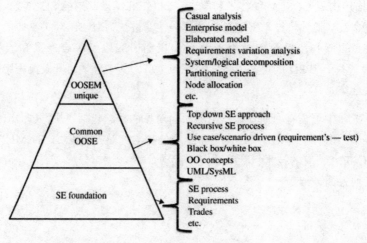

FIGURE 9.5   OOSEM builds on established SE foundations. Reprinted with permission from Howard Lykins. All other rights reserved.

9.4.2 方法概览

OOSEM 支持如图 9.6 所示的开发流程。

这个开发流程包括子流程，用以：

- *管理系统开发*——计划并控制技术工作，包括规划、风险管理、构型管理和测量
- *定义系统需求和设计*，包括规范系统需求、开发系统架构和将系统需求分配到系统元素
- *开发系统元素*——设计、实施和测试元素，以满足所分配的需求
- *综合并测试系统*——单独地和共同地综合系统元素并验证它们满足系统需求

本流程与典型的"V形"流程一致，如第 3 章所述。可在系统层级结构的每一层级进行递归、迭代地应用。例如，如果系统层级结构包括多个系统元素层级，则可在系统层级应用本流程，以规范系统元素需求的第一层级。然后，可以再次对处于第一层级的每个系统元素应用该流程，以规范对处于第二层级的系统元素的需求，并依次类推。

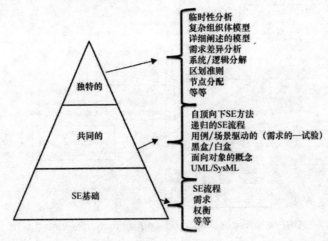

图 9.5 OOSEM 建立在既定的 SE 基础上。经 Howard Lykins 许可后转载。版权所有。

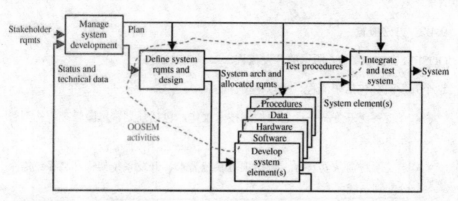

FIGURE 9.6 OOSEM activities in context of the system development process. Reprinted with permission from Howard Lykins. All other rights reserved.

To be effective, OOSEM development activities must be supported by systems engineers applying fundamental tenets of SE, including the use of multidisciplinary teams and disciplined management processes such as planning, risk management, configuration management, and measurement. OOSEM development activities and accompanying process flows are more fully described in *Object-Oriented Systems Engineering Method (OOSEM) Tutorial* (LMCO, 2008) and in *Tutorial Material— Model-Based Systems Engineering Using the OOSEM* (JHUAPL, 2011).

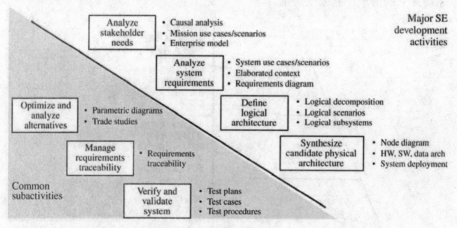

FIGURE 9.7 OOSEM activities and modeling artifacts. Reprinted with permission from Howard Lykins. All other rights reserved.

跨领域/学科系统工程方法

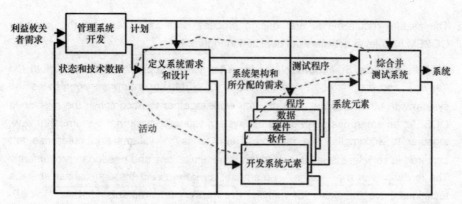

图 9.6　系统开发流程的背景环境中的 OOSEM 活动。经 Howard Lykins 许可后转载。版权所有。

为了高效，OOSEM 开发活动必须得到运用 SE 基本宗旨的系统工程师的支持，包括使用多学科团队和科律化的管理流程，例如规划、风险管理、构型管理和测度。在面向对象的系统工程方法（OOSEM）的教程（LMCO，2008）和指导资料——使用 OOSEM 的基于模型的系统工程（JHUAPL，2011）中更加全面地描述 OOSEM 开发活动和相关的流程流。

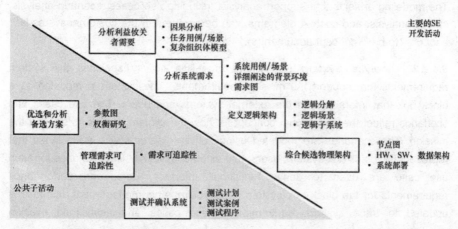

图 9.7　OOSEM 活动和建模制品。经 Howard Lykins 许可后转载。版权所有。

The system requirements and design process is decomposed into the following OOSEM high-level activities, as depicted in Figure 9.7.

9.4.2.1 Analyze Stakeholder Needs This activity supports analysis of both the "as-is" and the "to-be" enterprise. In OOSEM, an enterprise aggregates the system with other external systems that work together to accomplish the mission.in OOSEM, an enterprise aggregates the system with other external systems that work together to accomplish the mission. The "as-is" systems and enterprise are captured in sufficient detail to understand their limitations and needed improvements. The limitations of the "as-is" enterprise, as determined through causal analysis techniques, are the basis for deriving the mission requirements for the "to-be" enterprise.

OOSEM specifies the mission requirements for the "to-be" enterprise to reflect customer and other stakeholder needs. The mission requirements include definition of new and improved capabilities to address the limitations identified in the causal analysis. The capabilities for the "to-be" enterprise are represented as use cases with corresponding MOEs. The "to-be" enterprise sets the context for the system or system(s) to be developed.

The modeling artifacts that support analysis, including use cases, scenario analysis, causal analysis, and context diagrams, can be captured in the customer's "as-is" and/ or "to-be" concept document(s).

9.4.2.2 Analyze System Requirements This activity specifies the system requirements that support the mission requirements. The system is modeled as a black box that interacts with the external systems and users. The use cases and scenarios reflect the operational concept for how the system is used to support the mission. The scenarios are modeled using activity diagrams with swim lanes that represent the black box system, users, and external systems. The scenarios for each use case are used to derive functional, interface, data, and performance requirements for the black box system. The requirements management database is updated to trace system requirement to use cases and associated mission requirements.

Requirements may change as development proceeds.For example, a system's external interfaces may change, or its performance requirements may increase. Requirements variation is evaluated in terms of the probability that a requirement will change and the impact of such change on the mission. These factors are included in the risk assessments and later used to determine how to design the system to accommodate potential requirements changes.

系统需求和设计流程被分解成下述 OOSEM 高层级活动，如图 9.7 所描述。

9.4.2.1 分析利益攸关者的需要 这项活动支持对"当前"和"未来"复杂组织体的分析。在 OOSEM 中，复杂组织体将系统与共同工作完成任务的其他外部系统聚合。足够详细的捕获"当前"系统和复杂组织体，以理解其局限性和所需改进。通过因果分析技术所确定的"当前"复杂组织体的局限性是推演出"未来"复杂组织体的任务需求的基础。

OOSEM 规范"未来"复杂组织体的任务需求，以反映客户和其他利益攸关者的需要。任务需求包括定义新的和要改善的能力，以应对因果分析中识别的局限性。"未来"复杂组织体的能力被表示为具有与 MOEs 相应的用例。"未来"复杂组织体为要开发的一个或多个系统设置背景环境。

支持分析的建模制品，包括用例、场景分析、因果分析和背景环境图，可从客户的"当前"和/或"未来"概念文件中捕获。

9.4.2.2 分析系统需求 这项活动规范支持任务需求的系统需求。系统被建模为与外部系统和用户相互作用的"黑盒"。用例和场景反映如何利用系统支持该任务的运行概念。这些场景使用具有表达"黑盒"系统、用户和对外部系统的"泳道"的活动图建模。每种用例的场景用于推导"黑盒"系统的功能需求、接口需求、数据需求和性能需求。更新需求管理数据库，以跟踪系统的用例需求和相关的任务需求。

随着开发的进行，需求可能变化。例如，系统的外部接口可能变化，或者其性能需求可能增加。依照需求将会变化的概率以及这类变化对任务的影响来评价需求的变化差异。这些因素被包括在风险评估中，之后用于确定如何设计系统来适应潜在的需求变化。

9.4.2.3 Define Logical Architecture This activity includes decomposing and partitioning the system into logical elements, for example, a user interface that will be realized by a web browser or an environmental monitor that will be realized by an infrared sensor. The elements interact to satisfy system requirements and capture system functionality. Having a logical architecture/ design mitigates the impact of requirements and technology changes on system design.

OOSEM provides guidelines for decomposing the system into its logical elements. Functions for logical elements are derived from logical scenarios to support black box system functions. Logical element functionality and data may be repartitioned based on other criteria, such as cohesion, coupling, design for change, reliability, and performance.

9.4.2.4 Synthesize Candidate Physical Architectures The physical architecture describes relationships among physical system elements, including hardware, software, data, people, and procedures. Logical elements are allocated to physical elements. For distributed systems, OOSEM includes guidance for distributing the physical elements across the system nodes to address concerns, such as performance, reliability, and security. The system architecture continues to be refined to address concerns associated with the software, hardware, and data architectures. Requirements for each physical element are traced to the system requirements and maintained in the requirements management database.

9.4.2.5 Optimize and Evaluate Alternatives This activity is invoked throughout all other OOSEM activities to optimize the candidate architectures and conduct trade studies to select an architecture. Parametric models for modeling performance, reliability, availability, life cycle cost, human, and other specialty engineering concerns are used to analyze and optimize the candidate architectures to the level needed to compare the alternatives. Criteria and weighting factors used to perform the trade studies are traced to the system requirements and MOEs. TPMs are monitored, and potential risks are identified.

9.4.2.6 Manage Requirements Traceability This activity is performed throughout the other OOSEM activities to ensure traceability between requirements, architecture, design, analysis, and verification elements. Requirements relationships are established and maintained. Requirements in the system model are synchronized with the requirements management database. Traceability is continuously analyzed to assess and fill gaps or deficiencies. As requirements change, traceability is used to assess the impact of requirements changes on the system design, analysis, and verification elements.

9.4.2.3 定义逻辑架构　该活动包括将系统分解并区划成多个逻辑元素，例如，将通过网页浏览器实现的用户界面或者将通过红外传感器实现的环境监控器。元素相互作用，以满足系统需求并捕获系统功能性。逻辑架构/设计的存在缓解需求和技术变化对系统设计的影响。

OOSEM 提供用于将系统分解为逻辑元素的指导。逻辑元素的功能经由逻辑场景而导出，以支持"黑盒"系统功能。逻辑元素的功能性和数据可能基于诸如内聚、耦合、面向变化的设计、可靠性和性能等其他准则来进行重新区划。

9.4.2.4 综合候选物理架构　物理架构描述物理系统元素（包括硬件、软件、数据、人员和程序）之间的关系。将逻辑元素分配给物理元素。对于分布式系统，OOSEM 包括跨系统节点分布物理元素的指南，以应对诸如性能、可靠性和安保性等关注点。系统架构被继续细化，以应对与软件、硬件和数据架构相关的关注点。对每个物理元素的需求可追溯到系统需求，并保持在需求管理数据库中。

9.4.2.5 优化并评价备选方案　为优化候选架构并实施权衡研究以选择架构，贯穿于所有其他 OOSEM 活动调用该活动。为比较备选方案，利用建模性能、可靠性、可用性、生命周期成本、人员及其他专业工程关注点的参数模型来分析候选架构并优化到所需层级。用于实施权衡研究的准则和加权系数可追溯到系统需求和 MOE。监控 TPM，并识别潜在危险。

9.4.2.6 管理需求的可跟踪性　为确保需求、架构、设计、分析与验证元素之间的可追踪性，贯穿于其他 OOSEM 活动实施该项活动。建立并保持需求关系。系统模型中的需求与需求管理数据库同步。不断地分析可追溯性，以评估并填补空白或缺陷。随着需求的变化，利用可追踪性来评估需求变化对系统设计、分析和验证元素的影响。

9.4.2.7 Validate and Verify System This activity verifies that the system design satisfies its requirements and validates that those requirements meet the stakeholder needs. Verification plans, procedures, and methods (e.g., inspection, demonstration, analysis, and test) are developed. The primary inputs to the development of the test cases and associated verification procedures are system-level use cases, scenarios, and associated requirements. The verification system can be modeled using the same activities and artifacts described earlier for modeling the operational system. The requirements management database is updated during this activity to trace the system requirements and design information to the system verification methods, test cases, and results.

9.4.3 Applying OOSEM

OOSEM is an MBSE method used to specify and design systems. These include not only the operational system such as an aircraft or an automobile but also systems that enable the operational system throughout its life cycle, such as manufacturing, support, and verification systems. The method may also be applied to architect a system of systems or an enterprise, as well as to architect individual systems, or even system elements.

OOSEM should be tailored to support specific applications, project needs, and constraints. Tailoring may include varying the degree of emphasis on a particular activity and associated modeling artifacts and/or sequencing activities to suit a particular life cycle model.

The modeling artifacts can also be refined and reused in other applications to support product line and evolutionary development approaches. Product line modeling concerns the modeling of variability. Three refinements are to be considered in order to ensure what could be called as a "model-based product line SE":

- The modeling of the variability and the constraints between these variabilities for each type of artifacts: needs, requirements, architecture, tests, and others.
- The modeling of the variability for each type of SysML diagrams: Activity diagrams, use cases, and others.
- The modeling of the dependency links between these artifacts and the constraints between these variabilities to explain the relationship between them: As an example for a product line of water heaters, the variability of the electrical resistance component power depends, or not, on the variability of the water heater capacity.

OOSEM can be used in conjunction with principles and practices from various schools of thought, such as agile software development. Again, this requires adapting and tailoring how OOSEM is applied to integrate with the other approaches.

9.4.2.7 确认与验证系统 该活动验证系统设计满足其需求并确认那些需求满足利益攸关者的需要。开发验证计划、程序和方法（例如，检验、证实、分析和测试）。对测试案例和相关联的验证程序开发的初级输入是系统级用例、场景和相关联需求。验证系统可使用早期描述的用于对运行系统建模的相同活动和制品来建模。为了使系统需求和设计信息追溯到系统验证方法、测试用例和结果，在该活动期间，更新需求管理数据库。

9.4.3 应用 OOSEM

OOSEM 是用来规范和设计系统的 MBSE 方法。这些不仅包括诸如飞机或汽车等这样的运行系统，还包括使能运行系统贯穿于其生命周期的系统，诸如制造系统、支持系统和验证系统。该方法亦可用于架构系统之系统或复杂组织体，也可用于架构单独系统，乃至系统元素。

为支持特定的应用、项目需要和约束，应 OOSEM 进行剪裁。剪裁可以包括改变对特殊活动和相关建模制品和/或排序活动的强调程度，以适应特殊的生命周期模型。

为支持产品线和进化式开发方法，建模制品亦可在其他应用中被细化和复用。产品线建模考虑到差异性的建模。为了确保什么可以称为"基于模型的产品线 SE"，将考虑三种细化方式。

- 对每种类型的制品——需要、需求、架构、测试和其他——的差异性建模以及这些差异之间的约束。
- 对每种类型的 SysML 图——活动图、用例和其他——的差异性建模。
- 对这些制品之间的依赖链接以及这些差异性之间的约束建模，以解释它们之间的关系：以热水器的产品线为例，电阻部件的功率的差异性是否依赖于热水器容量的变异性。

OOSEM 可以与各种不同思想学派的原理和实践共同使用，诸如敏捷软件开发。此外，这要求调整和剪裁如何应用 OOSEM 来与其他方法综合。

9.5 PROTOTYPING

Prototyping is a technique that can significantly enhance the likelihood of providing a system that will meet the user's need. In addition, a prototype can facilitate both the awareness and understanding of user needs and stakeholder requirements.in addition, a prototype can facilitate both the awareness and understanding of user needs and stakeholder requirements. Two types of prototyping are commonly used: rapid and traditional.

Rapid prototyping is probably the easiest and one of the fastest ways to get user performance data and evaluate alternate concepts. A rapid prototype is a particular type of simulation quickly assembled from a menu of existing physical, graphical, or mathematical elements. Examples include tools such as laser lithography or computer simulation shells. They are frequently used to investigate form and fit, human–system interface, operations, or producibility considerations. Rapid prototypes are widely used and are very useful; but except in rare cases, they are not truly "prototypes."

Traditional prototyping is a tool that can reduce risk or uncertainty. A partial prototype is used to verify critical elements of the SOI. A partial prototype is used to verify critical elements of the SOI. A full prototype is a complete representation of the system. It must be complete and accurate in the aspects of concern.it must be complete and accurate in the aspects of concern. Objective and quantitative data on performance times and error rates can be obtained from these higher-fidelity interactive prototypes.

The original use of a prototype was as the first-of-a-kind product from which all others were replicated. However, prototypes are not "the first draft" of production entities. Prototypes are intended to enhance learning and should be set aside when this purpose is achieved. Once the prototype is functioning, changes will often be made to improve performance or reduce production costs. Thus, the production entity may require different behavior. The maglev train system (see Section 3.6.3) may be considered a prototype (in this case, proof of concept) for longer distance systems that will exhibit some but not all of the characteristics of the short line. Scientists and engineers are in a much better position to evaluate modifications that will be needed to create the next system because of the existence of a traditional prototype.

9.6 INTERFACE MANAGEMENT

Interface management is a proven set of activities that cut across the SE processes. Although some organizations treat interface management as a separate process, these are crosscutting activities of the technical and technical management processes that the project team should apply and track as a specific view of the system. When interface management is applied as a specific objective and focus of the technical processes, it will often help highlight underlying critical issues much earlier in the project than would otherwise be revealed. This would then impact upon project cost, schedule, and technical performance.

9.5 原型构建

原型构建是一种技巧，其可以显著地增强提供将会满足用户需要的系统的可能性。此外，原型可促进对用户需要和利益攸关者需求的认识和理解。两种常用的原型设计类型是：快速的和传统的。

快速原型构建可能是最容易的且可能是获取用户性能数据和评价备用概念的最快速的方式之一。快速原型是从现有物理元素、图形元素或数学元素的菜单中快速组合出的特殊类型的仿真。示例包括诸如激光成形或计算机仿真环境等工具。它们常用于调研外形和适配、人与系统接口、运行或生产性的考量。快速原型被广泛使用且非常有用；但除少数情况外，快速原型并非真是"原型"。

传统原型构建是一种可以降低风险或不确定性的工具。一个部分的原型用于验证 SOI 的关键元素。完整的原型是系统的一种完整表达。它必须在关注的方面完整、精确。性能时间和误差率的客观的和定量的数据可从这些保真度较高的交互式原型中获取。

原型最初的用途是作为复制出其他所有产品的同类的首个产品。然而，原型不是生产实体的"初稿"。原型旨在加强学习并且应在该目的实现后搁置于一旁。一旦该原型起作用，将会经常做出变更以提高性能或降低生产成本。因此，生产实体可能需要不同的行为。磁悬浮列车系统（见 3.6.3 节）可被看做较长距离系统的一个原型（在这种个案例中，为概念验证），它会展示一些但并非全部的短线特征。科学家与工程师处于更有利的地位来评价由于一个传统原型的存在而创建下一个系统所需的修改。

9.6 接口管理

接口管理是已被证明的跨领域/学科 SE 流程的活动集合。虽然一些组织将接口管理视为单独分开的流程，但这些是项目团队应该采用并作为系统的特定视图追踪的技术和技术管理流程的横向跨越活动。当接口管理用于技术流程的特定目的和聚焦点时，它往往有助于更早地强调项目中的底层关键问题，而不是等着问题被暴露。而这将影响项目成本、进度和技术性能。

CROSS-CUTTING SYSTEMS ENGINEERING METHODS

Interfaces are identified within the architecture definition process (Section 4.4), as the architecture models are developed. The interface requirements are defined through the system requirements definition process (Section 4.3). As the requirements are defined, the interface descriptions and definitions are defined to the extent needed for the architecture description within the architecture definition process (Section 4.4). Any further refinement and detail of the interface definition is provided by the design definition process (Section 4.5) as the details of the specific system implementation details are defined. The evolution of the system definition involves iteration between these processes, and the interface definition is an essential part of it. As the interface identification and definition evolve in the architecture definition process, there is an objective of keeping the interfaces as simple as possible (Fig. 4.8).

As part of the interface definition, many projects find the need or benefit to apply interface standards. In some cases, such as for plug-and-play elements or interfaces across open systems, it is necessary to strictly apply interface standards to ensure the necessary interoperation with systems for which the project team does not control. Examples of these standards include Internet Protocol standards and Modular Open Systems Architecture standards. Interface standards can also be beneficial for systems that are likely to have emergent requirements by enabling the evolution of capabilities through the use standard interface definitions that allow new system elements to be added.

Good communication is a vital part of ensuring the interface management of a system. Many projects incorporate the use of an Interface Control Working Group (ICWG), which include members responsible for each of the interfacing elements. The ICWG can be focused on internal interfaces within a single system or the external interfaces between interoperating or enabling systems. The use of ICWGs formalizes and enhances collaboration within project teams or between project teams/organizations. The use of ICWGs is an effective approach that helps to ensure adequate consideration of all aspects of the interfaces.

One of the objectives of the interface management activities is to facilitate agreements with other stakeholders. This includes roles and responsibilities, the timing for providing interface information, and the identification of critical interfaces early in the project through a structured process. This is done through the project planning process (Section 5.1). Through specific interface focus as a part of the risk management process (Section 5.4), early identification of issues, risks, and opportunities can be managed, avoiding potential impacts, especially during integration. Interface management also enhances relationships between the different organizations, giving an open communication system of issues and cooperation, where problems can be resolved more effectively.

开发架构模型时，在架构定义流程（4.4 节）内识别接口。通过系统需求定义流程（4.3 节）来定义接口需求。当定义需求时，在架构定义流程（4.4 节）内，将接口描述和定义限定到架构描述所需的范围。当定义特定系统实施细则的细节时，接口定义的任何进一步的细化和详述都应通过设计定义流程（4.5 节）提供。系统定义的演进涉及这些流程之间的迭代，接口定义是它的关键部分。当接口识别和定义在架构定义流程中演进时，目的是保持接口尽可能简单（图 4.8）。

作为接口定义的一部分，许多项目发现应用接口标准的需要或好处。在一些情况中，如跨开放系统的即插即用元件或接口，有必要严格地应用接口标准，以确保与项目团队未加控制的系统的有必要的互操作。这些标准的示例包括互联网协议标准和模块化开放系统架构标准。通过使能允许增加新系统元素的使用标准接口定义中能力演进的方式，接口标准还可以对可能有涌现需求的系统有益。

良好的沟通是确保系统的接口管理的一个关键部分。许多项目纳入接口控制工作组（ICWG）的使用，ICWG 包括负责每个接口元素的成员。ICWG 可聚焦于单一系统中的内部接口，或者互操作或使能系统之间的外部接口。ICWG 的使用使项目团队内的或项目团队/组织之间的协作得到正规化和加强。ICWG 的使用是一种帮助确保充分考虑接口的所有方面的有效途径。

接口管理活动的目的之一是促进与其他利益攸关者达成一致。这包括角色和职责，提供接口信息的时间安排，以及通过结构化流程在项目早期识别关键接口。这一点通过项目规划流程（5.1 节）完成。通过作为风险管理流程（5.4 节）的一部分的特定接口焦点，可以管理问题、风险和机会的早期识别，以避免潜在的影响，尤其是在综合期间。接口管理亦提高不同组织之间的关系，为问题及合作提供开放的沟通系统，在该系统中可以更有效地解决问题。

CROSS-CUTTING SYSTEMS ENGINEERING METHODS

Finally, after establishing baselines for requirements, architecture, and design artifacts, the configuration management process (Section 5.5) provides the ongoing management and control of the interface requirements and definitions, as well as any associated artifacts (such as interface control documents, interface requirement specifications, and interface description/definition documents). Interface management is intended to provide a simple but effective method to formally document and track the exchange of information as the life cycle processes are performed.

9.6.1 Interface Analysis Methods

There are several analysis methods and tools that aid the interface definition. These methods help to identify and understand the interfaces in the context of the system, the system elements, and/or the interfacing systems. Generally, the system analysis process (Section 4.6) is invoked by the system requirements definition, architecture definition, or design definition processes to perform the interface analysis.

N^2 diagrams (see Fig. 9.8) are a systematic approach to analyze interfaces. These apply to system interfaces, equipment (e.g., hardware) interfaces, or software interfaces. N^2 diagrams can also be used at later stages of the development process to analyze and document physical interfaces between system elements. For effective application, an N^2 diagram, which is a visual matrix, requires the systems engineer to generate complete definitions of all the system interfaces in a rigid bidirectional, fixed framework.

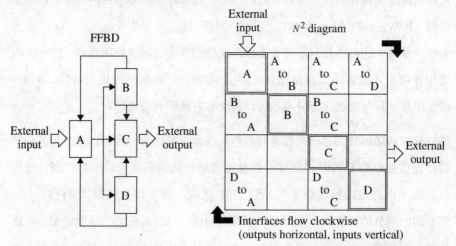

FIGURE 9.8 Sample FFBD and N^2 diagram. INCOSE SEH original figure created by Krueger and Forsberg. Usage per the INCOSE notices page. All other rights reserved.

最终，在为需求、架构和设计制品建立基线后，构型管理流程（5.5 节）提供对接口需求和定义以及任何相关的制品（例如，接口控制文件、接口需求规范和接口描述/定义文件）的持续管理和控制。接口管理旨在提供一种简单但有效的方法，以在实施生命周期流程时对信息的交换正规地文件化和追溯。

9.6.1 接口分析方法

辅助接口定义的分析方法和工具有多个。这些方法有助于识别和理解系统、系统元素和/或接口系统的背景环境中的接口。一般来说，系统分析流程（4.6 节）通过系统需求定义、架构定义或设计定义流程调用来实施接口分析。

N^2 图（见图 9.8）是分析接口的系统化方法。这些 N^2 图应用于系统接口、设备（如硬件）接口或软件接口。在开发流程的后期，N^2 图亦可用来分析和文件化系统元素之间的物理接口。为了有效的应用，N^2 图——一个可视化矩阵，要求系统工程师在严格的双向固定框架内产生所有系统接口的完整定义。

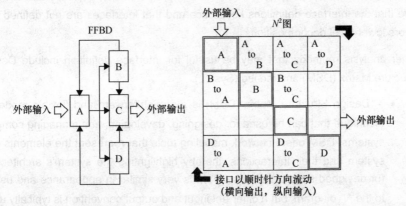

图 9.8　FFBD 和 N^2 图示例。由 Krueger 和 Forsberg 创建的 INCOSE SEH 原图。按照 INCOSE 通知页使用。版权所有。

The system functions or physical elements are placed on the chart diagonal. The rest of the squares in the $N \times N$ matrix represent the interface inputs and outputs. Interfaces between functions flow in a clockwise direction. The entity being passed from function A to function B, for example, can be defined in the appropriate square. When a blank appears, there is no interface between the respective functions. When all functions have been compared to all other functions, then the chart is complete. If lower-level functions are identified in the process with corresponding lower-level interfaces, then they can be successively described in expanded or lower level diagrams. Sometimes, characteristics of the entity passing between functions may be included in the box where the entity is identified. One of the main functions of the chart, besides interface identification, is to pinpoint areas where conflicts may arise between functions so that systems integration later in the development cycle can proceed efficiently (Becker et al., 2000; DSMC, 1983; Lano, 1977).

Alternatively, or in addition, FFBDs and DFDs can be used to characterize the flow of information among functions and between functions and the outside world. As the system architecture is decomposed to lower and lower levels, it is important to make sure that the interface definitions keep pace and that interfaces are not defined that ignore lower-level decompositions.

Other analysis methods that may be useful for interface definition include Design Structure Matrix (DSM) and the ibd (SysML).

- "Design Structure Matrix (DSM) is a straightforward and flexible modeling technique that can be used for designing, developing, and managing complex systems. DSM offers network modeling tools that represent the elements of a system and their interactions, thereby highlighting the system's architecture (or designed structure)." The DSM is very similar in appearance and usage to the N^2 diagram, but a differ-ent input and output convention is typically used (inputs on the horizontal rows and outputs on the vertical columns) (Eppinger and Browning, 2012).

- The ibd specifies the interconnection of parts of the system in SysML (see Section 9.1.9). They are used to describe the internal structure of a system in terms of its parts, ports, and connectors. The ibd provides the white box, or internal view, of a system block to represent the final assembly of all blocks within the main system block (Friedenthal et al., 2012).

系统功能或物理元素位于图表的对角线上。$N \times N$ 矩阵中其余方块表示接口输入和输出。功能之间的接口以顺时针方向流动。例如，经过功能 A 到功能 B 的能经过的实体可在适当的方块内定义。当出现空白时，对应的功能之间就没有接口。当所有功能已经与所有其他功能相对应时，图表则完整。若更低层级的功能在具有对应的更低层级接口的流程中被识别，则它们可在扩展的或更低层级图中被依次描述。有时，经过功能之间的实体的特性可包含在实体被识别的方块内。除了接口识别外，图表的主要功能之一是准确找出功能之间可能引发冲突的区域，以便使开发周期后期的系统综合能够高效地进行（Becker 等，2000；DSMC，1983；Lano，1977）。

可选择的或额外的，FFBD 和 DFD 还可用于描述各功能之间以及功能与外界之间的信息流的特征。随着系统架构分解到越来越低的层级，确保接口定义保持同步且确保忽略更低层级分解的接口不被定义则十分重要。

可用于接口定义的其他分析方法包括设计结构矩阵（DSM）和 ibd（SysML）。

- "设计结构矩阵（DSM）是一种直接的、灵活的建模技术，可以用来设计、开发和管理复杂系统。DSM 提供表达系统的元素及其相互作用的网络建模工具，从而强调系统的架构（或所设计的结构）。" DSM 在外观及使用方面与 N^2 图类似，但通常使用不同的输入和输出协定（水平行输入，竖列输出）（Eppinger 和 Browning，2012）。

- ibd 规定 SysML 中的系统的各个部分的互连（见 9.1.9 节）。它们被用来依照各个部分、端口和连接器来描述系统的内部结构。ibd 提供一个系统"块"的白盒或内部视图，以表达主要的系统"块"内的所有模块的最终汇集（Friedenthal 等，2012）。

9.7 INTEGRATED PRODUCT AND PROCESS DEVELOPMENT

Integrated Product Development (IPD) recognizes the need to consider all elements of the product life cycle, from conception through disposal, starting at the beginning of the life cycle. Important items to consider include quality, cost, schedule, user requirements, manufacturing, and support. IPD also implies the continuous integration of the entire product team, including engineering, manufacturing, verification, and support, throughout the product life cycle (DoD, 1998).

Risks inherent in concurrent product development are reduced by moving away from traditional hierarchical management structure and organized into integrated Product Teams (IPTs). Productivity gains come through decentralization of processes, avoidance of previous problems, and better integration between engineering and manufacturing. Traditional development with serial activities may be so lengthy such that the product becomes obsolete before it is completed. With good interface definition and control, IPD, involving the entire team, can speed up the development process.

IPPD further recognizes the importance of *process*. The following definitions apply to IPPD: The following definitions apply to IPPD:

- Integrated Product Development Team (IPDT)—A multidisciplinary group of people who are collectively responsible for delivering a defined product or process.

- *IPPD*—The process of using IPDTs to simultaneously develop the design for a product or system and the methods for manufacturing the product or system. The process verification may consist of review of a process description by an IPDT. It may also include a demonstration to an IPDT of a process.

- *Concurrent engineering*—is a management/ operational approach that aims to improve product design, production, operation, and maintenance by developing environments in which personnel from all disciplines (e.g., design, marketing, production engineering, process planning, and support) work together and share data throughout all stages of the product life cycle.

Integrated development has the potential to introduce *more* risk into a development program because down- stream activities are initiated on the *assumption* that upstream activities will meet their design and interface requirements. However, the introduction of a hierarchy of cross-functional IPDTs, each developing and delivering a product, can reduce risks and provide better products faster.

9.7 综合的产品和流程开发

综合产品开发（IPD） 认识到需要在生命周期一开始便考虑从概念形成到退出这一产品生命周期的所有元素。需要考虑的重要项包括质量、成本、进度、用户需求、制造和保障。IPD 还意味着，整个产品团队的持续综合，包括贯穿于产品生命周期的工程、制造、验证和维持（DoD, 1998）。

并行产品开发中固有的风险通过改变传统层级管理结构来缓解，并被组织到综合产品团队（IPT）中。通过分散流程，避免以前的问题并且在工程与制造之间更好地综合来提高生产率。具有串行活动的传统开发可能会如此冗长以至于产品在完成之前就作废了。使用良好的接口定义和控制，涉及整个团队的 IPD 可加速开发流程。

IPPD 进一步认可流程的重要性。下述定义适用于 IPPD：

- 综合的产品开发团队（IPDT）——一个共同负责交付定义的产品或流程的多学科人员群组。

- *IPPD*——使用 IPDT 同时开发产品或系统的设计以及产品或系统的制造方法的流程。流程验证可由 IPDT 进行的流程描述的评审组成，亦可包括流程的 IPDT 的验证。

- *并行工程*——是一种管理/运行方法，其目的在于通过对所有学科（如设计、市场、生产工程、流程规划和支持）人员共同工作并共享贯穿于产品生命周期所有阶段的数据环境的开发，来改善产品的设计、生产、运行和维护。

由于下游活动是基于上游活动将满足其设计和接口需求的假设而启动的，综合开发具有将更多的风险引入到开发计划中的潜在可能。然而，跨职能部门的 IPDT（各自开发和交付一种产品）的层级结构的引入可以降低风险并更快地提供更好的产品。

IPPD also improves team communications through IPDTs, implements a proactive risk process, makes decisions based on timely input from the IPDT, and improves customer involvement.

9.7.1 IPDT Overview

An IPDT is a process-oriented, integrated set of cross functional teams (i.e., an overall team comprising many smaller teams) given the appropriate resources and charged with the responsibility and authority to define, develop, produce, and support a product or process (and/ or service). Each team is staffed with the skills necessary to complete their assigned processes, which may include all or some of the development and production steps.

The general approach is to form cross-functional IPDTs for all products and services. The typical types of IPDTs are a Systems Engineering and integration Team (SEIT), a Product Integration Team (PIT), and a Product development Team (PDT). These teams each mimic a small, independent project focusing on individual elements and/or their integration into more complex system elements. The SEIT balances requirements between product teams, helps integrate the other IPDTs, focuses on the integrated system and system processes, and addresses systems issues, which, by their nature, the other IPDTs would most likely relegate to a lower priority. Although the teams are organized on a process basis, the organizational structure of the team of teams may approach a hierarchical structure for the product, depending upon the way the product is assembled and integrated.

The focus areas for these IPDT team types and their general responsibilities are summarized in Table 9.1. This arrangement often applies to large, multi-element, multiple subsystem programs but must be adapted to the specific project. For example, on smaller programs, the number of PIT teams can be reduced or eliminated. In service-oriented projects, the system hierarchy, focus, and responsibilities of the teams must be adapted to the appropriate services.

Team members' participation will vary throughout the product cycle, and different members may have primary, secondary, or minor support roles as the effort transitions from requirements development through the different stages of the life cycle. For example, the manufacturing and verification representatives may have minor, part-time advisory roles during the early product definition stage but will assume primary roles later, during manufacture and verification. Team members participate to the degree necessary from the outset to ensure their needs and requirements are reflected in overall project requirements and planning to avoid costly changes later. It is also good for some of the team to remain throughout the product cycle to retain the team's "project memory."

IPPD 还通过 IPDT 改善团队沟通，实施积极主动的风险流程，基于 IPDT 的及时输入做出决策，并提高客户的参与度。

9.7.1　IPDT 概览

IPDT 是流程导向的综合跨职能的团队集合（即由许多较小型团队组成的总体团队），得到合适的资源并具有定义、开发、生产和支持产品或流程（和/或服务）的职责和权限。每个团队拥有完成指派流程（可包括所有或部分开发和生产的步骤）所必需的技能。

一般方法是为所有产品和服务形成跨职能部门的 IPDT。IPDT 的典型类型是系统工程和综合团队（SEIT）、产品综合团队（PIT）和产品开发团队（PDT）。这些团队各自模仿一个小型的独立项目，聚焦于单个元素和/或其与较复杂的系统元素的综合。SEIT 平衡产品团队之间的需求，帮助与其他 IPDT 的综合，聚焦于综合系统和系统流程，并应对由于其本质属性而最有可能被其他 IPDT 降到较低优先级的系统问题。尽管团队是以流程为基础进行组织的，但团队之团队的组织结构可能接近产品的层级结构，这取决于产品汇集和综合的方式。

表 9.1 概括这些 IPDT 团队类型的关注领域及其一般职责。这种安排往往适用于多元素、多子系统的大型计划，但必须适应于特定项目。例如，就较小型的计划而言，可以减少 PIT 团队的数量或者撤销 PIT 团队。在面向服务的项目中，系统级结构、聚焦点和团队的职责必须适应于合适的服务。

由于工作从需求开发经过生命周期的不同阶段而进行转移，因此，团队成员的参与情况将贯穿于产品周期而变化，并且不同的成员可能扮演主要的、次要的或很少的支持角色。例如，在产品定义阶段的早期，制造和验证代表可能具有次要的兼职顾问角色，但在后来的制造和验证期间将担任主要角色。团队成员从一开始就以必要的程度来参与，从而确保他们的需要和需求都反映在整体项目需求和规划中，以避免后期产生代价高昂的变更。贯穿于产品周期保留团队中的某些成员是有好处的，以保留团队的"项目记忆"。

CROSS-CUTTING SYSTEMS ENGINEERING METHODS

TABLE 9.1 Types of IPDTS and their focus and responsibilities

System hierarchy	Team type + focus responsibilities
External interface and system	Systems engineering and integration team (SEIT)
	• Integrated system and processes • External and program issues • System issues and integrity • Integration and audits of teams
Upper-level elements	Product Integration Teams (PITs))
	• Integrated H/W and S/W • Deliverable item issues and integrity • Support to other teams (SEIT and PDTs)
Lower-level elements	Product development Teams (PDTs)Product development Teams (PDTs)
	• Hardware and software • Product issues and integrity • Primary participants (design and Mfg.) • Support to other teams (SEIT and PITs)

IPDTs must be empowered with full life cycle responsibility for their products and systems and with the authority to get the job done. They should *not* be looking to higher management for key decisions. They should, however, be required to justify their actions and decisions to others, including interfacing teams, the systems integration team, and project management.

表 9.1　IPDT 的类型及其聚焦点和职责

系统层级结构	团队类型+焦点职责
外部接口和系统	系统工程和综合团队（SEIT）
	• 综合的系统和流程
	• 外部的问题和计划问题
	• 系统问题和完整性
	• 团队的综合和审计
更高层级元素	产品综合团队（PIT）
	• 综合的硬件和软件
	• 可交付项的问题和完整性
	• 其他团队的支持（SEIT 和 PDT）
更低层级元素	产品开发团队（PDT）
	• 硬件和软件
	• 产品问题和完整性
	• 主要参与者（设计和制造）
	• 其他团队的支持（SEIT 和 PIT）

IPDT 必须被授予对产品和系统的全生命周期职责以及完成工作的权限。IPDT 不应希望从高级管理人员得到关键决策。然而，他们应该被要求向其他人员（包括交互的团队、系统综合团队和项目管理人员）证明他们的行动和决策是合理的。

9.7.2 IPDT Process

The basic principle of IPDT is to get all disciplines involved at the beginning of product development to ensure that needs and requirements are completely understood for the full life cycle of the product. Requirements are developed initially at the system level, then successively at lower levels as the requirements are flowed down. Teams, led by systems engineers, perform the up-front SE activities at each level.

IPDTs do their own internal integration in an IPPD environment. A SEIT representative belongs to each product team (or several) with internal and external team responsibilities. There is extensive iteration between the product teams and the SEIT to converge on requirements and design concepts, although this effort should slow down appreciably after the preliminary design review and as the design firms up.

Systems engineers participate heavily in the SEIT and PIT and to a much lesser extent in the PDT. Regardless, the iterative SE processes described in this handbook are just as applicable to all teams in the IPPD environment. It is even easier to apply the processes throughout the program because of the day-to-day presence of systems engineers on all teams.

IPDTs have many roles, and their integration roles overlap based on the type of team and the integration level. Figure 9.9 gives examples of program processes and system activities.

The three bars on the left show the roles of the types of product teams at different levels of the system. For example, the SEIT leads and audits in external integration and in systems integration activities, as indicated by the shaded bar. For program processes involving lower-level elements (e.g., parts, components, or subassemblies), the appropriate PDTs are the active lead and audit participants, supported by the SEIT and the PIT.

Basic system activities include system requirements derivation, system functional analysis, requirements allocation and flowdown, system trade-off analysis, system synthesis, systems integration, TPM definition, and system verification. The bars for system functions 1, 2, and 3 in the chart show that the SEIT leads and audits activities on different system activities while the element teams participate. The lower-level system element teams provide additional support, if requested.

The column at the right side of Figure 9.9 shows other integration areas where all teams have some involvement. The roles of the various teams must also be coordinated for these activities but should be similar to the example.

9.7.2.1 Organizing and Running a High-Performance IPDT
The basic steps and key activities to organize and run an IPDT are as follows:

9.7.2 IPDT 流程

IPDT 的基本原则是获得在产品开发开始时涉及的所有学科，以确保需要和需求在产品的全生命周期被完全理解。需求最初是在系统级上被开发的，然后随着需求的向下逐层细化依次在更低层级上开发。由系统工程师领导的团队执行每一层级前期的 SE 活动。

IPDT 在 IPPD 环境中进行他们自己的内部综合。SEIT 代表属于每个产品团队（或若干个团队），具有内部和外部团队的职责。尽管该工作应在初步设计评审后随着设计的确定明显放慢，产品团队和 SEIT 之间存在广泛的迭代，以汇集到需求和设计概念。

系统工程师广泛地加入 SEIT 和 PIT，却很少加入 PDT。无论如何，本手册中描述的 SE 迭代流程正好适用于 IPPD 环境中的所有团队。由于系统工程师日常都参与到所有团队，贯穿于整个计划应用流程更加容易。

IPDT 具有许多角色，并且他们的综合角色基于团队类型和综合水平重叠。图 9.9 提供项目流程和系统活动的示例。

左侧的三个条形区表明系统不同层级上各类产品团队的角色。例如，SEIT 在外部综合和系统综合活动中领导和审计，正如阴影条形区所指示。对于涉及更低层级元素（如零件、部件或子组件）的项目流程而言，合适的 PDT 是领导和审计的积极参与者，由 SEIT 和 PIT 支持。

基本的系统活动包括系统需求推导、系统功能分析、需求分配和细化、系统权衡分析、系统综合、系统综合、TPM 定义和系统验证。图中的系统功能条形 1、2、3 表明，SEIT 领导和审计不同系统活动中的活动而元素团队参与其中。若有要求，更低层级系统元素团队提供其他的支持。

图 9.9 右侧一栏表明所有团队中有一些团队参与的其他综合区域。为了这些活动，也必须协调各种不同团队的角色，但应与示例类似。

9.7.2.1 组织并运行高性能 IPDT　组织并运行 IPDT 的基本步骤和关键活动如下：

1. *Define the IPDT teams for the project*—develop IPDT teams that cover all project areas.

2. *Delegate responsibility and authority to IPDT leaders*—Select experienced team leaders early in the development process and avoid frequent budget changes throughout the life cycle.

3. *Staff the IPDT*—Candidates must work well in a team environment, communicate well, and meet their commitments:

  - Balance the competency, availability, and full time commitment of the core team.

  - Plan when competencies are needed and not needed.

  - identify issues where specialists are needed.

4. *Understand the team's operating environment*— recognize how the team directly or indirectly influences other teams and the project as a whole.

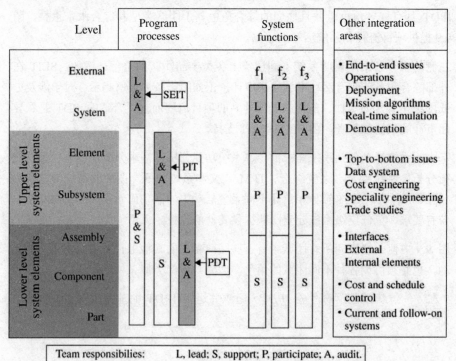

FIGURE 9.9 Examples of complementary integration activities of IPDTs. Adapted from Bob Lewis as INCOSE SEH v1 Figure 6.3. Usage per the INCOSE notices page. All other rights reserved.

1. *为项目定义 IPDT 团队*——发展涵盖所有项目区域的 IPDT 团队。
2. *授予 IPDT 领导者职责和权限*——在开发流程的早期选择有经验的团队领导并在贯穿于生命周期避免频繁的预算变更。
3. *为 IPDT 配备人员*——候选人必须在团队环境中工作优秀,沟通良好且履行承诺:

- 平衡核心团队的能力、可用性和全时投入。
- 计划何时需要/不需要能力。
- 识别需要专家解决的问题。

4. *理解团队的运行环境*——认识到团队如何直接或间接地影响其他团队和整个项目。

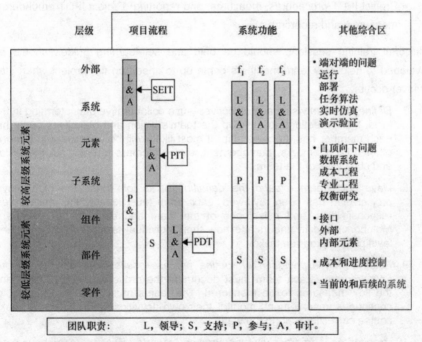

图 9.9 IPDT 互补的综合活动的示例。改编自 Bob Lewis 的 INCOSE SEH 第 1 版中的图 6.3。按照 INCOSE 通知页使用。版权所有。

5. *Plan and conduct the "kick-off meeting"*—recommend two kickoff meetings, one for the project as a whole and one for the individual IPDTs. Well-planned kick-off meetings will set the project off on the right foot.

6. *Train the team*—Training for the project is a critical element. The following recommended topics should be covered:

- Tailored SE process for the project

- Project description, stakeholders, purpose, mission, organization, schedule, and budget

- Terminology and nomenclature

- Access to project products

- Communications skills

- Project IPDT procedures, measures, and reporting• Project IPDT procedures, measures, and reporting

Additional training sessions should be held and self-learning guides should be developed to help new team members come up to speed on the project when staff turnover occurs.

7. *Define the team vision and objectives*—use collaborative brainstorming in the initial IPDT meetings to develop the team's vision and objectives such that each member has an ownership. It most likely will be necessary to bring in other IPDT members, management, and customers to flesh out the vision and objectives of the team.

8. *Have each team expand the definition of its job*—once the higher-level project plan has been reviewed, each team must identify the tasks, roles, responsibilities, and milestones of the team and each of the members. Members need to understand how their individual tasks fit into the higher-level project-program tasks.

9. *Establish an expectation of routine process assessment and continuous improvement*—Each team must document the process they are using and the key measures to be monitored. The teams must have the mindset of continuous improvement, monitor their own activities, and continually make course corrections along the way.

10. *Monitor team progress via measures and reports*—Each team will have a set of measures and reports to monitor its own progress. These reports and measures must be reviewed by the SEIT that coordinates among the other IPDTs. These measures may include an earned value report and technical measures, such as a defect rate report. The selected measures are dependent on the team's role on the project.

5. *计划并实施"启动会议"*——建议召开两次启动会议，一次用于整个项目，另一次用于单个的 IPDT。计划周详的启动会议将会使项目取得良好的开端。

6. *培训团队*——项目培训是一个关键元素。下述推荐主题应包含在内：

- 为项目剪裁的 SE 流程
- 项目描述、利益攸关者、目的、任务、组织、进度和预算
- 术语和名称
- 访问项目产品
- 沟通技巧
- 项目 IPDT 的程序、测度和报告

当发生人员流动时，应召开另外的培训会议并开发自学指南，来帮助团队的新成员跟上项目的进度。

7. *定义团队愿景和目的*——在初始的 IPDT 会议中使用协作式的头脑风暴法以开发团队愿景和目的，这样每个成员都拥有所有权。最有可能的是，为充实团队愿景和目的而引入其他 IPDT 成员、管理人员和客户。

8. *使每个团队扩展其工作的定义*——一旦更高层级的项目计划通过评审，每个团队必须识别团队及每个成员的任务、角色、职责和里程碑。成员需要理解他们自己的任务如何与更高层级项目计划任务适配。

9. *建立例行的流程评估和持续改进的期望值*——每个团队必须将他们正在使用的流程和需监控的关键测度文件化。团队必须具有持续改进的心态，监控他们自己的活动，并不断地修正前进方向。

10. *通过测度和报告监控团队进展*——每个团队将具有测度和报告集合，以监控团队的自身进展。这些报告和测度必须通过 SEIT 的评审，SEIT 在其他 IPDT 之间进行协调。这些测度可包括挣值报告和技术测度，如故障率报告。选定的测度依赖于团队在项目中的角色。

11. Sustain and evolve the team throughout the project—Personnel assignments to a team will vary as each team grows, shrinks, and changes skill mix over the project life cycle. As issues arise, technical specialists may need to join the team to help address these specific issues. Services such as marketing, program controls, procurement, finance, legal, and human resources gener-ally support the team at a steady, low level of effort, or as required.

12. Document team products—The team's products should be well defined. Because of the IPDT structure, the overhead of cross-organizational communication varies and should be reduced. When multiple documents are required, different team members with identified backups should be assigned as the respon-sible author with contributions from others. The IPDT should maintain a log of activities in addition to the mission, vision, objectives, deliverables, meeting minutes, decisions, tailored processes, agreements, team project information, and contact information.

13. Close the project and conduct follow-up activities—in conjunction with step 12, the IPDT should maintain records as though the project may be reengineered at some future time and all closeout products must be accessible. All IPDT logs should be organized the same way, when possible, such that they can be easily integrated into an overall project report. The closeout should include lessons learned, recommended changes, and a summary of measures for the team.

Project managers should review team staffing plans to ensure proper composition and strive for continuity of assignments. The advantages of a full-time contributor outweigh the work of many part-time team members. Similarly, the loss of a knowledgeable key team member can leave the team floundering. It is important to have people who can work well together and communicate, but team results may suffer without outstanding technical specialists and professionals who can make a difference. Recommended techniques for achieving high performance in an IPDT are as follows:

- *Carefully select the staff*—Excellent people do excellent work.

- *Establish and maintain positive team interaction dynamics*—All should know what is expected of the team and each individual and strive to meet commitments. Anticipate and surface potential problems quickly (internally and externally). Interactions should be informal but efficient and a "no blame" environment where problems are fixed and the team moves on. Acknowledge and reward good work.

- *Generate team commitment and buy-in*—Team alignment to the vision, objectives, tasks, and schedules. Maintain a team leader's notebook.

11. *团队贯穿于项目的维持和演进*——团队的人员分配随着项目生命周期内每个团队增长、缩小和变更技能的组合而变化。随着问题的出现,技术专家可能需要加入团队来帮助解决这些特定问题。如市场、计划控制、采购、财务、法律和人力资源等服务通常以稳定的、低投入水平或按要求来支持团队。

12. *将团队产品文件化*——团队的产品应被明确定义。由于 IPDT 结构,跨组织沟通的开销会发生变化并且应缩减。当要求多个文件时,具有备份身份的不同团队成员应被指定为责任编制者,且有来自其他成员的贡献。除了任务、愿景、目的、可交付物、会议记录、决策、剪裁后的流程、协议、团队项目信息和联系信息外,IPDT 还应保存活动日志。

13. *关闭项目并开展后续活动*——就好像项目也许在未来某一时间重建以及所有完工的产品一定易于获得一样,IPDT 应结合步骤 12 保存记录。若可能,应以同样的方式整理所有 IPDT 日志,使这些日志能够容易地综合到整体项目报告中。完工应包括经验教训、建议的变更以及团队测度的汇总。

项目管理者应评审团队的人员配备计划,以确保合适的人员构成和追求分配的持续性。全职贡献者的优势比许多兼职团队成员的工作更有价值。同样,知识渊博的团队关键成员的缺失可使团队举步维艰。重要的是,拥有可以很好地共同工作并沟通的人员,但团队可能深受其害的是,没有能够带来不同的杰出技术专家以及专业人员。在 IPDT 中达到高绩效所推荐技术如下:

- *仔细选择人员*——优秀的人员做优秀的工作。

- *建立并保持积极的团队互动活力*——所有人应明白团队和每个人的期望是什么,并且应努力兑现承诺。快速地预测和暴露潜在问题(内部和外部)。互动应是非正式的但却是高效的,在"没有指责的"环境中锁定问题且团队共同行动。认可和奖励出色的工作。

- *产生团队承诺及认同*——团队与愿景、目的、任务和进度的对齐。维持团队领导的工作记录。

- *Breakdown the job into manageable activities*— Those that can be accurately scheduled, assigned, and followed up on weekly.

- *Delegate and spread out routine administrative tasks among the team*—Free up the leader to participate in technical activities. Give every team member some administrative/managerial experience.

- *Schedule frequent team meetings with mandatory attendance for quick information exchanges*— Ensure everyone is current. Assign action items with assignee and due date.

9.7.3 Potential IPDT Pitfalls

There are ample opportunities to go astray before team members and leaders go through several project cycles in the IPDT framework and gain the experience of working together. Table 9.2 describes some pitfalls common to the IPDT environment that teams should watch out for.

9.8 LEAN SYSTEMS ENGINEERING

SE is regarded as an established, sound practice, but not always delivered effectively. Recent US government Accountability office (GAO, 2008) and NASA (2007a) studies of space systems document major budget and schedule overruns. Similarly, recent studies by the MIT based Lean Advancement Initiative (LAI) have identified a significant amount of waste in government programs, averaging 88% of charged time (LAI MIT, 2013; McManus, 2005; oppenheim, 2004; Slack, 1998). Most programs are burdened with some form of waste: poor coordination, unstable requirements, quality problems, and management frustration. This waste represents a vast productivity reserve in programs and major opportunities to improve program efficiency.

- *将工作分解为可管理的活动*——能够每周被精确地安排、指派和跟进的活动。

- *分派和传递团队之间的例行行政任务*——使领导有时间参加技术活动。为团队每个成员提供一些行政/管理经验。

- *安排频繁的团队会议，要求必须出席，以便进行快速的信息交换*——确保每个人都出席。向被分配人分配行动项和到期日。

9.7.3 潜在的 IPDT 陷阱

团队成员和领导在经历 IPDT 框架中的若干项目周期并获得共同工作经验之前，会出现偏离正轨的可能。表 9.2 描述团队应该提防的 IPDT 环境共有的一些陷阱。

9.8 精益系统工程

SE 被认为是一种已建立的、全面的实践，但并非总是能有效地交付。美国政府问责局（GAO，2008）和 NASA（2007a）的空间系统最新研究将主要预算超支和进度延期文件化。同样，麻省理工的"精益促进倡议行动"（LAI）的最新研究识别出政府计划中存在着巨大浪费，平均为所投入时间的 88%（"精益促进倡议行动"，麻省理工，2013；McManus，2005；oppenheim，2004；Slack，1998）。大部分计划都负担着某种形式的浪费：协调差、需求不稳定、质量问题和管理挫败。这种浪费表征在诸多计划中储备着的巨大生产力以及提高计划效率的极大机会。

TABLE 9.2 Pitfalls of using IPDT

IPDT pitfalls	What to do
Spending too much time defining the vision and objectives	Converge and move on
Insufficient authority—IPDT members must frequently check with management for approval	Give team leader adequate responsibility, or put the manager on the team
IPDT members are insensitive to management issues and overcommit or overspend	Team leader must remain aware of overall project objectives and communicate to team members
Teams are functionally oriented rather than cross-functionally process oriented	Review the steps in organizing and running an IPDT (see preceding text)
Insufficient continuity of team members throughout the project	Management should review staffing requirements
Transition to the next stage team specialists occurs too early or too late in the schedule	Review staffing requirements
Overlapping assignments for support personnel compromises their effectiveness	Reduce the number of team
Inadequate project infrastructure	Management involvement to resolve

Lean development and the broader methodology of lean thinking have their roots in the Toyota "just-in-time" philosophy, which aims at "producing quality products efficiently through the complete elimination of waste, inconsistencies, and unreasonable requirements on the production line" (Toyota, 2009). Lean SE is the application of lean thinking to SE and related aspects of organization and project management. SE is focused on the discipline that enables flawless development of complex technical systems. Lean thinking is a holistic paradigm that focuses on delivering maximum value to the customer and minimizing wasteful practices. A popular description of lean is "doing the right job right the first time" and "working smarter, not harder." lean thinking has been successfully applied in manufacturing, aircraft depots, administration, supply chain management, health-care, and product development, including engineering.

Lean SE is the area of synergy between lean thinking and SE, with the goal to deliver the best life cycle value for technically complex systems with minimal waste. The early use of the term lean SE is sometimes met with concern that this might be a "repackaged faster, better, cheaper" initiative, leading to cuts in SE at a time when the profession is struggling to increase the level and quality of SE effort in programs. Lean SE does not take away anything from SE and it does not mean *less* SE. It means more and better SE with higher responsibility, authority, and accountability, leading to better, waste free workflow with increased mission assurance. Under the lean SE philosophy, mission assurance is nonnegotiable, and any task that is legitimately required for success must be included, but it should be well planned and executed with minimal waste.

表 9.2 使用 IPDT 的陷阱

IPDT 陷阱	做什么
花费太多的时间来定义愿景和目的	收敛且行动
权限不够——IPDT 成员必须经常地与管理人员核对，以获审批	为团队领导者提供充分的职责，或使管理者加入团队
IPDT 成员对管理问题不敏感，过度承诺或超支	团队领导者必须始终认识到整体项目目的并与团队成员沟通
团队是以职能为导向的，而不是以跨职能流程为导向的	对组织和运行 IPDT 的步骤进行评审（见上文）
跨项目团队成员的连续性不够	管理者应审核人员配备需求
在进度安排中，向下一阶段团队专家的转移发生得过早或过晚	审核人员配备需求
支持人员的重叠指派损害其有效性	减少团队数量
项目基础设施不足	管理者参与解决

精益思考中的精益开发和更广泛的方法均起源于丰田"准时化"的哲学思想，其目标是"通过彻底消除生产线上的浪费、不一致性及不合理需求，高效率地生产优质产品"（Toyota，2009）。精益 SE 是将精益思考应用到 SE 中以及组织与项目管理的相关方面。SE 聚焦于促使复杂技术系统无缺陷开发的规程。精益思考是一种整体性的范式，聚焦于向客户交付最大价值并使浪费活动最小化。对精益的一般描述是"第一次就把正确的工作做正确"和"更聪明地工作，而不是更辛苦地工作"。精益思考已成功地应用于制造、飞机库管、行政管理、供应链管理、健康医疗和产品开发，包括工程。

精益 SE 是精益思考和 SE 之间协同的领域，目的是以最小的浪费向技术复杂系统提供生命周期的最大价值。术语精益 SE 的早期使用有时会遇到一个问题，其可能是"更快、更好、更低的成本的一揽子计划"的举措，从而在专业人员设法提高计划中 SE 的投入水平和质量时缩短 SE。精益 SE 并未减损 SE 中的任何事物，并且精益 SE 并不意味着更少的 SE。其意味着更多且更好的 SE，具有更高的职责、权限和责任，从而产生提高任务保证的更好的、无浪费的工作流。在精益 SE 理念下，任务保证是毋庸置疑的，并且为取得成功而合理要求的任何任务都必须被包括在内，但其应以最小的浪费被妥善地计划和执行。

CROSS-CUTTING SYSTEMS ENGINEERING METHODS

Lean thinking: *"Lean thinking is the dynamic, knowledge-driven, and customer-focused process through which all people in a defined enterprise continuously eliminate waste with the goal of creating value"* (Murman, 2002).

Lean SE: The application of lean principles, practices, and tools to SE to enhance the delivery of value to the system's stakeholders.

Three concepts are fundamental to the understanding of lean thinking: value, waste, and the process of creating value without waste (also known as lean principles).

9.8.1 Value

The value proposition in engineering programs is often a multiyear, complex, and expensive acquisition process involving numerous stakeholders and resulting in hundreds or even thousands of requirements, which, notoriously, are rarely stable. In lean SE, *"value"* is defined simply as mission assurance (i.e., the delivery of a flawless complex system, with flawless technical performance, during the product or mission development life cycle) and satisfying the customer and all other stakeholders, which implies completion with minimal waste, minimal cost, and the shortest possible schedule.

"Value is a measure of worth (e.g., benefit divided by cost) of a specific product or service by a customer, and potentially other stakeholders and is a function of (1) the product's usefulness in satisfying a customer need, (2) the relative importance of the need being satisfied, (3) the availability of the product relative to when it is needed, and (4) the cost of ownership to the customer" (McManus, 2004).

9.8.2 Waste in Product Development

The LAI classifies waste into seven categories: overprocessing, waiting, unnecessary movement, overproduction, transportation, inventory, and defects (McManus, 2005). lately, the eighth category is increasingly added: the waste of human potential.

Waste: *"The work element that adds no value to the product or service in the eyes of the customer. Waste only adds cost and time"* (Womack and Jones, 1996).

When applying lean thinking to SE and project planning, consider each waste category and identify areas of wasteful practice. The following illustrates some waste considerations for SE practice in each of the LAI waste classifications:

精益思考："精益思想是一个动态的、知识驱动的、以客户为中心的过程，由此使特定企业的所有人员以创造价值为目标不断地消除浪费"（Murman，2002）。

精益 SE：将精益原则、实践和工具应用到 SE，以提升面向系统利益攸关者的价值交付。

理解精益思想的三个基本概念是：价值、浪费以及无浪费创造价值的流程（亦称为精益原则）。

9.8.1 价值

工程计划中的价值主张常常是一个历经多年的、复杂的且昂贵的采办流程，涉及数千名利益攸关者并产生数百甚至数千个广为诟病的、难以稳定的需求。在精益 SE 中，"价值"被简单地定义为任务保证（即在产品或任务的开发生命周期中以无缺陷的技术实现交付无缺陷的复杂系统）并使客户和所有其他利益攸关者满意，这意味着以最少的浪费、最低的成本以及可能最短的进度完成任务。

"*价值*是客户和潜在的其他利益攸关者对特定产品或服务的意义（如收益与成本之比）的度量，是以下因素的函数：（1）产品满足客户需要的有用性，（2）所满足的需要的相对重要性，（3）产品相对于需要时间的可用性以及（4）客户的拥有成本"（McManus，2004）。

9.8.2 产品开发中的浪费

LAI 将浪费分为七类：过度处理、等待、不必要的移动、过度生产、运输、库存和缺陷。（McManus，2005）。最近，逐渐增加了第八类：人员潜能的浪费。

浪费："在客户眼中不能为产品或服务增加价值的工作单元。浪费只会增加成本和时间"（Womack 和 Jones，1996）。

当将精益思想应用于 SE 和项目规划时，要考虑每一类浪费并识别浪费活动所在的地方。以下阐明针对每类 LAI 浪费中 SE 实践的一些浪费因素：

- *Overprocessing*—Processing more than necessary to produce the desired output. Consider how projects "overdo it" and expend more time and energy than needed:
  - Too many hands on the "stuff" (material or information)
  - Unnecessary serial production
  - Excessive/custom formatting or reformatting
  - Excessive refinement, beyond what is needed for value
- *Waiting*—Waiting for material or information, or information or material waiting to be processed. Consider "things" that projects might be waiting for to complete a task:
  - Late delivery of material or information– late delivery of material or information
  - Delivery too early—leading to eventual rework– delivery too early—leading to eventual rework
- *Unnecessary movement*—Moving people (or people moving) to access or process material or information. Consider any unnecessary motion in the conduct of the task:
  - Lack of direct access—time spent finding what you need– lack of direct access—time spent finding what you need
  - Manual intervention
- *Overproduction*—Creating too much material or information. Consider how more "stuff" (e.g., material or information) is created than needed:
  - Performing a task that nobody needs or using a useless metric
  - Creating unnecessary data and information
  - Information overdissemination and pushing data
- *Transportation*—Moving material or information. Consider how projects move "stuff" from place to place:
  - Unnecessary hand-offs between people– unnecessary handoffs between people
  - Shipping "stuff" (pushing) when not needed

- *过度处理*——为产生所期望的输出而进行的不必要的处理。考虑项目是如何"做过头",且花费了更多不必要的时间和精力:

    - 过多的人经手"原料"(物料或信息)

    - 不必要的串行生产

    - 过度/自定格式或重新编排的格式

    - 超出价值所需的过度细化。

- *等待*——等待物料或信息,或信息或物料等待处理。考虑项目为完成某一任务需要等待的"事物":

    - 物料或信息的延迟交付

    - 过早交付——导致最终的返工

- *不必要的移动*——为了接近或处理物料或信息而移动人员(或人员的移动)。考虑执行任务时任何不必要的移动:

    - 缺乏直接获取——花费在寻找所需物品的时间

    - 人工干预

- *过度生产*——创造太多的物料或信息。考虑不必要的"原料"(例如物料或信息)是如何产生的:

    - 执行任何人不需要或使用没有用的指标的任务

    - 产生不必要的数据和信息

    - 信息过度传播和推送数据

- *运输*——移动物料或信息。考虑项目如何把"原料"从一处移到另一处:

    - 人员之间不必要的传递

    - 在不需要时运送"原料"(推送)

CROSS-CUTTING SYSTEMS ENGINEERING METHODS

- Incompatible communication—lost transportation through communication failures
- *Inventory*—Maintaining more material or information than is needed. Consider how projects stockpile information or materials:
  - Too much "stuff" buildup
  - Complicated retrieval of needed "stuff"
  - Outdated, obsolete information– outdated, obsolete information
- *Defects*—Errors or mistakes causing the effort to be redone to correct the problem. Consider how projects go back and do it again:
  - Lack of adequate review, verification, or validation
  - Wrong or poor information
- *Waste of human potential*—not utilizing or even suppressing human enthusiasm, energy, creativity, and ability to solve problems and general willing- ness to perform excellent work.

9.8.3 Lean Principles

Womack and Jones (1996) captured the process of creating value without waste into six lean principles. The principles (see Fig. 9.10) are abbreviated as value, value stream, flow, pull, perfection, and respect for people and are defined in detail in the following.

When applying lean thinking to SE, evaluate project plans, preparations of people, processes and tools, and organization behaviors using the lean principles. Consider how the customer defines *value* in the products and processes, then describe the *value stream* for creating products and processes, optimize *flow* through that value stream and eliminate waste, encourage *pull* from each node in that value stream, and strive to *perfect* the value stream to maximize value to the customer. These activities should all be conducted within a foundation of *respect* for customers, stakeholders, and project team members.

In 2009, the INCOSE lean SE Working group released a new online product entitled *Lean Enablers for Systems Engineering* (LEfSE), Version 1.0. It is a collection of practices and recommendations formulated as "do's" and "don'ts" of SE based on lean thinking. The practices cover a large spectrum of SE and other relevant enterprise management practices, with a general focus on improving program value and stakeholder satisfaction and reducing waste, delays, cost overruns, and frustrations. LEfSE are currently listed as 147 practices (referred to as subenablers) organized under 47 nonactionable topical headings called enablers and grouped into the six lean principles described below (oppenheim, 2011):

- 不协调的交流——由于交流失败而未传达

- *库存*——保留超出需要的物料或信息。考虑项目如何积压信息或物料：

  - "原料"积压太多

  - 所需"原料"检索复杂

  - 过时的、过期的信息

- *缺陷*——为纠正问题导致工作重做的错误或过失。考虑项目如何返工并重做：

  - 缺乏适当的评审、验证或确认

  - 错误或不足的信息

- *人员潜能的浪费*——未利用或甚至抑制人员的积极性、活力、创造力和解决问题的能力以及执行出色工作的意愿。

9.8.3 精益原则

Womack 和 Jones（1996）提出无浪费创造价值的流程分成六个精益原则。这些原则（见图 9.10）简称为：价值、价值流、流动、拉动、完美和对人的尊重，详细定义如下。

当将精益思维应用于 SE，使用精益原则来评估项目计划、人员准备、流程和工具以及组织行为。考虑客户如何定义产品和流程中的价值，然后为创造产品和流程对价值流进行描述，通过价值流优化流动且消除浪费，鼓励该价值流中每个节点的拉动，以及努力完善价值流，以最大化面向客户的价值。这些活动都应在尊重客户、利益攸关者和项目团队成员的基础上实施。

2009 年，INCOSE 精益 SE 工作组发布了一种名为"系统工程精益使能项"（LEfSE）V1.0 的全新在线产品。该产品基于精益思想收集了 SE 的实践和建议，并表述为"应做"和"不做"。这些实践涵盖了大范围的 SE 和其他相关的企业管理实践，总体上关注提高计划的价值和利益攸关者的满意度，并减少浪费、延期、成本超支和挫折。目前，LEfSE 在被称为使能项的 47 个非行动性的议题的标题之下，列出了 147 项实践（称为子使能项），并分组成为下述六个精益原则（oppenheim，2011）：

1. Under the *value principle*, subenablers promote a robust process of establishing the value of the end product or system to the customer with crystal clarity early in the program. The process should be customer focused, involving the customer frequently and aligning employees accordingly.

2. The subenablers under the *value stream principle* emphasize detailed program planning and waste preventing measures, solid preparation of the personnel and processes for subsequent efficient workflow, and healthy relationships between stakeholders (e.g., customer, contractor, suppliers, and employees); program frontloading; and use of leading indicators and quality measures. Systems engineers should prepare for and plan all end-to-end linked actions and processes necessary to realize streamlined value, after eliminating waste.

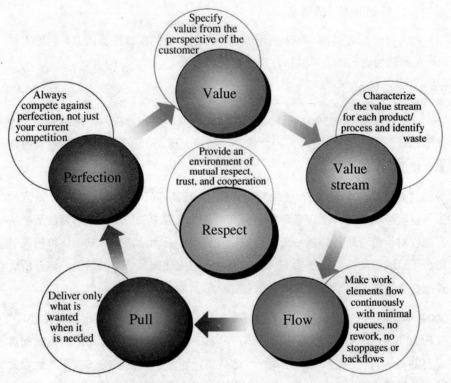

FIGURE 9.10 Lean development principles. Reprinted with permission from Bohdan Oppenheim. All other rights reserved.

1. 在*价值原则*下，子使能项以完全透明的方式在计划早期促成一个为客户建立最终产品或系统价值的鲁棒流程。该流程应聚焦客户，使客户频繁参与并相应地使员工协调一致。

2. 在价值流原则下的子使能项强调：详细的项目计划和浪费的预防措施、后续高效工作流中人员和流程的充分准备以及利益攸关者（如客户、承包商、供应商和员工）之间的良好关系；项目的前期准备；以及采用领先指标和质量度量。在消除浪费后系统工程师们应准备并计划实现简捷高效的价值所必需的所有端对端连接的行动和流程。

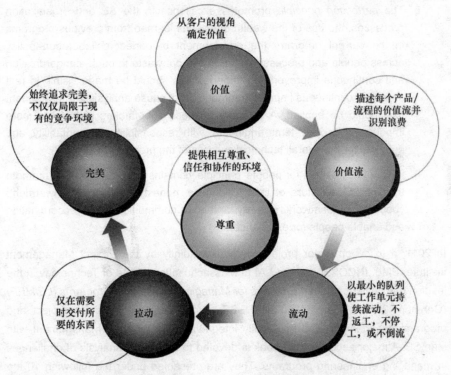

图 9.10　精益开发原则。经 Bohdan Oppenheim 许可后转载。版权所有。

3. The *flow principle* lists subenablers that promote the uninterrupted flow of robust quality work and first-time right products and processes, steady competence instead of hero behavior in crises, excellent communication and coordination, concurrency, frequent clarification of the requirements, and making program progress visible to all.

4. The subenablers listed under the *pull principle* are a powerful guard against the waste of rework and overproduction. They promote pulling tasks and outputs based on internal and external customer needs (including rejecting others as waste) and better coordination between the pairs of employees handling any transaction before their work begins so that the result can be first-time right.

5. The *perfection principle* promotes excellence in the SE and organization processes, the use of the wealth of lessons learned from previous programs in the current program, the development of perfect collaboration policy across people and processes, and driving out waste through standardization and continuous improvement. Imperfections should be made visible in real time, and continuous improvement tools (root cause analysis and permanent fix) should be applied. A category of these subenablers calls for a more important role of systems engineers, with responsibility, accountability, and authority for the overall technical success of the program.

6. Finally, the respect-for-people principle contains subenablers that promote the enterprise culture of trust, openness, honesty, respect, empowerment, cooperation, teamwork, synergy, and good communication and coordination and enable people for excellence.

In 2011, a follow-on major project undertaken jointly by the Project Management institute (PMI), INCOSE, and the LAI at Massachusetts institute of Technology in the leading role developed *Lean Enablers for Managing Engineering Programs (LEfMEP)* (Oehmen, 2012), incorporating all LEfSE, adding lean enablers for project and program management, and holistically integrating lean program management with lean SE. A major section of the book is devoted to a rigorous analysis of challenges in managing engineering programs. They are presented under the following 10 top challenge themes:

1. Firefighting—reactive program execution

2. Unstable, unclear, and incomplete requirements

3. Insufficient alignment and coordination of the extended enterprise

3. *流动原则*列出子使能项，促进高品质工作的不间断流动和第一次就做正确的产品和流程；稳定的能力而不是危机中的英雄行为；卓越的沟通和协调；并行性；经常地澄清需求；并使计划进展对所有人透明。

4. 在*拉动原则*下列出的子使能项能有力地防止返工和过度生产所造成的浪费。子使能项促进基于内部客户和外部客户的需要（包括被视为浪费而拒绝的其他需求）促进拉动任务和输出，并在开始工作前在处理各项事务的员工之间进行更好地协调，使结果第一次就是正确的。

5. *完美原则*促进 SE 和组织流程的卓越；将从先前计划中汲取宝贵的经验教训用于当前计划之中；开发人员与流程之间完美的协同策略；以及通过标准化和持续改进消除浪费。应使缺点实时可见，并且应使用持续改进工具（根本原因分析和永久地解决）。这一类子使能项呼吁系统工程师们扮演更重要角色，为了计划的全面技术成功承担职责、责任和权限。

6. 最后，在"对人的尊重"的原则中包含子使能项，其促进信任、公开、真诚、尊重、赋权、合作、团队工作、协同和良好沟通和协调的企业文化，并促使人员追求卓越。

2011 年，由项目管理协会（PMI）、INCOSE 以及起主导角色的麻省理工学院 LAI 共同承担的后续主要项目，开发管理工程计划的精益使能项（LEFMEP）(Oehmen, 2012)，纳入所有 LEFSE，增加项目和计划管理的精益使能项，并整体综合精益计划管理与精益 SE。本书的主要章节致力于精确分析管理工程计划中的挑战。在以下 10 项首要挑战主题下列出：

1. 救火式——反应式计划执行

2. 需求不稳定、不明确且不完整

3. 扩展企业的对准和协调不充分

CROSS-CUTTING SYSTEMS ENGINEERING METHODS

4. Processes that are locally optimized and not integrated for the entire enterprise
5. Unclear roles, responsibilities, and accountability
6. Mismanagement of program culture, team competency, and knowledge
7. Insufficient program planning
8. Improper metrics, metric systems, and key performance indicators
9. Lack of proactive program risk management
10. Poor program acquisition and contracting practices

The 326 lean enablers in Oehmen (2012) are listed in several convenient ways: under the six lean principles, under the 10 major challenge themes, under the SE processes used in this volume, and under the management performance domains defined in (PMI, 2013).

The LEfSE and LEfMEP are not intended to become mandatory practices. Instead, they should be used as a checklist of excellent holistic practices validated by the community of practice. Awareness of the enablers should improve the thinking at work and significantly improve program quality. Early feedback from the organizations practicing lean enablers indicates significant benefits (Oppenheim, 2011).

The *INCOSE Lean SE Working Group* public website (2011) contains a rich menu of publications and case studies related to both LEfSE and LEfMEP.

9.9 AGILE SYSTEMS ENGINEERING

Historically, agile software engineering processes came into awareness in 2001 with the declaration of the Agile Manifesto (Beck et al., 2001), which spawned interest in a number of methodologies, with names such as Scrum and Extreme Programming. But adoption of those meth-odologies and consideration of how they might inform nonsoftware engineering (Carson, 2013) has tended to focus on software-related specific practices rather than fundamental frameworks. In contrast, a cross-industry study in 1991 (Nagel, 1992) observed that technology and the environment in which it is deployed were coevolving at an increasing rate, outpacing the adaptation capabilities of most organized human endeavors. Agility, as a systemic characteristic, was thus identified, and subsequently studied to identify domain-independent metrics, architecture, and design principles (dove, 2001).

4. 局部优化且未综合的整个企业的流程

5. 不明确的角色、职责和责任

6. 项目文化、团队能力和知识的管理不当

7. 不充分的方案计划

8. 不正确的衡量标准、度量标准和关键性能指标

9. 缺乏前瞻性的计划风险管理

10. 不足的计划采办和合同实践

以多种简便的方式列出 Oehmen（2012）中的 326 项精益使能项：按照六项精益原则、按照 10 项主要挑战主题、按照本书中所使用的 SE 流程以及按照（PMI, 2013）定义的管理性能域。

LEfSE 和 LEfMEP 并非意图成为一种强制性实践。而相反，它应被当作经实践团体验证的良好完整实践的检查单。对使能项的认识应改进工作时的思维并明显提高计划质量。实践精益使能项的组织的早期反馈将会表明显著的效益（Oppenheim，2011）。

INCOSE 精益 SE 工作组公共网站（2011）包含涉及 LEfSE 和 lEfMEP 的各种出版物和案例研究的清单。

9.9 敏捷系统工程

历史上，随着敏捷宣言（Beck 等，2001）的宣告，敏捷软件工程流程在 2001 年开始被人们所认识，这引起了对诸多方法论的关注，例如，Scrum 和极限编程。但是，对这些方法论的采用以及对这些方法论如何贯穿于非软件工程的考虑（Carson，2013）已经倾向聚焦于与软件相关的特定实践而不是基本框架。相反，1991 年进行的跨行业研究（Nagel，1992）注意到技术及部署技术的环境正在快速地协同演进，正在速度上超过大多数有组织的人类工作的适应能力。敏捷性已作为一个系统性的特征被识别，随后的研究则是识别与独立于领域的指标、架构和设计原理（dove，2001）。

Agility is a capability exhibited by systems and processes that enables them to sustain effective operation under conditions of unpredictability, uncertainty, and change. The value proposition of an agile SE process is risk management, appropriate when development speed and customer satisfaction are likely to be affected by requirements understandings that evolve during system development.

Common causes for requirements evolution include insufficient initial understanding, new understandings revealed during development, and evolving knowledge of the deployment environment. If ignored, requirements evolution reduces or eliminates customer satisfaction. If unmitigated, requirements evolution causes rework and scrapped work, a principal source of time and cost overruns.

An agile system architecture incurs expense in infrastructure and modularity design, which should be weighed against probabilistic costs for requirements evolution. This is risk management. The purpose of an agile SE process is to reduce the technical, cost, and schedule risks associated with accommodating beneficial requirements evolution.

Agility is the ability to respond effectively to surprises—good or bad. Practices and techniques should be chosen for compatibility and synergy with the nature of the project (Carson, 2013; Sillitto, 2013) and the cultural environment in which they will be employed.

9.9.1 Agile SE Framework

Agile SE (Forsberg et al., 2005) is summarized as follows:

- Leverages an agile architecture for SE (process), enabling reconfiguration of goals, requirements, plans, and assets, predictably.

- Leverages an architecture for agile SE (product), enabling changes to the product (system) during development and fabrication, predictably.

- Leverages an empowered intimately involved "product owner" (chief systems engineer, customer, or equivalent responsible authority on product vision), enabling broad-level systems thinking to inform real-time decision making as requirements understanding evolve.

- Leverages human productivity factors that affect engineering, fabrication, and customer satisfaction in an unpredictable and uncertain environment.

9.9.2 Agile Metric Framework

Agility measures are enabled and constrained principally by architecture—in both the process and the product of development:

敏捷性是系统和流程所展示出的、能够使其在不可预见性、不确定性和变化的条件下保持有效运行的一种能力。敏捷 SE 流程的价值主张是风险管理，这在开发速度和客户满意度可能受到在系统开发期间演进的需求理解的影响时是适用的。

需求演进的常见原因包括不充分的初始理解、开发期间揭示的新理解，以及开发环境的演进知识。如果忽略这些原因，需求演进将降低或排除客户满意度。如果未缓解这些原因，需求演进将导致返工和工作前功尽弃，这是时间和成本超支的主要原因。

一个敏捷的系统架构会引起基础设施和模块化设计方面的开销，这应根据需求演进的可能成本进行权衡。这是风险管理。敏捷 SE 流程的目的是减少与适应有益的需求演进相关联的技术、成本和进度风险。

敏捷性是对意外（好的或坏的）有效响应的能力。选择的实践和技术应与项目的本质属性（Carson，2013；Sillitto，2013）以及采用这些实践和技术的文化环境兼容和协同。

9.9.1 敏捷 SE 框架

敏捷 SE（Forsberg 等，2005）如下文所概括：

- 更强有力地发挥敏捷 SE（流程）架构的作用，使能目标、需求、计划和资产的重构可预见。
- 更强有力地发挥敏捷 SE（产品）架构的作用，使能在开发和制造期间产品（系统）变更可预见。
- 更强有力地发挥经授权直接参与的"产品所有者"（首席系统工程师、客户或同等的产品愿景负责机制）的作用，使能广泛的系统思想随着需求理解而为实时决策提供依据。
- 在不可预见和不确定环境中更强有力地发挥影响工程、制造和客户满意度的人类生产力因素的作用。

9.9.2 敏捷测度框架

敏捷性测度主要由架构实现并受架构约束——在开发流程和产品方面：

- Time to respond, measured in both the time to understand a response is necessary and the time to accomplish the response
- Cost to respond, measured in both the cost of accomplishing the response and the cost incurred elsewhere as a result of the response
- Predictability of response capability, measured before the fact in architectural preparedness for response and confirmed after the fact in repeatable accuracy of response time and cost estimates
- Scope of response capability, measured before the fact in architectural preparedness for comprehensive response capability within mission and confirmed after the fact in repeatable evidence of broad response accommodation

9.9.3 Agile Architectural Framework

Agile SE and agile-systems engineering are two different things (Haberfellner and de Weck, 2005) with a shared common architecture that enables the agility in each (Dove, 2012). The architecture will be recognized in a simple sense as a drag-and-drop plug-and-play loosely coupled modularity, with some critical aspects not often called to mind with the general thoughts of a modular architecture.

There are three critical elements in the architecture: a roster of drag-and-drop encapsulated modules, a passive infrastructure of minimal but sufficient rules and standards that enable and constrain plug-and-play operation, and an active infrastructure that designates specific responsibilities that sustain agile operational capability:

- *Modules*—Modules are self-contained encapsulated units complete with well-defined interfaces that conform to the plug-and-play passive infrastructure. They can be dragged and dropped into a system of response capability with relationship to other modules determined by the passive infrastructure. Modules are encapsulated so that their interfaces conform to the passive infrastructure, but their methods of functionality are not dependent on the functional methods of other modules except as the passive infrastructure dictates.

- 响应时间，从必要的理解响应的时间和完成响应的时间两方面测量。
- 响应成本，从完成响应的成本和因响应而在其他地方所承担的成本两方面测量。
- 响应能力的可预测性，在响应的架构准备之前测量的以及在响应时间和成本估计的重复精度之后确认。
- 响应能力范围，在任务范围内对全面响应能力进行架构准备之前测量的以及在广泛的响应调节的重复迹象之后确认。

9.9.3 敏捷架构框架

敏捷 SE 和"敏捷系统工程"是两种不同的事物（Haberfellner 和 de Weck，2005），都具有一个使得敏捷性在其中能够实现的共享公共架构（Dove，2012）。从简单意义上讲，架构将被认为是一种拖—放和即插即用的松散耦合模块，具有常常超乎意料的某些关键方面的模块化架构的一般思想。

架构具有三个关键要素：拖放封装模块的注册表、实现和约束即插即用运行的最低但具有充分的规则和标准的被动式基础设施，以及指定多个保持敏捷运行能力的具体职责的主动式基础设施：

- *模块*——模块是具有完好定义的、符合即插即用被动式基础设施要求的接口的自包含封装单元。可按照与被动基础设施确定的其他模块的相互关系将该模块拖放至响应能力系统。模块是封装的，以便其接口符合被动基础设施，但是其功能性方法不取决于其他模块的功能方法，除非被动式基础设施有规定。

- *Passive infrastructure*—The passive infrastructure provides drag-and-drop connectivity between modules. Its value is in isolating the encapsulated modules so that unexpected side effects are minimized and new operational functionality is rapid. Selecting passive infrastructure elements is a critical balance between requisite variety and parsimony—just enough in standards and rules to facilitate module connectivity but not so much to overly constrain innovative system configurations.

- *Active infrastructure*—An agile system is not something designed and deployed in a fixed event and then left alone. Agility is most active as new system configurations are assembled in response to new requirements—something which may happen very frequently, even daily in some cases. In order for new configurations to be enabled when needed, four responsibilities are required: the collection of available modules must evolve to be always what is needed, the modules that are available must always be in deployable condition, the assembly of new configurations must be accomplished, and both the passive infrastructure and active infrastructure must have evolved when new configurations require new standards and rules. Responsibilities for these four activities must be designated and embedded within the system to ensure that effective response capability is possible at unpredictable times:

  - Module mix—Who (or what process) is responsible for ensuring that new modules are added to the roster and existing modules are upgraded in time to satisfy response needs?

  - Module readiness—Who (or what process) is responsible for ensuring that sufficient modules are ready for deployment at unpredictable times?

  - System assembly—Who (or what process) assembles new system configurations when new situations require something different in capability?

  - Infrastructure evolution—Who (or what process) is responsible for evolving the passive and active infrastructures as new rules and standards are anticipated and become appropriate?

9.9.4 Agile Architectural Design Principles

Ten reusable, reconfigurable, scalable design principles are briefly itemized in this section:

Reusable principles are as follows:

- *被动式基础设施*——被动式基础设施提供模块间的拖放连接性。其价值是隔离封装模块，以便未预料到的意外结果最小化并且新的运行功能性速度快。选择被动式基础设施要素是必要变化与简约性之间的关键平衡——标准和规则恰好足以促进模块连接性而不是过度约创新的系统配置。

- *主动式基础设施*——敏捷系统并不是在固定事件中设计和部署而后单独留下的某类事物。敏捷性是最主动的，在组装新系统构型以响应新需求时——发生频率非常高的事物，在某些情况下甚至是每天发生。为了在需要时能够启动新构型，需要四项职责：可用模块的集合必须演进，以达到总是要收集所需要的模块，可用模块必须总是处于可部署状态，新构型的组装必须完成，并且被动式基础设施和主动式基础设施必须在新构型要求新标准和规则时演进。必须在系统内指定和嵌入这四项活动的职责，以确保有效响应能力在不可预测的时间是可能的：

    - 模块混合——谁（或什么流程）负责确保新模块被添加至注册表，并且现有模块得到及时升级以满足响应需要？

    - 模块准备度——谁（或什么流程）负责确保已准备足够的模块，用于在不可预测的时间进行开发？

    - 系统组装——谁（或什么流程）在新情况要求不同能力时组装新系统构型？

    - 基础设施演进——谁（或什么流程）负责预备采用新规则和标准并在其适用时演进被动式基础设施和主动式基础设施？

9.9.4 敏捷架构设计原理

本节简要地分条列举了十个可复用、可重新配置且可扩展的设计原理：

可复用的原理如下：

- *Encapsulated modules*—Modules are distinct, separable, loosely coupled, independent units cooperating toward a shared common purpose.

- *Facilitated interfacing (plug compatibility)*—Modules share well-defined interaction and interface standards and are easily inserted or removed in system configurations.

- *Facilitated reuse*—Modules are reusable and replicable, with supporting facilitation for finding and employing appropriate modules.

Reconfigurable principles are as follows:

- *Peer-peer interaction*—Modules communicate directly on a peer-to-peer relationship; and parallel rather than sequential relationships are favored.

- *Distributed control and information*—Modules are directed by objective rather than method; decisions are made at point of maximum knowledge, and information is associated locally and accessible globally.

- *Deferred commitment*—requirements can change rapidly and continue to evolve. Work activity, response assembly, and response deployment that are deferred to the last responsible moment avoid costly wasted effort that may also preclude a subsequent effective response.

- *Self-organization*—Module relationships are self- determined where possible, and module interaction is self-adjusting or self-negotiated.

Scalable principles are as follows:

- *Evolving standards*—Passive infrastructure standardizes intermodule communication and interaction, defines module compatibility, and is evolved by designated responsibility for maintaining current and emerging relevance.

- *Redundancy and diversity*—Duplicate modules provide capacity right-sizing options and fail-soft tolerance, and diversity among similar modules employing different methods is exploitable.

- *Elastic capacity*—Modules may be combined in responsive assemblies to increase or decrease functional capacity within the current architecture.

- *封装模块*——模块是为共享的共同目标而协作的、与众不同的、可分离的、松散耦合的独立单元。
- *辅助的接口连接（接插兼容性）*——模块共享定义明确的交互和接口标准并易于插入系统构型或从中移除。
- *辅助的复用*——模块是可复用且可复制的，支持促进适当模块的发现和部署。

可重新配置原理如下：

- *对等交互*——模块直接以对等的关系沟通；并且更倾向于喜欢平行关系而不是顺序关系。
- *分布式控制和信息*——模块由目的指示而不是由方法指示；以最大知识点做出决策，信息是局部相关且可全局访问的。
- *遵从承诺*——需求可迅速变化并持续演进。工作活动、响应组装以及响应部署遵从最终的职责以避免过度浪费的工作还可能妨碍后续有效响应。
- *自组织*——模块关系是自主决定的（如有可能），并且模块交互是自调节或自协商的。

可扩展原理如下：

- *演进的标准*——被动式基础设施标准化模块间的通信和交互，定义模块兼容性，并且按照用于保持当前和新兴相关性的指定职责演进。
- *余度和多样性*——复制模块提供合理精简选项和性能弱化冗余的能力，并且采用不同方法的相似模块之间的多样性是可取的。
- *弹性能力*——模块可能组合为响应组件增加或减少当前架构内的功能能力。

10 SPECIALTY ENGINEERING ACTIVITIES

The objective of this chapter is to give enough information to systems engineers to appreciate the significance of various engineering specialty areas, even if they are not an expert in the subject. It is recommended that subject matter experts are consulted and assigned as appropriate to conduct specialty engineering analysis. The topics in this chapter are covered in alphabetical order by topic title to avoid giving more weight to one topic over another. More information about each specialty area can be found in references to external sources.

With a few exceptions, the forms of analysis presented herein are similar to those associated with SE. Most analysis methods are based on the construction and exploration of models that address specialized engineering areas, such as electromagnetic compatibility (EMC), reliability, safety, and security. Not every kind of analysis and associated model will be applicable to every application domain.

10.1 AFFORDABILITY/COST-EFFECTIVENESS/LIFE CYCLE COST ANALYSIS

As stated in Blanchard and Fabrycky (2011),

> Many systems are planned, designed, produced, and operated with little *initial* concern for affordability and the total cost of the system over its intended lifecycle...The technical [*aspects* are] usually considered first, with the economic [aspects] deferred until later.

This section addresses economic and cost factors under the general topics of affordability and cost-effectiveness. The concept of life cycle cost (LCC) is also discussed.[1]

10.1.1 Affordability Concepts

Improving design methods for affordability (Bobinis et al., 2013; Tuttle and Bobinis, 2013) is critical for all application domains. The INCOSE and the National Defense Industrial Association (NDIA) (the Military Operations Research Society (MORS) has also adapted these definitions) have addressed "affordability" through ongoing affordability working groups started in late 2009 and have defined system affordability through these ongoing working groups. Both organizations have defined system affordability as follows:

INCOSE Systems Engineering Handbook: A Guide for System Life Cycle Processes and Activities, Fourth Edition. Edited by David D. Walden, Garry J. Roedler, Kevin J. Forsberg, R. Douglas Hamelin and Thomas M. Shortell.

© 2015 John Wiley & Sons, Inc. Published 2015 by John Wiley & Sons, Inc.

10 专业工程活动

本章的目的是为系统工程师提供足够的信息，使其领会各种不同工程专业领域的重要性，即便他们并非这一领域的专家。建议主题领域专家得到咨询并在适当情况下被安排进行专业工程分析。本章中涉及的主题均按照主题标题的字母顺序排列，以避免给予一个主题相对于其他而言更高的重视度。有关每个专业领域的更多信息可在外部资源的参考中找到。

除少数例外情况，本文中提出的分析形式与 SE 相关形式相同。大多数的分析方法基于应对专业工程领域[如电磁兼容性（EMC）、可靠性、安全性和安保性]模型的构建与探索。并非每种分析和相关模型都会适用于各种应用领域。

10.1 可承受性/成本效能/生命周期成本分析

正如 Blanchard 和 Fabrycky（2011）中所述，

> 许多系统的规划、设计、生产和运行最初很少去关注可承受性以及系统在其预期生命周期的总体成本……。通常首先考虑[技术方面]，经济方面会推到之后。

本节涉及可承受性和成本效能一般性的主题之下的经济和成本因素。还论述生命周期成本（LCC）的概念。

10.1.1 可承受性的概念

改善可承受性的设计方法（Bobinis 等，2013；Tuttle 和 Bobinis, 2013）对于所有应用领域都很关键。INCOSE 和美国防务工业协会（NDIA）[美国军事运行学会（MORS）也已采纳这些定义]通过在 2009 年后期成立的持续可承受性工作组提出"可承受性"，并且通过这些持续工作组定义了系统的可承受性。两个组织均定义了系统的可承受性，如下所述：

INCOSE 系统工程手册：系统生命周期流程和活动指南，第 4 版。编撰：David D. Walden、GarryJ. Roedler、Kevin J. Forsberg、R. Douglas Hamelin 和 Thomas M. Shortell.
John Wiley & Sons 公司版权所有©2015。由 John Wiley & Sons 公司于 2015 年出版。

- INCOSE Affordability Working Group definitions (June 2011):

Affordability is the balance of system performance, cost and schedule constraints over the system life while satisfying mission needs in concert with strategic investment and organizational needs.

Design for affordability is the systems engineering practice of balancing system performance and risk with cost and schedule constraints over the system life satisfying system operational needs in concert with strategic investment and evolving stakeholder value.

- NDIA Affordability Working Group definition (June 2011):

Affordability is the practice of ensuring program success through the balancing of system performance (Kpps), total ownership cost, and schedule constraints while satisfying mission needs in concert with long-range investment, and force structure plans of the DOD.

The concept of affordability can seem straightforward. The difficulty arises when an attempt is made to specify and quantify the affordability of a system. This is significant when writing a specification or when comparing two affordable solutions to conduct an affordability trade study. Even though affordability has been defined by the INCOSE, NDIA, and MORS, in discussions at an MORS Special Meeting on *Affordability Analysis: How Do We Do It?* it was noted that all industry groups have discovered that affordability analysis is contextually sensitive, often leading to a misunderstanding and incompatible perspectives on what an "affordable system *is.*" The various industry working groups have recommended developing and formalizing affordability analysis processes, including recognizing the difference between cost and affordability analyses. As a result of these high-level discussions, the key affordability takeaways include:

- Affordability context, system(s), and portfolios (of systems capabilities) need to be consistently defined and included in any understanding of what an affordable system is.

- An affordability process/framework needs to be established and documented.

- Accountability (system governance) for affordability needs to be assigned across the life cycle, which includes stakeholders from the various contextual domains.

- INCOSE 可承受性工作组定义（2011年6月）：

可承受性是在整个系统生命中系统性能、成本和进度约束的平衡，同时与战略投资和组织需要协调一致并满足使命任务需要。

可承受性的设计是在系统生命中平衡系统性能和风险与成本和进度约束的系统工程实践，与战略投资和演进利益攸关者的价值协调一致。

- NDIA 可承受性工作组定义（2011年6月）：

可承受性是通过平衡系统性能（KPP）、总体拥有成本和进度约束以确保计划成功的实践，与长期投资以及 DOD 的军力结构规划协调一致并满足使命任务需要。

可承受性的概念看似很直观，但是，当试图规定并量化系统的可承受性时就会出现困难。在编写规范或者对比两个可承受的解决方案来进行可承受性权衡研究时，这一点十分显著。尽管 INCOSE、NDIA 和 MORS 已在 MORS 关于可承受性分析的特殊会议的讨论中对可承受性进行了定义：我们该怎么做？但显然，所有工业集团已经发现，可承受性分析是对背景环境敏感的，往往会造成对"什么是可承受的系统"的误解和互斥的观点。各种不同的工业工作组建议开发并规范可承受性分析流程，包括认识成本与可承受性分析之间的区别。作为这些高层级论述的结果，可承受性的关键重点包括：

- 可承受性的背景环境、系统和（系统能力的）全集需要一致地定义，并包括在关于什么是可承受的系统的任何理解之中。
- 可承受性流程/框架需要被建立并文件化。
- 对可承受性的责任追究（系统治理）需要跨生命周期分配，包括来自各种不同背景环境领域的利益攸关者。

SPECIALTY ENGINEERING ACTIVITIES

10.1.1.1 "Cost-Effective Capability" Is a Contextual Attribute As defined in "Better Buying power: Mandate for Restoring Affordability and productivity in Defense Spending" (Carter, 2011), "affordability means conducting a program at a cost constrained by the maximum resources the Department can allocate for that capability." Affordability includes acquisition cost and average annual operating and support cost. It is expanded to encompass additional elements required for the LCC of a system, as an outcome of various hierarchal contexts in which any system is embedded. Therefore, in the SE domain, affordability as an attribute must be determined both inside the boundaries of the system of interest (SOI) and outside (see Fig. 10.1). This defines, in practical terms, the link between system capability, cost, and what we call "affordability." Thus, the concept of affordability must encompass everything from a portfolio (e.g., family of automobiles) to an individual program (specific car model). Affordability as a design attribute of a system versus a program versus a domain remains contextually dependent on the stakeholder's context and the life cycle of the SOI under examination.

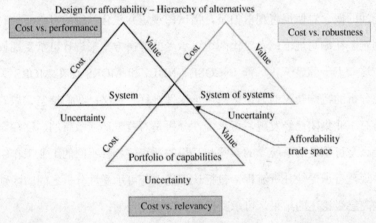

FIGURE 10.1 Contextual nature of the affordability trade space. Derived from Bobinis et al. (2013) Figure 1. Reprinted with permission from Joseph Bobinis. All other rights reserved.

10.1.1.2 Design Model for Affordability As previously stated, an affordability design model must be able to provide the ability to effectively manage and evolve systems over long life cycles. The derived requirements we will focus on in this section are as follows:

- Design perspective that assumes the system will change based on environmental influences, new uses, and disabling system causing performance deterioration

10.1.1.1 "成本效益能力"是一种背景环境的属性 按照"更好的购买力：恢复防务开支中的可承受性和生产率的委任"（Carter，2011）定义，"可承受性是指，以国防部可供给该能力的最大资源约束的成本开展项目"。可承受性包括采办成本和年平均运行及维持成本。可承受性又扩展到包括系统的 LCC 需要的附加元素，作为系统嵌入于各种不同层级背景环境的产出。因此，在 SE 领域中，作为属性的可承受性必须在所感兴趣之系统（SOI）边界内外确定（见图 10.1）。在实践中这个定义系统能力、成本及所谓的"可承受性"之间的联系。因此，可承受性的概念必须包括从项目群（如，汽车系列）到各个项目计划（特定的汽车型号）的所有事物。可承受性作为系统的设计属性与项目计划对比、与领域对比，从背景环境来看它仍然取决于利益攸关者的背景环境以及置于考查之下的 SOI 的生命周期。

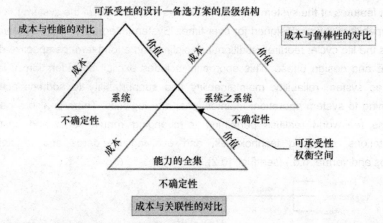

图 10.1 可承受性权衡空间的背景环境本质属性。来自 Bobinis 等（2013）图 1。经 Joseph Bobinis 许可后转载。版权所有。

10.1.1.2 可承受性的设计模型 如前面所述，可承受性设计模型必须能够提供在漫长生命周期内有效地管理并演进系统的能力。我们将在本节中聚焦的推演需求如下：

- 假设系统将基于环境影响、新的使用以及使性能退化的失效系统而变化的设计视角。

- Causes of system and system element life cycle differences (technology management)
- Feedback functions and measurement of system behaviors with processes to address emergence (life cycle control systems)
- A method to inductively translate system behaviors into actionable engineering processes (adaptive engineering)

One of the major assumptions for measuring the afford- ability of competing systems is that given two systems, which produce similar output capabilities, it will be the *nonfunctional* attributes of those systems that differentiate system value to its stakeholders. The affordability model is concerned with operational attributes of systems that determine their value and effectiveness over time, typically expressed as the system's "ilities" or specialty engineering as they are called in this handbook.

These attributes are properties of the system as a whole and as such represent the salient features of the system and are measures of the ability of the system to deliver the capabilities it was designed for over time. "System integration, and its derivatives across the life cycle, requires additional discipline and a long term perspective during the SE and design phase. This approach includes explicit consideration of issues such as system reliability, maintainability and supportability to address activities pertaining to system operation, maintenance, and logistics. There is also a need to address real-world realities pertaining to changing requirements and customer expectations, changing technologies, and evolving standards and regulations" (Gallios and verma, n.d.) (see Fig. 10.2).

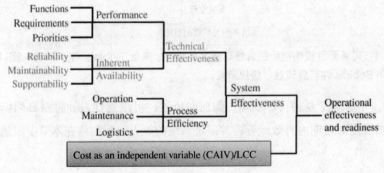

FIGURE 10.2 System operational effectiveness. Derived from Bobinis et al. (2013) Figure 4. Reprinted with permission from Joseph Bobinis. All other rights reserved.

- 系统和系统元素生命周期差异的原因（技术管理）
- 系统行为与流程反馈功能和测度以应对涌现性（生命周期控制系统）
- 将系统行为转化为可执行的工程流程的归纳推理方法（自适应的工程）

衡量竞争系统的可承受性主要假设之一是给定两个产生相似输出能力的系统，它将作为区别系统对利益攸关者的价值的那些系统的非功能属性。可承受性模型关注系统确定其随时间的价值和效能的运行属性，正如它们在本手册中的表达，通常表示为系统的"-性"或专业工程。

这些属性都是系统做为一个整体的特性，正因如此，它们代表系统的显著特征，并且作为系统随时间交付事先设计好的能力的测度。"系统综合及其跨生命周期的衍生物在 SE 和设计阶段需要其他的学科和长远的视角。该方法包括明确的考虑诸如系统可靠性、可维护性和保障性等问题，以应对与系统运行、维护和后勤相关的活动。还需要应对与需求和客户期望变化、技术变化以及标准和法规演进相关的真实世界的现实"（Gallios 和 verma，未注明日期）（见图 10.2）。

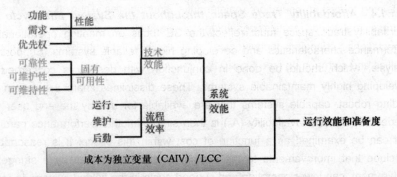

图 10.2 系统运行效能。来自 Bobinis 等（2013）图 4。经 Joseph Bobinis 许可后转载。版权所有。

SPECIALTY ENGINEERING ACTIVITIES

10.1.1.3 Impact to Affordability Managing a system within an affordability trade space means that we are concerned with the actual performance of the fielded system, defined in one or more appropriate metrics, bounded by cost over time. ("System performance" can be expressed in whatever way makes sense for the system under study.) The time dimension extends a specific "point analysis" (static) to a continuous life cycle perspective (dynamic). Quantifying a relationship between cost, performance, and time defines a functional space that can be graphed and analyzed mathematically. Then it becomes possible to examine how the output (performance, availability, capability, etc.) changes due to changes in the input (cost constraints or budget availability). Hence, utilizing this functional relationship between cost and outcome defines an affordability trade space. Done correctly, it is possible to analyze specifically the relationship between money spent and system performance and possibly determine the point of diminishing returns.

However, it is frequently necessary to estimate the value of one variable when the values of the others are known or specified. These kinds of point solutions (which in this sense are "coordinates" within the space) are useful in examining specific relationships, including making predictions of "average" behavior. Answering questions regarding "expected value" of some parameter in terms of others, often for specific points in time, frequently falls into this category. The overall trade space can then be thought of as the totality of all such point solutions.

10.1.1.4 Affordability Trade Space throughout the System Life Cycle The affordability trade space must reflect the SE focus on meeting operational and performance characteristics and developing highly reliable systems. Supportability analysis, which should be done in conjunction with design, is concerned with developing highly maintainable systems. These disciplines share a common goal: fielding robust, capable systems that are available for use by the end user when needed. Operational availability (A_o) is then simply another performance parameter that can be examined as a function of cost within this space. It is reasonable to conclude that improvements to the design, in this case driven by a stringent A_o requirement, can lower operation and support costs in the fielded system. In fact, A_o is an implicit performance measure used to calculate expected system effectiveness. That has to be applied, along with use/market size and operational requirements, to determine the overall cost-effective capability of the system under study.

Supporting a system throughout its life cycle requires systems engineers to account for changes in the system design as circumstances change over time, such as changes in the threat environment, diminishing material shortage (DMS) issues, and improvements in technology, SoS relationships, and impact of variables outside the system.

10.1.1.3 对可承受性的影响 在可承受性权衡空间内管理一个系统意指我们关注现场系统的实际性能，以一种或多种适当的衡量标准定义，随时间受成本的束缚。（"系统性能"可用任何一种使研究中的系统有意义的方式来表达。）时间维度将特定的"点分析"（静态）延伸到持续的生命周期视角（动态）。量化成本、性能和时间之间的关系可定义能在数学上用图表表示并分析的功能空间。于是，有可能检查输出（性能、可用性、能力等）如何由于输入（成本约束或预算可用性）的变化而变化。因此，利用成本与产出之间的这种功能关系能定义一个可承受性权衡空间。正确执行后，有可能规范地分析所花费用与系统性能之间的关系，并可能确定收益递减的点。

但是，在其他变量的值已知或特定时,往往有必要估计一个变量的值。这些种类的单点解决方案（从这层意义上说，是该空间内的"坐标"）在检查特定关系中是有用的，包括对"平均"行为进行预估。依照其他方面回答有关某些参数的"期望的价值"的问题，往往对于特定单点会及时地、频繁地划入该范畴。于是，总体权衡空间可被看做所有这类单点解决方案的总和。

10.1.1.4 贯穿于系统生命周期的可承受性权衡空间 可承受性权衡空间必须反映关于满足运行和性能特征并开发高可靠性系统的 SE 聚焦点。应对接设计开展维持性分析关注开发可维护性高的系统。这些学科共享一个共同的目标：现场鲁棒的、有能力的、在需要时可以由最终用户使用的系统。运行有效性（A_o）则只是另一个可以根据该空间内的成本检查的性能参数。在受严格约束的 A_o 需求驱动下，改善设计可降低现场系统的运行和维持成本，这一结论是合理的。事实上，A_o 是用于计算期望系统效能的固有的性能测度，与使用/市场规模及运行需求一道，确定研究中的系统的总体成本效益能力。

贯穿于生命周期来维持一个系统就要求系统工程师考虑系统设计随时间的环境变化而发生的变化，例如在威胁环境中的变化，减弱材料缺乏（DMS）问题，以及技术方面的提高、SoS 关系和系统外的变量影响。

Given that designers do not control the variables in the environment, the optimization problem for affordability is different or could be different at any point in the life cycle. This optimization problem can be managed by functional criticality, surge, and adaptive requirements or as a set of predetermined technology refresh cycles. It can be optimized for cost or performance whichever is of most *value*. The intention is to be able to measure both as a function of operational performance efficiency. The affordability model must enhance value engineering (VE) analysis through the ability to measure all functional contributions to operational performance. The designer must be provided with the ability to choose the range of functions to adjust for optimal impact and cost.

In all cases, the affordability trade space must be able to accommodate changes but always driven by the same key concepts: what does it cost to implement such a change, and what do we get for it. Consequently, identifying "cost-effective capability" is still the key to analysis within the affordability trade space across the system life cycle.

For instance, consider a system that is unique in the inventory but is suffering an unacceptably low A_o. Then the question arises, can we upgrade this system to improve field performance and do so in a cost-effective way? Is it possible to improve the existing reliability and maintainability characteristics of this system and increase A_o, given that the system itself has many years of service life remaining? So the trade space analysis must show that the upgrades will pay for themselves over time through lower operation and support costs, increased capability, or both.

In general, the factors that must be considered within the trade space across the system life cycle include:

- Cost versus benefits of different design solutions
- Cost versus benefits of different support strategies
- Methods and rationale used to develop these comparisons
- Ability to identify and obtain data required to analyze changes

From a programmatic perspective, analysis of these alternatives must also be done in conjunction with identification, classification, and analysis of associated risk and development of a costed plan for implementation.

10.1.1.5 Affordability Implementation When one considers the entire SOI, both the primary and enabling systems should be treated as an SoS. In the example in Figure 10.3, the mission-effectiveness affordability trade space brings together primary and enabling systems into an SoS. Note that the SoS is treated as a closed-loop system where requirements are modified as the mission needs evolve. These iterations allow for technology insertion as the design is updated. Design-to-cost (DTC) targets are set for the primary and enabling systems, which ensure that affordability throughout the system life cycle is considered, even as the system evolves. Affordability measurements that feed back into the system assessment are KPPs and operational availability, which ensures that the mission can be accomplished over time (e.g., those KPPs are met across the system life cycle; see Fig. 10.4).

如果设计师不能控制环境中的变量，可承受性的优化问题在生命周期的任一点是不同的或者可能是不同的。这种优化问题可通过功能关键性、"浪涌"的和自适应的需求来管理，或者是一系列事先确定的技术更新周期。它可以对成本或性能二者中的价值最大者进行优化。其目的是能够根据运行性能效率来衡量二者。可承受性模型必须通过衡量对运行性能的所有功能贡献的能力来加强价值工程（VE）分析。设计师必须具有选择功能范围，以为最优影响和成本而调整的能力。

在所有情况下，可承受性权衡空间必须能够适应变化，但始终由相同的关键概念所驱动：为了实施这种变化，它付出的代价是什么？以及我们从中得到了什么。因此，识别"成本效益能力"仍然是跨系统生命周期在可承受性权衡空间内分析的关键。

例如，考虑在库存中独一无二但却遭受低到不可接受的 A_o 的系统。然后，问题就出现了：我们可以升级该系统以提升现场性能吗？以及怎样以具有成本效益的方式做这件事？如果系统本身还有很多年的使用寿命，是否有可能改善该系统的现有可靠性和可维护性特征或者提高 A_o 吗？所以，权衡空间分析必须表明，升级将随时间通过降低运行及维持成本和/或提高能力来使其受益。

通常，跨系统生命周期在权衡空间内必须考虑的因素包括：

- 不同设计解决方案的成本与效益的对比
- 不同维持策略的成本与效益的对比
- 用于开发这些比较的方法和基本原理
- 识别并获得分析变化所需数据的能力

从程序化的视角来看，对这些备选方案的分析也必须结合对相关风险的识别、分类和分析以及有成本的实施计划的制定来进行。

10.1.1.5 可承受性的实施 当考虑整个 SOI 时，主要系统和使能系统都应被视为 SoS 来对待。在图 10.3 的示例中，任务效能可承受性权衡空间将主要系统和使能系统纳入到 SoS 之中。注意，SoS 被视为随着任务需要的演进而改进需求的闭环系统。当设计升级时，这些迭代允许技术导入。而为成本的设计（DTC）的目标是针对主要系统和使能系统设置的，以确保考虑到贯穿于系统生命周期的可承受性，正如系统的演进。反馈到系统评估中的可承受性衡量结果是 KPP 和运行有效性，以确保随时间可达成任务目标（如，跨系统生命周期满足那些 KPP；见图 10.4）。

SPECIALTY ENGINEERING ACTIVITIES

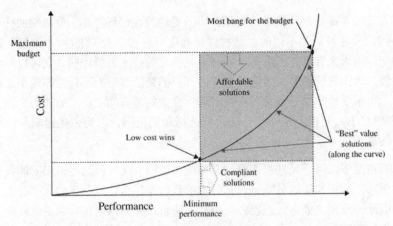

FIGURE 10.3 Cost versus performance. Reprinted with permission from Joseph Bobinis. All other rights reserved.

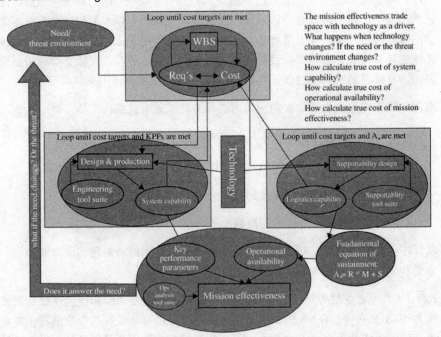

FIGURE 10.4 Affordability cost analysis framework. Derived from Bobinis et al. (2010). Reprinted with permission from Joseph Bobinis. All other rights reserved.

To define affordability for a particular program or system (see Fig. 10.3 as an example of an affordability range), we must define selected affordability components. As such, the following could be specified:

专业工程活动

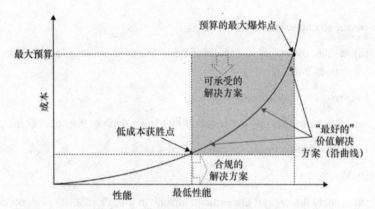

图 10.3 成本与性能的对比。经 Joseph Bobinis 许可后转载。版权所有。

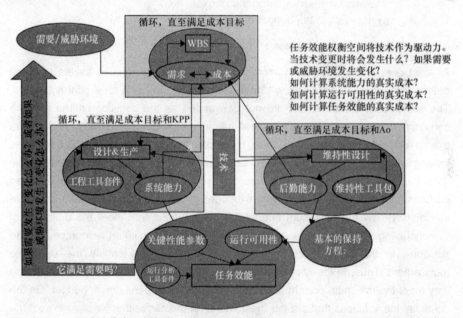

图 10.4 可承受性成本分析框架。来自 Bobinis 等。经 Joseph Bobinis 许可后转载。版权所有。

为定义特殊项目计划或系统的可承受性（可承受性范围的示例，见图 10.3），我们必须定义所选定的可承受性组成部分。正因如此，可以特定以下内容：

699

SPECIALTY ENGINEERING ACTIVITIES

1. Required capabilities

    (a) Identify the required capabilities and the time phasing for inclusion of the capabilities.

2. Required capabilities performance

    (a) Identify and specify the required MOEs for each of the capabilities.

    (b) Define time phasing for achieving the MOEs.

3. Budget

    (a) Identify the budget elements to include in the affordability evaluation.

    (b) Time-phased budget, either

    (i) for each of the budget elements or

    (ii) as the total budget.

At least one of the affordability elements needs to be designated as the decision criteria that will be used in either a trade study or as the basis for a contract award. The affordability elements that are not designated as the decision criteria become constraints, along with the constraints being specified. This is illustrated by an example depicted in Figure 10.3. Here, the capabilities and schedule have been fixed leaving either the cost or the performance to be the evaluation criteria, while the other becomes the constraint. This results in a relatively simple relationship between performance and cost. The maximum budget and the minimum performance are identified. Below the maximum budget line lie solutions that meet the definition of "...conducting a program at a cost constrained by the maximum resources...." The solutions to the right of the minimum performance line satisfy the threshold requirement. Thus, in the shaded rectangle lie the solutions to be considered since they meet the minimum performance and are less than the maximum budget. On the curve lay the solutions that are the "best value," in the sense that for a given cost the corresponding point on the curve is the maximum performance that can be achieved. (*Note: In the real world, the curve is rarely smooth or continuous.*) Similarly, for a given performance, the corresponding point on the curve is the minimum cost for which that performance can be achieved. Selecting the decision criterion as cost will result in achieving the threshold performance. Similarly, if the decision criterion is performance, all of the budget would be expended. Consequently, to specify affordability for a system or program requires determining which affordability element is the basis for the decision criteria and which elements are being specified as constraints.

1. 所需能力

    (a) 识别所需能力以及包含这些能力的时间段。

2. 所需能力性能

    (a) 识别并规定每个能力的所需 MOE

    (b) 定义完成 MOE 的时间段。

3. 预算

    (a) 识别预算元素，以包括在可承受性评价中

    (b) 分时间段的预算，或者

        （i） 针对每一个预算要素，或者

        （ii） 作为总体预算。

至少一个可承受性元素需将被指定为在权衡研究中使用或者用于签署合同的基础的决策准则。与制定的约束一同，未被指定为决策准则的可承受性元素成为约束。这一点通过图 10.3 所描述的示例来阐明。在此，能力和进度已经固定，使成本或性能成为评价准则，则另一个成为约束。在性能与成本之间产生了相对简单的关系。识别最大预算和最小性能。在最大预算直线下方，存在满足"…以最大资源约束的成本开展项目…"所定义的解决方案。最小性能直线右边的解决方案满足临界的要求。因此，在阴影矩形处存在将要考虑的解决方案，因为它们满足最小性能并且小于最大预算。位于曲线上的解决方案具有"最佳价值"，在一定意义上，对于给定成本而言，曲线上的对应点是可以达成的最大性能。（注意：在真实世界中，曲线很少平滑或者连续。）同样，对于给定的性能，曲线上的对应点是可以实现该性能的最低成本。选择成本作为决策准则将导致达成临界的性能。同样，如果决策准则是性能，将会耗费所有预算。因此，规定系统或项目的可承受性要求确定哪种可承受性元素是决策准则的基础，以及哪些元素正被规定为约束。

Affordability is the result of a disciplined decision making process—requiring systematic methodologies that support selection of the most affordable technologies and systems.

10.1.2 Cost-Effectiveness Analysis

As mentioned in the preceding section, a systems engineer can no longer afford the luxury of ignoring cost as an SE area of responsibility or as a major architectural driver. In essence, a systems engineer must be conversant in business and economics as well as engineering.

Cost-effectiveness analysis (CEA) is a form of business analysis that compares the relative costs and performance characteristics of two or more courses of action. At the system level, CEA helps derive critical system performance and design requirements and supports data-based decision making.

CEA begins with clear goals and a set of alternatives for reaching those goals. Comparisons should only be made for alternatives that have similar goals. A straightforward CEA cannot compare options with different goals and objectives.

Experimental or quasiexperimental designs can be used to determine effectiveness and should be of a quality capable of justifying reasonably valid conclusions. If not, there is nothing in the CEA method that will rescue the results. What CEA adds is the ability to consider the results of different alternatives relative to the costs of achieving those results. It does not change the criteria for what is a good effectiveness study. Alternatives being assessed should address a common specific goal where attainment of that goal can be measured such as miles per gallon, kill radius, or people served.

CEA is distinct from cost–benefit analysis (CBA), which assigns a monetary value to the measure of effect. The approach to measuring costs is similar for both techniques, but in contrast to CEA where the results are measured in performance terms, CBA uses monetary measures of outcomes. This approach has the advantage of being able to compare the costs and benefits in monetary values for each alternative to see if the benefits exceed the costs. It also enables a comparison among projects with very different goals as long as both costs and benefits can be placed in monetary terms.

Other closely related, but slightly different, formal techniques include cost–utility analysis, economic impact analysis, fiscal impact analysis, and social return on investment (SROI) analysis.

可承受性是科律化决策流程的结果——要求支持选择大多数的可承受技术和系统的系统性方法论。

10.1.2 成本效能分析

正如前面一节提到的，系统工程师可能不再承受得起忽视作为 SE 职责领域或作为主要架构驱动力的成本带来的"奢侈"。本质上，系统工程师必须像精通工程一样精通商业和经济学。

成本效能分析（CEA）是一种比较两个或多个行动路线的相对成本和性能特征的业务分析的一种形式。在系统层级，CEA 帮助推导关键系统性能和设计需求并支持基于数据的决策。

CEA 始于清晰的目标以及达到这些目标的备选方案集。应针对具有相似目标的备选方案进行比较。直接的 CEA 不能比较具有不同目标和目的的选项。

实验或准实验设计可以用来确定效能并且应具备证明合理有效的结论的高质量的能力。否则，在 CEA 方法中没有什么会获得这些结果。相对于实现那些结果的成本，CEA 增加的是考虑不同备选方案结果的能力。它不会改变良好效能研究的准则。正在评估的备选方案应该应对一个共同的特定目标，其中该目标的实现可以被衡量，诸如每加仑的英里数、杀伤半径或服务的人群。

CEA 与成本效益分析（CBA）截然不同，其向效应的测度赋予一种货币值。这两种技术的成本衡量方法相似，但与依据性能衡量结果的 CEA 完全不同，CBA 使用结果的货币测度。这种方法的优点是，能够以货币值比较每个备选方案的成本和效益，以审视效益是否超出成本。只要能够根据货币估定成本和效益，它还使能在极其不同目标的项目之间进行比较。

其他密切相关但稍有不同的正规技术包括成本—效用分析、经济影响分析、财政影响分析和社会投资回报（SROI）分析。

SPECIALTY ENGINEERING ACTIVITIES

In both CEA and CBA, the cost of risk and the risk of cost need to be included into the study. Risk is usually handled using probability theory. This can be factored into the discount rate (to have uncertainty increasing over time), but is usually considered separately. Particular consideration is often given to risk aversion—the irrational preference to avoid loss over achieving gain. Uncertainty in parameters (as opposed to risk of project failure) can be evaluated using sensitivity analysis, which shows how results and cost respond to parameter changes. Alternatively, a more formal risk analysis can be undertaken using Monte Carlo simulations. Subject matter experts in risk and cost should be consulted.

The concept of cost-effectiveness is applied to the planning and management of many types of organized activity. It is widely used in many aspects of life. Some examples are:

1. Studies of the desirable performance characteristics of commercial aircraft to increase an airline's market share at lowest overall cost over its route structure (e.g., more passengers, better fuel consumption)

2. Urban studies of the most cost-effective improvements to a city's transportation infrastructure (e.g., buses, trains, motorways, and mass transit routes and departure schedules)

3. In health services, where it may be inappropriate to monetize health effect (e.g., years of life, premature births averted, sight years gained)

4. In the acquisition of military hardware when competing designs are compared not only for purchase price but also for such factors as their operating radius, top speed, rate of fire, armor protection, and caliber and armor penetration of their guns

10.1.3 LCC Analysis

LCC refers to the total cost incurred by a system, or product, throughout its life. This "total" cost varies by circumstances, the stakeholders' points of view, and the product. For example, when you purchase an automobile, the major cost factors are the cost of acquisition, operation, maintenance, and disposal (or trade-in value). A more expensive car (acquisition cost) may have lower LCC because of lower operation and maintenance costs and greater trade-in value. However, if you are the manufacturer, other costs like development and production costs, including setting up the production line, need to be considered. The systems engineer needs to look at costs from several aspects and be aware of the stakeholders' perspectives. In some literature, LCC is equated to total cost of ownership (TCO) or total ownership cost (TOC), but many times, these measures only include costs once the systems is purchased or acquired.

在 CEA 和 CBA 中，风险的成本和成本的风险都需要包括在研究中。风险通常利用概率理论来应对，可以把折减比率考虑进去（不确定性随着时间而增加），但通常分开来考虑。通常会特别考虑风险规避——避免随着收益的实现而出现损失的非理性偏好。参数中的不确定性（与项目失败的风险截然相反）可以利用灵敏度分析来评价，以表明结果和成本如何响应参数的变更。或者，可以利用蒙特卡罗仿真进行更正规的风险分析。应咨询风险和成本方面的领域专家。

成本—效能的概念应用于许多种组织活动的规划和管理。在生活中的很多方面广泛使用。一些示例包括：

1. 用于按航线结构（如更多的乘客、更少的燃油消耗）的最低总体成本，增加航空公司的市场份额的商用飞机预期性能特征的研究

2. 最具成本效益的城市运输基础设施（如公共汽车、火车、高速公路以及公共交通路线和出发调度）改善的城市研究

3. 在健康服务中，把健康效应（如，寿命年限、避免早产、增加的目标年限）换算成金钱可能不是适当的

4. 在军用硬件的采办中，比较相对竞争的设计时不仅针对采购价而且还针对诸如运行半径、最大速度、发射速率、装甲防护以及喷枪的口径和护甲穿透等因素。

10.1.3 LCC 分析

LCC 指的是系统或产品贯穿于其生命周期所带来的总成本。这一"总"成本因环境、利益攸关者观点和产品而异。例如，当你购买一辆汽车时，主要的成本因素是购买、使用、维护和报废（或者以旧换新价值）的成本。一辆更为昂贵的汽车（购买成本）可能因为使用和维护成本更低及抵换价值更高而具有更低的 LCC。但是，如果你是制造商，需要考虑诸如开发及生产成本等其他成本，包括建立生产线的成本。系统工程师需要从若干个方面考虑成本，并需要知晓利益攸关者的视角。在一些文献中，LCC 等同于所有权总成本（TCO）或总拥有成本（TOC），但许多时候，一旦采购或采办了系统，则这些测度仅包括成本。

SPECIALTY ENGINEERING ACTIVITIES

Sometimes, it is argued that the LCC estimates are only to support internal program trade-off decisions and, therefore, must only be accurate enough to support the trade-offs (relative accuracy) and not necessarily realistic. By itself, this is usually a bad practice and, if done, is a risk element that should be tracked for some resolution of veracity. The analyst should always attempt to prepare as accurate cost estimates as possible and assign risk as required. These estimates are often reviewed by upper management and potential stakeholders. The credibility of results is significantly enhanced if reviewers sense the costs are "about right," based on their past experience. Future costs, while unknown, can be predicted based on assumptions and risk assigned. All assumptions when doing LCC analysis should be documented.

LCC analysis can be used in affordability and system cost-effectiveness assessments. The LCC is *not* the definitive cost proposal for a program since LCC "estimates" (based on future assumptions) are often prepared early in a program's life cycle when there is insufficient detailed design information. Later, LCC estimates should be updated with actual costs from early program stages and will be more definitive and accurate due to hands-on experience with the system. A major purpose of LCC studies is to help identify cost drivers and areas in which emphasis can be placed during the subsequent sub-stages to obtain the maximum cost reduction. Accuracy in the estimates will improve as the system evolves and the data used in the calculation is less uncertain.

LCC analysis helps the project team understand the total cost impact of a decision, compare between program alternatives, and support trade studies for decisions made *throughout* the system life cycle. LCC normally includes the following costs, represented in Figure 10.5:

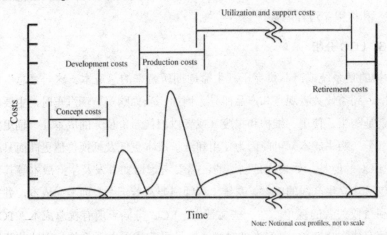

FIGURE 10.5 Life cycle cost elements (not to scale). Derived from INCOSE SEH v1 Figure 9.3. Usage per the INCOSE Notices page. All other rights reserved.

有时，人们争论，LCC 估算仅支持内部项目权衡决策，因此必须仅是足够精确以支持权衡（相对精确），而不必是现实的。通常情况下，这本身是一个不好的惯例，如果这样做，LCC 估算是应追踪一些真实解决方案的风险元素。分析师应不断地尽可能地尝试准备精确的成本估算，并根据要求分配风险。这些估算往往由高层管理人员和潜在的利益攸关者来评审。如果评审人员基于他们以往的经验认为这些成本"大致合适"，则结果的可信度显著提高。未来成本（在未知时）可基于假设以及所分配的风险来预计。进行 LCC 分析时的所有假设都应被文件化。

LCC 分析可用于可承受性和系统成本效能评估中。LCC 不是项目的明确成本建议，因为 LCC "估算"（基于未来的假设）往往在项目的生命周期早期当具有不充分的详细设计信息时准备。后期，LCC 估算应利用项目阶段早期实际成本而更新，并且将由于在系统方面的亲身实践经验更为明确和准确。LCC 研究的主要目的是，帮助识别在后续子阶段中为获得最大限度的成本降低可以重点强调的成本驱动因素和领域。随着系统的不断演进，估算的准确性将随之提高，并且用于计算的数据的不确定性减少。

LCC 分析帮助项目团队理解总体成本对于决策的影响，在项目备选方案之间进行比较，贯穿于系统生命周期支持决策的权衡研究。LCC 通常包括下列成本，如图 10.5 所示：

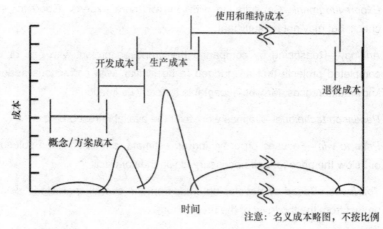

图 10.5 生命周期成本元素（不按比例）。来自于 INCOSE SEH 第 1 版中的图 9.3。按照 INCOSE 通知页使用。版权所有。

- *Concept costs*—Costs for the initial concept development efforts. Can usually be estimated based on average manpower and schedule spans and include overhead, general and administrative (G&A) costs, and fees, as necessary.
- *Development costs*—Costs for the system development efforts. Similar to concept costs, can usually be estimated based on average manpower and schedule spans and include overhead, G&A costs, and fees, as necessary.
- *Production costs*—Usually driven by tooling and material costs for large-volume systems. Labor cost estimates are prepared by estimating the cost of the first production unit and then applying learning curve formula to determine the reduced costs of subsequent production units. For an item produced with a 90% learning curve, each time the production lot size doubles (2, 4, 8, 16, 32, ... etc.) the average cost of units in the lot is 90% of the average costs of units in the previous lot. A production cost specialist is usually required to estimate the appropriate learning curve factor(s).
- *Utilization and support costs*—Typically based on future assumptions for ongoing operation and maintenance of the system, for example, fuel costs, manning levels, and spare parts.
- *Retirement costs*—The costs for removing the system from operation and includes an estimate of trade-in or salvage costs. Could be positive or negative and should be mindful of environmental impact to dispose.

Common methods/techniques for conducting LCC analysis are as follows:

- *Expert judgment*—Consultation with one or more experts. Good for sanity check, but may not be sufficient.
- *Analogy*—Reasoning by comparing the proposed project with one or more completed projects that are judged to be similar, with corrections added for known differences. May be acceptable for early estimations.
- *Parkinson technique*—Defines work to fit the available resources.
- *Price to win*—Focuses on providing an estimate, and associated solution, at or below the price judged necessary to win the contract.
- *Top focus*—Based on developing costs from the overall characteristics of the project from the top level of the architecture.
- *Bottoms up*—Identifies and estimates costs for each element separately and sums the contributions.

- *概念/方案成本*——初步概念/方案开发工作的成本。通常能够基于平均人力和进度跨度进行估算,必要时包括间接的、通用的和行政管理(G&A)的成本和费用。

- *开发成本*——系统开发工作的成本。与概念/方案成本相似,通常能够基于平均人力和进度跨度进行估算,必要时包括间接的、G&A 成本和费用。

- *生产成本*——对大生产量系统而言,通常由工具和材料成本驱动。人工成本估算通过先估算出首个生产单元的成本,之后应用学习曲线公式确定后续生产单元的降低成本的方式准备。对于所生产的学习曲线为 90% 的项,每当生产批量加倍(2、4、8、16、32、……等)时,批次中单元的平均成本是之前批次中单元平均成本的 90%。通常需要生产成本专家来估算适当的学习曲线系数。

- *使用及维持成本*——通常基于对系统的正进行中的运行和维护的未来假设,例如,燃料成本、人员配备水平和备件。

- *退役成本*——从运行中除去系统的成本,包括折价或利旧费的估算值。可能是正值或负值,并且应注意对处置的环境影响。

开展 LCC 分析的公共方法/技术如下:

- *专家判断*——向一个或多个专家咨询。适用于完整性检查,但是不一定是充分的。

- *类比*——通过将所提出项目与一个或多个已完成项目进行比较来进行推理,这些已完成项目与所提出项目相似,且针对已知的差异增加修正。对于早期估算可能是可接受的。

- *帕金森技术*——定义与可用资源适配的工作。

- *价格取胜*——聚焦于以等于或低于认为赢得合同所需价格来提供估算值和相关的解决方案。

- *顶层聚焦*——基于从架构的顶层来开发项目全部特征的成本。

- *自底向上*——分开识别和估算每个元素的成本并计算总贡献值。

SPECIALTY ENGINEERING ACTIVITIES

- *Algorithmic (parametric)*—Uses mathematical algorithms to produce cost estimates as a function of cost-driver variables, based on historical data. This technique is supported by commercial tools and models.

- *DTC*—Works on a design solution that meet a predetermined production cost.

- *Delphi techniques*—Builds estimates from multiple technical and domain experts. Estimates are only as good as the experts.

- *Taxonomy method*—Hierarchical structure or classification scheme for the architecture.

10.2 ELECTROMAGNETIC COMPATIBILITY

EMC is the engineering discipline concerned with the behavior of a system in an electromagnetic (EM) environment. A system is considered to be electromagnetically compatible when it can operate without malfunction in an EM environment together with other systems or system elements and when it does not add to that environment as to cause malfunction to other systems or system elements. When a system causes interference, the term electromagnetic interference (EMI) is often used. In EMC, the EM environment not only includes all phenomena and effects that are classically attributed to electromagnetics (such as radiation) but also electrical effects (conduction).

Successfully achieving EMC during system development requires a typical SE process, as shown in Figure 10.6.

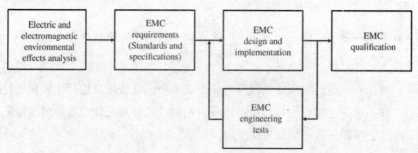

FIGURE 10.6 Process for achieving EMC. Reprinted with permission from Arnold de beer. All other rights reserved.

10.2.1.1 Electric and Electromagnetic Environmental Effects Analysis An electric and electromagnetic environmental effects (E^4) analysis describes all the threats (natural and man-made) that a system may encounter during its life cycle. MIL-STD-464C (DoD, 2010) can be used to guide this analysis, which should contain all the information needed to determine EMC requirements of the system.

- *算法（参数）*——基于历史数据，使用数学算法以成本驱动因素变量的函数生成成本估算。该技术得到商业工具和模型支持。

- *DTC*——致力于满足预先确定的生产成本的设计解决方案。

- *德尔菲技术*——由多个技术和领域专家构建估算。好的专家才能让估算更准确。

- *分类学方法*——架构的层级结构或分类方案。

10.2 电磁兼容性

EMC 是关注一个系统在电磁（EM）环境中的行为的工程学科。当系统可以在 EM 环境中连同其他系统或系统元素一同无故障地运行时，以及当系统造成其他系统或系统元素故障而并未附加额外影响环境时，系统被视为在电磁上是可兼容的。当系统造成干扰时，往往使用术语"电磁干扰（EMI）"。在 EMC 中，EM 环境不仅包括通常归因于电磁（诸如辐射）的所有现象和效应，而且还包括电效应（导电）。

在系统开发期间成功地实现 EMC 所要求的典型的 SE 流程，如图 10.6 所示。

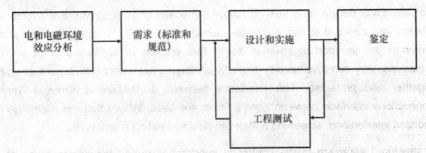

图 10.6 实现 EMC 的流程。经 Arnold de beer 许可后转载。版权所有。

10.2.1.1 电气和电磁环境效应分析 电气和电磁环境效应（E^4）的分析描述系统可能会在其生命周期期间遭遇的全部威胁（自然的和人造的）。MIL-STD-464C（DoD，2010）可用来指导这项分析，其应包含确定系统的 EMC 需求所需的全部信息。

SPECIALTY ENGINEERING ACTIVITIES

10.2.1.2 EMC Requirements (Standards and Specifications) EMC standards and specifications are used to regulate the EM environment in which a system is operating. It usually governs both the system's ability to function within its intended EM environment (sensitivity or susceptibility) and its contribution to that environment (emissions).

Standards and specifications are available for con-ducted emissions, conducted susceptibility, radiated emissions, and radiated susceptibility.

It is rare for a system to have custom-developed EMC requirements that do not follow existing standards and specifications. However, existing standards and specifications (whether commercial, military, avionic, automotive, or medical) classify a requirement into a class or category depending on the severity of the potential malfunction. It is a SE function to determine the correct EMC requirements with class or category according to the outcome of the E^4 analysis.

10.2.1.3 EMC Design and Implementation The EMC requirements are inputs to the concept and development stages. It is important that the EMC requirements are fixed at the beginning of physical design as the EMC design includes both mechanical and electrical/electronic hardware implementations that are not part of the other functional requirements.

For any EMC design, it is best to follow a process of zoning, where a system is divided into zones that can easily be managed for EMC. Typically, major system elements are grouped together in zones that are low in emissions or of similar emissions. Very sensitive circuits (typically analog/measurement circuits) are grouped together and protected. The interfaces between zones are controlled. Where connections interface between zones, filters are used. Where there is a change in radiated interference, screening and/or physical separation is employed.

A structured approach to the control of interference during the design stage of a system is to develop an EMI control plan. The EMI control plan typically includes all EMC requirements, zoning strategy, filtering and shielding, cabling, and detail mechanical and electrical design pertaining to EMC.

10.2.1.4 EMC Engineering Tests Prequalification tests may be required during the development stage. This is typically done on a system element level and even as low as single printed circuit board assembly level. Since EMC results are difficult to predict, the best way of ensuring design success is to test at lower system levels for compliance in order to maximize the probability of system compliance.

10.2.1.2 EMC 需求（标准和规范） EMC 标准和规范被用来监管系统运行的 EM 环境。它通常管控在其预期 EM 环境中系统功能的能力（灵敏性或易感性）以及其对该环境的贡献（发射）。

标准和规范可用于传导的发射、传导的易感性、辐射的发射和辐射的易感性。

对于系统而言，具有定制开发的不符合现有标准和规范的 EMC 需求是罕见的。但是，现有的标准和规范（不管是商用的、军用的、航空电子的、汽车的还是医疗的）依据潜在故障的严重程度将需求归为某一类或某一范畴。根据 E^4 分析的结果，采用类或范畴确定正确的 EMC 需求是一项 SE 功能。

10.2.1.3 EMC 设计与实施 EMC 需求是概念/方案和开发阶段的输入。重要的是，在物理设计开始时确定 EMC 需求，因为 EMC 设计包括机械的和电气/电子的硬件实施，并非其他功能需求一部分。

对于任何 EMC 设计，最好遵循一种分区流程，借助该流程，系统被分为可以很容易地管理 EMC 的区域。典型情况下，主要系统元素共同集合在发射率低或者有类似发射率的区域。非常敏感的电路（通常是模拟/测量电路）集合在一起进行保护。两个区域之间的接口是受控的。在两个区域之间连接的接口处，使用滤波器。在辐射干扰中有变化的地方，则采用筛查和/或物理隔离。

在系统的设计阶段控制干扰的一种结构化方法是开发 EMI 控制计划。EMI 控制计划通常包括所有 EMC 需求、分区策略、滤波和屏蔽、线缆布局以及关于 EMC 的详细机械和电气设计。

10.2.1.4 EMC 工程测试 在开发阶段可能需要预先鉴定测试。这通常在系统元素层级上进行，并甚至低到与单个印制电路板组件层级。因为 EMC 结果难以预计，所以确保设计成功的最佳方式是以较低的系统层级对合规性进行测试，以便使系统合规的概率的最大化。

SPECIALTY ENGINEERING ACTIVITIES

10.2.1.5 EMC Qualification EMC qualification tests are performed to verify the EMC design of a system against its requirements. The first part of this activity is to compile an EMC test plan, which maps each requirement to a test and test setup.

The EMC test setup is an integral part of EMC SE as test results can vary according to the setup. This setup can be challenging as the system or system element under test must be in operational mode during emission testing and it must be possible to detect malfunctions during susceptibility testing. Interfacing with the system or system element must be done while not compromising the EM zone in which it is tested. This may require special system-related EMC test equipment. When it is impractical to test a large system (such as a ship, aircraft, or complete industrial plant), the qualification tests of the system elements are used to qualify the larger system.

10.3 ENVIRONMENTAL ENGINEERING/IMPACT ANALYSIS

The European Union, the United States, and many other governments recognize and enforce regulations that control and restrict the environmental impact that a system may inflict on the biosphere. Such impacts include emissions to air, water, and land and have been attributed to cause problems such as eutrophication, acidification, soil erosion and nutrient depletion, loss of biodiversity, and damage to ecosystems (UNEP, 2012). The focus of environmental impact analysis is on potential harmful effects of a proposed system's development, production, utilization, support, and retirement stages. All governments that have legally expressed their concern for the environment restrict the use of hazardous materials (e.g., mercury, lead, cadmium, chromium 6, and radioactive materials) with a potential to cause human disease or to threaten endangered species through loss of habitat or impaired reproduction. Concern extends over the full life cycle of the system, from the materials used and scrap waste from the production process, operations of the system replacement parts, and consumables and their containers to final disposal of the system. These concerns are made evident by the European Union's 2006 resolution to adopt a legal restriction that system developers and their suppliers retain lifetime liability for decommissioning systems that they build and sell.

The ISO 14000 series of environmental management standards (ISO, 2004) are an excellent resource for organizations of methods to analyze and assess their operations and their impacts on the environment. Failure to comply with environmental protection laws carries penalties and should be addressed in the earliest phases of requirements analysis (Keoleian and Menerey, 1993). The Øresund Bridge (see Section 3.6.2) is an example of how early analysis of potential environmental impacts ensures that measures are taken in the design and construction to protect the environment with positive results. Two key elements of the success of this initiative were the continual monitoring of the environmental status and the integration of environmental concerns into the requirements from the owner.

10.2.1.5 EMC 鉴定 实施 EMC 鉴定测试，以针对系统的需求验证系统的 EMC 设计。该活动的第一部分是编制 EMC 测试计划，以将每个需求映射至测试和测试设置。

EMC 测试设置是 EMC SE 的主要部分，因为测试结果可以根据设置而变化。这种设置可能富有挑战性，因为正在测试中的系统或系统元素必须在进行发射测试期间处于运行模式，并且必须能在进行敏感性测试期间检测故障。与系统或系统元素的接口必须测试而不对进行测试的 EM 区域造成损害。这可能需要特殊的系统相关的 EMC 测试设备。当测试大型系统（诸如船舶、飞机或完整的工业厂房）不可实施时，系统元素的鉴定测试被用来对更大型系统进行鉴定。

10.3 环境工程/影响分析

欧盟、美国和许多其他政府认可并强行监管及限制系统可能对生物界产生的环境影响的规定。这些影响包括向空气、水和土地中放射，并且已归咎于造成诸如富营养化、酸化、土壤侵蚀和养分耗竭、生物多样性丧失及生态系统破坏等问题（UNEP，2012）。环境影响分析的聚焦点在于所提出系统的开发、生产、使用、维持和退役阶段的潜在不良影响。已经在法律上表达出对环境关注的所有政府都限制使用可能会导致人类疾病或以减少栖息地或以损害繁殖的方式威胁濒危物种的危险物质（例如水银、铅、镉、六价铬以及放射性物质）。这种关注扩展到系统的全生命周期，从所使用的材料和生产流程中的废品废料、系统更换件的运行、消耗品及其容器到系统的最终处置。欧盟 2006 年决议证实了这些关注点，以采用系统开发者及其供应商对他们所建造和出售的退役系统负有终身责任的法律限制。

ISO 14000 系列环境管理标准（ISO，2004）是组织分析并评估它们的运行及对环境的影响的方法的卓越资源。不遵守环境保护法会遭受处罚并且应在需求分析的最早阶段应对（Keoleian 和 Menerey，1993）。厄勒海峡大桥（见 3.6.2 节）是潜在环境影响的早期分析如何确保在设计和建造过程中采取措施以保护环境且产生积极效果的一个示例。该项举措取得成功的两个关键元素是环境状态的连续监控以及将环境关注点综合到所有者需求中。

Disposal analysis is a significant analysis area within environmental impact analysis. Traditional landfills for nonhazardous solid wastes have become less available within large city areas, and disposal often involves trans-porting the refuse to distant landfills at considerable expense. The use of incineration for disposal is often vigorously opposed by local communities and citizen committees and poses the problem of ash disposal since the ash from incinerators is sometimes classified as hazardous waste. Local communities and governments around the world have been formulating significant new policies to deal with the disposal of nonhazardous and hazardous wastes.

One goal of the architecture design is to maximize the economic value of the residue system elements and minimize the generation of waste materials destined for disposal. Because of the potential liability that accompanies the disposal of hazardous and radioactive materials, the use of these materials is carefully reviewed and alternatives used wherever and whenever possible. The basic tenet for dealing with hazardous waste is the "womb-to-tomb" control and responsibility for preventing unauthorized release of the material to the environment. This may include designing for reuse, recycling, or transformation (e.g., composing, biodegradation).

In accordance with the US and European Union laws, system developers and supporting manufacturers must analyze the potential impacts of the systems that they construct and must submit the results of that analysis to government authorities for review and approval to build the system. Failure to conduct and submit the environmental impact analysis can result in severe penalties for the system developer and may result in an inability to build or deploy the system. It is best when performing environmental impact analysis to employ subject matter experts who are experienced in conducting such assessments and submitting them for government review. Methods associated with life cycle assessment (LCA) and life cycle management (LCM) are increasingly sophisticated and supported by software (Magerholm et al., 2010). Government acquisitions are subject to legislation for Green public procurement (GPP) (Martin, 2010). Consumers of commercial products are offered assistance in their purchasing decisions through Environmental product Declarations and labeling, such as the Nordic Swan, and Blue Angel (Salzman, 1997).Another effort in the ISO community is the development of a standard for product carbon footprints as an indicator of the global environmental impact of a product expressed in carbon emission equivalents (Draucker et al., 2011).

处置分析是环境影响分析中的一个重要分析领域。存放无害固体废物的传统垃圾场在大城市地区已经变得较为少见,处置往往涉及以相当大的支出将废物运输至偏远垃圾场。采用焚化的方式进行处置往往会受到当地社区与公民委员会的强烈反对,并且会带来灰烬处置的问题,因为焚化炉中的灰烬有时被归类为有害废物。全世界范围内的当地社区和政府一直以来在制定有效的新策略,以应对无害废物和有害废物的处置。

架构设计的一个目标是最大化剩余系统元素的经济价值并即将处置的废料的生成最小化。由于伴随危险物质和放射性物质处置的潜在责任,这些物质的使用应经过仔细审查,并且无论何时何地,只要可能就使用备选方案。处理有害废物的基本宗旨是"从子宫到坟墓"的控制并负责防止在未经授权的情况下将该物质排入外界。这可能包括为复用、再循环或转换(例如,分解、生物降解)而设计。

根据美国和欧盟的法律,系统开发者以及支持制造商必须分析它们构建的系统的潜在影响,并且必须向政府当局提交该分析的结果,以为该系统的构建进行审查和审批。未能实施和提交环境影响分析可导致系统开发者受到严惩,并且可导致不能构建或部署这一系统。最好在执行环境影响分析时聘用领域专家,这些专家在进行这类评估并提交这类评估以备政府审查方面具有经验。与生命周期评估(LCA)和生命周期管理(LCM)相关的方法日益复杂并且由软件支持(Magerholm 等,2010)。政府采办须遵守绿色公共采购(GPP)法规(Martin,2010)。为商用产品的消费者提供通过环境产品声明和标签(诸如 Nordic Swan 和 Blue Angel)做出采购决策的帮助(Salzman,1997)。ISO 社区的另一项工作是,开发产品碳足迹的标准,作为以碳排放等效物表达的产品的全局环境影响的指标(Draucker 等,2011)。

10.4 INTEROPERABILITY ANALYSIS

Interoperability depends on the compatibility of elements of a large and complex system (which may be an SoS or a family of systems (FoS)) to work as a single entity. This feature is increasingly important as the size and complexity of systems continue to grow. Pushed by an inexorable trend toward electronic digital systems and pulled by the accelerating pace of digital technology invention, commercial firms and national organizations span the world in increasing numbers. As their spans increase, these commercial and national organizations want to ensure that their sunken investment in legacy elements of the envisioned new system is protected and that new elements added over time will work seamlessly with the legacy elements to form a unified system.

Standards have also grown in number and complexity over time, yet compliance with standards remains one of the keys to interoperability. The standards that correspond to the layers of the *ISO-OSI Reference Model* for peer-to-peer communication systems once fit on a single wall chart of modest size. Today, it is no longer feasible to identify the number of standards that apply to the global communications network on a wall chart of any size. Interoperability will increase in importance as the world grows smaller due to expanding communications networks and as nations continue to perceive the need to communicate seamlessly across international coalitions of commercial organizations or national defense forces.

The Øresund Bridge (see Section 3.6.2) demonstrates the interoperability challenges faced when just two nations collaborate on a project, for example, the meshing of regulations on health and safety and the resolution of two power supply systems for the railway.

10.5 LOGISTICS ENGINEERING

Logistics engineering (Blanchard and Fabrycky, 2011), which may also be referred to as product support engineering, is the engineering discipline concerned with the identification, acquisition, procurement, and provisioning of all support resources required to sustain operation and maintenance of a system. Logistics should be addressed from a life cycle perspective and be considered in all stages of a program and especially as an inherent part of system concept definition and development. The emphasis on addressing logistics in these stages is based on the fact that (through past experience) a significant portion of a system's LCC can be attributed directly to the operation and support of the system in the field and that much of this cost is based on design and management decisions made during early stages of system development. Furthermore, logistics should be approached from a system perspective to include all activities associated with design for supportability, the acquisition and procurement of the elements of support, the supply and distribution of required support material, and the maintenance and support of systems throughout their planned period of utilization.

10.4 互操作性分析

互操作性依赖于作为单一实体而运行的大型复杂系统[可能是系统之系统（SoS）或系统族（FoS）]中元素的兼容性。这一特征随着系统的规模和复杂性的不断增加而日益重要。在向电子数字系统发展的必然趋势的推动下，以及加速数字技术创新步伐的拉动下，向世界范围扩张的商业公司和国家组织越来越多。随着扩展不断增加，这些商业和国家组织希望确保它们在预想的新系统的遗产元素中吸收的投资受到保护，并且确保这些随时间演进导入的新元素会与遗产元素一起无缝工作，以形成一个统一的系统。

随着时间的演进，标准的数量和复杂性亦在增加，但是符合标准仍然是互操作性其中一个关键点。与对等通信系统的 ISO-OSI 参考模型的分层相对应的标准曾经可体现在一张大小适中的挂图上。如今，在一张任意尺寸的挂图上识别适用于全球通信网络的标准数量已不再可行。互操作性的重要性日益增加，这是因为世界由于通信网络的扩展而逐渐变小，同时各国也不断认识到需要跨商业组织或国防军队的国际联盟进行无缝通信。

厄勒海峡大桥（见 3.6.2 节）证明了只有两个国家在一个项目上展开合作时所面临的互操作性的挑战，例如健康和安全规定的合并以及铁路的两个供电系统的解决方案。

10.5 后勤工程

后勤工程（Blanchard 和 Fabrycky，2011）也可以称为产品支持工程，是关注维持系统运行与维护所需要的全部支持资源的识别、采办、采购和提供的工程学科。后勤应从一个生命周期的视角去应对，并且应在项目的全部阶段予以考虑，尤其作为系统概念定义和开发的固有部分。在这些阶段中应对后勤的重点基于（通过以往经验）系统 LCC 的重要部分可直接归因于系统在该领域的运行和支持，以及大量成本以系统开发早期做出的设计和管理决策为基础的事实。此外，后勤应从系统的视角接近，以包括与维持性的设计、维持元素的采办和采购、所需维持材料的供应和分发以及贯穿于计划使用阶段对系统的维护和维持相关的全部活动。

SPECIALTY ENGINEERING ACTIVITIES

The scope of logistics engineering is thus (i) to determine logistics support requirements, (ii) to design the system for supportability, (iii) to acquire or procure the support, and (iv) to provide cost-effective logistics support for a system during the utilization and support stages (i.e., operations and maintenance).Logistics engineering has evolved into a number of related elements such as supply chain management (SCM) in the commercial sector and integrated logistics support (ILS) in the defense sector. Further logistics engineering developments include acquisition logistics and performance based logistics. Logistics engineering is also closely related to reliability, availability, and maintainability (RAM) (refer to Section 10.8), since these attributes play an important role in the supportability of a system.

10.5.1 Support Elements

Support of a system during the utilization and support stages requires personnel, spares and repair parts, transportation, test and support equipment, facilities, data and documentation, computer resources, etc. Support planning starts with the definition of the support and maintenance concept (in the concept stage) and continues through supportability analysis (in the development stage) to the ultimate development of a maintenance plan. Planning, organization, and management activities are necessary to ensure that the logistics requirements for any given program are properly coordinated and implemented and that the following elements of support are fully integrated with the system:

- *Product support integration and management*—plan and manage cost and performance across the product support value chain, from the concept to retirement stages.

- *Design interface*—participate in the SE process to impact the design from inception throughout the life cycle. Facilitate supportability to maximize availability, effectiveness, and capability at the lowest LCC. Design interface evaluates all facets of the product from design to fielding, including the product's operational concept for support impacts and the adequacy of the support infrastructure. Prior to the establishment of logistics requirements, logistics personnel accomplish planning, trade-offs, and analyses to provide a basis for establishing support requirements and subsequent resources. These include support system effectiveness inputs to the system specifications and goals and integration of reliability and maintainability program requirements. Consideration of support alternatives and design into conceptual programs must be initiated at this early stage for effective problem identification and resolution. Logistics personnel conduct analyses to assist in identifying potential postproduction support problems and to contribute to possible LCC and support solutions.

因此，后勤工程的范围是：(i) 确定后勤支持需求，(ii) 设计系统的维持性，(iii) 采办或购买支持，以及 (iv) 在使用和维持阶段中（即，运行和维护）为系统提供成本效益后勤支持。后勤工程已经演进成许多相关的元素，诸如商业部门中的供应链管理（SCM）和防务部门中的综合后勤支持（ILS）。此外，后勤工程开发包括采办后勤和基于性能的后勤。后勤工程还与可靠性、可用性和可维护性（RAM）（见 10.8 节）密切相关，因为这些属性在系统的维持性中发挥着重要作用。

10.5.1 支持元素

在使用和支持阶段中支持系统需要人员、备件和维修配件、运输工具、测试及支持设备、设施、数据及文件、计算机资源等等。支持规划从定义支持和维护概念（在概念阶段）开始，然后经过维持性分析（在开发阶段）持续到维修计划的最终形成。为确保给定项目的后勤需求被适当地协调和实施，并确保下述支持元素与系统完全综合，规划、组织和管理活动是必不可少的：

- *产品支持综合与管理*——跨产品支持价值链（从概念阶段到退役阶段）计划并管理成本及性能。

- *设计接口*——参与 SE 流程，以从开始贯穿于生命周期来影响设计。促进维持性，从而以最低的 LCC 最大化可用性、效能和能力。设计接口评价产品的各个方面，从设计到现场，包括产品的支持影响的运行概念以及支持基础设施的适当性。在确立后勤需求之前，后勤人员完成规划、权衡和分析，以提供支持需求和后续资源的建立基础。这些包括对系统规范和目标的支持系统效能输入以及对可靠性和可维护性项目需求的综合。将支持备选方案和设计考虑到概念项目中都必须在早期启动，以便有效的识别和解决问题。后勤人员进行分析，以帮助识别潜在的生产后支持问题并有助于可能的 LCC 和支持解决方案。

SPECIALTY ENGINEERING ACTIVITIES

- *Sustaining engineering*—This effort spans those technical tasks (engineering and logistics investigations and analyses) to ensure continued operation and maintenance of a system with managed (i.e., known) risk. Technical surveillance of critical safety items, approved sources for these items, and the oversight of the design configuration baselines (basic design engineering responsibility for the overall configuration including design packages, maintenance procedures, and usage profiles) for the fielded system to ensure continued certification compliance are also part of the sustaining engineering effort. Periodic technical review of the in-service system performance against baseline requirements, analysis of trends, and development of management options and resource requirements for resolution of operational issues should be part of the sustaining effort.

- *Maintenance planning*—Identify, plan, fund, and implement maintenance concepts and requirements to ensure the best possible system capability is available for operations, when needed, at the lowest possible LCC. The support concept describes the support environment in which the system will operate. It also includes information related to the system maintenance concept: the support infrastructure, expected durations of support, reliability and maintainability rates, and support locations. The support concept is the foundation that drives the maintenance planning process. Establish general overall repair policies such as "repair or replace" criteria.

- *Operation and maintenance personnel*—Identify, plan, fund, and acquire personnel, with the training, experience, and skills required to operate, maintain, and support the system.

- *Training and training support*—plan, fund, and implement a strategy to train operators and maintainers across the system life cycle. As part of the strategy, plan, fund, and implement actions to identify, develop, and acquire Training Aids, Devices, Simulators, and Simulations (TADSS) to maximize the effectiveness of the personnel to operate and sustain the system equipment at the lowest LCC.

- *Supply support*—Consists of all actions, procedures, and techniques necessary to determine requirements to acquire, catalog, receive, store, transfer, issue, and dispose of spares, repair parts, and supplies. This means having the right spares, repair parts, and all classes of supplies available, in the right quantities, at the right place, at the right time, at the right price. The process includes provisioning for initial support, as well as acquiring, distributing, and replenishing inventories.

- *维持工程*——该项工作跨越那些技术任务（工程和后勤研究及分析），以确保具有受管理的（即，已知的）风险的系统继续运行和维护。关键安全项的技术监督、这些安全项的批准来源以及现场系统的设计构型基线的审视（对包括设计包装、维护程序和使用概况的总体构型的基本设计工程职责），以确保连续的认证合规也是维持工程工作的一部分。按照基线需求对在役系统性能的定期技术审查、趋势分析以及管理选项和解决运行问题的需求的开发是维持工作的一部分。

- *维护规划*——识别、计划、资助并实施维护概念和需求，以确保必要时以可能最低的 LCC 使可能最好的系统能力可用于运行。支持概念描述运行系统所在的支持环境。它还包括与系统维护概念相关的信息：支持基础设施、预期的支持持续时间、可靠性和可维护性比率以及支持场点。支持概念是驱动维护规划流程的基础。建立一般的总体维修方针，如"维修或更换"准则。

- *运行及维护人员*——识别、计划、资助和采办具有运行、维护和支持系统所需的培训、经验和技能的人员。

- *培训及培训支持*——跨系统生命周期计划、资助并实施培训操作人员和维护人员的策略。作为策略、计划、资助和实施行动的一部分，识别、开发和采办培训器材、装置、模拟器和仿真设备（TADSS），以最大化人员的效能，从而以最低的 LCC 运行和维持系统设备。

- *供应支持*——由确定需求所需的全部行动、程序和技术构成，以采办、分类、接收、存储、转移、发布和处置备件、维修零件和供给品。这意味着，可以在正确的地方，正确的时间，用正确的价格获得正确数量的备件、维修零件和所有类型的供给品。该流程包括对初始支持以及采办、分发和补充库存的供应。

SPECIALTY ENGINEERING ACTIVITIES

- *Computer resources (hardware and software)*—Computers, associated software, networks, and interfaces necessary to support all logistics functions. Includes resources and technologies necessary to support long-term data management and storage.

- *Technical data, reports, and documentation*—Represents recorded information of scientific or technical nature, regardless of form or character (such as equipment technical manuals and engineering drawings), engineering data, specifications, and standards. Procedures, guidelines, data, and checklists needed for proper operations and maintenance of the system, including:

    - System installation procedures
    - Operating and maintenance instructions
    - Inspection and calibration procedures
    - Engineering design data
    - Logistics provisioning and procurement data
    - Supplier data
    - System operational and maintenance data
    - Supporting databases

- *Facilities and infrastructure*—Facilities (e.g., buildings, warehouses, hangars, waterways, etc.) and infrastructure (e.g., IT services, fuel, water, electrical service, machine shops, dry docks, test ranges, etc.) required to support operation and maintenance.

- *Packaging, handling, storage, and transportation (PHS&T)*—The combination of resources, processes, procedures, design, considerations, and methods to ensure that all system, equipment, and support items are preserved, packaged, handled, and transported properly, including environmental considerations, equipment preservation for the short and long storage, and transportability. Some items require special environmentally controlled, shock-isolated containers for transport to and from repair and storage facilities via all modes of transportation (land, rail, sea, air, and space).

- *Support equipment*—Support equipment consists of all equipment (mobile or fixed) required to sustain the operation and maintenance of a system. This includes, but is not limited to, ground handling and maintenance equipment, trucks, air conditioners, generators, tools, metrology and calibration equipment, and manual and automatic test equipment.

- *计算机资源（硬件和软件）*——支持全部后勤功能所需的计算机、相关软件、网络和接口。包括支持长期数据管理和存储所需的资源和技术。

- *技术数据、报告和文件*——不考虑形式或特性，表示的科学本质属性或技术本质属性的存档信息（诸如，设备技术手册和工程图样）、工程数据、规范及标准。正确运行和维护系统所需的程序、指南、数据和检查单，包括：

  – 系统安装程序

  – 运行及维护说明

  – 检验和校准程序

  – 工程设计数据

  – 后勤规定及采购数据

  – 供应商数据

  – 系统运行及维护数据

  – 支持数据库

- *设施及基础建设*——支持运行和维护所需要的设施（如，建筑、仓库、机库、水道等）及基础建设（如，IT服务、燃料、水、电气服务、机器车间、干船坞、试验场等）。

- *包装、搬运、储存和运输（PHS&T）*——资源、流程、程序、设计、考虑和方法的组合，以确保恰当地保存、包装、搬运和运输全部系统、设备及支持项，包括环境考虑、短期和长期的设备保存以及可运输性。有些项需要专门的环境受控、隔振的容器，通过所有运输方式（陆地、铁路、海洋、空中和航天）往返于维修设施和储存设施运送。

- *支持设备*——支持设备由维持系统的运行及维护所需要的全部设备（移动的或固定的）组成。这包括但不限于地面搬运和维护设备、卡车、空调机、发电机、工具、计量和校准设备以及手动和自动试验设备。

10.5.2 Supportability Analysis

Supportability analysis is an iterative analytical process by which the logistics support requirements for a system are identified and evaluated. It uses quantitative methods to aid in (i) the initial determination and establishment of supportability requirements as an input to design; (ii) the evaluation of various design options; (iii) the identification, acquisition, procurement, and provisioning of the various elements of maintenance and support; and (iv) the final assessment of the system support infrastructure throughout the utilization and support stages.

Supportability analysis constitutes a design analysis process that is part of the overall SE effort. Functional analysis is used to define all system functions early in the concept stage and to appropriate levels in the system hierarchy. The resulting functional breakdown structure, together with the system requirements and the support and maintenance concept, provides the starting point of a supportability analysis as shown in Figure 10.7.

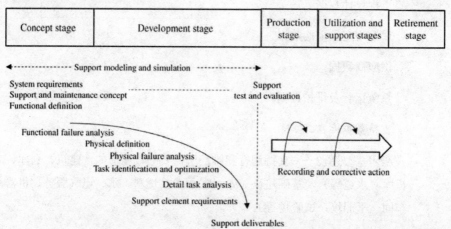

FIGURE 10.7 Supportability analysis. Reprinted with permission from Corrie Taljaard. All other rights reserved.

Supportability analysis may include analyses such as FMECA, fault tree analysis (FTA), reliability block diagram (RBD) analysis, Maintenance Task Analysis (MTA), RCM, and LORA. Products and activities identified in Figure 10.7 are described as follows:

- *Functional failure analysis*—A functional breakdown structure is used as reference to perform functional FMECA and/or FTA and RBD analysis. These analyses can be used to identify functional failure modes and to classify them according to criticality (i.e., severity of failure effects and probability of occurrence). The functional failure analysis can also provide valuable system design input (e.g., redundancy requirements).

10.5.2 维持性分析

维持性分析是一个迭代的分析流程，通过该流程识别并评价系统的后勤保障需求。它使用定量法来辅助（i）初步确定和建立作为设计输入的维持性需求；（ii）评价各种不同设计选项；（iii）各种不同维护和支持元素的识别、采办、采购和供给；以及（iv）贯穿于使用和支持阶段最终评估系统支持基础设施。

维持性分析构成了作为总体 SE 工作一部分的设计分析流程。在概念阶段早期利用功能分析来定义全部系统功能并达到系统层级结构中的适当层级。最终形成的功能分解结构连同系统需求和支持及维护概念一同提供维持性分析的起点，如图 10.7 所示。

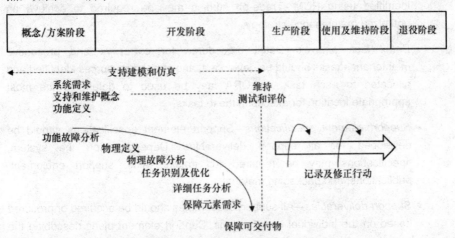

图 10.7 维持性分析。经 Corrie Taljaard 许可后转载。版权所有。

维持性分析可包括诸如 FMECA、故障树分析（FTA）、可靠性框图（RBD）分析、维护任务分析（MTA）、RCM 和 LORA 等分析。图 10.7 中识别的产品和活动描述如下：

- *功能故障分析*——功能分解结构用于作为参考，用来执行功能 FMECA 和/或 FTA 及 RBD 分析。这些分析可以用来识别功能故障模式并根据关键性（即，故障效应的严重程度以及发生故障的概率）来分类。功能故障分析还可以提供有价值的系统设计输入（如，冗余需求）。

SPECIALTY ENGINEERING ACTIVITIES

- *Physical definition*—During system design, the physical breakdown structure of a system should be developed to assist in identifying the actual location of items in the system. This breakdown structure is used as baseline at various stages throughout the life cycle of the system, and it should be continuously updated and refined to the required level of detail.

- *Physical failure analysis*—The physical breakdown structure is used as reference to perform hardware FMECA and/or FTA and RBD analysis with the objective of identifying all maintenance tasks for potential failure modes. The criticalities of failure modes are used to prioritize corrective and preventive maintenance task requirements.

- *Task identification and optimization*—Corrective maintenance tasks are primarily identified using FMECA, while preventive maintenance tasks are identified using RCM. Trade-off studies may be required to achieve an optimized maintenance strategy.

- *Detail task analysis*—Detail procedures for corrective and preventive maintenance tasks should be developed, and support resources identified and allocated to each task. A LORA may be used to determine the most appropriate location for executing these tasks.

- *Support element specifications*—Support element specifications should be developed for all support deliverables. Depending on the system, specifications may be required for training aids, support equipment, publications, and packaging material.

- *Support deliverables*—All support deliverables should be acquired or procured based on the individual specifications. Support element plans describing the management procedures for the support elements should also be developed.

- *Support modeling and simulation*—RAM and LCC modeling and simulation are integral parts of sup-portability analysis that should be initiated during the early stages to develop an optimized system design, maintenance strategy, and support system.

- *Support test and evaluation*—The support deliverables should be tested and evaluated against both support element specifications and the overall system requirements.

- *Recording and corrective action*—Failure recording and corrective action during the utilization and support stages form the basis for continuous improvement. System availability metrics should be used to continuously monitor the system in order to improve support where deficiencies are identified.

- *物理定义*——在系统设计期间,应开发系统的物理分解结构,以帮助识别系统中的各个项的实际位置。这种分解结构贯穿于系统的生命周期的各个不同阶段用作为基线,并且应不断地更新并细化到所需的详细层级。

- *物理故障分析*——物理分解结构作为参考来执行硬件 FMECA 和/或 FTA 及 RBD 分析,目标是识别全部维护任务的潜在故障模式。利用故障模式的关键性来对纠正性和预防性维护任务需求进行优先排列。

- *任务识别和优化*——纠正性维护任务主要利用 FMECA 来识别,而预防性维护任务利用 RCM 来识别。为实现最优化的维护策略,可能需要权衡研究。

- *详细任务分析*——应开发修正性和预防性维护任务的详细程序,识别支持资源,并分配给每个任务。LORA 可用来确定执行这些任务的最适合的位置。

- *支持元素规范*——应为全部支持交付项开发支持元素规范。根据系统,培训培训器材、支持设备、出版物和包装材料可能需要规范。

- *支持交付项*——所有支持交付项都应基于单独的规范采办或采购。还应形成描述支持元素的管理程序的支持元素计划。

- *支持建模和仿真*——RAM 和 LCC 建模和仿真是应该在早期启动以开发最优化的系统设计、维护策略和支持系统的维持性分析的固有部分。

- *支持试验和评价*——应按照支持元素规范和整个系统需求来测试并评价支持交付项。

- *记录和纠正行动*——在使用和维持阶段中的故障记录和修正行动形成持续改进的基础。应利用系统可用性度量标准来连续地监控系统,以便改善识别缺陷处的支持。

SPECIALTY ENGINEERING ACTIVITIES

10.6 MANUFACTURING AND PRODUCIBILITY ANALYSIS

The capability to manufacture or produce a system element is as essential as the ability to properly define and design it. A designed product that cannot be manufactured causes design rework and program delays with associated cost overruns. For this reason, producibility analysis and trade studies for each design alternative form an integral part of the architectural design process. One objective is to determine if existing proven processes are satisfactory since this could be the lowest risk and most cost-effective approach. The Maglev train contractor (see Section 3.6.3) experienced a steep learning curve to produce an unprecedented system from scientific theory.

Producibility analysis is a key task in developing low-cost, quality products. Multidisciplinary teams work to simplify the design and stabilize the manufacturing process to reduce risk, manufacturing cost, lead time, and cycle time and to minimize strategic or critical material use. Critical producibility requirements are identified during system analysis and design and included in the program risk analysis, if necessary. Similarly, long-lead-time items, material limitations, special processes, and manufacturing constraints are evaluated. Design simplification also considers ready assembly and disassembly for ease of maintenance and preservation of material for recycling. When production engineering requirements create a constraint on the design, they are communicated and documented. The selection of manufacturing methods and processes is included in early decisions.

Manufacturing analyses draw upon the production concept and support concept. Manufacturing test considerations are shared with the engineering team and are taken into account in built-in test and automated test equipment.

IKEA® is often used as an example of supply chain excellence. IKEA® has orchestrated a value creating chain that begins with motivating customers to perform the final stages of furniture assembly in exchange for lower prices and a fun shopping experience. They achieve this through designs that support low-cost production and transportability (e.g., the bookcase that comes in a flat package and goes home on the roof of a car).

10.7 MASS PROPERTIES ENGINEERING

Mass properties Engineering (MPE) ensures that the system or system element has the appropriate mass properties to meet the requirements (SAWE).Mass properties include weight, the location of center of gravity, inertia about the center of gravity, and product of the inertia about an axis.

10.6 制造及可生产性分析

制造或生产一个系统元素的能力与正确地定义并设计系统的能力同样重要。一个无法制造的设计产品会导致设计返工和项目推迟，同时带来相关的成本超支。为此，每一个设计备选方案的可生产性分析和权衡研究构成架构设计流程的不可缺少的部分。一个目的是，确定现有的已经证明的流程是否令人满意，因为这可能是风险最低且成本效益最高的方法。磁悬浮列车承包商（见 3.6.3 节）经历过陡峭的学习曲线阶段，生产了一个前所未有的来自科学理论的系统。

可生产性分析是开发低成本优质产品中的关键任务。跨学科团队工作是为了简化设计和稳定制造流程，进而降低风险、制造成本、前置时间和周期时间，并且最小化战略物资和关键材料的使用。在系统分析和设计期间识别关键的可生产性需求，并在必要时包括在项目风险分析中。同样，要对前置时间长的项、材料限制、特殊流程以及制造约束进行评价。设计简化也考虑现成的组装和拆卸，以便于维护和保存材料进行再循环。当生产工程需求对设计产生约束时，应对这些需要进行沟通和文件化。制造方法和流程的选择包括在早期决策中。

制造分析借鉴生产方案和支持方案。与工程团队共享制造试验考虑因素，并在自检测试和自动测试设备中予以考虑。

IKEA®经常被用作供应链的一个最佳示例。IKEA®精心设计价值创造链，其首先鼓励客户以较低的价格进行家具组装的最后阶段并获得有趣的购物体验。他们通过支持低成本生产和可运输性（例如，书架可以平板包装，放在车顶即可运回家）的设计获得这一结果。

10.7 质量特性工程

质量特性工程（MPE）确保系统或系统元素具有满足需求的合适的质量特性（SAWE）。质量特性包括重量、重心位置、重心惯性以及轴的惯性积。

SPECIALTY ENGINEERING ACTIVITIES

Typically, the initial sizing of the physical system is derived from other requirements, such as minimum payload, maximum operating weight, or human factor restrictions. Mass properties estimates are made at all stages of the system life cycle based on the information that is available at the time. This information may range from parametric equations to a three-dimensional product model to actual inventories of the product in service. A risk assessment is conducted, using techniques such as uncertainty analysis or Monte Carlo simulations, to verify that the predicted mass properties of the system will meet the requirements and that the system will operate within its design limits. MPE is conducted at the end of the production stage to assure all parties that the delivered system meets the requirements and then several times during the utilization stage to ensure the safety of the system, system element, or human operator. For a large project, such as oil platform or warship, the MPE level of effort is significant.

One trap in MPE is believing that three-dimensional modeling tools can be exclusively used to estimate the mass properties of the system or system element. This is problematic because (i) not all parts are modeled on the same schedule and (ii) most parts are modeled neat, that is, without such items as manufacturing tolerances, paint, insulation, fittings, etc., which can add from 10 to 100% to the system weight. For example, the liquid in piping and tanks can weigh more than the structural tank or metallic piping that contain it.

MPE usually includes a reasonableness check of all estimates using an alternative method. The simplest method is to justify the change between the current estimate and any prior estimates for the same system or the same system element on another project. Another approach is to use a simpler estimating method to repeat the estimate and then justify any difference.

10.8 RELIABILITY, AVAILABILITY, AND MAINTAINABILITY

To be reliable, a system must be robust—it must avoid failure modes even in the presence of a broad range of conditions including harsh environments, changing operational demands, and internal deterioration (Clausing and Frey, 2005). Reliability can thus be seen as the proper functioning of a system during its expected life under the full range of conditions experienced in the field.

Reliability engineering refers to the specialized engineering discipline that addresses the reliability of a system during its total life cycle. It includes related aspects such as availability and maintainability of a system. Therefore, reliability engineering is often used as collective term for the engineering discipline concerned with the RAM of a system.

通常,物理系统的初始尺度来自其他需求,例如最小有效载荷、最大运行重量或人为因素限制。在系统生命周期的所有阶段基于当时可用的信息进行质量特性估计。该信息的范围从参数方程到三维产品模型再到使用中产品的实际库存。使用诸如不确定性分析或蒙特卡罗模拟技术进行风险评估,以验证预测的系统质量特性是否满足需求并且验证系统是否会在其设计限制范围内运行。MPE 在生产阶段结束时进行,以向各方确保交付的系统满足需求,然后在使用阶段多次进行 MPE,以确保系统、系统元素或人工操作员的安全。对于诸如石油平台或军舰等大型项目而言,MPE 工作层级十分重要。

MPE 中的一个陷阱是相信三维建模工具可以专门用来估计系统或系统元素的质量特性。这是有问题的,因为(i)并非所有的零件均按照同一个进度建模,并且(ii)大部分零件被简化建模,即没有诸如制造公差、涂料、绝缘和配件等项,而这些项可使系统重量从 10%到 100%的增加。例如,管道和油箱中的液体比容纳液体的结构油箱或金属管道更重。

MPE 通常包括利用替代的方法对全部估计值进行合理性检查。最简单的方法是,证明现行估计值与另一个项目的同一系统或同一系统元素的任何预先估计值之间变化的合理性。另一个方法是,使用更简单的估计方法进行重复估计,然后证明任何差值的合理性。

10.8 可靠性、可用性和可维护性

要可靠,系统就必须是鲁棒的——必须避免故障模式,甚至在出现广泛的状况时,包括恶劣的环境、变化的运行需求和内部恶化(Clausing 和 Frey,2005)。因此,在该领域的全部状况下,可靠性在系统的期望寿命中可被看做系统功能的正常运行。

可靠性工程指的是,在系统的全生命周期中应对系统的可靠性的专门工程学科。它包括诸如系统的可用性和可维护性等的相关方面。因此,可靠性工程往往被用作关于系统 RAM 的工程学科的集体术语。

RAM are important attributes or characteristics of a given system. However, RAM should not actually be viewed as characteristics, but rather as nonfunctional requirements. It is therefore essential that SE processes should include RAM activities, selected, planned, and executed in an integrated manner with other technical processes.

Reliability engineering activities support other SE processes in two ways. Firstly, reliability engineering activities should be used to influence system design (e.g., the system architecture depends on reliability requirements). Secondly, reliability engineering activities should be used as part of system verification (e.g., system analysis or system test).

10.8.1 Reliability

The objectives of reliability engineering, in the order of priority, are (O'Connor and Kleyner, 2012):

- To apply engineering knowledge and specialist techniques to prevent or to reduce the likelihood or frequency of failures

- To identify and correct the causes of failures that do occur, despite the efforts to prevent them

- To determine ways of coping with failures that do occur, if their causes have not been corrected

- To apply methods for estimating the likely reliability of new designs and for analyzing reliability data

The priority emphasis is important, since proactive prevention of failure is always more cost-effective than reactive correction of failure. Timely execution of appropriate reliability engineering activities is of utmost importance in achieving the required reliability during operations.

Traditionally, reliability has been defined as *the probability that an item will perform a required function without failure under stated conditions for a stated period of time* (O'Connor and Kleyner, 2012). The emphasis on probability in the definition of reliability (to quantify reliability) resulted in a number of potentially misleading or even incorrect practices (e.g., reliability prediction and reliability demonstration of electronic systems).

RAM 是给定系统的重要属性或特征。但是，RAM 实际上不应该被视为特征，而是视为非功能需求。因此，最重要的是，SE 流程应包括以与其他技术流程综合的方式来选定、规划和执行的 RAM 活动。

可靠性工程活动以两种方式支持其他 SE 流程。首先，可靠性工程活动应当用来影响系统设计（如，系统架构依赖于可靠性需求）。其次，可靠性工程活动应被作为系统验证的一部分（如，系统分析或系统测试）。

10.8.1 可靠性

按照优先顺序，可靠性工程目标是（O'Connor 和 Kleyner，2012）：

- 应用工程知识和专家技术，防止或降低故障的可能性或频率
- 识别并纠正导致故障出现的原因，即使有防止故障的工作
- 如果尚未纠正故障的原因，要确定应对出现的故障的方式
- 采用估计新设计可能的可靠性并分析可靠性数据的方法

强调优先顺序很重要，因为具有前瞻性的故障预防总是比故障反应式的纠正更具成本效益。恰当的可靠性工程活动的及时执行在运行中实现所需可靠性方面极其重要。

传统上，可靠性已经被定义为某一项在规定条件下无故障地持续规定时间实施所需功能的概率（O'Connor 和 Kleyner，2012）。在可靠性定义中强调概率（量化可靠性）产生了大量的可能会误导乃至不正确的实践（如，电子系统的可靠性预计和可靠性演示验证）。

Modern approaches to reliability place more emphasis on the engineering processes required to prevent failure during the expected life of a system (i.e., failure-free operation). The concept of "design for reliability" has recently shifted the focus from a reactive "test–analyze– fix" approach to a proactive approach of designing reliability into the system. "Failure mode avoidance" approaches are aligned with other SE processes and attempt to improve reliability of a system early in the development stages (Clausing and Frey, 2005). It is performed by evaluating system functions, technology maturity, system architecture, redundancy, design options, etc. in terms of potential failure modes. The most significant improvements in system reliability can be achieved by avoiding physical failure modes in the first place and not by minor improvements after the system has been conceived, designed, and produced.

"Design for reliability" implies that reliability should be specified as a requirement in order to receive adequate attention during requirements analysis. Reliability requirements may be specified either in qualitative or quantitative terms, depending on the specific industry. Care should be taken with quantitative requirements, since verification of reliability is often not practical (especially for high reliability requirements). Also, the misuse of reliability metrics (e.g., mean time between failure (MTBF)) frequently results in "playing the numbers game" during system development, instead of focusing on the engineering effort necessary to achieve reliability (Barnard, 2008). For example, MTBF is often used as an indicator of "average life" of an item, which may be completely incorrect. It is therefore recommended that other reliability metrics be used for quantitative requirements (e.g., reliability (as success probability) at a specific time).

10.8.1.1 *Development of a Reliability Program Plan* Reliability engineering activities are often neglected during system development, resulting in a substantial increase in risk of project failure or customer dissatisfaction. It is therefore recommended that reliability engineering activities be formally integrated with other SE technical processes. A practical way to achieve integration is to develop a Reliability program plan at the start of the project.

Appropriate reliability engineering activities should be selected and tailored according to the objectives of the specific project. These activities should be captured in the Reliability program plan. The plan should indicate which activities will be performed, the planned timing of the activities, the level of detail required for the activities, and the persons responsible for execution of the activities.

可靠性的现代方法更加强调防止在系统的预期生命期间出现故障所需的工程流程（即，无故障运行）。"可靠性设计"的概念近来已经将聚焦点从反应式的"测试—分析—修补"方法转变为将可靠性设计到系统中的具有前瞻性的方法。"故障模式规避"方法与其他 SE 流程协调一致，并试图在开发阶段的早期提高系统的可靠性（Clausing 和 Frey，2005）。依照潜在的故障模式，通过评价系统功能、技术成熟度、系统架构、冗余、设计选项等来执行。系统可靠性的最明显改善首先可以通过避免物理故障模式来实现，而不是通过在已经构想、设计和生产系统后所做的较小的改进。

"面向可靠性的设计"意味着，可靠性应作为一项需求来规定，以便在需求分析期间得到充分的关注。可靠性需求可能根据特定的行业以定性的或定量的方式规定。应该慎重地考虑定量的需求，因为可靠性验证往往是难以切合实际的（尤其是高可靠性需求）。而且，可靠性度量标准的误用[如，平均故障间隔时间（MTBF）]往往导致在系统开发期间"玩数字游戏"，而不是聚焦于实现可靠性所必需的工程工作（Barnard，2008）。例如，MTBF 往往作为某一项的"平均寿命"的指标，可能完全不正确。因此，建议将其他可靠性度量标准用于定量的需求[如，在规定时间的可靠性（作为成功概率）]。

10.8.1.1　可靠性项目计划的开发　可靠性工程活动往往在系统开发期间被忽视，导致项目故障或客户不满意的风险显著增加。因此，建议可靠性工程活动正式地与其他 SE 技术流程相综合。实现综合的切实可行的方式是，在项目开始时开发可靠性项目计划。

适当的可靠性工程活动应按照特定项目的目标来选择和剪裁。这些活动应在可靠性项目计划中捕获。计划中应指出，将执行哪些活动、活动的计划的时序、活动需要的详细层级以及负责执行活动的人员。

SPECIALTY ENGINEERING ACTIVITIES

ANSI/GEIA-STD-0009-2008, *Reliability program standard for systems design, development, and manufacturing*, can be referenced for this purpose. This standard addresses not only hardware and software failures but also other common failure causes such as manufacturing, operator error, operator maintenance, training, quality, etc. "At the heart of the standard is a systematic 'design-reliability-in' process, which includes three elements:

- Progressive understanding of system-level operational and environmental loads and the resulting loads and stresses that occur throughout the structure of the system.

- Progressive identification of the resulting failure modes and mechanisms.

- Aggressive mitigation of surfaced failure modes."

ANSI/GEIA-STD-00092008, which supports a system life cycle approach to reliability engineering, consists of the following objectives:

- Understand customer/user requirements and constraints.

- Design and redesign for reliability.

- produce reliable systems/products.

- Monitor and assess user reliability.

The Reliability program plan thus provides a forward looking view on how to achieve reliability objectives. Complementary to the Reliability program plan is the Reliability Case that provides a retrospective (and documented) view on achieved objectives during the system life cycle.

Figure 10.8 indicates a few relevant questions that may be used to develop a Reliability program plan for a specific project.

10.8.1.2 Reliability Engineering Activities Reliability engineering activities can be divided into two groups, namely, engineering analyses and tests and failure analyses. These activities are supported by various reliability management activities (e.g., design procedures, design checklists, design reviews, electronic part derating guidelines, preferred parts lists, preferred supplier lists, etc.).

Engineering analyses and tests refer to traditional design analysis and test methods to perform, for example, load-strength analysis during design. Included in this group are finite element analysis, vibration and shock analysis, thermal analysis and measurement, electrical stress analysis, wear-out life prediction, highly accelerated life testing (HALT), etc.

为此，可以参考 ANSI/GEIA-STD-0009-2008，用于系统设计、开发和制造的可靠性项目标准。该标准应对的不仅仅是硬件和软件故障，还包括其他共同的故障原因，如，制造、操作者错误、操作者维护、培训、质量等。"该标准的核心之处是系统化的'设计可靠性'流程，包括三个元素：

- 逐渐理解系统级的运行和环境负荷以及遍及于系统的结构出现的载荷和应力。
- 逐渐识别产生的故障模式和机制。
- 积极缓解暴露出的故障模式。"

支持可靠性工程的系统生命周期方法—ANSI/GEIA-STD-00092008，包括以下目标：

- 理解客户/用户需求及约束。
- 可靠性的设计和再设计。
- 生产可靠的系统/产品。
- 监测并评估用户可靠性。

因此，可靠性项目计划提供关于如何实现可靠性目标的前瞻性视图。对可靠性项目计划的补充是可靠性用例，其提供系统生命周期期间关于所完成目标的有追溯性的（及文件化的）视图。

图10.8指出了可用来开发特定项目的可靠性项目计划的几个相关问题。

10.8.1.2　可靠性工程活动　可靠性工程活动可分为两组，即：工程分析和测试以及故障分析。这些活动由各种不同的可靠性管理活动（如，设计程序、设计检查清单、设计详审、电子零件额度值退化指南、首选的零件清单、首选的供应商清单等）来支持。

工程分析和测试是指，在设计期间执行例如载荷-强度分析的传统设计分析和测试方法。在这一组中包括的是，有限元分析、振动和冲击分析、热分析和测量、电应力分析、磨损寿命预测、高加速寿命试验（HALT）等。

SPECIALTY ENGINEERING ACTIVITIES

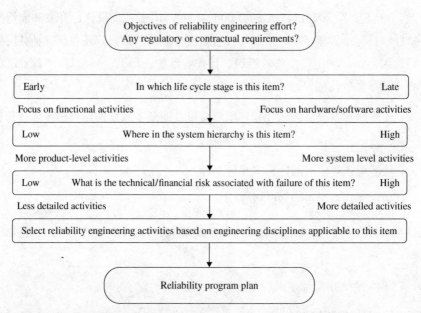

FIGURE 10.8 Reliability program plan development. Reprinted with permission from Albertyn Barnard. All other rights reserved.

Failure analyses refer to traditional RAM analyses to improve understanding of cause-and-effect relationships during design and operations. Included in this group are Failure Mode and Effects Analysis (FMEA), FTA, RBD analysis, systems modeling and simulation, root cause failure analysis, etc.

10.8.2 Availability

Availability is defined as the probability that a system, when used under stated conditions, will operate satisfactorily at any point in time as required. Availability is therefore dependent on the reliability and maintainability of the system, as well as the support environment during the utilization and support stages. It may be expressed and defined as inherent, achieved, or operational availability (Blanchard and Fabrycky, 2011):

- Inherent availability (A_i) is based only on the inherent reliability and maintainability of the system. It assumes an ideal support environment (e.g., readily available tools, spares, maintenance personnel) and excludes preventive maintenance, logistics delay time, and administrative delay time.

- Achieved availability (A_a) is similar to inherent availability, except that preventive (i.e., scheduled) maintenance is included. It excludes logistics delay time and administrative delay time.

专业工程活动

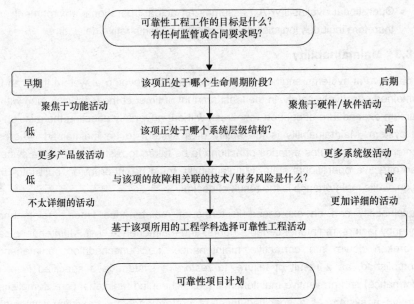

图 10.8 可靠性项目计划开发。经 Albertyn Barnard 许可后转载。版权所有。

故障分析是指,在设计和运行期间改善对因果关系的理解的传统 RAM 分析。在这一组中,包括故障模式与影响分析(FMEA)、FTA、RBD 分析、系统建模与仿真、根源故障分析等。

10.8.2 可用性

可用性被定义为系统在规定条件下使用时将根据需要及时地在任意地点良好运行的概率。因此,可用性取决于系统的可靠性和可维护性,以及使用和支持阶段中的支持环境。它可表达和定义为固有的、已实现的或运行的可用性(Blanchard 和 Fabrycky,2011):

- 固有可用性(A_i)仅仅基于系统的固有可靠性和可维护性。它假定一种理想的支持环境(如,随时可用的工具、备件、维护人员)并排除预防性维护、后勤延误时间及行政管理延误时间。

- 除了包括预防性(即,计划的)维护之外,已实现的可用性(A_a)与固有可用性类似。它排除后勤延误时间和行政管理延误时间。

SPECIALTY ENGINEERING ACTIVITIES

- Operational availability (A_o) assumes an actual operational environment and therefore includes logistics delay time and administrative delay time.

10.8.3 Maintainability

An objective in systems engineering is to design and develop a system that can be maintained effectively, safely, in the least amount of time, at the least cost, and with a minimum expenditure of support resources without adversely affecting the mission of that system. Maintainability is the *ability* of a system to be maintained, whereas maintenance constitutes a series of actions to be taken to restore or retain a system in an effective operational state. Maintainability must be inherent or "built into" the design, while maintenance is the result of design.

Maintainability can be expressed in terms of maintenance times, maintenance frequency factors, maintenance labor hours, and maintenance cost. Maintenance can be broken down into corrective maintenance (i.e., unscheduled maintenance accomplished, as a result of failure, to *restore* a system to a specified level of performance) and preventive maintenance (i.e., scheduled maintenance accomplished to *retain* a system at a specified level of performance by providing systematic inspection and servicing or preventing impending failures through periodic item replacements) (Blanchard and Fabrycky, 2011).

10.8.4 Relationship with Other Engineering Disciplines

Reliability engineering is closely related to other engineering disciplines, such as safety engineering and logistics engineering. The primary objective of reliability engineering is prevention of failure. The primary objective of safety engineering is prevention and mitigation of harm under both normal and abnormal conditions (see Section 10.10). The primary objective of logistics engineering is development of efficient logistics support (e.g., preventive and corrective maintenance; see Section 10.5).

These three disciplines not only have "failure" as common theme, but they may also use similar activities, albeit from different viewpoints. For example, an FMEA may be applicable to reliability, safety, and logistics engineering. However, a design FMEA will be different to a safety or logistics FMEA, due to the different objectives. Common to all disciplines is the necessity of early implementation during the system life cycle.

While reliability is concerned with failures (or rather the absence of failures), maintainability refers to the ability of a system to be maintained (or the ease of maintenance). Availability is a function of both reliability and maintainability and may include logistics aspects (as in the case of operational availability). The LCC of a system is highly dependent on reliability and maintainability, which are considered major drivers in support resources and related in-service costs.

- 运行可用性（A_o）一种假定实际的运行环境，因此包括后勤延误时间和行政管理延误时间。

10.8.3 可维护性

系统工程中的一个目标是以最少的时间，以最低的成本，以最少的支持资源开销来设计和开发一个能被有效、安全地维持的系统，且不会对系统的任务产生不利影响。可维护性是维护系统的能力，而维护构成一系列为恢复或保持系统的有效运行状态而采取的行动。可维护性必须是固有的，或者被"内置到"设计中，而维护是设计的结果。

可维护性可按照维护时间、维护频率因素、维护工时和维护成本来表示。维护可以分解为纠正性维护（即，由于故障，为将系统恢复到规定的性能水平所完成的非定期维护）和预防性维护（即，通过提供系统检验和检修或通过周期的物件更换防止即将发生的故障，以将系统保持在规定的性能水平上所完成的定期维护）（Blanchard 和 Fabrycky，2011）。

10.8.4 与其他工程学科的关系

可靠性工程与诸如安全性工程和后勤工程等其他工程学科密切相关。可靠性工程的首要目标是预防故障。安全性工程的首要目标是预防和缓解正常和异常情况下的伤害（见 10.10 节）。后勤工程的首要目标是开发高效的后勤支持（如，预防性和修正性维护；见 10.5 节）。

这三个学科不仅具有共同的主题——"故障"，而且还可能使用类似的活动，尽管来自于不同的视角。例如，FMEA 可能适用于可靠性、安全性和后勤工程。但是，一个设计 FMEA 由于目标不同而有别于一个安全性或后勤 FMEA。所有学科的共同之处是，在系统生命周期中尽早实施的必要性。

虽然可靠性关注的是故障（更应该说是不出现故障），但是可维护性指的是维护系统的能力（或方便于维护）。可用性是可靠性和可维护性的功能，可能包括后勤的各个方面（如运行可用性）。系统的 LCC 高度依赖于可靠性和可维护性，二者被视为支持资源和相关使用中成本的主要驱动力。

10.9 RESILIENCE ENGINEERING

10.9.1 Introduction

The general definition of resilience is "...the act of rebounding or springing back" (Little et al., 1973). For engineered systems, as defined in this handbook, resilience has taken on the following meaning (Haimes, 2012):

Resilience is the ability to prepare and plan for, absorb or mitigate, recover from, or more successfully adapt to actual or potential adverse events.

Although this definition can apply to the resilience of any engineered system including both physical assets and humans, early work (Hollnagel et al., 2006) focused on the resilience of organizational systems. Although this definition is widely used, some domains, for example, the military (Richards, 2009), define resilience to include only the recovery phase of a disruption.

Resilience has taken on a particular importance at a governmental level (NRC, 2012; The White House, 2010) with the resilience of infrastructure systems being of the highest priority. Infrastructure systems include fire protection, law enforcement, power, water, health- care, transportation, telecommunication, and other systems. The principles and practices outlined here can apply to any engineered system. Infrastructure systems are generally SoS as defined in Section 2.4 and pose a particular challenge to achieving resilience arising from the distinctive features of SoS. The SOIs are not limited to safety-critical systems. The resilience in question may apply to the restoration of a service, such as water, power, healthcare, and so forth. Water, power, and healthcare may most likely be safety-critical systems, contributing to safety functions, such as sprinkling systems (water), provision of life (health), and power (supporting safety-critical systems, such as grid, and critical infrastructure).

10.9.2 Description

Resilience pertains to the anticipation, survival, and recovery from a variety of disruptions caused by both human-made and natural threats. External human-originated threats include terrorist attacks. Internal human originated threats include operator and design error. Natural threats include extreme weather, geological events, wildfires, and so forth. Threats may be single or multiple. Threats confronted after the first in a multiple threat scenario may result from attempts to correct for the initial threat. Multiple threats may also result from cascading failures, which are common in infrastructure systems.

10.9 可恢复性工程

10.9.1 简介

一般的可恢复性定义是"……反弹或弹回的行为"（Little 等，1973）。对于工程系统，正如本手册中的定义，可恢复性有着下述含义（Haimes，2012）：

可恢复性是准备和计划、吸收或减缓、从中恢复或更成功地适应实际的或潜在的不利事件的能力。

尽管该定义可适用于任何工程系统（包括实物资产和人力）的可恢复性，但早期工作（Hollnagel 等，2006）聚焦于组织系统的可恢复性。尽管广泛使用了该定义，但举例来说，军队等一些领域（Richards，2009）规定可恢复性仅仅包括破坏的恢复阶段。

可恢复性在政府层面有着特殊的重要性（NRC，2012；白宫，2010），基础设施系统的可恢复性具有最高优先级。基础设施系统包括消防、执法、电、水、卫生保健、运输、通信和其他系统。在此概括的原则和实践可适用于任何工程系统。基础设施系统一般为按照 2.4 节定义的 SoS，并对实现 SoS 的区别性特征引起的可恢复性提出了特殊的挑战。SOI 不局限于安全性关键的系统。正在讨论的可恢复性可能适用于诸如水、电、卫生保健等服务的恢复。水、电和卫生保健最有可能是安全性关键的系统，有助于安全功能，诸如喷洒系统（水）、寿命保障（健康）和电（支持安全性关键的系统，如网格和关键的基础设施）。

10.9.2 描述

可恢复性属于来自于由人为的和自然的威胁造成的各种不同破坏的预期、生存和恢复。外部人员发起的威胁包括恐怖分子的攻击。内部人员发起的威胁包括操作人员和设计错误。自然的威胁包括极端天气、地质事件、野火等。威胁可能是单一的，也可能是多重的。在多重威胁场景中第一个威胁之后所面临的威胁可能是由试图修正初始威胁引起的。多重威胁还可能是由基础设施系统中共同的连锁性故障引起的。

229 Resilience is an emergent and nondeterministic property of a system (Haimes, 2012). It is emergent because it cannot be determined by the examination of individual elements of the system. The entire system and the interaction among the elements must be examined. It is nondeterministic because the wide variety of possible system states at the time of the disruption cannot be characterized either deterministically or probabilistically. Statistical data analysis (extreme amounts) may allow for probabilistic assessment. For example, in reference to Fukushima, data exist on earthquakes and tsunamis to make a quantitative prediction. Moreover, data are available on cooling system configuration and probability of failure under earthquake and tsunami conditions, making it possible to evaluate these events on a probabilistic basis. Because of these emergent and nondeterministic properties, resilience cannot be measured, nor outcomes of particular threats be accurately predicted except by iterative analytical trials of threats and system configurations.

The purpose of engineering a resilient system is to determine the architecture and/or other system characteristics that will anticipate, survive, and recover from a disruption or multiple disruptions. Figure 10.9 is a model of a disruption.

Figure 10.9 shows a disruption occurring in three states: the pre-event initial state, the intermediate states due to the event, and the post-event final state. The figure also shows a feedback loop that represents the multiple threat scenario. The ability of a system to accomplish these desirable outcomes depends on the application of one or more principles (Jackson and Ferris, 2013). These principles are abstract, allowing a system developer to design specific implementations that, in turn, will result in specific resilience characteristics. Principles can be either scientifically validated rules or heuristics. The characterization of these principles at the abstract level allows them to apply to any domain. The principles must be invoked during one or more of the phases in the diagram in Figure 10.9. The system developer can only determine which principles are preferred in a particular situation by proposing design solutions and modeling their effect. In addition, it has been determined (Jackson and Ferris, 2013) that resilience is achieved when the principles must be implemented in appropriate combinations. Hence, the following principle when implemented in the appropriate combination can be considered an integrated model of resilience. The system developer can develop concrete design proposals by following the reasoning that "an abstraction is a simplified replica of the concrete" (Lonergan, 1992). The top-level abstract principles and their associated dominant characteristics are described in the following text. Sub-principles to these principles can be found in the primary source (Jackson and Ferris, 2013).

可恢复性是系统的涌现的和不确定的性质（Haimes，2012）。涌现的原因在于，它不能通过检查系统的单个元素来确定。必须检查整个系统以及各元素之间的交互。非确定性的原因在于，在破坏时，不能确定地或概率地描述各式各样可能的系统状态。统计数据分析（极端数量）可能允许概率的评估。例如，在 Fukushima 的参考文献中，存在关于地震和海啸的数据，以进行定量预测。此外，数据对冷却系统构型以及在地震和海啸条件下的故障概率有效，使得有可能概率地评价这些事件。由于这些涌现性和不确定性的存在，无法衡量可恢复性，也不能准确地预测特殊威胁的后果，除非通过威胁和系统构型的迭代分析试验。

对一个可恢复系统进行工程设计的目的是，确定将预期、生存和从某一或多重破坏中恢复的架构和/或其他系统特征。图 10.9 是破坏的模型。

图 10.9 示出了在三种状态下出现的破坏：事件前的初始状态、因事件而产生的中间状态、事件后的最终状态。该图还示出了代表多重威胁场景的反馈回路。系统完成这些理想成果的能力取决于一个或多个原则的应用（Jackson 和 Ferris，2013）。这些原则是抽象的，允许系统开发者设计出反之会导致特殊弹性特征的特定实施方式。原则可以是在科学上得到确认的规则或启发法。这些原则在抽象层级的特征描述允许它们适用于任何领域。原则必须在图 10.9 所示的一个或多个阶段中调用。系统开发者只能通过提出设计解决方案和建模其效果来确定在特殊情况下首选哪些原则。此外，已经确定（Jackson 和 Ferris，2013）可恢复性在原则必须以适当组合实施时才能达成。因此，当以适当组合实施时，下述原则可被视为综合的可恢复性模型。系统开发者可以根据"抽象是具体的简化复制品"的推理制定具体的设计方案（Lonergan，1992）。以下述文本描述顶层抽象原则及其相关主导特征。这些原则的子原则可以在原始来源中找到（Jackson 和 Ferris，2013）。

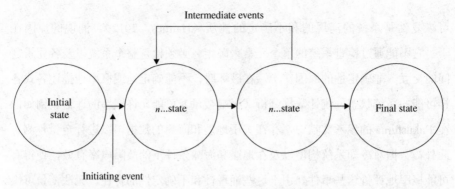

FIGURE 10.9 Resilience event model. Reprinted with permission from Scott Jackson. All other rights reserved.

The engineering of a resilient system is not a separate discipline. Its principles, listed in the following text, are recognized in other disciplines, for example, architecture design, reliability, and safety. Reliability is a key consideration in safety. They have one thing in common: the ability to enhance the resilience of an engineered system. The goal of each principle is to support a particular attribute or feature of the system that will enhance resilience. The following principles are listed according to the attribute they support:

- *Attribute: Capacity*—the ability to withstand a threat
  - Absorption: System capable of absorbing design threat level.
  - Physical redundancy: System consists of two or more identical and independent branches.
  - Functional redundancy: Also called layered diversity, system consists of two or more different and independent branches and is not vulnerable to common cause failure.
- *Attribute: Buffering*—the ability to maintain a distance from the boundary of unsafe operation or collapse
  - Layered defense: System does not have a single point of failure.
  - Reduce complexity: System capable of reducing the number of elements, interfaces, and/or variability among its elements.
  - Reduce hidden interactions: System capable of detecting undesirable interactions among its elements.
- *Attribute: Flexibility*—the ability to bend or restructure

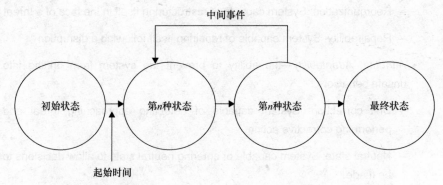

图 10.9　一个可恢复性事件模型。经 Scott Jackson 许可后转载。版权所有。

一个可恢复性系统的工程不是孤独的学科。下面列出的原则也在其他学科中得到认可，例如，架构设计、可靠性和安全性。可靠性是安全性的关键考虑因素。它们有一个共同之处：提高工程系统可恢复性的能力。每个原则的目标是，支持将提高一个可恢复性的系统的特殊属性或特征。根据它们支持的属性列出下述原则：

- *属性：能力*——承受一种威胁的能力
  - 吸收：能够吸收设定威胁层级的系统。
  - 物理冗余：系统由两个或更多个完全相同的、独立的分支组成。
  - 功能冗余：也称为分层多样性，系统由两个或更多个不同的、独立的分支组成并且不容易出现共因故障。
- *属性：缓冲*——与不安全运行或崩溃边缘保持一段距离的能力
  - 分层防卫：系统没有单点故障。
  - 减少复杂性：系统能够减少元素数量、接口和/或元素间的变化性。
  - 减少隐藏的交互：系统能够检测系统元素之间的不合需要的交互。
- *属性：灵活性*——屈服或重建的能力

- Reorganization: System capable of restructuring itself in the face of a threat
- Repairability: System capable of repairing itself following a disruption
- *Attribute: Adaptability*—the ability to prevent the system from drifting into unsafe behaviors
  - Drift correction: System capable of detecting approaching threat and performing corrective action.
  - Neutral state: System capable of entering neutral state to allow decisions to be made.
  - Human in the loop: System has human elements where needed.
  - Loose coupling: System resistant to cascading failure by slack and delays at the nodes.
- *Attribute: Tolerance*—the ability to degrade gracefully
  - Localized capacity: Individual elements of a system are capable of independent operation following failure of other elements.
- *Attribute: Cohesion*—the ability of the elements of a system to operate together as a system
  - Internode interaction: System has connections among all its nodes.

Key inputs for resiliency engineering are as follows:

- Threats: number, type, characteristics
- Objectives and priorities
- SOI: type and purpose
- Candidate principles: potentially appropriate for the SOI
- Solution proposals

Key outputs for resiliency engineering are as follows:

- preferred system characteristics
- System-predicted response to selected threats
- Loss and recovery of function, service, and financial impact
- Recovery time

- 重组：面对威胁能够重建自身的系统
- 可维修性：破坏之后能够修复自身的系统
- *属性：自适应性*——防止系统渐渐陷入不安全行为的能力
  - 偏差修正：系统能够检测出正在逼近的威胁并实施修正措施。
  - 中性状态：系统能输入中性状态以允许做出决策。
  - 人在回路中：系统在需要的地方具有人素。
  - 松耦合：系统通过在节点处松动和延迟防止连锁性故障。
- *属性：容忍性*——平稳退化的能力
  - 局部能力：能够在其他元素故障之后独立地运行的单独系统元素。
- *属性：内聚性*——系统的元素作为一个系统一起运行的能力
  - 节点间交互：系统在其所有节点之间有连接。

可恢复性工程的关键输入如下：

- 威胁：数量、类型、特征
- 目标和优先级
- SOI：类型和目的
- 候选原则：潜在地适合于 SOI
- 解决方案建议

可恢复性工程的关键输出如下：

- 首选的系统特征
- 对选定威胁的系统预计的响应
- 功能、服务和财务影响的损失和恢复
- 恢复时间

SPECIALTY ENGINEERING ACTIVITIES

Key activities of the resiliency engineering process are as follows:

- Create models, including system characteristics and threats.
- Select candidate resilience principles and combinations of principles appropriate to the relevant scenarios.
- Select a measure, or measures, of effectiveness.
- propose candidate solutions for each principle including inputs and outputs for each system element.
- Model threats for a selected range of types and magnitudes relevant to the scenarios:
  - Identify potential impacts of unanticipated threats.
- Execute the model for the range of threats and relevant system states.
- Conduct an impact analysis to determine the loss and recovery of function, service, or financial impact of the evaluated system.

10.10 SYSTEM SAFETY ENGINEERING

System safety engineering is an applied derivative of SE that builds upon the fundamentals of good systems thinking and applies them analytically through each of the system's life cycle phases. At the core of system safety engineering is the analysis of each of requirement, each system element, and each macro-to-micro behavior within the context of the system being developed, operated, or sustained to identify and eliminate or control safety risk potential. Safety risk potential is defined as any condition that would produce undesired conditions of the system resulting in damage to the system, harm to the humans involved in the operations and support of the system, or damage to the environment.

The primary objective of system safety engineering is to influence the design with safety-related requirements for the development, production, utilization, support, and retirement stages of a safe system. The benefits of a safe system are numerous and include, but not limited to, the reduction of risk associated with cost, schedule, operational effectiveness, system availability, and legal liability.

10.10.1 The Role of SE in System Safety

As today's systems increase in size and complexity, the SE approaches used become more critical. System-level properties such as safety must be designed into these systems. They cannot be added on afterward and expected to be safe. Systems are designed to achieve specific goals to satisfy the requirements and constraints. SE must develop a means to organize the engineering design process to ensure the goals of system safety are included.

可恢复性工程流程的关键活动如下：

- 创建模型，包括系统特征和威胁。
- 选择候选的可恢复性原则以及适于相关场景的原则组合。
- 选择效能的某一或某些测度。
- 针对每个原则提出候选的解决方案，包括每个系统元素的输入和输出。
- 对一个选定类型的范围和与场景相关的量级的威胁建模：
  – 识别未预料到的威胁的潜在影响。
- 执行威胁和相关系统状态范围内的模型。
- 进行影响分析，以确定所评价系统的功能、服务或财务影响的损失和恢复。

10.10 系统安全性工程

系统安全性工程是 SE 的应用推演，其建立在良好的系统思考的基础上，并通过分析在每个系统生命周期阶段中应用。系统安全性工程的核心是，在正在开发、运行或维持的系统的背景环境内分析每个需求、每个系统元素和每个宏观到微观的行为，以识别和消除或控制潜在的安全风险。潜在的安全风险被定义为会产生系统不希望有的情况以导致系统破坏、系统运行和支持中涉及的人员受伤害或环境受损的任何状况。

系统安全性工程的主要目标是，利用对安全系统的开发、生产、使用、维持和退役阶段的安全性相关需求来影响设计。安全系统的益处有很多，包括但不限于降低与成本、进度、运行效能、系统可用性和法律责任相关的风险。

10.10.1 SE 在系统安全中的作用

随着如今的系统规模越来越大，越来越复杂，所使用的 SE 方法变得更为关键。诸如安全性等系统级特性必须设计到这些系统中，不能以后再添加而预期它们是安全的。系统被设计成能实现特定的目标，以满足需求和约束。SE 必须开发一种组织工程设计流程的手段，以确保系统安全的目的包括在内。

SE must embed the system safety engineering effort into its engineering processes from the beginning such that safety can be designed into the system as engineering design decisions are made. With respect to system safety engineering, SE determines the goals of the system and participates in the identification and documentation of potential hazards to be avoided. From this information, a set of system functional requirements, safety requirements, and constraints can be identified and documented. These requirements will lay the foundation for design and operation of the system to ensure safety is designed into the system. SE must establish system safety engineering from the early concept stages and continue this process throughout the life cycle of the system. SE should ensure design decisions are guided by safety considerations while taking system requirements and constraints into account.

10.10.2 Identify and Integrate System Safety Requirements

System safety engineers review and identify applicable "best practice" system safety design requirements and guidelines from federal, military, national, and industry regulations, codes, standards, and other documents for the system to be designed and developed (e.g., Federal Motor vehicle Safety Standards (FMVSS), Military Standards (MIL-STDs), National Electrical Code (NEC), and Registration, Evaluation, Authorization, and Restriction of Chemicals (REACH)).These initial requirements are used to derive additional system safety design requirements that are provided to design engineers to eliminate or reduce hazard risks to acceptable levels. These requirements are integrated into the high-level system requirements and design documents. The system safety requirements are then related to the hazards identified in the system.

System safety engineering, along with SE, ensures that the system safety design requirements and guidelines are developed, refined, completely and correctly specified, and properly translated into system element requirements to ensure they are implemented in the design and development of the system hardware, software, and user interface.In addition, applicable safety requirements are identified for incorporation into procedures, processes, warnings, and cautions for use in the operator, user, and diagnostic manuals.

10.10.3 Identify, Analyze, and Categorize Hazards

Within the discipline of system safety engineering, there are numerous analytical methods, techniques, and products that are considered best practice and acceptable within the industry. For example, SAE International possesses specific methods defined in SAE ARP 4754 and 4761 that are commonly used within the aviation industry. The US Department of Defense MIL-STD-882 has also defined specific analysis techniques that are useful in the defense domain (DoD, 2010b). Regardless of the standard or guidance used, the following nonexhaustive list of analysis techniques and safety engineering artifacts reflects SE best practice:

SE 必须将系统安全性工程工作从一开始就嵌入到工程流程中,以便在做出工程设计决策时,可以将安全性设计到系统中。相对于系统安全性工程,SE 确定系统的目的,并参与将要避免的潜在危险的识别和文件化。从这一信息中,可以识别并文件化系统功能需求、安全需求和约束的集合。这些需求将为系统的设计和运行奠定基础,以确保将安全性设计到系统中。SE 必须从早期概念阶段开始确立系统安全性工程,并贯穿于系统的生命周期继续该流程。SE 应确保在考虑系统需求和约束的同时,通过安全考量来指导设计决策。

10.10.2 识别并综合系统安全需求

系统安全工程师根据联邦、军事、国家及行业的监管、规程、标准和其他用于设计和开发系统的文件来审查并识别可应用的"最佳实践"系统安全设计需求和指南[例如,联邦机动车安全标准(FMVSS)、军用标准(MIL-STD)、国家电气规程(NEC)以及化学品的登记、评价、授权和限制(REACH)]。利用这些初始需求来推导为设计工程师提供的附加系统安全设计需求,以消除危害风险或降低到可接受的水平。这些需求被综合到高层级系统需求和设计文件中。系统安全需求则与在系统中识别的危险有关。

系统安全性工程与 SE 一起确保开发、细化、完整和准确地规定系统安全性设计需求和指南,并恰当地转换成系统元素需求,以确保在系统硬件、软件和用户接口的设计和开发中予以实施。此外,为了加入到程序、流程、警告和警示中,还要识别适用的安全需求,以供操作人员、用户和诊断手册使用。

10.10.3 危险的识别、分析和分类

在系统安全性工程的学科内,有众多的解析方法、技巧和产品,它们在行业内被视为最佳实践并被接受。例如,"SAE 国际"拥有 SAE ARP 4754 和 4761 中定义的、在航空行业公共使用的特定方法。美国国防部 MIL-STD-882 也定义了在防务领域有用的特定分析技巧(DoD, 2010b)。不管使用什么标准或指南,以下分析技巧和安全性工程制品(非详尽列表)反映 SE 最佳实践:

SPECIALTY ENGINEERING ACTIVITIES

- Preliminary hazard analysis (PHA)
- Functional hazard analysis (FHA)
- System element hazard analysis (SEHA)
- System hazard analysis (SHA)
- Operations and support hazard analysis (O&SHA)
- Health hazard analysis (HHA)
- FTA
- Probabilistic risk assessment (PRA)
- Event tree analysis (ETA)

System safety engineers begin the identification of hazards at the beginning of system concept definition. A hazard analysis is initiated to identify potential hazards and their mishap potential that may be inherent in the concepts under consideration. The system safety engi-neer draws from safety experience on similar systems, including mishap/incident hazard tracking logs, safety lessons learned, and design guidelines to develop this list. As the system matures through the development cycle, the hazard analysis is updated to identify and analyze new hazards resulting from design changes.

An in-depth causal analysis is conducted for each identified hazard. This analysis identifies all contributions from hardware, software, and any/all control entities including the human operator, which could cause the hazard to occur. The identification of each causal factor allows the system safety engineer, along with the systems engineer, to identify specific system safety requirements/constraints necessary to mitigate the hazard and reduce the risk to an acceptable level.

Each hazard is analyzed to determine its severity and probability of occurrence. The hazard's severity and probability determine its risk categorization. Table A.I of MIL-STD-882E shows an example of mishap severity categories (DoD, 2010b), and Table A.II shows mishap probability levels. Mishap risk classification is per-formed by using a mishap risk assessment matrix. This matrix is used to rank the mishap risk potential of the hazards. Table A.III of MIL-STD-882E shows an example of the mishap risk assessment matrix (DoD, 2010b). This matrix is used to prioritize engineering efforts to mitigate the system hazards.

- 初步危险分析（PHA）
- 功能危险分析（FHA）
- 系统元素危险分析（SEHA）
- 系统危险分析（SHA）
- 运行及支持危险分析（O&SHA）
- 健康危险分析（HHA）
- FTA
- 概率风险评估（PRA）
- 事件树分析（ETA）

系统安全工程师从系统概念定义时就开始识别危险。启动危险分析，以识别潜在的危险及其在考虑之中的概念中可能固有的潜在事故。系统安全工程师在相似的系统上吸取安全经验，包括事故/事件危险跟踪日志、吸取的安全教训以及开发该列表的设计指南。随着系统在开发周期中成熟时，更新危险分析，以识别并分析由设计更改带来的新危险。

对识别的每个危险进行深入的原因分析。这项分析识别来自于硬件、软件和任何/所有控制实体（包括操作人员）的、可能会导致危险出现的全部"因素"。对每个原因因素的识别允许系统安全工程师以及系统工程师识别缓解危险并将风险降至可接受水平所需的特定系统安全需求/约束。

分析每个危险，以确定危险的严重程度及发生概率。危险的严重程度和发生概率决定它的风险分类。MIL-STD-882E 中的表 A.I 示出了事故严重程度分类的示例（DoD，2010b），表 A.II 示出了事故概率等级。事故风险分类利用事故风险评估矩阵执行。利用该矩阵对危险的潜在事故风险进行排序。MIL-STD-882E 中的表 A.III 示出了事故风险评估矩阵的示例（DoD，2010b）。利用该矩阵对工程工作进行优先顺序排列，以缓解系统危险。

SPECIALTY ENGINEERING ACTIVITIES

Hazards associated with software cannot rely solely on the hazard probability as identified in Table A.II of MIL-STD-882E (DoD, 2010b). Categorization of a software failure is generally determined by its hazard severity and the degree of command, control, and autonomy the software functionality exercises over the hardware. The software control categories include the following: (i) autonomous control, (ii) software exercises control or displays information allowing time for intervention by either an independent safety system or an operator, (iii) software issues commands or generates information requiring operator action to complete the control, and (iv) software does not control safety-critical hardware or provide safety-critical information. A soft-ware criticality index is then established using the severity categories and the control categories. This matrix can then be used to prioritize the level of rigor (LOR) assigned to the design, code, and test of the software. Safety-significant software is developed to a specific LOR to bring confidence that the software performs as expected functionally and does not perform unintended functions.

Hazards are prioritized so that corrective action efforts can be focused on the most serious hazards first. The goal of the system safety effort is to work with engineering to design systems that contain no hazards. Since it is impossible or impractical to design a system completely free from hazards, the effort is focused on developing a system design where there are no hazards with an unacceptable level of mishap risk. Each hazard identified is analyzed to determine the requirements to be incorporated into the design to reduce the risk associated with the hazard to an acceptable level. The system safety order of precedence is used to define the order followed for implementing system safety requirements to reduce the mishap risk. The order of precedence is as follows:

1. Eliminate hazards through design selection.
2. Reduce risk through design alteration.
3. Incorporate safety devices.
4. Provide warning devices.
5. Develop procedures and training.

Best practices dictate a function whose failure to operate or whose incorrect operation will directly result in a mishap of either catastrophic (death, permanent total disability, irreversible significant environmental impact, or monetary loss equal to or exceeding $10M) or critical (hospitalization of at least three personnel, reversible significant environmental impact, or monetary loss equal to or exceeding $1M but less than $10M). Severity cannot be mitigated solely through procedural mitigations.

与软件相关的危险不能完全依赖于在 MIL-STD-882E 的表 A.II 中识别的危险概率（DoD，2010b）。软件故障的分类通常通过其危险程度以及软件功能性运用于硬件的命令、控制及自主程度来确定。软件控制分类包括：（i）自主控制；（ii）软件实行控制或显示允许独立的安全系统或操作人员干预时间的信息；（iii）软件发布命令或产生要求操作者采取行动完成控制的信息；以及（iv）软件不控制安全性关键的硬件或提供安全性关键的信息。然后，利用严重程度分类和控制分类来建立软件的关键指数。其后，利用该矩阵对指定到软件的设计、规程和试验中的严苛层级（LOR）进行优先顺序排列。为特定的 LOR 开发安全性重要的软件，以为软件在功能上按照预期执行并且不执行计划外功能带来信心。

对危险进行优先排序，以便使纠正行动工作可以首先聚焦于最严重的危险。系统安全工作的目的是，研究设计不含危险的系统的工程。因为设计一个完全无危险的系统是不可能的或不切实际的，所以工作聚焦于开发一个没有不可接受的事故风险水平的危险的系统设计。分析所识别的每个危险，以确定将纳入设计中的需求，从而将该危险所涉及的风险降至可接受水平。系统安全先后顺序被用来定义实施系统安全需求所遵循的顺序以减少事故风险。先后顺序如下：

1. 通过设计选择消除危险。

2. 通过设计调整降低风险。

3. 纳入安全装置。

4. 提供警告装置。

5. 制定程序和培训。

众多实践记录，运行故障或不正确的运行将会直接导致毁灭性的（死亡、永久性完全残疾、不可逆转的重大环境影响或者大于等于 1 000 万美元的资金损失）或关键性的（至少三个人员住院治疗、可逆转的重大环境影响或者大于等于 100 万美元但小于 1 000 万美元的资金损失）事故的功能。严重程度不能仅仅通过程序上的缓和来减轻。

SPECIALTY ENGINEERING ACTIVITIES

The results of the hazard analysis activities are captured in a hazard tracking system (HTS). The HTS is used to track each hazard by documenting the implementation of the safety requirements, verification results, and residual mishap risk. The HTS is updated throughout the life cycle of the system.

10.10.4 Verify and Validate System Safety Requirements

System safety engineers, along with the systems engineers, provide input to all tests, demonstrations, models, and inspections to verify compliance of the system with the identified system safety requirements. This is done to ensure the safety of the design is adequately demonstrated for all hazards not eliminated by design. System safety engineers typically witness verification/validation activities for those hazards that were categorized as catastrophic and critical. Results from all test activities performed to validate/verify system safety requirements are reviewed by system safety engineers and captured in the HTS as well as the test and evaluation reports on hardware or software.

10.10.5 Assess Safety Risk

System safety engineers perform and document a comprehensive evaluation of the mishap risk being assumed prior to each key milestone of the program (preliminary design review, critical design review, program completion, etc.) as well as key test or operation activities. The safety assessment identifies all safety features of the hardware, software, and system design. It also identifies procedural, hardware, and software related hazards that may be present in the system at each milestone, test, or operation activity. Specific procedural controls and precautions are identified for those hazards still present in the system. The safety assessment also identifies and documents hazardous materials used in the design, operation, or maintenance of the system as well as those that will be generated by the system. An assessment as to why non- or less hazardous materials could not be used is developed and documented.

10.10.6 Summary

As an integral part of SE, system safety engineering focuses heavily on ensuring that the design engineers are provided a complete set of safety-related requirements to minimize the safety risk potential of the system during the development, production, utilization, support, and retirement stages. The end objective is to deploy, operate, and maintain a system that possesses an acceptable safety risk. The objective is ideally to be accident-free.

在危险跟踪系统（HTS）中捕获危险分析活动的结果。通过文件化安全需求、验证结果和残余事故风险的实施，HTS 被用来跟踪每个危险。贯穿于系统的生命周期更新 HTS。

10.10.4 验证并确认系统安全需求

系统安全工程师连同系统工程师一同向所有测试、验证、模型和检验提供输入，以验证系统与所识别的系统安全需求的合规性。这样做是为了确保针对通过设计未消除的全部危险充分证实设计的安全性。系统安全工程师通常见证用于被归类为灾难性和重大事故的那些危险的验证/确认活动。为确认/验证系统安全需求所执行的全部试验活动的结果由系统安全工程师来审查，并且在 HTS 以及关于硬件或软件的测试和评价报告中捕获。

10.10.5 评估安全风险

系统安全工程师实施并文件化项目计划中的每个关键里程碑（初步设计审查、关键设计审查、项目结束等）以及关键测试或运行活动之前假定的事故风险的综合性评价。安全性评估识别硬件、软件和系统设计的全部安全性特征。它还识别在每个里程碑、测试或运行活动可能存在于系统中的与程序、硬件和软件相关的危险。针对仍然存在于系统中的那些危险识别特定的程序控制和预防措施。安全性评估还识别并记录在系统的设计、运行或维护中使用的危险材料以及系统将产生的危险材料。开发并文件化关于为什么不能使用非危险材料或低危险材料的评估。

10.10.6 总结

作为 SE 的组成部分，系统安全性工程特别聚焦于确保为设计工程师提供一整套安全性相关的需求，以在开发、生产、使用、保障和退役阶段将系统的潜在安全风险降至最低程度。最终目标是，部署、运行和维护拥有可接受的安全风险的系统。理想的目标是无事故。

SPECIALTY ENGINEERING ACTIVITIES

10.11 SYSTEM SECURITY ENGINEERING

System security engineering is focused on ensuring a system can function under disruptive conditions associated with misuse and malicious behavior. System security engineering involves a disciplined application of SE principles in analyzing threats and vulnerabilities to systems and assessing and mitigating risk to the information assets of the system during its life cycle. It applies a blend of technology, management principles and practices, and operational rules to ensure sufficient protections are available to the system at all times.

System security engineering considers and accounts for the system and the associated environment. Sources of potential disruptive conditions (threats) are many and varied. They may be natural (e.g., weather) or man-made. They may emanate from external sources (e.g., political or power interruptions) or may be caused by internal forces (e.g., user or supporting systems). A disruption may be unintentional or intentional (malicious) in nature. The security capabilities, whether implemented through design, policy, or practice, must be usable from the user's perspective.

To be effective, system security engineering is applied throughout the life cycle of the system. System security engineering activities can be applied to each life cycle stage:

- *Concept*—System security engineering explores technology trends and advancements to identify potential technologies and promising security strategies to address current and future threats and support architectural and operational concepts that provide security to support and protect operational needs.

- *Development*—System security engineering ensures security concepts are translated into functional with verifiable requirements and defines security-level effectiveness during this stage.

- *Production*—System security engineering provides support during fabrication, build, and assembly to ensure that security settings are properly initialized and delivered with the final system and establishes security-level effectiveness during this stage.

- *Utilization*—System security engineering maintains security effectiveness during use by considering changes in operational, user, and threat environments.

- *Support*—System security engineering ensures that security features are updated and remain effective after maintenance and monitors security events to maintain security effectiveness.

10.11 系统安保性工程

系统安保性工程聚焦于确保系统可以在与误用和恶意行为相关的破坏性条件下运行。系统安保性工程涉及在分析系统的威胁和脆弱性中以及在系统的生命周期期间评估并降低系统信息资产的风险中严格地应用 SE 原则。它应用技术、管理原则和实践以及运行条例的混合,以确保可以随时对系统采取充分的保护。

系统安保性工程考虑并解释系统及相关的环境。潜在的破坏性条件(威胁)的来源多种多样。它们可能是自然的(如,天气),也可能是人为的。它们可能起源于外部来源(如,政治上的或动力中断),或者可能由内部力量(如,用户或支持系统)造成。破坏在本质上可能是非故意的,也可能是故意的(恶意的)。安保能力,不管是否通过设计、策略或实践来实施,都必须可从用户的视角是有用的。

为了有效,系统安保性工程贯穿于系统的生命周期应用。系统安保性工程活动可以适用于每个生命周期阶段:

- *方案*——系统安保性工程探索技术趋势和进展以识别潜在技术,并承诺安保策略以应对当前的和未来的威胁且支持架构和运行方案,为支持和保护运行需要提供安保。

- *开发*——系统安保性工程确保安保方案被转换成功能的、可验证的需求,并且在本阶段定义安保层级效能。

- *生产*——系统安保性工程在制造、构建和组装期间提供支持,以确保对安保设置进行适当地初始化并与最终系统一起交付,还在本阶段建立安保层级效能。

- *使用*——系统安保性工程在使用期间通过考虑运行、用户和威胁环境中的变化来保持安保效能。

- *支持*——系统安保性工程确保安保特征被更新,且在维护后仍然有效,并监控安保事件,以保持安保效能。

SPECIALTY ENGINEERING ACTIVITIES

- *Retirement*—System security engineering ensures that effective security practices are employed during retirement of the system and associated information.

10.11.1 Systems Engineer and System Security Engineer Roles and Responsibilities

System security engineering brings security-focused disciplines, technology, and concerns into the SE process to ensure that protection considerations are given the proper weight in decisions concerning the system (Dove et al., 2013). Systems engineers must employ security subject matter experts and security engineers on a timely basis to perform system security analysis and provide effective security recommendations.

System security engineering consists of subspecialties such as anti-tamper, supply chain risk management, hardware assurance, information assurance, software assurance, system assurance, and others that the system security engineer needs to trade off to provide a balanced system security engineering view. An example of the relationship of SE and system security engineering to the security expert domain is illustrated by the joint contribution to defining and deploying an engineering solution to meet user needs. Some of these joint work products include:

- System security plan
- Vulnerability assessment plan
- Security risk management plan
- System security architecture views
- Security test plan
- Deployment plan
- Disaster recovery and continuity plan

10.11.2 System Security Engineering Activities for Requirements

Stakeholder security interests include intellectual property, information assurance, security laws, supply chain compliance, and security standards. Examples of standards include ISO/IEC 27002, Information security standard (2013); Chapter 13 of the Defense Acquisition Guide (DAU, 2010); and the Engineering for Systems Assurance Guide (NDIA, 2008). Systems engineers need to consider stakeholder security interests during the stakeholder needs and requirements definition process and should be captured in the stakeholder requirements.

- *退役*——系统安保性工程确保在系统及相关信息退役期间采用有效的安保实践。

10.11.1 系统工程师和系统安保性工程师的作用及职责

系统安保性工程将聚焦安保的学科、技术和关注点汇集到 SE 流程中，以确保在做出系统相关决策时适当地给予保护考虑（Dove 等，2013）。系统工程师必须采用安保领域专家和安保工程师及时地执行系统安保性分析和提供有效的安保性建议。

系统安保性工程包括诸如防篡改、供应链风险管理、硬件保证、信息保证、软件保证、系统保证和系统安保性工程师需要权衡以提供平衡的系统安保性工程视图的其他方面等子专业。通过定义和部署满足用户需要的解决方案的共同贡献，阐明了在安保专家领域内的 SE 和系统安保性工程的关系示例。这些共同工作产品中的一些是：

- 系统安保计划
- 脆弱性评估计划
- 安保风险管理计划
- 系统安保架构视图
- 安保测试计划
- 部署计划
- 灾难恢复和连续性计划

10.11.2 面向需求的系统安保性工程活动

利益攸关者安保利益包括知识产权、信息保证、安保法、供应链合规性和安保标准。标准示例包括 ISO/IEC 27002 信息安保标准（2013）；防务采办指南的第 13 章（DAU，2010）；以及系统保证工程指南（NDIA，2008）。系统工程师需要在利益攸关者需要及需求定义流程中考虑利益攸关者安保利益，并且应在利益攸关者需求中予以捕获。

During system requirements definition, the critical functions and data are identified that are most in need of protection. A risk-based analysis pattern of criticality analysis, threat assessment, vulnerability assessment, and identification of potential protection controls and mitigations are used as inputs to a CBA. The CBA takes into account impacts to system performance, afford- ability, and usage compatibility. Security scenarios are developed for both normal security processing and misuse/abuse situations to assist with system requirements definition. These scenarios are also important for use during verification and validation.

System requirements definition needs to consider system security protections. System security protections can be grouped into three categories: prevention, detection, and response. Prevention includes access control and critical function isolation and separation. Detection includes functions that monitor and log security-related behavior. Response includes mitigations that switch to a degraded mode when the primary function or data has been compromised.

The results of these activities are a set of system security requirements (including process requirements), security OpsCon and support concept, and a set of scenarios to be used for verification and validation.

10.11.3 System Security Engineering for Architecture Definition and Design Definition

Engagement of system security engineering in the architecture definition and design definition processes is important because system architecture enables or impedes system security. Adversaries learn system protective measures and change their methods rapidly. Architecture should enable rapid change of protective measures, facilitating operational-time engineering intervention. This is achieved by modifying architectural structure and security functional detail and is enabled by an agile architectural strategy articulated in the system OpsCon. Evolving threats can be countered with a security architecture composed of loosely coupled encapsulated security functional system elements that can be replaced, augmented with additional functionality, and reconfigured for different interconnections (Dove et al., 2013).

Long system life expectancies are especially vulnerable to unadaptable, rigid security architectures.

Architecture focuses on the high-level allocation of responsibilities between different system elements of the SOI and defines the interactions and connectivity between those system elements. Responsibility for security requirements established during the requirements processes is allocated to functional system elements and security-specialty system elements as appropriate during the architectural design process.

在系统需求定义期间，识别需要保护的大多数的关键功能和数据。关键性分析、威胁评估、脆弱性评估和潜在的保护控制及缓解的识别的基于风险的分析模式被用作 CBA 的输入。CBA 考虑到对系统性能、可承受性和使用兼容性的影响。针对正常的安保处理和误用/滥用情况开发安保场景，以有助于系统需求定义。这些场景对于在验证和确认期间使用而言也很重要。

系统需求定义需要考虑系统安保保护。系统安保保护可以组合成三类：预防、检测和响应。预防包括访问控制和关键的功能隔离及分离。检测包括监控并记录安保相关行为的功能。响应包括在损害了主要功能或数据时切换到降级模式的缓解。

这些活动的结果是系统安保需求（包括流程需求）、安保运行概念 OpsCon 和支持概念的集合以及用于验证和确认的方案的集合。

10.11.3 针对架构定义和设计定义的系统安保性工程

系统安保性工程参与架构定义和设计定义流程很重要，因为系统架构使能或阻碍系统安保。攻击对手将学习系统防护措施并迅速地改变他们的方法。架构应使防护措施能速度改变，以辅助运行时的工程干涉。这一点通过修改架构结构和安保功能细节的方式实现，并且由系统运行概念中链接的敏捷架构策略促动。演进中的威胁可以用安保架构反击，安保架构由松耦合的封装的安保功能系统元素构成，这些元素可以被附加的功能性取代、增加并重新配置用于不同的互连（Dove 等，2013）。

长系统生命期望尤其易受非适应的、刚性的安保架构的影响。

架构聚焦于在 SOI 的两个不同系统元素之间高层级职责分配，并定义这些系统元素之间的交互和连通。在需求流程期间建立的安保需求职责在架构设计流程中被酌情分配给功能系统元素和安保专业系统元素。

SPECIALTY ENGINEERING ACTIVITIES

Resilience engineering works closely with system security engineering. System resilience permits a system to operate while under, and recover after, attack, perhaps with degraded performance but with continued delivery of critical functionality. Resilience is an architectural feature that is difficult to provide later in the development life cycle and is very costly after deployment.

Long life systems will have functional upgrades and system element replacements throughout their life. Insider threat and supply chain threat may manifest as system elements developed with embedded malicious capabilities, which may lie dormant until activated on demand. This suggests self-protective system elements that distrust communications and behaviors of interconnected system elements, rather than relying on system perimeter protection or trusted environment expectations.

10.11.4 System Security Engineering Activities for Verification and Validation

Verification and validation processes develop strategies for verifying and validating the SOI. System security engineering's involvement is needed for verifying security-related requirements and security impacts to enabling systems. Verification methods are identified for each security requirement with objective pass/fail criteria for inspection, analysis, demonstration, and test. Milestone reviews are conducted to confirm that measures have been planned, threat and vulnerability assessments are current, and system security requirements under test map to a comprehensive criticality analysis. validation is performed on systems using assessment scenarios that focus on evolving threats in operational environments; risk evaluations and criticality assessments are updated based on current threats and vulnerabilities; and end-to-end scenarios are updated to address new vulnerabilities and threats.

10.11.5 System Security Engineering Activities for Maintenance and Disposal

System security engineering responsibilities do not end when the system is delivered. As part of maintenance, work instructions need to indicate authorized maintenance activities to prevent vulnerability insertion. As part of sustainment operations, threats and security features need to be reevaluated to determine if there are new vulnerabilities in the existing systems. As part of any capability upgrade or technology refresh, criticality analysis must be repeated in the context of new threats, end-to-end scenarios reevaluated to confirm effective protection, and supply chain vulnerabilities evaluated for updated equipment. Prior to system disposal, hardware and software must be brought to a state that cannot be reverse engineered.

可恢复性工程与系统安保性工程密切合作。系统弹性使系统在受到攻击和恢复后，运行中也许性能降低但关键功能性还能持续交付。可恢复性是在开发生命周期后期很难提供并且在部署后极其昂贵的架构特征。

长生命系统将贯穿于其生命周期进行功能升级和系统元素更换。内部人员威胁和供应链威胁可能表示用嵌入的恶意能力开发的系统元件，可以潜伏，直到需要时激活。这表明，自我保护系统元素不相信互连系统元素的沟通和行为，而不是依靠系统周边保护或可信的环境期望。

10.11.4 针对验证和确认的系统安保性工程活动

验证和确认流程开发 SOI 的验证及确认策略。为了验证安保相关需求以及安保对使能系统的影响，需要系统安保性工程的加入。利用目标通过/失效的准则针对每个安保要求识别验证方法，以进行检验、分析、验证和测试。进行里程碑审查，以确认已经计划了测度，威胁和脆弱性评估是最新的，处于测试阶段的系统安保需求映射到综合的关键性分析。利用聚焦于在运行环境中演进威胁的评估场景对系统进行确认；基于最新的威胁和脆弱性更新风险评价和关键性评估；以及更新端对端场景，以应对新的脆弱性和威胁。

10.11.5 针对维护和退出的系统安保性工程活动

系统安保性工程职责并不在交付系统时结束。作为维护的一部分，工作说明需求表明授权的维护活动防止脆弱性插入。作为维持运行的一部分，需要重新评价威胁和安保特征，以确定现有系统中是否有新的脆弱性。作为任何能力升级或技术更新的一部分，必须在新威胁的背景环境中重复关键性分析，重新评价端对端场景以确认有效防护，评价所更新设备的供应链脆弱性。在系统退出之前，必须使硬件和软件处于不能被逆向工程的状态。

SPECIALTY ENGINEERING ACTIVITIES

10.11.6 System Security Engineering Activities for Risk Management

A mission criticality analysis, a threat assessment, and a vulnerability assessment can be used as security inputs to the risk management process to improve the objectivity of the risk identification and risk-level determination. Balancing security risk reduction within the context of the overall system performance, cost, and schedule requires an objective CBA. The earlier the security risk identification and analysis is done, the more effectively security can be built into the designs, development process, and supply chain. Because of the dynamic innovative nature of threats and the continuous discovery of vulnerabilities, the risk identification and analysis needs to be repeated often throughout the system life cycle.

10.11.7 System Security Engineering Activities for Configuration and Information Management

Configuration and information management ensures that the state of the system is known in total only to those authorized to understand its configuration and capability. Configuration and information management should allow members of the design team to interact with the portions of the system they are authorized to see in order to maintain a consistent and accurate view of the system as it evolves from concept to delivery. Changes to the configuration must be controlled to restrict access for changing the system and documented to provide insight for forensic analysis in the event of attack or discovery of vulnerabilities.

10.11.8 System Security Engineering Activities for Acquisition and Supply

System security considerations go beyond the SOI. The enabling systems and supply chain must be protected, and all aspects of assembling a solution must be known. System security analysis, evaluation, protection implementation, and updates should be part of the request for proposal to ensure that the acquired system or system element is delivered with acceptable risk.

Vulnerability assessments must be made and updated throughout the life cycle. Hardware and software products that are COTS, while often considered for affordability, may have limited assurance that malicious insertion has not occurred. COTS products may also be used outside the SOI, which allows vulnerability discovery that could be exploited against the SOI.

The system security engineer evaluates the consequence and likelihood of losing business/mission capability in order to select system elements that have secure supply chains. In some cases, the best-performing COTS system element may not be selected if the supply chain risk cannot be reduced through countermeasures to an acceptable level.

10.11.6 针对风险管理的系统安保性工程活动

任务关键性分析、威胁评估和脆弱性评估可以被用作风险管理流程的安保输入，以改善风险识别和风险层级测定的客观性。在总体系统性能、成本和进度的背景环境内平衡安保风险降低需要目标 CBA。安保风险识别和分析做得越早，安保越可以更有效地构建到设计、开发流程和供应链中。由于威胁的动态创新型本质属性和脆弱性的持续发现，往往需要贯穿于系统生命周期重复风险识别和分析。

10.11.7 针对构型和信息管理的系统安保性工程活动

构型和信息管理确保只有那些被授权知悉系统构型和能力的人员知道系统的状态。构型和信息管理应允许团队的成员与他们被授权观察的系统各个部分交互，以便在从概念演变为交付时保持系统的一致、准确的视图。必须控制构型的变更，以限制改变系统的机会，并进行文件化，以在攻击或发现脆弱性的情况下提供对司法分析的理解。

10.11.8 针对采办和供应的系统安保性工程活动

系统安保考虑要超出 SOI 的范围。必须保护使能系统和供应链，并且必须知道组合一个解决方案的所有方面。系统安保分析、评价、保护实施以及更新应为建议请求的一部分，以确保所采办的系统或系统元素以可接受的风险交付。

必须进行脆弱性评估并贯穿于生命周期进行更新。属于 COTS 但往往考虑到可承受性的硬件和软件产品可能具有未出现恶意插入的有限保证。COTS 产品还可能用于 SOI 以外，这就允许针对 SOI 能利用的脆弱性发现。

系统安保性工程师评价丧失业务/任务能力的后果和可能性，以便选择具有安保供应链的系统元素。在一些情况下，如果供应链风险无法通过对策降低到可接受水平，则可以选择表现最好的 COTS 系统元素。

SPECIALTY ENGINEERING ACTIVITIES

10.12 TRAINING NEEDS ANALYSIS

Training needs analyses support the development of products and processes for training the users, maintainers, and support personnel of a system. Training analysis includes the development of personnel capabilities and proficiencies to accomplish tasks at any point in the system life cycle to the level they are tasked. These analyses address initial and follow-on training necessary to execute required tasks associated with system use and maintenance. An effective training analysis begins with a thorough understanding of the concept documents and the requirements for the SOI. A specific list of functions or tasks can be identified from these sources and represented as learning objectives for operators, maintainers, administrators, and other system users. The learning objectives then determine the design and development of the training modules and their means of delivery.

Important considerations in the design of training include who, what, under what conditions, how well each user must be trained, and what training will meet the objectives. Each of the required skills identified must be transformed into a positive learning experience and mapped onto an appropriate delivery mechanism. The formal classroom environment is rapidly being replaced with or augmented by simulators, computer-based training, Internet-based distance delivery, and in-systems electronic support, to name a few. Updates to training content use feedback from trainees after they have some experience to improve training effectiveness.

10.13 USABILITY ANALYSIS/HUMAN SYSTEMS INTEGRATION

Human systems integration (HSI) is the interdisciplinary technical and management process for integrating human considerations within and across all system elements. HSI focuses on the human, an integral element of every system, over the system life cycle. It is an essential enabler to SE practice as it promotes a "total system" approach that includes humans, technology (e.g. hardware, software), the operational context, and the necessary interfaces between and among the elements to make them all work in harmony (Bias and Mayhew, 1994; Blanchard and Fabrycky, 2011; Booher, 2003; Chapanis, 1996; ISO 13407, 1999; Rouse, 1991). The "human" in HSI includes all personnel who interact with the system in any capacity, such as:

- System owners
- Operators
- Maintainers
- Trainers
- Users/customers

10.12 培训需要分析

培训需要分析支持用于培训系统用户、维护人员和保障人员的产品和流程的开发。培训分析包括在系统生命周期任何一点上按照任务委派级别完成任务的人员能力和熟练程度的开发。这些分析应对执行与系统使用和维护相关的规定任务所需的初期和后续培训。有效的培训分析开始于对概念文件以及 SOI 的需求的彻底理解。功能或任务的特定列表可根据这些来源确定，并可表达为操作人员、维护人员、行政管理人员和其他系统用户的学习目标。学习目标则决定培训模块及其交付手段的设计和开发。

培训设计的重要考虑因素包括：何人、何事、在何条件下、每个用户必须被培训至何种程度以及何种培训将满足目标。识别后的每个所要求的技能必须被转换为积极的学习经验并映射到适当的交付机制上。正规的课堂环境将很快被模拟器、基于计算机的培训、基于互联网的远程交付以及系统内置的电子支持（仅列举几个）所替换或增强。在受训人员具备一定经验后，使用受训人员的反馈更新培训内容，以提升培训的有效性。

10.13 可用性分析/人与系统综合

人与系统综合（HSI）是在所有系统元素内且跨所有系统元素综合人为因素的跨学科的技术和管理流程。HSI 在整个系统生命周期中聚焦于人，即每个系统的组成元素。它对于 SE 实践而言是一种基本的使能项，因为它提倡一种"全系统"方法，这种方法包括人员、技术（即硬件、软件）、运行背景环境和位于两个和更多个元素之间使其都协调工作的所需接口。Blanchard 和 Fabrycky，2011；Booher，2003；Chapanis，1996；ISO 13407，1999；Rouse，1991）。HSI 中的"人员"包括以任何能力与系统交互的所有人员，诸如：

- 系统所有者

- 操作人员

- 维护人员

- 培训人员

- 用户/客户

SPECIALTY ENGINEERING ACTIVITIES

- Decision makers
- Support personnel
- Peripheral personnel

Humans are an element of most systems, so many systems benefit from HSI application. HSI establishes human-centered disciplines and concerns into the SE process to improve the overall system design and performance. The primary objective of HSI is to ensure that human capabilities and limitations are treated as critical system elements, regardless of whether humans in the system operate as individuals, crews, teams, units, or organizations. The technology elements of the system have inherent capabilities; similarly, humans possess particular knowledge, skills, and abilities (KSAs), expertise, and cultural experiences. Deliberate design effort is essential to ensure development of quality interfaces between technology elements and the system's intended users, operators, maintainers, and support personnel in operational environments. It is also important to acknowledge that humans outside the system may be affected by its operation.

While many systems and design engineers intuitively understand that the human operator and maintainer are part of the system under development, they often lack the expertise or information needed to fully specify and incorporate human capabilities. HSI brings this technical expertise into the SE process and serves as the focal point for human considerations in the system concept, development, production, utilization, support, and retirement stages. The comprehensive application of HSI to system development, design, and acquisition is intended to optimize total system performance (e.g., humans, hardware, and software) while accommodating the characteristics of the population that will use, operate, maintain, and support the system and also support efforts to reduce LCC.

A key method of HSI is trade studies and analyses. HSI analyses, especially requirements analyses that include human issues and implications, often result in insights not otherwise realized. Trade studies that include human-related issues are critical to determining the design that is the most effective, efficient, suitable (including useful and understandable), usable, safe, and affordable.

HSI helps systems engineers focus on long-term costs since much of that cost is directly related to human element areas. One example (unfortunately, of many) is the Three Mile Island power plant nuclear incident in the United States:

The accident was caused by a combination of personnel error, design deficiencies, and component failures. The problems identified from careful analysis of the events during those days have led to permanent and sweeping changes in how NRC regulates its licensees—which, in turn, has reduced the risk to public health and safety. (NRC, 2005)

- 决策人员
- 支持人员
- 外围人员

人是大多数系统的一种元素，因此，很多系统都从 HSI 应用中受益。HSI 建立以人为中心的规程，并关注 SE 流程，以改善总体系统设计和性能。HSI 的主要目标是确保人的能力和局限性被作为关键系统元素处理，无论系统中的人员是作为个体、群体、团队、单位还是组织运行。系统的技术元素具有固有的能力；同样地，人也拥有特殊的知识、技能和能力（KSA）、专长和文化阅历。周密设计工作对于在运行环境中确保技术元素与系统预期用户、操作人员、维修人员和保障人员之间的高品质接口的开发非常重要。对于确认位于系统外部的人员会受到其影响也非常重要。

虽然许多系统和设计工程师直观地了解到人类操作者和维护者是正在开发的系统的一部分，但他们通常缺乏充分规定并纳入人员能力所需的专业知识或信息。HSI 将这种技术专长带入 SE 流程，并充当在系统方案、开发、生产、使用、保障和退役阶段中的人为因素的聚焦点。将 HIS 全面应用于系统开发、设计和采办的目的是，在适应使用、运行、维护和保障系统的人员特点的同时，优化全系统性能（即人、硬件和软件），并支持减少 LCC 的工作。

HSI 的关键方法是权衡研究及分析。HSI 分析，尤其是包括人员问题和隐含的需求分析，往往导致在其他情况无法实现的洞察。包括人员相关问题的权衡研究对于确定最有效的、高效率的、适当的（包括有用的和可理解的）、可用的、安全的和可承受的设计很关键。

HSI 帮助系统工程师聚焦于长期成本，因为大量这类成本与人素域直接相关。一个示例（很不幸，有很多）是美国三里岛的发电厂核事件。

该事故是由人为过失、设计缺陷和部件故障综合造成的。在那些日子根据事件的仔细分析识别的问题已经引起了关于 NRC 永久地和彻底地改变如何监管其被许可方的方式，这反过来也降低了公共健康和安全的风险。（NRC，2005）

SPECIALTY ENGINEERING ACTIVITIES

Failure to include HSI within a comprehensive SE frame-work resulted in loss of confidence in nuclear power in the United States and delayed progress in the field for almost 30 years. The cleanup costs, legal liability, and significant resources associated with responding to this near catastrophe trace directly to lack of attention to the human element of a highly complex system, resulting in flawed operations technology and work methods. It also emphasized that while human performance includes raw efficiency in terms of tasks accomplished per unit time and accuracy, human performance directly impacts the overall system performance.

10.13.1 HSI is Integral to the SE Process

The foundation for program success is rooted in requirements development. Human performance requirements are derived from and bounded by other performance requirements within the system. Front-end analyses (FEA) generate system requirements and incorporate HIS-related requirements. Effective FEA start with a thorough understanding of the mission of the new system and the work to be performed, successes or problems with any predecessor systems, and the KSAs and training associated with the people who are likely to interact with the proposed system technology. HSI modeling views the system as a collection of interrelating elements that behave according to a shared organizing principle. This perspective underpins models and simulations with mathematical rigor approaching that of other engineering disciplines (SEBoK, 2014). It also highlights the essential HSI truth—there are no unpopulated systems. Simulation early in the development process, particularly before system hardware and software elements are developed, ensures all the interfaces are captured for requirements definition, trade space analysis, and iterative design activities. HSI analyses allocate human-centered functions within the system and identify potential human (or system) capability gaps. For example, humans excel at solving induction problems, and machines excel at deduction (Fitts, 1954). The requirement for inductive or deductive decision making is inherent in the structure of the system design.

It is critical to include HSI early in system development (during stakeholder requirements generation) and continuously through the development process to realize the greatest benefit to the final system solution and substantial LCC savings. Fully utilizing HSI in IPPD helps ensure that systems will not require expensive "train arounds" or latestage fixes to correct ineffective usability and operational inefficiencies driven by bad, unspecified, or undefined human interfaces. The systems engineer plays a critical role in engaging HSI expertise to support the IPDTs. A knowledgeable, interdisciplinary HSI team is generally required to address the full spectrum of human considerations, and the systems engineer is key to ensuring that HSI is included throughout the system's life cycle. Program managers, chief engineers, and systems engineers should ensure that HSI practitioners are actively participating in design reviews, working groups, and IPDTs. Consistent involvement and communications with customers, users, developers, scientists, testers, logisticians, engineers, and designers (human, hardware, and software) are essential.

未能将 HSI 包括在一个广泛的 SE 框架内导致了美国丧失核能方面的信心，使其在该领域的进步推迟了近 30 年。回应这次近期灾难相关的清理成本、法律责任以及重要资源，可直接追溯到对高度复杂系统中人员元素关注的缺乏，从而造成了有缺陷的运行技术和工作方法。还应强调，当人的效能包括每单位时间完成的任务方面的原始效率和精确度时，人的效能直接地影响着系统总体性能。

10.13.1 HSI 对 SE 流程不可或缺

项目成功的基础源于需求开发。人的效能需求来自于系统内的其他性能需求并受这些需求的限制。前端分析（FEA）产生系统需求并纳入 HIS 相关需求。有效的 FEA 开始于彻底了解：新系统的任务以及待执行的工作；任何先前系统的成功或问题；以及与有可能和所提出的系统技术交互的人员有关的 KSA 和培训。HSI 建模将系统视为按照共享组织原则行动的互相关连的元素的集合。这种视角利用属于其他工程学科的数学严格方法支撑模型和仿真（SEBoK，2014）。它还强调基本的 HSI 事实——不存在不含人的系统。在开发流程早期的仿真，特别是在开发系统硬件和软件元素之前，确保捕获了所有接口来进行需求定义、权衡空间分析和迭代的设计活动。HSI 分析将以人为中心的功能分配到系统内，并识别潜在的人员（或系统）能力缺口。例如，人擅长于解决归纳的问题，机器擅长于解决推演的问题（Fitts，1954）。对归纳的或推演的决策的需求是系统设计结构中固有的。

在系统开发（在利益攸关者需求产生期间）中提早包括 HSI，并持续通过开发过程来实现最终系统解决方案的最大收益，以及最大化节省 LCC 是非常关键的。在 IPPD 中全面利用 HSI 有助于确保系统不会要求昂贵的"培训围绕"或后期修理，以纠正由不好的、未规定的或未定义的人员接口导致的无效的使用性和运行低效率。系统工程师在使用 HSI 专业知识支持 IPDI 中扮演关键角色。通常需要知识渊博的、跨学科的 HSI 团队应对全方位的人素考量，并且系统工程师对于确保 HSI 被包含在整个系统生命周期内是非常关键的。项目经理、总工程师以及系统工程师应确保 HSI 实践者能够积极参与到设计评审、工作组和 IPDT 中。与客户、用户、开发者、科学家、测试人员、后勤人员、工程师和设计师（人、硬件和软件）的一致参与和沟通是基本要求。

10.13.2 Technical and Management HSI Processes

HSI must be addressed by all program-level IPDTs and at program, technical, design, and decision reviews throughout the life of the system. HSI influences the design and acquisition of all systems and system modifications and makes explicit the role that the human plays in system performance and cost and how these factors are shaped by design decisions. In addition, HSI is one of the essential components of engineering practice for system development, contributing technical and management support to the development process itself.

10.13.2.1 HSI Domains HSI processes facilitate trade-offs among interdependent, human-centered domains without replacing individual domain activities, responsibilities, or reporting channels. The human-centered domains typically cited by various organizations may differ in what they are called or in number, but the human considerations addressed are quite similar. The following human-centered domains with recognized application to HSI serve as a good foundation of human considerations that need to be addressed in system design and development, but clearly are not all inclusive:

- *Manpower*—Addresses the number and type of personnel and the various occupational specialties required and potentially available to train, operate, maintain, and support the deployed system.

- *Personnel*—Considers the type of KSAs, experience levels, and aptitudes (e.g., cognitive, physical, and sensory capabilities) required to operate, maintain, and support a system and the means to provide (e.g., recruit and retain) such people.

- *Training*—Encompasses the instruction and resources required to provide personnel with requisite KSAs to properly operate, maintain, and support systems. The training community develops and delivers individual and collective qualification training programs, placing emphasis on options that:

    - Enhance user capabilities to include operator, maintainer, and support personnel
    - Maintain skill proficiencies through continuation training and retraining
    - Expedite skill and knowledge attainment
    - Optimize the use of training resources

Training systems, such as simulators and trainers, should be developed in conjunction with the emerging system technology.

10.13.2 技术和管理 HSI 流程

HSI 必须由所有项目计划层级 IPDT 来处理，并且必须贯穿于系统生命期的项目、技术、设计和决策评审阶段来处理。HSI 影响所有系统和系统改进的设计和采办，并使人员在系统性能和成本中起的作用以及如何通过设计决策形成这些因素变得明确。此外，HSI 是系统开发工程实践的基本部分之一，为开发流程本身提供技术和管理支持。

10.13.2.1 HSI 领域 HSI 流程促进在相互依存的、以人为中心的领域间权衡，无须取代单个的领域活动、职责或报告通道。通常由各种不同组织提出的以人为中心的领域，其在称谓或数目方面会有所不同，但是所应对的人素考量是基本相似的。认可在 HSI 中应用的下述以人为中心的领域充当系统设计和开发中应对的人素考量的良好基础，但并未清晰地包括全部内容：

- *人力资源*——涉及人员的数量和类型以及所要求的且可能适用于培训、运行、维护和保障已部署系统的各种不同专业。

- *人员*——考虑运行、维护和保障一个系统所要求的 KSA、经验水平以及天资（如，认知的、体能的和感知的能力）的类型以及提供（即招聘和留用）该类人员的手段。

- *培训*——包括提供给人员能够正确地操作、维护和保障系统的必要 KSA 所需的指导及资源。培训团体开发并交付个人的和集体的鉴定培训项目，重点强调的选项是：

    – 提高用户能力，包含操作人员、维护人员和支持人员

    – 通过持续培训和再培训保持技能熟练程度

    – 加速技能和知识的获得

    – 优化培训资源的使用

应结合新兴的系统技术开发培训系统，例如模拟器和训练器。

SPECIALTY ENGINEERING ACTIVITIES

- *Human factors engineering (HFE)*—Involves an understanding of human capabilities (e.g., cognitive, physical, sensory, and team dynamic) and comprehensive integration of those capabilities into system design beginning with conceptualization and continuing through system disposal. A key objective for HFE is to clearly characterize the actual work to be performed and then use this information to create effective, efficient, and safe human/hardware/software interfaces to achieve optimal total system performance. This "optimal performance" is the achievement of the following:

    – Conducting task analyses and design trade-off studies to optimize human activities, creating workflow

    – Making the human goals and performance the design driver to assure an intuitive system for humans who will use, operate, maintain, and support it

    – providing deliberately designed primary, secondary, backup, and emergency tasks and functions

    – Meeting or exceeding performance goals and objectives established for the system

    – Conducting analyses to eliminate/minimize the performance and safety risks leading to task errors and system mishaps across all expected operational, support, and maintenance environments

HFE uses task and function analyses (including cognitive task analysis) supported by increasingly functions and interfaces. These efforts should recognize the increasing complexity of technology and the associated demands on people, giving careful consideration to the capabilities and limitations of humans. The design should not demand unavailable or unachievable skills. HFE seeks to maximize usability for the targeted range of users/customers and to minimize design characteristics that induce frequent or critical errors. HSI/HFE tools provide data that can be directly incorporated into system design elements such as the information architecture.

HFE works with the IPDTs to ensure that representative personnel are tested in situations to determine whether the human can operate, maintain, and support the system in adverse environments while working under the full range of anticipated mission stress and endurance conditions.

- *Environment*—In the context of HSI, this domain involves environmental considerations that can affect operations and requirements, particularly human performance.

- *人素工程学（HFE）*——涉及对人员能力（如认知的、体能的、感知的和团队动态的）的了解，以及将那些从概念化开始持续到系统退出的能力全面综合到系统设计之中。HFE 的关键目标是清晰地描述待实施的实际工作，然后使用该信息创建有效的、高效率的和安全的人员/硬件/软件接口，以实现最优的总体系统性能。这一"最优性能"是通过下述内容实现的：

  – 进行任务分析和设计权衡研究，以优化产生工作流程的人员活动

  – 制定人员目标，并执行设计驱动力，以保证使用、运行、维护和保障系统的人员使用直观的系统

  – 提供精心设计的主要的、次要的、备用的和应急的任务与功能

  – 满足或超过性能目标以及为该系统建立的目标

  – 进行分析，以消除/最小化性能及安全风险，这些风险会导致跨全部预期运行、保障和维护环境出现任务差错和系统事故

HFE 使用受越来越多的功能和接口支持的任务与功能分析（包括认知任务分析）。这些工作应认识到技术复杂性以及对人员相关要求的增加，以对人员能力和局限性进行仔细的考虑。设计不应要求不可用的或不可实现的技能。HFE 力图使用户/客户目标范围的可用性最大化，并使导致常见错误和关键性错误的设计特征最小化。HSI/HFE 工具提供可以直接纳入诸如信息架构等系统设计元素的数据。

HFE 与 IPDT 共同确保在一些情况下测试有代表性的人员，以在各种预期的任务压力和耐久性条件下工作的同时，确定人员能否在环境恶劣中运行、维护和保障系统。

- *环境*——在 HSI 背景环境中，该领域涉及可影响运行和需求（特别是人员效能）的环境因素。

SPECIALTY ENGINEERING ACTIVITIES

- *Safety*—promotes system design characteristics and procedures to minimize the risk of accidents or mishaps that cause death or injury to operators, maintainers, and support personnel; threaten the operation of the system; or cause cascading failures in other systems. Prevalent issues include the following:
  - Factors that threaten the safety of personnel and their operation of the system
  - Walking/working surfaces, emergency egress pathways, and personnel protection devices
  - Pressure and temperature extremes
  - Prevention/control of hazardous energy releases (e.g., mechanical, electrical, fluids under pressure, ionizing or nonionizing radiation, fire, and explosions)
- *Occupational health*—promotes system design features and procedures that serve to minimize the risk of injury, acute or chronic illness, and disability and to enhance job performance of personnel who operate, maintain, or support the system. Prevalent issues include the following:
  - Noise and hearing protection
  - Chemical exposures and skin protection
  - Atmospheric hazards (e.g., Confined space entry and oxygen deficiency)
  - Vibration, shock, acceleration, and motion protection
  - Ionizing/nonionizing radiation and personnel protection
  - Human factors considerations that can result in chronic disease or discomfort (e.g., repetitive motion injuries or other ergonomic-related problems)
- *Habitability*—Involves characteristics of system living and working conditions, such as the following:
  - Lighting and ventilation
  - Adequate space
  - Availability of medical care, food, and/or drink services
  - Suitable sleeping quarters, sanitation and personal hygiene facilities, and fitness/recreation facilities

- *安全*——促使系统设计特性和程序将意外或事故的风险降至最低,这些意外或事故可导致操作人员、维修人员以及保障人员死亡或受伤;威胁系统的运行;或导致其他系统的连锁性故障。普遍存在的问题包括下述几个方面:

  – 威胁人员安全及其系统运行的因素

  – 行走面/工作面、应急疏散通道和人员防护设备

  – 压力和温度极限

  – 危险能量释放(例如机械的、电气的、受到压力的液体、电离或非电离辐射、起火和爆炸)的预防/控制。

- *职业健康*——改善系统设计特性和程序以使受伤、急性病或慢性病、残疾的风险最小化,并有助于提高运行、维护或保障系统人员工作效能。普遍存在的问题包括下述几个方面:

  – 噪声和听力保护

  – 化学暴露及皮肤防护

  – 大气危害(例如,进入密闭空间和缺氧)

  – 振动、冲击、加速度和运动防护

  – 电离/非电离性辐射和人员防护

  – 导致慢性病或不适(如重复运动损伤或其他与人体工程学相关的问题)的人素考量

- *适居性*——涉及系统生活和工作条件的特性,比如以下:

  – 照明和通风

  – 充足的空间

  – 医疗护理、食品和/或饮水服务的可用性

  – 适宜的卧室、卫生和个人卫生设施以及健身/娱乐设施

SPECIALTY ENGINEERING ACTIVITIES

- *Survivability*—Addresses human-related characteristics of a system (e.g., life support, body armor, helmets, plating, egress/ejection equipment, air- bags, seat belts, electronic shielding, alarms, etc.) that reduce susceptibility of the total system to mission degradation or termination, injury or loss of life, and partial or complete loss of the system or any of its elements.

The domains outlined above must be considered simultaneously since decisions made in one domain can have significant impacts on other domains. Each individual domain decision generates the need to concurrently assess HSI issues across all the domains and against mission performance, prior to making formal programmatic decisions. This approach mitigates the potential for unintended, adverse consequences, including increased technical risk and cost.

10.13.2.2 *Key HSI Activities and Tenets* HSI programs have distilled the following HSI activities and associated key actionable tenets:

- *Initiate HSI early and effectively*—HSI should begin in early system concept development with FEA and requirements definition.

HSI-related requirements include not just those of the individual domains but also those that arise from interactions among the HSI domains.

HSI requirements must be developed in consonance with other system requirements and consider any constraints or capability gaps and must be reconsidered, refined, and revised as program documents and system requirements are updated. The human considerations identified in the requirements must address the capabilities and limitations of users, operators, maintainers, and other personnel as they interact with and within the system. Doing this early, continuously, and comprehensively as part of the SE process provides the opportunity to identify risks and costs associated with program decisions.

HSI should be conducted by professionals who are trained in the domains outlined earlier, who are resourced with appropriate tools, and who have access to data from other concurrent IPDT activities.

- *Identify issues and plan analysis*—projects/ programs must identify HIS-related issues that require analysis to ensure thorough consideration.

The systems engineer must have a comprehensive plan for HSI early in the acquisition process and summarize HSI planning in the acquisition strategy. This plan can be stand-alone or integrated into the SEMP and include details associated with the analyses in SE language to support HSI trade-offs. Systems engineers should address HSI throughout the entire acquisition cycle as part of the SE process.

- *生存性*——应对系统中与人员相关的特性（如生命保障、防弹衣、防护帽、护板、撤离设备/弹射设备、安全气囊、安全带、电子防护罩和警报器等），以降低整个系统对任务降级或终止、人员伤亡以及系统或其系统元素的部分或全部损失的敏感性。

必须同时考虑上述领域，因为在一个领域做出的决策可能对其他领域具有明显的影响。在做出正式的计划性决策前，每个单独领域决策均会产生跨所有领域并针对任务性能同时评估 HSI 问题的需要。这种方法会降低出现意想不到的、不利后果的可能性，包括增加的技术风险和成本。

10.13.2.2 关键 HSI 活动和宗旨　HSI 项目精选出下述 HSI 活动以及相关的关键可行的宗旨：

- *尽早并有效地启动 HSI*——在早期系统概念开发中，HSI 应始于 FEA 和需求定义。

与 HSI 相关的需求不仅包括这些单个领域的活动，亦包括由 HSI 领域间的相互作用产生的活动。

HSI 需求的开发必须与其他系统需求一致，并且考虑全部约束或能力缺口，在更新项目文件和系统需求时还必须重新考虑、改进和修正。在需求中识别的人素考量，必须在用户、操作人员、维护人员以及其他工作人员与系统交互或在系统内交互时应对其能力和局限性的问题。在部分 SE 流程提供识别与项目决策相关的风险和成本的机会时，尽早持续而全面地开展这项工作。

HSI 应由专业人员进行，这些专业人员在前述领域受过培训，配备充足的专用工具，并且可以使用来自其他并行 IPDT 活动的数据。

- *识别问题和计划分析*——项目/项目计划必须识别需要分析的 HSI 相关问题，以确保考虑全面性。

系统工程师必须在采办流程中提早地为 HSI 做出全面计划，并总结采办策略中的 HSI 规划。这一计划可以是独立的，也可以综合到 SEMP 中，并且包括与分析相关的用 SE 语言编写的详细说明，以支持 HSI 权衡。系统工程师应当贯穿于整个采办周期应对 HSI，作为 SE 流程一部分。

SPECIALTY ENGINEERING ACTIVITIES

Efficient, timely, and effective planning/replanning and FEA are the cornerstones of HSI efforts, ensuring human considerations are effectively integrated into capability requirements, system concept development, and acquisition. HSI FEA establishes and assesses criteria for success and helps determine when a system design change is required.

- *Document HSI requirements*—Systems engineers derive HSI requirements, as needed at each level of the system hierarchy, using HSI plans, analyses, and reports as sources (or rationale) for the derived requirements. HSI requirements should be cross-referenced with other documents, plans, and reports and captured in the requirements traceability documents and maintained in system requirements databases in the same manner as all other system requirements.

- *Make HSI a factor in source selection for contracted development efforts*—HSI requirements must be explicit in source selection planning and implementation with adequate priority in the SOW.

- *Execute integrated technical processes*—HSI domain integration begins in early system concept development with FEA and requirements definition and continues through development, operations, sustainment, modification, and all the way to eventual system disposal.

HSI activities and considerations must be included in each key project planning document (e.g., acquisition strategy, SEMP, test plan, verification, etc.) and in the system architectural framework.

Systems engineers and HSI personnel must be prepared to exchange technical data and to present accurate, integrated cost data whenever possible to demonstrate reduced LCC, thereby justifying trade off decisions that may increase design and acquisition costs.

- *Conduct proactive trade-offs*—In conducting trade off analyses both within HSI domains and across the system, the primary goal is to ensure the system meets or exceeds the performance requirements without compromising survivability, environment, safety, occupational health, and habitability.

- *Conduct HSI assessments*—The purpose of the HSI assessment process is to evaluate the application of HSI principles throughout the system life cycle. The process enables cross-discipline HSI collaboration in acquisition program evaluations and pro-vides an integral avenue for HSI issue identification and resolution.

高效的、及时的和有效的规划/再规划以及 FEA 是 HSI 工作的基础，以确保人为因素被有效地综合到能力需求、系统概念开发和采办之中。HSI 的 FEA 建立和评估成功准则，并帮助确定需要系统设计变更的时间。

- *将 HSI 需求文件化*——按每一层系统层级结构的需要，系统工程师推演 HSI 需求，将 HSI 计划、分析以及报告用作所推演需求的来源（或根本原因）。HSI 需求应与其他文件、计划以及报告相互参照并在需求可追溯性文件中捕获，且以与所有其他系统需求相同的方式保留在系统需求数据库中。

- *使 HSI 成为合同开发工作中资源选择的因素*——在资源选择规划中 HSI 需求必须是清楚的，并且在 SOW 中应优先实施。

- *执行综合的技术流程*——在早期系统概念开发中，HSI 领域综合始于 FEA 和需求定义，并持续经过开发、运行、维持、改进，直到最终系统退出。

HSI 活动和因素必须包含在每个关键项目规划文件（例如采办策略、SEMP、试验计划、验证等）以及系统架构框架中。

系统工程师和 HSI 人员必须在任何可能的时候做好交换技术数据和提出精确的、综合的成本数据的准备，以展示已减少的 LCC，从而证明可能增加设计和采办成本的权衡决策的合理性。

- *进行具有主动性的权衡*——在 HSI 领域内并且跨系统进行权衡分析的过程中，主要目标是确保系统在不损害可生存性、环境、安全、职业健康和可居住性情况下满足或超过性能需求。

- *开展 HSI 评估*——HSI 评估流程的目的是贯穿于系统生命周期评价 HSI 原则的应用情况。该流程使能在采办项目评价过程中进行跨学科的 HSI 协作，并提供识别和解决 HSI 问题的整体路径。

HSI assessment should be initiated early in the system life cycle and addressed throughout the system development process, particularly in SE technical reviews and during working group meetings, design reviews, logistical assessments, verification, and validation.

HSI assessments should be based on sound data collection and analyses. Deficiencies should be captured and include a detailed description of the deficiency, its operational impact, recommended corrective action, and current status.

10.14 VALUE ENGINEERING

VE, value management (VM), and value analysis (VA) are all terms that pertain to the application of the VE process (Bolton et al., 2008; Salvendy, 1982; SAVE, 2009).

This chapter discusses how VE supports, compliments, and "adds value" to the SE discipline.

VE originated in studies of product changes resulting from material shortages during World War II. During that period, materials identified in designs were substituted without sacrificing quality and performance. Lawrence D. Miles noticed this success and developed a formal methodology to utilize teams to examine the function of products manufactured by General Electric.

VE uses a systematic process (e.g., a formal job plan), VE-certified facilitators/team leads, and a multidisciplinary team approach to identify and evaluate solutions to complex problems in the life cycle of a project, process, or system. The VE process utilizes several industry standard problem-solving/decision-making techniques in an organized effort directed at independently analyzing the functions of programs, projects, organizations, processes, systems, equipment, facilities, services, and supplies. The objective is to achieve the essential functions at the lowest LCC consistent with required performance, reliability, availability, quality, and safety. VE is not a cost reduction activity but a function-oriented method to improve the value of a product. There is no limit to the field in which VE may be applied.

The objective of performing VE is to improve the economical value of a project, product, or process by reviewing its elements to accomplish the following:

- Achieve the essential functions and requirements
- Lower total LCC (resources)
- Attain the required performance, safety, reliability, quality, etc.
- Meet schedule objectives

HSI 评估应在系统生命周期早期启动，并贯穿于系统开发流程应对，特别是在 SE 技术评审中和工作团队会议、设计评审、后勤评估、验证以及确认期间。

HSI 评估应基于合理的数据收集和分析。不足之处应被捕获并包含对不足之处及其运行影响、建议的纠正措施和最新状态的详细描述。

10.14 价值工程

VE、价值管理（VM）和价值分析（VA）都是属于 VE 流程的应用的术语（Bolton 等，2008；Salvendy，1982；SAVE，2009）。

本章论述 VE 如何对 SE 学科进行支持、倡导和"增值"。

VE 起源于第二次世界大战期间由于物料短缺而导致对产品变更的研究。那个时期，在不牺牲质量和性能的情况下对设计中识别的物料进行了替换。Lawrence D.Miles 注意到了这种成功，开发了一个正式的方法论，以利用团队来检查通用电气公司制造的产品的功能。

VE 使用系统化的流程（即正规的工作计划）、经 VE 认证的推动者/团队领导以及多学科团队方法来识别和评价项目、流程或系统生命周期中复杂问题的解决方案。VE 流程在以独立分析项目群、项目、组织、流程、系统、设备、设施、服务和供给功能为目的的有组织的工作中，利用若干行业标准的解决问题/决策技术。目的是以符合所要求性能、可靠性、可用性、质量和安全性要求的最低 LCC 实现必要的功能。VE 并非降低成本的活动，而是一个提高产品价值的面向功能的方法。可能应用 VE 的可应用领域没有限制。

实施 VE 的目的是通过审查其元素的方式提高项目、产品或流程的经济价值以达到下述要求：

- 达成最基本的功能和需求
- 降低总的 LCC（资源）
- 获得所需性能、安全性、可靠性、质量等
- 满足进度目标

Value is defined as a fair return or the equivalent in goods, services, or money for something exchanged. In other words, value is based on "what you get" relative to "what it cost." It is represented by the relationship:

Value=function / cost

Function is measured by the requirements from the stakeholder. Cost is calculated in the materials, labor, price, time, etc., required to accomplish that function.

According to the Society of American value Engineers (SAVE) International, to qualify as a value study, the following conditions must be satisfied:

- The value study team follows an organized, six phase VE job plan (see following text) and performs function analysis.

- The value study team is a multidisciplinary group, chosen based on their expertise.

- The value study team leader (i.e., facilitator) is trained in the value methodology.

10.14.1 Systems Engineering Applicability

VE supports the various aspects and techniques of SE as well as being a comprehensive approach to SE project initiation. VE is implemented through a systematic, rational process consisting of a series of team-based techniques, including:

- Mission definition, strategic planning, or problem solving techniques to determine current and future states. High-level requirements definition can be initiated in these processes.

- Functional analysis to define what a system or system element does or its reason for existence. This technique serves as the basis or structure for technical, functional, and/or operational requirements.

- Innovative, creative, or speculative techniques to generate new alternatives. Trade studies are often defined in VE workshops from the more "viable" alternatives. Time can be given to conduct in-depth analyses, such as feasibility studies, cost estimates, etc. The VE workshop can then be reconvened (e.g., 1–6 months later) to determine the preferred alternative.

- Evaluation techniques to select preferred alternatives. These range in complexity from completely subjective to quantitative, depending on the LOR and justification required.

价值被定义为商品、服务或用于交换某物的货币的合理收益或等价物。换言之，价值是"你之所得"与"物之成本"之比。可通过下述关系表示：

价值=功能/成本

功能通过利益攸关者的需求来衡量。成本是依据实现该功能所需的物料、劳动、价格、时间等计算出的。

根据美国价值工程师协会（SAVE）国际部，欲取得价值研究资格，必须满足下列条件：

- 价值研究团队遵守有组织的、分六阶段的 VE 工作计划（见下述文本）并实施功能分析。

- 价值研究团队是一个根据其专门知识选择的多学科小组。

- 价值研究团队领导（即推动者）要在价值方法论方面受培训。

10.14.1 系统工程适用性

VE 支持 SE 的各个不同方面和技术，并作为 SE 项目启动的综合方法。VE 通过由一系列基于团队的技术所组成的系统合理的流程来实现，包括：

- 确定当前状态和未来状态的任务定义、战略规划或解决问题的技术。可以在这些流程中启动高层级的需求定义。

- 定义系统或系统元素是什么或其存在原因的功能分析。这项技术充当技术、功能和/或运行需求的基础或结构。

- 产生新备选方案的创新性的、创造性的或推测性的技术。在 VE 研讨会上权衡研究通常可根据更"可行的"备选方案来定义。可以投入一定时间进行深入分析，例如可行性研究、成本估算等。然后可重新召集 VE 研讨会（例如，1～6 个月之后），以确定首选的备选方案。

- 选择首选的备选方案的评估技术。这些技术根据所要求的严格程度及证明的合理性从完全的主观扩展到定量的复杂性。

SPECIALTY ENGINEERING ACTIVITIES

VE can be used to effectively and efficiently initiate an SE project. The team approach brings the customer and other stakeholders together along with engineering, planning, finance, marketing, etc. so that the strategic elements of the project can be discussed or developed. Stakeholder needs can be identified and established as requirements. Functions are brainstormed and alternatives developed. In general, the boundaries of the scope of work are identified and clarified with all affected groups. The SE effort can then proceed with initial planning already established and stakeholder input integrated. Some of the uses and benefits of VE include:

- Clarify objectives
- Solve problems (obtain "buy-in" to solution)
- Improve quality and performance
- Reduce costs and schedule
- Assure compliance
- Identify and evaluate additional concepts and options
- Streamline and validate processes and activities
- Strengthen teamwork
- Reduce risks
- Understand customer requirements

10.14.2 VE Job Plan

The VE process uses a formal job plan, which is a scientific method of problem solving that has been optimized over the last 50 years. The premise of VE is to use a beginning-to-end process that analyzes functions in such a way that creates change or value-added improvement in the end. A typical six-phase VE job plan is as follows:

- *Phase 0: Preparation/planning*—plan the scope of the VE effort.

- *Phase 1: Information gathering*—Clearly identify the problem(s) to be solved or the objective of the work scope. This includes gathering background information pertinent to meeting the objective and identifying and understanding customer/stakeholder requirements and needs.

- *Phase 2*: Functional analysis—Define the project functions using an active verb and measurable noun and then review and analyze the functions to determine which are critical, need improvements, or are unwanted. Several techniques can be used to enhance functional analysis. Functions can be organized in a WBS, in a Function Analysis System Technique (FAST) diagram (discussed in the following text), or in a flow diagram.

VE 可用来有效地且高效率地启动一个 SE 项目。团队方法可让客户和其他利益攸关者同时进行工程、计划、财务和市场等工作,以便讨论或提出项目的战略元素。利益攸关者的需要可以被识别和建立为需求。功能是头脑风暴出来的,并开发了备选方案。通常,可与所有受到影响的团队共同识别和阐明工作范围的边界。然后,SE 工作可以根据已经确定好的初步计划和整合的利益攸关者输入继续进行。VE 的一些用途和益处包括:

- 阐明目标
- 解决问题(获得"认可"解决方案)
- 提高质量和性能
- 降低成本并缩短进度
- 确保合规性
- 识别和评价附加概念和选项
- 简化和确认流程及活动
- 加强团队合作
- 降低风险
- 理解客户需求

10.14.2 VE 工作计划

VE 流程采用正式的工作计划,这种正式工作计划是一种解决问题的科学方法,在过去的 50 年中已不断得到优化。VE 的前提是使用一种自始至终的流程,其以在最终产生变更或者增值改进这样一种方式分析功能。典型的六步 VE 工作计划如下:

- *阶段 0*:准备/规划——计划 VE 工作的范围。
- *阶段 1*:信息收集——明确识别待解决的问题或工作范围目标。这包括收集与满足目标有关的背景信息,并识别和理解客户/利益攸关者的需求和需要。
- *阶段 2*:功能分析——使用主动动词和可数名词定义项目功能,然后评审和分析功能,以确定哪些是关键的,哪些需要改进,或者哪些是不想要的。可使用多种技术改进功能分析。功能可在 WBS 中组织,可在功能分析系统技术(FAST)图(在下述文本中论述)中组织,或可在流程图中组织。

SPECIALTY ENGINEERING ACTIVITIES

Functions can be assigned a cost. Depending on the objective of the VE study, determining the cost/ function relationships is one method to identify where unnecessary costs exist within the scope of the study. Other criteria to identify those areas needing improvement are personnel, environmental safety, quality, reliability, construction time, etc. Cost/function relationships provide direction for the team related to areas that provide the greatest opportunity for cost improvement and the greatest benefits to the project. This can be captured in a cost/ function worksheet.

- *Phase 3: Creativity*—Brainstorm different ways to accomplish the function(s), especially those that are high cost or low value.

- *Phase 4: Evaluation*—Identify the most promising functions or concepts. The functions can also be used in this phase to generate full alternatives that accomplish the overall objective. Then the alternatives can be evaluated using the appropriate structured evaluation technique.

- *Phase 5: Development*—Develop the viable ideas into alternatives that are presented to decision makers to determine the path forward. Alternatively, the full alternatives that were evaluated in phase 4 can be further developed into proposals, which may include rough-order-of-magnitude cost, estimated schedule, resources, etc.

- *Phase 6: Presentation/implementation*—It may be desirable to present the alternatives/proposals to management or additional stakeholder for final direction. This is where the implementation plan concepts are identified and a point of contact assigned to manage the actions.

10.14.3 FAST Diagram

The FAST diagramming technique was developed in 1964 by Charles W. Bytheway to help identify the dependencies and relationships between functions. A FAST diagram (see Fig. 10.10) is not time oriented like a program Evaluation Review Technique (PERT) chart or flowchart. It is a function-oriented model that applies intuitive logic to verify the functions that make up the critical path.

There are several different types of FAST diagrams as well as varying levels of complexity. The purpose is to develop a statement of the function (verb and noun) for each part or element of the item or process under study. Functions are classified as basic and secondary. The critical path is made up of the main functions within the scope of the activity or project. The basic function is typically the deliverable or end state of the effort. The higher-order function is the ultimate goal and is the answer to "why" the basic function is performed. Functions that "happen at the same time" or "are caused by" some function on the critical path are known as "when" functions and are placed below the critical path functions. Functions that happen "all the time," such as safety, aesthetic functions, etc., are placed above the critical path functions.

功能可被分配一个成本。根据 VE 研究的目标，确定成本/功能关系是一种在研究范围内识别哪里存在不必要成本的方法。其他识别那些需要改进的领域的准则是人员、环境安全性、质量、可靠性和建造时间等。成本/功能关系为与这些领域相关的团队提供指导，这些领域为项目提供最大成本改进机会并提供最大利益。可将以上内容收集到成本/功能工作表单中。

- *阶段 3：创造性*——"头脑风暴"得出不同的方式以实现这些功能，特别是那些高成本或低价值的功能。

- *阶段 4：评价*——识别最有希望的功能或概念。这些功能也可用于这一阶段以产生实现总体目标的完整备选方案。然后，可使用适当的结构化评价技术评价备选方案。

- *阶段 5：开发*——将可行的想法开发成提供给决策者的备选方案，以确定前行的路径。或者，可使在阶段 4 中评价的完整备选方案进一步形成建议，可包括粗略量级估算的成本、估计的进度和资源等。

- *阶段 6：展现/实施*——可能可取的是，为管理人员或额外的利益攸关者提供备选方案/建议用作最终指导。在这一阶段识别实施计划概念并指定一个接触点来管理行动。

10.14.3　FAST 图

FAST 图技术由 Charles W. Bytheway 于 1964 年提出，以便帮助识别功能间的依赖性与关系。FAST 图（见图 10.10）不是以时间为导向，而是像项目评价审查技术（PERT）图或流程图一样。它是一种功能导向模型，该模型可使用直觉性逻辑来验证构成关键路径的功能。

存在多种不同类型的 FAST 图以及不同层级的复杂性。目的是为研究中的项目或流程的每个部分或元素开发一种功能说明（动词和名词）。功能可分为基本的和次要的两类。关键路径由活动或项目范围内的主功能组成。基本功能通常是工作的可交付状态或最终状态。高阶功能是最终目标，可用于回答"为什么"执行基本功能。"在同一时间发生"的功能或由关键路径上的某个功能"引起"的功能被称为"同时发生"的功能，并且位于关键路径功能的下方。"在所有时间"发生的功能，例如安全性、美学功能等，位于关键路径功能的上方。

SPECIALTY ENGINEERING ACTIVITIES

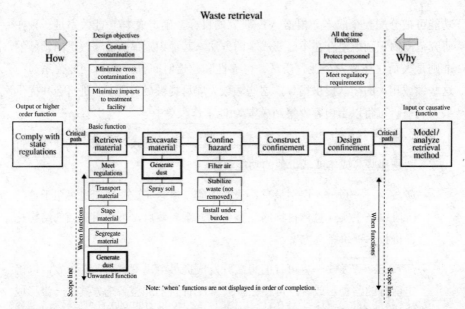

FIGURE 10.10 Sample Function Analysis System Technique (FAST) diagram. Reprinted with permission from Doug Hamelin. All other rights reserved.

Unwanted functions are noted by a double lined box. They are not essential to the performance of the scope, but are a consequence of the selected design solution. Limiting unwanted functions and minimizing the cost of basic/critical path functions result in an item of "best value" that is consistent with all performance, reliability, quality, maintainability, logistics support, and safety requirements.

There is no "correct" FAST model; team discussion and consensus on the final diagram is the goal. Using a team to develop the FAST diagram is beneficial for several reasons:

- Applies intuitive logic to verify functions
- Displays functions in a diagram or model
- Identifies dependence between functions
- Creates a common language for team
- Tests validity of functions

10.14.4 VE Certification

A Certified value professional should be used to facilitate VE workshops. They have the training and experience to manage a team, implement the methodology, and maximize the benefits to the customer.

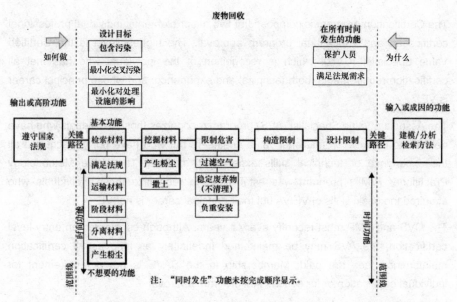

图 10.10 功能分析系统技巧（FAST）示例图。经 Doug Hamelin 许可后转载。版权所有。

使用双线框指出不想要的功能。它们对于该范围的性能并不重要，但却是选定的设计解决方案的结果。限制不想要的功能和最小化基本/关键的路径功能的成本会产生一个符合所有性能、可靠性、质量、可维护性、后勤保障以及安全性需求的"最佳值"。

不存在"正确的"FAST 模型；目标是就对最终图的团队讨论和共识。使用一个团队开发 FAST 图是有益的，原因如下：

- 使用直觉逻辑验证功能
- 以图或模型显示功能
- 识别功能间的依赖性
- 创建团队的一种公共语言
- 测试功能的有效性

10.14.4 VE 认证

应利用注册价值专业人员来推动 VE 研讨会。他们经过培训并在管理团队、实施方法论以及使客户利益最大化方面具有经验。

SPECIALTY ENGINEERING ACTIVITIES

The Certification program is composed of two major elements: individual professional certification and educational program approval. The highest level is the Certified Value Specialist (CVS), which is recognition of the individual who has met all certification requirements, both technical and experience, and whose principal career is VE.

The Associate Value Specialist (AVS) program recognizes those individuals who have decided to become professional value engineers but who have not yet acquired all the experience or technical skills expected of a CVS. The Value Methodology Practitioner (VMP) program was established to recognize those individuals who acquired the basic skills of VE/VA but their principal career is not VE.

The CVS and VMP must recertify every 4 years. Although considered an entry-level certification, the AVS may be maintained indefinitely as long as all certification maintenance fees are paid. Membership in the SAVE is not a requirement for individual certification or for educational program approval.

10.14.5 Conclusion

VE is a best business practice. Projects that use VE in the early life cycle stages have shown high rates of success. A VE study is more rigorous than the typical project review. Each VE study brings together an impartial and independent team of technical experts with a common purpose of improving and optimizing the project's value. VE is seen as a systematic and creative approach for increasing the "return on investment" in systems, facilities, and other products. For more information on VE, consult the website of the SAVE International (2009).

认证计划包括两个主要元素：个人专业认证以及教育计划批准。最高层级的是注册价值评估专家（CVS），这是对已经在技术和经验方面满足所有认证要求且主要职业为 VE 的个人的认可。

助理价值评估专家（AVS）计划认可那些决定成为专业价值工程师但是还未获得 CVS 期望具有的所有经验或技术技能的个人。价值方法论实践者（VMP）计划被建立用于认可那些获得 VE/VA 的基本技能，但是其主要职业并非 VE 的个人。

CVS 和 VMP 必须每隔四年重新认证一次。尽管被认为是入门级的认证，但是只要支付了全部认证维持费，AVS 便可无限期地保持。SAVE 的会员资格并非个人认证或教育计划批准的要求。

10.14.5 结论

VE 是一种最佳的商业实践。在早期的生命周期阶段中使用 VE 的项目已经表明了高度的成功。VE 研究比典型的项目评审更加严格。每项 VE 研究会汇聚一个以提高和优化项目价值为共同目的的公正且独立的技术专家团队。VE 被视为一种用于增加系统、设施及其他产品"投资回报"的系统化和创造性的方法。关于 VE 的更多信息，登录 SAVE 国际部（2009）的网站咨询。

APPENDIX A: REFERENCES （参考文献）

Achenbach, J. (2009). Mars Mission Has Some Seeing Red. *The Washington Post*, 11 February 2009.

Adams, J. L. (1990). *Conceptual Blockbusting*, 3rd Ed. San Francisco, CA: San Francisco Book Company, Inc.

Alavi, M., & Leidner, D. E. (1999). Knowledge Management Systems: Issues, Challenges, and Benefits. *Communications of the AIS*, 1(2).

Albright, D., Brannan, P., & Walrond, C. (2010). *Did Stuxnet Take Out 1,000 Centrifuges at the Natanz Enrichment Plant?* Washington, DC: Institute for Science and International Security. Retrieved from http://isis-online.org/isis-reports/detail/did-stuxnet-take-out-1000-centrifuges-at-the-natanz-enrichment-plant/ (accessed January 26, 2015).

Albright, D., Brannan, P., & Walrond, C. (2011). Stuxnet Malware and Natanz: Update of ISIS December 22, 2010 Report. Institute for Science and International Security. Retrieved from http://isis-online.org/uploads/isis-reports/documents/-stuxnet_update_15Feb2011.pdf (accessed October 24, 2014).

ANSI/AIAA G-043A. (2012). *Guide to the Preparation of Operational Concept Documents*. Reston, VA: American National Standards Institute/American Institute of Aeronautics and Astronautics.

ANSI/EIA 632. (2003). *Processes for Engineering a System*. Arlington, VA: American National Standards Institute/Electronic Industries Association.

ANSI/EIA 649B. (2011). Configuration Management Standard. TechAmerica.

ANSI/GEIA-STD-0009. (2008). Reliability Program Standard for Systems Design, Development, and Manufacturing. American National Standards Institute/Government Electronic Industries Association.

Arnold, S., & Lawson, H. (2003). Viewing Systems from a Business Management Perspective: The ISO/IEC 15288 Standard. *Systems Engineering*, 7(3), 229-42.

ARP 4754A. (2010). *Guidelines for Development of Civil Aircraft and Systems*. Warrendale, PA: SAE International.

INCOSE Systems Engineering Handbook: A Guide for System Life Cycle Processes and Activities, Fourth Edition. Edited by David D. Walden, Garry J. Roedler, Kevin J. Forsberg, R. Douglas Hamelin and Thomas M. Shortell.

© 2015 John Wiley & Sons, Inc. Published 2015 by John Wiley & Sons, Inc.

REFERENCES

Ashby, W. R. (1956). *Introduction to Cybernetics*. London, UK: Methuen.

ASQ. (2007). Quality Progress. In *Quality Glossary*. Milwaukee, WI: American Society for Quality Control.

AT&T. (1993). *AT&T Engineering Guides for Managing Risk*. TX: McGraw-Hill.

Barnard, R. W. A. (2008). What Is Wrong with Reliability Engineering? *Proceedings of the 18th Annual INCOSE International Symposium*. Utrecht, the Netherlands: International Council on Systems Engineering.

Barton, T. L., Shenkir, G., & Walker, P. L. (2002). *Making Enterprise Risk Management Pay Off: How Leading Companies Implement Risk Management*. Upper Saddle River, NJ: Financial Times/Prentice Hall PTR/Pearson Education Company.

BBC. (2002). China's Supertrain Takes to Tracks. Retrieved from BBC News World Edition: Asia-Pacific: http://news.bbc.co.uk/2/hi/asia-pacific/2182975.stm (accessed May 29, 2003).

Beck, K., Beedle, M., van Bennekum, A., Cockburn, A., Cunningham, W., Fowler, M., ... Thomas, D. (2001). Manifesto for Agile Software Development. Retrieved from http://agilemanifesto.org (accessed October 24, 2014).

Becker, O., Ben-Ashe, J., & Ackerman, I. (2000). A Method for Systems Interface Reduction Using N2 Charts. *Systems Engineering, Systems Engineering*, 3(1), 27-37.

Beer, S. (1959). *Cybernetics and Management*. New York, NY: John Wiley & Sons, Inc.

Bellinger, G. (2013). Systems Thinking World. Retrieved from http://www.systemswiki.org/ (accessed October 24, 2014).

von Bertalanffy, L. (1950). The Theory of Open Systems in Physics and Biology. *Science*, 111(2872), 23-9.

von Bertalanffy, L. (1968). *General System Theory: Foundations, Development, Applications*. New York, NY: Braziller.

Bias, R. G., & Mayhew, D. J. (1994). *Cost Justifying Usability*. Boston, MA: Academic Press.

Blanchard, B., & Fabrycky, W. (2011). *Systems Engineering and Analysis*, 5th Ed. Boston, MA: Prentice Hall.

Boardman, J., & Sauser, B. (2008). *Systems Thinking—Coping with 21st Century Problems*. Boca Raton, FL: CRC Press.

REFERENCES

Bobinis, J., Dean, E., Mitchell, T., & Tuttle, P. (2010). INCOSE Affordability Working Group Mission. *2010 ISPA/SCEA Joint Annual Conference & Training Workshop Proceedings*. Society for Cost Estimating and Analysis.

Bobinis, J., Haimowitz, J., Tuttle, P., Garrison, C., Mitchell, T., & Klingberg, J. (2013). Affordability Considerations: Cost Effective Capability. *Proceedings of the 23rd Annual INCOSE International Symposium*. Philadelphia, PA: International Council on Systems Engineering.

Boehm, B. (1986), A Spiral Model of Software Development and Enhancement. *ACM SIGSOFT Software Engineering Notes, ACM*, 11(4), 14-24.

Boehm, B. (1996). Anchoring the Software Process. *Software*, 13(4), 73-82.

Boehm, B., & Lane, J. (2007). Using the Incremental Commitment Model to Integrate System Acquisition, Systems Engineering, and Software Engineering. *CrossTalk*, 20, 4-9.

Boehm, B., & Turner, R. (2004). *Balancing Agility and Discipline*. Boston, MA: Addison-Wesley.

Boehm, B., Lane, J., Koolmanojwong, S., & Turner, R. (2014). *The Incremental Commitment Spiral Model: Principles and Practices for Successful Systems and Software*. Boston, MA: Addison-Wesley Professional.

Bogdanich, W. (2010, January 23). Radiation Offers New Cures, and Ways to Do Harm. *The New York Times*.

Bolton, J. D., Gerhardt, D. J., Holt, M. P., Kirk, S. J., Lenaer, B. L., Lewis, M. A., ... Vicers, J. R. (2008). *Value Methodology: A Pocket Guide to Reduce Cost and Improve Value through Function Analysis*. Salem, NH: GOAL/QPC.

Booher, H. R. (Ed.). (2003). *Handbook of Human Systems Integration*. New York, NY: John Wiley & Sons, Inc.

Brenner, M. J. (n.d.). TQM, ISO 9000, Six Sigma: Do Process Management Programs Discourage Innovation? Retrieved from Knowledge@Wharton, University of Pennsylvania: http://knowledge.wharton.upenn.edu/article/tqm-iso-9000-six-sigma-do-process-management-programs-discourage-innovation/ (accessed January 26, 2015).

Briedenthal, J., & Forsberg, K. (2007). *Organization of Systems Engineering Plans According to Core and Off-Core Processes*. Pasadena, CA: California Institute of Technology, Jet Propulsion Laboratory.

REFERENCES

Brykczynski, D. B., & Small, B. (2003). Securing Your Organization's Information Assets. *CrossTalk*, 16(5), 12-6.

Buede, D. M. (2009). *The Engineering Design of Systems: Models and Methods* (2nd Ed). Hoboken, NJ: John Wiley & Sons, Inc.

Carpenter, S., Delugach, H., Etzkorn, L., Fortune, J., Utley, D., & Virani, S. (2010). The Effect of Shared Mental Models on Team Performance. *Industrial Engineering Research Conference*. Cancun, Mexico: Institute of Industrial Engineers.

Carson, R. (2013). Can Systems Engineering be Agile? Development Lifecycles for Systems, Hardware, and Software. *Proceedings of the 23rd Annual INCOSE International Symposium*. Philadelphia, PA: International Council on Systems Engineering.

Carter, A. (2011). Better Buying Power: Mandate for Restoring Affordability and Productivity in Defense Spending. *Memorandum for Defense Acquisition and Logistics Professionals*. Under Secretary of Defense for Acquisition, Technology, and Logistics.

Chang, C. M. (2010). *Service Systems Management and Engineering: Creating Strategic Differentiation and Operational Excellence*. Hoboken, NJ: John Wiley & Sons, Inc.

Chapanis, A. (1996). *Human Factors in Systems Engineering*. New York, NY: John Wiley & Sons, Inc.

Chase, W. P. (1974). *Management of System Engineering*. New York, NY: John Wiley & Sons, Inc.

Checkland, P. (1975). The Origins and Nature of Hard Systems Thinking. *Journal of Applied Systems Analysis*, 5(2), 99-110.

Checkland, P. (1998). *Systems Thinking, Systems Practice*. Chichester, UK: John Wiley & Sons, Ltd.

Christensen, C. M. (2000). *The Innovator's Dilemma*. New York, NY: HarperCollins Publishers.

Churchman, C. W., Ackoff, R. L., & Arnoff, E. L. (1950). *Introduction to Operations Research*. New York, NY: John Wiley & Sons, Inc.

CJCS. (2012). *CJCSI 3150.25E, Joint Lessons Learned Program*. Washington, DC: Office of the Chairman of the Joint Chiefs of Staff.

Clausing, D., & Frey, D. D. (2005). Improving System Reliability by Failure Mode

REFERENCES

Avoidance including Four Concept Design Strategies. *Systems Engineering*, 8(3), 245-61.

Cloutier, R., DiMario, M., & Pozer, H. (2009). Net Centricity and Systems of Systems. In M. Jamshidi, (Ed.), *Systems of Systems Engineering*. Wiley Series in Systems Engineering. Boca Raton, FL: CRC Press/Taylor & Francis Group.

CMMI Product Team. (2010). Capability Maturity Model Integration, Version 1.3 (CMU/SEI-2010-TR-033). Software Engineering Institute, Carnegie Mellon University. Retrieved from CMMI Institute: http://cmmiinstitute.com (accessed October 24, 2014).

Cockburn, A. (2000). Selecting a Project Methodology. *IEEE Software*, 7(4), 64-71.

Conrow, E. H. (2003). *Effective Risk Management*, (2 Ed). Reston, VA: American Institute of Aeronautics and Astronautics, Inc.

Conway, M. E. (1968). How Do Committees Invent? *Datamation*, 14(5), 28-31. Retrieved from http:www.melconway.com/research/committees.html (accessed October 24, 2014).

Cook, M. (2004). Understanding the Potential Opportunities provided by Service-Oriented Concepts to Improve Resource Productivity. In T. Bhamra, & B. Hon (Eds.), *Design and Manufacture for Sustainable Development* (pp. 123-34). Suffolk, UK: Professional Engineering Publishing Limited.

Crosby, P. B. (1979). *Quality Is Free*. New York, NY: New American Library.

Dahmann, J. (2014). System of Systems Pain Points. *24th Annual INCOSE International Symposium*. Las Vegas, NV: International Council on Systems Engineering.

Daskin, M. S. (2010). *Service Science*. New York, NY: John Wiley & Sons, Inc.

DAU. (1993). *Committed Life Cycle Cost against Time. 3.1*. Fort Belvoir, VA: Defense Acquisition University.

DAU. (2010). *Defense Acquisition Guidebook*. Fort Belvoir, VA: Defense Acquisition University. Retrieved from https://acc.dau.mil/CommunityBrowser.aspx?id=22907&lan=en-US (accessed October 24, 2014).

Deming, W. E. (1986). *Out of the Crisis*. Cambridge, MA: MIT Center for Advanced Engineering Study.

DeRosa, J. K. (2005). Enterprise Systems Engineering. *Air Force Association, Industry Day*. Danvers, MA.

REFERENCES

DeRosa, J. K. (2006). An Enterprise Systems Engineering Model. *16th Annual INCOSE International Symposium*. Orlando, FL: International Council on Systems Engineering.

DoD. (1998). Integrated Product and Process Development Handbook. August. Retrieved from http://www.acq.osd.mil/se/docs/DoD-IPPD-Handbook-Aug98.pdf (accessed October 24, 2014).

DoD 5000.59. (2007). *Directive:DoD Modeling and Simulation (M&S) Management*. Washington, DC: U.S. Department of Defense.

DoD. (2010a). *MIL-STD-464C. Electromagnetic Environmental Effects, Requirements for Systems*. Washington, DC: U.S. Department of Defense.

DoD. (2010b). *MIL-STD-882. System Safety Program Require-ments*. Washington, DC: US Department of Defense.

DoDAF. (2010). DoD Architecture Framework, Version 2.02. Retrieved from http://dodcio.defense.gov/dodaf20.aspx (accessed October 24, 2014).

DoD and US Army. (2003). PSM Guide V4.0c, Practical Software and Systems Measurement: A Foundation for Objective Project Management. Picatinny Arsenal, NJ.

Domingue, J., Fensel, D., Davies, J., Gonzalez-Cabero, R., & Pedrinaci, C. (2009). The Service Web: A Web of Billions of Services. In G. Tselentis, J. Domingue, A. Galis, A. Gavras, D. Hausheer, S. Krco, ... T. Zeheriadis (Eds.), *Toward the Future Internet—A European Research Perspective*. Amsterdam, the Netherlands: IOS Press.

Dove, R. (2001). *Response Ability: The Language, Structure, and Culture of the Agile Enterprise*. New York, NY: John Wiley & Sons, Inc.

Dove, R. (2012). Agile Systems and Processes: Necessary and Sufficient Fundamental Architecture (Agile 101). *INCOSE Webinar*. Retrieved from http://www.parshift.com/s/Agile Systems-101.pdf (accessed September 19).

Dove, R., Popick, P., & Wilson, E. (2013). The Buck Stops Here: Systems Engineering is Responsible for System Security. *Insight*, 16(2), 6-10.

Draucker, L., Kaufman, S., Kuile, R. T., & Meinrenken, C. (2011). Moving Forward on Product Carbon Footprint Standards. *Journal of Industrial Ecology*, 15(2), 169-71.

DSMC. (1983). *Systems Engineering Management Guide*. Fort Belvoir, VA: Defense Systems Management College.

Edwards, W., Miles Jr., R. F., & Von Winterfeldt, D. (2007). *Advances in Decision

REFERENCES

Analysis: From Foundations to Applications. New York, NY: Cambridge University Press.

Eisner, H. (2008). *Essentials of Project and Systems Engineering Management*. Hoboken, NJ: John Wiley & Sons, Inc.

Elm, J., & Goldenson, D. (2012). The Business Case for Systems Engineering Study: Results of the Systems Engineering Effectiveness Study. Software Engineering Institute, Carnegie Mellon University. Retrieved from http://resources.sei.cmu.edu/library/asset-view.cfm?assetID=34061 (accessed January 26, 2015).

Engel, A. (2010). *Verification, Validation, and Testing of Engineered Systems*. Hoboken, NJ: John Wiley & Sons, Inc. Retrieved from INCOSE Systems Engineering Center of Excellence: http://www.incose.org/secoe/0105.htm (accessed September 23).

Eppinger, S., & Browning, T. (2012). *Design Structure Matrix Methods and Applications*. Cambridge, MA: MIT Press.

Estefan, J. (2008). Survey of Model-Based Systems Engineering (MBSE) Methodologies, Rev. B, Section 3.2. NASA Jet Propulsion Laboratory.

FAA. (2006). Systems Engineering Manual, Version 3.1. Federal Aviation Administration.

Failliere, N. L. (2011). W32.Stuxnet Dossier Version 1.4. Wired. Retrieved from http://www.wired.com/images_blogs/threatlevel/2011/02/Symantec-Stuxnet-Update-Feb-2011.pdf (accessed October 24, 2014).

Fairley, R. E. (2009). *Managing and Leading Software Projects*. Los Alamitos, CA: IEEE Computer Society; John Wiley & Sons, Inc.

Fitts, P. M. (1954). The Information Capacity of the Human Motor System in Controlling the Amplitude of Movement. *Journal of Experimental Psychology*, 47(6), 381-91.

Flood, R. L. (1999). *Rethinking the Fifth Discipline: Learning with the Unknowable*. London, UK: Routledge.

Flood, R. L., & Carson, E. R. (1993). *Dealing with Complexity: An Introduction to the Theory and Application of Systems Science* (2 Ed). New York, NY: Plenum Press.

Forrester, J. W. (1961). *Industrial Dynamics*. Waltham, MA: Pegasus Communications.

Forsberg, K. (1995). If I Could Do That, Then I Could ...: Systems Engineering in a

REFERENCES

Research and Development Environment. *Proceedings of the Fifth Annual INCOSE Symposium*. St. Louis, MO: International Council on Systems Engineering.

Forsberg, K., & Mooz, H. (1991), The Relationship of System Engineering to the Project Cycle. *Proceedings of the National Council for Systems Engineering (NCOSE) Conference*, Chattanooga, TN, pp. 57-65. October.

Forsberg, K., Mooz, H., & Cotterman, H. (2005). *Visualizing Project Management*, (3 Ed). Hoboken, NJ: John Wiley & Sons, Inc.

Fossnes, T. (2005). Lessons from Mt. Everest Applicable to Project Leadership. *Proceedings of the Fifteenth Annual INCOSE Symposium*. Rochester, NY: International Council on Systems Engineering.

Friedenthal, S. (1998). Object Oriented Systems Engineering. *Process Integration for 2000 and Beyond: Systems Engineering and Software Symposium*. New Orleans, LA: Lockheed Martin Corporation.

Friedenthal, S., Moore, A., & Steiner, R. (2012). *A Practical Guide to SysML: The Systems Modeling Language*, (2 Ed). New York, NY: Morgan Kaufmann Publishers, Inc.

Gallios, B., & Verma, D. (n.d.). System Design and Operational Effectiveness (SDOE): Blending Systems and Supportability Engineering Education. *Partnership in RMS Standards*, 5(1), 2.

GAO. (2008). Best Practices—Increased Focus on Require-ments and Oversight Needed to Improve DOD's Acquisition Environment and Weapon System Quality. Washington, DC: U.S. Government Accountability Office. Retrieved from http://www.gao.gov/new.items/d08294.pdf (accessed October 24, 2014).

Giachetti, R. E. (2010). *Designing of Enterprise Systems: Theory, Architecture, and Methods*. Boca Raton, FL: CRC Press.

Gilb, T. (2005). *Competitive Engineering*. Philadelphia, PA: Elsevier.

Gilb, T., & Graham, D. (1993). *Software Inspection*. Reading, MA: Addison- Wesley-Longman.

Grady, J. O. (1994). *System Integration*. Boca Raton, FL: CRC Press.

Gupta, J., & Sharma, S. (2004). *Creating Knowledge Based Organizations*. Boston, MA: Ida Group Publishing.

Haberfellner, R., & de Weck, O. (2005). Agile SYSTEMS ENGINEERING versus AGILE SYSTEMS Engineering. *Proceedings of the 15th Annual INCOSE*

REFERENCES

International Symposium. Rochester, NY: International Council on Systems Engineering.

Haimes, Y. (2012). Modelling Complex Systems of Systems with Phantom System Models. *Systems Engineering*, 15(3), 333-46.

Hall, A. (1962). *A Methodology for Systems Engineering*. Princeton, NJ: Van Nostrand.

Heijden, K., Bradfield, R., Burt, G., Cairns, G., & Wright, G. (2002). *The Sixth Sense: Accelerating Organizational Learning with Scenarios*. Chichester, UK: John Wiley & Sons, Ltd.

Herald, T., Verma, D., Lubert, C., & Cloutier, R. (2009). An Obsolescence Management Framework for System Baseline Evolution—Perspectives through the System Life Cycle. *Systems Engineering*, 12, 1-20.

Hipel, K., Jamshidi, M., Tien, J., & White, C. (2007). The Future of Systems, Man, and Cybernetics: Application Domains and Research Methods. *IEEE Transactions on Systems, Man and Cybernetics—Part C: Applications and Reviews*, 13(5), 726-43.

Hitchens, D. K. (2003). *Advanced Systems Thinking, Engineer-ing, and Management*. Boston, MA: Artech House.

Hollnagel, E., Woods, D. D., & Leveson, N. (Eds.). (2006). *Resilience Engineering: Concepts and Precepts*. Aldershot, UK: Ashgate Publishing Limited.

Honour, E. (2013). Systems Engineering Return on Investment. Ph.D. Thesis, Defense and Systems Institute, University of South Wales. Retrieved from http://www.hcode.com/seroi/index.html (accessed January 26, 2015).

Hughes, T. (1998). In *Rescuing Prometheus* (pp. 141-95). New York, NY: Pantheon Books.

Hybertson, D. (2009). *Model-Oriented Systems Engineering Science: A Unifying Framework for Traditional and Complex Systems*. Boca Raton, FL: Auerback/CRC Press.

IEEE Std 828. (2012). *IEEE Standard for Configuration Management in Systems and Software Engineering*. New York, NY: Institute of Electrical and Electronics Engineers.

IEEE 1012. (2012). *IEEE Standard for System and Software Verification and Validation*. New York, NY: Institute of Electrical and Electronics Engineers.

REFERENCES

IIBA. (2009). *A Guide to the Business Analysis Body of Knowledge (BABOK Guide)*. Toronto, ON: International Institute of Business Analysis.

INCOSE. (2004). What Is Systems Engineering? Retrieved from http://www.incose.org/practice/whatissystemseng.aspx (accessed June 14, 2004).

INCOSE. (2006). INCOSE Code of Ethics. International Council on Systems Engineering. Retrieved from http://www.incose.org/about/ethics.aspx (accessed February 15).

INCOSE. (2007). Systems Engineering Vision 2020. International Council on Systems Engineering, Technical Operations. Retrieved from http://www.incose.org/Products Pubs/pdf/SEVision2020_20071003_v2_03.pdf (accessed October 24, 2014).

INCOSE. (2010a). Model-Based Systems Engineering (MBSE) Wiki, hosted on OMG server. Retrieved from http://www.omgwiki.org/MBSE/doku.php (accessed October 24, 2014).

INCOSE. (2010b). Systems Engineering Measurement Primer, TP-2010-005, Version 2.0. Measurement Working Group, San Diego, CA: International Council on Systems Engineering. Retrieved from http://www.incose.org/ProductsPubs/pdf/INCOSE_SysEngMeasurementPrimer_2010-1205.pdf (accessed October 24, 2014).

INCOSE. (2011). Lean Enablers for Systems Engineering. Lean Systems Engineering Working Group. Retrieved from INCOSE Connect https://connect.incowse.org/tb/leansw/ (accessed February 1).

INCOSE. (2012). Guide for the Application of Systems Engineer-ing in Large Infrastructure Projects. TP-2010-007-01. INCOSE Infrastructure Working Group. San Diego, CA. June.

INCOSE RWG. (2012). *Guide for Writing Requirements*. San Diego, CA: International Council on Systems Engineering, Requirements Working Group.

INCOSE & PSM. (2005). Technical Measurement Guide, Version 1.0, December 2005, http://www.incose.org and http://www.psmsc.com (accessed October 24, 2014).

INCOSE UK. (2010). Systems Engineering Competencies Framework. INCOSE United Kingdom, Ltd. Retrieved from https://connect.incose.org/products/SAWG%20Shared%20 Documents/Other%20Technical%20Products/FRAMEWORK_WEB_

REFERENCES

Jan2010_Issue3.pdf (accessed October 24, 2014).

ISM. (n.d.). Institute for Supply Management. Retrieved from http://www.ism.ws/ (accessed October 24, 2014).

ISO 9001. (2008). *Quality management systems—Requirements*. Geneva, Switzerland: International Organization for Standardization.

ISO 10007. (2003). *Quality Management Systems—Guidelines for Configuration Management*. Geneva, Switzerland: International Organization for Standardization.

ISO 10303-233. (2012). *Industrial Automation Systems and Integration—Product Data Representation and Exchange—Part 233: Application Protocol: Systems Engineering*. Geneva, Switzerland: International Organization for Standardization.

ISO 13407. (1999). *Human-Centered Design Processes for Interactive Systems*. Geneva, Switzerland: International Organization for Standardization.

ISO 14001. (2004). *Environmental Management Systems — Requirements with Guidance for Use*. Geneva, Switzerland: International Organization for Standardization. Retrieved from http://14000store.com (accessed October 24, 2014).

ISO 14971. (2007) *Medical Devices—Application of Risk Management to Medical Devices*. Geneva, Switzerland: International Organization for Standardization.

ISO 17799. (2005) *Information Technology—Security Techniques: Code of Practice for Information Security Management*. Geneva, Switzerland: ISO/IEC.

ISO 26262. (2011). *Road Vehicles — Functional Safety*. Geneva, Switzerland: International Organization for Standardization.

ISO 31000. (2009). *Risk Management — Principles and Guidelines*. Geneva, Switzerland: International Organization for Standardization.

ISO 31010. (2009). *Risk Management—Risk Assessment Techniques*. Geneva, Switzerland: International Organization for Standardization.

IOS Guide 73. (2002). Risk Management—Vocabulary. International Organization for Standardization.

ISO Guide 73. (2009). *Risk Management — Vocabulary*. Geneva, Switzerland: International Organization for Standardization.

ISO/IEC 16085. (2006). *Systems and Software Engineering—Life Cycle Processes: Risk Management*. Geneva, Switzerland: International Organization for

REFERENCES

Standardization.

ISO/IEC 27002. (2013). *Information Technology—Security Techniques: Code of Practice for Information Security Controls*. Geneva, Switzerland: International Organization for Standardization.

ISO/IEC Guide 51. (1999). *Safety Aspects—Guidelines for the Inclusion in Standards*. Geneva, Switzerland: International Organization for Standardization.

ISO/IEC TR 19760. (2003) *Systems Engineering—A Guide for the Application of ISO/IEC 15288*. Geneva, Switzerland: International Organization for Standardization.

ISO/IEC TR 24748-1. (2010). *System and Software Engineering — Life Cycle Management—Part 1: Guide for Life Cycle Management*. Geneva, Switzerland: International Organization for Standardization. Retrieved from http://standards.iso.org/ittf/PubliclyAvailableStandards/index.html (accessed October 24, 2014).

ISO/IEC TR 24748-2. (2010). *Systems and Software Engineering—Life Cycle Management—Part 2: Guide to the Application of ISO/IEC 15288*. Geneva, Switzerland: International Organization for Standardization.

ISO/IEC/IEEE 15288. (2015). *Systems and Software Engineering— System Life Cycle Processes*. Geneva, Switzerland: International Organization for Standardization.

ISO/IEC/IEEE 15939, 2007. *Systems and Software Engineering— Measurement Process, ISO/IEC 2007*. Geneva, Switzerland: International Organization for Standardization.

ISO/IEC/IEEE 16326. (2009). *Systems and Software Engineering— Life Cycle Processes: Project Management*. Geneva, Switzerland: International Organization for Standardization.

ISO/IEC/IEEE 24748-4. (2014). *Systems and Software Engineering—Life Cycle Management — Part 4: Systems Engineering Planning*. Geneva, Switzerland: International Organization for Standardization.

ISO/IEC/IEEE 24765. (2010). *Systems and Software Engineering— Vocabulary*. Geneva, Switzerland: International Organiza-tion for Standardization. Retrieved from http://pascal.-computer.org/sev_display/index.action (accessed October 24, 2014).

ISO/IEC/IEEE 29110 Series (2014) *Systems and Software Engineering—Lifecycle Profiles for Very Small Entities (VSEs)*. International Organization for

REFERENCES

Standardization.

ISO/IEC/IEEE 29119. (2013). *Software Testing Standard*. Geneva, Switzerland: International Organization for Standardization.

ISO/IEC/IEEE 29148. (2011). *Systems and Software Engineering— Life Cycle Processes: Requirements Engineering*. Geneva, Switzerland: International Organization for Standardization.

ISO/IEC/IEEE 42010. (2011). *Systems and Software Engineering—Recommended Practice for Architectural Descriptions of Software-Intensive Systems*. Geneva, Switzerland: International Organization for Standardization.

Jackson, M. (1989). Which Systems Methodology When? Initial Results from a Research Program. In R. Flood, M. Jackson, & P. Keys (Eds.), *Systems Prospects: The Next Ten Years of Systems Research*. New York, NY: Plenum Press.

Jackson, S., & Ferris, T. (2013). Resilience Principles for Engineered Systems. *Systems Engineering* 19(2), 152-164.

Jacky, J. (1989). Programmed for Disaster. *The Sciences*, 29, 22-7.

Jensen, J. (2014). The Øresund Bridge — Linking Two Nations. Retrieved from http://www.cowi.dk (accessed October 24, 2014).

JHUAPL. (2011). Tutorial Material—Model-Based Systems Engineering Using the Object-Oriented Systems Engineer-ing Method (OOSEM). The Johns Hopkins University Applied Physics Laboratory. Retrieved from APL Technology Transfer http://www.jhuapl.edu/ott/technologies/copyright/sysml.asp (accessed October 24, 2014).

Johnson, S. (2010). *Where Good Ideas Come From: The Natural History of Innovation*. New York, NY: Riverhead Books.

Juran, J. M. (Ed.). (1974). *Quality Control Handbook*, 3rd Ed. New York, NY: McGraw-Hill.

Kaposi, A., & Myers, M. (2001). *Systems for All*. London, UK: Imperial College Press.

Katzan, H. (2008). *Service Science*. Bloomington, IN: -iUniverse Books.

KBS. (2010). IDEF Family of Methods. Knowledge Based Systems, Inc. Retrieved from IDEF: Integrated DEFinition Methods http://www.idef.com/ (accessed October 24, 2014).

Keeney, R. L. (2002). Common Mistakes in Making Value Trade-Offs. *Operations*

REFERENCES

Research, 50(6), 935-45.

Keeney, R., & Gregory, R. (2005). Selecting Attributes to Measure the Achievement of Objectives. *Operations Research*, 15(1), 1-11.

Keoleian, G. A., & Menerey, D. (1993). Life Cycle Design Guidance Manual: Environmental Requirements and the Product System, EPA/600/R-92/226. Environmental Protection Agency, Risk Reduction Engineering Laboratory. Retrieved from http://css.snre.umich.edu/css_doc/CSS93-02.pdf (accessed October 24, 2014).

Klir, G. (1991). *Facets of Systems Science*. New York, NY: Plenum Press.

Kotter, J. P. (2001). *What Leaders Really Do*. Boston, MA: Harvard Business Review: Best of HBR.

Kruchten, P. (1999). The Software Architect and the Software Architecture Team. In P. Donohue (Ed.), *Software Architecture* (pp. 565-583). Boston, MA: Kluwer Academic Publishers.

LAI MIT. (2013). Lean Enterprise Value Phase 1. Retrieved from MIT Lean Advancement Initiative http://lean.mit.edu/about/history/phase-i?highlight= WyJwaGFzZSIsMV0= (accessed October 24, 2014).

LAI, INCOSE, PSM, and SEARI. (2010). Systems Engineering Leading Indicators Guide, Version 2.0, January 29. Retrieved from http://www.incose.org and http://www.psmsc.com (accessed October 24, 2014).

Langley, M., Robitaille, S., & Thomas, J. (2011). Toward a New Mindset: Bridging the Gap Between Program Management and Systems Engineering. *PM Network*, 25(9), 24-6. Retrieved from http://www.pmi.org (accessed October 24, 2014).

Langner, R. (2012). Stuxnet Deep Dive. Miami Beach, FL: SCADA Security Scientific Symposium (S4). Retrieved from http://vimeopro.com/user10193115/s4-2012#/video/35806770 (accessed October 24, 2014).

Lano, R. (1977). *The N2 Chart*. Euclid, OH: TRW, Inc.

Larman, C., & Basili, V. (2003, June). Iterative and Incremental Development: A Brief History. *IEEE Software* 36(6): 47-56.

Lawson, H. (2010). *A Journey Through the Systems Landscape*. Kings College, UK: College Publications.

Lefever, B. (2005). ScSCE Methodology. Retrieved from SeCSE Service Centric Systems Engineering: http://www.secse-project.eu/ (accessed January 26, 2015).

REFERENCES

Leveson, N., & Turner, C. S. (1993). An Investigation of the Therac-25 Accidents. *IEEE Computer*, 26(7), 18-41.

Lewin, K. (1958). *Group Decision and Social Change*. New York, NY: Holt, Rinehart and Winston.

Lin, F., & Hsieh, P. (2011). A SAT View on New Service Development. *Service Science*, 3(2), 141-57.

Lindvall, M., & Rus, I. (2000). Process Diversity in Software Development. *IEEE Software*. 17(4), 14-8.

Little, W., Fowler, H. W., Coulson, J., & Onions, C. T. (1973). *The Shorter Oxford English Dictionary on Historical Principles*, 3rd Ed. Oxford, UK: Oxford University Press.

LMCO. (2008). Object-Oriented Systems Engineering Method (OOSEM) Tutorial, Version 3.11. Bethesda, MD: Lockheed Martin Corporation and San Diego, CA: INCOSE OOSEM Working Group.

Lombardo, M. M., & Eichinger, R. W. (1996). *The Career Architect Development Planner*, 1st Ed. Minneapolis, MN: Lominger.

Lonergan, B. (1992). Insight: A Study of Human Understanding. In F. E. Crowe, & R. M. Doran (Eds.), *Collected Works of Bernard Lonergan* (5 Ed, Vol. 3). Toronto, ON: University of Toronto Press.

Long, D. (2013). The Holistic Perspective — Systems Engineering as Leaders. *INCOSE Great Lakes Regional Conference Key Note Address*. West Lafayette, IN.

Luzeaux, D., & Ruault, J. R. (Eds.). (2010). *Systems of Systems*. New York, NY: John Wiley & Sons, Inc.

Lykins, H., Friedenthal, S., & Meilich, A. (2000). Adapting UML for an Object-Oriented Systems Engineering Method (OOSEM). *Proceedings of the 20th Annual INCOSE International Symposium*. Chicago, IL: INCOSE.

M&SCO. (2013). Verification, Validation, & Accreditation (VV&A) Recommended Practices Guide (RPG). Retrieved from U.S. DoD Modeling & Simulation Coordination Office: http://www.msco.mil/VVA_RPG.html (accessed October 24, 2014).

Magerholm, F. A., Shau, E. M., & Haskins, C. (2010). A Framework for Environmental Analyses of Fish Food Production Systems Based on Systems Engineering Principles. *Systems Engineering*, 13, 109-18.

REFERENCES

Maglio, P., & Spohrer, J. (2008). Fundamentals of Service Science. *Journal of the Academy of Marketing Science*, 36(1), 18-20.

Maier, M. W. (1998). Architecting Principles for Systems of Systems. *Systems Engineering*, 1(4), 267-84.

Maier, M. W., & Rechtin, E. (2009). *The Art of Systems Architecting*, 3rd Ed. Boca Raton, FL: CRC Press.

Martin, B. (2010). New Initiatives in the Army Green Procure-ment Program. Presentation, U.S. Army Public Health Command. Retrieved from http://www.dtic.mil/dtic/tr/-fulltext/u2/a566679.pdf (accessed October 24, 2014).

Martin, J. N. (1996). *Systems Engineering Guidebook: A Process for Developing Systems and Products*. Boca Raton, FL: CRC Press.

Martin, J. N. (2011). Transforming the Enterprise Using a Systems Approach. *Proceedings of the 21st Annual INCOSE International Symposium*. Denver, CO: International Council on Systems Engineering.

McAfee, A. (2009). *Enterprise 2.0: New Collaborative Tools for Your Organizations Toughest Challenges*. Boston, MA: Harvard Business School Press.

McConnel, S. (1998, May). The power of process. *IEEE Computer* 31(5), 100-102.

McDonough, W. (2013). McDonough Innovations: Design for the Ecological Century. Retrieved from http://www.mcdonough.com/ (accessed October 24, 2014).

McGarry, J., Card, D., Jones, C., Layman, B., Clark, E., Dean, J., & Hall, F. (2001). *Practical Software Measurement: Objective Information for Decision Makers*, Boston, MA: Addison-Wesley.

McGrath, D. (2003). China Awaits High Speed Maglev. Wired News. Retrieved from http://archive.wired.com/science/discoveries/news/2003/01/57163 (accessed January 26, 2015).

McManus, H. L. (2004). *Product Development Value Stream Mapping Manual, LAI Release Beta*. Boston, MA: MIT Lean Advancement Initiative.

McManus, H. L. (2005). *Product Development Transition to Lean (PDTTL) Roadmap, LAI Release Beta*. Boston, MA: MIT Lean Advancement Initiative.

Michel, R. M., & Galai, D. (2001). *Risk Management*. New York, NY: McGraw-Hill.

Miller, G. A. (1956). The Magical Number Seven, Plus or Minus Two: Some Limits on our Capacity for Processing Information. *Psychological Review*, 63(2), 81.

REFERENCES

MoDAF. (n.d.). The Website for MODAF Users and Implementers. UK Secretary of State for Defence (run by Model Futures, Ltd.). Retrieved from http://www.modaf.org.uk/ (accessed October 24, 2014).

Murman, E. M. (2002). *Lean Enterprise Value*. New York, NY: Palgrave.

Nagel, R. N. (1992). 21st Century Manufacturing Enterprise Strategy Report. Prepared for the Office of Naval Research. Retrieved from http://www.dtic.mil/cgi-bin/GetTRDoc?AD=ADA257032 (accessed October 24, 2014).

NASA. (2007a). NASA Pilot Benchmarking Initiative: Exploring Design Excellence Leading to Improved Safety and Reliability. Final Report. National Aeronautic and Space Administration.

NASA. (2007b). NASA Systems Engineering Handbook. National Aeronautic and Space Administration. Retrieved from http://www.es.ele.tue.nl/education/7nab0/2013/doc/NASA-SP-2007-6105-Rev-1-Final-31Dec2007.pdf (accessed October 24, 2014).

NDIA. (2008). Engineering for Systems Assurance, Version 1.0. National Defense Industrial Association, Systems Assurance Committee. Retrieved from http://www.acq. osd.mil/se/docs/SA-Guidebook-v1-Oct2008.pdf (accessed October 24, 2014).

NDIA. (2011). System Development Performance Measure-ment Report. Arlington, VA: National Defense Industrial Association, December.

Nissen, J. (2006). The Øresund Link. *The Arup Journal*, 31(2), 37-41.

NIST. (2012). National Vulnerability Database, Version 2.2. National Institute of Standards and Technology, Computer Science Division. Retrieved from http://nvd.nist.gov/ (accessed December 19)

Norman, D. (1990). *The Design of Everyday Things*. New York, NY: Doubleday.

NRC (Nuclear Regulatory Commission). (2005). Fact Sheets, Three Mile Island Accident. U.S. Nuclear Regulatory Commission. Report NUREG/BR-0292. Retrieved from http://pbadupws.nrc.gov/docs/ML0825/ML082560250.pdf (accessed October 24, 2014).

NRC (National Research Council). (2008). Pre-Milestone A and Early-Phase Systems Engineering. National Research Council of the National Academies. Washington, DC: The National Academies Press. Retrieved from http://www.nap.edu/ (accessed October 24, 2014).

REFERENCES

NRC (National Resilience Coalition). (2012). Definition of Resilience. National Resilience Coalition. Washington, DC: The Infrastructure Security Partnership.

O'Connor, P. D., & Kleyner, A. (2012). *Practical Reliability Engineering*, 5th Ed. Hoboken, NJ: John Wiley & Sons, Inc.

Oehmen, J. (Ed.). (2012). The Guide to Lean Enablers for Managing Engineering Programs, Version 1.0. Cambridge, MA: Joint MIT-PMI-INCOSE Community of Practice on Lean in Program Management. Retrieved from http://hdl.handle.net/1721.1/70495/ (accessed October 24, 2014).

Office of Government Commerce. (2009). *ITIL Lifecycle Publication Suite Books*. London, UK: The Stationery Office.

OMG. (2013a). Documents Associated with Unified Profile for DoDAF and MODAF (UPDM), Version 2.1. Object Management Group, Inc. Retrieved from http://www.omg.org/spec/UPDM/Current (accessed October 24, 2014).

OMG. (2013b). OMG Systems Modeling Language (SysML). Object Management Group, Inc. Retrieved from http://www.omgsysml.org (accessed October 24, 2014).

Oppenheim, B. W. (2004). Lean Product Development Flow. *Systems Engineering*, 7(4), 352-76.

Oppenheim, B. W. (2011). *Lean for Systems Engineering with Lean Enablers for Systems Engineering*. Hoboken, NJ: John Wiley & Sons, Inc.

Parnell, G. S., Bresnick, T., Tani, S., & Johnson, E. (2013). *Handbook of Decision Analysis*. Hoboken, NJ: John Wiley & Sons, Inc.

Patanakul, P., & Shenhar, A. (2010). Exploring the Concept of Value Creation in Program Planning and Systems Engineering Processes. *Systems Engineering*, 13, 340-52.

Pineda, R. (2010). Understanding Complex Systems of Systems Engineering. *Fourth General Assembly Cartagena Network of Engineering*, Metz, France.

Pineda, Martin, & Spoherer, (2014), in SEBoK, Part 3, Value of Service Systems Engineering.

PLCS. (2013). Product Life Cycle Support (PLCS). Retrieved from http://www.plcs-resources.org/ (accessed October 24, 2014)..

PMI. (2000). *A Guide to the PMBOK*. Newton Square, PA: Project Management Institute.

REFERENCES

PMI. (2009). *Practice Standard for Project Risk Management*. Newton Square, PA: Project Management Institute.

PMI. (2013). *The Standards for Program Management*, 3rd Ed. Newton Square, PA: Project Management Institute.

Porrello, A. M. (n.d.). Death and Denial: The Failure of the THERAC-25, A Medical Linear Accelerator. California Polytechnic State University, San Luis Obispo, CA. Retrieved from http://users.csc.calpoly.edu/~jdalbey/SWE/Papers/THERAC25.html (accessed October 24, 2014).

Qiu, R. (2009). Computational Thinking of Service Systems: Dynamics and Adaptiveness Modeling. *Service Science*, 1(1), 42-55.

Rebovich, G. (2006). Systems thinking for the enterprise: new and emerging perspectives. *Proceedings 2006 IEEE /SMC International Conference on System of Systems Engineering*. Los Angeles, CA: Institute of Electrical and Electronics Engineers.

Rebovich, G., & White, B. E. (Eds.). (2011). *Enterprise Systems Engineering: Advances in the Theory and Practice*. Boca Raton, FL: CRC Press.

Richards, M. G. (2009). *Multi-Attribute Tradespace Exploration for Survivability*. Boston: Massachusetts Institute of Technology.

Roedler, G. (2010). Knowledge Management Position. *Proceedings of the 20th Annual INCOSE International Symposium*. Chicago, IL: International Council on Systems Engineering.

Roedler, G. J., & Jones, C. (2006). Technical Measurement: A Collaborative Project of PSM, INCOSE, and Industry. INCOSE Measurement Working Group. INCOSE TP-2003-020-01.

Roedler, G., Rhodes, D. H., Schimmoler, H., & Jones, C. (Eds.). (2010). Systems Engineering Leading Indicators Guide, v2.0. Boston: MIT; INCOSE; PSM.

Ross, A. M., Rhodes, D. H., & Hastings, D. E. (2008). Defining Changeability: Reconciling Flexibility, Adaptability, Scalability, Modifiability, and Robustness for Maintaining System Lifecycle Value. *Systems Engineering*, 11, 246-62.

Rouse, W. B. (1991). *Design for Success: A Human-Centered Approach to Designing Successful Products and Systems*. New York, NY: John Wiley & Sons, Inc.

Rouse, W. B. (2005). Enterprise as Systems: Essential Challenges and Enterprise Transformation. *Systems Engineering*, 8(2), 138-50.

Rouse, W. B. (2009). Engineering the Enterprise as a System. In A. P. Sage, & W. B. Rouse (Eds.), *Handbook of Systems Engineering and Management*, 2nd Ed. New York, NY: John Wiley & Sons, Inc.

Royce, W. W. (1970). Managing the Development of Large Software Systems, *Proceedings, IEEE WESCON*, pp. 1-9. August.

Ryan, A. (2008). What Is a Systems Approach? *Journal of Non-Linear Science*, arXiv, 0809.1698.

Ryan, M. (2013). An Improved Taxonomy for Major Needs and Requirements Artifacts. *Proceedings of the 23rd Annual INCOSE International Symposium*. Philadelphia, PA: International Council on Systems Engineering.

SAE Aerospace Quality Standard AS9100:C. (2009). *Quality Management Systems — Requirements for Aviation, Space, and Defense Organizations*. Warrendale, PA: Society of Automotive Engineers.

SAE ARP 4754. (2010). *Guidelines for Development of Civil Aircraft and Systems*. Warrendale, PA: Society of Automotive Engineers.

SAE JA 1011 (2009). Evaluation criteria for Reliability-Centered Maintenance (RCM) processes. Warrendale, PA: Society of Automotive Engineers.

Salter, K. (2003). Presentation Given at the Jet Propulsion Laboratory. Pasadena, CA.

Salvendy, G. (Ed.). (1982). *Handbook of Industrial Engineer-ing*. New York, NY: John Wiley & Sons, Inc.

Salzman, J. (1997). Informing the Green Consumer: The Debate Over the Use and Abuse of Environmental Labels. *Journal of Industrial Ecology*, 1(2), 11-21.

SAVE. (2009). Welcome to SAVE International. SAVE International. Retrieved from http://www.value-eng.org/ (accessed October 24, 2014).

SAWE. (n.d.). Standards and Practices. Retrieved from Society of Allied Weight Engineers, Inc.: http://www.sawe.org/rp (accessed October 24, 2014).

Scholtes, P. R. (1988). *The Team Handbook: How the Use Teams to Improve Quality*. Madison, WI: Joiner Associates, Inc.

SE VOCAB. (2013). Software and Systems Engineering Vocabulary. Retrieved from http://pascal.computer.org/sev_display/index.action (accessed October 24, 2014).

SEBoK. (2014). BKCASE Editorial Board. The Guide to the Systems Engineering Body of Knowledge (SEBoK), version 1.3. R.D. Adcock (EIC). Hoboken, NJ: The

REFERENCES

Trustees of the Stevens Institute of Technology. www.sebokwiki.org. BKCASE is managed and maintained by the Stevens Institute of Technology Systems Engineering Research Center, the International Council on Systems Engineering, and the Institute of Electrical and Electronics Engineers Computer Society.

Software Engineering Institute. (2010). CMMI® (Measurement and Quantitative Management Process Areas), Version 1.3, November. Retrieved from http://www.sei.cmu.edu (accessed October 24, 2014).

Senge, P. (1990). *The Fifth Discipline: The Art & Practice of the Learning Organization*. New York, NY: Crown Business.

Shaw, T. E., & Lake, J. G. (1993). Systems Engineering: The Critical Product Development Enabler. *36th APICS International Conference Proceedings*. American Produc-tion and Inventory Control Society.

Sheard, S. (1996). Twelve Systems Engineering Roles. *Proceedings of the 6th Annual INCOSE International Symposium*. Boston, MA: International Council on Systems Engineering.

Shewhart, W. A. (1939). *Statistical Method from the Viewpoint of Quality Control*. New York, NY: Dover.

Sillitto, H. G. (2012). Integrating Systems Science, Systems Thinking, and Systems Engineering: Understanding the Differences and Exploiting the Synergies. *Proceedings of the 22nd Annual INCOSE International Symposium*. Rome, Italy: International Council on Systems Engineering.

Sillitto, H. G. (2013). Composable Capability—Principles, Strategies, and Methods for Capability Systems Engineering. *Proceedings of the 23rd Annual INCOSE International Symposium*. Philadelphia, PA: International Council on Systems Engineering.

Skanska. (2013). Øresund Bridge: Improving Daily Life for Commuters, Travelers, and Frogs. Retrieved from Skanska: Øresund Consortium: http://www.group.skanska.com/Campaigns/125/Oresund-Bridge/ (accessed October 24, 2014).

Skyttner, L. (2006). *General Systems Theory: Perspectives, Problems, Practice*, 2nd Ed. Singapore: World Scientific Publishing Company.

Slack, R. A. (1998). Application of Lean Principles to the Military Aerospace Product Development Process. Master of Science—Engineering and Management Thesis, Massachusetts Institute of Technology.

SMTDC. (2005). Shanghai Maglev Project Background. Retrieved from Shanghai

REFERENCES

Maglev Train: http://www.smtdc.com/en/ (accessed January 26, 2015).

Sols, A., Romero, J., & Cloutier, R. (2012). Performance-based Logistics and Technology Refreshment Programs: Bridging the Operational-Life Performance Capability Gap in the Spanish F-100 Frigates. *System Engineering*, 15, 422-32.

Spath, D., & Fahnrich, K. P. (Eds.). (2007). *Advances in Services Innovations*. Berlin/Heidelberg, Germany: Springer-Verlag.

Spohrer, J. C. (2011). Service Science: Progress & Direction. International Joint Conference on Service Science. Taipei, Taiwan.

Spohrer, J. C., & Maglio, P. P. (2010). Services Science: Toward a Smarter Planet. In G. Slavendy, & W. Karwowski (Eds.), *Introduction to Service Engineering*. Hoboken, NJ: John Wiley & Sons, Inc.

Srinivansan, J. (2010). Towards a Theory Sensitive Approach to Planning Enterprise Transformation. 5th EIASM Workshop on Organizational Change and Development. Vienna, Austria: European Institute for Advanced Studies in Management.

Stoewer, H. (2005). Modern Systems Engineering: A Driving Force for Industrial Competitiveness. Presentation to Members of the Japan INCOSE Chapter. Tokyo, Japan.

The White House. (2010). *National Security Strategy*. Washington, DC: The White House.

Theilmann, W., & Baresi, L. (2009). Multi-level SLAs for Harmonized Management in the Future Internet. In G. Tselentix, J. Dominque, A. Galis, A. Gavras, D. Hausheer, S. Krco, . . . T. Zehariadis (Eds.), *Towards the Future Internet—A European Research Perspective*. Amsterdam, the Netherlands: IOS Press.

Tien, J., & Berg, D. (2003). A Case for Service Systems Engineering. *Journal Systems Science and Systems Engineering*, 12(1), 13-38.

Toyota. (2009). Toyota Production System: Just-in-Time—Productivity Improvement. Retrieved from http://www.-toyota-global.com/company/vision_philosophy/toyota _ production_system/ (accessed October 24, 2014).

Transrapid. (2003, May 23). Transrapid International. Retrieved from http://www.transrapid.de/cgi/en/basics.prg?session=86140b5952535725_590396 (accessed October 24, 2014).

Tuttle, P., & Bobinis, J. (2013). Specifying Affordability. *Proceedings of the 23rd*

REFERENCES

Annual INCOSE International Symposium. Philadelphia, PA: International Council on Systems Engineering.

UNEP. (2012). GEO-5: Global Environmental Outlook. Retrieved from United National Environment Programme: http://www.unep.org/geo (accessed October 24, 2014).

Urwick, L. E. (1956). The Manager's Span of Control. Harvard Business Review. May/June.

Vargo, S. L., & Akaka, M. A. (2009). Service-Dominant Logic as a Foundation for Service Science: Clarifications. *Service Science*, 1(1), 32-41.

Walden, D. (2007). YADSES: Yet Another Darn Systems Engineering Standard. *Proceedings of the 17th Annual INCOSE International Symposium*. San Diego, CA: International Council on Systems Engineering.

Warfield, J. (2006). *An Introduction to Systems Science*. Hackensack, NJ: World Scientific Publishing Company.

Waters Foundation. (2013). Systems Thinking. Retrieved from Systems Thinking in Schools: http://watersfoundation.org/systems-thinking/overview/ (accessed October 24, 2014).

Wideman, R. M. (2002). Comparative Glossary of Project Management Terms, Version 3.1. Retrieved from http://-maxwideman.com/pmglossary/ (accessed October 24, 2014).

Wideman, R. M. (Ed.). (2004). *Project and Program Risk Management: A Guide to Managing Project Risks and Opportunities*. Newtown Square, PA: Project Management Institute.

Wiener, N. (1948). *Cybernetics or Control and Communication in the Animal and the Machine*. (Hermann, & Cie, Eds.) New York, NY: John Wiley & Sons, Inc.

Wild, J. P., Jupp, J., Kerley, W., Eckert, W., & Clarkson, P. J. (2007). Towards a Framework for Profiling of Products and Services. 5th International Conference on Manufacturing Research (ICMR). Leicester, UK.

Womack, J. P., & Jones, D. T. (1996). *Lean Thinking*. New York, NY: Simon & Schuster.

Wymore, A.W. (1967). *A Mathematical Theory of Systems Engineering: The Elements*. Malabar, FL: Krieger Publication Co.

REFERENCES

Wymore, A. W. (1993). *Model-Based Systems Engineering*. Boca Raton, FL: CRC Press.

Yourdon, E. (1989). *Modern Structured Analysis*. Upper Saddle River, NJ: Yourdon Press.

Zachman, J. A. (1987). A Framework for Information Systems Architecture. *IBM Systems Journal*, 26(3), 276-92. Retrieved from http://www.zifa.com (accessed October 24, 2014).

APPENDIX B: ACRONYMS

A_a	Achieved availability
A_i	Inherent availability
A_o	Operational availability
act	Activity diagrams [SysML™]
AECL	Atomic Energy Commission Limited [Canada]
AIAA	American Institute of Aeronautics and Astronautics [United States]
ANSI	American National Standards Institute [United States]
API	Application programming interface
ARP	Aerospace Recommended Practice
AS	Aerospace Standard
ASQ	American Society for Quality
ASAM	Association for Standardization of Automation and Measuring Systems
ASEP	Associate Systems Engineering Professional [INCOSE]
AUTOSAR	AUTomotive Open System Architecture
AVS	Associate Value Specialist [SAVE]
bdd	Block definition diagram [SysML™]
BRS	Business Requirements Specification
CAIV	Cost as an Independent Variable
CBA	Cost-benefit analysis
CBM	Condition-based maintenance
CCB	Configuration Control Board
CE	Conformité Européenne [EU]
CEA	Cost effectiveness analysis
CFR	Code of Federal Regulations [United States]
CI	Configuration item
CMMI®	Capability Maturity Model® Integration [CMMI Institute]
CMP	Configuration management plan
ConOps	Concept of operations
COTS	Commercial off-the-shelf
CSEP	Certified Systems Engineering Professional [INCOSE]
CVS	Certified Value Specialist [SAVE]
DAU	Defense Acquisition University [United States]

INCOSE Systems Engineering Handbook: A Guide for System Life Cycle Processes and Activities, Fourth Edition. Edited by David D. Walden, Garry J. Roedler, Kevin J. Forsberg, R. Douglas Hamelin and Thomas M. Shortell.

© 2015 John Wiley & Sons, Inc. Published 2015 by John Wiley & Sons, Inc.

附录 B：首字母缩写词

A_a	已实现的可用性
A_i	固有可用性
A_o	运行可用性
act	活动图 [SysML™]
AECL	原子能委员会有限公司[加拿大]
AIAA	美国国家航空航天管理局[美国]
ANSI	美国国家标准协会[美国]
API	应用编程接口
ARP	航空航天所推荐的实践
AS	航空航天标准
ASQ	美国质量学会
ASAM	自动化及测量系统标准协会
ASEP	助理级系统工程专业人员[INCOSE]
AUTOSAR	汽车开放系统架构
AVS	助理价值评估专家
bdd	块定义图[SysML™]
BRS	业务需求规范
CAIV	成本作为独立变量
CBA	成本收益分析
CBM	基于状态的维护
CCB	构型控制委员会
CE	欧洲统一[欧洲]
CEA	成本效能分析
CFR	联邦法规汇编[美国]
CI	配置项
CMMI®	能力成熟度模型综合[CMMI研究所]
CMP	构型管理计划
ConOps	运行意图
COTS	商用货架产品
CSEP	认证的系统工程专业人员 [INCOSE]
CVS	认证价值评估专家[SAVE]
DAU	国防采办大学 [美国]

INCOSE 系统工程手册：系统生命周期流程和活动指南，第 4 版。编撰：David D. Walden, Garry J. Roedler, Kevin J. Forsberg, R. Douglas Hamelin 和 Thomas M. Shortell。
John Wiley & Sons 公司版权所有©2015。由 John Wiley & Sons 公司于 2015 年出版。

ACRONYMS

	DFD	Data flow diagrams
	DMS	Diminishing material shortages
	DoD	Department of Defense [United States]
	DoDAF	Department of Defense Architecture Framework [United States]
	DSM	Design Structure Matrix
	DTC	Design to cost
	ECP	Engineering Change Proposal
257	ECR	Engineering change request
	EIA	Electronic Industries Alliance
	EM	Electromagnetic
	EMC	Electromagnetic compatibility
	EMI	Electromagnetic interference
	EN	Engineering notice
	ER	Entity relationship diagram
	ESEP	Expert Systems Engineering Professional [INCOSE]
	ETA	Event tree analysis
	FAST	Function Analysis System Technique
	FBSE	Functionsbased systems engineering
	FEA	Frontend analyses
	FEP	Fuel enrichment plant
	FFBD	Functional flow block diagram
	FHA	Functional hazard analysis
	FMEA	Failure Mode and Effects Analysis
	FMECA	Failure modes, effects, and criticality analysis
	FMVSS	Federal Motor Vehicle Safety Standards [United States]
	FoS	Family of systems
	FTA	Fault tree analysis
	G&A	General and administrative
	GAO	Government Accountability Office [United States]
	GEIA	Government Electronic Industries Alliance
	GENIVI	Geneva In-Vehicle Infotainment Alliance
	GNP	Gross national product
	GPP	Green Public Procurement
	HALT	Highly accelerated life testing

DFD	数据流图	
DMS	减弱材料短缺	
DoD	国防部 [美国]	
DoDAF	美国国防部架构框架[美国]	
DSM	设计结构矩阵	
DTC	为成本设计	
ECP	工程变更建议书	
ECR	工程变更请求	
EIA	电子工业联盟	
EM	电磁	
EMC	电磁兼容性	
EMI	电磁干扰	
EN	工程通知	
ER	实体关系图	
ESEP	专家级系统工程专业人员[INCOSE]	
ETA	事件树分析	
FAST	功能分析系统技术	
FBSE	基于功能的系统工程	
FEA	前端分析	
FEP	燃料浓缩工厂	
FFBD	功能流块图	
FHA	功能危险分析	
FMEA	失效模式和影响分析	
FMECA	失效模式、影响和危害性分析	
FMVSS	联邦机动车安全标准[美国]	
FoS	系统族	
FTA	故障树分析	
G&A	通用的和行政管理	
GAO	政府问责局[美国]	
GEIA	政府电子工业联盟	
GENIVI	车载信息娱乐联盟	
GNP	国民生产总值	
GPP	绿色公共采购	
HALT	高加速寿命试验	

ACRONYMS

HFE	Human factors engineering
HHA	Health hazard analysis
HSI	Human systems integration
HTS	Hazard tracking system
ibd	Internal block diagram [SysML™]
IBM	International Business Machines
ICD	Interface control document
ICS	Industrial control system
ICSM	Incremental Commitment Spiral Model
ICWG	Interface Control Working Group
IDEF	Integrated definition for functional modeling
IEC	International Electrotechnical Commission
IEEE	Institute of Electrical and Electronics Engineers
IFWG	Interface Working Group
IID	Incremental and iterative development
ILS	Integrated logistics support
INCOSE	International Council on Systems Engineering
IPAL	INCOSE Product Asset Library [INCOSE]
IPD	Integrated Product Development
IPDT	Integrated Product Development Team
IPO	Input-process-output
IPPD	Integrated Product and Process Development
IPT	Integrated Product Team
ISO	International Organization for Standardization
IT	Information technology
IV&V	Integration, verification, and validation
JSAE	Japan Society of Automotive Engineers [Japan]
JPL	Jet Propulsion Laboratory [United States]
KDR	Key driving requirement
KM	Knowledge management
KPP	Key Performance Parameter
KSA	Knowledge, skills, and abilities
LAI	Lean Advancement Initiative
LCA	Life cycle assessment

HFE	人因工程学
HHA	健康危害分析
HSI	人机综合
HTS	危险跟踪系统
ibd	内部块图 [SysML™]
IBM	国际商业机器公司
ICD	接口控制文档
ICS	工业控制系统
ICSM	渐进承诺螺旋模型
ICWG	接口控制工作组
IDEF	综合的功能建模定义
IEC	国际电工委员会
IEEE	电气与电子工程师协会
IFWG	接口工作组
IID	渐进和迭代开发
ILS	综合后勤支持
INCOSE	国际系统工程委员会
IPAL	INCOSE产品资产库[INCOSE]
IPD	综合产品开发
IPDT	综合产品开发团队
IPO	输入—流程—输出
IPPD	综合产品和流程开发
IPT	综合产品团队
ISO	国际标准化组织
IT	信息技术
IV&V	综合、验证与确认
JSAE	日本汽车工程师协会[日本]
JPL	喷气推进实验室[美国]
KDR	关键驱动需求
KM	知识管理
KPP	关键性能参数
KSA	知识、技能和能力
LAI	精益促进倡议行动
LCA	生命周期评估

ACRONYMS

LCC	Life cycle cost
LCM	Life cycle management
LEFMEP	Lean Enablers for Managing Engineering Programs
LEFSE	Lean Enablers for Systems Engineering
LINAC	Linear accelerator
LOR	Level of rigor
LORA	Level of Repair Analysis
MBSE	Model-based systems engineering
MIT	Massachusetts Institute of Technology
MoC	Models of computation
MODA	Multiple objective decision analysis
MoDAF	Ministry of Defense Architecture Framework [United Kingdom]
MOE	Measure of effectiveness
MOP	Measure of performance
MORS	Military Operations Research Society
MOS	Measure of suitability
MPE	Mass Properties Engineering
MTA	Maintenance Task Analysis
MTBF	Mean time between failure
MTBR	Mean time between repair
MTTR	Mean time to repair
N^2	N squared diagram
NASA	National Aeronautics and Space Administration [United States]
NEC	National Electrical Code [United States]
NCOSE	National Council on Systems Engineering (pre - 1995)
NCS	Network - Centric Systems
NDI	Nondevelopmental item
NDIA	National Defense Industrial Association [United States]
O&SHA	Operations and support hazard analysis
OAM&P	Operations, administration, maintenance, and provisioning
OEM	Original Equipment Manufacturer
OMG	Object Management Group
OOSEM	Object - Oriented Systems Engineering Method
OpEMCSS	Operational Evaluation Modeling for Context - Sensitive Systems

LCC	生命周期成本
LCM	生命周期管理
LEFMEP	工程计划管理精益使能项
LEFSE	系统工程精益使能项
LINAC	线性加速器
LOR	严苛层级
LORA	修复层级分析
MBSE	基于模型的系统工程
MIT	麻省理工学院
MoC	计算模型
MODA	多目标决策分析
MoDAF	国防部架构框架[英国]
MOE	有效性测度
MOP	性能测度
MORS	军事运行学会
MOS	适用性测量
MPE	质量特性工程
MTA	维护任务分析
MTBF	平均故障间隔时间
MTBR	平均修复间隔时间
MTTR	平均修复时间
N^2	N^2 图
NASA	国家航空航天局[美国]
NEC	国家电气规程[美国]
NCOSE	国家系统工程委员会（1995 年前）
NCS	网络中心系统
NDI	非开发项
NDIA	国家防务工业协会[美国]
O&SHA	运行及支持危险分析
OAM&P	运行、管理、维护与供应
OEM	原始设备制造商
OMG	对象管理组
OOSEM	面向对象的系统工程方法
OpEMCSS	对背景环境敏感的系统的运行评价建模

ACRONYMS

OPM	Object - Process Methodology
OpsCon	Operational concept
OSI	Open System Interconnect
par	Parametric diagram [SysML™]
PBL	Performance - based logistics
PBS	Product breakdown structure
PDT	Product Development Team
PERT	Program Evaluation Review Technique
PHA	Preliminary hazard analysis
PHS&T	Packaging, handling, storage, and transportation
PIT	Product Integration Team
pkg	Package diagram [SysML™]
PLC	Programmable logic controller
PLCS	Product Life Cycle Support
PLM	Product line management
PMI	Project Management Institute
PRA	Probabilistic risk assessment
PSM	Practical Software and Systems Measurement
QA	Quality assurance
QM	Quality management
R&D	Research and development
RAM	Reliability, availability, and maintainability
RBD	Reliability block diagram
RCM	Reliability - centered maintenance
REACH	Registration, Evaluation, Authorization, and Restriction of Chemical Substances
req	Requirement diagram [SysML™]
RFC	Request for change
RFP	Request for proposal
RFQ	Request for quote
RFV	Request for variance
RMP	Risk management plan
ROI	Return on investment
RUP	Rational Unified Process [IBM]

OPM	对象流程方法论
OpsCon	运行方案
OSI	开放系统互连
par	参数图[SysML™]
PBL	基于绩效的后勤
PBS	产品分解结构
PDT	产品开发团队
PERT	计划评估的评审技术
PHA	初步危险分析
PHS&T	包装、装卸、储存和运输
PIT	产品综合团队
pkg	包图[SysML™]
PLC	可编程逻辑控制器
PLCS	产品生命周期保障
PLM	产品线管理
PMI	项目管理举措
PRA	概率风险评估
PSM	实用软件和系统测量
QA	质量保证
QM	质量管理
R&D	研究与开发
RAM	可靠性、可用性和可维护性
RBD	可靠性框图
RCM	以可靠性为中心的维护
REACH	化学品的登记、评价、授权和限制
req	需求图[SysML™]
RFC	更改请求
RFP	建议请求
RFQ	报价申请书
RFV	偏差请求
RMP	风险管理计划
ROI	投资回报
RUP	Rational 统一流程[IBM]

ACRONYMS

RUP - SE	Rational Unified Process for Systems Engineering [IBM]
RVTM	Requirements Verification and Traceability Matrix
SA	State Analysis [JPL]
SAE	SAE International [formerly the Society of Automotive Engineers]
SAR	Safety Assessment Report
SAVE	Society of American Value Engineers
SCM	Supply chain management
SCN	Specification change notice
sd	Sequence diagram [SysML™]
SE	Systems engineering
SEARI	Systems Engineering Advancement Research Institute
SEBoK	Guide to the Systems Engineering Body of Knowledge
SEH	Systems Engineering Handbook [INCOSE]
SEHA	System element hazard analysis
SEIT	Systems Engineering and Integration Team
SEMP	Systems engineering management plan
SEMS	Systems Engineering Master Schedule
SEP	Systems engineering plan
SHA	System hazard analysis
SLA	Service - level agreement
SOI	System of interest
SoS	System of systems
SOW	Statement of work
SROI	Social return on investment
SRR	System Requirements Review
SSDP	Service system design process
STEP	Standard for the Exchange of Product Model Data
stm	State machine diagram [SysML™]
StRS	Stakeholder Requirements Specification
SWOT	Strength–weakness–opportunity–threat
SysML™	Systems Modeling Language [OMG]
SyRS	System Requirements Specification
SySPG	Systems Engineering Process Group
TADSS	Training Aids, Devices, Simulators, and Simulations

RUP-SE	系统工程的 Rational 统一流程[IBM]
RVTM	需求验证和可追溯性矩阵
SA	状态分析[JPL]
SAE	国际汽车工程师学会[之前称为美国汽车工程师学会]
SAR	安全评估报告
SAVE	美国价值工程师协会
SCM	供应链管理
SCN	规范变更通知单
sd	顺序图[SysML™]
SE	系统工程
SEARI	系统工程促进研究所
SEBoK	系统工程知识主体指南
SEH	系统工程手册[INCOSE]
SEHA	系统元素危险分析
SEIT	系统工程和综合团队
SEMP	系统工程管理计划
SEMS	系统工程主进度计划
SEP	系统工程计划
SHA	系统危险分析
SLA	服务等级协议
SOI	所感兴趣之系统
SoS	系统之系统
SOW	工作说明
SROI	社会投资回报
SRR	系统需求评审
SSDP	服务系统设计流程
STEP	产品模型数据交换标准
stm	状态机图[SysML™]
StRS	利益攸关者需求规范
SWOT	优势-劣势-机会-威胁
SysML™	系统建模语言[OMG]
SyRS	系统需求规范
SySPG	系统工程流程组
TADSS	培训器材、装置、模拟器和仿真设备

ACRONYMS

	TCO	Total cost of ownership
259	TOC	Total ownership cost
	TOGAF	The Open Group Architecture Framework
	TPM	Technical performance measure
	TRL	Technology readiness level
	TRP	Technology refreshment program
	TQM	Total quality management
	TR	Technical report
	uc	Use case diagram [SysML™]
	UIC	International Union of Railways
	UK	United Kingdom
	UL	Underwriters Laboratory [United States and Canada]
	UML™	Unified Modeling Language™ [OMG]
	US	United States
	USB	Universal Serial Bus
	USD	US dollars [United States]
	V&V	Verification and validation
	VA	Value analysis
	VE	Value engineering
	VM	Value management
	VMP	Value Methodology Practitioner [SAVE]
	VSE	Very small entities
	VSME	Very small and micro enterprises
	VV&A	Verification, validation, and accreditation
	WBS	Work breakdown structure
260	WG	Working group

TCO	所有权总成本
TOC	总拥有成本
TOGAF	The Open Group 架构框架
TPM	技术性能测度
TRL	技术成熟度
TRP	技术更新计划
TQM	全面质量管理
TR	技术报告
uc	用例图[SysML™]
UIC	国际铁路联盟
UK	英国
UL	保险商实验室[美国和加拿大]
UML™	统一建模语言™ [OMG]
US	美国
USB	通用串行总线
USD	美元[美国]
V&V	验证与确认
VA	价值分析
VE	价值工程
VM	价值管理
VMP	价值方法论实践者[SAVE]
VSE	极小型实体
VSME	极小型和微型复杂组织体
VV&A	验证、确认和鉴定
WBS	工作分解结构
WG	工作组

APPENDIX C: TERMS AND DEFINITIONS

Words not included in this glossary carry meanings consistent with general dictionary definitions. Other related terms can be found in SE VOCAB (2013).

Acquirer	The stakeholder that acquires or procures a product or service from a supplier
"-ilities"	The developmental, operational, and support requirements a program must address (named because they typically end in "ility"—availability, maintainability, vulnerability, reliability, supportability, etc.)
Acquisition logistics	Technical and management activities conducted to ensure supportability implications are considered early and throughout the acquisition process to minimize support costs and to provide the user with the resources to sustain the system in the field
Activity	A set of cohesive tasks of a process
Agile	Project execution methods can be described on a continuum from "adaptive" to "predictive." Agile methods exist on the "adaptive" side of this continuum, which is not the same as saying that agile methods are "unplanned" or "undisciplined"
Agreement	The mutual acknowledgment of terms and conditions under which a working relationship is conducted
Architecture	(System) fundamental concepts or properties of a system in its environment embodied in its elements, relationships, and in the principles of its design and evolution (see ISO 42010)
[1]*Baseline*	The gate-controlled step-by-step elaboration of business, budget, functional, performance, and physical characteristics, mutually agreed to by buyer and seller, and under formal change control. Baselines can be modified between formal decision gates by mutual consent through the change control process. An agreed-to description of the attributes of a product at a point in time, which serves as a basis for defining change (ANSI/EIA-649-1998)

INCOSE Systems Engineering Handbook: A Guide for System Life Cycle Processes and Activities, Fourth Edition. Edited by David D. Walden, Garry J. Roedler, Kevin J. Forsberg, R. Douglas Hamelin and Thomas M. Shortell.
© 2015 John Wiley & Sons, Inc. Published 2015 by John Wiley & Sons, Inc.

附录C：术语和定义

未包含在本词汇表中的词汇含有一般字典中定义的解释。其他相关术语可在 SE 词汇表（2013）中找到。

采办方	从供应商处采办或采购产品或服务的利益攸关者
"-性"	一个项目计划必须应对的开发、运行和维持需求（这样命名是因为其通常以"性"结尾——可用性、可维护性、易损性、可靠性、可维持性等）
采办后勤	技术活动和管理活动被实施确保在采办流程早期以及在整个采办流程期间考虑维持性的影响，以尽可能减少维持成本并向用户提供现场维持系统的资源
活动	一个流程中紧密衔接的任务集合
敏捷	项目执行方法可以用从"适应"到"预测"的连续统一体来描述。敏捷方法存在于该连续统一体的"适应性"一侧，但这并不等同于敏捷方法"无计划"或"无规矩"
协议	相互认可的条款和条件，并据此使工作关系得以实施
架构	系统中的元素、关系及其设计和演进的原则所体现的系统在其所处环境中的（系统）基本概念或特性（见 ISO 42010）
基线	在正式变更的控制下，买方与卖方双方均同意的业务、预算、功能、性能和物理特性"控制门"的逐步详细阐明。可在正式"决策门"之间通过变更控制流程共同一致地对基线进行修改。
	在某一个时间点商定的产品属性的描述作为定义变化的基础（ANSI/EIA-649-1998）

INCOSE 系统工程手册：系统生命周期流程和活动指南，第 4 版。编撰：David D.Walden、Garry J.Roedler、Kevin J.Forsberg、R.Douglas Hamelin 和 Thomas M.Shortell。

John Wiley & Sons 公司版权所有©2015。由 John Wiley & Sons 公司于 2015 年出版。

TERMS AND DEFINITIONS

Black box/ white box	Black box represents an external view of the system (attributes). White box represents an internal view of the system (attributes and structure of the elements)
Capability	An expression of a system, product, function, or process ability to achieve a specific objective under stated conditions
Commercial off-the-shelf (COTS)	Commercial items that require no unique acquirer modifications or maintenance over the life cycle of the product to meet the needs of the procuring agency
Commonality	(Of a product line) refers to functional and nonfunctional characteristics that can be shared with all member products within a product line (ISO 26550 2nd CD)
Configuration	A characteristic of a system element, or project artifact, describing their maturity or performance
Configuration item (CI)	A hardware, software, or composite item at any level in the system hierarchy designated for configuration management. (The system and each of its elements are individual CIs.) CIs have four common characteristics: 1. Defined functionality 2. Replaceable as an entity 3. Unique specification 4. Formal control of form, fit, and function
Decision gate	A decision gate is an approval event (often associated with a review meeting). Entry and exit criteria are established for each decision gate; continuation beyond the decision gate is contingent on the agreement of decision makers
Derived requirements	Detailed characteristics of the system of interest (SOI) that typically are identified during elicitation of stakeholder requirements, requirements analysis, trade studies, or validation
Design constraints	The boundary conditions, externally or internally imposed, for the SOI within which the organization must remain when executing the processes during the concept and development stages

黑盒/白盒	黑盒表示系统的外部视图（属性）。白盒表示系统的内部视图（元素的属性和结构）
能力	在申明的条件下达成特定目标的系统、产品、功能或流程能力的表述
商用货架产品（COTS）	为满足采购机构需要，在产品的生命周期内，无须独特的采办方维护和修改的商品
共通性	（产品线的共通性）指的是产品线范围内可与所有成员产品共享的功能特征和非功能特征（ISO 26550 2nd CD）
构型（配置）	系统元素或项目制品的一种特征，描述其成熟度或性能
构型配置项（CI）	指定用于构型管理的系统层级结构中任意级别的硬件、软件或复合项。（系统和每个元素都是独立的 CI。）CI 具有四个共同特点： 1. 定义的功能性 2. 可替换的实体 3. 唯一的规范 4. 形式、适配和功能的正式控制
决策门	决策门是一个审批事件（常常与评审会议相关）。为每个决策门建立了入口和出口准则；决策门以外的延续取决于决策者们的协议
导出的需求	所感兴趣之系统（SOI）的详细特点，通常在利益攸关者需求的提出、需求分析、权衡研究或确认过程中进行识别
设计约束	外部或内部施加给 SOI 的边界条件，在概念和开发阶段中实施流程时，组织必须保持这些边界条件

TERMS AND DEFINITIONS

Domain asset	Is the output of a subprocess of domain engineering that is reused for producing two or more products in a product line. A domain asset may be a variability model, an architectural design, a software component, a domain model, a requirements statement or specification, a plan, a test case, a process description, or any other element useful for producing products and services.Syn: domain artifact (ISO 26550 2nd CD)
	Note: In systems engineering, domain assets may be subsystems or components to be reused in further system designs. Domain assets are considered through their original requirements and technical characteristics.Domain assets include, but are not limited to, use cases, logical principles, environmental behavioral data, and risks or opportunities learned from the previous projects
	Domain assets are not physical products available off-the-shelf and ready for commissioning. Physical products (e.g., mechanical parts, electronic components, harnesses, optic lenses) are stored and managed according to the best practices of their respective disciplines
	Note: In software engineering, domain assets can include source or object code to be reused during the implementation
	Note: Domain assets have their own life cycles. ISO/I EC/IEEE 15288 may be used to manage a life cycle
Domain scoping	Identifies and bounds the functional domains that are important to an envisioned product line and provide sufficient reuse potential to justify the product line creation. Domain scoping builds on the definitions of the product scoping (ISO 26550 2nd CD)
Element	See system element
Enabling system	A system that supports a SOI during its life cycle stages but does not necessarily contribute directly to its function during operation
Enterprise	A purposeful combination of interdependent resources that interact with each other to achieve business and operational goals (Rebovich and White, 2011)

领域资产	是被重新用于在一个产品线中生产两个或多个产品的领域工程子流程的输出。领域资产可能是可变性模型、架构设计、软件部件、领域模型、需求声明或规范、计划、试验案例、流程描述或可用于生产产品和服务的任何其他元素。同义词：领域制品（ISO 26550 2nd CD）
	注：在系统工程中，领域资产可能是在未来系统设计中复用的子系统或部件。通过领域资产的原始需求和技术特征考虑领域资产。领域资产包括但不限于用例、逻辑原则、环境性能数据以及从之前的项目中获悉的风险或机会。
	领域资产不是可用的商用货架和准备投入试运行的物理产品。物理产品（例如，机械零件、电子部件、导线束、光学透镜）是根据其各自学科的最佳实践贮存和管理的。
	注：在软件工程中，领域资产可包括实施中复用的源代码或目标代码。
	注：领域资产具有其自己的生命周期。ISO/I EC/IEEE 15288可用于管理生命周期
领域范围界定	识别和约束对于预想的产品线非常重要并提供足够的复用潜力以证明产品线创建正确的功能领域。领域范围界定是建立于产品范围界定的定义之上（ISO 26550 2nd CD）
元素	见系统元素
使能系统	一种系统，在其生命周期阶段中支持 SOI，但未必直接致力于运行中的功能
复杂组织体	为实现业务和运行目标而彼此交互的、相互依赖的资源的有目的的组合（Rebovich 和 White, 2011）

TERMS AND DEFINITIONS

Environment	The surroundings (natural or man-made) in which the SOI is utilized and supported or in which the system is being developed, produced, and retired
Facility	The physical means or equipment for facilitating the performance of an action, for example, buildings, instruments, and tools
Failure	The event in which any part of an item does not perform as required by its specification. The failure may occur at a value in excess of the minimum required in the specification, that is, past design limits or beyond the margin of safety
Functional configuration audit	An evaluation to ensure that the product meets baseline functional and performance capabilities (adapted from ISO/EC/IEEE 15288)
Human factors	The systematic application of relevant information about human abilities, characteristics, behavior, motivation, and performance. It includes principles and applications in the areas of human-related engineering, anthropometries, ergonomics, job performance skills and aids, and human performance evaluation
Human systems integration	The interdisciplinary technical and management processes for integrating human considerations within and across all system elements; an essential enabler to SE practice
Interface	A shared boundary between two functional units, defined by functional characteristics, common physical interconnection characteristics, signal characteristics, or other characteristics, as appropriate (ISO 2382-1)
Integration definition for functional modeling (IDEF)	A family of modeling languages in the fields of systems and software engineering that provide a multiple-page (view) model of a system that depicts functions and information or product flow. Boxes illustrate functions and arrows illustrate information and product flow (KBS, 2010). Alphanumeric coding is used to denote the view: • IDEF0—functional modeling method • IDEF1—information modeling method • IDEF1X—data modeling method • IDEF3—process description capture method • IDEF4—object-oriented design method • IDEF5—ontology description capture method

术语和定义

环境	周围的环境（天然的或人造的），SOI 在该环境中使用并得到支持或者系统在该环境中开发、生产和退役
设施	用于辅助行动执行的物理手段或设备，如建筑物、仪器和工具
失效	物品的任意部分没按规范的要求而实施的事件。失效可能在数值超过规范的最低要求（即超过设计限制或超出安全裕度）时出现
功能构型审核	一种评价，以确保产品满足基线功能能力和性能能力（改编自 ISO/EC/IEEE 15288）
人素	有关人员能力、特征、行为、动机和绩效等相关信息的系统化应用。它包括与人员相关的工程、人体测量学、人机工程学、工作岗位技能和辅助手段及人员绩效评估等领域的原理和应用
人员系统综合	在所有系统元素内部或跨所有系统元素综合人素考量的跨学科技术和管理流程；SE 实践的基本使能项
接口	由功能特征、共同物理互联特征、信号特征或其他特征适当定义的两个功能单元之间的共享边界（ISO 2382-1）
功能建模综合定义（IDEF）	系统和软件工程领域中的建模语言系列，可提供系统的多页（视图）模型，描述功能和信息或产品流。盒表示功能，箭头表示信息和产品流（KBS, 2010）。字母数字编码用于表示视图： • IDEF0——功能建模方法 • IDEF1——信息建模方法 • IDEF1X——数据建模方法 • IDEF3——流程描述捕获方法 • IDEF4——面向对象设计方法 • IDEF5——本体论描述捕获方法

TERMS AND DEFINITIONS

IPO diagram	Figures in this handbook that provide a high-level view of the process of interest. The diagram summarizes the process activities and their inputs and outputs from/to external actors; some inputs are categorized as controls and enablers. A control governs the accomplishments of the process; an enabler is the means by which the process is performed
Life cycle cost (LCC)	The total cost of acquisition and ownership of a system over its entire life. It includes all costs associated with the system and its use in the concept, development, production, utilization, support, and retirement stages
Life cycle model	A framework of processes and activities concerned with the life cycle, which also acts as a common reference for communication and understanding
Measures of effectiveness	Measures that define the information needs of the decision makers with respect to system effectiveness to meet operational expectations
Measures of performance	Measures that define the key performance characteristics the system should have when fielded and operated in its intended operating environment
N^2 diagrams	Graphical representation used to define the internal operational relationships or external interfaces of the SOI
Operator	An individual who, or an organization that, contributes to the functionality of a system and draws on knowledge, skills, and procedures to contribute the function
Organization	Person or a group of people and facilities with an arrangement of responsibilities, authorities, and relationships (adapted from ISO 9001:2008)
Performance	A quantitative measure characterizing a physical or functional attribute relating to the execution of a process, function, activity, or task; performance attributes include quantity (how many or how much), quality (how well), timeliness (how responsive, how frequent), and readiness (when, under which circumstances)
Physical configuration audit	An evaluation to ensure that the operational system or product conforms to the operational and configuration documentation (adapted from ISO/IEC7 IEEE 15288)

IPO 图	本手册中提供了对受关注流程的高层级视角的图。该图总结了流程活动及其来自/去到外部参与者的输入和输出；有些输入被归类为"控制项"和"使能项"。"控制项"支配着流程的完成；"使能项"是实施流程的手段
生命周期成本（LCC）	在系统的全生命周期内采办和拥有系统的总成本。它包括与系统及其在概念、开发、生产、使用、保障和退役阶段中的使用相关的所有成本
生命周期模型	与生命周期相关的流程和活动的框架，亦可用作沟通和理解的公共参照
有效性测度	定义决策者对满足运行期望的系统有效性的信息需要的测度
性能测度	定义系统在其预期运行环境现场和运行时应具有的关键性能特征的测度
N^2 图	用于定义 SOI 的内部运行关系或外部接口的图形表示
操作人员	致力于系统的功能性并利用知识、技能和程序致力于该功能的个人或组织
组织	具有职责、权限和关系设置安排的个人或群体与设施（改编自 ISO 9001:2008）
性能	与流程、功能、活动或任务执行相关的物理或功能属性的定量测度；性能属性包括数量（多少）、品质（怎么样）、时间性（响应如何、频度如何）及准备度（何时，在何种情况下）
物理构型审核	一种评价，以确保运行的系统或产品符合运行文件和构型文件的要求（改编自 ISO/IEC7 IEEE 15288）

TERMS AND DEFINITIONS

Process	A set of interrelated or interacting activities that transforms inputs into outputs (adapted from ISO 9001:2008)
Product line	1. Group of products or services sharing a common, managed set of features that satisfy specific needs of a selected market or mission. ISO/ I EC/IEEE 24765 (2010), Systems and software engineering vocabulary
	2. A collection of systems that are potentially derivable from a single-domain architecture. IEEE 1517-1999 (R2004) IEEE standard for information technology—Software life cycle processes—Reuse processes (3.14) (ISO/IEC FCD 24765.5)
Product line scoping	Defines the products that will constitute the product line and the major (externally visible) common and variable features among the products, analyzes the products from an economic point of view, and controls and schedules the development, production, and marketing of the product line and its products. Product management is primarily responsible for this process (ISO 26550 2nd CD)
Project	An endeavor with defined start and finish criteria undertaken to create a product or service in accordance with specified resources and requirements
Proof of concept	A naive realization of an idea or technology to demonstrate its feasibility
Prototype	A production-ready demonstration model developed under engineering supervision that is specification compliant and represents what manufacturing should replicate
Qualification limit	Proving that the design will survive in its intended environment with margin. The process includes testing and analyzing hardware and software configuration items to prove that the design will survive the anticipated accumulation of acceptance test environments, plus its expected handling, storage, and operational environments, plus a specified qualification margin
Requirement	A statement that identifies a system, product, or process characteristic or constraint, which is unambiguous, clear, unique, consistent, stand-alone (not grouped), and verifiable, and is deemed necessary for stakeholder acceptability

流程	将输入转变为输出的相互关联或相互作用的活动的集合（改编自 ISO 9001:2008）
产品线	1. 产品或服务的群组，这些产品或服务共享一组受管理的、满足所选市场或任务的特定需要的公共特点的集合，ISO/ I EC/IEEE 24765（2010），系统和软件工程词汇表
2. 可能从单域架构潜在地可导出的系统的一个汇集。IEEE 1517-1999（R2004）IEEE 信息技术标准——软件生命周期流程——复用流程（3.14）（ISO/IEC FCD 24765.5）
产品线范围界定	定义将构成产品线和产品间主要（外部可见）的共同特征和可变特征的产品，从经济观点分析该产品，以及控制与安排该产品线及其产品的开发、生产和销售。产品管理人员主要负责此流程（ISO 26550 2nd CD）
项目	按照特定的资源和需求创建产品或服务所付出的具有明确定义的开始与完成准则的努力
概念验证	某个想法或技术的非常初级地实现，以验证其可行性
原型	一种在工程监督下开发的可量产演示模型，其符合规范并表示应重复什么样的制造
鉴定限制	证明设计将会在其意图的环境中具有裕度的生存。该流程包括试验和分析硬件和软件构型配置项，以证明该设计将会在经历预期积累的验收试验环境、其期望的处理环境、贮存环境和运行环境以及指定的鉴定裕度后继续生存。
需求	识别一个系统、产品或流程的特征或约束的说明，是明确的（无歧义的）、清晰的、唯一的、一致的、独立的（不是成组的）和可验证的，这对利益攸关者的可接受性十分必要

TERMS AND DEFINITIONS

Resource	An asset that is utilized or consumed during the execution of a process
Return on investment	Ratio of revenue from output (product or service) to development and production costs, which determines whether an organization benefits from performing an action to produce something (ISO/IEC 24765.5 FCD; ISO/IEC/IEEE 24765, 2010)
Reuse	1. The use of an asset in the solution of different problems. [IEEE 1517-1999 (R2004)]
	2. Building a software system at least partly from existing pieces to perform a new application.[ISO/IEC/ IEEE 24765 (2010)]2.
Specialty engineering	Analysis of specific features of a system that requires special skills to identify requirements and assess their impact on the system life cycle
Stage	A period within the life cycle of an entity that relates to the state of its description or realization
Stakeholder	A party having a right, share, or claim in a system or in its possession of characteristics that meet that party's needs and expectations
Supplier	An organization or an individual that enters into an agreement with the acquirer for the supply of a product or service
System	An integrated set of elements, subsystems, or assemblies that accomplish a defined objective.These elements include products (hardware, software, firmware), processes, people, information, techniques, facilities, services, and other support elements (INCOSE)
	A combination of interacting elements organized to achieve one or more stated purposes (ISO/IEC/IEEE 15288)
System element	Member of a set of elements that constitutes a system
System life cycle	The evolution with time of a SOI from conception to retirement
System of interest	The system whose life cycle is under consideration

资源	在流程执行期间被使用或消耗的资产
投资回报	来自输出（产品或服务）的收入与开发和生产成本的比率，该比率确定组织是否受益于实施产生某些事物的活动（ISO/IEC 24765.5 FCD；ISO/IEC/IEEE 24765，2010年）
复用	1. 在不同问题的解决方案中使用一种资产。[IEEE 1517-1999（R2004）] 2. 建立至少部分依据现有资产的软件系统以实施一个新的应用。[ISO/IEC/IEEE 24765（2010）]
专业工程	对系统特定的特点的分析，其要求特殊技能以识别系统需求并评估对系统生命周期的影响的专门技能
阶段	实体生命周期内的某一时期，与其描述或实现状态相关
利益攸关者	在满足其需要和期望某一系统或其拥有的特征中具有权利、份额或要求权的一方
供应商	与采办方签订协议以供应产品或服务的组织或个人
系统	完成某一确定目标的一组综合的元素、子系统或组件的集合。这些元素包括产品（硬件、软件和固件）、流程、人员、信息、技术、设施、服务和其他支持元素（INCOSE） 实现一个或多个指定目的的相互作用的元素组织起来的一个组合，（ISO/IEC/IEEE 15288）
系统元素	构成系统的一个元素集合中的成员
系统生命周期	SOI从概念到退役随时间的演进
所感兴趣之系统	全生命周期都在考虑之中的系统

TERMS AND DEFINITIONS

System of systems	A SOI whose system elements are themselves systems; typically, these entail large-scale interdisciplinary problems with multiple, heterogeneous, distributed systems	
Systems engineering	Systems engineering (SE) is an interdisciplinary approach and means to enable the realization of successful systems. It focuses on defining customer needs and required functionality early in the development cycle, documenting requirements, and then proceeding with design synthesis and system validation while considering the complete problem: operations, cost and schedule, performance, training and support, test, manufacturing, and disposal. SE considers both the business and the technical needs of all customers with the goal of providing a quality product that meets the user needs (INCOSE)	
Systems engineering effort	Systems engineering effort integrates multiple disciplines and specialty groups into a set of activities that proceed from concept to production and to operation. SE considers both the business and the technical needs of all stakeholders with the goal of providing a quality system that meets their needs	
Systems engineering management plan (SEMP)	Structured information describing how the systems engineering effort, in the form of tailored processes and activities, for one or more life cycle stages, will be managed and conducted in the organization for the actual project	
Tailoring	The manner in which any selected issue is addressed in a particular project. Tailoring may be applied to various aspects of the project, including project documentation, processes and activities performed in each life cycle stage, the time and scope of reviews, analysis, and decision making consistent with all applicable statutory requirements	
Technical performance measures	Measures that define attributes of a system element to determine how well a system or system element is satisfying or expected to satisfy a technical requirement or goal	
Trade-off	Decision-making actions that select from various requirements and alternative solutions on the basis of net benefit to the stakeholders	

系统之系统	系统元素本身也是系统的 SOI；这些系统之系统通常带来大规模的跨学科问题，包括多重的与异构的分布式的系统
系统工程	系统工程（SE）是一种使系统能够成功实现的跨学科的方法和手段。系统工程专注于：在开发周期的早期阶段定义客户的需要与所要求的功能性，将需求文件化，然后进行设计综合和系统确认同时考虑完整问题，亦即运行、成本、进度、性能、培训、保障、试验、制造和退出。SE 以提供满足用户需要的高质量产品为目的，同时考虑了所有客户的业务和技术需要（INCOSE）
系统工程的努力	系统工程的努力是工作把多种学科和专业组综合到从概念到生产再到运行的一系列活动中。SE 以提供满足需要的高质量系统为目的，考虑了所有利益攸关者的业务和技术需要
系统工程管理计划（SEMP）	描述在一个或多个生命周期阶段，以剪裁的流程和活动的形式描述系统工程工作将在实际项目的组织中如何被管理和实施的结构化信息
剪裁	在特殊项目中应对任何选定项目的方式。剪裁可适用于项目的各个方面，包括项目文件编制、在生命周期每个阶段执行的流程和活动以及符合所有适用法定要求的评审、分析和决策的时间与范围
技术性能测度	定义系统元素的属性以确定系统或系统元素满足或期望满足技术需求或目标的程度的测度
权衡	决策行动，根据利益攸关者的净利益从各种要求和备选解决方案中进行选择

TERMS AND DEFINITIONS

User	Individual who or group that benefits from a system during its utilization
Validation	Confirmation, through the provision of objective evidence, that the requirements for a specific intended use or application have been fulfilled (ISO/IEC/IEEE 15288)
	Note: Validation is the set of activities ensuring and gaining confidence that a system is able to accomplish its intended use, goals, and objectives (i.e., meet stakeholder requirements) in the intended operational environment
Value	A measure of worth (e.g.. benefit divided by cost) of a specific product or service by a customer, and potentially other stakeholders and is a function of (i) the product's usefulness in satisfying a customer need, (ii) the relative importance of the need being satisfied, (iii) the availability of the product relative to when it is needed, and (iv) the cost of ownership to the customer (McManus. 2004)
Variability	Of a product line refers to characteristics that may differ among members of the product line (ISO 26550 2nd CD)
Variability constraints	Denotes constraint relationships between a variant and a variation point, between two variants, and between two variation points (ISO 26550 2nd CD)
Verification	Confirmation, through the provision of objective evidence, that specified requirements have been fulfilled (ISO/ IEC/IEEE 15288)
	Note: Verification is a set of activities that compares a system or system element against the required characteristics. This may include, but is not limited to, specified requirements, design description, and the system itself
Waste	Work that adds no value to the product or service in the eyes of the customer (Womack and Jones. 1996)

术语和定义

用户	在系统使用期间从中获益的个人或群体
确认	通过客观证据的提供，证实某一特定的使用或应用的需求已经得到实现（ISO/IEC/IEEE 15288）
	注：确认是确保和获得信心的活动集，即系统能够在意图的操作环境中完成其意图的使用、目标和目的（即，满足利益攸关者需求）
价值	"客户和潜在的其他利益攸关者对特定产品或服务的意义（如收益除以成本）的度量，是以下因素的函数：（i）产品满足客户需要的有用性，（ii）所满足的需要的相对重要性，（iii）产品相对于需要时间的可用性以及（iv）客户的拥有成本（McManus.，2004）
可变性	产品线的可变性指的是在产品线成员间可能有所不同的特征（ISO 26550 2nd CD）
可变性约束	表示变体与变化点之间、两个变体之间以及两个变化点之间的约束关系（ISO 26550 2nd CD）
验证	通过客观证据的提供，证实规定的需求已经得到实现（ISO/IEC/IEEE 15288）
	注：验证是比较系统或系统元素与所需特征的一组活动。这可能包括但不限于指定的需求、设计描述以及系统本身
浪费	在客户眼中是不能为产品或服务增加价值的工作（Womack 和 Jones，1996）

APPENDIX D: N^2 DIAGRAM OF SYSTEMS ENGINEERING PROCESSES

Figure D.1 illustrates the input/output relationships between the various SE processes presented in the handbook and shows the interactions depicted on the IPO diagrams throughout this handbook. The primary flows represent a typical system development program.

The individual processes are placed on the diagonal by abbreviation to the process names, as follows:

EXT	External inputs and outputs
BMA	Business or mission analysis
SNRD	Stakeholder needs and requirements definition
SRD	System requirements definition
AD	Architecture definition
DD	Design definition
SA	System analysis
IMPL	Implementation
INT	Integration
VER	Verification
TRAN	Transition
VAL	Validation
OPER	Operation
MAINT	Maintenance
DISP	Disposal
PP	Project planning
PAC	Project assessment and control
DM	Decision management
RM	Risk management
CM	Configuration management
INFOM	Information management
MEAS	Measurement

NCOSE Systems Engineering Handbook: A Guide for System Life Cycle Processes and Activities, Fourth Edition. Edited by David D. Walden, Garry J. Roedler, Kevin J. Forsberg, R. Douglas Hamelin and Thomas M. Shortell.
© 2015 John Wiley & Sons, Inc. Published 2015 by John Wiley & Sons, Inc.

附录 D：系统工程流程的 N^2 图

图 D.1 阐明本手册提供的各种不同 SE 流程之间的输入/输出关系，并表明贯穿于本手册的 IPO 图上所描述的交互。主要流程表示典型的系统开发计划。

单个流程以其名称的缩写形式置于斜线上，如下：

EXT	外部输入和输出
BMA	业务或使命任务分析
SNRD	利益攸关者需要和需求定义
SRD	系统需求定义
AD	架构定义
DD	设计定义
SA	系统分析
IMPL	实施
INT	综合
VER	验证
TRAN	转移
VAL	确认
OPER	运行
MAINT	维护
DISP	退出
PP	项目规划
PAC	项目评估和控制
DM	决策管理
RM	风险管理
CM	构型配置管理
INFOM	信息管理
MEAS	测量/度

INCOSE 系统工程手册：系统生命周期流程和活动指南，第 4 版。编撰：David D.Walden、Garry J.Roedler、Kevin J.Forsberg、R.Douglas Hamelin 和 Thomas M.Shortell。

John Wiley & Sons 公司版权所有©2015。由 John Wiley & Sons 公司于 2015 年出版。

N^2 DIAGRAM OF SYSTEMS ENGINEERING PROCESSES

QA	Quality assurance
ACQ	Acquisition
SUP	Supply
LCMM	Life cycle model management
INFRAM	Infrastructure management
PM	Portfolio management
HRM	Human resource management
QM	Quality management
KM	Knowledge management
TLR	Tailoring

The off-diagonal squares represent the inputs/outputs interface shared by the processes that intersect at a given square. Outputs flow horizontally; inputs flow vertically and can be read in a clockwise fashion.

Note 1: The absence of an *x* in an intersection does not preclude tailoring to create a relationship between any two processes.

Note 2: This is the result of one possible instance of the life cycle processes. Other instances of the process relationships are possible.

FIGURE D.1 Input/output relationships between the various SE processes. INCOSE SEH original figure created by David Walden. Usage per the INCOSE Notices page. All other rights reserved.

QA	质量保证	
ACQ	采办	
SUP	供应	
LCMM	生命周期模型管理	
INFRAM	基础设施管理	
PM	项目群管理	
HRM	人力资源管理	
QM	质量管理	
KM	知识管理	
TLR	剪裁	

非对角线方块表示给定方块上相交流程所共享的输入/输出接口。输出水平流动；输入垂直流动，并能够按顺时针方向来读取。

注 1：相交处不存在 x，并不排除通过剪裁在任意两个流程之间创建一种关系。

注 2：这是生命周期流程的一种可能实例的结果。流程关系的其他实例是可能的。

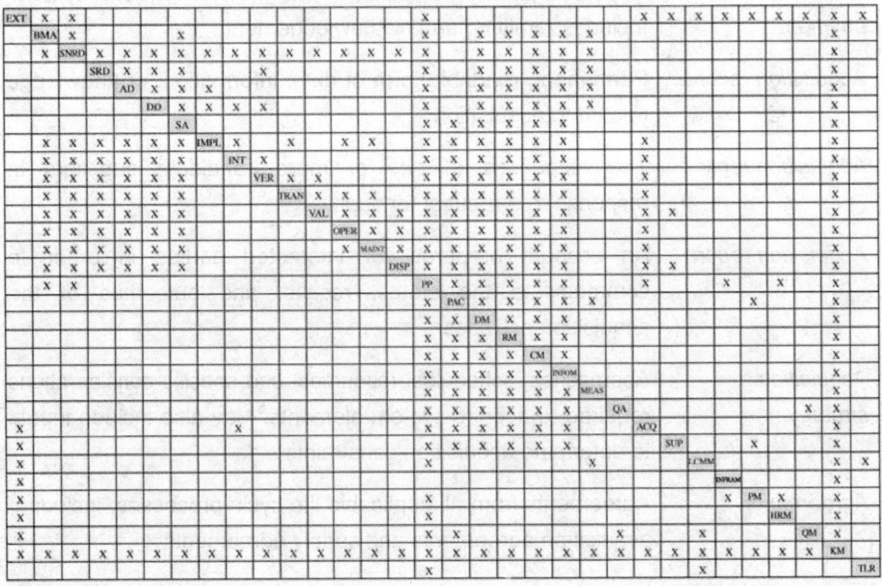

图 D.1　各种不同 SE 流程之间的输入/输出关系。INCOSE SEH 原始图由 David Walden 创建。按照 INCOSE 通知页使用。版权所有。

APPENDIX E: INPUT/OUTPUT DESCRIPTIONS

Accepted system or system element	System element or system is transferred from supplier to acquirer and the product or service is available to the project
Acquired system	The system or system element (product or service) is delivered to the acquirer from a supplier consistent with the delivery conditions of the acquisition agreement
Acquisition agreement	An understanding of the relationship and commitments between the project organization and the supplier. The agreement can vary from formal contracts to less formal interorganizational work orders. Formal agreements typically include terms and conditions
Acquisition need	The identification of a need that cannot be met within the organization encountering the need or a need that can be met in a more economical way by a supplier
Acquisition payment	Payments or other compensations for the acquired system. Includes remitting and acknowledgement
Acquisition record	Permanent, readable form of data, information, or knowledge related to acquisition
Acquisition reply	The responses of one or more candidate suppliers in response to a request for supply
Acquisition report	An account prepared for interested parties in order to communicate the status, results, and outcomes of the acquisition activities
Acquisition strategy	Approaches, schedules, resources, and specific considerations required to acquire system elements. May also include inputs to determine acquisition constraints
Agreements [1]	Agreements from all applicable life cycle processes, including acquisition agreements and supply agreements

---

INCOSE Systems Engineering Handbook: A Guide for System Life Cycle Processes and Activities, Fourth Edition. Edited by David D. Walden, Garry J. Roedler, Kevin J. Forsberg, R. Douglas Hamelin and Thomas M. Shortell.
© 2015 John Wiley & Sons, Inc. Published 2015 by John Wiley & Sons, Inc.

附录 E：输入/输出描述

已接受的系统或系统元素	系统元素或系统从供应商转移到采办方且产品或服务可用于项目
采办的系统	系统或系统元素（产品或服务）由供应商交付给与采办协议的交付条件一致的采办方
采办协议	对项目组织和供应商之间的关系和承诺的理解。协议可根据正式合同或不太正式的组织间的工作指令而变化。正式协议通常包括条款和条件
采办需要	识别组织内遇到的但不能被满足的一个需要或能以更经济的方式由供应商满足的一个需要
采办支付	采办的系统的付款或其他补偿。包括汇款和确认
采办记录	与采办有关的永久可读取形式的数据、信息或知识
采办回复	一个或多个候选供应商对供应请求的响应
采办报告	为有关方准备一份记述报告，以便沟通采办活动的状态、结果和成果
采办策略	采办系统元素所要求的途径、进度、资源和特定的考量。亦可包括确定采办约束的输入
协议	所有来自适用生命周期流程的协议，包括采办协议和供应协议

2

INCOSE 系统工程手册：《系统生命周期流程和活动指南》，第 4 版。编撰：David D.Walden、Garry J.Roedler、Kevin J.Forsberg、R.Douglas Hamelin 和 Thomas M.Shortell。

John Wiley & Sons 公司版权所有©2015。由 John Wiley & Sons 公司于 2015 年出版。

INPUT/OUTPUT DESCRIPTIONS

Alternative solution classes	Identifies and describes the classes of solutions that may address the problem or opportunity
Analysis situations	The context information for the analysis including life cycle stage, evaluation drivers, cost drivers, size drivers, team characteristics, project priorities, or other characterization information and parameters that are needed to understand analysis and represent the element being analyzed. Relevant information from the process that invokes the analysis. Any existing models related to the element being analyzed. Any data related to the element being analyzed, including historical, current, and projected data. Can originate from any life cycle process
Applicable laws and regulations	International, national, or local laws or regulations
Architecture definition record	Permanent, readable form of data, information, or knowledge related to architecture definition
Architecture definition strategy	Approaches, schedules, resources, and specific considerations required to define the selected system architecture that satisfies the requirements
Architecture traceability	Bidirectional traceability of the architecture characteristics
Business or mission analysis record	Permanent, readable form of data, information, or knowledge related to business or mission analysis
Business or mission analysis strategy	Approaches, schedules, resources, and specific considerations required to conduct business or mission analysis and ensure business needs are elaborated and formalized into business requirements
Business requirements	Definition of the business framework within which stakeholders will define their requirements. Business requirements govern the project, including agreement constraints, quality standards, and cost and schedule constraints. Business requirements may be captured in a Business Requirements Specification (BRS), which is approved by the business leadership Note: Business requirements may not always be formally captured in the system life cycle

备选解决方案类别	识别和描述可应对问题或机会的解决方案的类别
分析情况	用于分析的背景环境信息包括理解分析和表示正在分析的元素所需要的生命周期阶段、评价驱动因素、成本驱动因素、规模驱动因素、团队特征、项目优先级或其他特征描述信息和参数。流程的相关信息调用该分析。任何现有模型均与正在分析的元素有关。任何数据均与正在分析的元素有关,包括历史数据、当前数据以及预计的数据。可起源于任何生命周期流程
适用法律和法规	国际、国家或当地法律或法规
架构定义记录	与架构定义有关的永久可读取形式的数据、信息或知识
架构定义策略	定义满足需求的所选系统架构所要求的途径、进度、资源和特定的考量
架构可追溯性	架构特征的双向可追溯性
业务或任务分析记录	与业务或任务分析有关的永久可读取形式的数据、信息或知识
业务或任务分析策略	进行业务或任务分析并确保业务需要在业务需求中详细阐述和形成所需的途径、进度、资源和特定考虑因素
业务需求	利益攸关者将在其中定义其需求的业务框架的定义。业务需求治理项目,包括协议约束、质量标准以及成本和进度约束。业务需求可在经业务领导层批准的业务需求规范(BRS)中捕获
	注:业务需求可能不会总是在系统生命周期中正式捕获

INPUT/OUTPUT DESCRIPTIONS

Business requirements traceability	Bidirectional traceability of the business requirements
Candidate configuration items (CIs)	Items for configuration control. Can originate from any life cycle process
Candidate information items	Items for information control. Can originate from any life cycle process
Candidate risks and opportunities	Risks and opportunities that arise from any stakeholder. In many cases, risk situations are identified during the project assessment and control process. Can originate from any life cycle process
Concept of operations (ConOps)	The ConOps is a verbal and/or graphic statement prepared for the organization's leadership that describes the assumptions or intent regarding the overall operation or series of operations of the enterprise, to include any new capability (ANSI/ AIAA, 2012; ISO/IEC/IEEE 29148, 2011)
Configuration baselines	Items placed under formal change control. The required configuration baseline documentation is developed and approved in a timely manner to support required systems engineering (SE) technical reviews, the system's acquisition and support strategies, and production
Configuration management record	Permanent, readable form of data, information, or knowledge related to configuration management
Configuration management report	An account prepared for interested parties in order to communicate the status, results, and outcomes of the configuration management activities.
	Documents the impact to any process, organization, decision (including any required change notification), products, and services affected by a given change request

业务需求可追溯性	业务需求的双向可追溯性
候选构型项（CI）	构型配置控制项。可起源于任何生命周期流程
候选信息项	信息控制项。可起源于任何生命周期流程
候选风险和机会	来源于任何利益攸关者的风险和机会。许多情况下，风险情况在项目评估和控制流程中得以识别。可起源于任何生命周期流程
运行方案（ConOps）	ConOps 是一个为组织的领导层准备的口头和/或图形说明，描述有关复杂组织体总体运行或一系列运行的假设或意图，以包括任何新的能力（ANSI/AIAA, 2012；ISO/IEC/IEEE 29148, 2011）
构型配置基线	置于正式变更控制之下的构型项。及时地开发和审批所要求的构型基线文档，以支持所要求的系统工程（SE）技术评审、系统的采办和支持策略以及生产
构型配置管理记录	与构型管理有关的永久可读取形式的数据、信息或知识
构型配置管理报告	为有关方准备一份记述报告，以便沟通构型管理活动的状态、结果和成果。
	文件化给定变更请求对任何流程、组织、决策（包括任何所要求的变更通知）、产品和服务造成的影响。

INPUT/OUTPUT DESCRIPTIONS

Configuration management strategy	Approaches, schedules, resources, and specific considerations required to perform configuration management for a project. Describes and documents how to make authorized changes to established baselines in a uniform and controlled manner
Customer satisfaction inputs	Responses to customer satisfaction surveys or other instruments
Decision management strategy	Approaches, schedules, resources, and specific considerations required to perform decision management for a project
Decision record	Permanent, readable form of data, information, or knowledge related to decision management
Decision report	An account prepared for interested parties in order to communicate the status, results, and outcomes of the decision management activities. Should include a recommended course of action, an associated implementation plan, and key findings through effective trade space visualizations underpinned by defendable rationale grounded in analysis results that are repeatable and traceable. As decision makers seek to understand root causes of top-level observations and build their own understanding of the trade-offs, the ability to rapidly drill down from top-level trade space visualizations into lower-level analyses supporting the synthesized view is often beneficial
Decision situation	Decisions related to decision gates are taken on a prearranged schedule; other requests for a decision may arise from any stakeholder, and initial information can be little more than broad statements of the situation. Can originate from any life cycle process
Design definition record	Permanent, readable form of data, information, or knowledge related to design definition

构型配置管理策略	执行项目构型管理所需的途径、进度、资源和特定考虑因素。描述和文件化如何以统一和受控的方式对已建立的基线进行授权的变更
客户满意度输入	对客户满意度调查或其他手段的响应
决策管理策略	对一个项目实施决策管理所要求的途径、进度、资源和特定考量
决策记录	与决策管理有关的永久可读取形式的数据、信息或知识
决策报告	为有关方准备的一份记述报告,以便沟通决策管理活动的状态、结果和成果。应包括一个推荐的行动路线、一份相关的实施计划以及在有效权衡空间可视化中由基于可重复且可追溯的分析结果的经得起推敲的基本原理所支撑的关键发现。在决策者设法理解顶层观察数据的根本原因并构建其自身对权衡的理解时,由顶层权衡空间可视化迅速深入到那些支持综合视角的较低层级分析的能力往往是有益。
决策情况	基于预先安排的进度做出与决策门有关的决策;针对一个决策的其他请求可能来源于任何利益攸关者,并且初始信息可能比该情况的概括性申明略多。可起源于任何生命周期流程
设计定义记录	与设计定义有关的永久可读取形式的数据、信息或知识

INPUT/OUTPUT DESCRIPTIONS

Design definition strategy	Approaches, schedules, resources, and specific considerations required to define the system design that is consistent with the selected system architecture and satisfies the requirements
Design traceability	Bidirectional traceability of the design characteristics, the design enablers, and the system element requirements
Disposal constraints	Any constraints on the system arising from the disposal strategy including cost, schedule, and technical constraints
Disposal enabling system requirement	Requirements for any systems needed to enable disposal of the system of interest
Disposal procedure	A disposal procedure that includes a set of disposal actions, using specific disposal techniques, performed with specific disposal enablers
Disposal record	Permanent, readable form of data, information, or knowledge related to disposal
Disposal report	An account prepared for interested parties in order to communicate the status, results, and outcomes of the disposal activities. May include an inventory of system elements for reuse/storage and any documentation or reporting required by regulation or organization standards
Disposal strategy	Approaches, schedules, resources, and specific considerations required to ensure the system or system elements are deactivated, disassembled, and removed from operations
Disposed system	Disposed system that has been deactivated, disassembled, and removed from operations
Documentation tree	Defines the hierarchical representation of the set of system definition products for the system under development. Based on the evolving system architecture
Enabling system requirements	Enabling system requirements from all applicable life cycle processes, including implementation enabling system requirements, integration enabling system requirements, verification enabling system requirements, transition enabling system requirements, validation enabling system requirements, operation enabling system requirements, maintenance enabling system requirements, and disposal enabling system requirements

设计定义策略	定义与所选系统架构一致并满足需求的系统设计所需的途径、进度、资源和特定考量
设计可追溯性	设计特征、设计使能项以及系统元素需求的双向可追溯性
退出的约束	退出策略引发的对系统的任何约束,包括成本约束、进度约束和技术约束
退出的使能系统需求	使能所感兴趣之系统退出所需的任何系统的需求
退出的程序	一个退出程序包括退出行动的一个集合,使用特定退出技术,由特定的退出使能项实施
退出的记录	与退出有关的永久可读取形式的数据、信息或知识
退出的报告	为有关方准备的一份记述报告,以便沟通退出活动的状态、结果和成果。可包括用于复用/储存的系统元素的详细目录,以及规章或组织标准所要求的任何文档或报告
退出的策略	确保从运行中停用、拆解并移除系统或系统元素所需的途径、进度、资源和特定考量
已退出的系统	已经从运行中停用、拆解并移除的已退出的系统
文档树	定义用于开发中的系统的系统定义产品集合的层级表征。基于演进的系统架构
使能系统需求	所有适用生命周期流程的使能系统需求,包括实施使能系统需求、综合使能系统需求、验证使能系统需求、转移使能系统需求、确认使能系统需求、运行使能系统需求、维护使能系统需求以及退出使能系统需求

INPUT/OUTPUT DESCRIPTIONS

Final Requirements Verification and Traceability Matrix (RVTM)	Final list of requirements, their verification attributes, and their traces. Includes any proposed changes to the system requirements due to the verification actions
Human resource management plan	Approaches, schedules, resources, and specific considerations required to identify the skill needs of the organization and projects. Includes the organizational training plan needed to develop internal personnel and the acquisition of external personnel
Human resource management record	Permanent, readable form of data, information, or knowledge related to human resource management
Human resource management report	An account prepared for interested parties in order to communicate the status, results, and outcomes of the human resource management activities
Implementation constraints	Any constraints on the system arising from the implementation strategy including cost, schedule, and technical constraints
Implementation enabling system requirements	Requirements for any systems needed to enable implementation of the system of interest
Implementation record	Permanent, readable form of data, information, or knowledge related to implementation
Implementation report	An account prepared for interested parties in order to communicate the status, results, and outcomes of the implementation activities
Implementation strategy	Approaches, schedules, resources, and specific considerations required to realize system elements to satisfy system requirements, architecture, and design
Implementation traceability	Bidirectional traceability of the system elements
Information management record	Permanent, readable form of data, information, or knowledge related to information management

最终需求验证和可追溯性矩阵（RVTM）	需求及其验证属性、跟踪文件的最终列表。包括因验证行动而对系统需求做出的任何建议性更改
人力资源管理计划	识别组织和项目的技能需要所需的途径、进度、资源和特定考虑因素。包括培养内部人员并采办外部人员所需的组织培训计划
人力资源管理记录	与人力资源管理有关的永久可读取形式的数据、信息或知识
人力资源管理报告	为有关方准备一份记述报告，以便沟通人力资源管理活动的状态、结果和成果
实施约束	实施策略引发的对系统的任何约束，包括成本约束、进度约束和技术约束
实施使能系统需求	使所感兴趣之系统能够实施所需的任何系统的需求
实施记录	与实施有关的永久可读取形式的数据、信息或知识
实施报告	为有关方准备一份记述报告，以便沟通实施活动的状态、结果和成果
实施策略	实现系统元素以满足系统需求、架构和设计所需的途径、进度、资源和特定考量
实施可追溯性	系统元素的双向可追溯性
信息管理记录	与信息管理有关的永久可读取形式的数据、信息或知识

INPUT/OUTPUT DESCRIPTIONS

Information management report	An account prepared for interested parties in order to communicate the status, results, and outcomes of the information management activities
Information management strategy	Approaches, schedules, resources, and specific considerations required to perform information management for a project
Information repository	A repository that supports the availability for use and communication of all relevant project information artifacts in a timely, complete, valid, and, if required, restricted manner
Infrastructure management plan	Approaches, schedules, resources, and specific considerations required to define and sustain the organizational and project infrastructures
Infrastructure management record	Permanent, readable form of data, information, or knowledge related to infrastructure management
Infrastructure management report	An account prepared for interested parties in order to communicate the status, results, and outcomes of the infrastructure management activities. Includes cost, usage, downtime/response measures, etc. These can be used to support capacity planning for upcoming projects
Initial RVTM	A preliminary list of requirements, their verification attributes, and their traces
Installation procedure	An installation procedure that includes a set of installation actions, using specific installation techniques, performed with specific transition enablers
Installed system	Installed system ready for validation
Integrated system or system element	Integrated system element or system ready for verification. The resulting aggregation of assembled system elements
Integration constraints	Any constraint on the system arising from the integration strategy including cost, schedule, and technical constraints

信息管理报告	为有关方准备一份记述报告，以便沟通信息管理活动的状态、结果和成果
信息管理策略	执行项目信息管理所需的途径、进度、资源和特定考虑因素
信息库	支持以及时、完整、有效和限制（如果需要）的方式保障所有相关项目信息制品的使用和沟通的有效性的库
基础设施管理计划	定义和保持组织与项目基础设施所需的途径、进度、资源和特定考量
基础设施管理记录	与基础设施管理有关的永久可读取形式的数据、信息或知识
基础设施管理报告	为有关方准备一份记述报告，以便沟通基础设施管理活动的状态、结果和成果。包括成本、用法、停工时间/响应测度等。这些因素可用来支持即将到来的项目的能力规划
初始RVTM	需求及其验证属性、跟踪文件的初始列表
安装程序	安装程序包括安装行动集，使用特定安装技术，由特定的转移使能项实施
已安装的系统	准备好要确认的已安装的系统
综合的系统或系统元素	准备好要验证的综合的系统元素或系统。产生的已组装系统元素的集合
综合约束	综合策略引发的对系统的任何约束，包括成本约束、进度约束和技术约束

INPUT/OUTPUT DESCRIPTIONS

Integration enabling system requirements	Requirements for any systems needed to enable integration of the system of interest
Integration procedure	An assembly procedure that groups a set of elementary assembly actions to build an aggregate of implemented system elements, using specific integration techniques, performed with specific integration enablers
Integration record	Permanent, readable form of data, information, or knowledge related to integration
Integration report	An account prepared for interested parties in order to communicate the status, results, and outcomes of the integration activities. Includes documentation of the integration testing and analysis results, areas of nonconformance, and validated internal interfaces
Integration strategy	Approaches, schedules, resources, and specific considerations required to integrate the system elements
Interface definition	The logical and physical aspects of internal interfaces (between the system elements composing the system) and external interfaces (between the system elements and the elements outside the system of interest)
Interface definition update identification	Identification of updates to interface requirements and definitions, if any
Knowledge management plan	Establishes how the organization and projects within the organization will interact to ensure the right level of knowledge is captured to provide useful knowledge assets. Includes a list of applicable domains; plans for obtaining and maintaining knowledge assets for their useful life; characterization of the types of assets to be collected and maintained along with a scheme to classify them for the convenience of users; criteria for accepting, qualifying, and retiring knowledge assets; procedures for controlling changes to the knowledge assets; and definition of a mechanism for knowledge asset storage and retrieval
Knowledge management report	An account prepared for interested parties in order to communicate the status, results, and outcomes of the knowledge management activities

综合使能系统需求	使所感兴趣之系统能够综合所需的任何系统的需求
综合程序	一个汇集程序将基本汇集行动集分组以构建已实施的系统元素集合，该程序使用特定综合技术，由特定的综合使能项实施
综合记录	与综合有关的永久可读取形式的数据、信息或知识
综合报告	为有关方准备一份记述报告，以便沟通综合活动的状态、结果和成果。包括综合试验和分析结果、不符合项的领域以及确认的内部接口的文档
综合策略	综合系统元素所需的途径、进度、资源和特定考虑因素
接口定义	内部接口（构成系统的系统元素之间）和外部接口（系统元素与所感兴趣之系统外部的元素之间）的逻辑和物理方面
接口定义更新识别	对接口需求和定义的更新的识别，如果有的话
知识管理计划	确定组织和该组织内的项目将如何相互作用，以确保捕获适当水平的知识来提供有用的知识资产。包括适用领域的列表；用于在其使用寿命内获得和保留知识资产的计划；将与方案一同收集和保留的资产类型的特征描述，以将资产分类，方便用户使用；接受、限定和退役知识资产的准则；控制知识资产变更的程序；以及知识资产的存储与检索机制的定义
知识管理报告	为有关方准备一份记述报告，以便沟通知识管理活动的状态、结果和成果

INPUT/OUTPUT DESCRIPTIONS

Knowledge management system	Maintained knowledge management system. Project suitability assessment results for application of existing knowledge. Lessons learned from execution of the organizational SE processes on projects. Should include mechanisms to easily identify and access the assets and to determine the level of applicability for the project considering its use. Can be used by any life cycle process
Life cycle concepts	Articulation and refinement of the various life cycle concepts consistent with the business needs in the form of life cycle concept documents on which the system of interest is based, assessed, and selected. The architecture is based on these concepts, and they are essential in providing context for proper interpretation of the system requirements. Typical concepts include: • Acquisition concept • Deployment concept • Operational concept (OpsCon) • Support concept • Retirement concept
Life cycle constraints	Constraints from all applicable life cycle processes, including implementation constraints, integration constraints, verification constraints, transition constraints, validation constraints, operation constraints, maintenance constraints, and disposal constraints
Life cycle model management plan	Approaches, schedules, resources, and specific considerations required to define a set of organizational life cycle models. Includes identification of new needs and the evaluation of competitiveness from the perspective of the organization strategy. Includes criteria for assessments and approvals/ disapprovals
Life cycle model management record	Permanent, readable form of data, information, or knowledge related to life cycle model management

知识管理系统	保留的知识管理系统。对现有知识的应用的项目适用性评估结果来自于组织对项目执行组织的 SE 流程的教训。应包括易于识别和访问资产的机制以及确定考虑使用该资产的项目的适用性水平的机制。可由任何生命周期流程使用
生命周期概念/方案	以建立、评估和选择所感兴趣之系统所依据的生命周期概念文件的形式对符合业务需要的各种不同的生命周期概念/方案的阐明和细化。架构基于这些概念/方案,并且这些概念/方案在为系统需求的适当解释提供背景环境方面是十分重要的。典型概念/方案包括: • 采办方案 • 部署方案 • 运行方案（OpsCon） • 保障方案 • 退役方案
生命周期约束	所有适用生命周期流程的约束,包括实施约束、综合约束、验证约束、转移约束、确认约束、运行约束、维护约束以及退出约束
生命周期模型管理计划	定义组织生命周期模型集所需的途径、进度、资源和特定考虑因素。包括新需要的识别以及从组织策略视角对竞争力评价。包括用于评估和批准/不批准的准则
生命周期模型管理记录	与生命周期模型管理有关的永久可读取形式的数据、信息或知识

INPUT/OUTPUT DESCRIPTIONS

Life cycle model management report	An account prepared for interested parties in order to communicate the status, results, and outcomes of the life cycle model management activities
Life cycle models	Life cycle model or models appropriate for the project. Includes definition of the business and other decision-making criteria regarding entering and exiting each life cycle stage. The information and artifacts are collected and made available to be used and reused
Maintenance constraints	Any constraints on the system arising from the maintenance strategy including cost, schedule, and technical constraints
Maintenance enabling system requirements	Requirements for any systems needed to enable operation of the system of interest
Maintenance procedure	A maintenance procedure that includes a set of maintenance actions, using specific maintenance techniques, performed with specific maintenance enablers
Maintenance record	Permanent, readable form of data, information, or knowledge related to maintenance
Maintenance report	An account prepared for interested parties in order to communicate the status, results, and outcomes of the maintenance activities
Maintenance strategy	Approaches, schedules, resources, and specific considerations required to perform corrective and preventive maintenance in conformance with operational availability requirements
Major stakeholder identification	List of legitimate external and internal stakeholders with an interest in the solution. Major stakeholders are also derived from analysis of the ConOps
Measurement data	Measurement data from all applicable life cycle processes, including measure of effectiveness (MOE) data, measure of performance (MOP) data, technical performance measures (TPM) data, project performance measures data, and organizational process performance measures data

输入/输出描述

生命周期模型管理报告	为有关方准备一份记述报告，以便沟通生命周期模型管理活动的状态、结果和成果
生命周期模型	适于该项目的一个或多个生命周期模型。包括进入和退出每个生命周期阶段相关的业务准则和其他决策准则的定义。收集信息和制品并使之可被使用或复用
维护约束	维护策略引发的对系统的任何约束，包括成本约束、进度约束和技术约束
维护使能系统需求	使所感兴趣之系统能够运行所需的任何系统的需求
维护程序	维护程序包括维护行动集，使用特定维护技术，由特定的维护使能项执行
维护记录	与维护有关的永久可读取形式的数据、信息或知识
维护报告	为有关方准备一份记述报告，以便沟通维护活动的状态、结果和成果
维护策略	实施与运行有效性需求一致的纠正和预防性维修所需的途径、进度、资源和特定考量
主要利益攸关者识别	对该解决方案感兴趣的合法外部利益攸关者和内部利益攸关者的列表。主要利益攸关者亦来自 ConOps 分析
测量数据	所有适用生命周期流程的测量数据，包括有效性测度（MOE）数据、性能测度（MOP）数据、技术性能测度（TPM）数据、项目性能测度数据和组织流程性能测度数据

INPUT/OUTPUT DESCRIPTIONS

	Measurement needs	Measurement needs from all applicable life cycle processes, including MOE needs, MOP needs, TPM needs, project performance measures needs, and organizational process performance measures needs
	Measurement record	Permanent, readable form of data, information, or knowledge related to measurement
	Measurement report	An account prepared for interested parties in order to communicate the status, results, and outcomes of the measurement activities. Includes documentation of the measurement activity results, the measurement data that was collected and analyzed and results that were communicated, and any improvements or corrective actions driven by the measures with their supporting data
	Measurement repository	A repository that supports the availability for use and communication of all relevant measures in a timely, complete, valid, and, if required, confidential manner
	Measurement strategy	Approaches, schedules, resources, and specific considerations required to perform measurement for a project. Addresses the strategy for performing measurement: describing measurement goals, identifying information needs and applicable measures, and defining performance and evaluation methodologies
	MOE data	Data provided for the identified measurement needs
	MOE needs	Identification of the MOEs (Roedler and Jones, 2006), which define the information needs of the decision makers with respect to system effectiveness to meet operational expectations
	MOP data	Data provided for the identified measurement needs
	MOP needs	Identification of the MOPs (Roedler and Jones, 2006), which define the key performance characteristics the system should have when fielded and operated in its intended operating environment
	Operation constraints	Any constraints on the system arising from the operational strategy including cost, schedule, and technical constraints

测量需要	所有适用生命周期流程的测量需要，包括 MOE 需要、MOP 需要、TPM 需要、项目性能测度需要和组织流程性能测度需要
测量记录	与测量有关的永久可读取形式的数据、信息或知识
测量报告	为有关方准备一份记述报告，以便沟通测量活动的状态、结果和成果。包括测量活动结果的文档，收集、分析的测量数据和传达的结果以及具有支持数据的测度所驱动的任何改进或纠正措施
测量库	支持以及时、完整、有效和保密（如果需要）的方式保障所有相关测度的使用和沟通的有效性的库
测量策略	执行项目测量所需的途径、进度、资源和特定考量。涉及执行测量的策略：描述测量目标、识别信息需要和适用的测度并定义性能和评价方法论
MOE 数据	为识别的测量需要提供的数据
MOE 需要	定义决策者对满足运行期望的系统有效性的信息需要的 MOE（Roedler 和 Jones，2006）的识别
MOP 数据	为识别的测量需要提供的数据
MOP 需要	定义系统在其预期运行环境中部署和运行时应具有的关键性能特征的 MOP（Roedler 和 Jones，2006）的识别
运行约束	运行策略引发的对系统的任何约束，包括成本约束、进度约束和技术约束

INPUT/OUTPUT DESCRIPTIONS

Operation enabling system requirements	Requirements for any systems needed to enable operation of the system of interest
Operation record	Permanent, readable form of data, information, or knowledge related to operation
Operation report	An account prepared for interested parties in order to communicate the status, results, and outcomes of the operation activities
Operation strategy	Approaches, schedules, resources, and specific considerations required to perform system operations
Operator/ maintainer training materials	Training capabilities and documentation
Organization infrastructure	Resources and services that support the organization. Organizational-level facilities, personnel, and resources for hardware fabrication, software development, system implementation and integration, verification, validation, etc.
Organization infrastructure needs	Specific requests for infrastructure products or services from the organization, including commitments to external stakeholders
Organization lessons learned	Organizational-related lessons learned. Results from an evaluation or observation of an implemented corrective action that contributed to improved performance or increased capability. A lesson learned also results from an evaluation or observation of a positive finding that did not necessarily require corrective action other than sustainment
Organization portfolio direction and constraints	Organization business objectives, funding outlay and constraints, ongoing research and development (R&D), market tendencies, etc., including cost, schedule, and solution constraints
Organization strategic plan	The overall organization strategy, including the business mission or vision and strategic goals and objectives

运行使能系统需求	使所感兴趣之系统能够运行所需的任何系统的需求
运行记录	与运行有关的永久可读取形式的数据、信息或知识
运行报告	为有关方准备一份记述报告,以便沟通运行活动的状态、结果和成果
运行策略	执行系统操作所需的途径、进度、资源和特定考虑因素
操作者/维护人员培训资料	培训能力和文档
组织基础设施	支持组织的资源和服务。用于硬件制造、软件开发、系统实施以及综合、验证、确认等的组织层级的设施、人员和资源
组织基础设施需要	组织对基础设施产品或服务的特定请求,包括外部利益攸关者的承诺
吸取的组织教训	吸取的与组织有关的教训。来自对已实施的有助于改进性能或提高能力的纠正措施的评价或观察。吸取的教训亦来自对未必要求除维持以外的纠正措施的积极结果的评价或观察。
组织项目组合方向与约束	组织业务目的、资金支出和约束、持续研发(R&D)、市场趋势等,包括成本、进度和解决方案约束。
组织策略计划	总体组织策略,包括业务任务或愿景以及战略目标和目的

INPUT/OUTPUT DESCRIPTIONS

Organization tailoring strategy	Approaches, schedules, resources, and specific considerations required to incorporate new or updated external standards into the organization's set of standard life cycle processes
Organizational policies, procedures, and assets	Items related to the organization's standard set of life cycle processes, including guidelines and reporting mechanisms. Organization process guidelines in the form of organization policies, procedures, and assets for applying the system life cycle processes and adapting them to meet the needs of individual projects (e.g., templates, checklists, forms). Includes defining responsibilities, accountability, and authority for all SE processes within the organization
Organizational process performance measures data	Data provided for the identified measurement needs
Organizational process performance measures needs	Identification of the organizational process performance measures, which measure how well the organization is satisfying it objectives
Portfolio management plan	Approaches, schedules, resources, and specific considerations required to define a project portfolio
Portfolio management record	Permanent, readable form of data, information, or knowledge related to portfolio management
Portfolio management report	An account prepared for interested parties in order to communicate the status, results, and outcomes of the portfolio management activities
Preliminary interface definition	The preliminary logical and physical aspects of internal interfaces (between the system elements composing the system) and external interfaces (between the system elements of the system and the elements outside the system of interest)

组织剪裁策略	将新的或更新的外部标准纳入组织的标准生命周期流程集所需的途径、进度、资源和特定考虑因素
组织方针、程序和资产	与组织的标准生命周期流程集有关的项，包括指南和报告机制。以组织方针、程序和资产的形式用于应用系统生命周期流程并使其适应个别项目（例如模板，检查单、表格）需要的组织流程指南。包括定义组织内所有SE流程的职责、责任和权限
组织流程性能测度数据	为识别的测量需要提供的数据
组织流程性能测度需要	测量组织满足其目的的程度的组织流程性能测度的识别
项目组合管理计划	定义项目组合所需的途径、进度、资源和特定考虑因素
项目群管理记录	与项目组合管理有关的永久可读取形式的数据、信息或知识
项目群管理报告	为有关方准备一份记述报告，以便沟通项目组合管理活动的状态、结果和成果
初步接口定义	内部接口（构成系统的系统元素之间）和外部接口（系统的系统元素与所感兴趣之系统外部的元素之间）的初始逻辑方面和物理方面

INPUT/OUTPUT DESCRIPTIONS

Preliminary life cycle concepts	Preliminary articulation of the various life cycle concepts consistent with the business needs in the form of life cycle concept documents on which the system of interest is based, assessed, and selected. The architecture is based on these concepts, and they are essential in providing context for proper interpretation of the system requirements. Typical concepts include: • Acquisition concept • Deployment concept • OpsCon • Support concept • Retirement concept
Preliminary MOE data	Preliminary data provided for the identified measurement needs
Preliminary MOE needs	Preliminary identification of the MOEs (Roedler and Jones, 2006), which define the information needs of the decision makers with respect to system effectiveness to meet operational expectations
Preliminary TPM data	Preliminary data provided for the identified measurement needs
Preliminary TPM needs	Preliminary identification of the TPM (Roedler and Jones, 2006), which measure attributes of a system element to determine how well a system or system element is satisfying or expected to satisfy a technical requirement or goal
Preliminary validation criteria	The preliminary validation criteria (the measures to be assessed), who will perform validation activities, and the validation environments of the system of interest
Problem or opportunity statement	Description of the problem or opportunity. Should be derived from the organization strategy and provide enough detail to understand the gap or new capability that is being considered
Procedures	Procedures from all applicable life cycle processes, including integration procedure, verification procedure, installation procedure, validation procedure, maintenance procedure, and disposal procedure

初始生命周期概念/方案	以建立、评估和选择所感兴趣之系统所依据的生命周期概念文件的形式对符合业务需要的各种不同的生命周期概念的初步阐明。架构基于这些概念，并且这些概念在为系统需求的适当解释提供背景环境方面是十分重要的。典型概念包括： • 采办方案 • 部署方案 • 运行方案 • 保障方案 • 退役方案
初始 MOE 数据	为识别的测量需要提供的初始数据
初始 MOE 需要	定义决策者对满足运行期望的系统有效性的信息需要的 MOE（Roedler 和 Jones，2006）的初步识别
初始 TPM 数据	为确定的测量需要提供的初始数据
初始 TPM 需要	测量系统元素的属性以确定系统或系统元素满足或期望满足技术需求或目标的程度的 TPM（Roedler 和 Jones，2006）的初步识别
初始确认准则	用于指定谁将执行确认活动以及所感兴趣之系统的确认环境的初始确认准则（待评估的测度）
问题或机会说明	问题或机会的描述。应来自组织策略并提供足够详细的说明以理解考虑中的差距或新能力
程序	所有适用生命周期流程的程序，包括综合程序、验证程序、安装程序、确认程序、维护程序以及退出程序

INPUT/OUTPUT DESCRIPTIONS

Project assessment and control record	Permanent, readable form of data, information, or knowledge related to project assessment and control
Project assessment and control strategy	Approaches, schedules, resources, and specific considerations required to perform assessment and control for a project
Project budget	A prediction of the costs associated with a particular project. Includes labor, infrastructure, acquisition, and enabling system costs along with reserves for risk management
Project change requests	Requests to update any formal baselines that have been established. In many cases, the need for change requests is identified during the project assessment and control process. Can originate from any life cycle process
Project constraints	Any constraints on the system arising from the technical management strategy including cost, schedule, and technical constraints
Project control requests	Internal project directives based on action required due to deviations from the project plan. New directions are communicated to both project team and customer, when appropriate. If assessments are associated with a decision gate, a decision to proceed, or not to proceed, is taken
Project direction	Organizational direction to the project. Includes sustainment of projects meeting assessment criteria and redirection or termination of projects not meeting assessment criteria
Project human resource needs	Specific requests for human resources needed by the project, including commitments to external stakeholders
Project infrastructure	Resources and services that support a project. Project-level facilities, personnel, and resources for hardware fabrication, software development, system implementation and integration, verification, validation, etc.
Project infrastructure needs	Specific requests for infrastructure products or services needed by the project, including commitments to external stakeholders
Project lessons learned	Project-related lessons learned. Results from an evaluation or observation of an implemented corrective action that contributed to improved performance or increased capability. A lesson learned also results from an evaluation or observation of a positive finding that did not necessarily require corrective action other than sustainment (CJCS, 2012)

项目评估和控制记录	与项目评估和控制有关的永久可读取形式的数据、信息或知识
项目评估和控制策略	执行项目评估和控制所需的途径、进度、资源和特定考虑因素
项目预算	对与特殊项目相关联的成本的预测。包括人工、基础设施、采办和使能系统成本,以及风险管理的储备金
项目变更请求	更新已经建立的任何正式基线的请求。许多情况下,变更请求的需要在项目评估和控制流程中得以识别。可起源于任何生命周期流程
项目约束	技术管理策略引发的对系统的任何约束,包括成本约束、进度约束和技术约束
项目控制请求	因偏离项目计划而要求的基于行动的内部项目指令。若适合,将新的指导传达给项目团队和客户。若评估与决策门有关,则做出继续进行或不继续进行的决策。
项目指导	组织对项目的指导包括达到评估准则的项目的维持和未达到评估准则的项目的重定向或终止
项目人力资源需要	项目对所需的人力资源的特定请求,包括外部利益攸关者的承诺
项目基础设施	支持项目的资源和服务。用于硬件制造、软件开发、系统实施以及综合、验证、确认等的项目层级的设施、人员和资源
项目基础设施需要	项目对所需的基础设施产品或服务的特定请求,包括外部利益攸关者的承诺
吸取项目的教训	吸取的与项目有关的教训。来自对已实施的有助于改进性能或提高能力的纠正措施的评价或观察。吸取的教训亦来自对未必要求除维持以外的纠正措施的积极结果的评价或观察(CJCS,2012)

INPUT/OUTPUT DESCRIPTIONS

Project performance measures data	Data provided for the identified measurement needs
Project performance measures needs	Identification of the project performance measures, which measure how well the project is satisfying it objectives
Project planning record	Permanent, readable form of data, information, or knowledge related to project planning
Project portfolio	The necessary information for all of the organizations' projects. The initiation of new projects or the setting up of a product line management approach. Includes the project goals, resources, budgets identified and allocated to the projects, and clearly defined project management accountability and authorities
Project schedule	A linked list of a project's milestones, activities, and deliverables with intended start and finish dates. May include a top-level milestone schedule and multiple levels (also called tiers) of schedules of increasing detail and task descriptions with completion criteria and work authorizations
Project status report	An account prepared for interested parties in order to communicate the status, results, and outcomes of the overall project activities. Includes status on meeting the objectives set out for the project, information on the health and maturity of the project work effort, status on project tailoring and execution, and status on personnel availability and effectiveness for the project
Project tailoring strategy	Approaches, schedules, resources, and specific considerations required to incorporate and tailor the organization's set of standard life cycle processes for a given project
Quality management (QM) corrective actions	Actions taken when quality goals are not achieved. Resulting from project-related and process-related reviews and audits
Qualified personnel	The right people with the right skills are assigned at the right time to projects per their skill needs and timing

| 项目性能测度数据 | 为识别的测量需要提供的数据 |

| 项目性能测度需要 | 测量项目满足其目的的程度的项目性能测度的识别 |

| 项目规划记录 | 与项目规划有关的永久可读取形式的数据、信息或知识 |

| 项目群 | 组织所有项目的必要信息。新项目的启动或产品线管理方法的建立。包括识别并分配给项目的项目目标、资源和预算,以及定义明确的项目管理责任和权限。 |

| 项目进度 | 带有预期的开始时间和结束时间的项目里程碑、活动和可交付物的链接表。可包括顶层里程碑进度和使用完成准则和工作授权来增加细节和任务描述的多层级(亦称为多层)进度 |

| 项目状态报告 | 为有关方准备一份记述报告,以便沟通总体项目活动的状态、结果和成果。包括关于满足为项目制定的诸多目的的状态、关于项目工作的良好运行状态和成熟度方面的信息、关于项目剪裁和执行的状态以及关于项目的个人可用性和有效性的状态 |

| 项目剪裁策略 | 纳入和剪裁组织中用于给定项目的标准生命周期流程集所需的途径、进度、资源和特定考虑因素 |

| 质量管理(QM)纠正措施 | 当未达到质量目标时,采取措施。产生于项目相关和流程相关的评审和审计 |

| 有资质的人员 | 根据技能需要和时机在适当时间为项目指派具有适当技能的适当人员 |

INPUT/OUTPUT DESCRIPTIONS

Quality assurance evaluation report	An account prepared for interested parties in order to communicate evidence of whether the project's quality assurance activities are effective. Includes the assessment of all the project-related process and any suggested improvements or necessary corrective actions. Provides constructive input for improvements to an organization's life cycle model implementation
Quality assurance plan	The set of project quality assurance activities, tailored to the project, designed to monitor development and SE processes. Describes the quality assurance organization and applicable audit, evaluation, and monitoring activities. This includes the set of policies and procedures, including specific methods and techniques that apply to quality assurance practices within the organization and within individual projects. It also includes quality objectives for processes and systems that are measurable, along with linkages to the assigned accountability and authority for QM within the organization. The plan also references activities performed by other organizations or functions that are monitored or audited by the quality assurance organization
Quality assurance record	Permanent, readable form of data, information, or knowledge related to quality assurance
Quality assurance report	An account prepared for interested parties in order to communicate the status, results, and outcomes of the quality assurance activities. Includes information on deviations from nominal conditions during the product life cycle and actions to be taken when quality assurance goals and objectives are not achieved
Quality management evaluation report	An account prepared for interested parties in order to communicate evidence of whether the organization's QM activities are effective. Includes the assessment of all the organizational-related process and any suggested improvements or necessary corrective actions. Provides constructive input for improvements to an organization's life cycle model implementation

质量保证评价报告	为有关方准备一份记述报告，以便沟通有关项目质量保证活动是否有效的证据。包括对所有与项目有关的流程以及任何建议的改进或必要的纠正措施的评估。为组织生命周期模型实施的改善提供建设性输入
质量保证计划	设计用于监控开发和 SE 流程的、按项目剪裁的项目质量保证活动集描述质量保证组织和适用的审计、评价和监控活动。这包括方针和程序集，包括适用于组织内和个别项目内质量保证实践的特定方法和技术。亦包括可测量的流程和系统的质量目的以及指定的组织内质量管理的责任和权限的链接。该计划亦参考受质量保证组织监控或审计的其他组织或功能所执行的活动
质量保证记录	与质量保证有关的永久可读取形式的数据、信息或知识
质量保证报告	为有关方准备一份记述报告，以便沟通质量保证活动的状态、结果和成果。包括关于在产品生命周期期间与正常条件的偏差以及在未达到质量保证目标和目的时将要采取的措施的信息
质量管理评价报告	为有关方准备一份记述报告，以便沟通有关组织质量管理活动是否有效的证据。包括对所有与组织有关的流程以及任何建议的改进或必要的纠正措施的评估。为组织生命周期模型实施的改善提供建设性输入

INPUT/OUTPUT DESCRIPTIONS

Quality management guidelines	Guidelines for quality practices within the organization, within individual projects, and as part of the execution of system life cycle processes
Quality management plan	The overarching guidance that explains the organization's quality philosophy and quality organization. Describes the QM organization and applicable audit, evaluation, and monitoring activities. This includes the set of policies and procedures, including specific methods and techniques that apply to QM practices within the organization. It also includes quality objectives for processes and systems that are measurable, along with the assigned accountability and authority for QM within the organization. The set of project QM activities form the basis of the project quality assurance
Quality management record	Permanent, readable form of data, information, or knowledge related to QM
Quality management report	An account prepared for interested parties in order to communicate the status, results, and outcomes of the QM activities. Includes the results of any customer satisfaction surveys and any issues that need to be addressed
Records	Records from all applicable life cycle processes, including business or mission analysis record, stakeholder needs and requirements definition record, system requirements definition record, architecture definition record, design definition record, system analysis record, implementation record, integration record, verification record, transition record, validation record, operation record, maintenance record, disposal record, project planning record, project assessment and control record, decision record, risk record, configuration management record, information management record, measurement record, quality assurance record, acquisition record, supply record, life cycle model management record, infrastructure management record, portfolio management record, human resource management record, and QM record

质量管理指南	组织内和个别项目内的质量实践以及作为系统生命周期流程执行的部分的指南
质量管理计划	解释组织质量哲学和质量组织的至关重要的指南。描述 QM 组织和适用的审计、评价和监控活动。这包括方针和程序集，包括适用于组织内 QM 实践的特定方法和技巧。亦包括可测量的流程和系统的质量目标以及指定的组织内 QM 的责任和权限。项目 QM 活动集形成项目质量保证的基础
质量管理记录	与 QM 有关的永久可读取形式的数据、信息或知识
质量管理报告	为有关方准备一份记述报告，以便沟通 QM 活动的状态、结果和成果。包括任何客户满意度调查的结果以及需要处理的任何问题
记录	所有适用生命周期流程的记录，包括业务或任务分析记录、利益攸关者需要和需求定义记录、系统需求定义记录、架构定义记录、设计定义记录、系统分析记录、实施记录、综合记录、验证记录、转移记录、确认记录、运行记录、维护记录、退出记录、项目规划记录、项目评估和控制记录、决策记录、风险记录、构型管理记录、信息管理记录、测量记录、质量保证记录、采办记录、供应记录、生命周期模型管理记录、基础设施管理记录、项目组合管理记录、人力资源管理记录和 QM 记录

INPUT/OUTPUT DESCRIPTIONS

Reports	Project reports from all applicable life cycle processes, including system analysis report, implementation report, integration report, verification report, transition report, validation report, operation report, maintenance report, disposal report, decision report, risk report, configuration management report, information management report, measurement report, quality assurance report, acquisition report, and supply report (other reports go to other process areas and are not aggregated here)
Request for supply	A request to an external supplying organization to propose a solution to meet a need for a system element or system (product or service). The organization can identify candidate suppliers that could meet this need. Inputs are received from the project personnel in the organization with the need
Risk management strategy	Approaches, schedules, resources, and specific considerations required to perform risk management for a project
Risk record	Permanent, readable form of data, information, or knowledge related to risk management
Risk report	An account prepared for interested parties in order to communicate the status, results, and outcomes of the risk management activities. The risks are documented and communicated along with rationale, assumptions, treatment plans, and current status. For selected risks, an action plan is produced to direct the project team to update the project plan and properly respond to the risks. If appropriate, change requests are generated to mitigate technical risk. Risk profiles and/or risk matrices summarize the risks and contain the findings of the risk management process
SEMP	Systems engineering management plan. The top-level plan for managing the SE effort. It defines how the project will be organized, structured, and conducted and how the total engineering process will be controlled to provide a product that satisfies stakeholder requirements. Includes identification of required technical reviews and their completion criteria, methods for controlling changes, risk and opportunity assessment and methodology, and identification of other technical plans and documentation to be produced for the project

报告	所有适用生命周期流程的项目报告，包括系统分析报告、实施报告、综合报告、验证报告、转移报告、确认报告、运行报告、维护报告、退出报告、决策报告、风险报告、构型管理报告、信息管理报告、测量报告、质量保证报告、采办报告以及供应报告（属于其他流程领域且本书中未收集的其他报告）
供应请求	对外部供应组织的某一要求，提出一种满足系统元素或系统（产品或服务）需要的解决方案。组织能够识别可满足此需要的候选供应商。从具有需要的组织内的项目人员中接收输入
风险管理策略	执行项目风险管理所需的途径、进度、资源和特定考虑因素
风险记录	与风险管理有关的永久可读取形式的数据、信息或知识
风险报告	为有关方准备一份记述报告，以便沟通风险管理活动的状态、结果和成果。将风险文件化，并与依据、假设、处理计划和当前状态一起传递下去。对于所选择的风险，产生一个行动计划，以指导项目团队更新项目计划并正确地应对风险。若适当，就生成变更要求，以减轻技术风险。风险概况和/或风险矩阵总结风险并包含风险管理流程的结果
SEMP	系统工程管理计划。管理 SE 工作的顶层计划。其定义如何组织、结构化和实施项目以及总体工程流程将如何被控制以提供符合利益攸关者需求的产品。包括识别所需技术评审及其达成准则、控制变更的方法、风险与机遇的评估和方法论，以及识别要为项目所产生的其他技术计划和文档

INPUT/OUTPUT DESCRIPTIONS

	Source documents	External documents relevant to the particular stage of procurement activity for the system of interest. Includes the written directives embodied in the source documents relevant to organizational strategies and policies
	Stakeholder needs	Needs determined from communication with external and internal stakeholders in understanding their expectations, needs, requirements, values, problems, issues, and perceived risks and opportunities
	Stakeholder needs and requirements definition record	Permanent, readable form of data, information, or knowledge related to stakeholder needs and requirements definition
	Stakeholder needs and requirements definition strategy	Approaches, schedules, resources, and specific considerations required to reflect consensus among the stakeholder classes to establish a common set of acceptable requirements. Includes the approach to capture the stakeholder needs, transform them into stakeholder requirements, and manage them through the life cycle
	Stakeholder requirements	Requirements from various stakeholders that will govern the project, including required system capabilities, functions, and/or services; quality standards; system constraints; and cost and schedule constraints. Stakeholder requirements may be captured in the Stakeholder Requirements Specification (StRS)
	Stakeholder requirements traceability	Bidirectional traceability of the stakeholder requirements
	Standards	This handbook and relevant industry, country, military, acquirer, and other specifications and standards. Includes new knowledge from industry-sponsored knowledge networks
	Strategy documents	Strategies for all applicable life processes, including business or mission analysis strategy, stakeholder needs and requirements definition strategy, system requirements definition strategy, architecture definition strategy, design definition strategy, system analysis strategy, implementation strategy, integration strategy, verification strategy, transition strategy, validation strategy, operation strategy, maintenance strategy, disposal strategy, project assessment and control strategy, decision management strategy, risk management strategy, configuration management strategy, information management strategy, measurement strategy, acquisition strategy, and supply strategy

源文件	与所感兴趣之系统的采购活动的特殊阶段有关的外部文件。与组织策略和方针相关的源文件中包含的书面指令
利益攸关者需要	根据在理解其期望、需要、需求、价值、问题、议题以及意识到的风险和机遇方面与外部和内部利益攸关者的沟通所确定的需要
利益攸关者需要和需求定义记录	与利益攸关者需要和需求定义有关的永久可读取形式的数据、信息或知识
利益攸关者需要和需求定义策略	反映诸多利益攸关者级别间为建立共同的可接受需求集达成共识所需的途径、进度、资源和特定考虑因素包括捕获利益攸关者需要、将该需要转换为利益攸关者需求以及在生命周期中管理这些需要的方法
利益攸关者需求	各种不同的利益攸关者的需求将支配项目,包括:所需系统的能力、功能和/或服务;质量标准;系统约束;以及成本约束和进度约束。可在利益攸关者需求规范(StRS)中捕获利益攸关者需求
利益攸关者需求可追溯性	利益攸关者需求的双向可追溯性
标准	本手册和相关的行业、国家、军用、采办方规范和标准以及其他规范和标准。包括来自于行业资助知识网络的新知识
策略文件	所有适用生命周期流程的策略,包括业务或任务分析策略、利益攸关者需要和需求定义策略、系统需求定义策略、架构定义策略、设计定义策略、系统分析策略、实施策略、综合策略、验证策略、转移策略、确认策略、运行策略、维护策略、退出策略、项目评估和控制策略、决策管理策略、风险管理策略、构型管理策略、信息管理策略、测量策略、采办策略和供应策略

INPUT/OUTPUT DESCRIPTIONS

Supplied system	The system or system element (product or service) is delivered from the supplier to the acquirer consistent with the delivery conditions of the supply agreement
Supply agreement	An understanding of the relationship and commitments between the project organization and the acquirer. The agreement can vary from formal contracts to less formal interorganizational work orders
	Formal agreements typically include terms and conditions
Supply payment	Payments or other compensations for the supplied system. Includes receipt and acknowledgement
Supply record	Permanent, readable form of data, information, or knowledge related to supply
Supply report	An account prepared for interested parties in order to communicate the status, results, and outcomes of the supply activities
Supply response	The organization response to the request for supply
Supply strategy	Approaches, schedules, resources, and specific considerations required to identify candidate projects for management consideration. May also include inputs to determine supply constraints. Should also include the identification of potential acquirers
System analysis record	Permanent, readable form of data, information, or knowledge related to system analysis
System analysis report	An account prepared for interested parties in order to communicate the status, results, and outcomes of the system analysis activities. Includes the results of costs analysis, risks analysis, effectiveness analysis, and other critical characteristics analysis. Also includes all models or simulations that are developed for the analysis
System analysis strategy	Approaches, schedules, resources, and specific considerations required to accomplish the various analyses to be carried out, including methods, procedures, evaluation criteria, or parameters

所供应系统	系统或系统元素（产品或服务）由供应商交付给与供应协议的交付条件一致的采办方
供应协议	对项目组织和采办方之间的相互关系和承诺的理解。协议可根据正式合同或不太正式的组织间的工作指令而变化。正式协议通常包括条款和条件
供应款项	供应的系统的款项或其他补偿。包括收据和回单
供应记录	与供应有关的永久可读取形式的数据、信息或知识
供应报告	为有关方准备一份记述报告，以便沟通供应活动的状态、结果和成果
供应响应	组织对供应请求的响应
供应策略	识别供管理考虑的候选项目所需的途径、进度、资源和特定考虑因素。亦包括确定供应约束的输入。亦应包括潜在采办方的识别
系统分析记录	与系统分析有关的永久可读取形式的数据、信息或知识
系统分析报告	为有关方准备一份记述报告，以便沟通系统分析活动的状态、结果和成果。包括成本分析、风险分析、有效性分析以及其他关键特征分析的结果。亦包括开发用于该分析的所有模型或仿真
系统分析策略	完成待实施的各种不同分析所需的途径、进度、资源和特定考虑因素，包括方法、程序、评价标准或参数

INPUT/OUTPUT DESCRIPTIONS

System architecture description	Description of the selected system architecture, typically presented in a set of architectural views (e.g., views from architecture frameworks), models (e.g., logical and physical models, although there are other kinds of models that might be useful), and architectural characteristics (e.g., physical dimensions, environment resistance, execution efficiency, operability, reliability, maintainability, modularity, robustness, safeguard, understandability, etc.) (ISO/IEC/IEEE 42010, 2010). Architecturally significant system elements are identified and defined to some degree in this artifact. (Other system elements might need to be added during the design definition process as the design is fleshed out)
System architecture rationale	Rationale for architecture selection, technological/technical system element selection, and allocation between system requirements and architectural entities (e.g., functions, input/output flows, system elements, physical interfaces, architectural characteristics, information/data elements, containers, nodes, links, communication resources)
System design description	Description of the selected system design. System elements are identified and defined
System design rationale	Rationale for design selection, system element selection, and allocation between system requirements and system element. Includes rationale of major selected implementation options and enablers
System element descriptions	Design characteristics description of the system elements contained in the system; the description depends on the implementation technology (e.g., data sheets, databases, documents, exportable data files)
System element documentation	Detailed drawings, codes, and material specifications. Updated design documentation, as required by corrective action or adaptations caused by acquisition or conformance to regulations
System elements	System elements implemented or supplied according to the acquisition agreement

系统架构描述	选定系统架构的描述，通常以架构视图（例如，从架构框架的视图）、模型（例如，逻辑模型和物理模型，尽管存在可能有用的其他种类的模型）以及架构特征（例如物理维度、环境阻力、执行效率、可操作性、可靠性、可维护性、模块化、鲁棒性、安全措施、可理解性等）一个集合的形式来表达（ISO/IEC/IEEE 42010, 2010）。本制品中在某种程度上识别和定义在架构方面重要的系统元素。（其他系统元素可能需要随着设计的更加具体化在设计定义流程期间增加）
系统架构基本原理	用于架构选择、工业技术/工艺技术系统元素选择以及系统需求与架构实体（例如功能、输入/输出流、系统元素、物理接口、架构特征、信息/数据元素、容器、节点、链接和通信资源）之间的分配的基本原理
系统设计描述	选定系统设计的描述。识别和定义系统元素
系统设计基本原理	用于设计选择、系统元素选择以及系统需求和系统元素之间的分配的基本原理。包括主要选定实施选项和使能项的基本原理
系统元素描述	包含在系统中的系统元素的设计特征描述；取决于实施技术（例如数据表、数据库、文件、可输出的数据文件）的描述
系统元素文档	详细的图样、编码和材料规范。纠正措施所需的或由采办或规则一致性引起的调整所需的更新后的设计文档
系统元素	根据采办协议实施或供应的系统元素

INPUT/OUTPUT DESCRIPTIONS

System function definition	Definition of the functional boundaries of the system and the functions the system must perform
System function identification	Identification of the system functions
System functional interface identification	Identification and documentation of the functional interfaces with systems external to the boundaries and the corresponding information exchange requirements
System requirements	What the system needs to do, how well, and under what conditions, as required to meet project and design constraints. Includes types of requirements such as functional, performance, interface, behavior (e.g., states and modes, stimulus responses, fault and failure handling), operational conditions (e.g., safety, dependability, human factors, environmental conditions), transportation, storage, physical constraints, realization, integration, verification, validation, production, maintenance, disposal constraints, and regulation. System requirements may be captured in a document called the System Requirements Specification (SyRS) or just System Specification. This includes the requirements at any level in the system hierarchy
System requirements definition record	Permanent, readable form of data, information, or knowledge related to system requirements definition
System requirements definition strategy	Approaches, techniques, resources, and specific considerations required to be used to identify and define the system requirements and manage the requirements through the life cycle
System requirements traceability	Bidirectional traceability of the system requirements
TPM data	Data provided for the identified measurement needs
TPM needs	Identification of the TPM, which measure attributes of a system element to determine how well a system or system element is satisfying or expected to satisfy a technical requirement or goal

系统功能定义	系统的功能边界以及系统必须实施的功能的定义
系统功能的识别	系统功能的识别
系统功能接口识别	功能接口与该边界以外的系统以及相应信息交换需求的识别和文件化
系统需求	为按要求满足项目和设计约束，系统需要做什么、达到何种程度以及处于何种条件下包括诸多需求类型，如功能、性能、接口、行为（例如状态和模式、激励响应、错误和故障处理）、运行条件（例如，安全性、可依赖性、人为因素、环境条件）、运输约束、储存约束、物理约束、实现约束、综合约束、验证约束、确认约束、生产约束、维护约束、退出约束和法规。可在被称为系统需求规范（SyRS）或只是系统规范的文件中捕获系统需求。这包括系统层级结构中任何级别的需求
系统需求定义记录	与系统需求定义有关的永久可读取形式的数据、信息或知识
系统需求定义策略	用来识别和定义系统需求并管理生命周期中的需求所需的方法、技术、资源和特定考量
系统需求可追溯性	系统需求的双向可追溯性
TPM 数据	为识别的测量需要提供的数据
TPM 需要	测量系统元素的属性以确定系统或系统元素满足或期望满足技术需求或目标的程度的 TPM 的识别

INPUT/OUTPUT DESCRIPTIONS

Trained operators and maintainers	Trained humans that will operate and maintain the system
Transition constraints	Any constraints on the system arising from the transition strategy including cost, schedule, and technical constraints
Transition enabling system requirements	Requirements for any systems needed to enable transition of the system of interest
Transition record	Permanent, readable form of data, information, or knowledge related to transition
Transition report	An account prepared for interested parties in order to communicate the status, results, and outcomes of the transition activities. Includes documentation of the transition results and a record of any recommended corrective actions, such as limitations, concessions, and ongoing issues. Should also include plans to rectify any problems that arise during transition
Transition strategy	Approaches, schedules, resources, and specific considerations required to transition the systems into its operation environment
Updated RVTM	An updated list of requirements, their verification attributes, and their traces
Validated requirements	Confirmation that the various requirements will satisfy the business and stakeholder requirements
Validated system	Validated system ready for supply and operation. Also informs maintenance and disposal
Validation constraints	Any constraint on the system arising from the validation strategy including cost, schedule, and technical constraints
Validation criteria	The validation criteria (the measures to be assessed), who will perform validation activities, and the validation environments of the system of interest
Validation enabling system requirements	Requirements for any systems needed to enable validation of the system of interest

经培训的操作者和维护者	将操作和维护该系统的经培训的人员
转移约束	转移策略引发的对系统的任何约束,包括成本约束、进度约束和技术约束
转移使能系统需求	使所感兴趣之系统能够转移所需的任何系统的需求
转移记录	与转移有关的永久可读取形式的数据、信息或知识
转移报告	为有关方准备一份记述报告,以便沟通转移活动的状态、结果和成果。包括转移结果的文档,以及对任何建议性纠正措施的记录,如限制、妥协和持续存在的问题。亦应包括纠正转移期间出现的任何问题的计划。
转移策略	将系统转移至其运行环境所需的途径、进度、资源和特定考虑因素
已更新的RVTM	需求及其验证属性、跟踪文件的已更新列表
已确认的需求	证实关于各种不同需求将满足业务和利益攸关者需求
已确认的系统	准备用于供应和运行的已确认的系统。亦通告维护和退出
确认约束	确认策略引发的对系统的任何约束,包括成本约束、进度约束和技术约束
确认准则	用于指定谁将执行确认活动以及所感兴趣之系统的确认环境的确认准则(待评估的测度)
确认使能系统需求	使所感兴趣之系统能够确认所需的任何系统的需求

INPUT/OUTPUT DESCRIPTIONS

Validation procedure	A validation procedure that includes a set of validation actions, using specific validation techniques, performed with specific validation enablers
Validation record	Permanent, readable form of data, information, or knowledge related to validation
Validation report	An account prepared for interested parties in order to communicate the status, results, and outcomes of the validation activities. Includes validation results and the objective evidence confirming that the system satisfies its stakeholder requirements and business requirements or not. Should also communicate an assessment of the confidence level of the findings or results
Validation strategy	Approaches, schedules, resources, and specific considerations required to accomplish the selected validation actions that minimize costs and risks while maximizing operational coverage of system behaviors
Verification constraints	Any constraint on the system arising from the verification strategy including cost, schedule, and technical constraints
Verification criteria	The verification criteria (the measures to be assessed), who will perform verification activities, and the verification environments of the system of interest
Verification enabling system requirements	Requirements for any systems needed to enable verification of the system of interest
Verification procedure	A verification procedure that includes a set of verification actions, using a specific verification method/technique, performed with specific verification enablers
Verification record	Permanent, readable form of data, information, or knowledge related to verification
Verification report	An account prepared for interested parties in order to communicate the status, results, and outcomes of the verification activities. Includes verification results and the objective evidence confirming that the system fulfills its requirements, architectural characteristics, and design properties or not. Should also communicate an assessment of the confidence level of the findings or results

确认程序	确认程序包括确认行动集，使用特定确认技术，由特定的确认使能项执行
确认记录	与确认有关的永久可读取形式的数据、信息或知识
确认报告	为有关方准备一份记述报告，以便沟通确认活动的状态、结果和成果。包括确认结果以及证实该系统是否满足其利益攸关者需求以及业务需求的客观证据。亦应传达对调查结果或结果的置信度的评估
确认策略	完成选定的确认行动所需的途径、进度、资源和特定考虑因素，该确认行动在使系统行为的运行覆盖范围最大化的同时最小化成本和风险
验证约束	验证策略引发的对系统的任何约束，包括成本约束、进度约束和技术约束
验证准则	验证准则（待评估的测度）、将执行验证活动的人员，以及所感兴趣之系统的验证环境
验证使能系统需求	使所感兴趣之系统能够验证所需的任何系统的需求
验证程序	验证程序包括验证行动集，使用特定验证方法/技术，由特定的验证使能项执行
验证记录	与验证有关的永久可读取形式的数据、信息或知识
验证报告	为有关方准备一份记述报告，以便沟通验证活动的状态、结果和成果。包括验证结果以及证实该系统是否实现其需求、架构特征以及设计特性的客观证据。亦应传达对调查结果或结果的置信度的评估

INPUT/OUTPUT DESCRIPTIONS

Verification strategy	Approaches, schedules, resources, and specific considerations required to accomplish the selected verification actions that minimize costs and risks while maximizing operational coverage of system behavior
Verified system	Verified system (or system element) ready for transition
WBS	The work breakdown structure is the decomposition of a project into smaller components and provides the necessary framework for detailed cost estimating and control. Includes a data dictionary. The costs for and description of the physical end products (hardware and software) may be captured in a product breakdown structure (PBS). The PBS supports bottoms up and algorithmic (parametric) cost estimating (see 10.1.3). The PBS is a key ingredient of commercial cost estimating tools

验证策略	完成选定的验证行动所需的途径、进度、资源和特定考虑因素，该验证行动在使系统行为的运行覆盖范围最大化的同时最小化成本和风险
已验证的系统	准备用于转移的已验证的系统（或系统元素）
工作分解结构	工作分解结构是将项目分解成诸多较小部分并提供用于详细的成本估计和控制的必要框架。包括一个数据字典。可在产品分解结构（PBS）中捕获物理成品（硬件和软件）的成本和描述。PBS 支持逆向和算法（参数的）成本估计（见 10.1.3）。PBS 是商业成本估计工具的关键组成部分

APPENDIX F: ACKNOWLEDGEMENTS

SEH V4 CONTRIBUTIONS

The *INCOSE Systems Engineering Handbook* version 4 editorial team owes a debt of gratitude to all the contributors to prior editions (versions 1, 2, 2A, and 3). Tim Robertson led the effort to create version 1 of the hand- book. Version 2 was led by James Whalen (ESEP) and Richard Wray (ESEP). Version 3 was led at various times by Kevin Forsberg (ESEP), Terje Fossnes (ESEP), Douglas Hamelin, Cecilia Haskins (ESEP), Michael Krueger (ESEP), and David Walden (ESEP). The framework they provided gave a solid basis for moving ahead with this version.This revision reflects changes to the previous version based on three primary objectives: first, to reflect the updated ISO/IEC/IEEE 15288:2015 standard; second, to reflect the state of the practice based on inputs from the relevant INCOSE Working Groups (WGs); and third, to be consistent with the Systems Engineering Body of Knowledge (SEBoK) wherever possible. Version 4 also corrected several minor issues identified by the INCOSE community.

A great deal of effort and enthusiasm was provided by the section leads and key authors, most of whom also serve as INCOSE WG Chairs or SEBoK authors. We acknowledge them in alphabetical order: Erik Aslaksen (CSEP), Albertyn Barnard, Joe Bobinis, Barry Boehm, Ed Casey, Dan Cernoch, Hugo Chale Gongora, Matthew Cilli, John Clark (CSEP), Bjorn Cole, Judith Dahmann, Arnold de Beer, Charles Dickerson, Rick Dove, Joe Elm (ESEP), Tom Fairlie, Alain Faisandier, Gauthier Fanmuy, Paul Frenz (CSEP), Sandy Friedenthal, Katri Hakola, Alan Harding, Cecilia Haskins (ESEP), Mimi Heisey, Eric Honour (CSEP), Scott Jackson, Ken Kepchar (ESEP), Alain Kouassi, Gary Langford, Claude Laporte, Alain LePut, Howard Lykins, Ray Madachy, James Martin, Jen Narkevicius, Warren Naylor, Bohdan Oppenheim, Ricardo Pineda, Paul Popick, Derek Price, Melinda Reed, Kevin Robinson, JeanClaude Roussel (ESEP), Mike Ryan, Frank Salvatore (CSEP), Hillary Sillitto (ESEP), Jack Stein, Richard Swanson (ASEP), Corrie Taljaard, Chris Unger, Beth Wilson (ESEP), and Mark Wilson (ESEP).

INCOSE Systems Engineering Handbook: A Guide for System Life Cycle Processes and Activities, Fourth Edition. Edited by David D. Walden, Garry J. Roedler, Kevin J. Forsberg, R. Douglas Hamelin and Thomas M. Shortell.

© 2015 John Wiley & Sons, Inc. Published 2015 by John Wiley & Sons, Inc.

附录 F：致谢

SEHV4 的贡献

INCOSE 系统工程手册第 4 版的编辑团队要向所有为手册之前版本（第 1 版、第 2 版、第 2A 版以及第 3 版）做出贡献的人们表示感谢。Tim Robertson 领导第 1 版手册的创建工作。第 2 版手册由 James Whalen（ESEP）和 Richard Wray（ESEP）领导。第 3 版手册在各个不同时期分别由 Kevin Forsberg（ESEP）、Terje Fossnes（ESEP）、Douglas Hamelin、Cecilia Haskins（ESEP）、Michael Krueger（ESEP）和 David Walden（ESEP）领导。他们所提供的框架为推进本版本提供了一个坚实的基础。本修订版反映了基于以下三个主要目标对之前版本所做的更改：第一，反映更新的 ISO/IEC/IEEE 15288:2015 标准；第二，反映基于相关 INCOSE 工作组（WG）意见的实践的状态；第三，将在可能情况下符合系统工程知识主体（SEBoK）的要求。第 4 版本亦纠正了由 INCOSE 领域识别出的若干次要问题。

章节负责人和主要作者付出了大量努力并投入了极大热情，他们中的很多人都担任 INCOSE WG 主席或 SEBoK 作者。我们对下列人员表示感谢（按名字首字母顺序排列）：Erik Aslaksen（CSEP）、Albertyn Barnard、Joe Bobinis、Barry Boehm、Ed Casey、Dan Cernoch、Hugo Chale Gongora、Matthew Cilli、John Clark（CSEP）、Bjorn Cole、Judith Dahmann、Arnold de Beer、Charles Dickerson、Rick Dove、Joe Elm（ESEP）、Tom Fairlie、Alain Faisandier、Gauthier Fanmuy、Paul Frenz（CSEP）、Sandy Friedenthal、Katri Hakola、Alan Harding、Cecilia Haskins（ESEP）、Mimi Heisey、Eric Honour（CSEP）、Scott Jackson、Ken Kepchar（ESEP）、Alain Kouassi、Gary Langford、Claude Laporte、Alain LePut、Howard Lykins、Ray Madachy、James Martin、Jen Narkevicius、Warren Naylor、Bohdan Oppenheim、Ricardo Pineda、Paul Popick、Derek Price、Melinda Reed、Kevin Robinson、JeanClaude Roussel（ESEP）、Mike Ryan、Frank Salvatore（CSEP）、Hillary Sillitto（ESEP）、Jack Stein、Richard Swanson（ASEP）、Corrie Taljaard、Chris Unger、Beth Wilson（ESEP）和 Mark Wilson（ESEP）。

INCOSE 系统工程手册：系统生命周期流程和活动指南，第 4 版。编撰：David D.Walden、Garry J.Roedler、Kevin J.Forsberg、R.Douglas Hamelin 和 Thomas M.Shortell。
John Wiley & Sons 公司版权所有©2015。由 John Wiley & Sons 公司于 2015 年出版。

ACKNOWLEDGEMENTS

The INCOSE Technical Operations review team led by Quoc Do generated excellent comments that significantly improved the handbook. Other individual reviewers also generated useful review comments. We acknowledge them in alphabetical order: Aaron Chia, Stephen Cook, Judith Dahmann, Bruce Douglass, Nick Dutton (ASEP), Jeff Grady (ESEP), Robert Halligan, Alan Harding, Cecilia Haskins (ESEP), Ray Hentzschel, Charles Homes, Rainer Ignetik, Vernon Ireland, Julian Johnson (ASEP), Mario Kossmann (CSEP), Michael Krueger (ESEP), Paul Logan, Bill Miller, Kevin Patrick, Jack Ring, Steve Saunders, Paul Schreinemakers, David Schultz, Zane Scott, Despina Tramoundanis, Joyce van den Hoek Ostende, Charles Wilson (CSEP), and Kenneth Zemrowski (ESEP). We would also like to thank the certification beta exam participants and the specific and anonymous reviewers who provided comments on v3.2, v3.2.1, and v3.2.2. Their inputs were much appreciated.

The editors wish to acknowledge the Idaho National Laboratory Systems Analysis Department for their significant support of this update and thank them for their contribution of technical editor Douglas Hamelin. We also wish to acknowledge Lockheed Martin Corporation for their significant support of this update and thank them for their contribution of Thomas Shortell (CSEP), who was the lead for the handbook figures and the electronic versions. The editors wish to thank Vitech Corporation for the use of their CORE tool, which was used to create an underlying process model that helped ensure consistency in the handbook IPO diagrams.

Any errors introduced as part of the editorial process rest with the editors, not the contributors.

We apologize if we unintentionally omitted anyone from these lists.

Gratefully, David Walden (ESEP), Garry Roedler (ESEP), and Kevin Forsberg (ESEP).

Quoc Do 领导的 INCOSE 技术操作评审团队形成了显著改进本手册的极佳意见。其他个人评审者亦形成了有用的评审意见。我们对下列人员表示感谢（按名字首字母顺序排列）：Aaron Chia、Stephen Cook、Judith Dahmann、Bruce Douglass、Nick Dutton（ASEP）、Jeff Grady（ESEP）、Robert Halligan、Alan Harding、Cecilia Haskins（ESEP）、Ray Hentzschel、Charles Homes、Rainer Ignetik、Vernon Ireland、Julian Johnson（ASEP）、Mario Kossmann（CSEP）、Michael Krueger（ESEP）、Paul Logan、Bill Miller、Kevin Patrick、Jack Ring、Steve Saunders、Paul Schreinemakers、David Schultz、Zane Scott、Despina Tramoundanis、Joyce van den Hoek Ostende、Charles Wilson（CSEP）和 Kenneth Zemrowski（ESEP）。我们亦要向为 3.2、3.2.1 和 3.2.2 版本提出意见的那些认证测试考试的参与者以及特定的和不知姓名的评审者表示感谢。非常感谢他们为此提供的意见。

编者要为爱达荷国家实验室系统分析部在此次版本更新过程所提供的大力支持表示感谢，并感谢他们对技术编辑 Douglas Hamelin 的贡献。我们亦要为洛克希德—马丁公司在此次版本更新过程所提供的大力支持表示感谢，并感谢他们对此手册插图和电子版本的领导 Thomas Shortell（CSEP）的贡献。编者要为使用 Vitech 公司的 CORE 工具向其表示感谢，该工具被用于创建一个有助于确保本手册 IPO 图的一致性的基本流程模型。

被作为编辑流程的一部分引入的任何错误均由编辑负责，而不是贡献者。

如果我们无意中漏掉了这些列表中的任何一个人，在此致以歉意。

敬上，David Walden（ESEP），Garry Roedler（ESEP）和 Kevin Forsberg（ESEP）。

APPENDIX G: COMMENT FORM

Reviewed document: INCOSE SE Handbook v4.0

Name of submitter: Given FAMILY (given name and family name)

Date submitted: DD-MMM-YYYY

Contact info: john.doe@anywhere.com (email address)

Type of submission: Group (individual/group)

Group name and number of contributors: INCOSE XYZ WG (if applicable)

Comments: Detailed comments with reference to document section, paragraph, etc. Please include detailed recommendations, as shown in the table below

Send comments to info@incose.org

Comment ID	Category (TH, TL, E, G)	Section number (e.g., 3.4.2.1, no alpha)	Specific reference (e.g., paragraph, line, figure, table)	Issue, comment, and rationale (rationale must make comment clearly evident and supportable)	Proposed change/new text-mandatory entry (must be substantial to increase the odds of acceptance)

E, editorial; G, general; TH, technical high; TL, technical low.

INCOSE Systems Engineering Handbook: A Guide for System Life Cycle Processes and Activities, Fourth Edition. Edited by David D. Walden, Garry J. Roedler, Kevin J. Forsberg, R. Douglas Hamelin and Thomas M. Shortell.

© 2015 John Wiley & Sons, Inc. Published 2015 by John Wiley & Sons, Inc.

附录 G：意见表

审核的文档：　　　　　INCOSE SE 手册 v4.0
提交者姓名：　　　　　姓名（名字和姓氏）
提交日期：　　　　　　DD-MMM-YYYY
联系方式：　　　　　　john.doe@anywhere.com（电子邮件地址）
提交形式：　　　　　　团体（个人/团体）
团体名称和贡献者数量：　INCOSE XYZ 工作组（如果适用）
意见：　　　　　　　　关于文档章节、段落等的详细意见，请提供详细建议，如下表所示。

将意见发送至 info@incose.org

意见 ID	类别 （TH, TL, E, G）	章节号 （如 3.4.2.1,α）	特定参考（如段、行、图、表）	问题、意见和基本原理 （基本原理的阐述必须是清晰、明显和言之有据的）	建议性变更/新文本强制性输入（对于提高接受机会是非常重要的）

E：编辑的；G：通用的；TH：技术等级高；TL：技术等级低。

INCOSE 系统工程手册：系统生命周期流程和活动指南，第 4 版。编撰：David D.Walden、Garry J.Roedler、Kevin J.Forsberg、R.Douglas Hamelin 和 Thomas M.Shortell。

John Wiley & Sons 公司版权所有©2015。由 John Wiley & Sons 公司于 2015 年出版。

INDEX

acquirer, 13, 51, 88-93, 123-4, 134, 139-44, 261
acquisition, 2, 51, 54-5, 72, 98, 106, 139-42, 237, 241, 261, 267-8
affordability, 27, 30, 32, 64, 74, 98, 107, 134, 211-19
aggregate, 69, 80-82
agile, 27, 106, 110, 207-10, 261
agreement, 2, 139-44, 261, 269-70, 280
allocate/allocation, 29, 57-60, 66-72, 77, 124, 135, 150-155, 188, 190-196, 201, 224, 236, 238
analysis, 9-10, 18-19, 32, 41-2, 48-51, 54-5, 59-67, 72, 74-7, 86, 92, 98-101, 107-8, 112-13, 120, 132, 138, 164, 170-171, 195-9, 211-45
architecture, 6, 19, 30-31, 64-70, 81-2, 85, 92, 110, 159, 181, 190-199, 208-10, 261
architecture definition, 2, 33, 48, 64-70, 73, 182, 235-6, 267-8
assessment, 22, 70, 74, 108-10, 113, 135-8, 147-9, 152-8, 171, 202
associate systems engineering professional (ASEP), 23
attribute, 6, 60-63, 87, 113, 119, 173-4, 212-13, 230-231
audits, 77-9, 106, 109-10, 124-7, 129, 137-8, 148-9, 167, 200-2, 263, 264
availability, 96, 128-30, 226-9

baseline, 26-7, 34-8, 60, 64, 107, 122-7, 262, 270
behavioral architecture *see* functional architecture
benchmark, 148-9
black box, 6-7, 60-61, 194-6, 262
boundary, 6, 20, 44, 57, 59-60, 66, 73-4
brainstorming, 116, 118, 120, 202, 242-3
business or mission analysis, 2, 33, 47-52, 182, 267-8
business requirements, 33, 47-52, 182, 270

case studies, 39-46
certified systems engineering professional (CSEP), 23

INCOSE Systems Engineering Handbook: A Guide for System Life Cycle Processes and Activities, Fourth Edition. Edited by David D. Walden, Garry J. Roedler, Kevin J. Forsberg, R. Douglas Hamelin and Thomas M. Shortell.
© 2015 John Wiley & Sons, Inc. Published 2015 by John Wiley & Sons, Inc.

索引

（原书页码，见偶数页左侧标注）

采办方, 13, 51, 88-93, 123-4, 134, 139-44, 261
采办, 2, 51, 54-5, 72, 98, 106, 139-42, 237, 241, 261, 267-8
可承受性, 27, 30, 32, 64, 74, 98, 107, 134, 211-19
聚集, 69, 80-82
敏捷, 27, 106, 110, 207-10, 261
协议, 2, 139-44, 261, 269-70, 280
分配（动词）/分配（名词）, 29, 57-60, 66-72, 77, 124, 135, 150-155, 188, 190-196, 201, 224, 236, 238
分析, 9-10, 18-19, 32, 41-2, 48-51, 54-5, 59-67, 72, 74-7, 86, 92, 98-101, 107-8, 112-13, 120, 132, 138, 164, 170-171, 195-9, 211-45
架构, 6, 19, 30-31, 64-70, 81-2, 85, 92, 110, 159, 181, 190-199, 208-10, 261
架构定义, 2, 33, 48, 64-70, 73, 182, 235-6, 267-8
评估, 22, 70, 74, 108-10, 113, 135-8, 147-9, 152-8, 171, 202
助理级系统工程专业人员（ASEP）, 23
属性, 6, 60-63, 87, 113, 119, 173-4, 212-13, 230-231
审计, 77-9, 106, 109-10, 124-7, 129, 137-8, 148-9, 167, 200-2, 263, 264
可用性, 96, 128-30, 226-9

基线, 26-7, 34-8, 60, 64, 107, 122-7, 262, 270
行为架构 见功能架构
基准, 148-9
黑盒, 6-7, 60-61, 194-6, 262
边界, 6, 20, 44, 57, 59-60, 66, 73-4
头脑风暴, 116, 118, 120, 202, 242-3
业务或任务分析, 2, 33, 47-52, 182, 267-8
业务需求, 33, 47-52, 182, 270

案例研究, 39-46
认证的系统工程专业人员（CSEP）, 23

INCOSE 系统工程手册：系统生命周期流程和活动指南，第 4 版。编撰：David D.Walden、GarryJ.Roedler、Kevin J.Forsberg、R.Douglas Hamelin 和 Thomas M.Shortell。
John Wiley & Sons 公司版权所有©2015。由 John Wiley & Sons 公司于 2015 年出版。

INDEX

change control, 34, 126-7, 186

commercial off-the-shelf (COTS), 10, 72, 79, 99, 161, 167, 237, 262

complexity, ix, 8-10, 12-14, 17-18, 21, 44, 127, 148, 165-70, 189, 221, 231

concept, 2, 5-8, 11, 13-14, 17-18, 20, 25-36, 38-46, 48-51, 53-8, 60-61, 64
- documents, 13, 54-7, 99, 195, 274, 276
- stage, 13, 25-31, 41-3, 45-6, 224

concept of operations (ConOps), 48, 50-51, 52, 270 *see also* operational concept (OpsCon)

configuration control board (CCB), 124-6

configuration item (CI), 123-7, 262, 270

configuration management, 2, 27, 122-7, 267-8

consensus, 55, 165, 244, 280

constraint, 26, 49-50, 53-5, 58-62, 65-8, 71-2, 74-5, 77-80, 83-4, 90-91, 95-103, 105-7, 173, 215-16, 262, 266, 271-82

context, 6, 9-10, 13, 20-22, 47, 49-50, 55-7, 66-7, 74, 86-7, 104, 121, 139, 178, 182-3, 195, 198

contract(s)/subcontract(s), 41-2, 55, 61, 80, 84, 89, 96, 98, 101-2, 126, 130-131, 136-7, 139-40, 142-4, 215, 219, 269, 280

contractual, 84, 90, 114, 121, 124, 127, 169, 228

cost effectiveness, 107, 211-19

cost estimating, 168, 208, 217-19, 242-3

coupling, 196, 231

coupling matrix, 69, 81-2 *see also* N^2 diagram

customer, 11, 22, 26, 34, 39, 47, 53-6, 60-61, 63, 95-9, 107-8, 121, 125-8, 133, 142-3, 147-8, 153, 156-8, 165, 170-175, 195, 204-8, 227, 237, 242-4 *see also* stakeholder

decision gates, 25-9, 93, 106-8, 123-4, 130, 135, 142, 147, 153, 262

decision management, 2, 33, 67, 110-14, 164, 267-8

decisions, 7, 12-14, 21, 110-14, 130-135, 163-4, 200-201, 271

demonstration, 86

derivation, 63, 195-6, 201, 225, 241

derived requirement, 56, 59, 62-4, 66-7, 201, 213, 232, 238, 241, 262

design, 14, 60, 64, 73-4, 113, 134, 181-2, 209-10, 281

design definition, 2, 29, 31, 70-4, 267-8

索引

变更控制, 34, 126-7, 186
商用货架产品（COTS）, 10, 72, 79, 99, 161, 167, 237, 262
复杂性, ix, 8-10, 12-14, 17-18, 21, 44, 127, 148, 165-70, 189, 221, 231
概念, 2, 5-8, 11, 13-14, 17-18, 20, 25-36, 38-46, 48-51, 53-8, 60-61, 64
    文件, 13, 54-7, 99, 195, 274, 276
    阶段, 13, 25-31, 41-3, 45-6, 224
运行概念（ConOps）, 48, 50-51, 52, 270 *还见运行概念*
构型控制委员会（CCB）, 124-6
配置项（CI）, 123-7, 262, 270
构型配置管理, 2, 27, 122-7, 267-8
共识, 55, 165, 244, 280
约束, 26, 49-50, 53-5, 58-62, 65-8, 71-2, 74-5, 77-80, 83-4, 90-91, 95-103, 105-7,
    173, 215-16, 262, 266, 271-82
背景环境, 6, 9-10, 13, 20-22, 47, 49-50, 55-7, 66-7, 74, 86-7, 104, 121, 139, 178,
    182-3, 195, 198
合同/分包合同 41-2, 55, 61, 80, 84, 89, 96, 98, 101-2, 126, 130-131, 136-7, 139-40,
    142-4, 215, 219, 269, 280
合同的, 84, 90, 114, 121, 124, 127, 169, 228
成本效益, 107, 211-19
成本估算, 168, 208, 217-19, 242-3
耦合, 196, 231
耦合矩阵, 69, 81-2 *还见 N^2 图*
客户, 11, 22, 26, 34, 39, 47, 53-6, 60-61, 63, 95-9, 107-8, 121, 125-8, 133, 142-3, 147-
    8, 153, 156-8, 165, 170-175, 195, 204-8, 227, 237, 242-4 *还见利益攸关者*

决策门, 25-9, 93, 106-8, 123-4, 130, 135, 142, 147, 153, 262
决策管理, 2, 33, 67, 110-14, 164, 267-8
决策, 7, 12-14, 21, 110-14, 130-135, 163-4, 200-201, 271
证明, 86
推演, 63, 195-6, 201, 225, 241
需求推导, 56, 59, 62-4, 66-7, 201, 213, 232, 238, 241, 262
设计, 14, 60, 64, 73-4, 113, 134, 181-2, 209-10, 281
设计定义, 2, 29, 31, 70-4, 267-8

INDEX

design structure matrix (DSM), 199 *see also* N^2 diagram
design to cost (DTC), 215, 219
development models, 32-9
development stage, 25-9, 31, 33, 93, 107, 224
disposal, 2, 14, 30, 32, 51, 101-3, 110, 124, 217-18, 220-221, 236, 267-8
documentation tree, 59, 272
domain, 6, 19, 49, 60, 73, 124, 152, 159-61, 165-71, 179, 184-6, 239-41, 262-3

effectiveness, 14-15, 22, 74-6, 98, 106-7, 109, 133, 147-9, 204, 213, 216
electromagnetic compatibility (EMC), 219-20
emergent properties/behaviors, 6, 9-10, 12, 20-21, 56, 64, 68, 73, 229-30
enabling system, 10-11, 27, 31, 49, 52, 59, 66, 72, 75, 78, 80, 84, 89, 91, 96, 98, 100-102, 106, 145, 198, 215, 236-7, 272
enterprise, 48, 69, 145, 175-9, 263 *see also* organization/organizational
environment, 6, 10-11, 20-21, 40-42, 47, 49, 53-7, 60, 66-70, 87, 91, 96, 100-103, 125, 165-70, 218, 220-221, 223, 233, 240, 263
environmental engineering, 220-221
estimating, 119, 146, 168, 218, 226, 283
ethics, 23, 141, 143
evaluation criteria, 66-7, 75, 215
evolutionary development, 8, 36-7, 122, 196
expert systems engineering professional (ESEP), 24

failure modes, effects, and criticality analysis (FMECA), 101, 117, 224
family of systems (FoS), 159-61, 221 *see also* system of systems (SoS)
flowdown, 60, 181, 193, 201
functional analysis, 107, 190-193, 201, 224, 242-3
functional architecture, 190-197
functional breakdown structure, 224
functional flow block diagram (FFBD), 56, 192, 198-9
functions-based systems engineering (FBSE), 190-193

gates *see* decision gate

hardware, 5, 20, 31, 53, 69, 78-9, 82, 126-7, 177, 182, 185, 194, 200
hazard, 231-4

设计结构矩阵（DSM），199 还见 N^2 图
按成本设计（DTC），215, 219
开发模型，32-9
开发阶段，25-9, 31, 33, 93, 107, 224
退出，2, 14, 30, 32, 51, 101-3, 110, 124, 217-18, 220-221, 236, 267-8
文档树，59, 272
领域，6, 19, 49, 60, 73, 124, 152, 159-61, 165-71, 179, 184-6, 239-41, 262-3

有效性，14-15, 22, 74-6, 98, 106-7, 109, 133, 147-9, 204, 213, 216
电磁兼容性（EMC），219-20
涌现特性/行为，6, 9-10, 12, 20-21, 56, 64, 68, 73, 229-30
使能系统，10-11, 27, 31, 49, 52, 59, 66, 72, 75, 78, 80, 84, 89, 91, 96, 98, 100-102, 106, 145, 198, 215, 236-7, 272
复杂组织体，48, 69, 145, 175-9, 263 还见组织/组织的
环境，6, 10-11, 20-21, 40-42, 47, 49, 53-7, 60, 66-70, 87, 91, 96, 100-103, 125, 165-70, 218, 220-221, 223, 233, 240, 263
环境工程，220-221
估算，119, 146, 168, 218, 226, 283
伦理，23, 141, 143
评价准则，66-7, 75, 215
演进式开发，8, 36-7, 122, 196
专家级系统工程专业人员（ESEP），24

失效模式、影响与危害性分析（FMECA），101, 117, 224
系统族（FoS），159-61, 221 还见系统之系统（SoS）
细化，60, 181, 193, 201
功能分析，107, 190-193, 201, 224, 242-3
功能架构，190-197
功能分解结构，224
功能流程方块图（FFBD），56, 192, 198-9
基于功能的系统工程（FBSE），190-193

门 见决策门

硬件，5, 20, 31, 53, 69, 78-9, 82, 126-7, 177, 182, 185, 194, 200
危害，231-4

INDEX

hierarchy, 7-8, 32-3, 59, 61, 92, 173, 187, 192-3, 200, 228, 241
human resource management, 2, 154-6, 267-8
human systems integration (HSI), 237-41

ICWG/IFWG, 56, 198
IDEF, 192, 263
ilities *see* specialty engineering
implementation, 2, 70-74, 77-9, 267-8
INCOSE, iii, ix, vii, 5, 11-12, 23-4, 132, 206-7, 211-12, 284-6
incremental, 14, 32, 36-8, 64, 80-82, 101, 122, 182
incremental commit spiral model (ICSM), 36-8
information management, 2, 27, 128-30, 236, 267-8
infrastructure management, 2, 149-51, 267-8
inspection, 78, 86, 99, 109, 136-7, 223, 229
integrated product and process development (IPPD), 199-204, 238
integrated product development team (IPDT), 78, 106, 155, 199-204, 239-41
integration, 2, 7, 9, 30-34, 69, 79-82, 87, 182, 185-6, 198-202, 213, 267-8
interface, 10, 20, 30-31, 47, 52-7, 59-63, 66-70, 72, 78-82, 94, 107, 133, 184, 191, 197-202, 209, 237, 263
interoperability, 68, 134, 175, 186, 221
ISO/IEC/IEEE 15288, vii-viii,1-4, 12-13, 29, 162
iteration, 28, 32-3, 35-6, 58, 60, 65, 191-2, 197, 201, 215

key performance parameters (KPPs), 134, 212, 215-16
knowledge management, 2, 158-61, 267-8

leadership, 9, 21-2, 51, 111, 133, 137
leading indicators, 132-3, 207
lean, 203-7
lessons learned, 105, 109, 116, 119, 142, 144, 146-9, 152, 159-61, 203, 207, 233, 273, 275, 277
life cycle, 1-3, 13-14, 20, 25-39, 48-51, 54-7, 98, 104, 110, 121, 145-9, 181-2, 186, 199-203, 220-225, 227-8, 234-5
life cycle cost (LCC), 13-14, 62, 76, 95, 99, 107, 125, 211-19, 222-3, 229, 242, 263
life cycle model, 104-5, 145-9, 181-2, 186, 263, 274

索引

层级结构, 7-8, 32-3, 59, 61, 92, 173, 187, 192-3, 200, 228, 241
人力资源管理, 2, 154-6, 267-8
人员系统综合（HSI）, 237-41

ICWG/IFWG, 56, 198
IDEF, 192, 263
性 见专业工程
实施, 2, 70-74, 77-9, 267-8
INCOSE, iii, ix, vii, 5, 11-12, 23-4, 132, 206-7, 211-12, 284-6
增量, 14, 32, 36-8, 64, 80-82, 101, 122, 182
增量承诺螺旋模型（ICSM）, 36-8
信息管理, 2, 27, 128-30, 236, 267-8
基础设施管理, 2, 149-51, 267-8
检验, 78, 86, 99, 109, 136-7, 223, 229
综合产品与流程开发（IPPD）, 199-204, 238
综合产品开发团队（IPDT）, 78, 106, 155, 199-204, 239-41
综合, 2, 7, 9, 30-34, 69, 79-82, 87, 182, 185-6, 198-202, 213, 267-8
接口, 10, 20, 30-31, 47, 52-7, 59-63, 66-70, 72, 78-82, 94, 107, 133, 184, 191, 197-202, 209, 237, 263
互操作性, 68, 134, 175, 186, 221
ISO/IEC/IEEE 15288, vii-viii,1-4, 12-13, 29, 162
迭代, 28, 32-3, 35-6, 58, 60, 65, 191-2, 197, 201, 215

关键性能参数（KPP）, 134, 212, 215-16
知识管理, 2, 158-61, 267-8

领导, 9, 21-2, 51, 111, 133, 137
领先指标, 132-3, 207
精益, 203-7
经验教训, 105, 109, 116, 119, 142, 144, 146-9, 152, 159-61, 203, 207, 233, 273, 275, 277
生命周期, 1-3, 13-14, 20, 25-39, 48-51, 54-7, 98, 104, 110, 121, 145-9, 181-2, 186, 199-203, 220-225, 227-8, 234-5
生命周期成本（LCC）, 13-14, 62, 76, 95, 99, 107, 125, 211-19, 222-3, 229, 242, 263
生命周期模型, 104-5, 145-9, 181-2, 186, 263, 274

INDEX

life cycle model management, 2, 145-9, 267-8
logical architecture *see* functional architecture
logistics, 97-101, 222-5, 261

maintainability, 45-6, 97-101, 213, 222-9
maintainer, 31, 77, 88, 95, 97-8, 100, 182, 223, 237-41, 275, 282 *see also* stakeholder
maintenance, 2, 97-101, 110, 213, 222-5, 228-9, 236, 267-8
manufacturing, 225
margin, 70-71, 81, 94, 263-4
mass properties, 188, 225-6
measures/measurement, 2, 6, 110-13, 117, 127, 130-135, 149, 158, 193, 203, 208, 267-8
measures of effectiveness (MOEs), 54, 59, 74, 131, 133-4, 215, 263, 275-6
measures of performance (MOPs), 59, 74, 131, 133-4, 264, 275-6
measures of suitability (MOSs), 54, 59, 74
mission analysis *see* business or mission analysis
model, 66-70, 74-7, 180-189, 192-7, 224-5, 231
model-based systems engineering (MBSE), 189-97

N^2 diagram, 69, 198-9, 264, 267-8
nondevelopmental item (NDI), 38, 72, 161 *see also* commercial off-the-shelf (COTS)

object-oriented systems engineering method (OOSEM), 190, 193-7
operation, 2, 95-7, 267-8
operational concept (OpsCon), 30, 48-51, 55-7, 60, 64, 74, 79, 91-2, 96, 235, 274, 276 *see also* concept of operations (ConOps)
operator, 6, 31, 40, 49, 52, 56, 69, 77, 86, 88-9, 92, 95-7, 99, 170, 181-2, 185, 223, 226-7, 229, 232, 233, 237-41, 264, 275, 282 *see also* stakeholder
opportunity, 49-52, 110, 114-22, 153, 177-8
organization/organizational, 36-39, 51, 88, 133, 137, 139-40, 145, 163-5, 176-8, 209, 264 *see also* enterprise

peer reviews, 76, 78, 86
performance, ix, 14-16, 59, 63-4, 96, 99-101, 108-10, 113, 134, 155, 174, 190-193, 212-17, 222-3, 239-41, 264

生命周期模型管理, 2, 145-9, 267-8
逻辑架构 见功能架构
后勤, 97-101, 222-5, 261

可维护性, 45-6, 97-101, 213, 222-9
维护人员, 31, 77, 88, 95, 97-8, 100, 182, 223, 237-41, 275, 282 *see also* stakeholder
维护, 2, 97-101, 110, 213, 222-5, 228-9, 236, 267-8
制造, 225
裕度, 70-71, 81, 94, 263-4
质量特性, 188, 225-6
测度/测量, 2, 6, 110-13, 117, 127, 130-135, 149, 158, 193, 203, 208, 267-8
有效性测度（MOE）, 54, 59, 74, 131, 133-4, 215, 263, 275-6
性能测度（MOP）, 59, 74, 131, 133-4, 264, 275-6
适用性测度（MOS）, 54, 59, 74
任务分析 见业务或任务分析
模型, 66-70, 74-7, 180-189, 192-7, 224-5, 231
基于模型的系统工程（MBSE）, 189-97

N² 图, 69, 198-9, 264, 267-8
非开发项（NDI）, 38, 72, 161 还见商用货架产品（COTS）

面向对象的系统工程方法（OOSEM）, 190, 193-7
运行, 2, 95-7, 267-8
运行概念（OpsCon）, 30, 48-51, 55-7, 60, 64, 74, 79, 91-2, 96, 235, 274, 276 还见运行方案（ConOps）
操作人员, 6, 31, 40, 49, 52, 56, 69, 77, 86, 88-9, 92, 95-7, 99, 170, 181-2, 185, 223, 226-7, 229, 232, 233, 237-41, 264, 275, 282 还见利益攸关者
机会, 49-52, 110, 114-22, 153, 177-8
组织/组织的, 36-39, 51, 88, 133, 137, 139-40, 145, 163-5, 176-8, 209, 264 还见复杂组织体

同行评审, 76, 78, 86
性能, ix, 14-16, 59, 63-4, 96, 99-101, 108-10, 113, 134, 155, 174, 190-193, 212-17, 222-3, 239-41, 264

INDEX

physical architecture, 64-70, 81, 195-6
physical breakdown structure, 69, 224
physical model, 66, 68, 75, 183-5, 188
planning, 21, 31-32, 104-8, 120, 178, 207, 223, 241
portfolio management, 2, 142, 144, 151-4, 177, 267-8, 277
process, ix, 1-4, 6, 11-14, 28, 38-9, 145-9, 162-5, 199-203, 205-7, 264
producibility, 225
product breakdown structure (PBS), 283
product line management (PLM), 19, 63, 68, 152-3, 160-161, 166, 170-172, 196-7, 262-4, 266
production stage, 28-9, 31, 110, 205-7, 218-19, 224, 225-6
professional development, 22-4, 175
project, ix, 27, 36-9, 47, 104-14, 151-4, 165, 264
project assessment and control, 2, 108-10, 267-8
project planning, 2, 104-8, 267-8
prototyping, 14-15, 30-31, 42-3, 45-6, 67, 70, 85, 101, 183, 197, 264

qualification (system), 94, 158, 166, 220, 264
qualification margin, 94, 264
qualification/qualified (person), 27, 91, 98, 154-6, 239, 278
quality assurance, 2, 135-8, 267-8
quality management, 2, 156-8, 267-8

recursion, 28, 32-3, 58 *see also* recursion
reliability, 98-101, 213, 226-9
requirements, 9-10, 28-31, 47-64, 83-7, 89-95, 124-7, 133-5, 139-44, 213, 219-20, 235, 262
requirements analysis, 59-64, 220
requirements verification and traceability matrix (RVTM), 84, 91, 272, 273, 282
resilience, 22, 229-31, 236
resource, 20, 79, 105-8, 118, 132, 149-51, 154-6, 176, 223, 264
retirement stage, 25-9, 32, 48, 51, 101-3, 110, 121, 159, 218, 224, 235, 274, 276
return on investment (ROI), 14-17, 22, 26, 170-172, 217, 245, 264
reviews, 26-7, 31, 76-8, 86, 93-4, 105-9, 124, 147-9, 153, 155, 164, 236, 239

物理架构, 64-70, 81, 195-6
物理分解结构, 69, 224
物理模型, 66, 68, 75, 183-5, 188
规划, 21, 31-32, 104-8, 120, 178, 207, 223, 241
项目组合管理, 2, 142, 144, 151-4, 177, 267-8, 277
流程, ix, 1-4, 6, 11-14, 28, 38-9, 145-9, 162-5, 199-203, 205-7, 264
可生产性, 225
产品分解结构（PBS）, 283
产品线管理（PLM）, 19, 63, 68, 152-3, 160-161, 166, 170-172, 196-7, 262-4, 266
生产阶段, 28-9, 31, 110, 205-7, 218-19, 224, 225-6
专业开发, 22-4, 175
项目, ix, 27, 36-9, 47, 104-14, 151-4, 165, 264
项目评估与控制, 2, 108-10, 267-8
项目规划, 2, 104-8, 267-8
原型构建, 14-15, 30-31, 42-3, 45-6, 67, 70, 85, 101, 183, 197, 264

鉴定（体系）, 94, 158, 166, 220, 264
鉴定裕度, 94, 264
鉴定/鉴定的（人员）, 27, 91, 98, 154-6, 239, 278
质量保证, 2, 135-8, 267-8
质量管理, 2, 156-8, 267-8

递归, 28, 32-3, 58 *还见递归*
可靠性, 98-101, 213, 226-9
需求, 9-10, 28-31, 47-64, 83-7, 89-95, 124-7, 133-5, 139-44, 213, 219-20, 235, 262
需求分析, 59-64, 220
需求验证和可追溯性矩阵（RVTM）, 84, 91, 272, 273, 282
可恢复性, 22, 229-31, 236
资源, 20, 79, 105-8, 118, 132, 149-51, 154-6, 176, 223, 264
退出阶段, 25-9, 32, 48, 51, 101-3, 110, 121, 159, 218, 224, 235, 274, 276
投资回报（ROI）, 14-17, 22, 26, 170-172, 217, 245, 264
评审, 26-7, 31, 76-8, 86, 93-4, 105-9, 124, 147-9, 153, 155, 164, 236, 239

INDEX

risk, ix,13, 22, 25-7, 30, 33-8, 56, 59, 63, 70, 74-6, 82, 85, 104, 106, 108, 113, 114-22, 125, 132-5, 142-4, 153, 162-7, 169, 174-5, 179, 185, 189-90, 197-200, 207-8, 217, 225, 228, 231-7, 239-43, 270
risk management, 2, 33-8, 114-22, 267-8, 279

safety, 19, 39-40, 53, 121, 231-4, 240
scenario, 54, 56, 78, 94, 194-6, 235
security, 10, 19, 43-5, 128-30, 174, 234-7
sensitivity analysis, 112-13, 217
services, 1, 10, 20, 23, 32, 79, 171-5, 177
similarity, 86-7
simulation, 19, 70, 76, 78, 87, 113, 180-189, 192, 223-5, 238
software, 6, 19-20, 53, 69, 73, 79, 82, 85, 106, 177, 194-5, 200, 223, 228, 233, 262
specialty engineering, 211-45, 261
specification, 48, 52, 55, 57, 59, 64, 127, 141, 219-20, 224-5, 270, 280-281
spiral, 32, 36-8
stages, 13-14, 25-32, 93, 95, 121, 146, 162, 218, 224
stakeholder, 9, 22-23, 25, 27-8, 30-34, 36-8, 47-78, 85, 89-96, 106, 111-19, 128, 131-2, 142, 144, 147-8, 151-3, 156-9, 165, 170-178, 180-182, 189, 195-8, 265
  needs and requirements definition, 2, 52-7, 267-8
  requirements, 28-31, 48, 52-7, 89-95, 280
standards, 1, 12-13, 60, 93-4, 107, 146-9, 163-70, 186, 197-8, 210, 219, 221, 235
state, 6, 20-21, 96, 188, 192, 228, 230-231
supplier, 13, 93, 139-44, 165, 265
supply, 2, 142-4, 267-8
support stage, 25-9, 32, 48, 51, 56, 95, 97-101, 110, 123, 144, 218, 222-5, 228, 234, 274, 276
survivability, 240
SysML *see* Systems Modeling Language (SysML)
system(s), 1, 5-8, 25, 265
  analysis, 2, 74-7, 267-8
  element, 5-8, 10, 20, 48, 58, 64-74, 77-103, 110, 113, 122-7, 134-5, 140-141, 160-161, 177, 181-8, 192-6, 198-203, 219-22, 225-6, 231, 235-8, 265
  engineer, ix, 4, 21-5, 29-30, 52, 54-5, 70, 104, 106-7, 112, 124, 130, 139, 159, 201, 216-17, 235

索引

风险, ix,13, 22, 25-7, 30, 33-8, 56, 59, 63, 70, 74-6, 82, 85, 104, 106, 108, 113, 114-22, 125, 132-5, 142-4, 153, 162-7, 169, 174-5, 179, 185, 189-90, 197-200, 207-8, 217, 225, 228, 231-7, 239-43, 270
风险管理, 2, 33-8, 114-22, 267-8, 279

安全性, 19, 39-40, 53, 121, 231-4, 240
场景, 54, 56, 78, 94, 194-6, 235
安保性, 10, 19, 43-5, 128-30, 174, 234-7
灵敏度分析, 112-13, 217
服务, 1, 10, 20, 23, 32, 79, 171-5, 177
相似性, 86-7
仿真, 19, 70, 76, 78, 87, 113, 180-189, 192, 223-5, 238
软件, 6, 19-20, 53, 69, 73, 79, 82, 85, 106, 177, 194-5, 200, 223, 228, 233, 262
专业工程, 211-45, 261
规范, 48, 52, 55, 57, 59, 64, 127, 141, 219-20, 224-5, 270, 280-281
螺旋, 32, 36-8
阶段, 13-14, 25-32, 93, 95, 121, 146, 162, 218, 224
利益攸关者, 9, 22-23, 25, 27-8, 30-34, 36-8, 47-78, 85, 89-96, 106, 111-19, 128, 131-2, 142, 144, 147-8, 151-3, 156-9, 165, 170-178, 180-182, 189, 195-8, 265
  需要和需求定义, 2, 52-7, 267-8
  需求, 28-31, 48, 52-7, 89-95, 280
标准, 1, 12-13, 60, 93-4, 107, 146-9, 163-70, 186, 197-8, 210, 219, 221, 235
状态, 6, 20-21, 96, 188, 192, 228, 230-231
供应商, 13, 93, 139-44, 165, 265
供应, 2, 142-4, 267-8
维持阶段, 25-9, 32, 48, 51, 56, 95, 97-101, 110, 123, 144, 218, 222-5, 228, 234, 274, 276
生存性, 240
SysML 见系统建模语言（SysML）系统, 1, 5-8, 25, 265
系统，1，5-8,25,265
  分析, 2, 74-7, 267-8
  元素, 5-8, 10, 20, 48, 58, 64-74, 77-103, 110, 113, 122-7, 134-5, 140-141, 160-161, 177, 181-8, 192-6, 198-203, 219-22, 225-6, 231, 235-8, 265
    工程师, ix, 4, 21-5, 29-30, 52, 54-5, 70, 104, 106-7, 112, 124, 130, 139, 159, 201, 216-17, 235

INDEX

science, 17-21
thinking, 17-21
system of interest (SOI), 6-8, 10-11
system of systems (SoS), 8-10, 13, 36, 172, 215, 229
system requirements, 28-31, 48, 57-64, 83-7, 224, 235, 281
definition, 2, 57-64, 267-8
systems engineering (SE), ix, 1, 11-13, 25, 53, 265
systems engineering and integration team (SEIT), 200-203
systems engineering body of knowledge (SEBoK), guide to, vii, 1, 12-13, 17, 19-20, 132, 176, 186, 284
systems engineering management plan (SEMP), 104-8, 135, 241, 265
systems engineering plan (SEP), 106 *see also* systems engineering management plan (SEMP)
Systems Modeling Language (SysML), 187-8, 193-4, 199

tailoring, ix, 1-3, 25, 105-8, 110, 145-9, 158, 162-79, 196-7, 202-3, 265, 267-8
taxonomy, 160, 183-4, 219
team, 11, 21-2, 33, 36-9, 69, 78, 88, 106-10, 120, 155-6, 176-7, 199-204
technical performance measures (TPMs), 59, 74, 106-7, 131, 134-5, 265, 276, 282
test, 86
testing, 10, 78, 82, 85, 108, 120, 182, 227
tools, 38, 55-6, 66, 100, 148, 150-151, 173, 188, 192
traceability, 33, 50, 54, 56-60, 62, 64, 67, 72, 78, 80, 84, 89, 91-2, 96, 99, 114, 119, 126-7, 142, 187, 191-2, 195-6
trade study, 25, 107, 110-14, 120, 130, 212-4, 266
training, 22-3, 51, 77-9, 88-9, 96, 98, 100, 108, 120, 149, 151, 155, 159, 182, 202, 223, 233, 237, 239, 275
transition, 2, 88-9, 267-8

usability, 237-41
utilization stage, 25-9, 32, 88, 93, 95, 110, 121, 123, 218, 222-6, 228, 234

validation, 2, 21, 89-95, 267-8
value, 6, 13-17, 22, 36, 99, 103, 106, 111-14, 122, 132, 145-6, 170-178, 180, 204-8, 212-17, 221, 225, 241-5, 266

科学, 17-21
思考, 17-21
所感兴趣之系统（SOI）, 6-8, 10-11
系统之系统（SoS）, 8-10, 13, 36, 172, 215, 229
系统需求, 28-31, 48, 57-64, 83-7, 224, 235, 281
    定义, 2, 57-64, 267-8
系统工程（SE）, ix, 1, 11-13, 25, 53, 265
系统工程和综合团队（SEIT）, 200-203
系统工程知识体指南（SEBoK）, vii, 1, 12-13, 17, 19-20, 132, 176, 186, 284
系统工程管理计划（SEMP）, 104-8, 135, 241, 265
系统工程计划（SEP）, 106 还见系统工程管理计划（SEMP）
系统建模语言（SysML）, 187-8, 193-4, 199

剪裁, ix, 1-3, 25, 105-8, 110, 145-9, 158, 162-79, 196-7, 202-3, 265, 267-8
分类法, 160, 183-4, 219
团队, 11, 21-2, 33, 36-9, 69, 78, 88, 106-10, 120, 155-6, 176-7, 199-204
技术性能测度（TPM）, 59, 74, 106-7, 131, 134-5, 265, 276, 282
试验, 86
测试, 10, 78, 82, 85, 108, 120, 182, 227
工具, 38, 55-6, 66, 100, 148, 150-151, 173, 188, 192
可追溯性, 33, 50, 54, 56-60, 62, 64, 67, 72, 78, 80, 84, 89, 91-2, 96, 99, 114, 119, 126-7, 142, 187, 191-2, 195-6
权衡研究, 25, 107, 110-14, 120, 130, 212-4, 266
培训, 22-3, 51, 77-9, 88-9, 96, 98, 100, 108, 120, 149, 151, 155, 159, 182, 202, 223, 233, 237, 239, 275
转移, 2, 88-9, 267-8

可用性, 237-41
使用阶段, 25-9, 32, 88, 93, 95, 110, 121, 123, 218, 222-6, 228, 234

确认, 2, 21, 89-95, 267-8
价值, 6, 13-17, 22, 36, 99, 103, 106, 111-14, 122, 132, 145-6, 170-178, 180, 204-8, 212-17, 221, 225, 241-5, 266

INDEX

value engineering, 241-5
Vee model, 32-6, 81, 169, 193
verification, 2, 21, 27, 30-31, 33, 56-63, 69, 78, 81, 83-8, 267-8
very small and micro enterprises (VSME), 179
view, 6-8, 22, 28-9, 36, 48, 51, 55-7, 64-70, 72, 104, 117, 119, 175-6, 197, 199, 227, 262

waste, 27, 101-2, 131, 133, 159, 173-4, 204-7, 220-221, 266
white box, 6, 78, 194, 199, 262
work breakdown structure (WBS), 105-7, 109, 168, 216, 243, 283

价值工程, 241-5
V 形模型, 32-6, 81, 169, 193
验证, 2, 21, 27, 30-31, 33, 56-63, 69, 78, 81, 83-8, 267-8
极小型和微型复杂组织体（VSME）, 179
视图, 6-8, 22, 28-9, 36, 48, 51, 55-7, 64-70, 72, 104, 117, 119, 175-6, 197, 199, 227, 262

浪费, 27, 101-2, 131, 133, 159, 173-4, 204-7, 220-221, 266
白盒, 6, 78, 194, 199, 262
工作分解结构（WBS）, 105-7, 109, 168, 216, 243, 283